Identification and Geographical Distribution
of the Mosquitoes of North America, North of Mexico

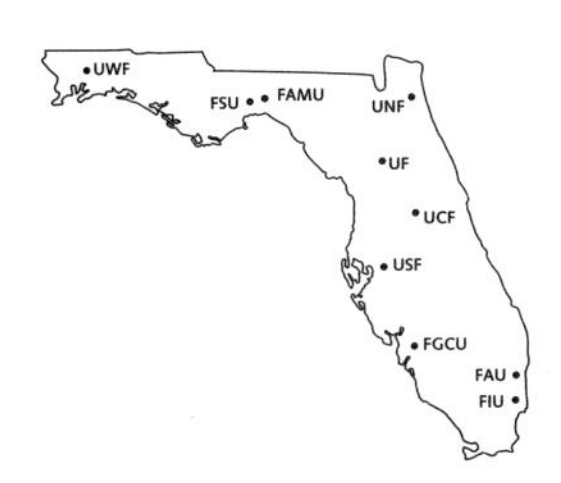


UNIVERSITY PRESS OF FLORIDA / STATE UNIVERSITY SYSTEM

Florida A&M University, Tallahassee
Florida Atlantic University, Boca Raton
Florida Gulf Coast University, Ft. Myers
Florida International University, Miami
Florida State University, Tallahassee
University of Central Florida, Orlando
University of Florida, Gainesville
University of North Florida, Jacksonville
University of South Florida, Tampa
University of West Florida, Pensacola

University Press of Florida

Gainesville · Tallahassee · Tampa · Boca Raton · Pensacola · Orlando · Miami · Jacksonville · Ft. Myers

# Identification and Geographical Distribution of the

# of North America, North of Mexico

Richard F. Darsie Jr. and Ronald A. Ward

With illustrations by Chien C. Chang and Taina Litwak

Printed in the United States of America on acid-free paper


10 09 08 07 06 05 6 5 4 3 2 1

Library of Congress Cataloging-in-Publication Data
Darsie, Richard F.
Identification and geographical distribution of the mosquitoes of North America, north of Mexico / Richard F. Darsie, Jr., and Ronald A. Ward;
with illustrations by Chien C. Chang and Taina Litwak.
p. cm.
Includes bibliographical references (p.) and index.
ISBN 0-8130-2784-5 (alk. paper)
1. Mosquitoes—United States—Identification. 2. Mosquitoes—Canada—Identification. 3. Mosquitoes—United States—Geographical distribution.
4. Mosquitoes—Canada—Geographical distribution. 5. Insects—United States.
6. Insects—Canada. I. Ward, Ronald A., 1929– II. Title.
QL536.D34 2004
595.77'2—dc22 2004058007

The University Press of Florida is the scholarly publishing agency for the State University System of Florida, comprising Florida A&M University, Florida Atlantic University, Florida Gulf Coast University, Florida International University, Florida State University, University of Central Florida, University of Florida, University of North Florida, University of South Florida,and University of West Florida.

University Press of Florida
15 Northwest 15th Street
Gainesville, FL 32611-2079
http://www.upf.com

To the memory of John N. Belkin
for his outstanding contributions
to mosquito systematics

# Contents

# Preface to the First Edition

This publication was conceived by Harold D. Chapman in 1975 while he was serving as President of the American Mosquito Control Association. Since that time it has been supported by Presidents D. Bruce Francy, Lewis T. Nielsen, Paul A. Hunt, Glenn W. Stokes and Robert K. Washino, their respective Boards of Directors, the publication committees and Executive Directors Thomas A. Mulhern and W. Donald Murray. The publication Editorial Board was composed of William E. Bickley, John D. Edman, Lewis T. Nielsen and the authors.

We are greatly indebted to the Department of Entomology, U.S. National Museum, Smithsonian Institution, its head, Don R. Davis and Oliver S. Flint, Jr., the principal investigator of the Medical Entomology Project, for granting the use of the national collection and space in which to work during the development and illustration of the keys. Many others have provided specimens to be used in preparing illustrations. Especially do we thank the late John N. Belkin, Harold D. Chapman, Lewis T. Nielsen, Sunthorn Sirivanakarn, Stephen M. Smith, and Thomas J. Zavortink.

The following persons kindly shared with us unpublished records on the occurrence of mosquito species in their political units: Peter Belton (BC), Richard L. Berry (OH), John F. Burger (NH), Charles F. Burr (VA), Robert J. Dicke (WI), James S. Haeger (FL), Fred W. Knapp (KY), Robert W. Lake (DE), Harold D. Newson (MI), Margaret A. Parsons (OH), Aileen Pucat (ALTA ) and William Wills (PA). California records were provided by Gail Grodhaus. D. M. Wood allowed us to see parts of his unpublished manuscript on the mosquitoes of Canada. For this we give him thanks.

We are particularly grateful to those who have participated in the review process and contributed their expertise to improve the various sections of the publication. We especially wish to thank the late John N. Belkin, Michael E. Faran, Ralph E. Harbach, Bruce A. Harrison, Lewis T. Nielsen, Charles H. Porter, Sunthorn Sirivanakarn and D. M. Wood.

The respective agencies to which we belong are due our gratitude for their support in many ways without which the project could never have been completed. In this regard special thanks go to the late Robert L. Kaiser and Ladene Newton of the Bureau of Tropical Diseases, Center for Disease Control, to the several directors of the Central America Research Station, San Salvador, El Salvador, Central America, where the senior author was stationed during much of the preparation and to Marta Ambrogi, who was responsible for typing most of the manuscript. The Walter Reed Army Institute of Research permitted the junior author to devote considerable time during a two year period for this work with the necessary facilities provided by the Smithsonian Institution.

We would like to acknowledge the assistance of the Defense Pest Management Information and Analysis Center which readily supplied literature references needed for the bibliography.

High praise must go to the artist, Chien C. Chang, for her superior quality work in preparing the illustrations at considerable sacrifice, and to the Department of Entomology, Kansas State University, its former chairman, Richard J. Sauer and staff members Herbert C. Knutson and H. Derrick Blocker, for their support of the art work by providing space, instruments, technical assistance, use of their reproduction equipment and mailing facilities. We also are deeply indebted to the late Kenneth L. Knight who arranged for the artist to work at Kansas State University and who has helped sustain this project in many ways. We also thank him for permission to use the illustrations on Plates 2, 3, 4, 5, 7, 8.

Thanks need also to be extended to Mr. and Mrs. George L. Gattoni who gave refuge to the senior author during his several visits to the U.S. National Museum and offered assistance and counsel during the publication's production.

# Preface to the Second Edition

The revised edition was prepared at the behest of many officials and members of the American Mosquito Control Association. Support was provided from two sources, the first a loan from the Florida Mosquito Control Association through the American Mosquito Control Association, and two additional grants from the latter. Preparation of this edition would not have been possible without the support and cooperation of the Florida Medical Entomology Laboratory, Institute of Food and Agricultural Sciences, University of Florida, and its directors Richard H. Baker and Walter J. Tabachnick. They provided office and laboratory space for the senior author and the office facilities and supplies so necessary for a project like this.

The authors are indebted to Taina Litwak for preparing illustrations for the twelve species added to the book since the first edition and to James M. Newman, graphic artist at the Florida Medical Entomology Laboratory, for preparing the thirty-one map plates and injecting his ideas for improving them over those in the first edition. Also thanks to the supervisor of the Graphics Laboratory, C. C. Lord, and to J. F. Day for their support of the project, and to L. P. Lounibos, B. F. Eldridge, and D. Strickman for reviewing the manuscript.

## Abbreviations of the United States of America and the Provinces of Canada

*United States*

| | |
|---|---|
| AL | Alabama |
| AK | Alaska |
| AR | Arkansas |
| AZ | Arizona |
| CA | California |
| CO | Colorado |
| CT | Connecticut |
| DC | District of Columbia |
| DE | Delaware |
| FL | Florida |
| GA | Georgia |
| IA | Iowa |
| ID | Idaho |
| IL | Illinois |
| IN | Indiana |
| KS | Kansas |
| KY | Kentucky |
| LA | Louisiana |
| MA | Massachusetts |
| MD | Maryland |
| ME | Maine |
| MI | Michigan |
| MN | Minnesota |
| MO | Missouri |
| MS | Mississippi |
| MT | Montana |
| NE | Nebraska |
| NC | North Carolina |
| ND | North Dakota |
| NH | New Hampshire |
| NJ | New Jersey |
| NM | New Mexico |
| NV | Nevada |
| NY | New York |
| OH | Ohio |
| OK | Oklahoma |
| OR | Oregon |
| PA | Pennsylvania |
| RI | Rhode Island |
| SC | South Carolina |
| SD | South Dakota |
| TN | Tennessee |
| TX | Texas |
| UT | Utah |
| VA | Virginia |
| VT | Vermont |
| WA | Washington |
| WI | Wisconsin |
| WV | West Virginia |
| WY | Wyoming |

*Canada*

| | |
|---|---|
| ALTA | Alberta |
| BC | British Columbia |
| LAB | Labrador* |
| MAN | Manitoba |
| NB | New Brunswick |
| NFLD | Newfoundland |
| NS | Nova Scotia |
| NUN | Nunavut |
| NWT | Northwest Territories |
| ONT | Ontario |
| PEI | Prince Edward Island |
| PQ | Quebec |
| SASK | Saskatchewan |
| YUK | Yukon |

*Although Labrador is now a part of Newfoundland, mosquito records are here listed separately.

# Introduction

In 1955 Carpenter and LaCasse published a monograph titled *Mosquitoes of North America North of Mexico* (146). They included 143 species and subspecies in 11 genera and 19 subgenera, identification keys to genera and species, and descriptions of the known adult female, male, and larval stages. Their distribution information consists of lists of states of the United States and provinces of Canada in which each taxon was collected, with substantiating references.

There are now 174 known species and subspecies in 14 genera and 29 subgenera. The additions and changes in the names of the North American mosquito fauna have been reviewed by Carpenter (131, 135, 144), Darsie (204, 206), Darsie and Ward (217), and Reinert (599, 802).

The principal objective of this volume is to revise the identification keys to adult female and larval stages in order to incorporate all 174 taxa. Each key is preceded by a detailed description of the morphology of the stage, which is needed to use the key successfully. In addition, every couplet in the two keys is illustrated, to assist the user in interpreting the characters employed.

A second purpose is to present up-to-date information on the geographical distribution of the mosquito taxa. We are continuing the arrangement used by Carpenter and LaCasse (146), listing the states and provinces from which each species has been reported with confirming references. In addition, we are depicting the distribution on maps (plates 9–42); plate captions contain the specific states/provinces in which each taxon is found. Using Carpenter and LaCasse (146) as a starting point and listing the state/province data given by them, we are adding a total of 556 new records that encompass the twelve species added since 1955. Detailed also are the fifteen instances in which species once reported as occurring in particular states/provinces have been deleted.

The morphological terms employed in this volume are substantially changed from those used by Carpenter and LaCasse (146). Harbach and Knight (308) published *Taxonomists' Glossary of Mosquito Anatomy* in 1980, and the morphological terms proposed by them and used in this volume take into consideration homology and phylogeny and use of those terms generally in the field of dipterous insects.

Another modification from the 1955 monograph is the adoption of the chaetotaxical nomenclature espoused by John N. Belkin (1950, 1952, 1953, 1954, 1960, 1962) and the abbreviations he used to designate parts of the body and setae borne on them, especially in the immature stages (e.g., T for metathorax and 6-T for seta 6 on that segment). This practice has been used by Belkin and his associates (1, 10, 11, 13, 42, 469, 511, 514) and by many other taxonomists (250, 258, 647, 778).

Let us reiterate that the user should study the applicable sections on morphology before starting to identify specimens. Wherever possible we have used adult characters that are the least disturbed by the mechanical light trapping process, but in some couplets, especially in the genus *Ochlerotatus*, the use of traits disrupted by trapping was unavoidable. The user should be familiar with the proper method of preserving mosquito larvae because the presence of a full complement of the appendages and setae is essential for their identification in the larval keys. We also tried to quantify insofar as practicable all characters, to reduce the guesswork in dealing with relative terms.

Beside each species named in the keys is a plate number where its distribution is shown. The user can immediately ascertain if an identified species has been reported from the locality where it was collected.

References cited by year are found in the Selected Bibliography of Mosquito Morphology and those cited by number are found in the Bibliography of Mosquito Taxonomy and Geographical Distribution, both at the end of this volume. Abbreviations of generic and subgeneric names used throughout the work follow Reinert (812).

An appendix provides locality data for the voucher specimens selected for illustration in the keys. These mosquitoes are located by figure number. Most are from the Walter Reed Biosystematics Unit, U.S. National Museum, Smithsonian Institution mosquito collection.

# Systematics

Mosquitoes belong to the phylum Arthropoda, class Insecta, order Diptera. They are bilaterally symmetrical insects; adults are covered with an exoskeleton and bear jointed legs and two functional wings. A second pair of wings is represented by knobbed halteres. Mosquitoes may be distinguished from other dipterous insects by the presence of scales on the wing veins and by mouthparts in the form of an elongated proboscis, adapted for piercing and sucking. They are holometabolous; that is, they have four dissimilar stages in their life cycle: egg, larva, pupa, and adult. This volume deals with adult females and fourth instar larvae, which are so different in appearance that they seem not to be related.

It is assumed that the user can already recognize species belonging to the order Diptera and family Culicidae. If not, general references such as Romoser and Stoffolano (1998) should be consulted.

In this volume we follow the classification of the family Culicidae as given by Knight and Stone (385) and Knight (384). We do not deal with suprageneric categories except in relation to certain morphological structures that belong to anophelines, referring to members of the subfamily Anophelinae, or culicines, meaning members of the subfamily Culicinae, as interpreted by Knight and Stone (385) and Harbach and Kitchings (813). Also no infrasubspecies are considered in this work.

A systematic index of the species of Culicidae now known from North America, north of Mexico, is given in table 1, page 4. After each taxon the zoogeographical region, area, or specific country where each is found, outside the region being considered, is given. Those marked as indigenous are confined to the region.

The following are the mosquito species added to the fauna of the United States and Canada since Carpenter and LaCasse (1955); numbers refer to confirming references in the bibliography of mosquito taxonomy at the back of the book.

**New Species**

*Ochlerotatus brelandi* Zavortink—793
*Ochlerotatus burgeri* Zavortink—79
*Ochlerotatus churchillensis* Ellis & Brust—250
*Ochlerotatus clivis* Lanzaro & Eldridge—394
*Ochlerotatus deserticola* Zavortink—789
*Ochlerotatus monticola* Belkin & McDonald—46
*Ochlerotatus nevadensis* Chapman & Barr—159
*Ochlerotatus papago* Zavortink—792
*Ochlerotatus washinoi* Lanzaro & Eldridge—394
*Anopheles diluvialis* Reinert—599
*Anopheles hermsi* Barr & Guptavanij—26
*Anopheles inundatus* Reinert—599
*Anopheles judithae* Zavortink—788
*Anopheles maverlius* Reinert—599
*Anopheles smaragdinus* Reinert—599
*Culex biscaynensis* Zavortink & O'Meara—795
*Culiseta minnesotae* Barr—19
*Deinocerites mathesoni* Belkin & Hogue—43

### Resurrected from Synonymy

*Aedes epactius* Dyar & Knab—7
*Aedes euedes* Howard, Dyar & Knab—783
*Aedes hendersoni* Cockerell—98
*Aedes mercurator* Dyar—201
*Aedes tahoensis* Dyar—108
*Anopheles perplexens* Ludlow—49
*Culiseta particeps* (Adams)—698

### Exotic Species Introduced into the United States and Canada

*Aedes albopictus* (Skuse)—691
*Ochlerotatus bahamensis* Berlin—554
*Ochlerotatus j. japonicus* (Theobald)—587
*Ochlerotatus togoi* (Theobald)—53
*Deinocerites pseudes* Dyar & Knab—43
*Haemagogus equinus* Theobald—727
*Orthopodomyia kummi* Edwards—471
*Psorophora mexicana* (Bellardi)—372
*Toxorhynchites moctezuma* Dyar & Knab—794

In order for the user to have a better understanding of our position on certain taxa included herein, the following comments are offered.

*Aedes*. We do not recognize *Aedes hemiteleus* Dyar as separate from *Ae. cinereus* Meigen as proposed by Bohart and Washino (82). Bickley (72) and to a lesser extent Wood et al. (784) present convincing evidence that the characters used to separate the two taxa are so variable as to preclude considering them distinct.

The now well-established exotic species *Aedes albopictus* (Skuse) was first detected in Harris County, Texas, in 1985 (691). From that start it spread throughout the eastern United States. The latest summary of its distribution by Moore (499) lists 919 counties in 26 states.

*Aedes vexans* has several subspecies, including *Ae. vexans nipponii* (Theobald), reported by Reinert (799) in Delaware, New Jersey, and Ohio. Recently it has been reported from Connecticut (Andreadis 2000, personal communication).

>*Ochlerotatus*. This genus was recently raised to generic rank from a subgenus under *Aedes* by Reinert (802), who presented evidence for the change based mainly on the structure of the female and male genitalia. It affects the generic assignment of a number of species formerly in *Aedes* and adds parentheses to the authors of many species because of the change from the originally assigned genus. The new genus is divided into two sections based on differences in larval morphology, delegating subgenera *Finlaya*, *Ochlerotatus*, *Protomacleaya*, and *Rusticoides* to section I and subgenera *Abraedes*, *Howardina*, and *Kompia* to section II. Section I has 77 species and subspecies while Section II has only 3 species; see table 1. This action will entail significant changes in the nomenclature of many common pest species. The species *Oc. bicristatus* and *Oc. provocans* are newly assigned to subgenus *Rusticoides* by Reinert (800).

A study of thirty-three populations of *Oc. increpitus* by Lanzaro and Eldridge (394) has resulted in the recognition of two new sibling species, *Oc. clivis* and *Oc. washinoi*. The former occurs only in California and the latter in California and Oregon (249).

Morphometric and electrophoretic studies of nine populations of *Oc. communis* complex by Brust and Munstermann (108) has validated a species that was formerly a synonym of *Oc. communis* (237), namely, *Oc. tahoensis*.

An imported species, *Oc.* (*Howardina*) *bahamensis*, was reported by Pafume et al. (554) from rural Dade County, Florida. Its spread into Broward County was noted by O'Meara et al. (550) and it is the first species in the subgenus *Howardina* to occur in the United States.

The exotic species *Oc. japonicus japonicus* was first collected in Suffolk County, New York, and Ocean County, New Jersey, and reported by Peyton et al. (567). It is a rock pool and container breeder and since its introduction has spread to Connecticut (808) and Pennsylvania (S. Spichiger, pers. comm. 2000) and invaded Washington State (F. Maloney, pers. comm. 2002).

>*Anopheles*. *Anopheles quadrimaculatus* was considered a single species until 1988 when detection of the existence of sibling species was determined by molecular testing (375–377, 508–512). Five sibling species are recognized, formerly given letter designations. Reinert et al. (599) have named them, viz. *An. quadrimaculatus* (A), *An. smaragdinus* (B), *An. diluvialis* ($C_1$), *An. inundatus* ($C_2$), and *An. maverlius* (D). Each is included in the keys to adult females and larvae along with appropriate illustrations. Although molecular markers are available, the five species are recognized morphologically in all life stages.

A new species separated from *An. freeborni* by Barr and Guptavanij (26) is *An. hermsi*. It was later discovered in New Mexico (266) and in Arizona and Colorado (806).

Fritz et al. (811), studying *An. punctipennis* in California, found that the wing characters used to separate it from *An. perplexens*—i.e., the subcostal pale spot and the subapical dark spot—were not entirely useful. They concluded that the only reliable way to distinguish them is by egg characters reported by Linley and Kaiser (412). We continue to use the wing spots until a more definitive study of the *An. punctipennis* taxa provides a better solution.

*Culex*. The first edition of this book included *Cx.* (*Tinolestes*) *latisquama* (Coquillett) on the strength of one male collected in Estero, Lee County, Florida, by J. B. Van Duzee in 1906 and confirmed by Stone (704). However, Berlin and Belkin (62) contended that the specimen was erroneously labeled and does not occur in Florida. It is distributed from Honduras south to Colombia, South America, and usually breeds in crabholes.

Another new discovery is *Cx.* (*Micraedes*) *biscaynensis* Zavortink and O'Meara (795), the first in this subgenus to occur in the United States (61, 547). It was first collected from bromeliad leaf axils in Fairchild Tropical Garden, Miami, on May 15, 1996. It is not clear whether this new species was recently introduced or had existed in Florida undetected.

A new species, *Cx. cedecei*, in the *taeniopus* group of subgenus *Melanoconion*, was described by Stone and Hair (707). In the first edition it was called *Cx. opisthopus* Komp because of the placement of *Cx. cedecei* in synonymy with it by Belkin (38). Subsequently Sirivanakarn and Belkin (810) delegated *Cx. opisthopus* to synonmy of *Cx. taeniopus* Dyar and Knab. However, following results of transmission studies with *Cx. cedecei* and *Cx. taeniopus* by Weaver et al. (754), striking differences in their vector competence for strains of Venezuelan equine encephalitis virus led to the conclusion that *Cx. cedecei* should be resurrected as a valid species. Indeed this was confirmed by Sallum and Forattini (809).

>*Deinocerites*. The possibility that the genus *Deinocerites* may be a subgenus of *Culex* has been suggested by Navarro and Liria (815). In fact they propose the reduction of *Deinocerites* to a subgenus of *Culex*. Their cladistic study included only one species of *Deinocerites* and 7 of 24 subgenera of *Culex*. Therefore their taxonomic action is premature and not accepted here.

>*Psorophora*. Belkin et al. (42) proposed changing the name of the common pest *Ps. confinnis* (Lynch Arribalzaga) to *Ps. columbiae*, applying it to populations in eastern and southern United States. In our first edition the identification of the southwestern populations was in question but is now mostly solved by Bohart and Washino (82), who declared that in California it is recognized as *Ps. columbiae*. The larva of *Psorophora mexicana* is unknown.

*Toxorhynchites*. In the first edition an unidentified species of *Toxorhynchites* was mentioned as first reported from southern Arizona by Zavortink (791). Subsequently he has identified the species as *Tx. moctezuma* (794).

*Wyeomyia*. The first edition included *Wy. haynei* Dodge. Experiments conducted by Bradshaw and Lounibos (86) compared this taxon with *Wy. smithii* in aspects of photoperiodism, larval diapause, and morphology of the anal papillae in different latitudes and altitudes and included cross-mating. They concluded that only one species exists. Therefore, *Wy. haynei* was declared a junior synonym of *Wy. smithii*.

**Table 1. Systematic index of the Culicidae of North America, north of Mexico, and distribution in other regions, areas, or specific countries**

| Taxon | Extralimital Distribution |
|---|---|
| Genus ANOPHELES Meigen | |
| Subgenus *Anopheles* Meigen | |
| *atropos* Dyar & Knab | Caribbean |
| *barberi* Coquillett | Indigenous |
| *bradleyi* King | Mexico |
| *crucians* Wiedemann | Neotropical |
| *diluvialis* Reinert | Indigenous |
| *earlei* Vargas | Indigenous |
| *franciscanus* McCracken | Mexico |
| *freeborni* Aitken | Mexico |
| *georgianus* King | Indigenous |
| *hermsi* Barr & Guptavanij | Indigenous |
| *inundatus* Reinert | Indigenous |
| *judithae* Zavortink | Mexico |
| *maverlius* Reinert | Indigenous |
| *occidentalis* Dyar & Knab | Indigenous |
| *perplexens* Ludlow | Indigenous |
| *pseudopunctipennis* Theobald | Neotropical |
| *punctipennis* (Say) | Mexico |
| *quadrimaculatus* Say | Mexico |
| *smaragdinus* Reinert | Indigenous |
| *walkeri* Theobald | Mexico |
| Subgenus *Nyssorhynchus* Blanchard | |
| *albimanus* Wiedemann | Neotropical |
| Genus AEDES Meigen | |
| Subgenus *Aedes* Meigen | |
| *cinereus* Meigen | Palearctic |
| Subgenus *Aedimorphus* Theobald | |
| *vexans* (Meigen) | Worldwide |
| Subgenus *Stegomyia* Theobald | |
| *aegypti* (Linnaeus) | Cosmotropical |
| *albopictus* (Skuse) | Cosmopolitan |
| Genus OCHLEROTATUS Lynch Arribalzaga | |
| **Section I** | |
| Subgenus *Finlaya* Theobald | |
| *japonicus japonicus* (Theobald) | Palearctic |
| *togoi* (Theobald) | Palearctic |
| Subgenus *Ochlerotatus* Lynch Arribalzaga | |
| *aboriginis* (Dyar) | Indigenous |
| *abserratus* (Felt & Young) | Indigenous |
| *aloponotum* (Dyar) | Indigenous |
| *atlanticus* (Dyar & Knab) | Indigenous |
| *atropalpus* (Coquillett) | Indigenous |
| *aurifer* (Coquillett) | Indigenous |
| *bimaculatus* (Coquillett) | Neotropical |
| *campestris* (Dyar & Knab) | Mexico |
| *canadensis canadensis* (Theobald) | Mexico |
| *canadensis mathesoni* (Middlekauff) | Indigenous |
| *cantator* (Coquillett) | Indigenous |
| *cataphylla* (Dyar) | Palearctic |
| *churchillensis* (Ellis & Brust) | Indigenous |
| *clivis* (Lanzaro & Eldridge) | Indigenous |
| *communis* (DeGeer) | Palearctic |
| *decticus* (Howard, Dyar & Knab) | Indigenous |
| *deserticola* (Zavortink) | Indigenous |
| *diantaeus* (Howard, Dyar & Knab) | Palearctic |
| *dorsalis* (Meigen) | Palearctic<br>Mexico |
| *dupreei* (Coquillett) | Mexico |
| *epactius* (Dyar & Knab) | Neotropical |
| *euedes* (Howard, Dyar & Knab) | Palearctic |
| *excrucians* (Walker) | Palearctic |
| *fitchii* (Felt & Young) | Palearctic |
| *flavescens* (Müller) | Palearctic |
| *fulvus pallens* (Ross) | Cuba |
| *grossbecki* (Dyar & Knab) | Indigenous |
| *hexodontus* Dyar | Palearctic |
| *impiger* (Walker) | Palearctic |
| *implicatus* (Vockeroth) | Palearctic |
| *increpitus* (Dyar) | Indigenous |
| *infirmatus* (Dyar & Knab) | Mexico |
| *intrudens* (Dyar) | Palearctic |
| *melanimon* (Dyar) | Indigenous |

| Taxon | Extralimital Distribution |
|---|---|
| *mercurator* (Dyar) | Palearctic |
| *mitchellae* (Dyar | Mexico |
| *monticola* (Belkin & McDonald) | Mexico |
| *muelleri* (Dyar) | Mexico |
| *nevadensis* (Chapman & Barr) | Indigenous |
| *nigripes* (Zellerstedt) | Palearctic |
| *nigromaculis* (Ludlow) | Mexico |
| *niphadopsis* (Dyar & Knab) | Indigenous |
| *pionips* (Dyar) | Palearctic |
| pullatus (Coquillett) | Palearctic |
| *punctodes* (Dyar) | Palearctic |
| *punctor* (Kirby) | Palearctic |
| *rempeli* (Vockeroth) | Palearctic |
| *riparius* (Dyar & Knab) | Palearctic |
| *scapularis* (Rondani) | Neotropical |
| *schizopinax* (Dyar) | Indigenous |
| *sierrensis* (Ludlow) | Indigenous |
| *sollicitans* (Walker) | Caribbean Mexico |
| *spencerii idahoensis* (Theobald) | Indigenous |
| *spencerii spencerii* (Theobald) | Indigenous |
| *squamiger* (Coquillett) | Mexico |
| *sticticus* (Meigen) | Palearctic |
| *stimulans* (Walker) | Indigenous |
| *taeniorhynchus* (Wiedemann) | Neotropical |
| *tahoensis* (Dyar) | Indigenous |
| *thelcter* (Dyar) | Mexico |
| *thibaulti* (Dyar & Knab) | Palearctic |
| *tormentor* (Dyar & Knab) | Neotropical |
| *tortilis* (Theobald) | Neotropical |
| *trivittatus* (Coquillett) | Mexico |
| varipalpus (Coquillett) | Indigenous |
| *ventrovittis* (Dyar) | Indigenous |
| *washinoi* (Lanzaro & Eldridge) | Indigenous |
| Subgenus *Protomacleaya* Theobald | |
| *brelandi* (Zavortink) | Mexico? |
| *burgeri* (Zavortink) | Mexico |
| *hendersoni* (Cockerell) | Indigenous |
| *triseriatus* (Say) | Mexico |
| *zoosophus* (Dyar & Knab) | Mexico |
| Subgenus *Rusticoides* Shevchenko & Prudkina | |
| *bicristatus* (Thurman & Winkler) | Indigenous |
| *provocans* (Walker) | Indigenous |
| **Section II** | |
| Subgenus *Abraedes* Zavortink | |
| *papago* (Zavortink) | Indigenous |
| Subgenus *Howardina* Theobald | |
| *bahamensis* (Berlin) | Caribbean |
| Subgenus *Kompia* Aitken | |
| *purpureipes* (Aitken) | Mexico |
| Genus HAEMAGOGUS Williston | |
| Subgenus *Haemagogus* Williston | |
| *equinus* Theobald | Neotropical |
| Genus PSOROPHORA Robineau-Desvoidy | |
| Subgenus *Grabhamia* Theobald | |
| *columbiae* (Dyar & Knab) | Caribbean |
| Mexico | |
| *discolor* (Coquillett) | Mexico |
| *pygmaea* (Theobald) | Caribbean |
| *signipennis* (Coquillett) | Mexico |
| Subgenus *Janthinosoma* Lynch Arribalzaga | |
| *cyanescens* (Coquillett) | Neotropical |
| *ferox* (von Humboldt) | Neotropical |
| *horrida* (Dyar & Knab) | Indigenous |
| *johnstonii* (Grabham) | Caribbean |
| *longipalpus* Randolph & O'Neill | Indigenous |
| *mathesoni* Belkin & Heinemann | Indigenous |
| *mexicana* (Bellardi) | Mexico |
| Subgenus *Psorophora* Robineau-Desvoidy | |
| *ciliata* (Fabricius) | Neotropical |
| *howardii* Coquillett | Neotropical |
| Genus CULEX Linnaeus | |
| Subgenus *Culex* Linnaeus | |
| *bahamensis* Dyar & Knab | Caribbean |
| *chidesteri* Dyar | Neotropical |
| *coronator* Dyar & Knab | Neotropical |
| *declarator* Dyar & Knab | Neotropical |
| *erythrothorax* Dyar | Mexico |
| *interrogator* Dyar & Knab | Neotropical |
| *nigripalpus* Theobald | Neotropical |
| *pipiens* Linnaeus | Palearctic Neotropical S Subsahara S |
| *quinquefasciatus* Say | Cosmotropical |
| *restuans* Theobald | Mexico |
| *salinarius* Coquillett | Mexico |
| *stigmatosoma* Dyar | Neotropical |
| *tarsalis* Coquillett | Mexico |
| *thriambus* Dyar | Neotropical |
| Subgenus *Melanoconion* Theobald | |
| *abominator* Dyar & Knab | Indigenous |
| *anips* Dyar | Mexico |
| *atratus* Theobald | Neotropical |
| *cedecei* Stone & Hair | Indigenous |
| *erraticus* (Dyar & Knab) | Neotropical |
| *iolambdis* Dyar | Neotropical |
| *mulrennani* Basham | Caribbean |
| *peccator* Dyar & Knab | Caribbean Mexico |
| *pilosus* Dyar & Knab | Neotropical |
| Subgenus *Micraedes* Coquillett | |

| Taxon | Extralimital Distribution |
|---|---|
| *biscaynensis* Zavortink & O'Meara | Indigenous |
| Subgenus *Neoculex* Dyar | |
| *apicalis* Adams | Mexico |
| *arizonensis* Bohart | Mexico |
| *boharti* Brookman & Reeves | Indigenous |
| *reevesi* Wirth | Mexico |
| *territans* Walker | Palearctic |
| Genus DEINOCERITES Theobald | |
| *cancer* Theobald | Neotropical |
| *mathesoni* Belkin & Hogue | Mexico |
| *pseudes* Dyar & Knab | Neotropical |
| Genus CULISETA Felt | |
| Subgenus *Climacura* Howard, Dyar & Knab | |
| *melanura* (Coquillett) | Indigenous |
| Subgenus *Culicella* Felt | |
| *minnesotae* Barr | Indigenous |
| *morsitans* (Theobald) | Palearctic |
| Subgenus *Culiseta* Felt | |
| *alaskaensis* (Ludlow) | Palearctic |
| *impatiens* (Walker) | Indigenous |
| *incidens* (Thomson) | Mexico |
| *inornata* (Williston) | Mexico |
| *particeps* (Adams) | Neotropical |
| Genus COQUILLETTIDIA Dyar | |
| Subgenus *Coquillettidia* Dyar | |
| *perturbans* (Walker) | Mexico |
| Genus MANSONIA Blanchard | |
| Subgenus *Mansonia* Blanchard | |
| *dyari* Belkin, Heinemann & Page | Neotropical |
| *titillans* (Walker) | Neotropical |
| Genus ORTHOPODOMYIA Theobald | |
| *alba* Baker | Indigenous |
| *kummi* Edwards | Neotropical |
| *signifera* (Coquillett) | Caribbean<br>Mexico |
| Genus WYEOMYIA Theobald | |
| Subgenus *Wyeomyia* Theobald | |
| *mitchellii* (Theobald) | Caribbean<br>Mexico |
| *smithii* (Coquillett) | Indigenous |
| *vanduzeei* Dyar & Knab | Caribbean |
| Genus URANOTAENIA Lynch Arribalzaga | |
| Subgenus *Pseudoficalbia* Theobald | |
| *anhydor anhydor* Dyar | Mexico |
| *anhydor syntheta* Dyar & Shannon | Mexico |
| Subgenus *Uranotaenia* Lynch Arribalzaga | |
| *lowii* Theobald | Neotropical |
| *sapphirina* (Osten Sacken) | Mexico |
| Genus TOXORHYNCHITES Theobald | |
| Subgenus *Lynchiella* Lahille | |
| *moctezuma* Dyar & Knab | Neotropical |
| *rutilus rutilus* (Coquillett) | Indigenous |
| *rutilus septentrionalis* Dyar & Knab | Indigenous |

# Morphology of Adult Female Mosquitoes

The morphological descriptions below deal mostly with the structures used in the keys. For a more detailed account of mosquito anatomy, consult the references listed in the bibliography of mosquito morphology on page 341.

## Basic Structures

The body of the adult mosquito is composed of hardened plates, called **sclerites**, separated from each other by lines, known as **sutures**, or by membranes of various sizes. These structures comprise the integument, or outer covering of the body; those important in identification of the female will be discussed below.

Since **scales** are common on adult females and indeed constitute one of the principal structures of recognition, they must be distinguished from setae. **Setae** (hairs, hair tufts, bristles, and spiniforms) are usually round in cross section, tapering from base to apex, and arise from a relatively large movable socket, called an **alveolus** (pl. alveoli). Scales, on the other hand, are flat in cross section, widening from base to apex, with longitudinal ridges, attached to minute alveoli on the integument. They occur in three basic forms: broad and flat, narrow and curved, and erect and apically forked. The scales on the fringe of the mosquito wing are fusiform (see Harbach and Knight 1980).

The color of scales varies from black and brown to golden, shades of yellow, such as dingy yellow in *Cx. salinarius*, to white and silvery. The white color can be brownish white, as in *Cs. minnesotae*, to grayish white. The colors tend to fade as the pinned adult ages, so in the keys herein, pale has been used to mean shades of white, and dark to mean black or brown. It is important to adjust the lighting to observe the true color of scales.

The body of the adult female is divided into three principal regions: head, thorax, and abdomen (see plate 1).

## Head

The structure of the head is shown in plates 1 and 2C. It is ovoid and a large proportion is occupied by the **compound eyes** (CE). They are composed of circular morphological units called **corneal facets** (CoF). The **antennae** arise between the eyes. The sclerite ventrad to their bases is the convex **clypeus** (Clp). Dorsad is a sclerite between and above the antennae, the **frons** (Fr), above which is the dorsum of the head, made up of the **vertex** (V) anteriorly and the **occiput** (Occ) posteriorly. Since there is no dividing suture between them, it is customary to refer to the whole dorsum simply as the occiput. The anterior border along the dorsal edge of the compound eye is known as the **ocular line** (OL).

The head bears the following five appendages: two antennae, two palpi, and the proboscis (plates 2A and 2B). The two antennae are composed of a narrow, basal ring, the **scape** (Sc), the bulbous **pedicel** (= torus) (Pc), and the **flagellum** (Fl), which contains 13 or 14 flagellomeres (= flagellar segments) (Flm), each bearing a whorl of setae. A pair of **maxillary palpi** (Mplp), called simply palpi (sing. palpus), is located ventrolateral to the clypeus and each consists of five **palpomeres** (Plp); however, in some females the basal palpomere is small or rudimentary so that the palpi appear to be 4-segmented. The **proboscis** (P) extends forward from the anteroventral base of the head. Normally, only the outer scaled covering of the proboscis, known as the **labium** (Lb), and the two terminal lobes, the **labella** (La) (sing. labellum), can be seen. Inside the labium in most species are thin stylets for piercing the host's skin.

Nine characters of the head are used in the keys as follows: (1) Shape of proboscis. It is usually nearly straight, but in genus *Toxorhynchites*, it is decidedly curved downward (fig. 1). (2) Scales on proboscis. Sometimes the proboscis has a definite pale-scaled ring near the middle, as in *Oc. sollicitans* (fig. 49), or it is variously marked with pale scales; however, in most species it is dark-scaled throughout. (3) Length of palpi. This character is used to differentiate anopheline and culicine females. In the former, the palpi are as long as the proboscis while in the latter they are not more than 0.4 as long. Within the culicine species, *Ps. longipalpus* (fig. 523) has rather long palpi; i.e., somewhat more than 0.33 as long as the proboscis, and in some species of subgenus *Neoculex* the length is compared to the length of flagellomere 4 of the antenna (fig. 430). (4) Scales of palpi. Apices of some or all of segments 2–5 may have pale-scaled rings, as in *An. walkeri* (fig. 355), scattered pale among dark scales, or only dark scales. (5) Scales on antennal pedicel. The number and color are diagnostic (e.g., *Oc. fitchii*, fig. 125). (6) Length of antennae and flagellomere 1. Flagellomere 1 is unusually long in genus *Deinocerites* (fig. 39); in addition, antennae are longer than the proboscis. (7) Width of frons. The width of the frons between the eyes, called the interocular distance, can be measured by comparing it with the diameter of a corneal facet (e.g., *Oc. epactius*,fig. 167). (8) Interocular setae (IS). These are located on the dorsal part of the frons and medioanterior area of the vertex and are long and usually dark, but in some species they are pale (e.g., *An. freeborni*, fig. 353). (9) Scales on dorsum of head. Posteriorly the scales are erect, usually forked, while anteriorly and laterally they are decumbent and either narrow and curved (e.g, subgenus *Culex*, fig. 382), or broad and flat (e.g., subgenus *Melanoconion,* fig. 384).

## Thorax

The thorax (plates 3 and 4), the body region between the head and abdomen, is divided into three segments: **prothorax**, **mesothorax**, and **metathorax**. Each bears a pair of legs; in addition, the mesothorax has a pair of functional wings, and the metathorax, a pair of knobbed **halteres** (Hl). The dipterous mesothorax is typically greatly enlarged to accommodate the flight muscles associated with the functional wings. The pro- and metathorax are correspondingly reduced in size.

In dorsal view (plates 3A and 3B) proceeding from anterior to posterior, the **antepronota** (= anterior pronotal lobes) (Ap), part of the prothorax, are found laterally just posterior to the head. The size and scales are used in the keys. Two genera, *Haemagogus* and *Wyeomyia*, have enlarged antepronota that approach each other middorsally (fig. 31).

The next three structures are mesothoracic, starting with the **scutum** (Scu), the largest sclerite of the mosquito body and rather spheroid. Anterolateral depressions in the sphere are known as the **scutal fossae** (SF) and the slightly depressed, usually unscaled area posteromedially is the **prescutellar space** (PrA). The scutum has setae arranged in three somewhat irregular longitudinal rows in the middle third. The central one is composed of **acrostichal setae** (AcS), and the rows on either side are the **dorsocentral setae** (DS). In addition, there is a group in front of and superior to the wing root, the **supraalar setae** (SaS). Those anterolateral setae occurring around and in the scutal fossa are the **scutal fossal setae** (SFS) (plates 3A and 4A). In some species the scutal setae are quite numerous and long (e.g., *An. barberi*, fig. 348), while in others they are shorter and fewer. In the subgenus *Melanoconion* (fig. 383) the acrostichal setae are absent, and in some species the acrostichal and dorsocentral setae are absent anteriorly, a condition termed the "acrostichal gap" and the "dorsocentral gap" by Lunt and Nielsen (1972). The color of some of these setae, particularly the supraalars, is diagnostic for several species (e.g., *Oc. hexodontus*, fig. 319).

The scutal integument may have spots or be a distinctive color (e.g., *Cx. erythrothorax*,fig. 401). The patterns made by the scutal scales are extensively employed in culicine mosquito identification (see *Oc. atlanticus*, fig. 189), and usually have the same names as the setae just described when they occur in the same location. One difficulty commonly encountered is rubbed specimens in which the scutum is devoid of scales and setae. This is particularly true of those collected in mechanical light traps. When such specimens are examined under high power of a stereoscopic microscope the color of some few scales still attached may give a clue about the pattern of that species. Likewise the presence of alveoli will indicate the prior location of setae in the specimen.

Posterior to the scutum is a narrow transverse sclerite, the **scutellum** (Stm). In the subfamily Anophelinae (fig. 5) it is arcuate and bears an even row of setae, the **scutellar setae** (MSS, LSS). In the subfamily Culicinae the scutellum is trilobate, with a group of setae on each lobe (fig. 7). In addition, the kind and color of scales and setae on this sclerite may be important.

The shiny, dome-shaped structure posterior to the scutellum is the **mesopostnotum** (Mpn). In most species it is bare, but in the sabethine mosquitoes (e.g., *Wyeomyia*) a group of setae occurs near its attachment to the **metanotum** (Mtn) and **abdominal tergum I** (Ab-1) (fig. 9) and is known as the **mesopostnotal setae** (MpnS).

Posteriorly is the metanotum, a thin sclerite that enlarges laterally and there bears the **halteres**, the organs of balance. Next the intersegmental cleft separates the thorax from abdominal segment I, then there is a second, very thin, metathoracic element, the **metapostnotum** (Mtpn). It actually adheres to the first abdominal tergum, but extends lateroventrally as a thin strip to touch the metameron (see plate 4A). The halteres are usually dark scaled, but have pale scales in *An. walkeri* (fig. 375).

The three thoracic segments are also represented by the structures of the thoracic pleuron (plate 4A). Two of the visible sclerites, the **antepronotum** (Ap) and the **postpronotum** (Ppn), are components of the tergum of the prothorax, not of its pleuron. Starting anteriorly, the prothoracic elements consist of the antepronotum, which is connected ventrally by a straplike piece to the **proepisternum** (Ps); both of these bear setae, i.e., **antepronotal setae** (ApS) and **upper proepisternal setae** (PeSU), and sometimes scales. The proepisternum bends around medially to cover the ventroanterior face of the thorax below the head and cervix (see plate 3A) and lobes from each side extend ventrally between the forecoxae. This anterior face of the proepisternum is sometimes covered with scales, the **lower proepisternal scales** (Pscl) (e.g., *Oc. hexodontus*, fig. 310). The last prothoracic sclerite, the postpronotum, is found posterior to the antepronotum and lateral to the scutum at the level of the scutal fossa. It bears scales that sometimes have a distinctive pattern; and a number of setae (PpS), usually confined to the posterior margin, but sometimes scattered over the posterior 0.5 (e.g., *Oc. impiger*, fig. 298).

The mesothoracic pleuron has five large and important sclerites. Just posterior to the postpronotum is an opening in the thorax, the **mesothoracic spiracle** (MS). It is surrounded by a large sclerite, the **anterior mesanepisternum** (Amas), and divided into four areas: (1) The **prespiracular area** (PsA), a small triangle dorsoanterior to the spiracle. It adjoins the posterior border of the postpronotum, and sometimes bears setae, the **prespiracular setae** (PsS) (e.g., genus *Culiseta*, fig. 18). (2) The **postspiracular area** (PA), a rather large space posterior to the spiracle with or without setae and scales; when present these are the **postspiracular setae** (PS) (e.g., genus *Psorophora*, fig. 17), and **postspiracular scales** (PoSc) (e.g., *Oc. brelandi*, fig. 200). (3) The **hypostigmal area** (HyA), immediately ventral to the spiracle and at times with **hypostigmal scales** (HySc) (e.g., *Oc. pullatus*, fig. 241), or a dark integumental spot, as in *Oc. fulvus pallens* (fig. 183). (4) The **subspiracular area** (SA), a depression ventral to the hypostigmal area, adjoining the mesokatepisternum ventrally, with or without **subspiracular setae** (SaS) and **scales** (Ssc) (e.g., *Oc. varipalpus*, fig. 175).

The largest of the mesopleural sclerites, the **mesokatepisternum** (Mks) is rather pear shaped, bulging ventroanteriorly. It is united with a dorsal narrow linear area, the **posterior mesanepisternum** (Pmas), containing the prealar area (Pa) with its prealar knob (PK) that bears a group of setae, the

**prealar setae** (PaS). The mesokatepisternum has two groups of setae, the **upper** (MkSU) and **lower** (MkSL) **mesokatepisternal setae.** These are often combined into a single line of setae, the **mesokatepisternal setae** (MkS). The **mesokatepisternal scales** (MkSc) are sometimes arranged in distinct patterns, e.g., narrow lines of scales as in *Oc. papago* (fig. 64), or more frequently an extensive patch that may or may not reach the anterior angle, as in *Oc. provocans* (fig. 253). Between the forecoxa and the ventroanterior border of the mesokatepisternum there is a membrane, the **postprocoxal membrane** (PM). In some species of *Ochlerotatus*, it bears a small patch of scales, the **postprocoxal scales** (Psc) (e.g., *Oc. punctor*, fig. 268).

The rectangular sclerite just posterior to the mesokatepisternum and ventral to the wing root is the **mesepimeron** (Mam). It bears a group of setae in the dorsoposterior corner, the **upper mesepimeral setae** (MeSU). Sometimes another group, the **lower mesepimeral setae** (MeSL), usually with not more than 1–6 setae in a single row, occurs along the anteroventral border. These are often used to separate groups of species in the genus *Ochlerotatus* (e.g., *Oc. riparius*, fig. 105, vs. *Oc. stimulans*, fig. 106).

The mesepimeron may also have varying amounts of scaling. In some species of the subgenus *Melanoconion* the mesepimeron has a definite pale spot or light and dark colored integumental areas that provide species differentiation (fig. 440).

Just ventral to the mesepimeron is the fifth and smallest, mesopleural sclerite, the **mesomeron** (MsM). It is triangular and situated between the mid- and hindcoxae. The relation of the base of the mesomeron to the base of the hindcoxa is a generic character. Usually the base of the hindcoxa is distinctly ventral to the base of the mesomeron, but in the sabethine females the base of the hindcoxa is about even with the base of the mesomeron (see figs. 10, 12).

The metathoracic pleuron is much reduced (plate 4A). The largest element is the **metepisternum** (Mts), located posterior to the mesepimeron. It is strap shaped with a dorsoventral axis and surrounds the **metathoracic spiracle** (MtS), the other opening in the thorax, in its dorsal half. Ventral to the metepisternum is a small sclerite, the **metameron** (Mem), articulating with the hindcoxa posteriorly and with the ventroposterior border of the mesepimeron. Rarely it bears scales (see fig. 248). Dorsoposterior to the metepimeron is the metanotum, already discussed under the mesopostnotum.

The sternal elements of the thorax are not included in this discussion since they have not been used as identifying characters—except for one, the intersegmental membrane connecting the metepisternum with abdominal sternum I. It sometimes bears **postmetasternal scales** (MscP) (e.g., *Oc. pionips*, fig. 318).

### *Appendages of the Thorax*

**Wings.** The two functional wings (W) of adult mosquitoes are attached to the mesothorax (see plate 3C). Each is composed of a network of longitudinal thickenings called **veins.** Between the veins are stretched membranes known as **cells.** The veins are clothed with scales dorsally and ventrally. The apical and posterior margin of the wing is bordered by long, fusiform scales, the **wing fringe** (FS), which may have pale and dark sections, best exemplified in *Ps. signipennis* (fig. 497), or a coppery or silvery, apical spot (e.g., *An. earlei*, fig. 328).

The veins and cells have names as shown in plate 3C, which follows the Comstock-Needham system of nomenclature. There are six major longitudinal veins: costa (C), subcosta (Sc), radius (R), media (M), cubitus (Cu), and anal (A). If the veins are traced from base to apex, several have one or more subdivisions. For example, the radius has a basal vein R, with primary branches $R_1$ and $R_s$. The latter further divides into $R_{2+3}$ and $R_{4+5.}$ The $R_{2+3}$ separates into $R_2$ and $R_3$ apically. There are several crossveins, short connectors between major veins. The humeral crossvein (h) joins the costa and subcosta, the radiomedial crossvein (r-m), the radius with the media, and the mediocubital (m-cu), the media with the cubital veins.

The cells likewise have names, per plate 3C (letters in italics). Each cell derives its name from the vein just anterior to it. An important one to know is cell $R_2$ because it is shortened in the genus *Uranotaenia* (fig. 13). In the key character its length is compared to the length of the vein $R_{2+3}$, a portion of vein $R_s$ between the branching of $R_{4+5}$ and the junction of veins $R_2$ and $R_3$. This section of the vein is called the "petiole" by some authors.

The wing scales provide many useful characters. They can be broad and numerous (e.g., *Cq.*

*perturbans*, fig. 37), triangular shaped (e.g., *Oc. grossbecki*,fig. 91), or narrow and filiform (e.g., *Cx. pipiens*, fig. 15 ). Colors are important, too. Many species have the wing scales entirely dark, or they may vary in number of pale scales, from a small patch at the base of the costa (e.g., *Oc. atropalpus*, fig. 155), to scattered pale scales on the anterior veins (e.g., *Oc. cataphylla*, fig. 236), to generally intermixed pale and dark scales (e.g., *Oc. sollicitans*, fig. 54), to alternating mostly dark- with mostly pale-scaled veins (e.g., *Oc. s. idahoensis*, fig. 221), to mostly pale scales (e.g., *Oc. dorsalis* fig. 145).

Furthermore, there are wings with unicolorous spots produced by dense clusters of scales along some veins (e.g., *An. quadrimaculatus*, fig. 328). The costa, subcosta, and radial veins in some anophelines possess spots of pale scales that are named (Wilkerson and Peyton 1990). The area of pale scales at or near the apex of the wing is called the apical spot and the subcostal spot is found where the subcostal vein joins the costal vein. Although they are called "spots" they are really patches of pale scales sometimes extending over several veins (e.g., *An. punctipennis*, fig. 332). Most mosquito wings do not bear prominent setae, but in the genus *Culiseta* (fig. 28), a row occurs ventrally near the base of the subcosta.

**Legs.** There are three pairs of legs, one attached to each thoracic segment. The leg consists of five main parts: **coxa** (CI, CII, CIII), **trochanter** (Tr), **femur** (Fe), **tibia** (Ti), and **tarsus** (Ta) (plate 2D). The tarsus is composed of five segments known as **tarsomeres.** The fifth tarsomere ($Ta_5$) bears two **unguis** (U) (claws, Cl) that, in most species, have a secondary element, the tooth. The tarsal claws are used frequently in the *Ochlerotatus* key (e.g., *Oc. excrucians*, fig. 103). They can be studied under the stereoscopic microscope best by shining the light on the stage below the specimen and viewing the claws in silhouette. Tarsomere 4 ($Ta_4$) is unusually small in the fore- and midlegs of the genus *Orthopodomyia* (fig. 34).

Scale patterns on the various segments of the legs are extensively employed as key characters. The scales on coxa I can be brown or pale (e.g., *Ae. cinereus*, fig. 275). The femora may have the basal half pale (e.g., *Oc. zoosophus*, fig. 75), or with subapical pale rings (e.g., *Ps. columbiae*, fig. 488), or with apical pale rings (= knee spots) (e.g., *Oc. implicatus*, fig. 252). The foretibia sometimes has a complete line of pale scales separate from the pale-scaled ventral half of the segment (e.g., *Cx. tarsalis*, fig. 393). The femora and tibiae of some *Psorophora* species have long, erect scales apically, giving them a shaggy appearance (fig. 499). The tarsomeres, especially on the hindleg, may have basal pale rings, which are narrow, as in *Ae. vexans* (fig. 77), or broad as in *Oc. excrucians* (fig. 45), both apical and basal pale rings, creating the effect of appearing to be very wide bands, as in *Oc. c. canadensis* (fig. 48), or with tarsomeres 4, 5 and part of 3 all pale, as in *Ps. ferox* (fig. 501). In some cases it is necessary to distinguish shades of the pale scales. For example, the pale band on the hindtarsomere 1 in *Oc. sollicitans* is yellow scaled, while in *Oc. nigromaculis*, it is white scaled (see figs. 59, 61).

## Abdomen

The abdomen is composed of ten segments, of which the first seven are quite similar in external structure. The three terminal segments are specialized for reproduction and excretion. It has become customary to refer to the abdominal segments by Roman numerals (e.g., abdominal segment III) and they are referred to in the keys by just the Roman numeral.

Each of the seven segments has a dorsal sclerite, the **tergum** (Te), and a ventral sclerite, the **sternum** (S) (see plate 4B). Laterally, they are connected by expandable, elastic tissue, the **pleural membrane** (Pme). A similar intersegmental membrane separates the terga dorsally and the sterna ventrally. These membranes permit the abdomen to expand during blood feeding and when the female becomes gravid.

Segments VIII–X are shortened and modified. In some genera, e.g., *Culex*, *Culiseta*, and *Mansonia* (fig. 19), these segments are mostly telescoped inside the terminal segments, making the apex of the abdomen appear bluntly rounded. In other genera, e.g., *Ochlerotatus* and *Psorophora*, part of these segments protrude posteriorly, giving the abdominal terminus a pointed appearance. Also in those with blunt abdomens, segment VII is almost the same width as VI, for in the pointed abdomens, VII is decidedly smaller than VI. Abdominal segment VIII usually has a larger sternum than tergum.

Posterior to tergum VIII can be seen two elongated lobes, the **cerci** (sing. cercus). These structures are long, straight, and visible in the genera with pointed abdomens, but are shorter, usually curved medially, and not so visible in the genera with blunt abdomens. Ventrally, posterior to sternum VIII is a smaller lobe lying ventral to the cerci, the **postgenital lobe** (PGL). Both of these terminal organs are parts of the female genitalia.

No attempt will be made to describe completely the female genitalia, since their parts are rarely used in the keys, but some elements are described above because they should be recognized. For an account of the female genitalia, consult Reinert (1974).

The anopheline abdomen, with the exception of *An. albimanus*, is devoid of scales, although it bears a number of tergal and sternal setae. In the other genera, both setae and scales are present on the abdomen. The patterns of dark and pale scales are very important in identification. Sometimes the pale scales are located basally on the tergum, i.e., the part nearest the base of the abdomen, where it is attached to the thorax (e.g., *Oc. intrudens*, fig. 256), or sometimes on the apical part, i.e., nearest to the distal end of the abdomen (e.g., *Cx. territans*, fig. 386). Likewise, the scales on the sterna may be unicolorous or have distinctive patterns (e.g., *Cx. tarsalis*, fig. 394).

In *Mansonia* there are special spiniforms along the posterior border of tergum VII (*Ma. titillans*, fig. 473), and thick peglike spiniforms on tergum VIII of all species. The cerci of *De. cancer* (fig. 471) have specialized, spatulate setae.

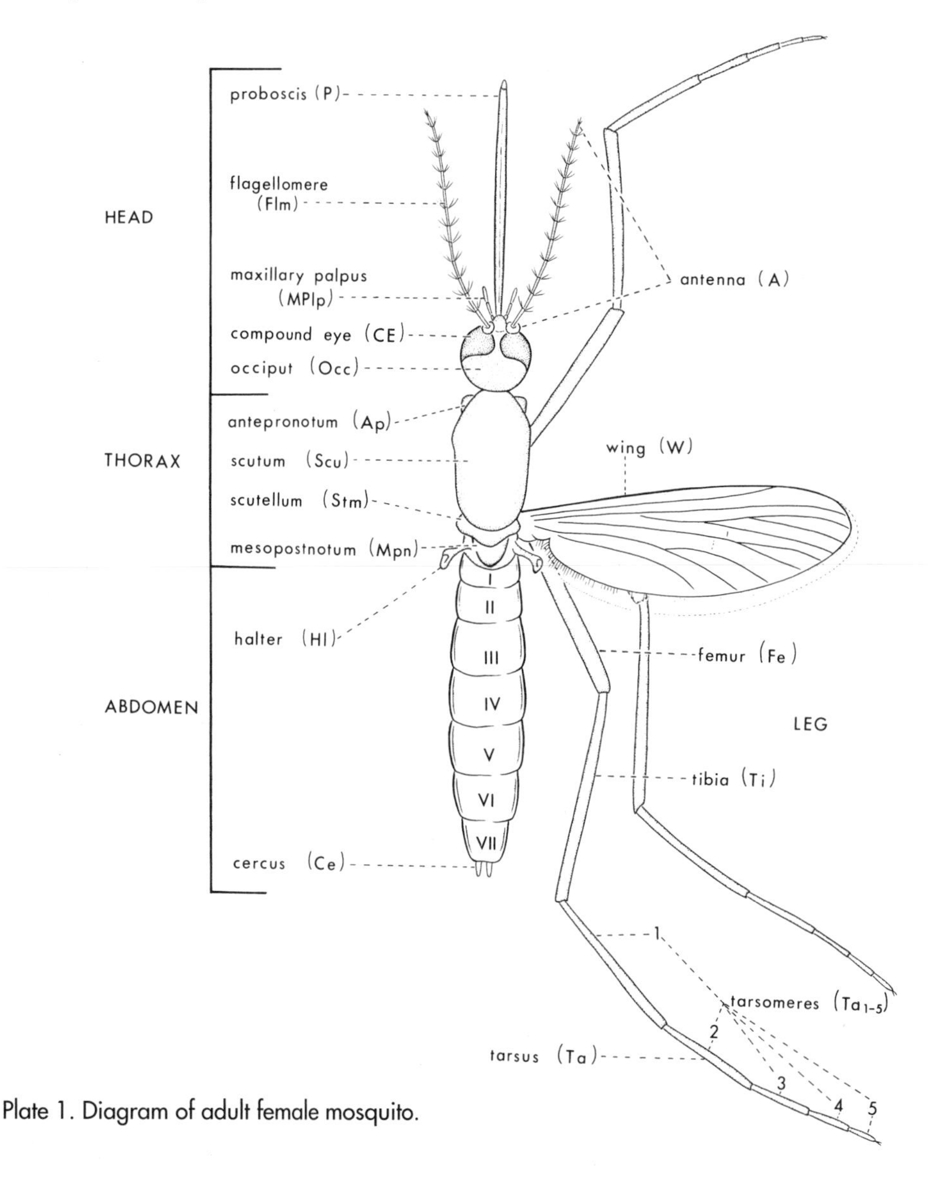


Plate 1. Diagram of adult female mosquito.

## Abbreviations of Adult Female Morphology in Plates

*Plate 2*

A antenna
C coxa
CE compound eye
Cl claw
Clp clypeus
CoF corneal facet
Fe femur
Fl flagellum
Flm flagellomere
Fr frons
IS interocular space
La labellum
Lb labium
Mplp maxillary palpus
Occ occiput
OL ocular line
P proboscis
Pe pedicel
Plp palpomere
Sc scape
Ta tarsus
$Ta_{1-5}$ tarsomere
Ti tibia
Tr trochanter
V vertex

*Plate 3, Illustrations A and B*

AcS acrostichal
Ap antepronotum
ApS antepronotal setae
C-I forecoxa
Cv cervix
DS dorsocentral setae
LSS lateral scutellar setae
Mpn mesopostnotum
MSS median scutellar setae
Mtn metanotum
PeSU upper proepisternal setae
Ppn postpronotum
PpS postpronotal setae
PrA prescutellar area
Ps proepisternum
SaS supraalar setae
Scu scutum
Stm scutellum
SF scutal foss
SFS scutal fossal setae
W wing

*Plate 3, Illustration C (Wing)*

A anal vein
*A* anal cell
C costal vein
C costal cell
Cu cubital vein
$Cu_1$ cubital cell
$Cu_2$ posterior branch of cubital vein
$Cu_2$ $cubital_2$ cell
FS fringe scales
h humeral crossvein
M medial vein
*M* medial cell
$M_{1+2}$ anterior branch of medial vein
$M_2$ $medial_2$ cell
$M_{3+4}$ posterior branch of medial vein
$M_4$ $medial_4$ cell
m-cu mediocubital crossvein
R radial vein
*R* radial cell
$R_1$ anteriormost branch of radial vein
$R_1$ $radial_1$ cell
$R_s$ radial sector vein
$R_2$ anterior branch of radial sector vein
$R_2$ $radial_2$ cell
$R_{2+3}$ connector vein of radial sector vein
$R_3$ median branch of radial sector vein
$R_3$ $radial_3$ cell
$R_{4+5}$ posterior branch of radial sector vein
$R_5$ $radial_5$ cell
r-m radiomedial crossvein
Sc subcostal vein
*Sc* subcostal cell

*Plate 4*

Ab-I abdominal segment I
Amas anterior mesanepisternum
Ap antepronotum
ApS antepronotal setae
C-I forecoxa
C-II midcoxa
C-III hindcoxa
Ce cercus
Cv cervix
DS dorsocentral setae
H head
Hl halter
HyA hypostigmal area
LSS lateral scutellar setae

Mam mesepimeron
Mem metameron
MeSL lower mesepimeral setae
MeSU upper mesepimeral setae
Mks mesokatepisternum
MkSL lower mesokatepisternal setae
MkSU upper mesokatepisternal setae
Mpn mesopostnotum
MS mesothoracic spiracle
Msm mesomeron
MSS medial scutellar setae
Mtm metepimeron
Mtn metanotum
Mtpn metapostnotum
Mts metepisternum
MtS metathoracic spiracle
PA postspiracular area
PaS prealar setae
PeSU upper proepisternal setae
PGL postgenital lobe
PM postprocoxal membrane
Pmas posterior mesanepisternum
Ppn postpronotum
PpS postpronotal setae
Ps proepisternum
PS postspiracular setae
PsS prespiracular setae
PsA prespiracular area
S sternum
SA subspiracular area
SaS supraalar setae
Scu scutum
SF scutal fossa
SFS scutal fossal setae
Stm scutellum
Te tergum of abdomen
W wing

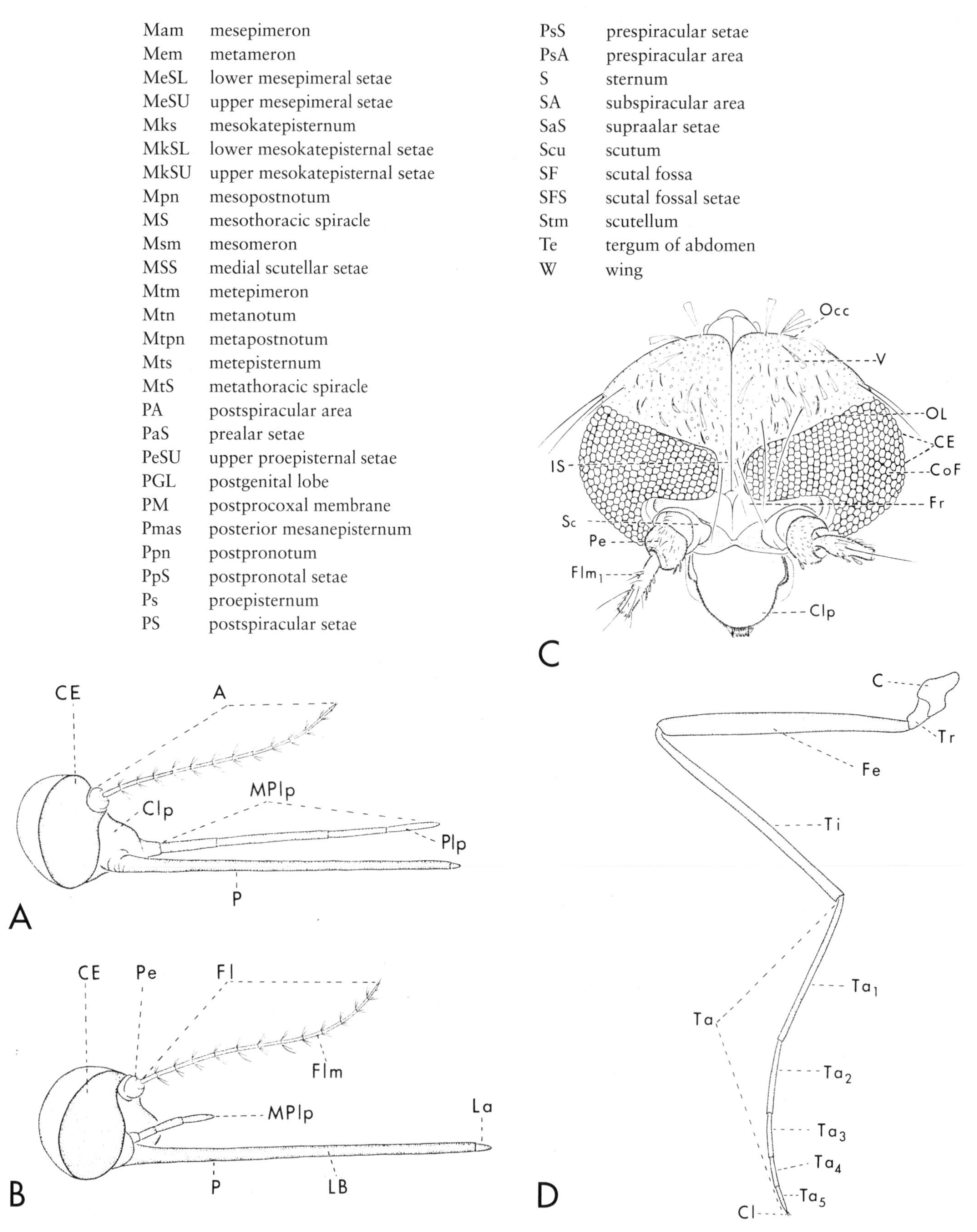


Plate 2. Head and leg of adult female mosquito. *A*. Lateral view of anopheline head; *B*. Lateral view of culicine head; *C*. Dorsal view of culicine head; *D*. Lateral view of leg.

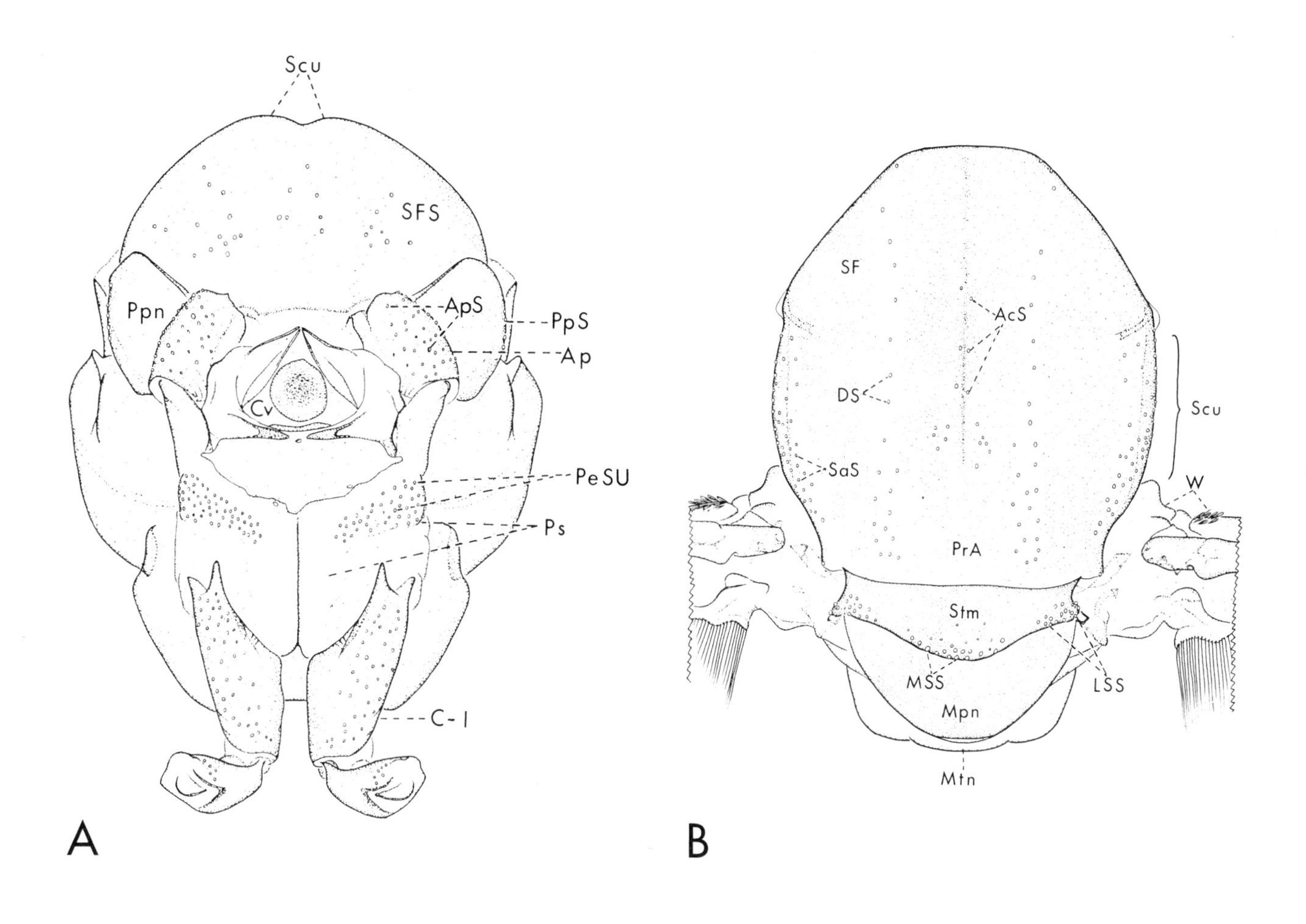


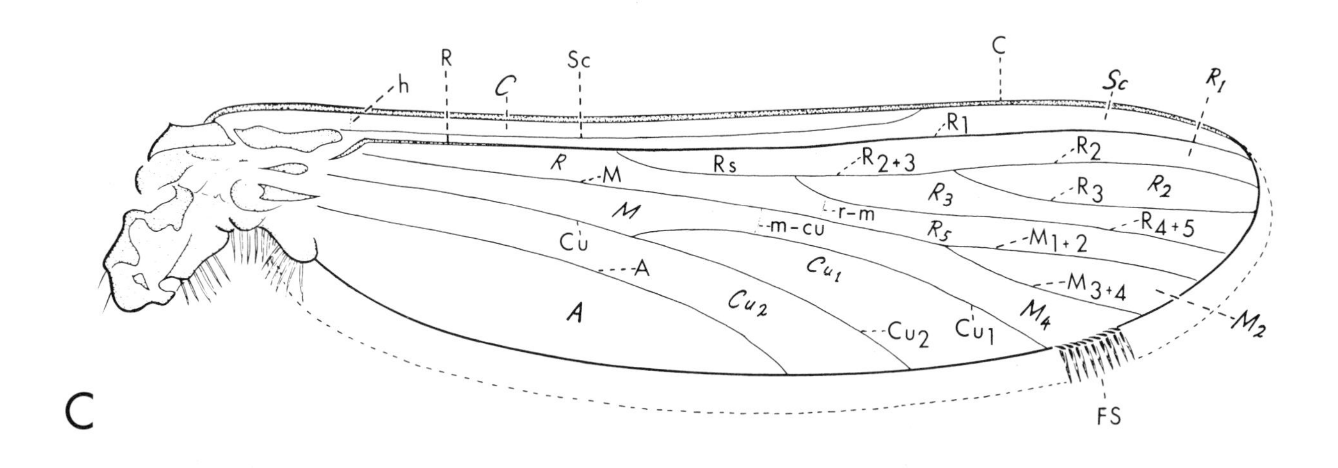


Plate 3. Thorax and wing of adult female mosquito. *A*. Anterior view of thorax; *B*. Dorsal view of thorax; *C*. Dorsal view of wing: longitudinal veins designated by gothic letters, cells by italics.

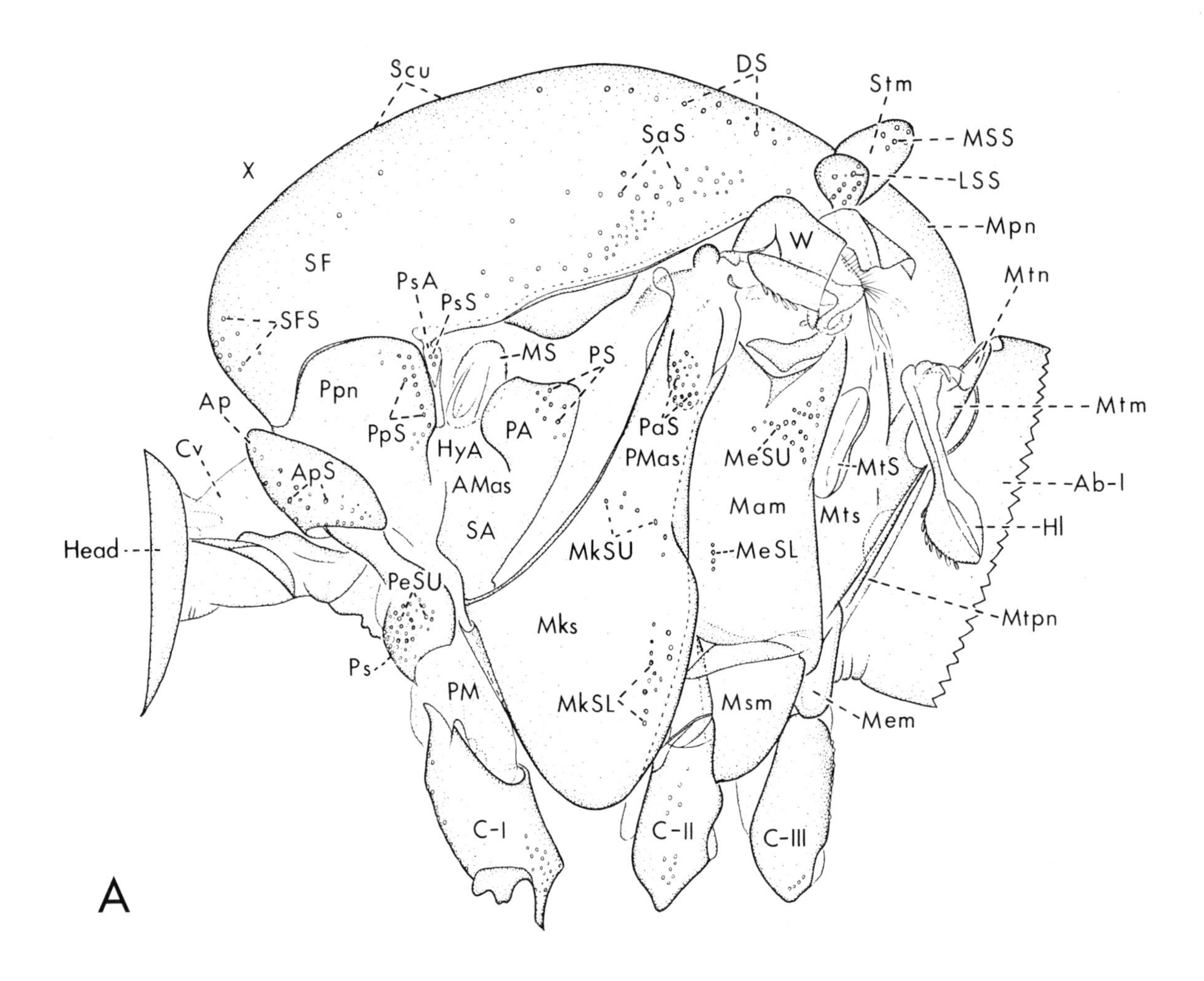


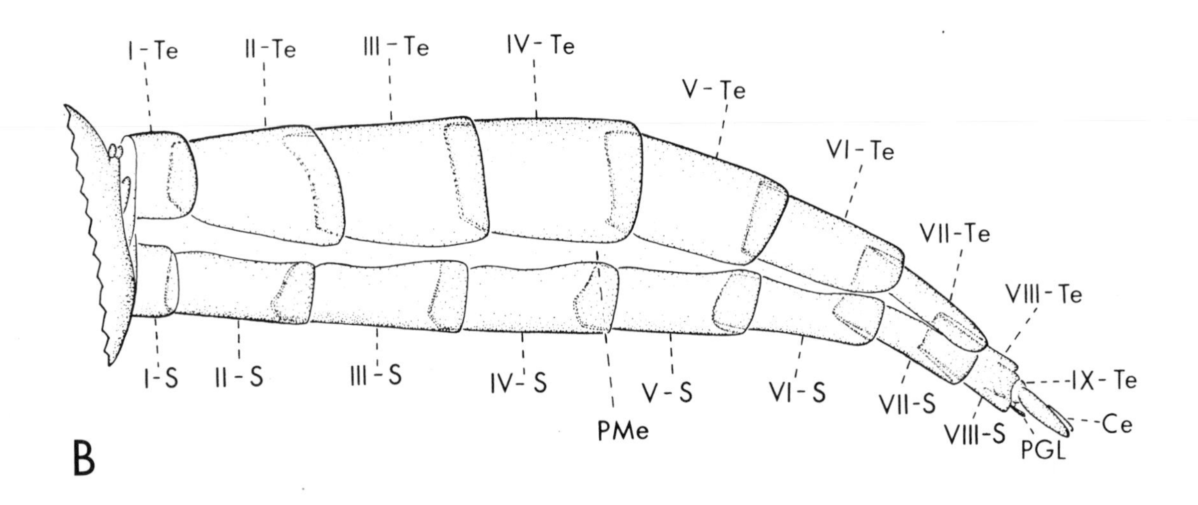


Plate 4. Thorax and abdomen of adult female mosquito. *A*. Lateral view of thorax; *B*. Lateral view of abdomen.

# Keys to the Adult Female Mosquitoes of North America, North of Mexico

## Key to the Genera of Adult Females

1. Proboscis long and strongly recurved (fig. 1); posterior border of wing distinctly emarginate at apex of $Cu_2$ (fig. 2) ................................................ *Toxorhynchites r. rutilus* (plate 42D)
*Toxorhynchites r. septentrionalis* (plate 37B)
*Toxorhynchites moctezuma* (plate 42C)

   Proboscis not so long and only slightly recurved, if at all (fig. 3); wing border rounded or slightly emarginate at apex of vein $Cu_2$ (fig. 4) ................................................................ 2

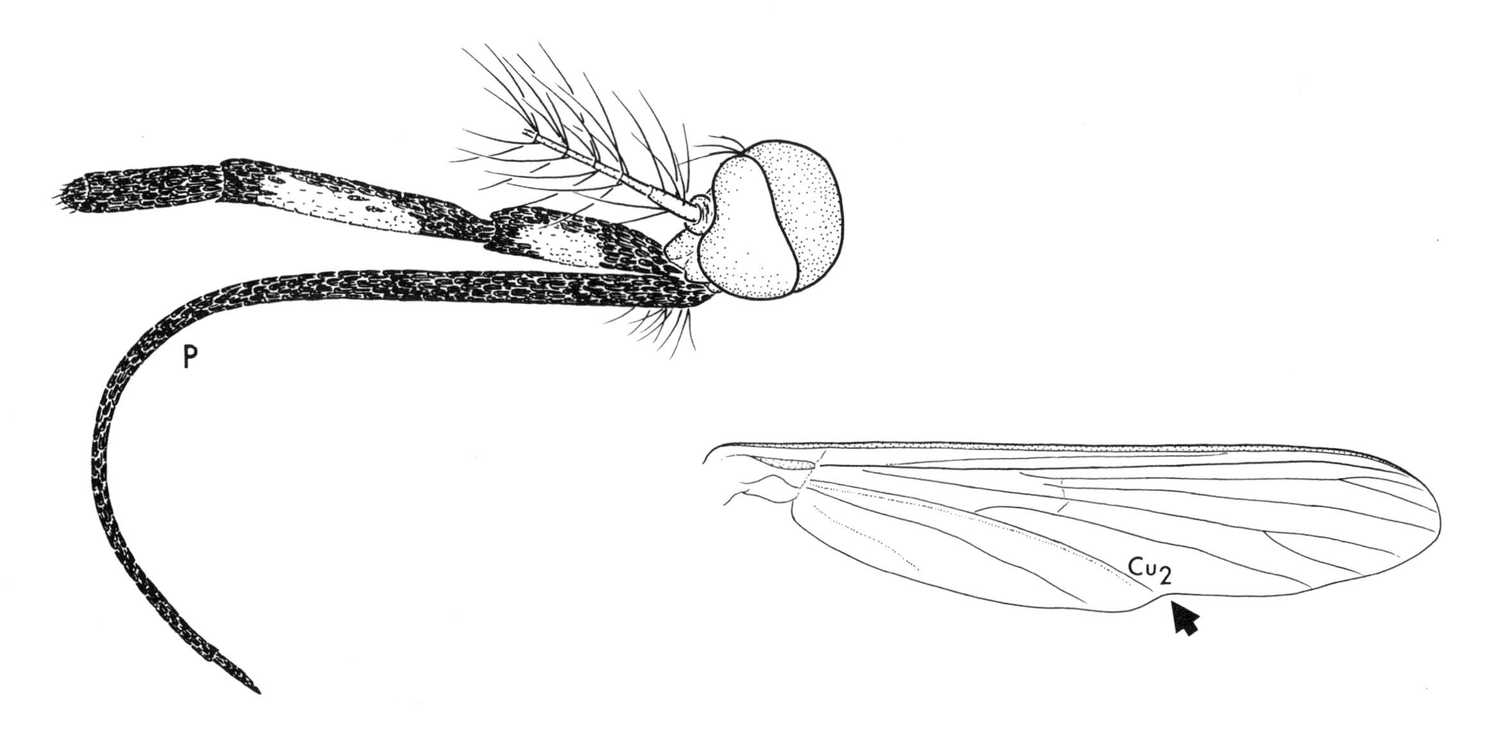


Fig. 1. Lateral view of head: *Tx. r. septentrionalis*

Fig. 2. Dorsal view of wing: *Tx. r. septentrionalis*

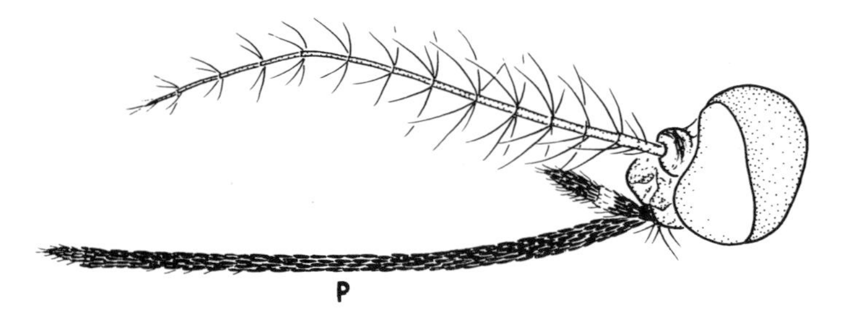


Fig. 3. Lateral view of head: *Ae. vexans*

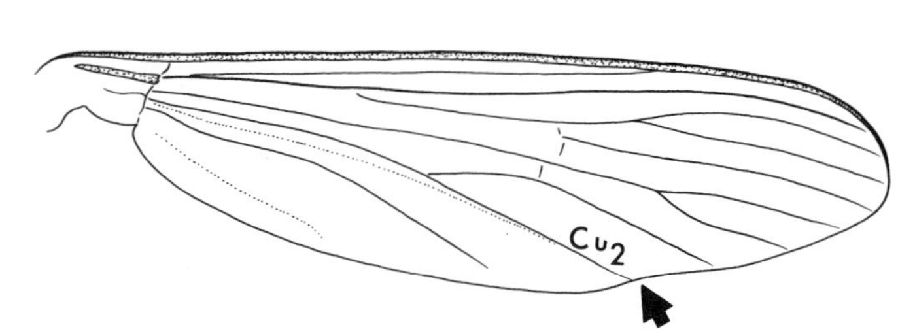


Fig. 4. Dorsal view of wing: *Ae. vexans*

2(1). Scutellum evenly rounded, with setae more or less evenly distributed (fig. 5); maxillary palpus about as long as proboscis (fig. 6) .......... *Anopheles*

Scutellum trilobed, with setae in 3 distinct groups (fig. 7); maxillary palpus shorter than proboscis (fig. 8) .......... 3

3(2). Mesopostnotum with setae (fig. 9); base of hindcoxa in line with base of mesomeron or slightly dorsad (fig. 10) .......... *Wyeomyia*

Mesopostnotum without setae (fig. 11); base of hindcoxa distinctly ventral to base of mesomeron (fig. 12) .......... 4

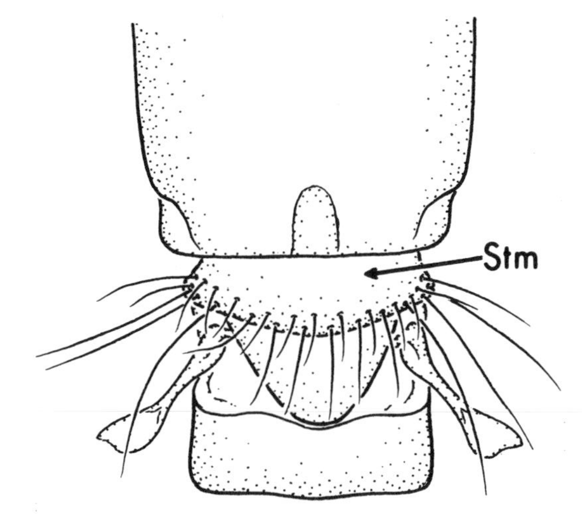


Fig. 5. Posterior dorsal view of thorax: *An. quadrimaculatus*

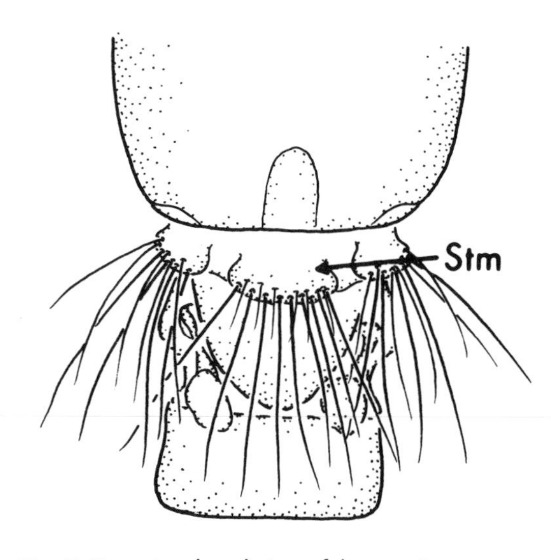


Fig. 7. Posterior dorsal view of thorax: *Ae. vexans*

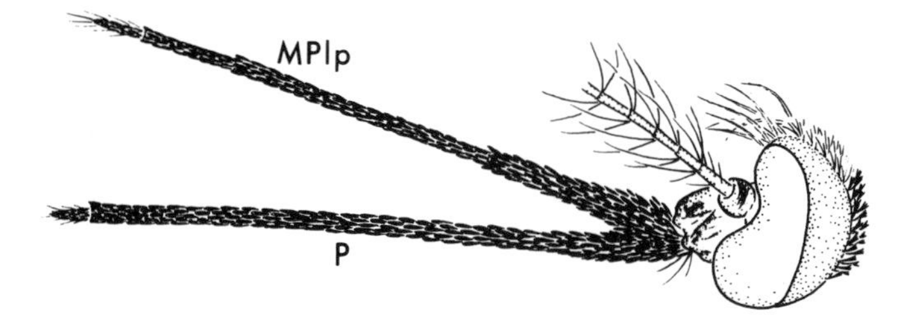


Fig. 6. Lateral view of head: *An. quadrimaculatus*

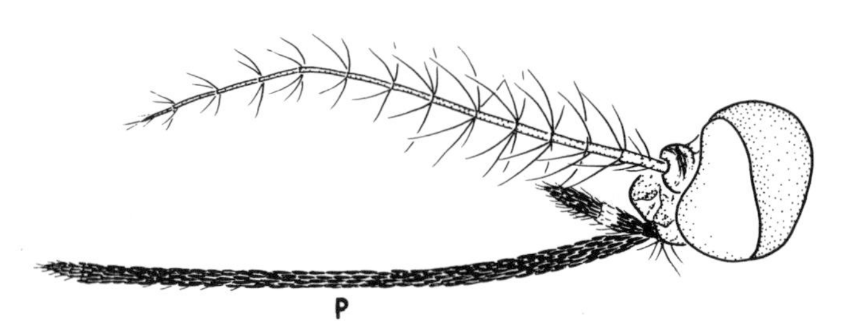


Fig. 8. Lateral view of head: *Ae. vexans*

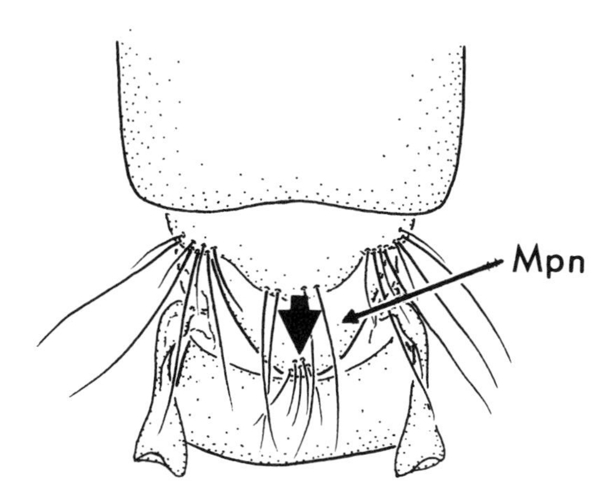


Fig. 9. Posterior dorsal view of thorax: *Wy. smithii*

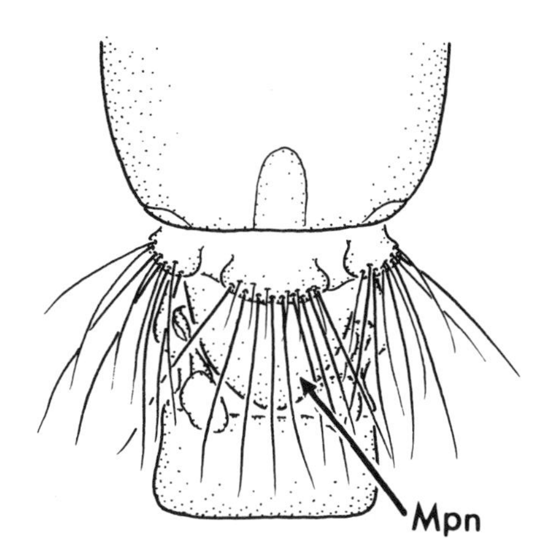


Fig. 11. Posterior dorsal view of thorax: *Ae. vexans*

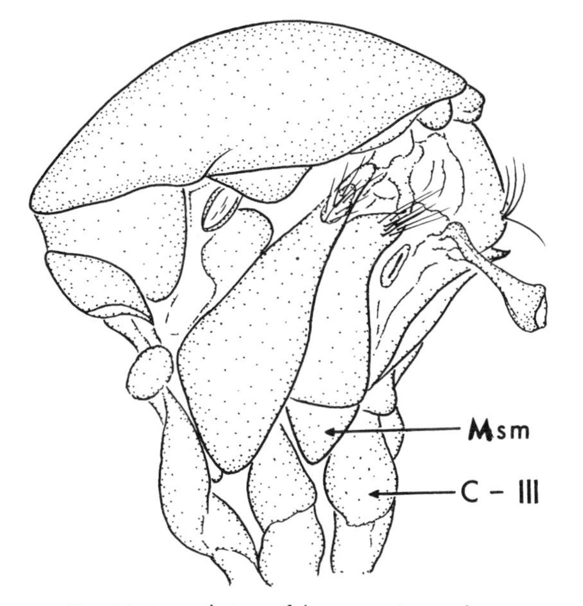


Fig. 10. Lateral view of thorax: *Wy. smithii*

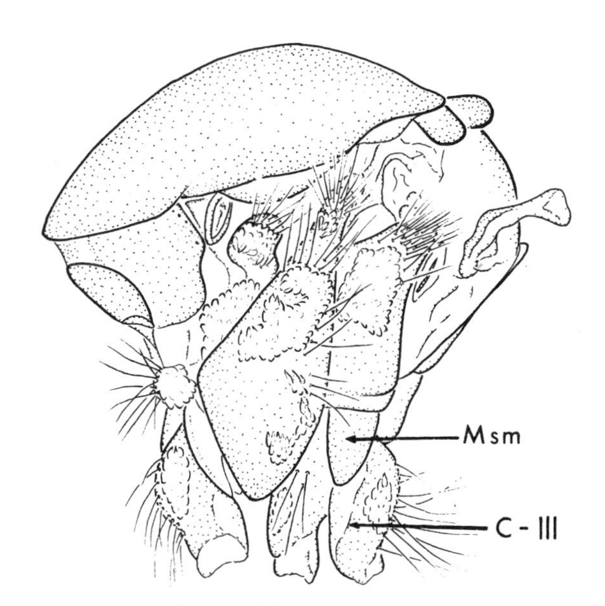


Fig. 12. Lateral view of thorax: *Ae. vexans*

4(3). Cell $R_2$ of wing shorter than vein $R_{2+3}$ (fig. 13); thorax usually with lines of iridescent blue scales (fig. 14) .......... *Uranotaenia*

Cell $R_2$ at least as long as vein $R_{2+3}$ (fig. 15); iridescent blue scales absent on thorax (fig. 16) .......... 5

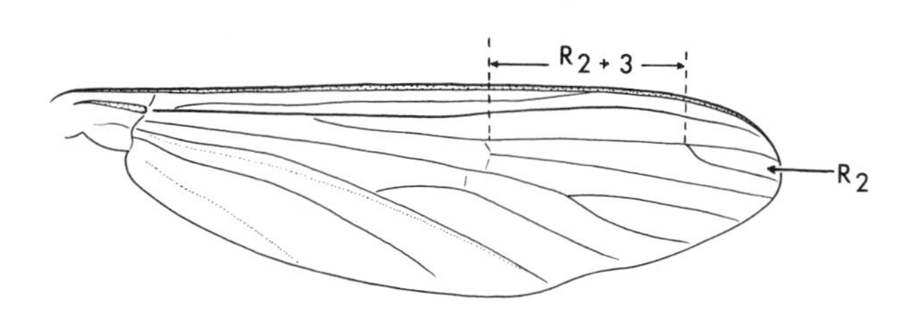


Fig. 13. Dorsal view of wing: *Ur. sapphirina*

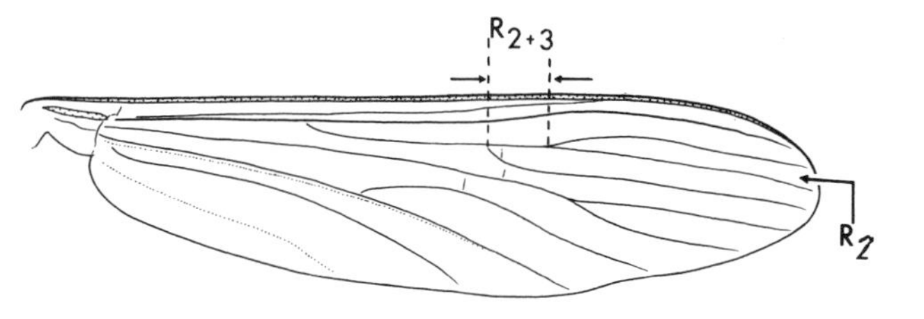


Fig. 15 Dorsal view of wing: *Cx. pipiens*

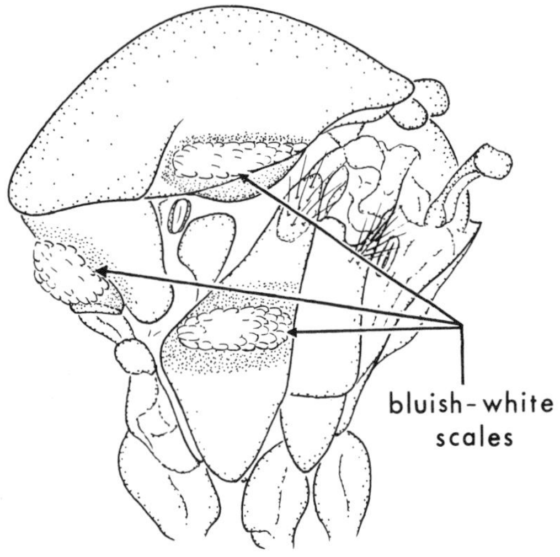

Fig. 14. Lateral view of thorax: Ur. *sapphirina*

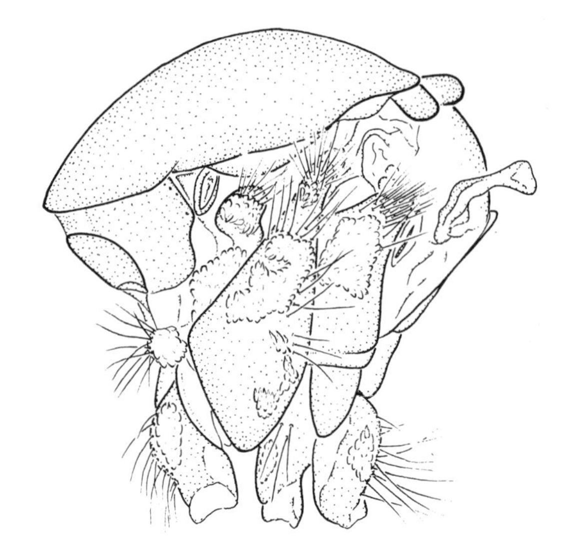

Fig. 16. Lateral view of thorax: *Ae. vexans*

5(4). Postspiracular setae present (fig. 17) ........................................................................... 6

Postspiracular setae absent (fig. 18) ........................................................................... 8

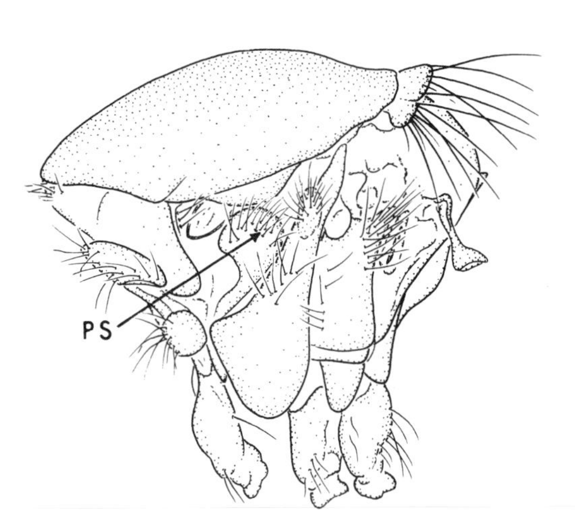


Fig. 17. Lateral view of thorax: *Ps. ciliata*

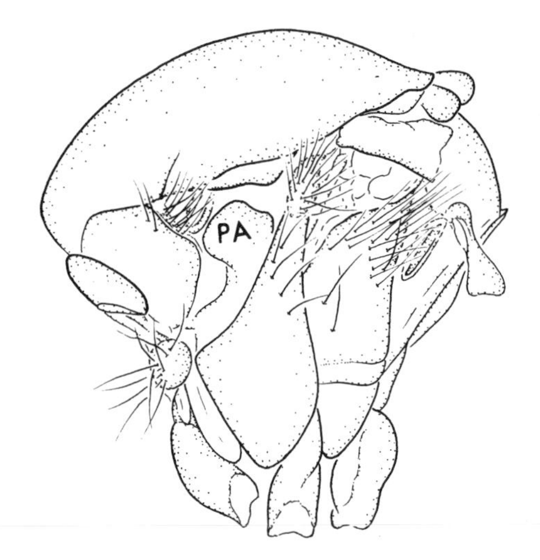


Fig. 18. Lateral view of thorax: *Cs. inornata*

6(5). Apex of abdomen bluntly rounded in dorsal view (fig. 19); most scales on dorsal surface of wing very broad (fig. 20) ........................................................................... *Mansonia*

Apex of abdomen tapering to a point in dorsal view, segment VII markedly narrower than VI (fig. 21); dorsal wing scales long and slender, at least in veins $R_s$ and M (fig. 22) ............. 7

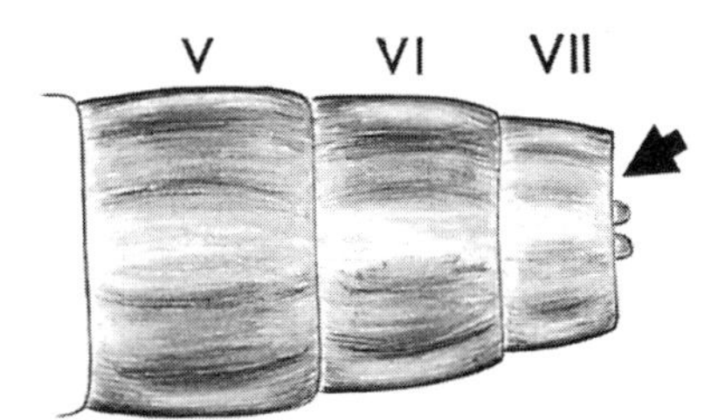


Fig. 19. Dorsal view of abdomen: *Ma. titillans*

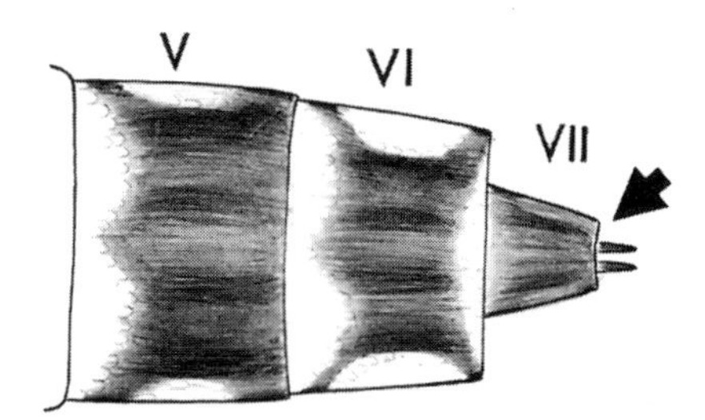


Fig. 21. Dorsal view of abdomen: *Ae. vexans*

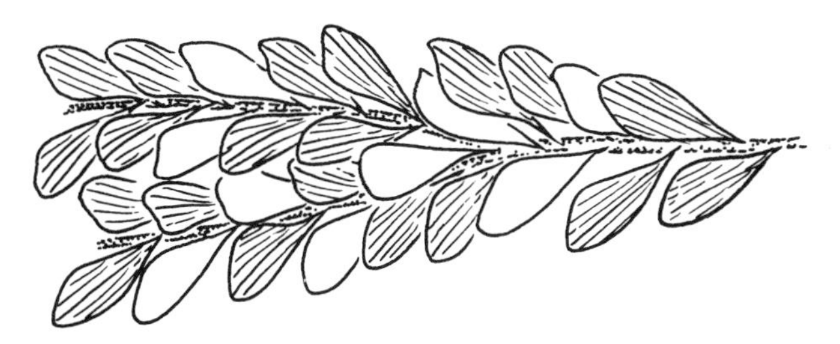

Fig. 20. Dorsal view of some veins: *Ma. titillans*

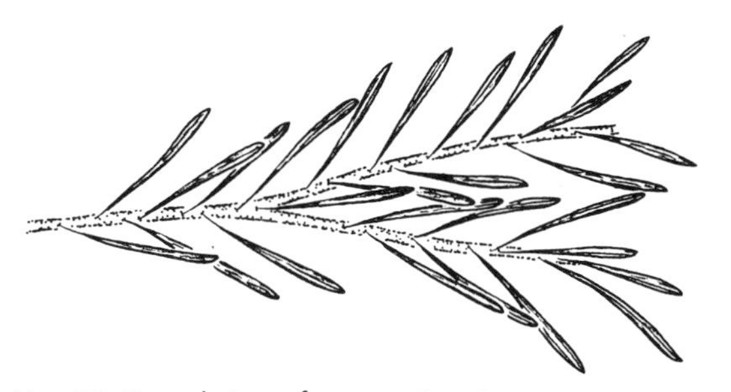

Fig. 22. Dorsal view of some veins: *Ae. vexans*

7(6). Prespiracular setae present (fig. 23); pale transverse bands or lateral patches, when present, apical on abdominal terga (fig. 24) .................................................................. *Psorophora*

Prespiracular setae absent (fig. 25); pale transverse bands or lateral patches basal on abdominal terga (fig. 26) .............................................................................................. *Aedes* (in part) *Ochlerotatus*

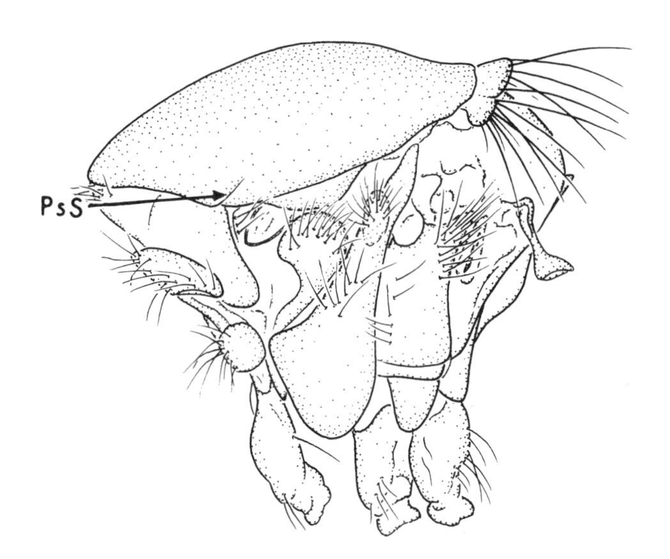


Fig. 23. Lateral view of thorax: *Ps. ciliata*

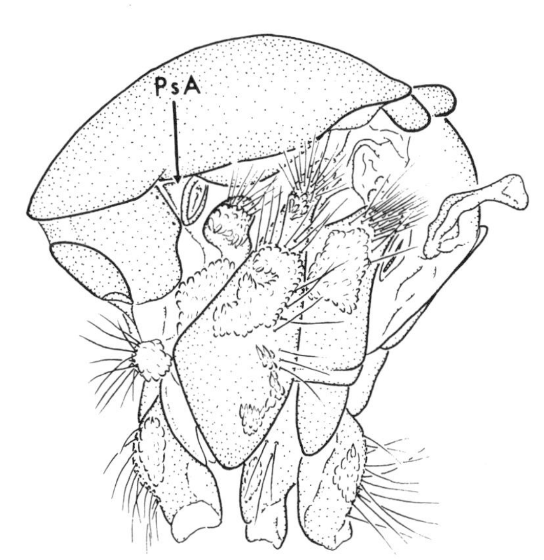


Fig. 25. Lateral view of thorax: *Ae. vexans*

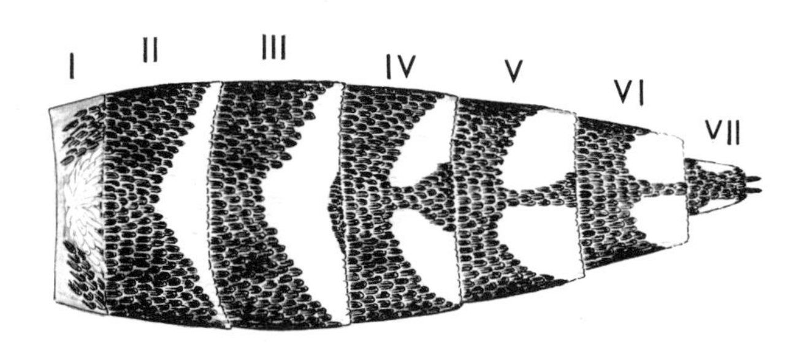


Fig. 24. Dorsal view of abdomen: *Ps. cyanescens*

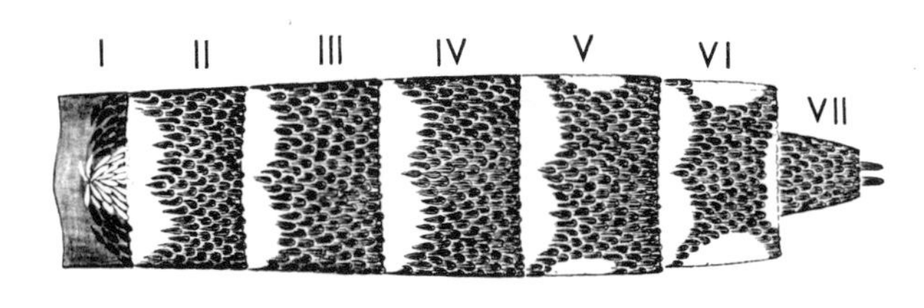


Fig. 26. Dorsal view of abdomen: *Ae. vexans*

8(5). Prespiracular setae present (fig. 27); base of wing vein Sc with row of setae ventrally (fig. 28) .................................................................................................................. *Culiseta*

Prespiracular and vein Sc setae absent (figs. 29, 30) ........................................................ 9

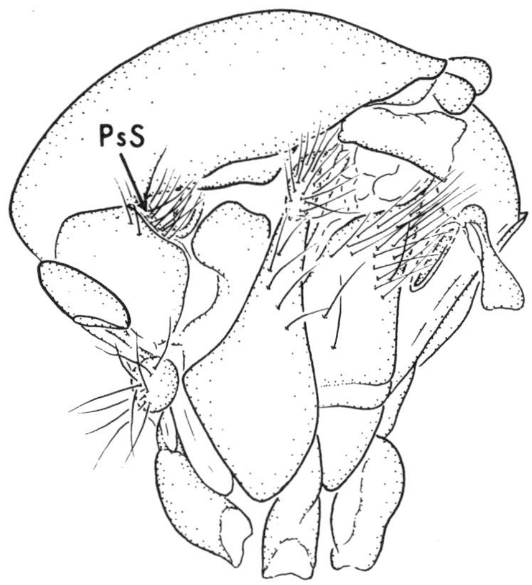

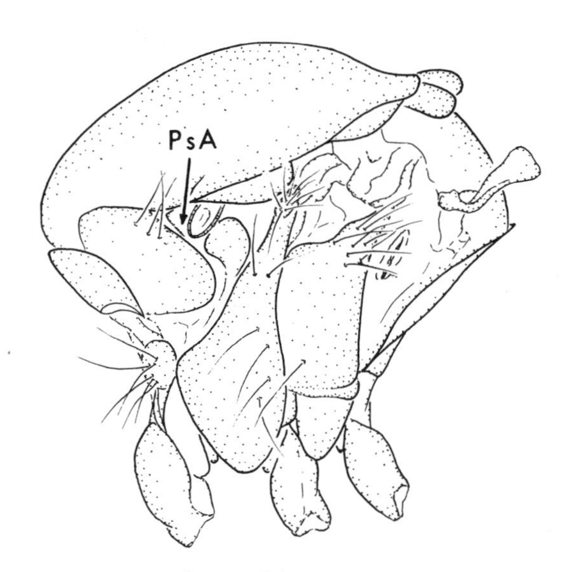


Fig. 27. Lateral view of thorax: *Cs. inornata*

Fig. 29. Lateral view of thorax: *Cx. pipiens*

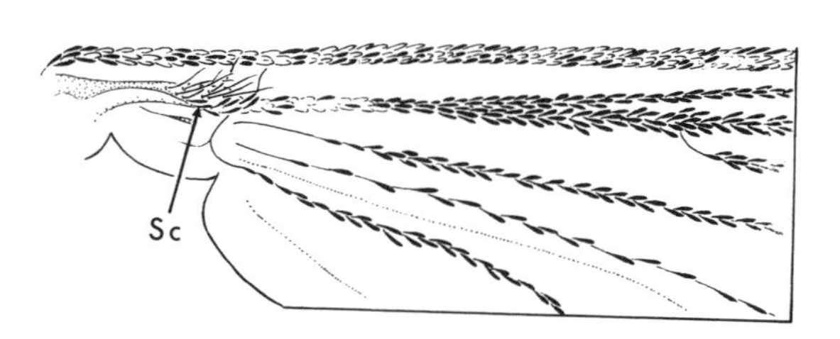


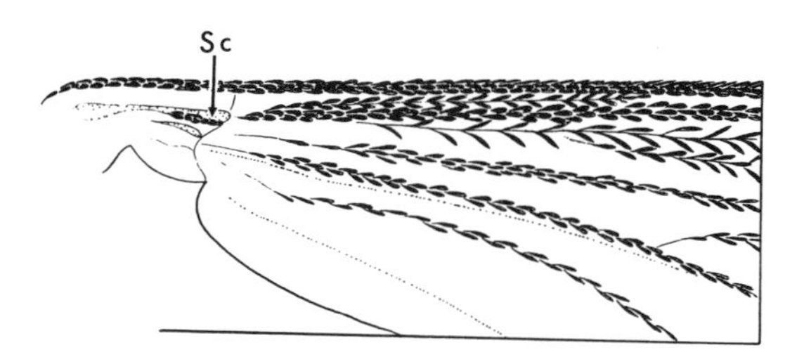


Fig. 28. Ventral view of basal half of wing: *Cs. inornata*

Fig. 30. Ventral view of basal half of wing: *Cx. pipiens*

9(8). Scutum covered with broad flat metallic scales; antepronotum large, approaching middorsally (fig. 31) .................................................................. *Haemagogus equinus* (plate 40D)

Scutal ornamentation not of broad flat metallic scales; antepronotum small, not approaching middorsally (fig. 32) .......................................................................................................... 10

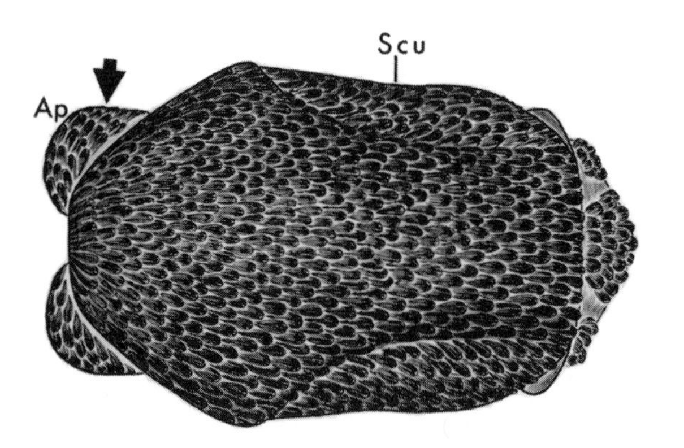


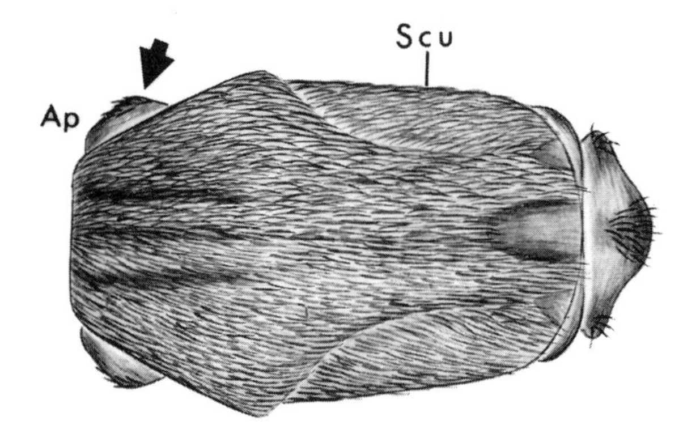


Fig. 31. Dorsal view of thorax: *Hg. equinus*

Fig. 32. Dorsal view of thorax: *Cx. pipiens*

10(9). Scutum with narrow lines of pale scales (fig. 33); tarsomere 1 of fore- and midlegs longer than other 4 tarsomeres combined, tarsomere 4 very short, about as long as wide (fig. 34) ........................................................................................................ *Orthopodomyia*

Scutum without narrow lines of pale scales (fig. 35); tarsomere 1 of fore- and midlegs shorter than other 4 combined, tarsomere 4 much longer than wide (fig. 36) ............ 11

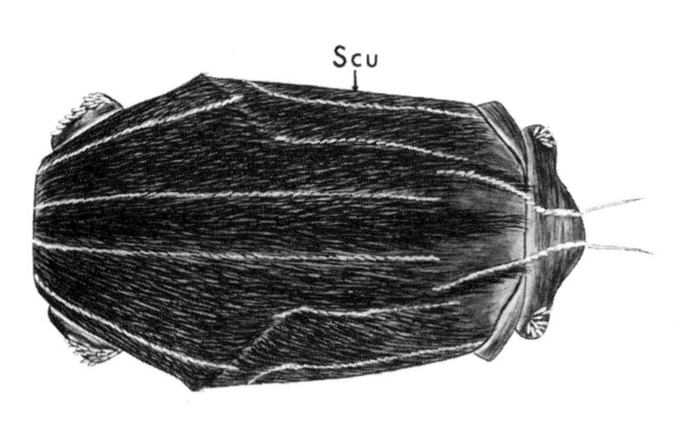


Fig. 33. Dorsal view of scutum: *Or. signifera*

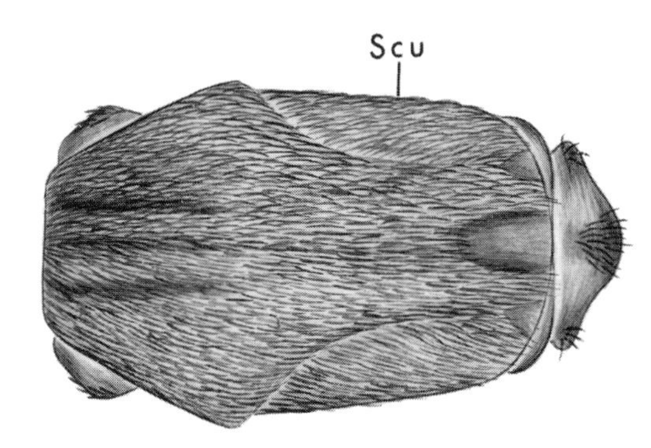


Fig. 35. Dorsal view of thorax: *Cx. pipiens*

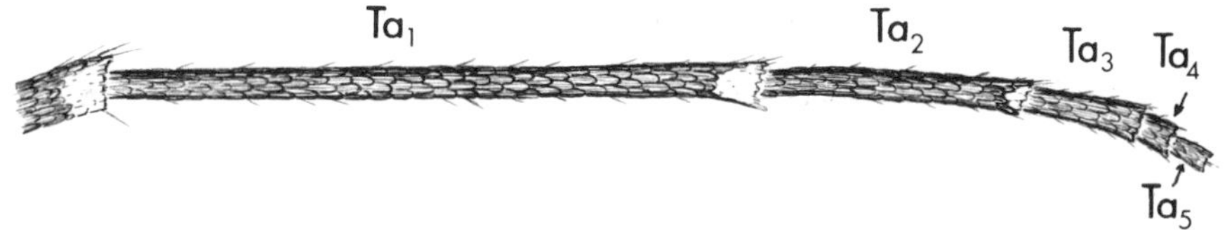


Fig. 34. Enlargement of tarsal segments of midleg: *Or. signifera*

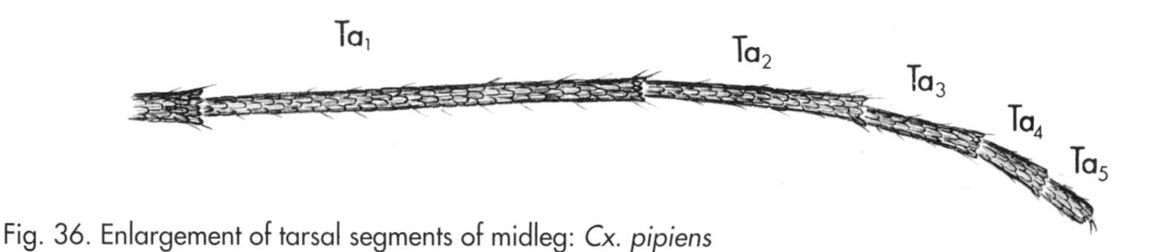


Fig. 36. Enlargement of tarsal segments of midleg: *Cx. pipiens*

11(10). Most scales on dorsal surface of wing very broad (fig. 37) .......................................... ........................................................................ *Coquillettidia perturbans* (plate 32A)

Scales on dorsal surface of wing long and narrow, at least on veins $R_s$ and M (fig. 38) ... ........................................................................................................................ 12

Fig. 37. Dorsal view of some wing veins: *Cq. perturbans*

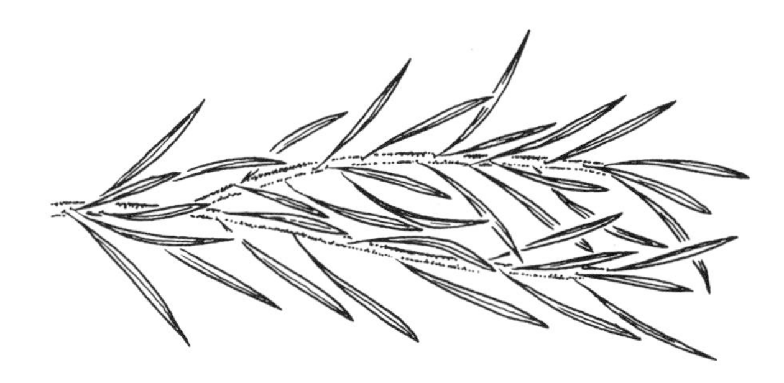

Fig. 38. Dorsal view of some wing veins: *Cx. pipiens*

12(11). Antenna longer than proboscis, flagellomere 1 longer than Flm 2 (fig. 39) .................... ........................................................................................................... *Deinocerites*

Antenna subequal to, or shorter than, proboscis, flagellomere 1 about as long as Flm 2 (fig. 40) ........................................................................................................... 13

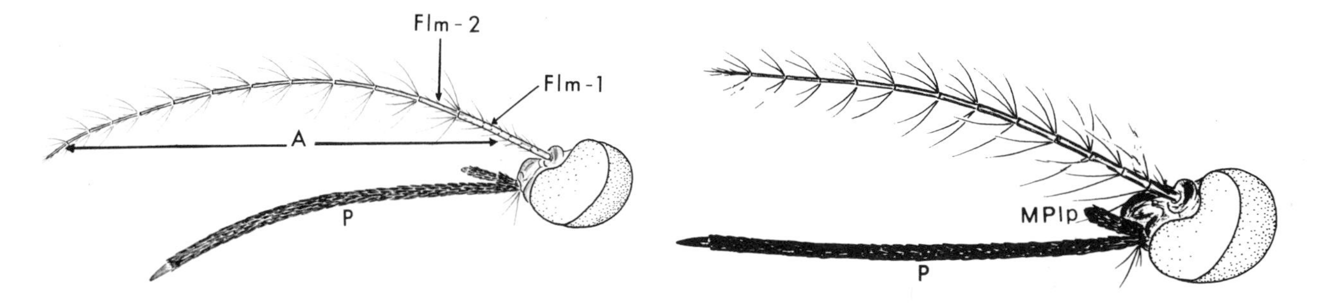


Fig. 39. Lateral view of head: *De. pseudes*

Fig. 40. Lateral view of head: *Cx. pipiens*

13(12). Apex of abdomen tapering to point in dorsal view, terga with basolateral patches of silvery scales (fig. 41); scutum with pattern of black, brown, and golden scales (fig. 42) (subgenus *Kompia*) (in part) ........................................................................................ *Ochlerotatus*

Apex of abdomen bluntly rounded in dorsal view, terga with baso- or apicolateral patches of pale white or dingy yellow scales, never silvery (fig. 43); scutum with other pattern of scales (fig. 44) ........................................................................................................ *Culex*

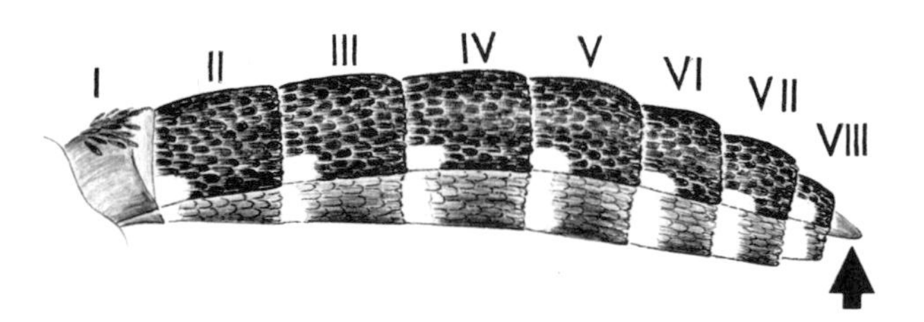


Fig. 41. Lateral view of abdomen: *Oc. purpureipes*

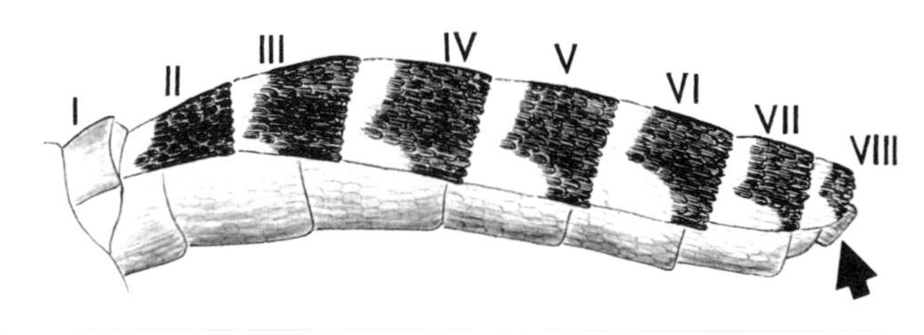


Fig. 42. Dorsal view of thorax: *Cx. pipiens*

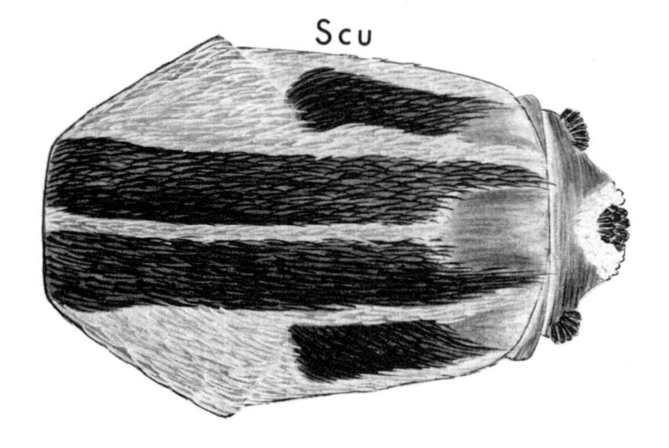


Fig. 43. Lateral view of abdomen: *Oc. purpureipes*

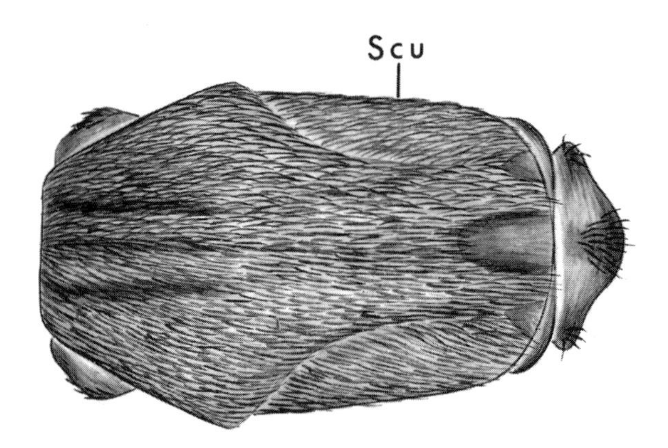


Fig. 44. Dorsal view of thorax: *Cx. pipiens*

## Key to Adult Females of Genera *Aedes* (*Ae.*) and *Ochlerotatus* (*Oc.*)

1. Hindtarsomeres with pale bands (fig. 45) ........................................ 2

   Hindtarsomeres entirely dark scaled (fig. 46) ........................................ 39

Fig. 45. Hindleg: *Oc. excrucians*

Fig. 46. Hindleg: *Oc. triseriatus*

2(1). Hindtarsomeres pale scaled only on basal part of segment (fig. 47) ........................................ 3

   Hindtarsomeres with pale bands both basally and apically, at least on some segments (fig. 48) ........................................ 28

Fig. 47. Hindleg: *Oc. excrucians*

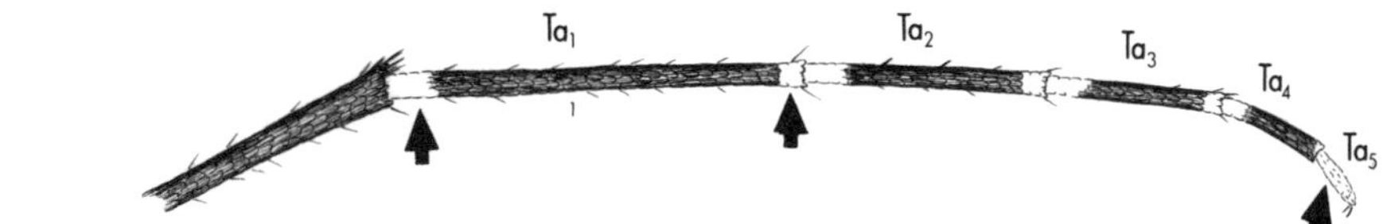


Fig. 48. Hindleg: *Oc. c. canadensis*

3(2). Proboscis with definite pale-scaled band near middle (fig. 49) ........................................ 4

   Proboscis lacking definite pale-scaled band near middle (fig. 50) ........................................ 7

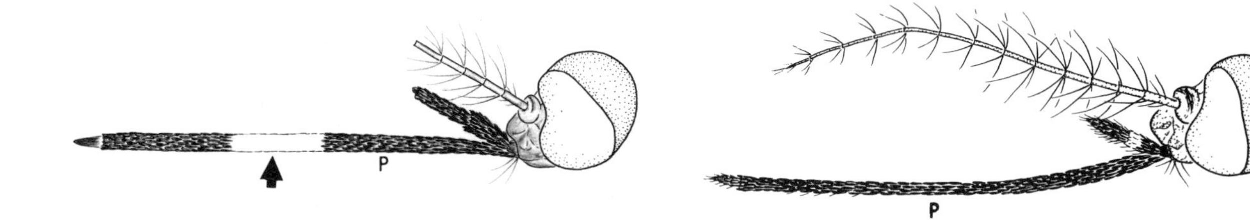


Fig. 49. Lateral view of head: *Oc. sollicitans*

Fig. 50. Lateral view of head: Ae. vexans

4(3). Abdominal terga with transverse basal pale bands, but lacking median longitudinal stripe of pale scales (fig. 51); wing dark scaled (fig. 52) .................... *Oc. taeniorhynchus* (plate 22B)

Abdominal terga with pale-scaled median longitudinal stripe or row of disconnected spots (fig. 53); wing scales either all dark or intermixed dark and pale (fig. 54) ........................ 5

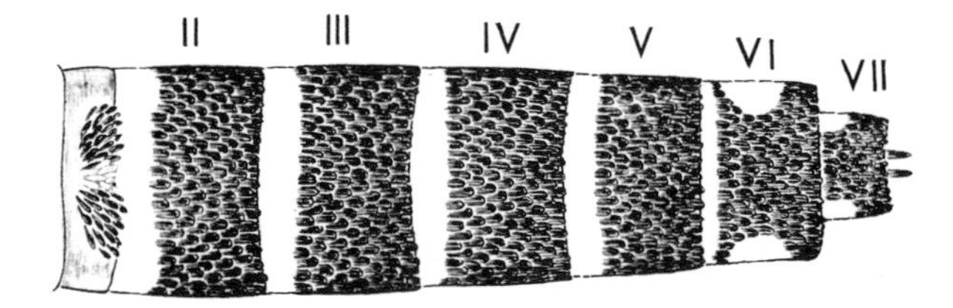


Fig. 51. Dorsal view of abdomen: *Oc. taeniorhynchus*

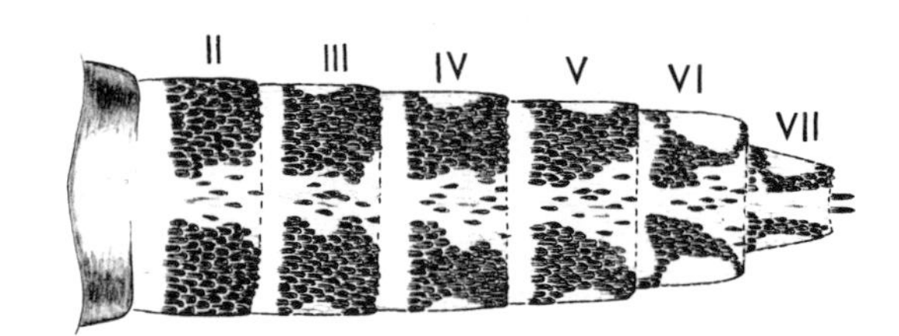


Fig. 53. Dorsal view of abdomen: *Oc. sollicitans*

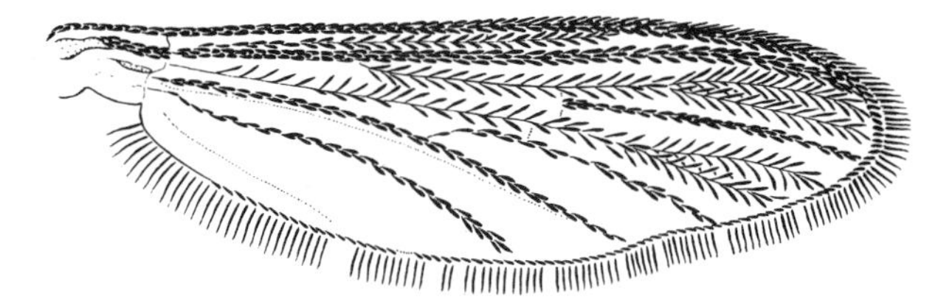

Fig. 52. Dorsal view of wing: *Oc. taeniorhynchus*

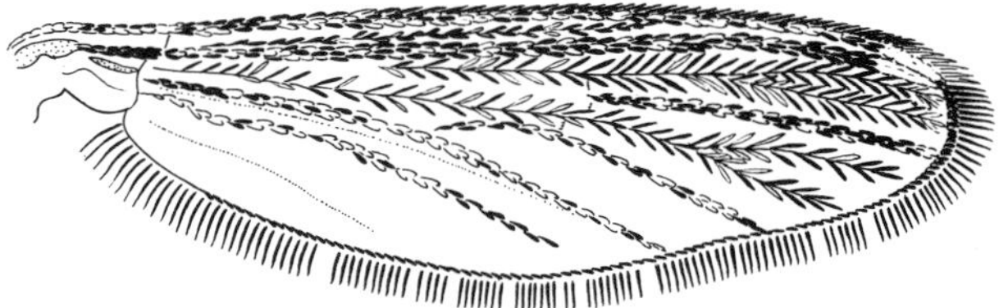

Fig. 54. Dorsal view of wing: *Oc. sollicitans*

5(4). Wing with scales all dark (fig. 55); hypostigmal scales absent (fig. 56) .................................................................................................................... *Oc. mitchellae* (plate 17D)

Wing with dark and pale scales intermixed (fig. 57); hypostigmal scales present .............. 6

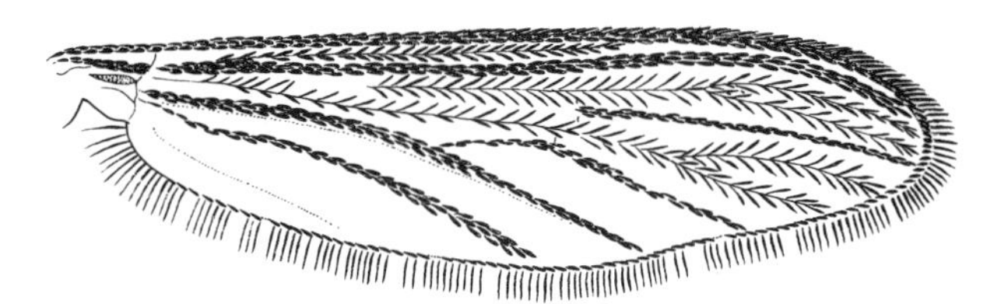

Fig. 55. Dorsal view of wing: *Oc. mitchellae*

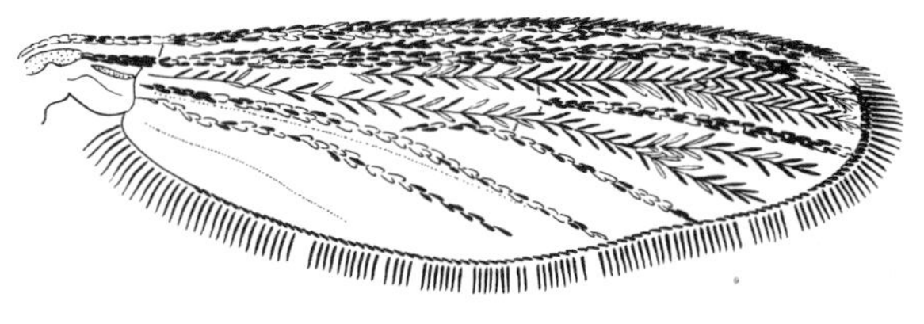

Fig. 57. Dorsal view of wing: *Oc. sollicitans*

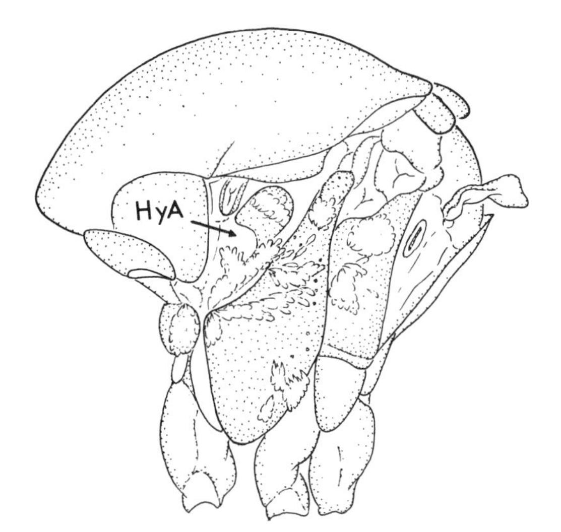


Fig. 56. Lateral view of thorax: *Oc. mitchellae*

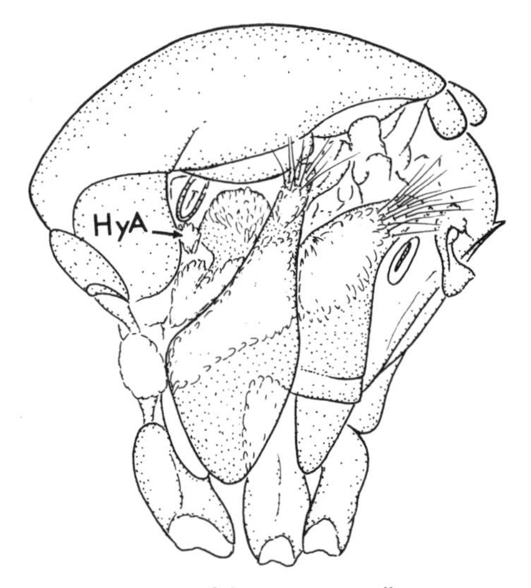


Fig. 58. Lateral view of thorax: *Oc. sollicitans*

6(5). Hindtarsomere 1 with definite yellow-scaled median band (fig. 59); basolateral patches on abdominal terga whitish (fig. 60) .................................................. *Oc. sollicitans* (plate 21)

Hindtarsomere 1 usually without median pale band; if present, then scales whitish (fig. 61); basolateral patches on abdominal terga yellowish scaled (fig. 62) ........................................ .......................................................................... (in part) *Oc. nigromaculis* (plate 18)

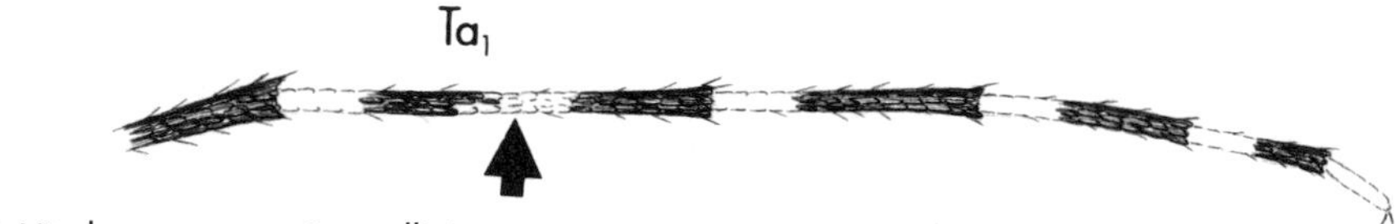


Fig. 59. Hindtarsomeres: *Oc. sollicitans*

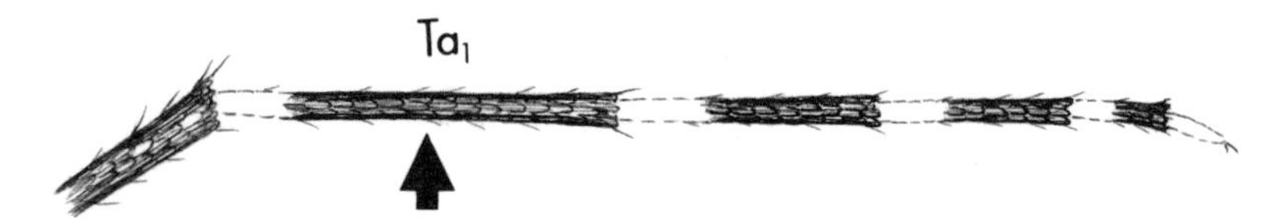


Fig. 61. Hindtarsomeres: *Oc. nigromaculis*

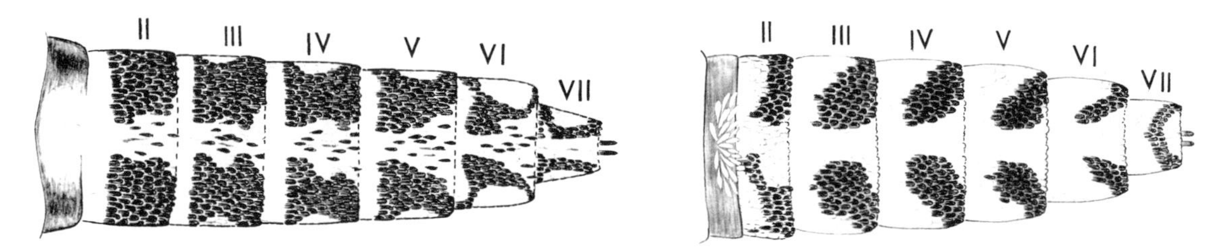


Fig. 60. Dorsal view of abdomen: *Oc. sollicitans*

Fig. 62. Dorsal view of abdomen: *Oc. nigromaculis*

7(3). Abdominal terga VI, VII with large submedian scaleless areas containing setae (fig. 63); mesokatepisternum with 2 narrow diagonal lines of silvery scales (fig. 64) ........................ ............................................................................................... *Oc. papago* (plate 40B)

Abdominal terga VI–VII fully scaled (fig. 65); mesokatepisternum variously scaled, never with 2 narrow diagonal lines of scales (fig. 66) ................................................................ 8

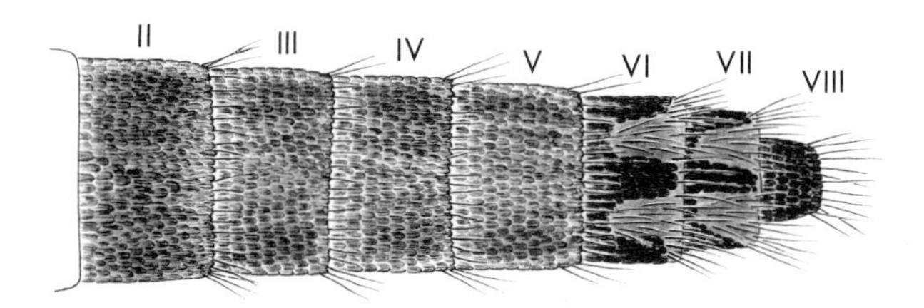


Fig. 63. Dorsal view of abdomen: *Oc. papago*

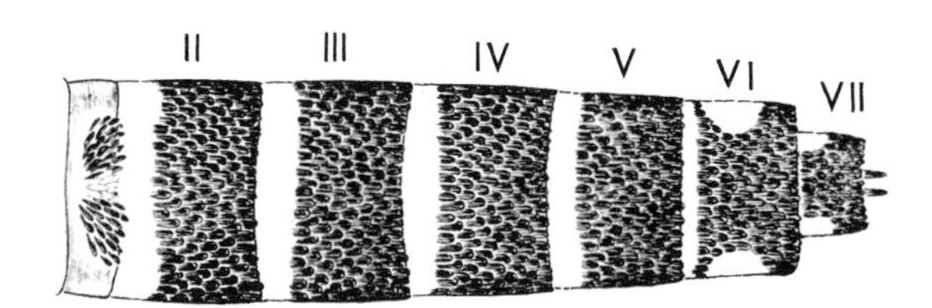


Fig. Dorsal view of abdomen: *Oc. taeniorhynchus*

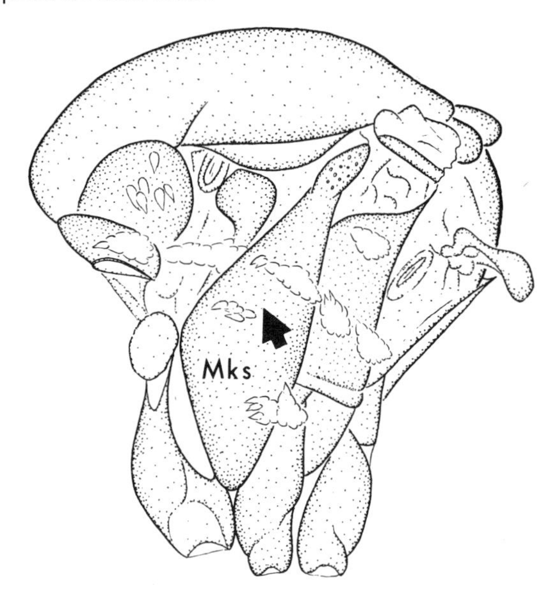


Fig. 64. Lateral view of thorax: *Oc. papago*

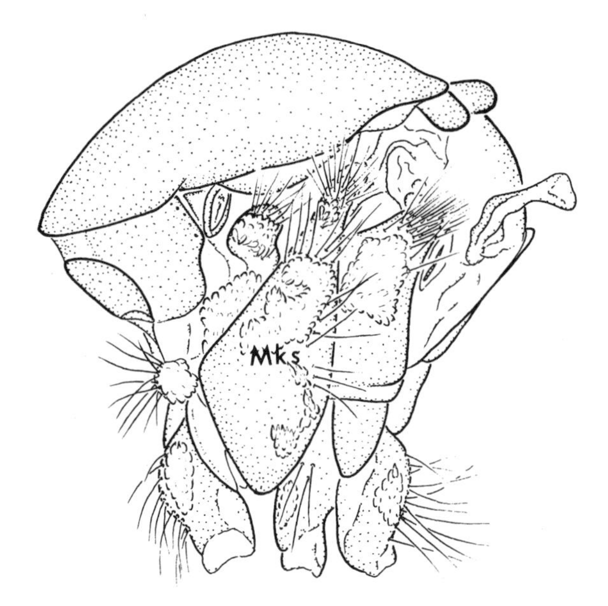


Fig. 66. Lateral view of thorax: *Ae. vexans*

8(7). Scutum with lyre- or modified lyre-shaped marking of silvery or yellow scales on dark-scaled background (fig. 67) .......... 9

Scutum with other scale markings (fig. 68) .......... 10

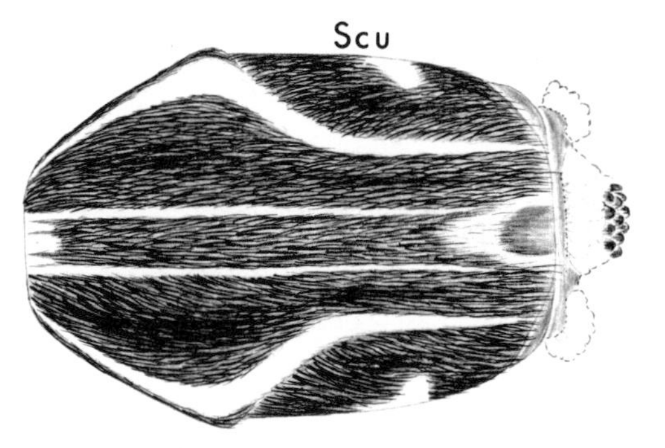


Fig. 67. Dorsal view of scutum: *Ae. aegypti*

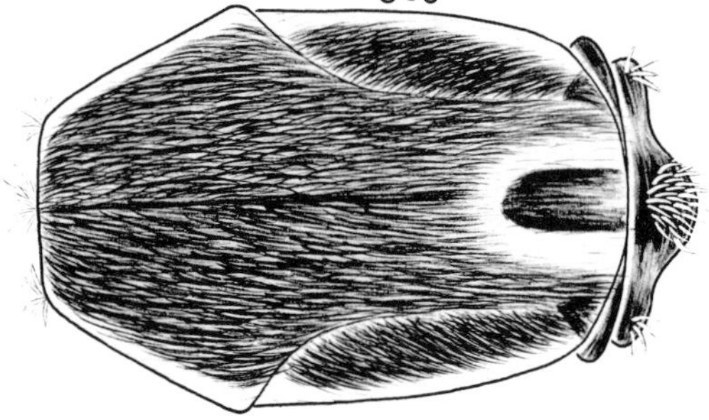

Fig. 68. Dorsal view of scutum: *Oc. c. canadensis*

9(8). Scutum with median longitudinal stripe of yellow scales (fig. 69): abdominal terga III–VII without dorsal pale bands (fig. 70); hindtarsomere 5 dark scaled (fig. 71) .......... *Oc. j. japonicus* (plate 39D)

Scutum without median longitudinal stripe (fig. 72); abdominal terga III–VII with basal transverse pale bands (fig. 73); hindtarsomere 5 pale scaled (fig. 74) ... *Ae. aegypti* (plate 9B)

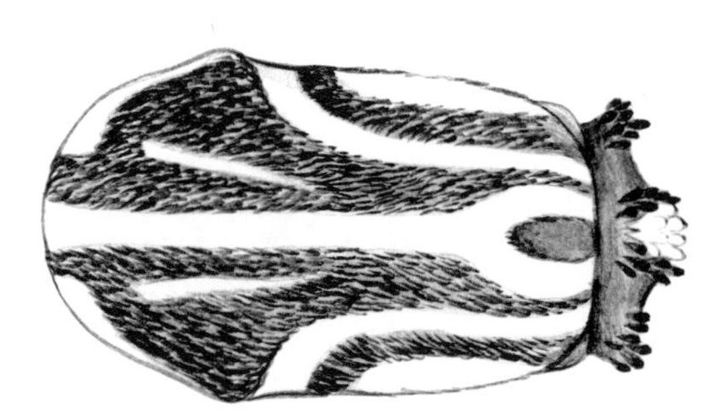

Fig. 69. Dorsal view of scutum: *Oc. j. japonicus*

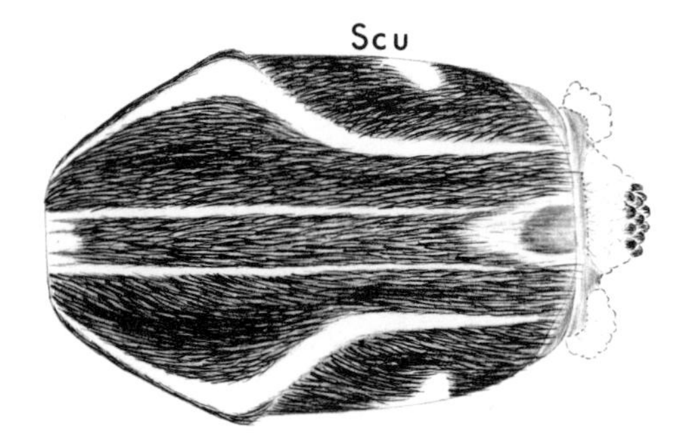


Fig. 72. Dorsal view of scutum: *Ae. aegypti*

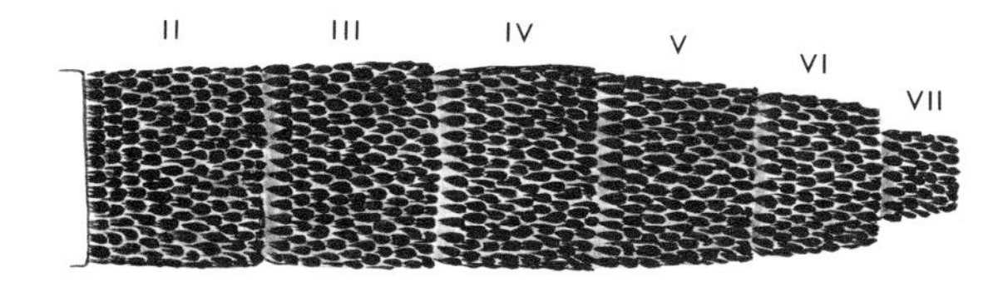


Fig. 70. Dorsal view of abdominal segments III–VII: *Oc. j. japonicus*

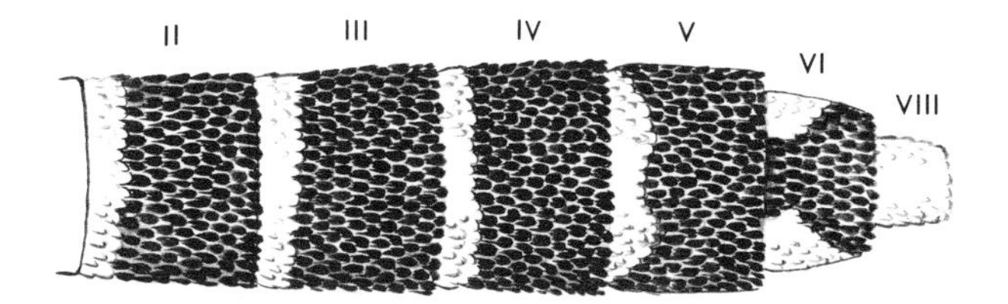


Fig. 73. Dorsal view of abdominal segments III–VII: *Ae. aegypti*

Fig. 71. Hindtarsomeres: *Oc. j. japonicus*

Fig. 74. Hindtarsomeres: *Ae. aegypti*

10(8). Basal 0.5 of hindfemur entirely pale scaled (fig. 75) .................. *Oc. zoosophus* (plate 24)

Basal 0.5 of hindfemur with anterior surface dark scaled or with dark and pale scales intermixed (fig. 76) .......................................................................................................... 11

Fig. 75. Hindleg: *Oc. zoosophus*

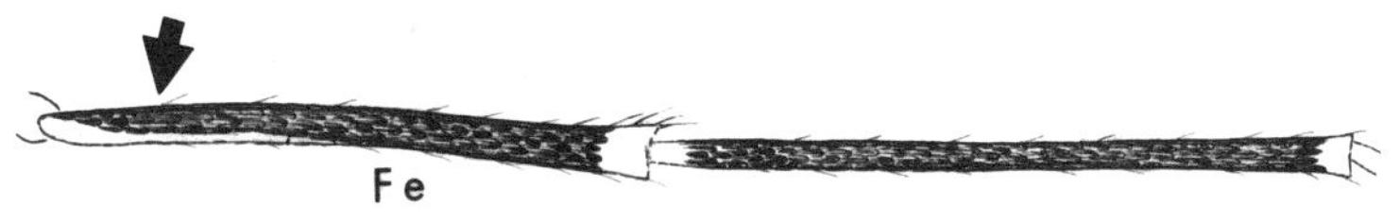


Fig. 76. Hindleg: *Oc. epactius*

11(10). Basal pale bands of hindtarsomeres narrow, 0.2 or less than length of segment 2 (fig. 77) .......................................................................................................... 12

Basal pale bands of hindtarsomeres broad, that on tarsomere 2 more than 0.3 length of segment (fig. 78) .......................................................................................................... 13

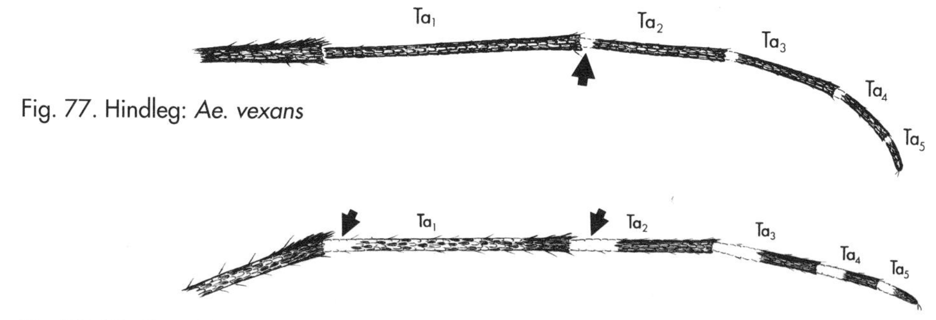


Fig. 77. Hindleg: *Ae. vexans*

Fig. 78. Hindleg: *Oc. excrucians*

12(11). Basal pale bands on abdominal terga II–VI with 2 posterior lobes, tergum VII mostly dark scaled (fig. 79); lower mesepimeral setae absent (fig. 80) ............ *Ae. vexans* (plate 24A)

Basal pale bands on terga II–VI not bilobed nor clearly defined, tergum VII mostly pale scaled (fig. 81); lower mesepimeral setae present (fig. 82) ......... *Oc. cantator* (plate 11C)

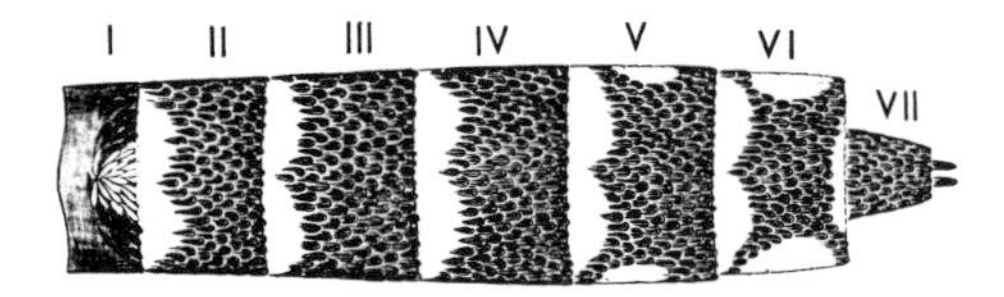


Fig. 79. Dorsal view of abdomen: *Ae. vexans*

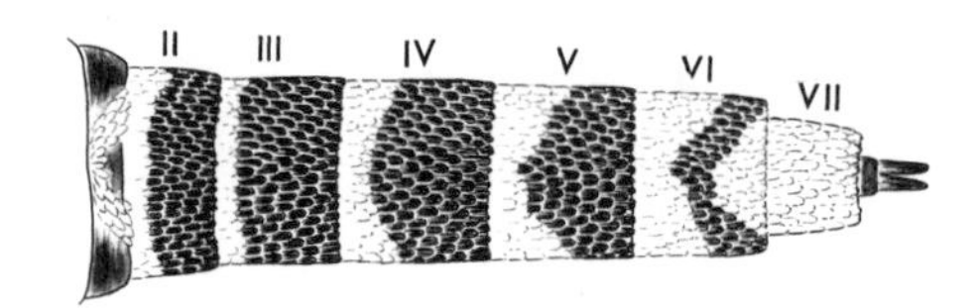


Fig. 81. Dorsal view of abdomen: *Oc. cantator*

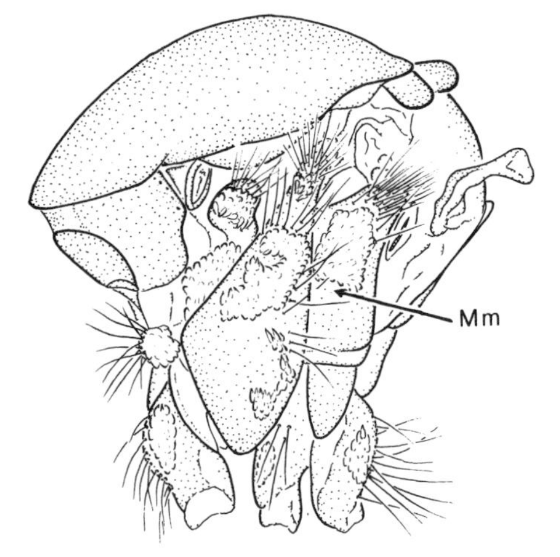


Fig. 80. Lateral view of thorax: *Ae. vexans*

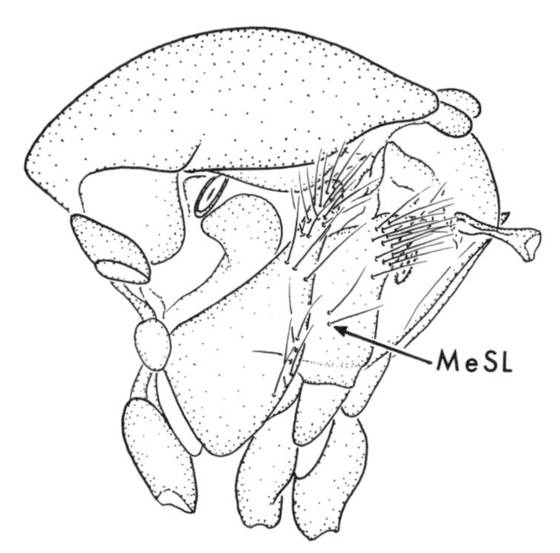


Fig. 82. Lateral view of thorax: *Oc. cantator*

13(11). Wings entirely dark scaled (fig. 83); hindtarsomeres 1–3 with broad pale bands (fig. 84) ........................................................................................................................ 14

Wings with pale and dark scales intermixed (fig. 85), if all dark, then at least hindtarsomeres 1–4 with broad pale bands (fig. 86) ............................................................... 15

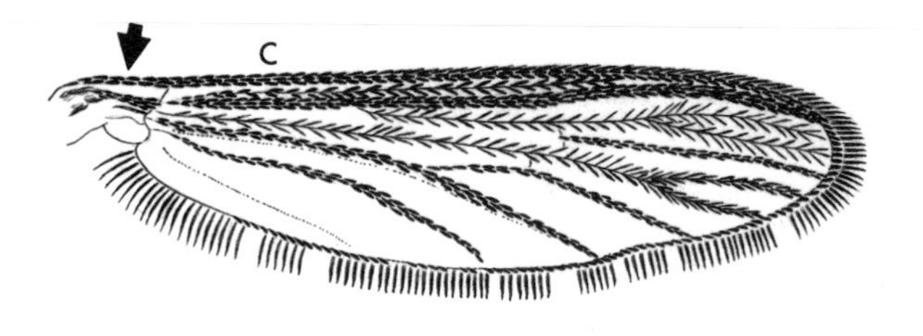


Fig. 83. Dorsal view of wing: *Oc. c. canadensis*

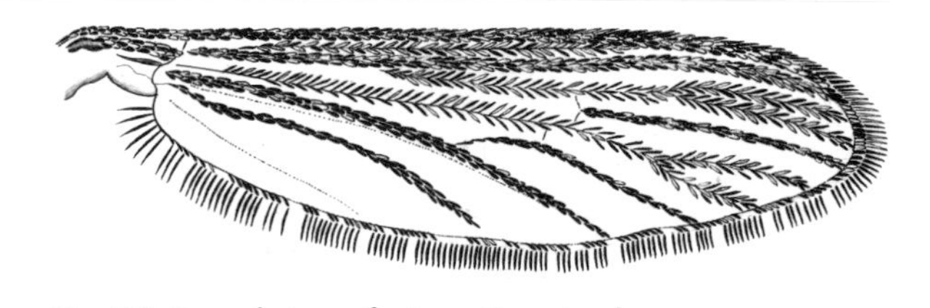

Fig. 85. Dorsal view of wing: *Oc. stimulans*

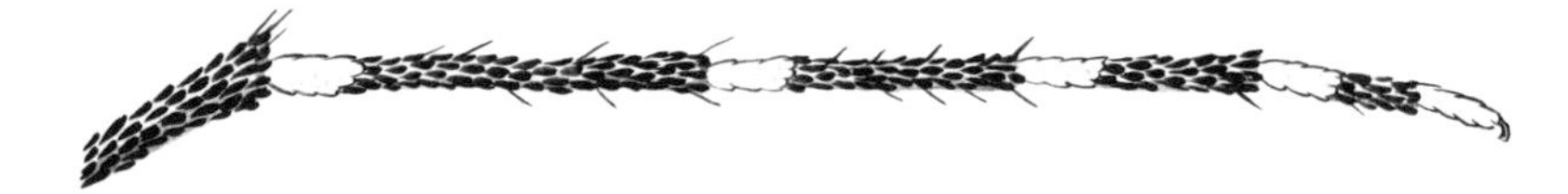

Fig. 84. Hindleg: *Ae. albopictus*

Fig. 86. Hindleg: *Oc. excrucians*

14(13). Scutum with narrow lines of white, creamy, and golden scales (fig. 87); hindtarsomeres 4,5 dark scaled (fig. 88) ........................................................ *Oc. bahamensis* (plate 39A)

Scutum with median narrow stripe of white scales (fig. 89); hindtarsomere 4 with basal white band, 5 entirely white (fig. 90) ...................................... *Ae. albopictus* (plate 9D)

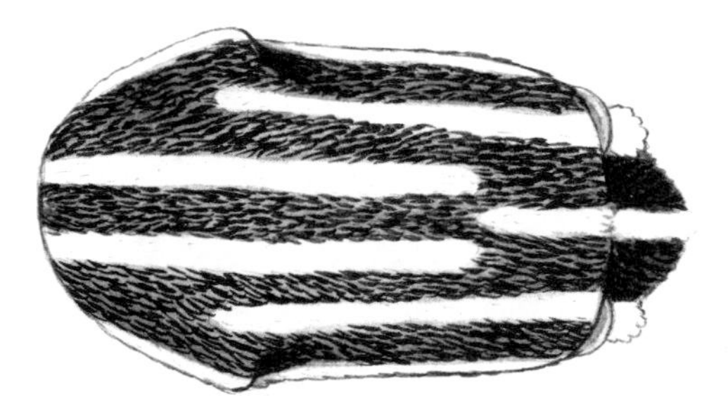

Fig. 87. Dorsal view of scutum: *Oc. bahamensis*

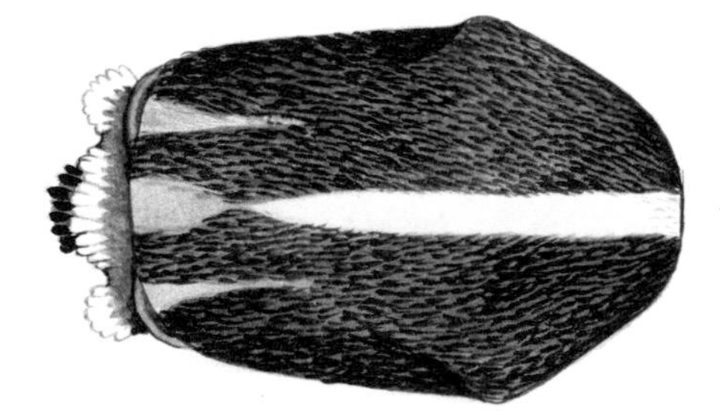

Fig. 89. Dorsal view of scutum: *Ae. albopictus*

Fig. 88. Hindleg: *Oc. bahamensis*

Fig. 90. Same as fig. 84 (Hindleg: *Ae. albopictus*)

15(13). Wing with broad triangular-shaped dark and pale scales, rather evenly intermixed dorsally (fig. 91) .................................................................................................... 16

At least some dorsal wing scales narrow, with dark and pale scales; usually unevenly distributed (fig. 92) ........................................................................................ 17

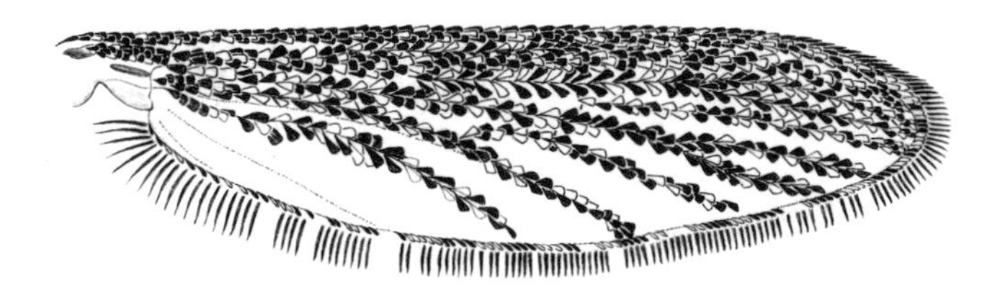

Fig. 91. Dorsal view of wing: *Oc. grossbecki*

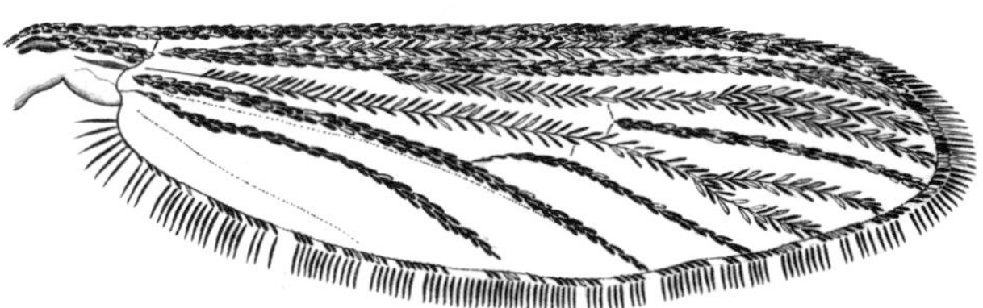

Fig. 92. Dorsal view of wing: *Oc. stimulans*

16(15). Proboscis with many dark and pale scales intermixed (fig. 93); scutum with mixed brown and pale scales laterally (fig. 94) ........................................... *Oc. squamiger* (plate 40B)

Proboscis with few scattered pale scales on basal 0.5 (fig. 95); scutum with mostly pale scales laterally (fig. 96) ......................................................... *Oc. grossbecki* (plate 15B)

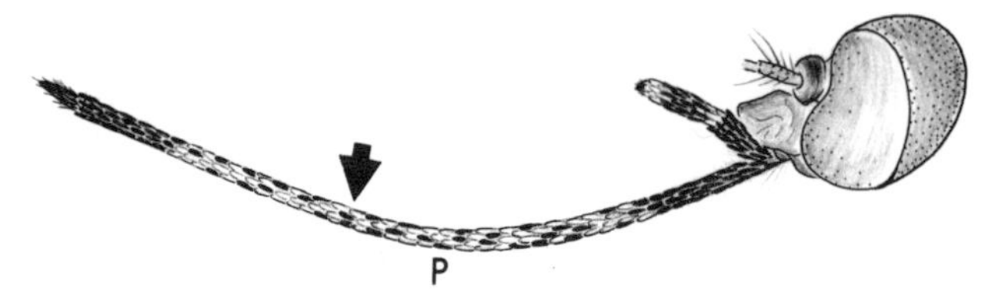


Fig. 93. Lateral view of head: *Oc. squamiger*

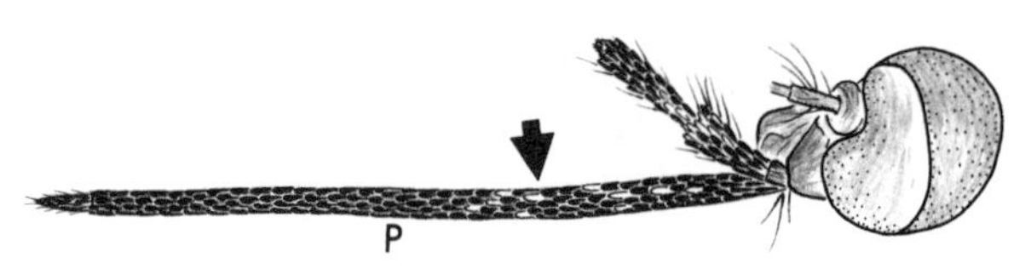


Fig. 95. Lateral view of head: *Oc. grossbecki*

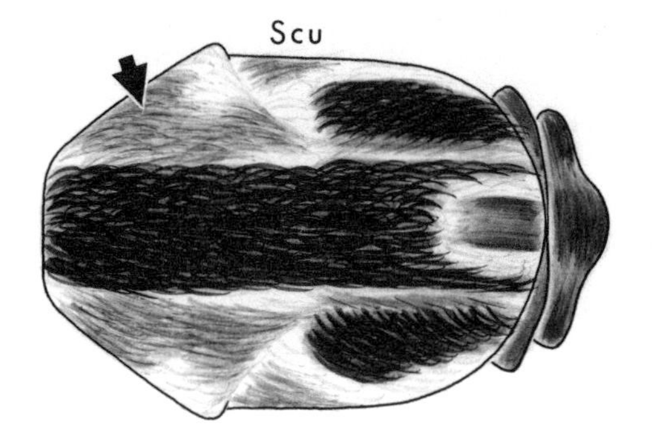


Fig. 94. Dorsal view of scutum: *Oc. squamiger*

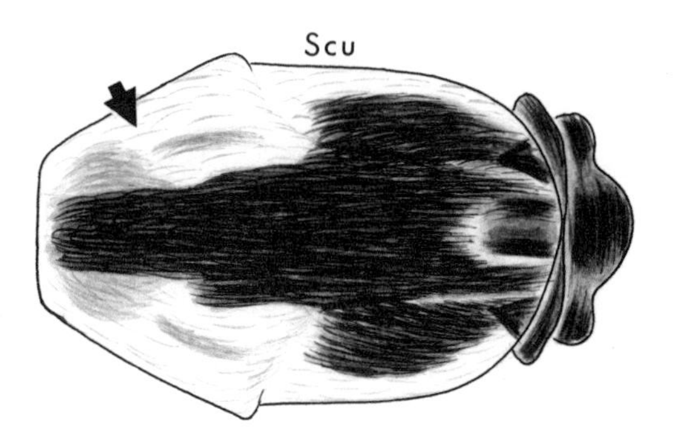


Fig. 96. Dorsal view of scutum: *Oc. grossbecki*

17(15). Palpus dark scaled (fig. 97); abdominal terga with yellowish scales forming median longitudinal stripe (fig. 98) ......................................... (in part) *Oc. nigromaculis* (plate 18C)

Palpus with some pale scales (fig. 99); pale scales on abdominal terga never forming distinct and complete median longitudinal stripe (fig. 100) ............................................ 18

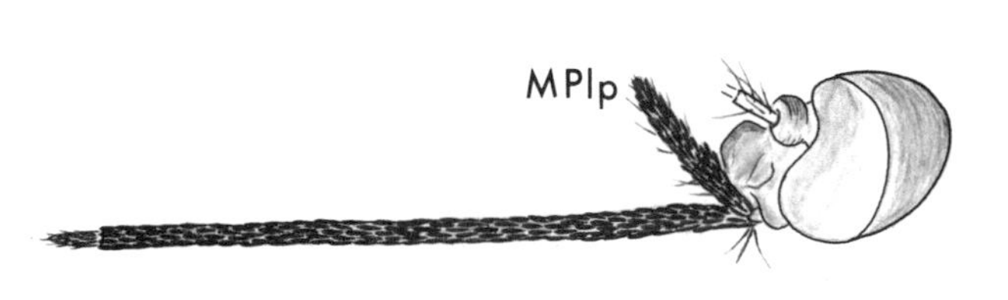


Fig. 97. Lateral view of head: *Oc. nigromaculis*

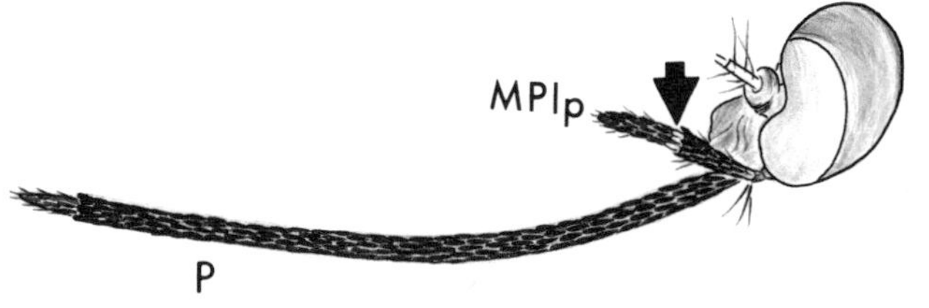


Fig. 99. Lateral view of head: *Oc. increpitus*

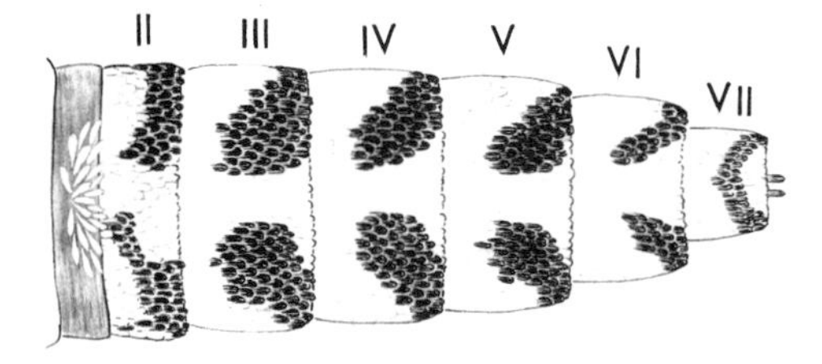


Fig. 98. Dorsal view of abdomen: *Oc. nigromaculis*

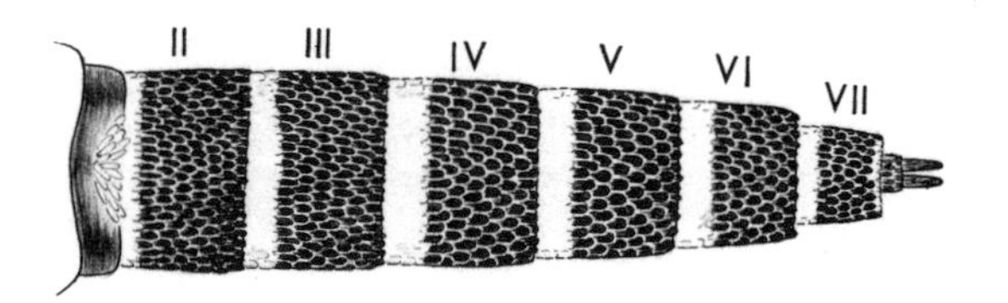


Fig. 100. Dorsal view of abdomen: *Oc. increpitus*

18(17). Abdominal terga entirely clothed with yellow scales (fig. 101) ........................................ ........................................................................................ *Oc. flavescens* (plate 14D)

Abdominal terga with some dark scales, usually with pale-scaled basal bands on some segments (fig. 102) ........................................................................................ 19

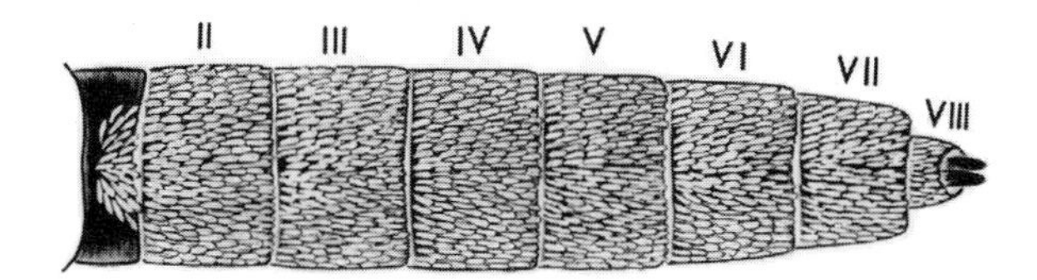


Fig. 101. Dorsal view of abdomen: *Oc. flavescens*

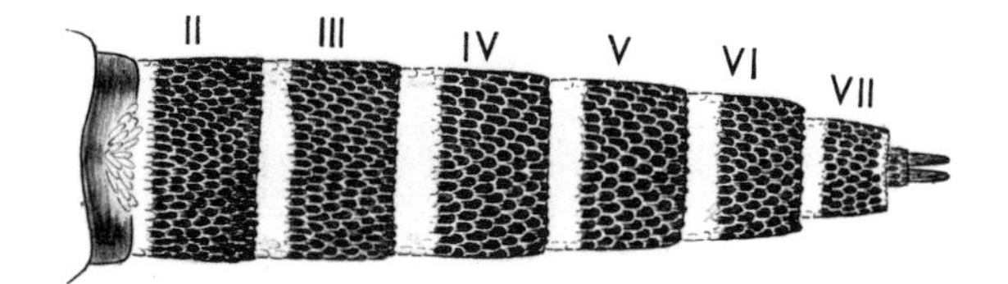


Fig. 102. Dorsal view of abdomen: *Oc. increpitus*

19(18). Foreclaw sharply bent and subparallel to long tooth (fig. 103) ........................................ *Oc. excrucians* (plate 14B)

Foreclaw not sharply bent nor nearly parallel to shorter tooth (fig. 104) .................... 20

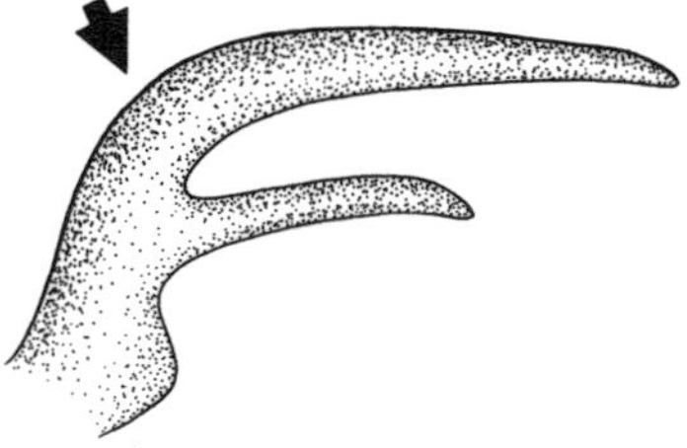

Fig. 103. Foreclaw: *Oc. excrucians* enlarged

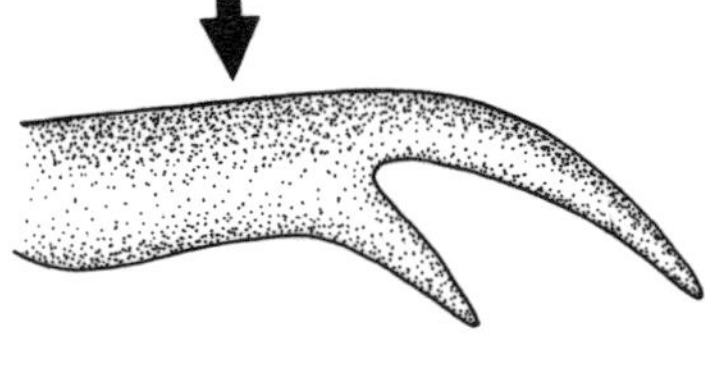

Fig. 104. Foreclaw: *Oc. increpitus* enlarged

20(19). Lower mesepimeral setae absent; mesomeron bare (fig. 105) .................................... 21

Lower mesepimeral setae present, mesomeron with few scales on dorsoposterior corner (fig. 106) ........................................................................................................ 24

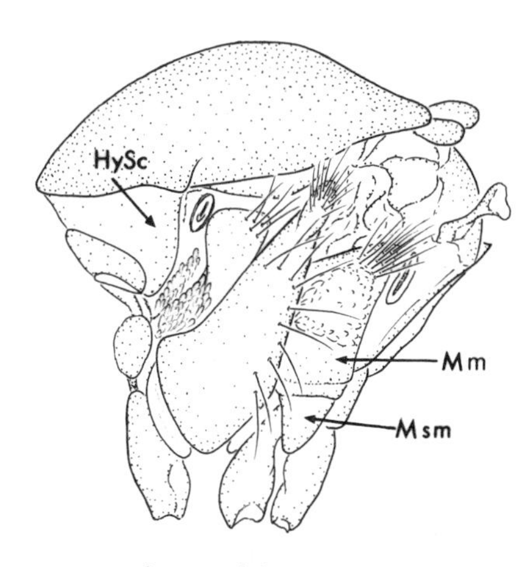


Fig. 105. Lateral view of thorax: *Oc. riparius*

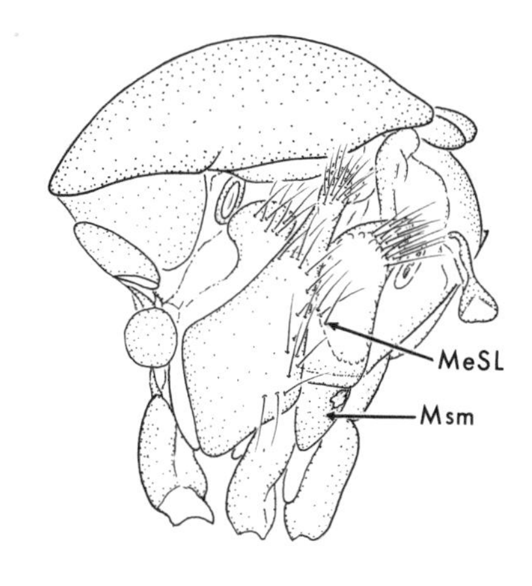


Fig. 106. Lateral view of thorax: *Oc. stimulans*

21(20). Tooth of foreclaw short, blunt, less than 0.5 length of the usually elongate claw (fig. 107) ........................................................................................................ 22

Tooth of foreclaw long, thin, 0.5 or more length of claw, which is markedly curved just distal to attachment of tooth ........................................................................ 23

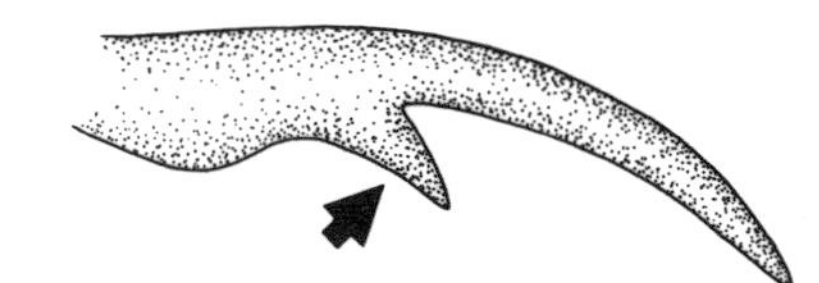

Fig. 107. Foreclaw: *Oc. riparius*

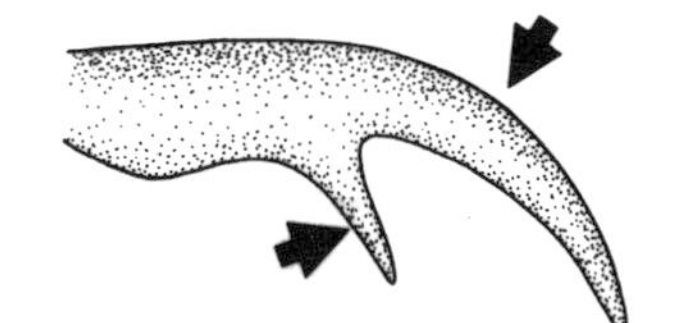

Fig. 108. Foreclaw: *Oc. fitchii*

22(21). Hypostigmal area scaled (fig. 109); abdominal terga with many scattered yellowish scales in dark-scaled areas (fig. 110) .................................................... *Oc. riparius* (plate 20B)

Hypostigmal area bare (fig. 111); abdominal terga with few or no pale scales in dark-scaled areas (fig. 112) .................................................................. Oc. *aloponotum* (plate 10A)

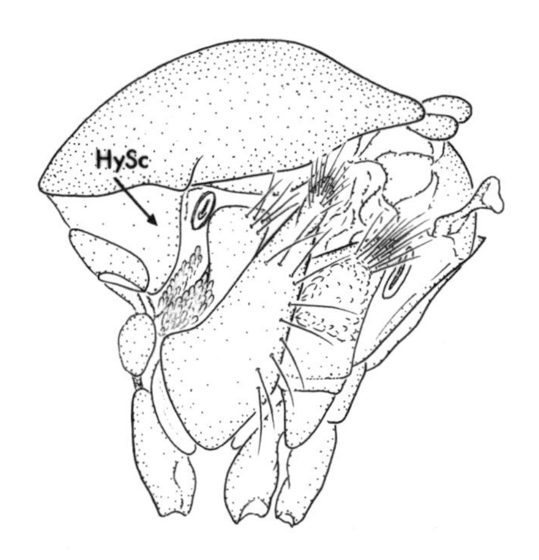


Fig. 109. Lateral view of thorax: *Oc. riparius*

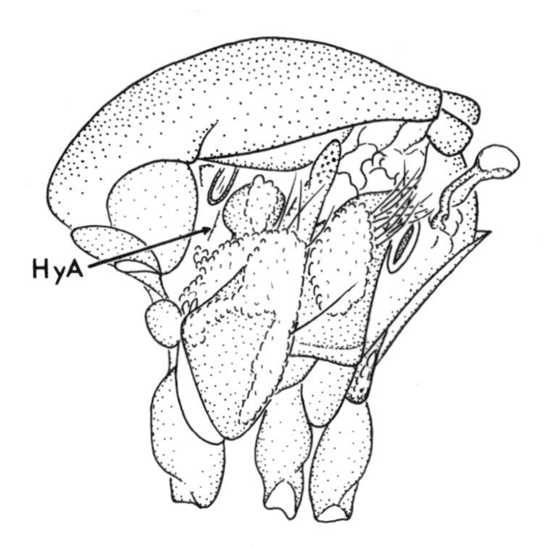


Fig. 111. Lateral view of thorax: *Oc. aloponotum*

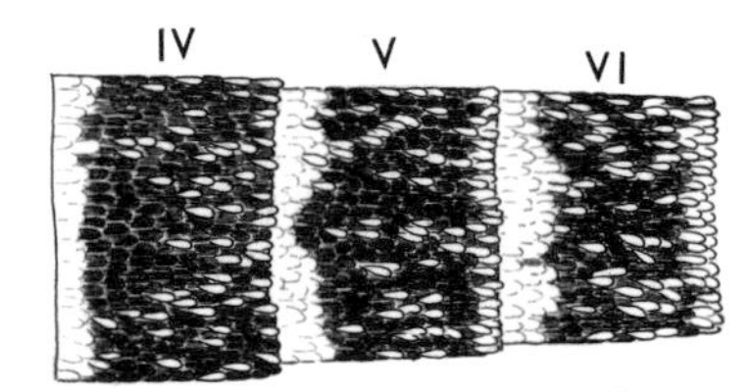


Fig. 110. Dorsal view of abdominal segments IV–VI: *Oc. riparius*

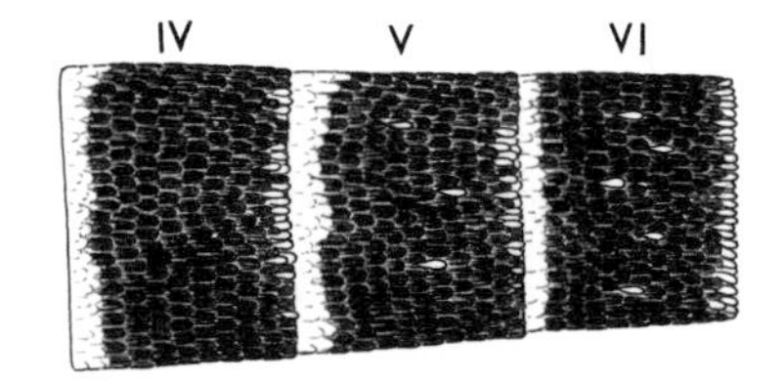


Fig. 112. Dorsal view of abdominal segments IV–VI: *Oc. aloponotum*

23(21). Proboscis, cercus, and tarsomere 1 of all legs with numerous pale scales (figs. 113, 114, 115); foreclaw long, straight distal to attachment of tooth (fig. 116) ............................. ........................................................................................ (in part) *Oc. euedes* (plate 14A)

Proboscis, cercus, and tarsomere 1 distal to basal ring, usually dark scaled (figs. 117, 118, 119); foreclaw shorter and more strongly curved distal to attachment of tooth (fig. 120) ........................................................................................ (in part) *Oc. fitchii* (plate 14C)

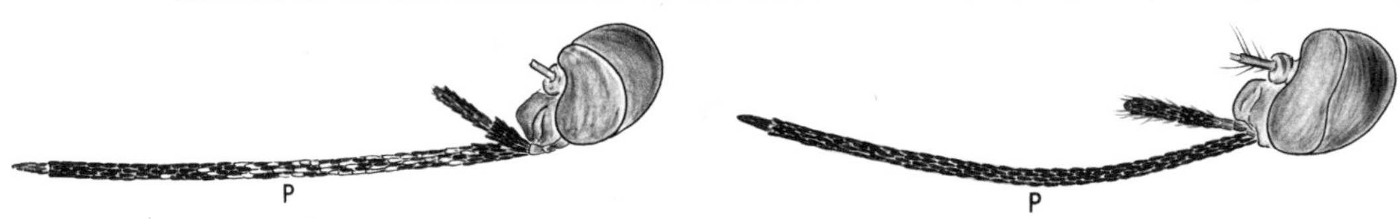


Fig. 113. Lateral view of head: *Oc. euedes*

Fig. 117. Lateral view of head: *Oc. fitchii*

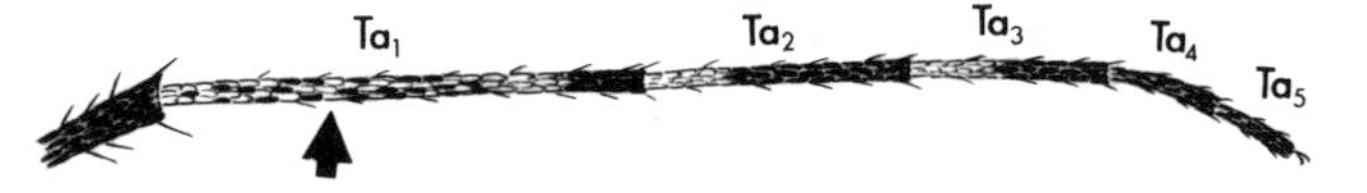


Fig. 115. Lateral view of hindleg: *Oc. euedes*

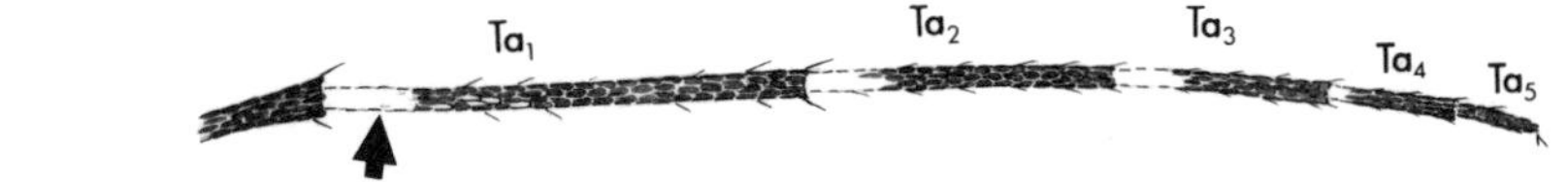


Fig. 119. Lateral view of hindleg: *Oc. fitchii*

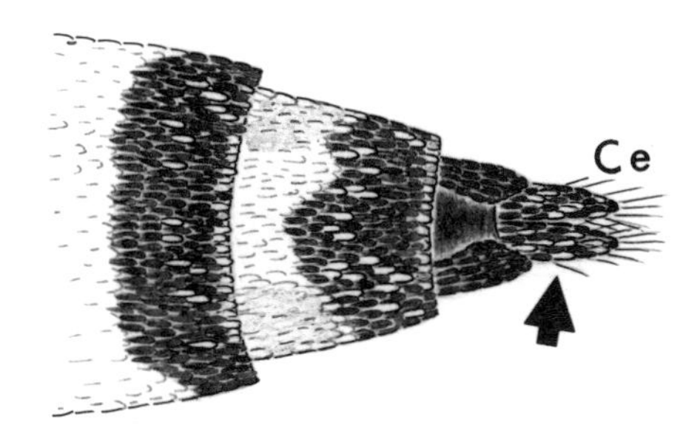


Fig. 114. Dorsal view of abdominal segments VII–X: *Oc. euedes*

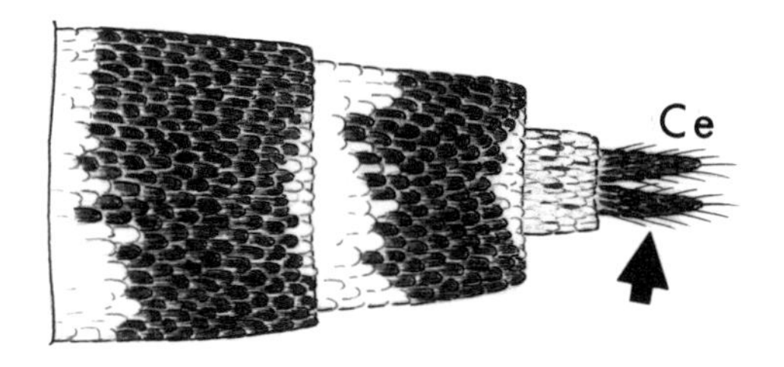


Fig. 118. Dorsal view of abdominal segments VII–X: *Oc. fitchii*

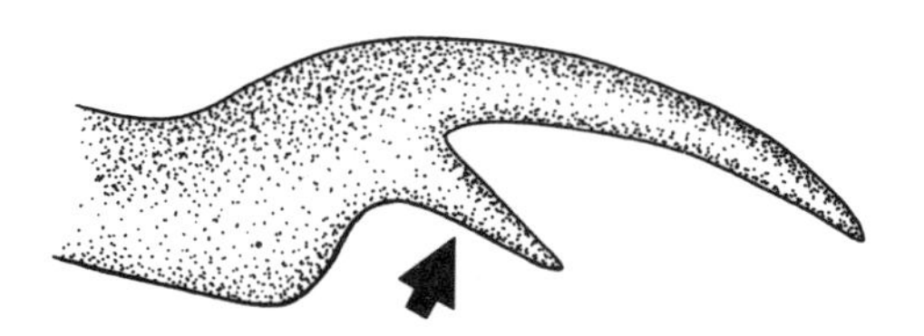

Fig. 116. Foreclaw enlarged: *Oc. euedes*

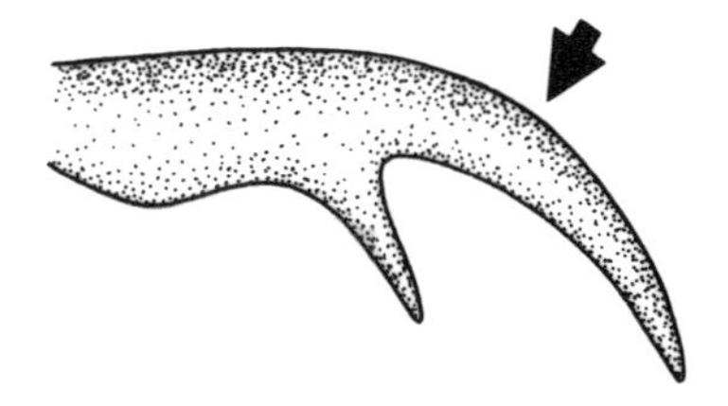

Fig. 120. Foreclaw: *Oc. fitchii*

24(20). ... Segments 2,3 of palpus dark scaled with apical pale-scaled rings (fig. 121); abdominal sterna IV,V with lateral patches of dark scales (fig.122); proboscis dark scaled (fig. 121)
*Oc. increpitus* (plate 16) ........................................................................ *clivis* (plate 39C)
*washinoi* (plate 41A)

Segments 2,3 of palpus with scattered pale scales (fig. 123); abdominal sterna IV,V pale scaled; if dark scales present, not laterally (fig. 124); proboscis usually with some pale scales (fig. 123) ........................................................................................................25

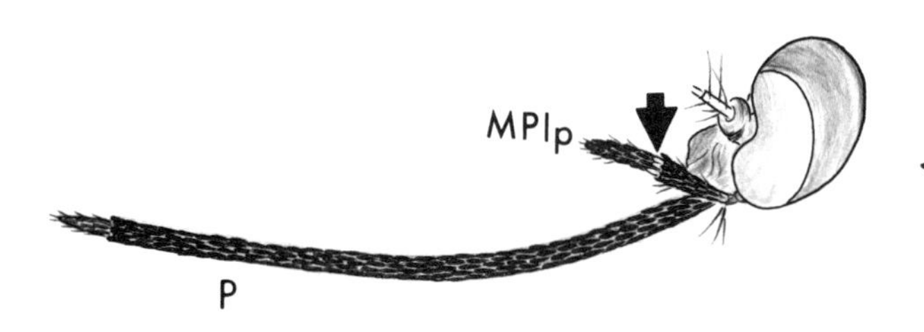


Fig. 121. Lateral view of head: *Oc. increpitus*

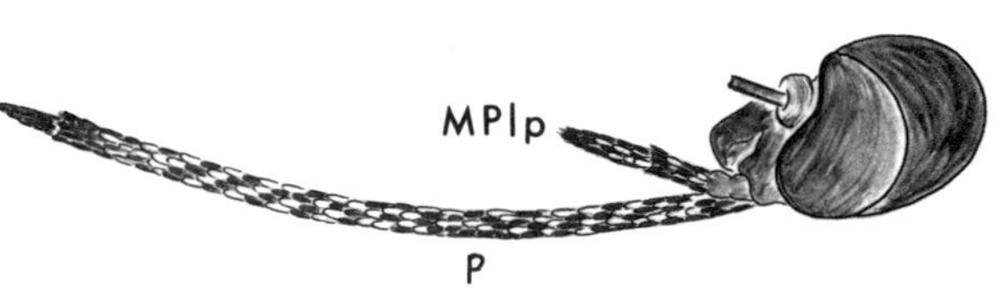


Fig. 123. Lateral view of head: *Oc. stimulans*

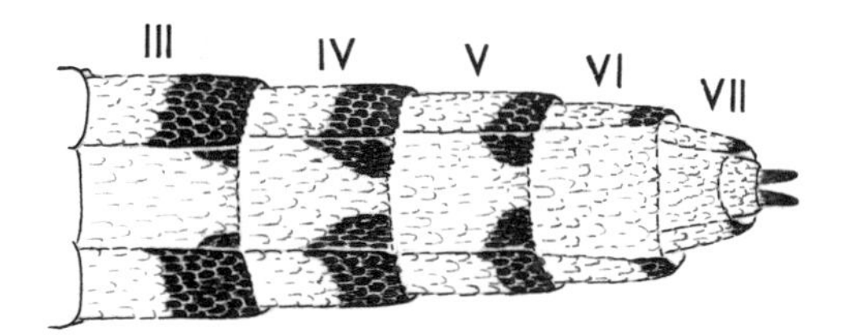


Fig. 122. Ventral view of abdomen: *Oc. increpitus*

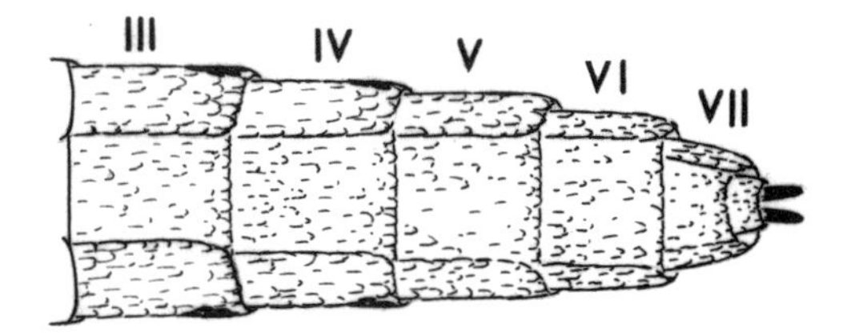


Fig. 124. Ventral view of abdomen: *Oc. stimulans*

25(24).Scales on antennal pedicel numerous, mostly pale (fig. 125); scutum with medium to dark brown longitudinal stripe (fig. 126) .................................................................... 26

Scales on antennal pedicel few, mostly dark (fig. 127); scutum with reddish brown scales medially, sometimes with stripe of light scales (fig. 128) ..............................................27

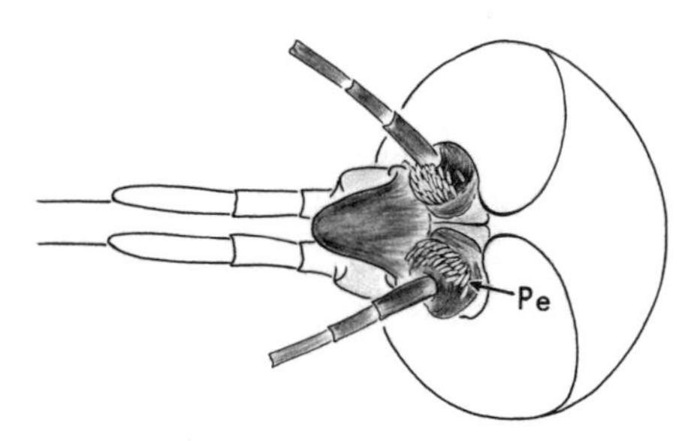

Fig. 125. Anterior view of head: *Oc. fitchii*

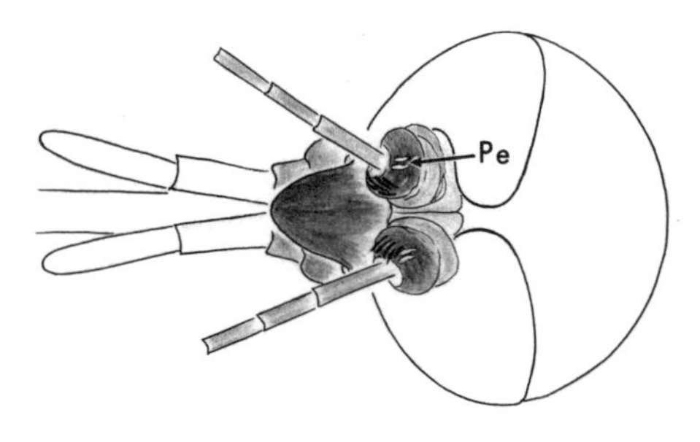

Fig. 127. Anterior view of head: *Oc. stimulans*

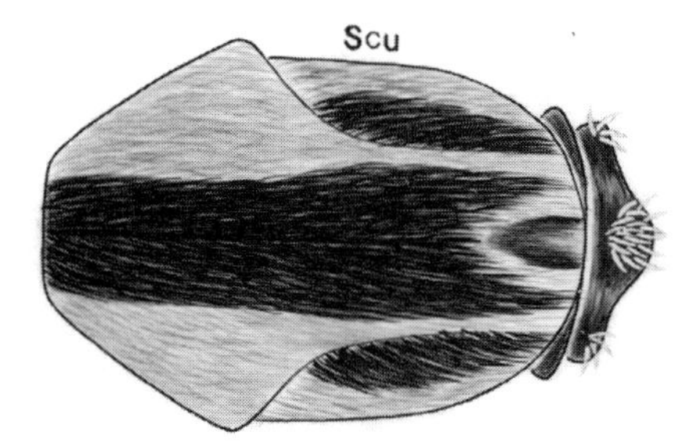

Fig. 126. Dorsal view of thorax: *Oc. mercurator*

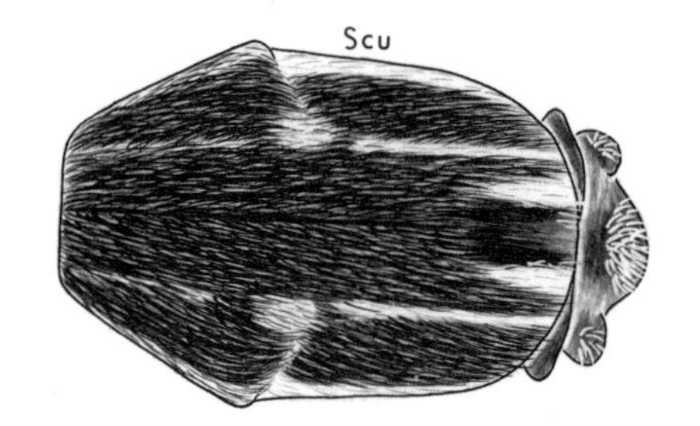

Fig. 128. Dorsal view of thorax: *Oc. euedes*

26(25). Scutum with pale yellowish scales laterally (fig. 129); dorsal brown-scaled area of postpronotum at most 0.5 as large as ventral pale-scaled area (fig. 130); foretarsomere 3 with incomplete basal pale ring (fig. 131) .................................... *Oc. mercurator* (plate 17C)

Scutum with pale white scales, often mixed with yellow or light brown scales, laterally (fig. 132); dorsal brown-scaled area of postpronotum equal to or larger than ventral pale-scaled area (fig. 133); foretarsomere 3 with complete basal pale ring (fig. 134) ........................ ........................................................................................ (in part) *Oc. fitchii* (plate 14C)

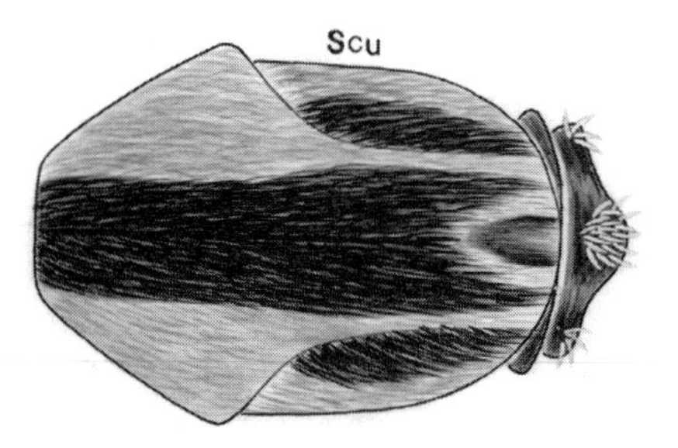

Fig. 129. Dorsal view of thorax: *Oc. mercurator*

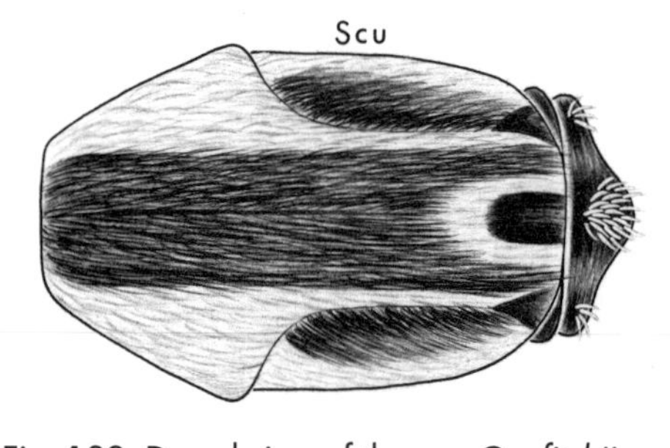

Fig. 132. Dorsal view of thorax: *Oc. fitchii*

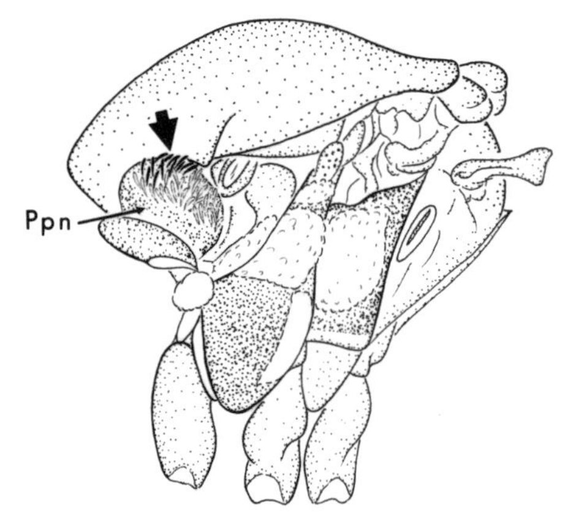

Fig. 130. Lateral view of thorax: *Oc. mercurator*

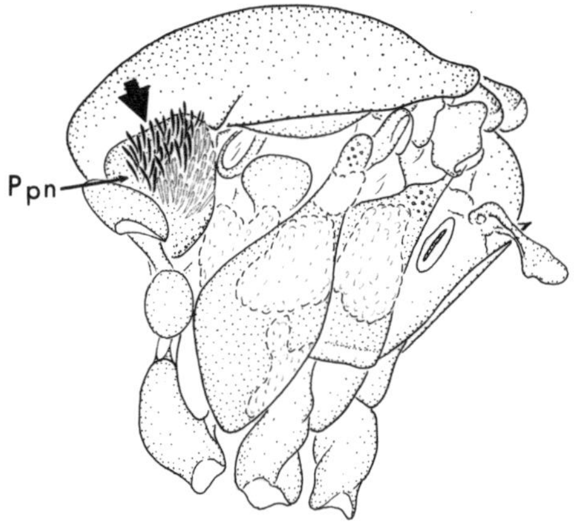

Fig. 133. Lateral view of thorax: *Oc. fitchii*

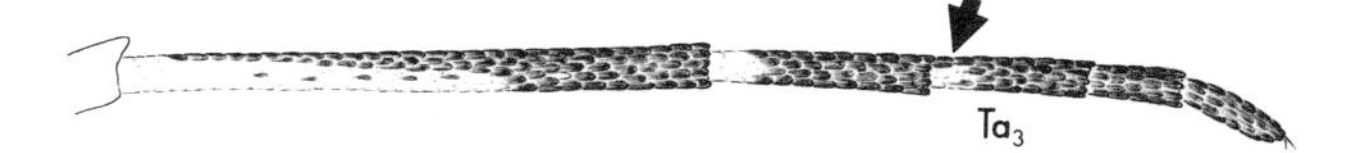

Fig. 131. Lateral view of foretarsi: *Oc. mercurator*

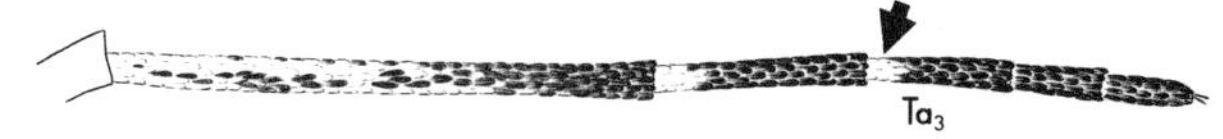

Fig. 134. Lateral view of foretarsi: *Oc. fitchii*

27(25). Foreclaw markedly bent just distad to tooth (fig. 135); abdominal sterna VI–VIII pale scaled or with few dark scales only (fig. 136) ........................ *Oc. stimulans* (plate 22A)

Foreclaw evenly curved distad to tooth (fig. 137); abdominal sterna VI–VIII pale scaled with rather broad medioapical dark-scaled patches (fig. 138) ........................................ ........................................................................................ (in part) *Oc. euedes* (plate 14A)

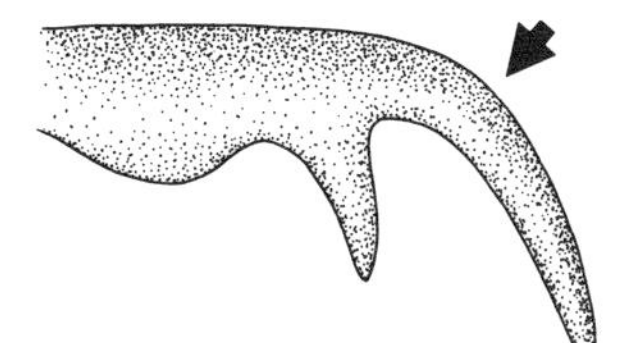
Fig. 135. Foreclaw: *Oc. stimulans*

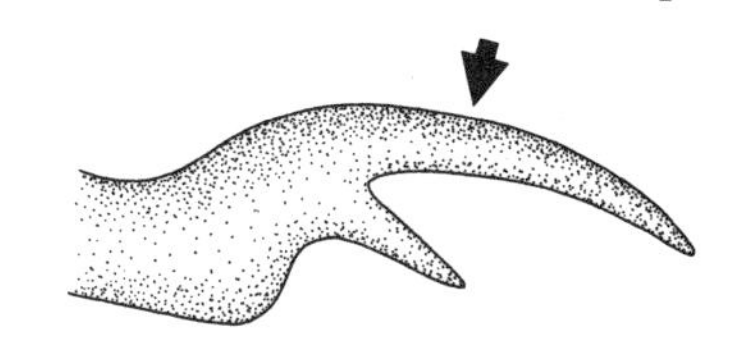
Fig. 137. Foreclaw: *Oc. euedes*

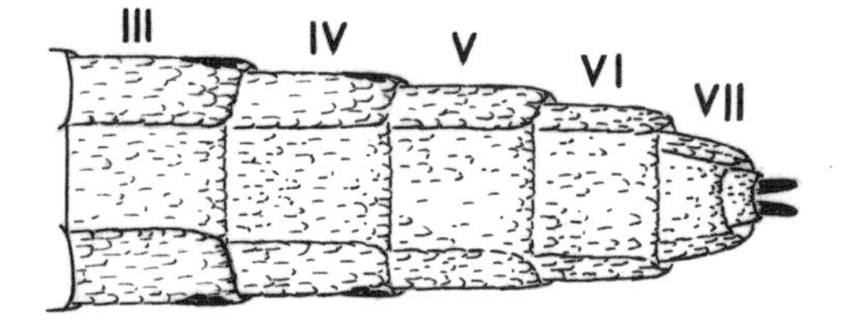

Fig. 136. Ventral view of abdomen: *Oc. stimulans*

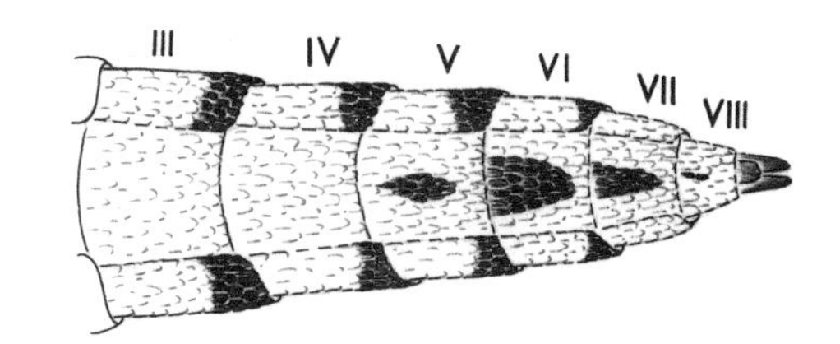

Fig. 138. Ventral view of abdomen: *Oc. euedes*

28(2). Wing with dark and pale scales intermixed, usually pale scaled (fig. 139); postprocoxal scale patch present (fig. 140) ........................................................................................ 29

Wing entirely dark scaled or with some pale scales on anterior veins (fig. 141); postprocoxal scale patch absent (fig. 142) ........................................................................... 31

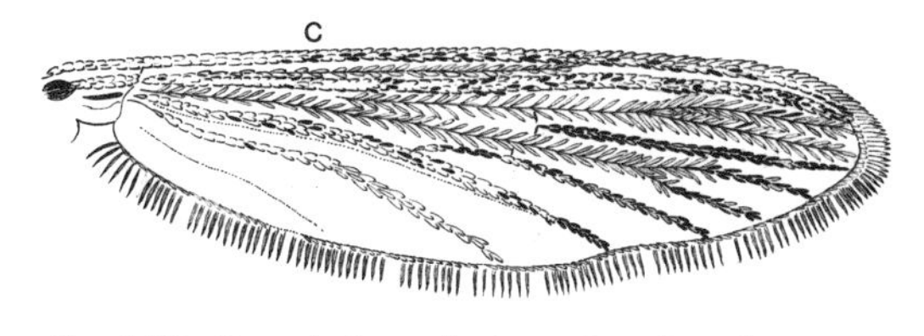

Fig. 139. Dorsal view of wing: *Oc. dorsalis*

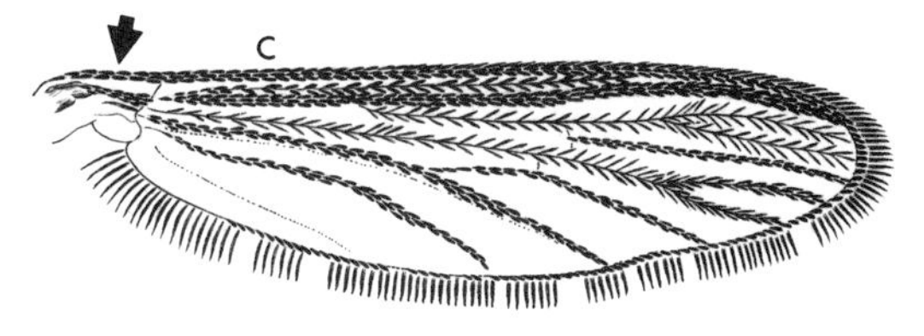

Fig. 141. Dorsal view of wing: *Oc. c. canadensis*

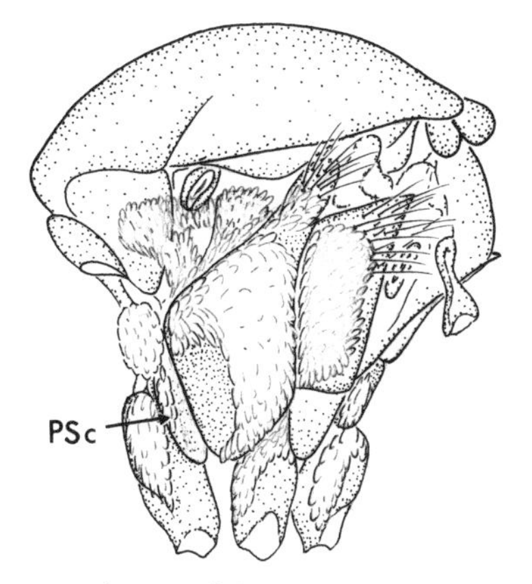

Fig. 140. Lateral view of thorax: *Oc. dorsalis*

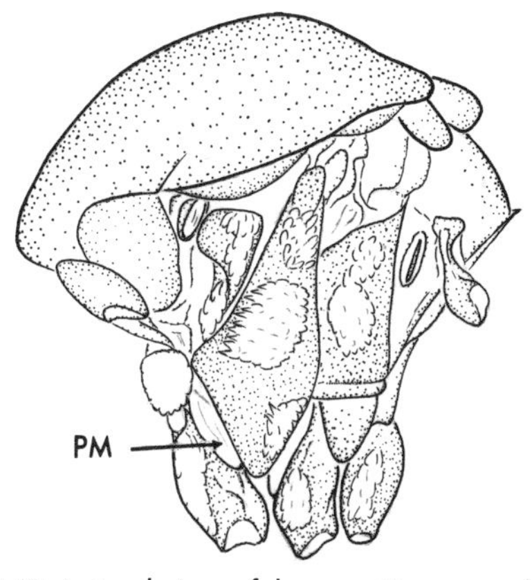

Fig. 142. Lateral view of thorax: *Oc. atropalpus*

29(28). Wing vein C mostly dark scaled (fig. 143); abdominal tergum VII usually with more dark than pale scales (fig. 144) ..................................................... *Oc. melanimon* (plate 17B)

Wing vein C mostly pale scaled (fig. 145); tergum VII with more pale than dark scales (fig. 146) ............................................................................................................ 30

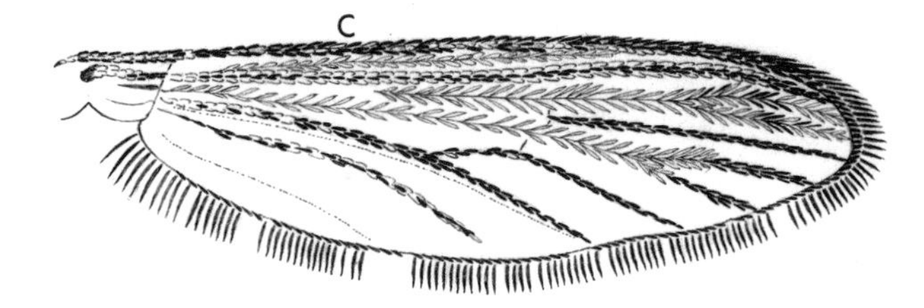


Fig. 143. Dorsal view of wing: *Oc. melanimon*

Fig. 145. Dorsal view of wing: *Oc. dorsalis*

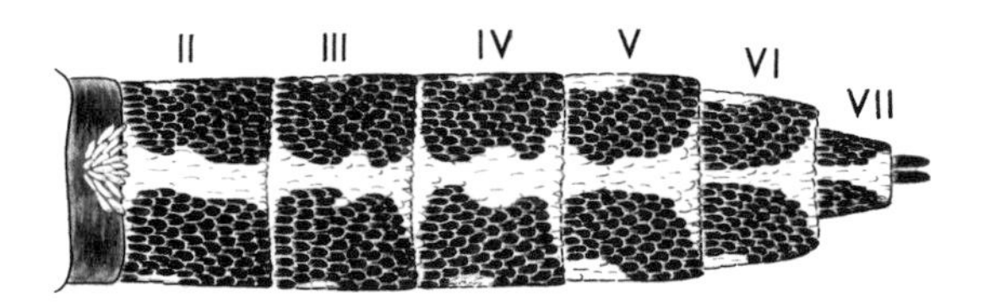


Fig. 144. Dorsal view of abdomen: *Oc. melanimon*

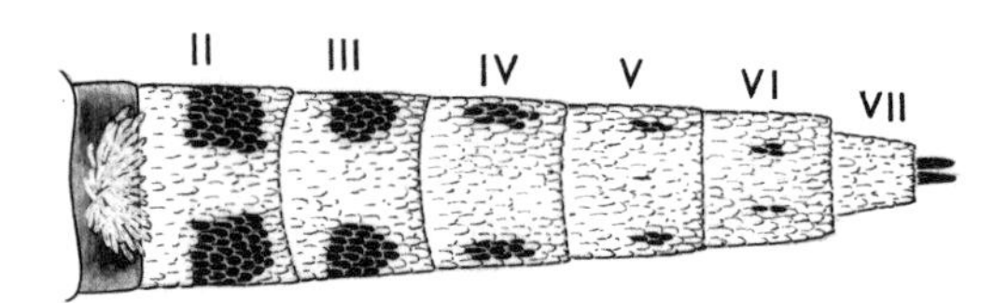


Fig. 146. Dorsal view of abdomen: *Oc. dorsalis*

30(29). Wing vein $R_{4+5}$ with more dark scales than veins $R_2$ and $R_3$ (fig. 147); foreclaw almost straight in middle (fig. 148) ....................................................... *Oc. dorsalis* (plate 13B)

Wing vein $R_{4+5}$ with as many dark scales as $R_2$ and $R_3$ (fig. 149); foreclaw abruptly curving near attachment of tooth (fig. 150) ....................................... *Oc. campestris* (plate 14A)

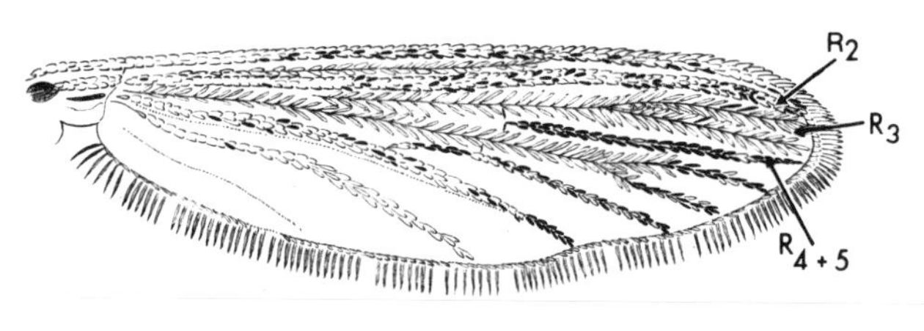


Fig. 147. Same as fig. 139 (Dorsal view of wing: *Oc. dorsalis*)

Fig. 149. Dorsal view of wing: *Oc. campestris*

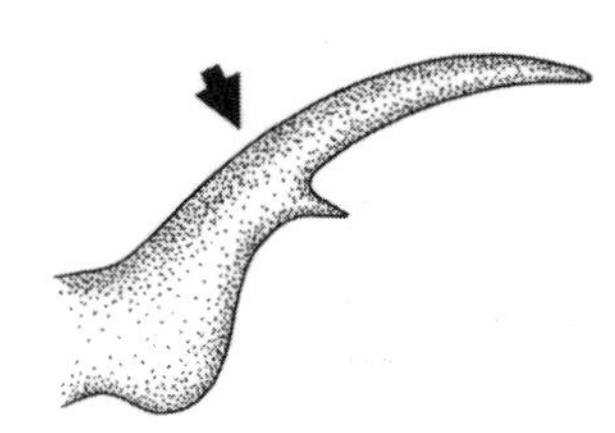

Fig. 148. Foreclaw: *Oc. dorsalis*

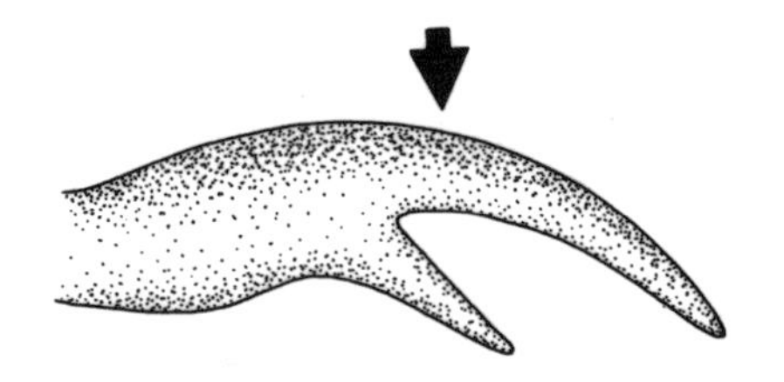

Fig. 150. Foreclaw: *Oc. campestris*

31(28). Scutum with lyre-shaped pattern of golden scales, usually with 4 median stripes (fig. 151) ............................................................................................... *Oc. togoi* (plate 10B)

Scutum without such a lyre-shaped pattern (fig. 152) ................................................ 32

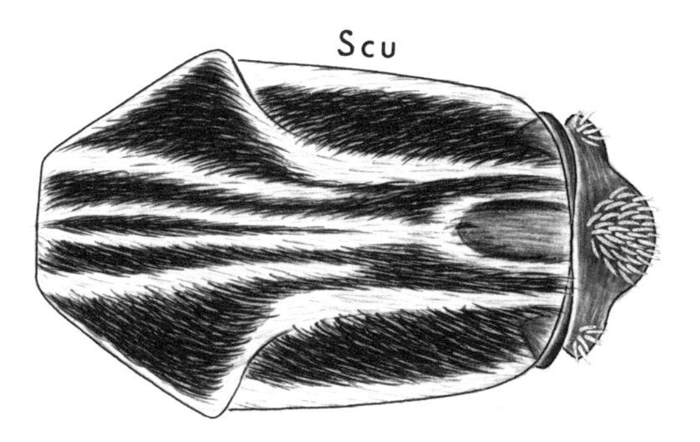

Fig. 151. Dorsal view of scutum: *Oc.* togoi

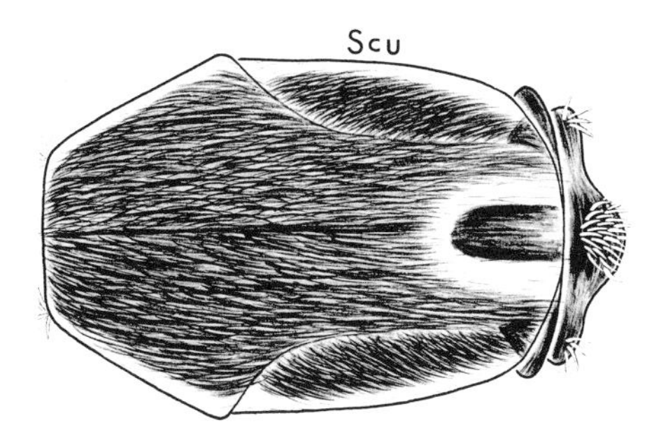


Fig. 152. Dorsal view of scutum: *Oc. c. canadensis*

32(31). Wing entirely dark scaled (fig. 153); scutum without dark median stripe, usually rather evenly reddish or golden brown (fig. 154) .................................................................. 33

Wing with prominent patch of pale scales on base of vein C (fig. 155); scutum with broad dark brown or golden median longitudinal stripe (fig. 156) ........................................ 34

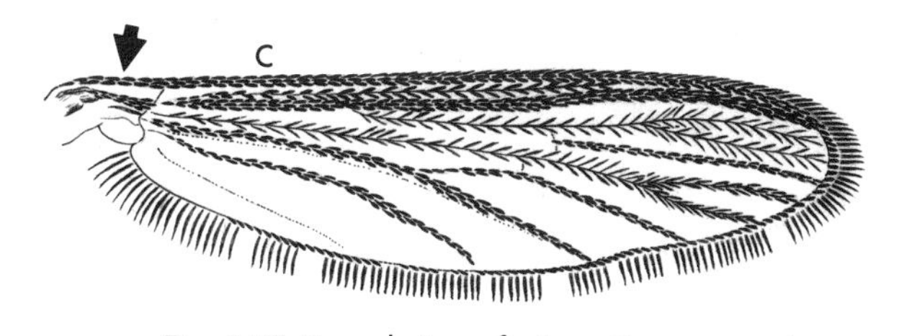


Fig. 153. Dorsal view of wing: *Oc. c. canadensis*

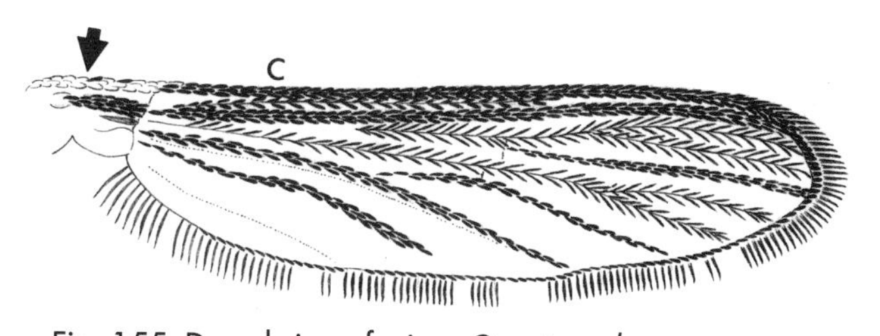


Fig. 155. Dorsal view of wing: *Oc. atropalpus*

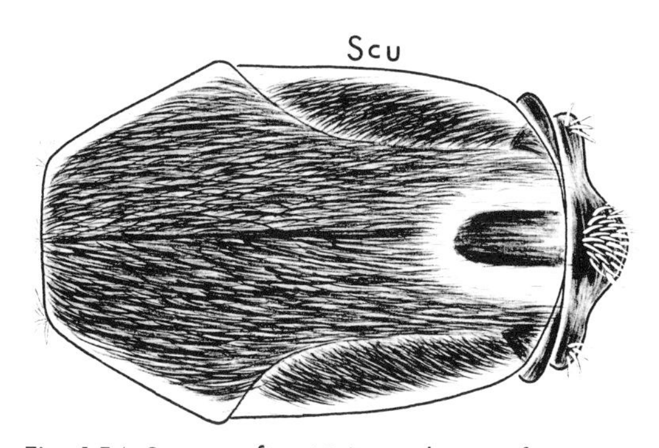


Fig. 154. Same as fig. 68 (Dorsal view of scutum: *Oc. c. canadensis*)

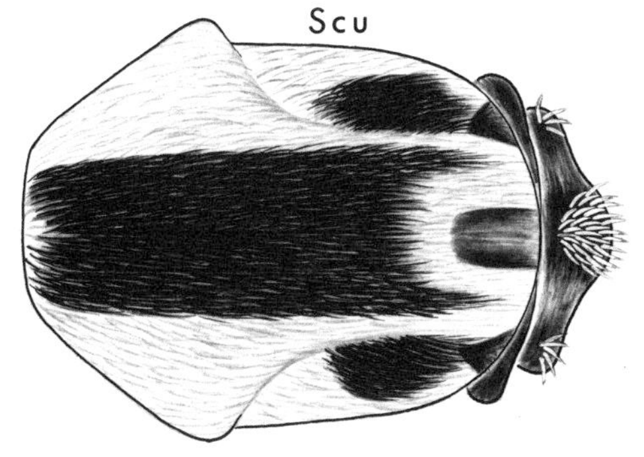


Fig. 156. Dorsal view of scutum: *Oc. atropalpus*

33(32). Hindtarsomeres 1–4 with broad pale basal and apical rings, tarsomere 5 entirely pale scaled (fig. 157); scutum with golden brown scales (fig. 158) ........................................ .............................................................................. *Oc. canadensis canadensis* (plate 11B)

Hindtarsomeres with narrow pale basal and apical rings on 1,2, basally only on 3,4, and tarsomere 5 dark scaled (fig. 159); scutum with scales mostly dark brown, with indefinite median stripe of paler scales (fig. 160) .................. *Oc. canadensis mathesoni* (plate 39B)

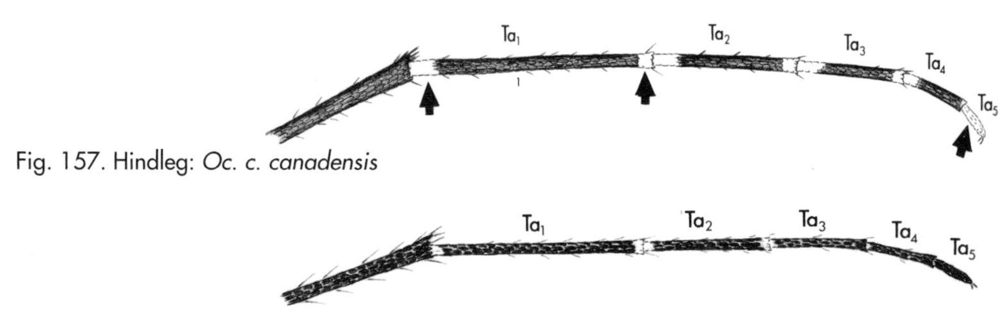


Fig. 157. Hindleg: *Oc. c. canadensis*

Fig. 159. Hindleg: *Oc. c. mathesoni*

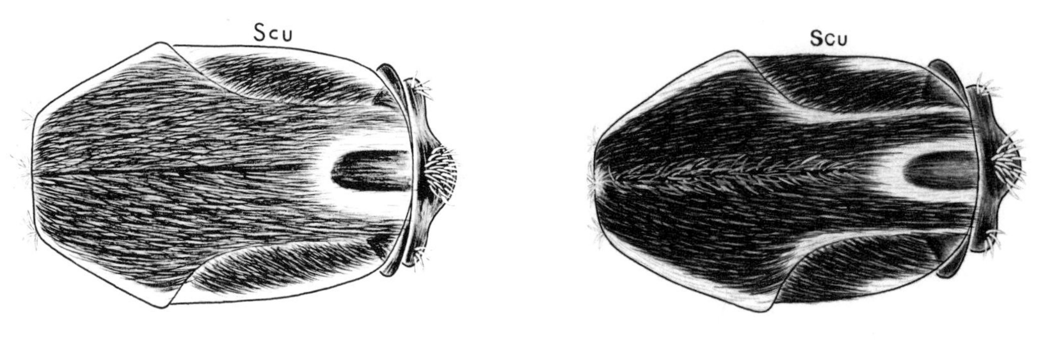


Fig. 158. Dorsal view of scutum: *Oc. c. canadensis*

Fig. 160. Dorsal view of scutum: *Oc. c. mathesoni*

34(32). Palpi almost entirely dark scaled (fig. 161); pale rings on hindtarsomere joint 1,2 subequal (fig. 162); scutellum with narrow yellow to brown scales (fig. 163) ........................... 35

Palpi with bands of pale scales (fig. 164); pale rings on hindtarsomere joint 1,2 longer on 1 than on 2 (fig. 165); scutellum with broad pale scales (fig. 166) ............................. 36

Fig. 161. Lateral view of head: *Oc. atropalpus*

Fig. 164. Lateral view of head: *Oc. sierrensis*

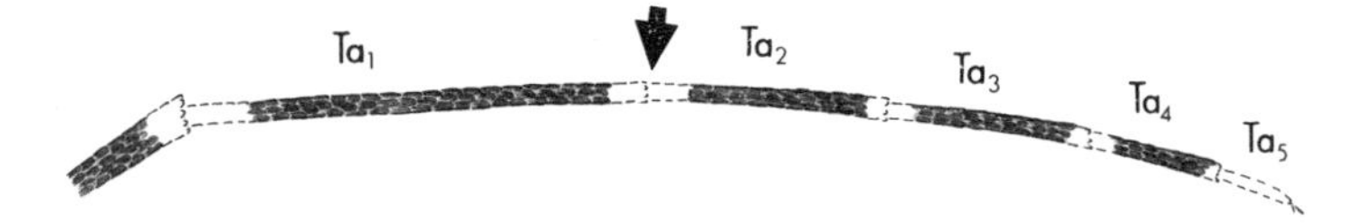


Fig. 162. Lateral view of hindtarsi: *Oc. atropalpus*

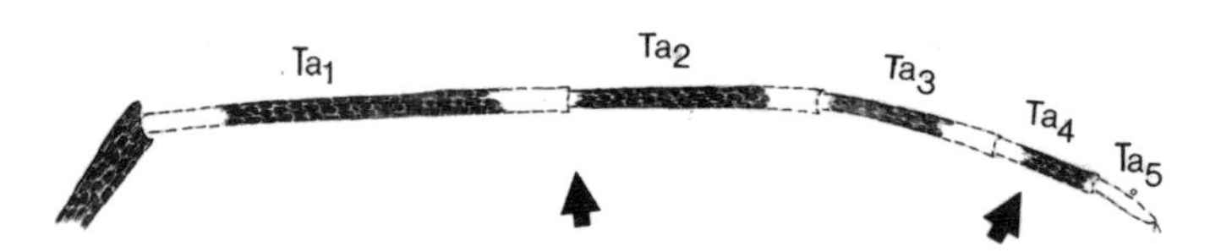


Fig. 165. Lateral view of hindtarsi: *Oc. sierrensis*

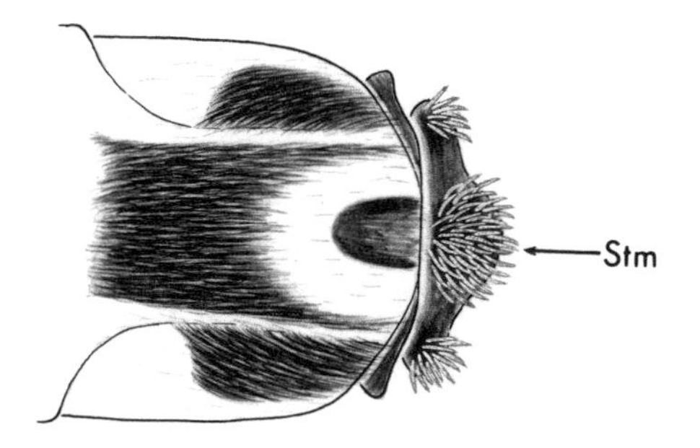


Fig. 163. Posterior dorsal view of thorax: *Oc. atropalpus*

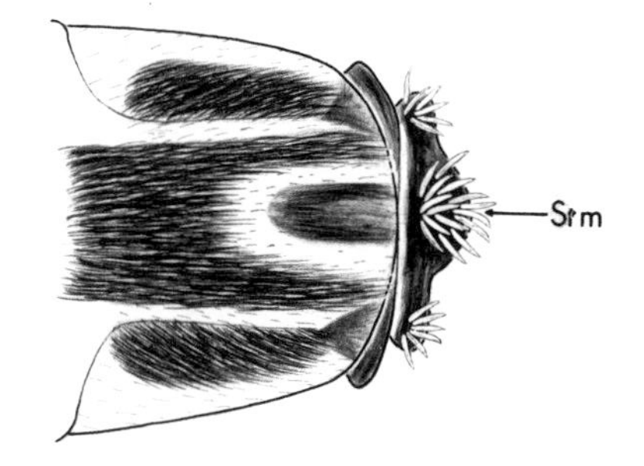


Fig. 166. Posterior dorsal view of thorax: *Oc. sierrensis*

35(34). Interocular space no wider than 2.0 diameters of single corneal facet (fig. 167); hindfemur with dark scales to near base dorsally (fig. 168); scutal fossa with 1 or more strong posterior setae (fig. 169) ........................................................ *Oc. epactius* (plate 13D)

Interocular space at least 2.5 diameters of single corneal facet (fig. 170); hindfemur entirely pale in basal 0.3–0.5 (fig. 171); scutal fossa without posterior setae (fig. 172) ........................................................ *Oc. atropalpus* (plate 10C)

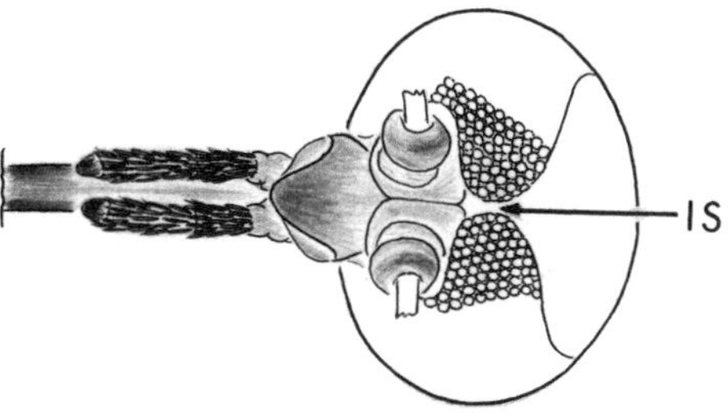


Fig. 167. Front view of head: *Oc. epactius*

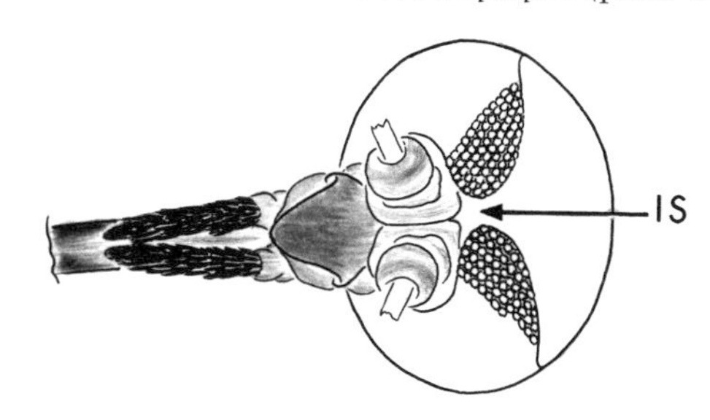


Fig. 170. Front view of head: *Oc. atropalpus*

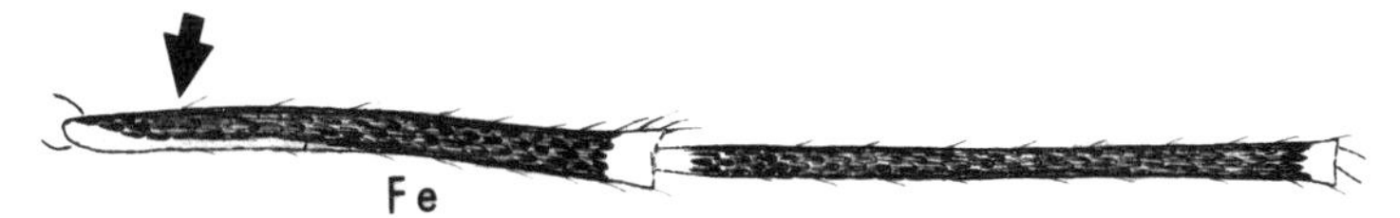


Fig. 168. Hindleg: *Oc. epactius*

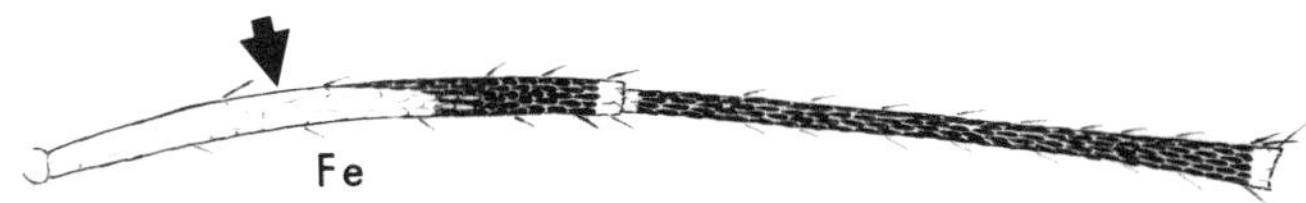


Fig. 171. Hindleg: *Oc. atropalpus*

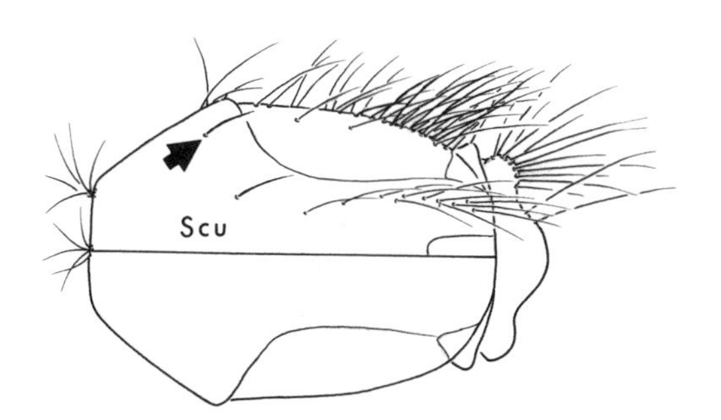


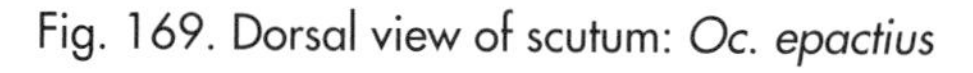

Fig. 169. Dorsal view of scutum: *Oc. epactius*

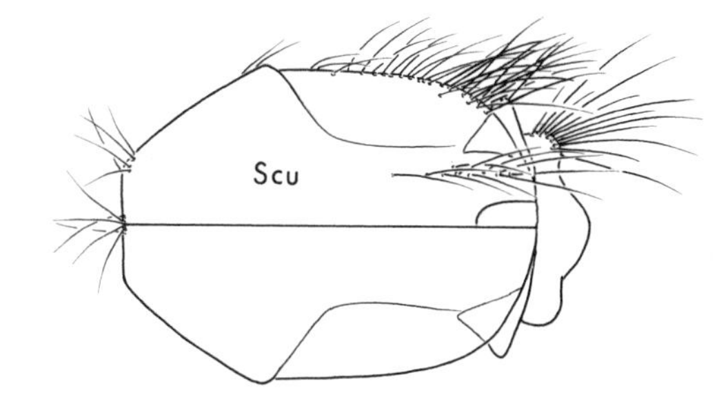


Fig. 172. Dorsal view of scutum: *Oc. atropalpus*

36(34). Postprocoxal scale patch present (fig. 173) ........................... *Oc. monticola* (plate 40A)

Postprocoxal scale patch absent (fig. 174) ........................................................ 37

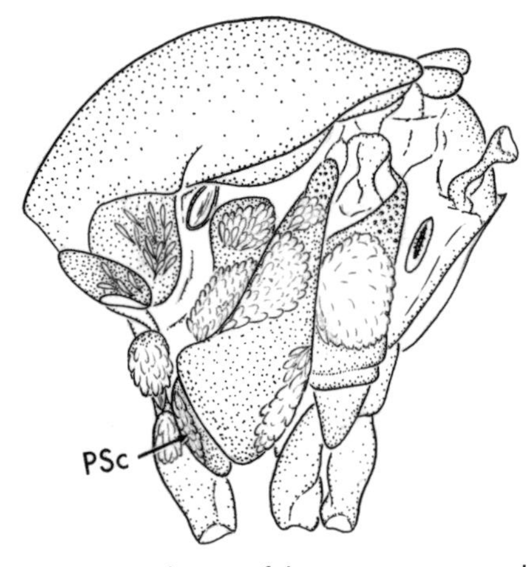


Fig. 173. Lateral view of thorax: *Oc. monticola*

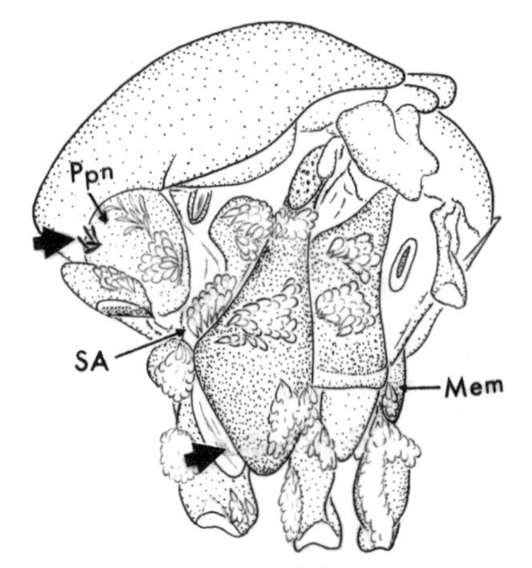


Fig. 174. Lateral view of thorax: *Oc. sierrensis*

37(36). Subspiracular area with several light colored setae arising from scale patch (fig. 175) .... ............................................................ *Oc. varipalpus* (plate 40D)

Subspiracular area without setae (fig. 176) ............................................................ 38

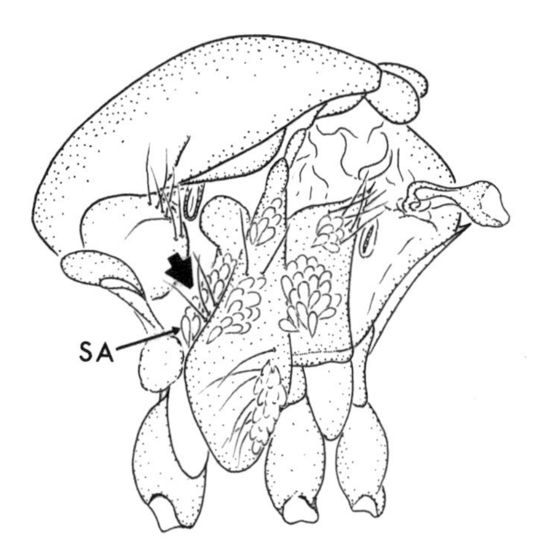


Fig. 175. Lateral view of thorax: *Oc. varipalpus*

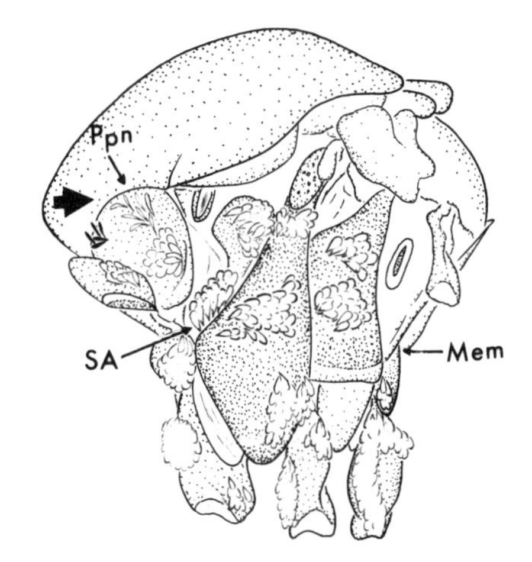


Fig. 176. Lateral view of thorax: *Oc. sierrensis*

38(37). Base of hindtarsomere 4 with broad band of pale scales (fig. 177); metameron with scales; anterodorsal border of postpronotum with dark scales (fig. 178) .................................. ............................................................ *Oc. sierrensis* (plate 20D)

Base of hindtarsomere 4 with at most very narrow band of pale scales (fig. 179); metameron bare; postpronotum entirely pale scaled (fig. 180) ..... *Oc. deserticola* (plate 39D)

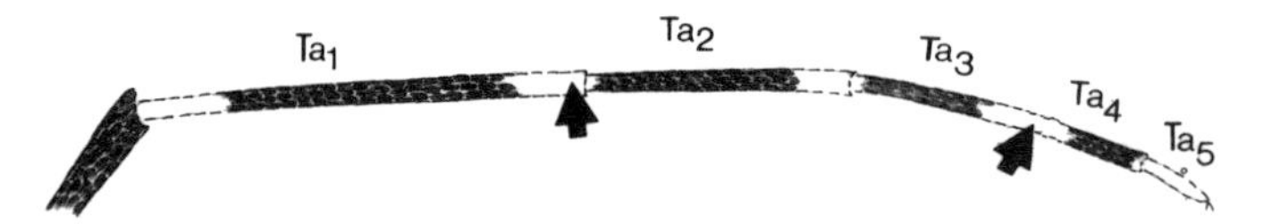


Fig. 177. Lateral view of hindtarsi: *Oc. sierrensis*

Fig. 179. Lateral view of hindtarsi: *Oc. deserticola*

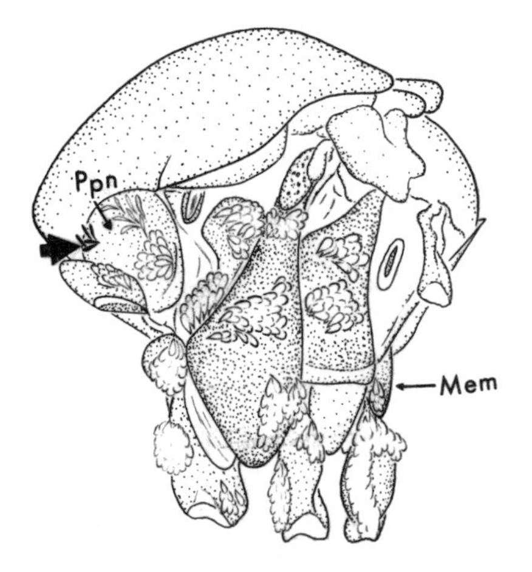


Fig. 178. Lateral view of thorax: *Oc. sierrensis*

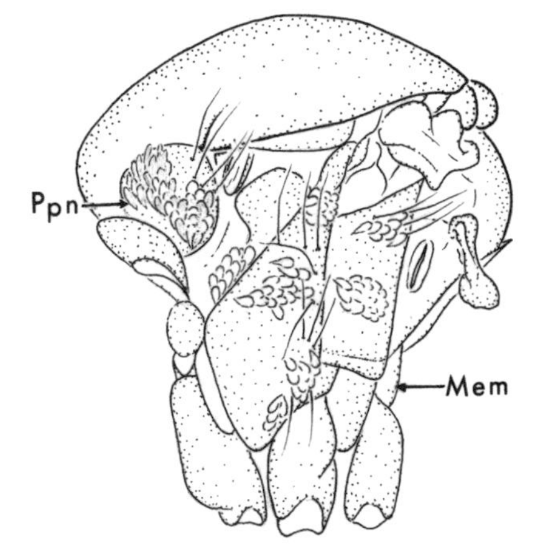


Fig. 180. Lateral view of thorax: *Oc. deserticola*

39(1). Scutal integument with pair of dark posterolateral spots (fig. 181) ............................ 40

Scutal integument lacking dark posterolateral spots (fig. 182) .................................... 41

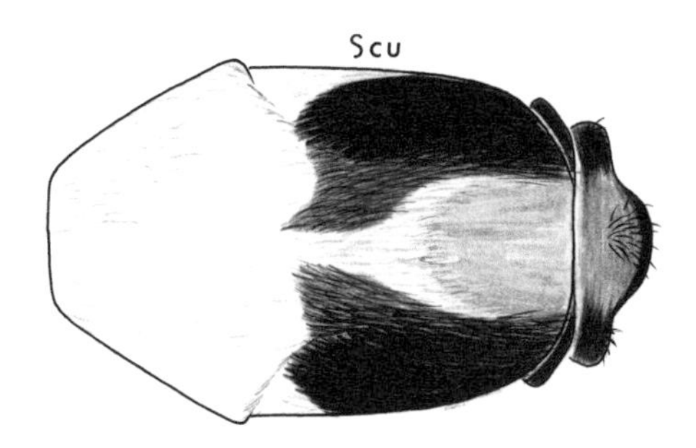


Fig. 181. Dorsal view of scutum: *Oc. fulvus pallens*

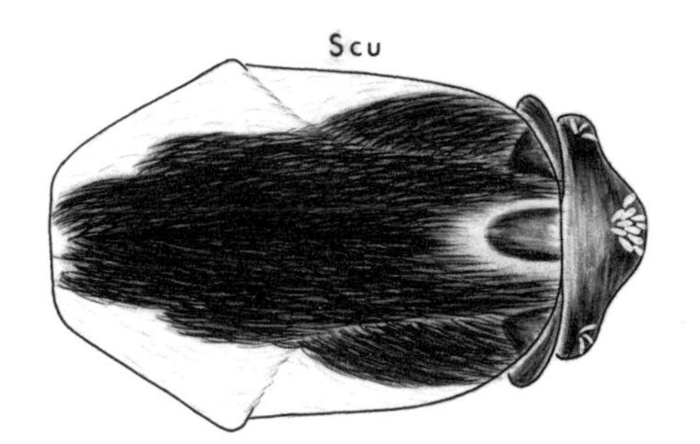


Fig. 182. Dorsal view of scutum: *Oc. triseriatus*

40(39). Hypostigmal area with dark integumental spot (fig. 183); abdominal terga II–VI basally yellow scaled, apically dark scaled (fig. 184) .................... *Oc. fulvus* pallens (plate 15A)

Hypostigmal area without dark spot (fig. 185); abdominal terga II–VI entirely yellow scaled (fig. 186) ................................................................ *Oc. bimaculatus* (plate 39A)

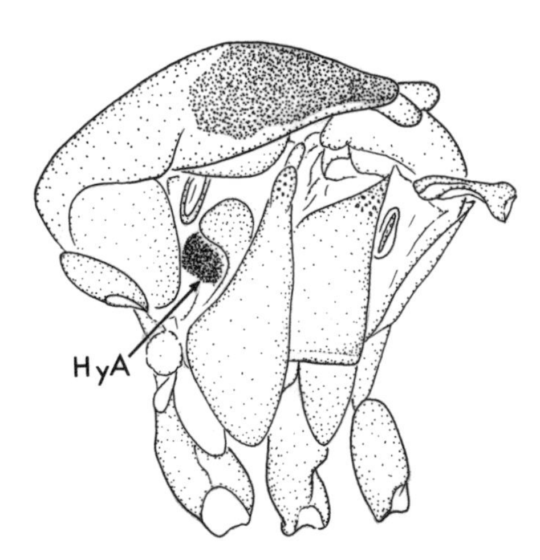


Fig. 183. Lateral view of thorax: *Oc. fulvus pallens*

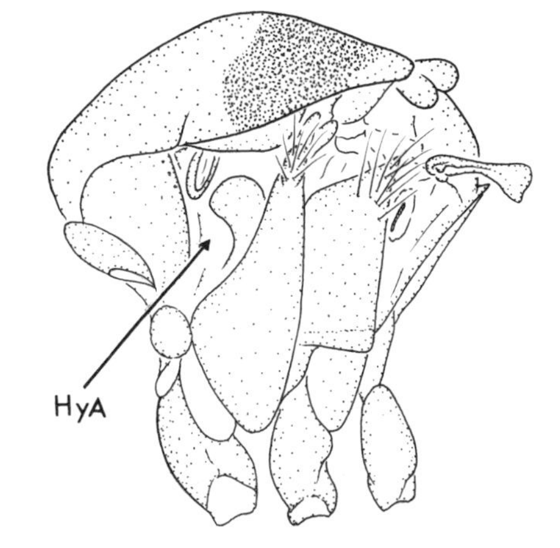


Fig. 185. Lateral view of thorax: *Oc. bimaculatus*

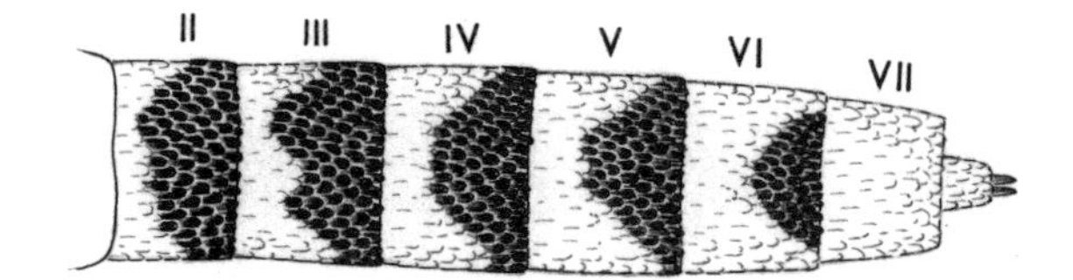


Fig. 184. Dorsal view of abdomen: *Oc. fulvus pallens*

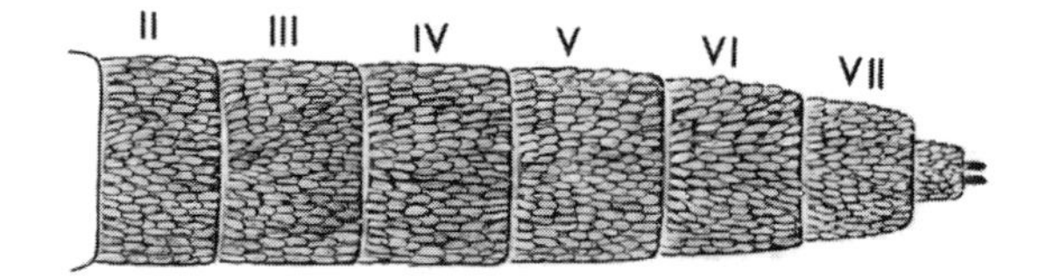


Fig. 186. Dorsal view of abdomen: *Oc. bimaculatus*

41(39). Postspiracular setae absent (fig. 187) .................................. *Oc. purpureipes* (plate 40C)

Postspiracular setae present (fig. 188) .......................................................................... 42

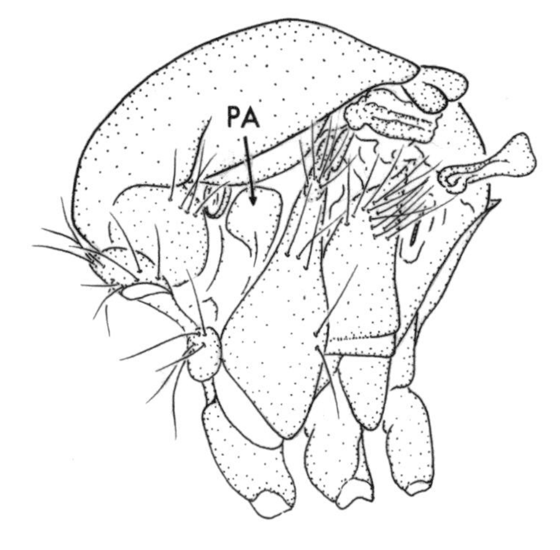


Fig. 187. Lateral view of thorax: *Oc. purpureipes*

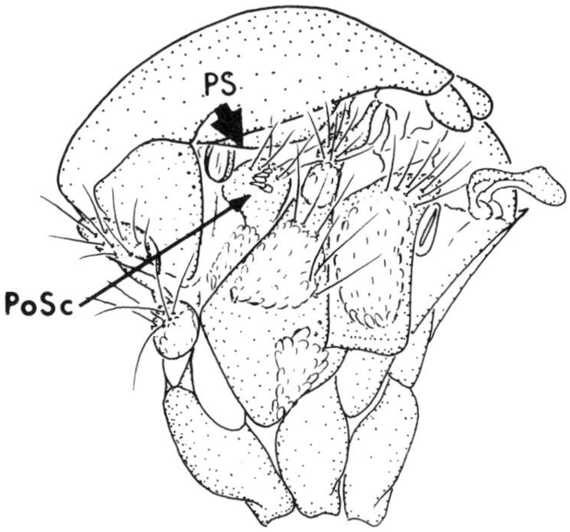

Fig. 188. Lateral view of thorax: *Oc. hendersoni*

42(41). Scutum with submedian stripes or median patch of silvery white, pale white, or pale yellow scales, or with silvery white scales laterally (figs. 189, 190) ....................................... 43

Scutum without silvery white scales medially or laterally nor pale white or pale yellow scales medially (fig. 191) ............................................................................................ 52

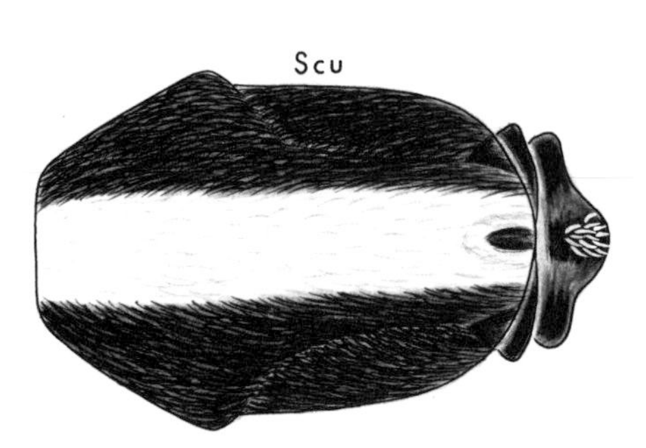


Fig. 189. Dorsal view of scutum: *Oc. atlanticus*

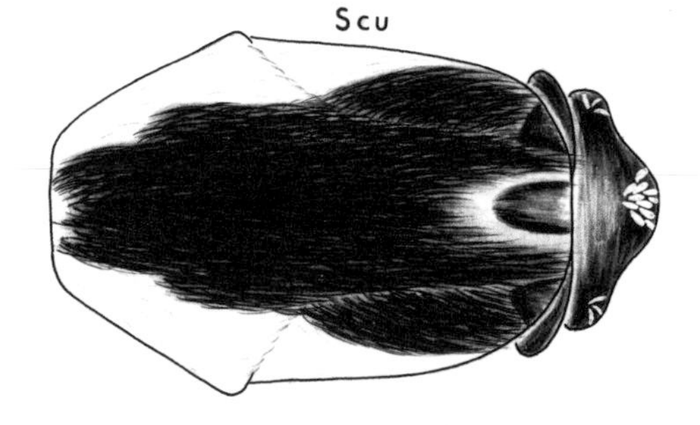


Fig. 190. Dorsal view of scutum: *Oc. triseriatus*

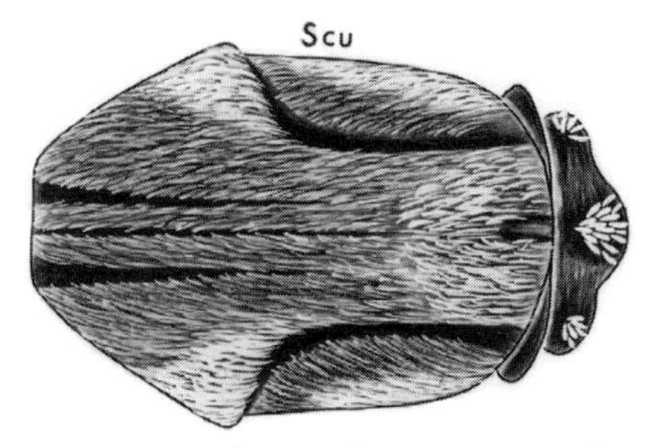


Fig. 191. Dorsal view of scutum: *Oc. pullatus*

43(42). Scutum with median longitudinal stripe of dark brown scales and silvery white scales laterally (fig. 192) .......... 44

Scutum with broad patch or 1 or 2 stripes of silvery white, pale white, or sometimes pale yellow scales medially (fig. 193) .......... 46

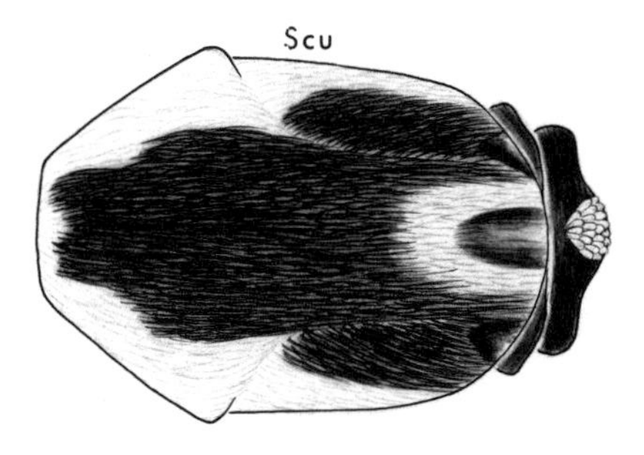


Fig. 192. Dorsal view of thorax: *Oc. triseriatus*

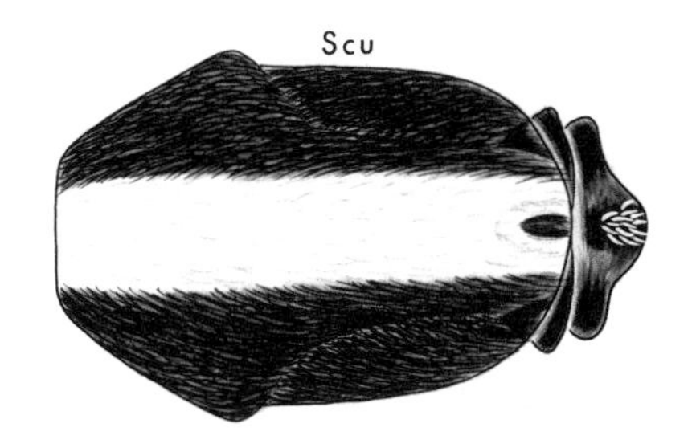


Fig. 193. Dorsal view of scutum: *Oc. atlanticus*

44(43). Setae of anterior portion of scutum relatively few and weak; silvery scaling of scutal fossa usually restricted to lateral and posterior portions (fig. 194); fore- and midclaws evenly curved, tooth less than 0.3 length of claw (fig. 195) .......... *Oc. triseriatus* (plate 23B)

Setae of anterior portion of scutum numerous and well developed; silvery scaling usually covering entire scutal fossa (fig. 196); claws of fore- and midlegs abruptly curving, tooth 0.3–0.5 length of claw (fig. 197) .......... 45

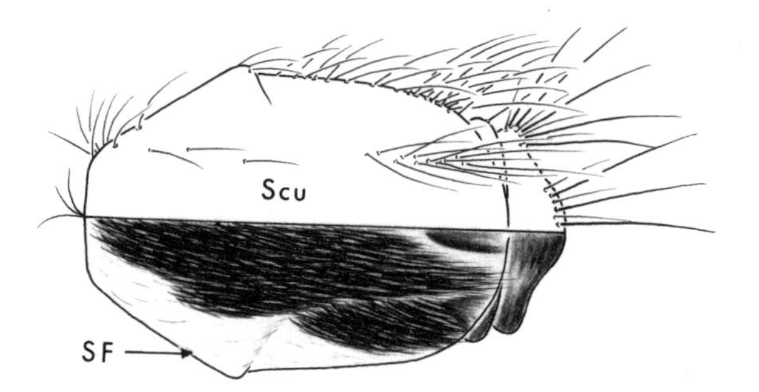


Fig. 194. Dorsal view of thorax: *Oc. triseriatus*

Fig. 196. Dorsal view of thorax: *Oc. hendersoni*

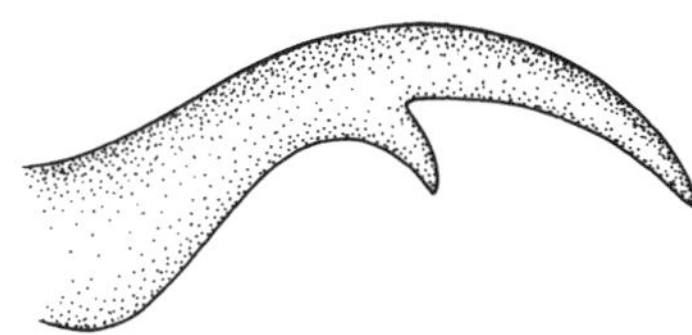

Fig. 195. Foreclaw: *Oc. triseriatus*

Fig. 197. Foreclaw: *Oc. hendersoni*

45(44). Postspiracular scale patch small or absent (fig. 198); setae around scutal fossa lightly to moderately pigmented (fig. 199) .......... *Oc. hendersoni* (plate 15C)

Postspiracular scale patch large (fig. 200); setae of scutal fossa darkly pigmented (fig. 201) .......... *Oc. brelandi* (plate 39B)

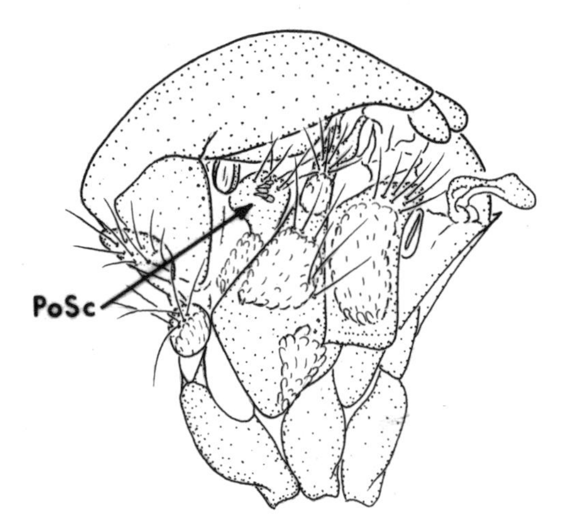

Fig. 198. Lateral view of thorax: *Oc. hendersoni*

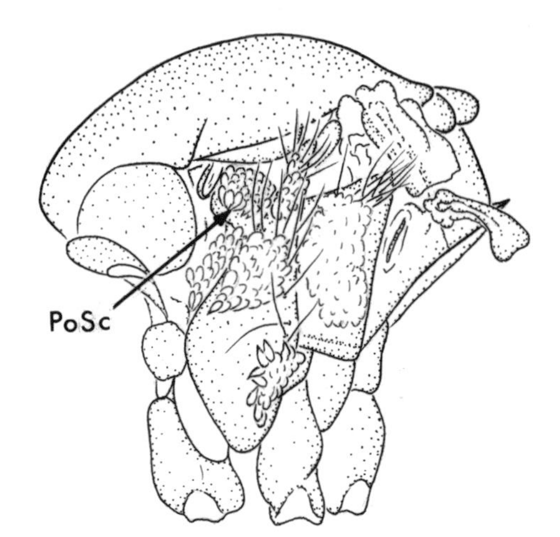

Fig. 200. Lateral view of thorax: *Oc. brelandi*

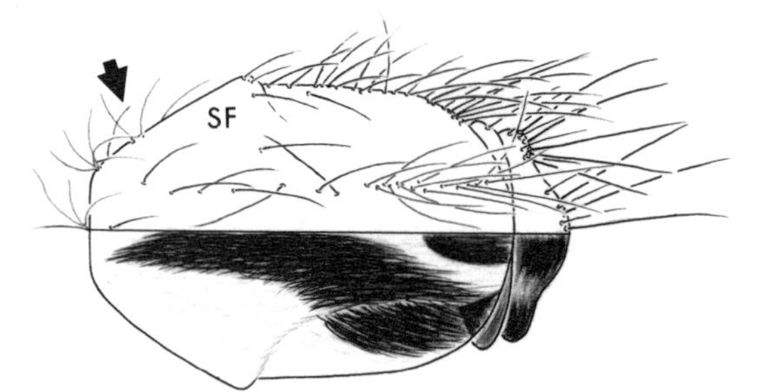

Fig. 199. Dorsal view of thorax: *Oc. hendersoni*

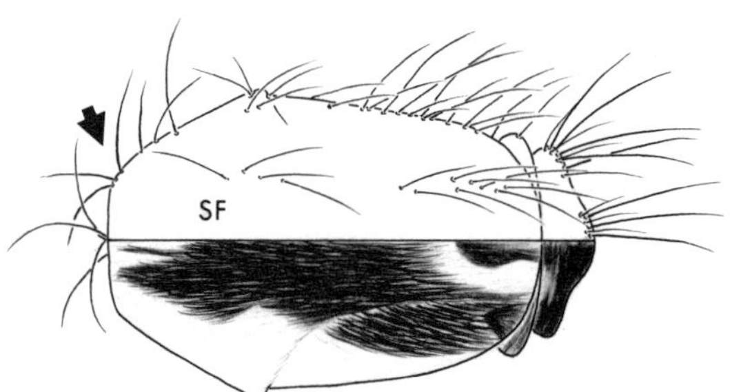

Fig. 201. Dorsal view of thorax: *Oc. brelandi*

46(43). Scutum with pair of submedian pale-scaled stripes, separated by dark stripe of about same width (fig. 202) ........ *Oc trivittatus* (plate 23C)

Scutum without pair of submedian pale-scaled stripes (fig. 203) ........ 47

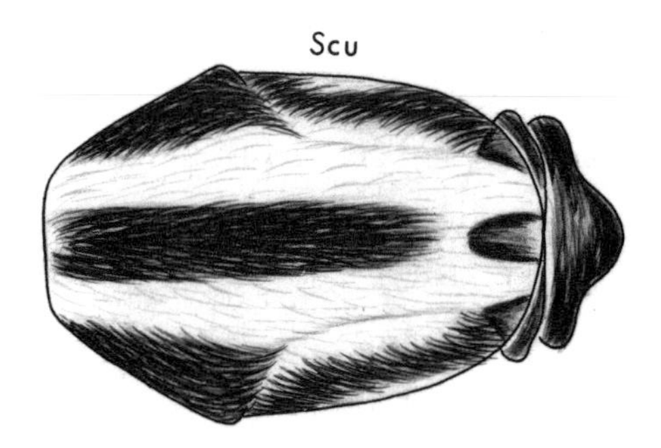

Fig. 202. Dorsal view of scutum: *Oc. trivittatus*

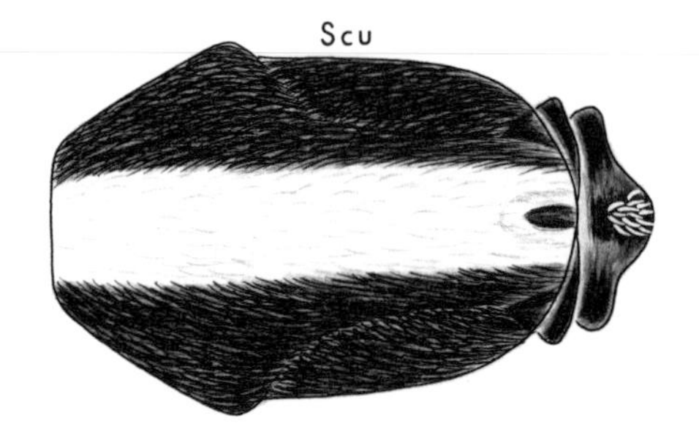

Fig. 203. Dorsal view of scutum: *Oc. atlanticus*

47(46). Scutum with anteromedian patch of silvery white or pale yellow scales, extending to middle or a little beyond, much broader than lateral dark-scaled areas (fig. 204) ........ 48

Scutum with median longitudinal stripe of silvery scales extending full length, usually narrower than lateral dark-scaled areas (fig. 205) ........ 49

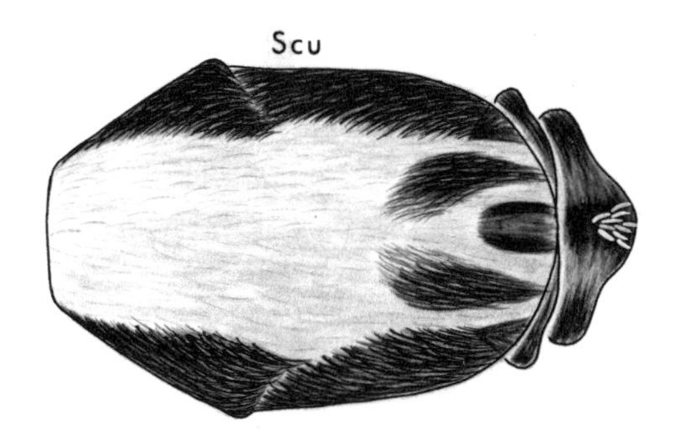


Fig. 204. Dorsal view of scutum: *Oc. infirmatus*

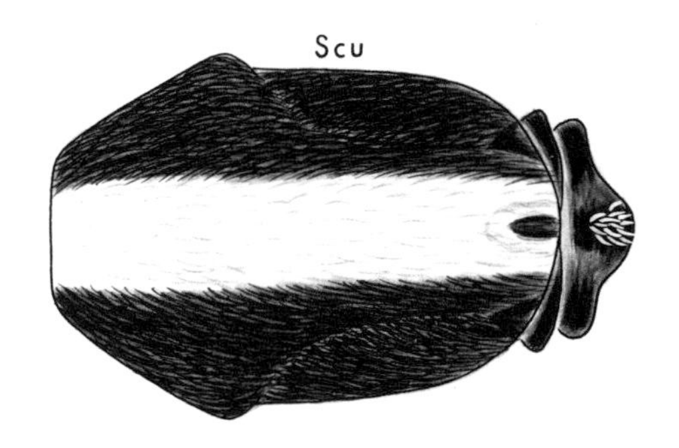


Fig. 205. Dorsal view of scutum: *Oc. atlanticus*

48(47). Hindtibia with basal and apical dark-scaled bands (fig. 206); abdominal terga VI–VIII with lighter colored scales medially (fig. 207) ................................ *Oc. scapularis* (plate 23D)

Hindtibia with dark scales from base to apex (fig. 208); abdominal terga VI–VIII dark scaled medially (fig. 209) ....................................................... *Oc. infirmatus* (plate 16D)

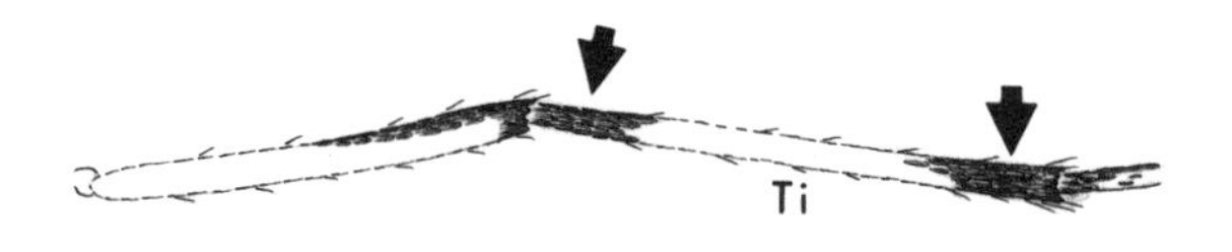


Fig. 206. Hindleg: *Oc. scapularis*

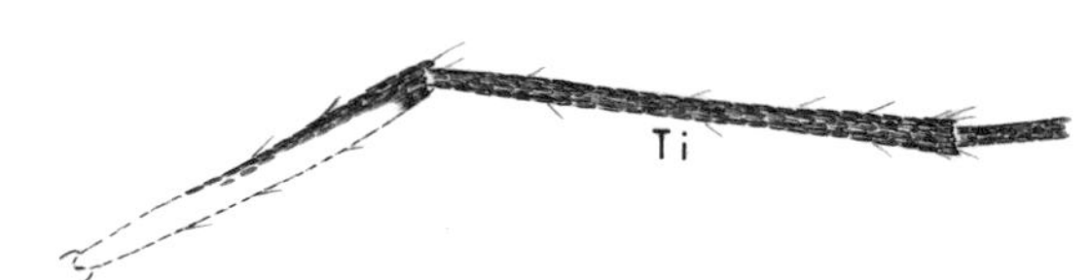


Fig. 208. Hindleg: *Oc. infirmatus*

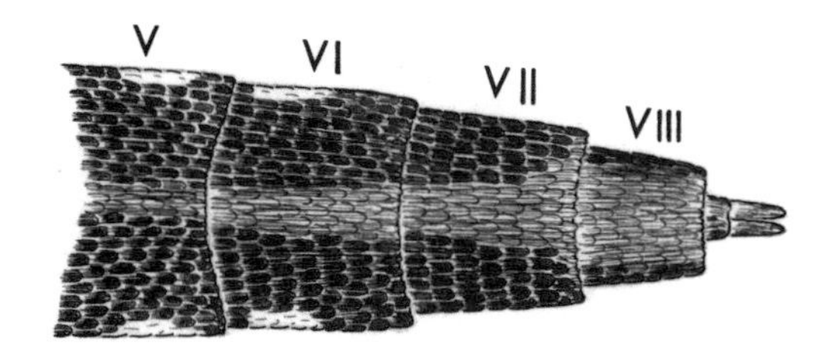


Fig. 207. Dorsal view of abdomen: *Oc. scapularis*

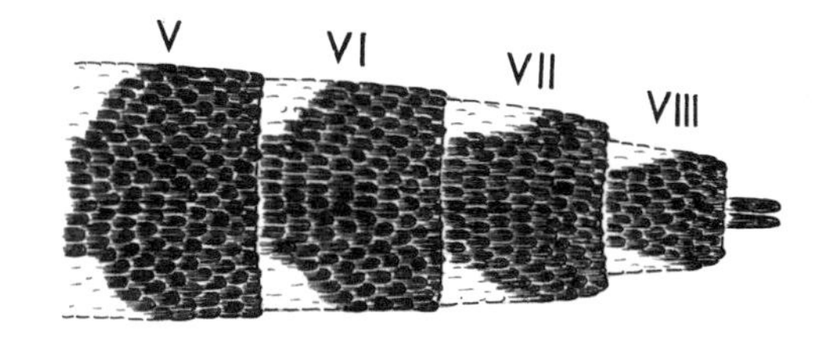


Fig. 209. Dorsal view of abdomen: *Oc. infirmatus*

49(47). Midtarsomere 1 with broad pale band (fig. 210); foretarsomere 1 with pale-scaled basal patch (fig. 211) ........................................................................... *Oc. burgeri* (Plate 39B)

Tarsomere 1 of mid- and forelegs dark scaled (figs. 212, 213) ................................... 50

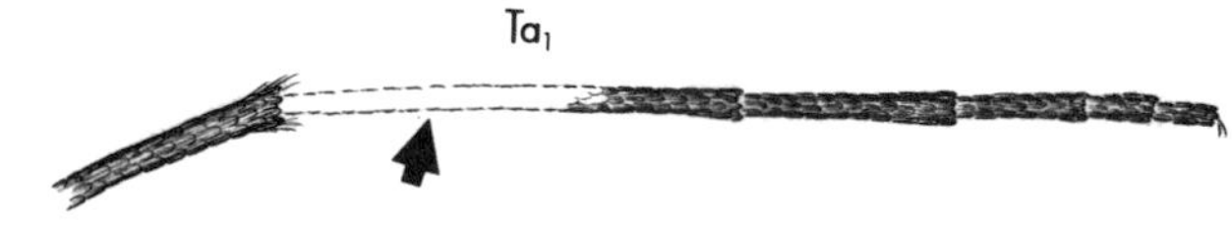


Fig. 210. Midleg: *Oc. burgeri*

Fig. 212. Midleg: *Oc. atlanticus*

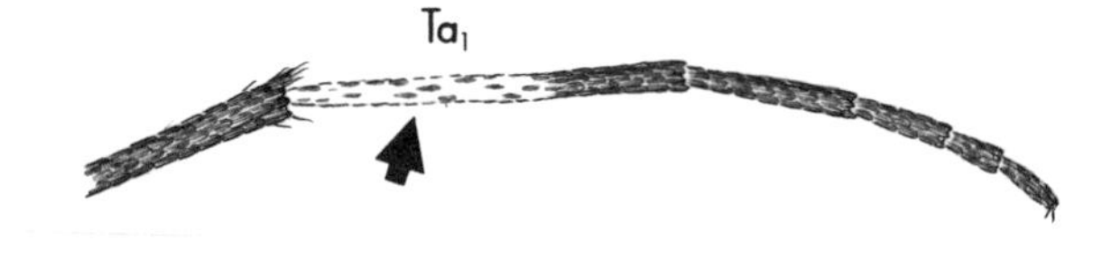


Fig. 211. Foreleg: *Oc. burgeri*

Fig. 213. Foreleg: *Oc. atlanticus*

50(49). Abdominal terga with basal pale bands (fig. 214); scutum with submedian dark-scaled longitudinal stripes (fig. 215) .................................................... *Oc. muelleri* (plate 39C)

Abdominal terga with pale-scaled basolateral patches only (fig. 216); scutum without submedian dark- scaled longitudinal stripes (fig. 217) .................................................... 51

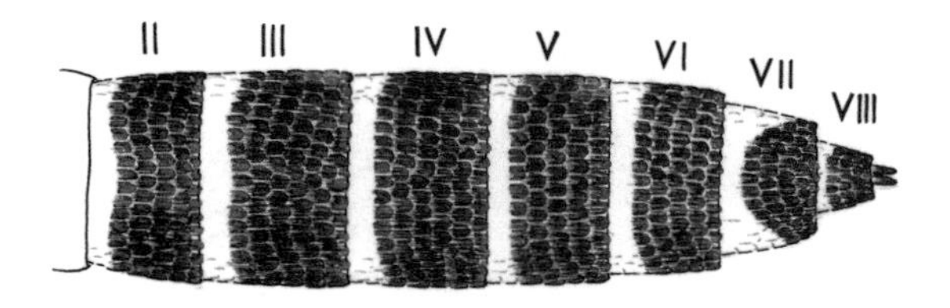


Fig. 214. Dorsal view of abdomen: *Oc. muelleri*

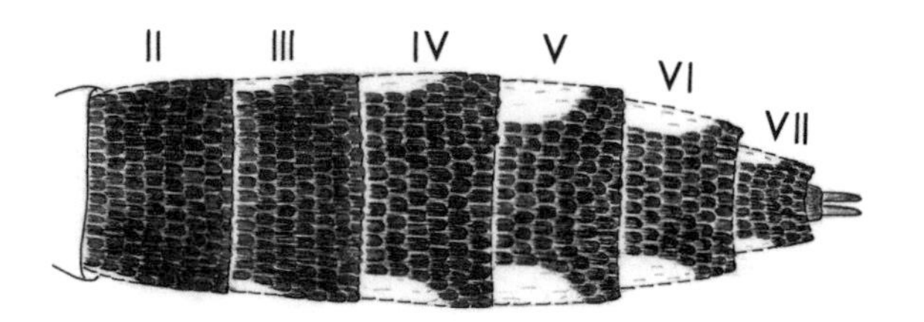


Fig. 216. Dorsal view of abdomen: *Oc. atlanticus*

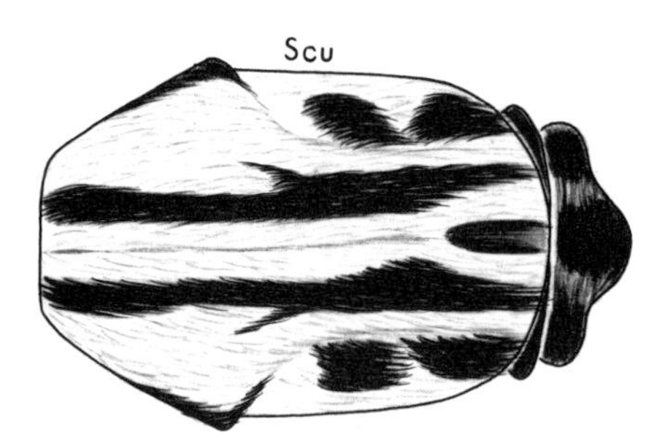


Fig. 215. Dorsal view of scutum: *Oc. muelleri*

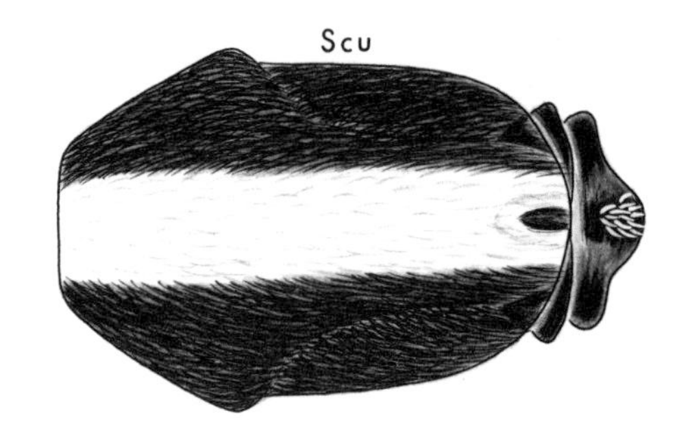


Fig. 217. Dorsal view of scutum: *Oc. atlanticus*

51(50). Occiput with few or no dark scales laterally (fig. 218); small species, wing length about 2.5 mm .................................................... *Oc. dupreei* (plate 13C)

Occiput with prominent spots of dark appressed scales laterally (fig. 219); medium sized species, wing length 3.0–4.0 mm .......................................... *Oc. atlanticus* (plate 10B)
*Oc. tormentor* (plate 23A)

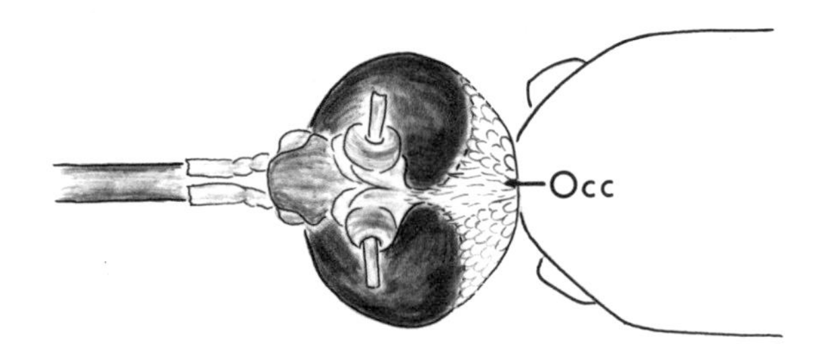


Fig. 218. Dorsal view of head: *Oc. dupreei*

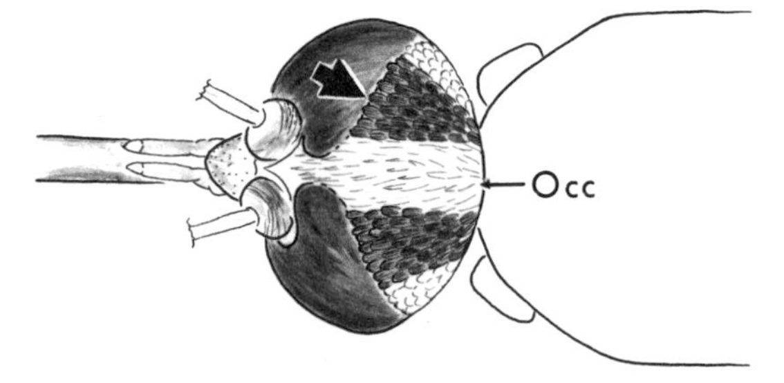


Fig. 219. Dorsal view of head: *Oc. atlanticus*

52(42). Wing with many pale scales either confined to anterior veins, with some on all veins, or on alternating veins dark and pale (figs. 220, 221) .................................................... 53

Wing veins entirely dark or with pale scales at base of vein C and sometimes Sc and R (fig. 222) .................................................... 58

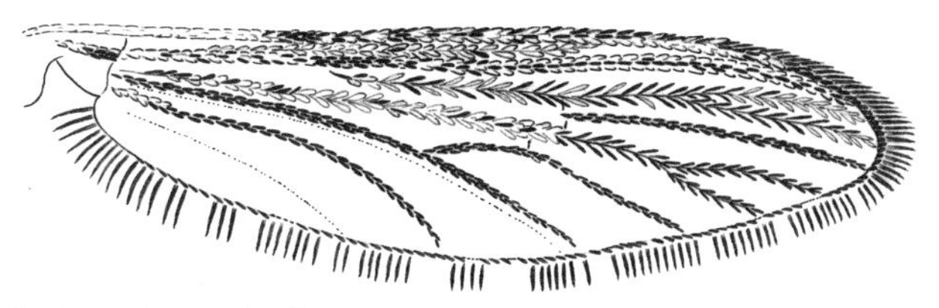

Fig. 220. Dorsal view of wing: *Oc. niphadopsis*

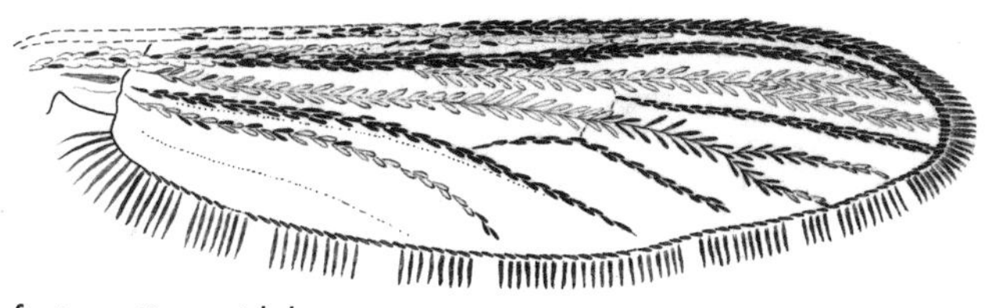

Fig. 221. Dorsal view of wing: *Oc. s. idahoensis*

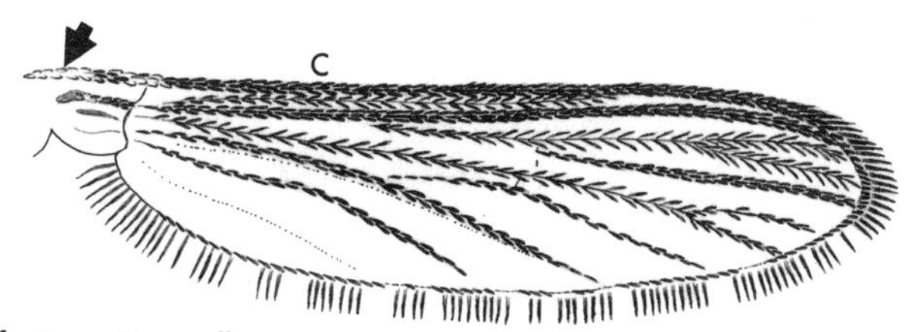


Fig. 222. Dorsal view of wing: *Oc. pullatus*

53(52). Wing with veins alternating dark and pale scales, $R_1$, $R_{4+5,}$ and Cu dark, others pale (fig. 223) .......... 54

Wing with pale scales scattered over all veins or confined to anterior veins (fig. 224) .... 55

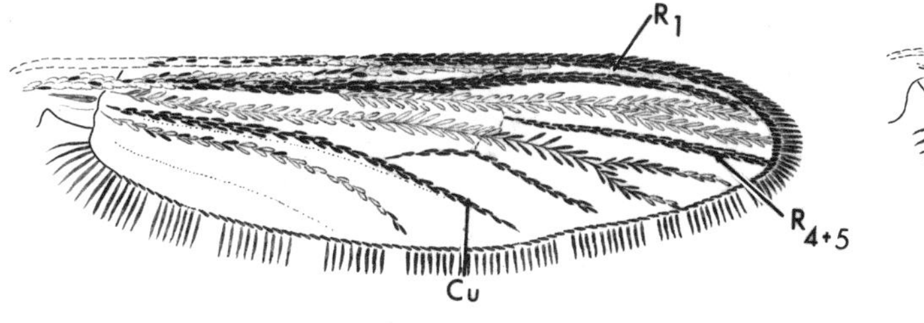


Fig. 223. Dorsal view of wing: *Oc. s. idahoensis*

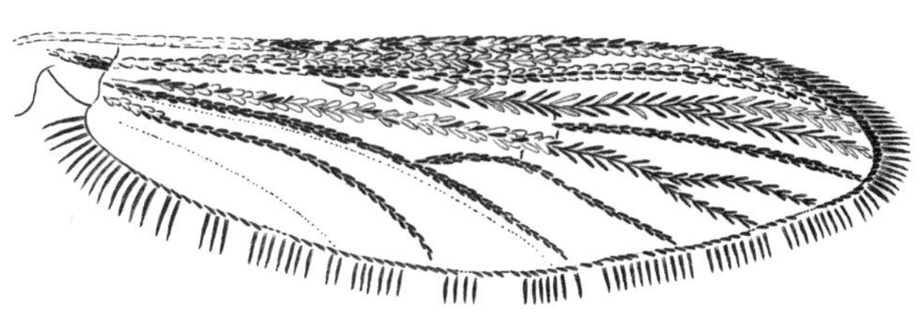

Fig. 224. Dorsal view of wing: *Oc. niphadopsis*

54(53). Abdominal terga with dorsal median longitudinal stripe of pale scales, or almost entirely pale scaled (fig. 225); scales on dorsal 0.5 of postpronotum brown (fig. 226) .......... *Oc. spencerii spencerii* (plate 21C)

Abdominal terga with only basal bands of pale scales (fig. 227); dorsal 0.5 of postpronotum with some pale scales (fig. 228) .......... *Oc. spencerii idahoensis* (plate 21B)

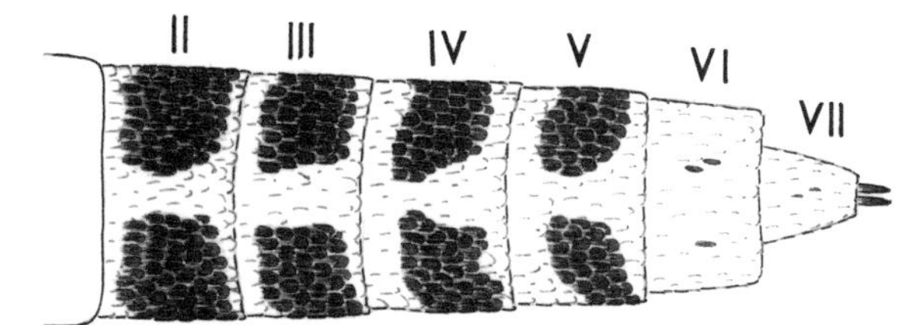


Fig. 225. Dorsal view of abdomen: *Oc. s. spencerii*

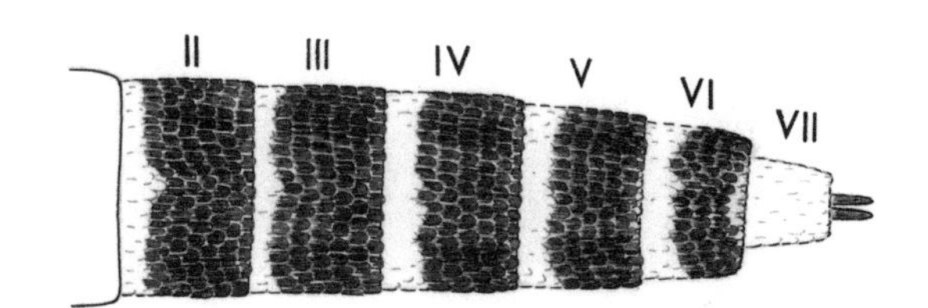


Fig. 227. Dorsal view of abdomen: *Oc. s. idahoensis*

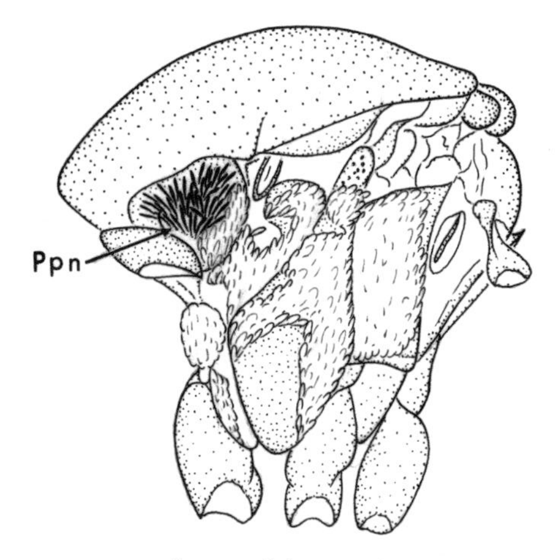


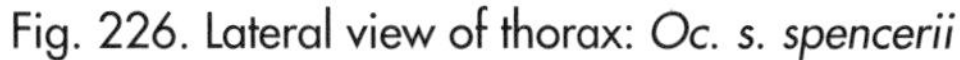

Fig. 226. Lateral view of thorax: *Oc. s. spencerii*

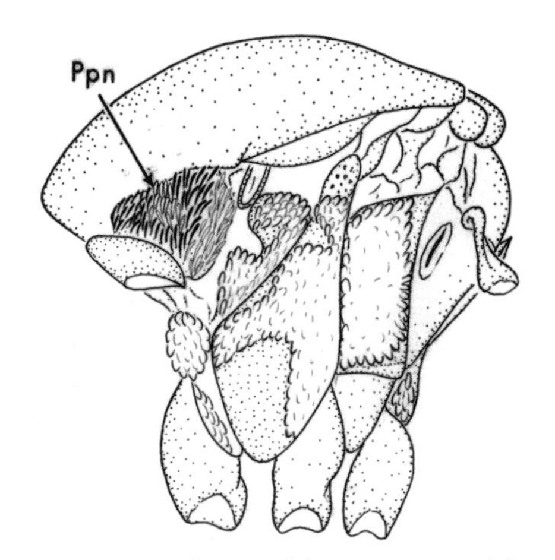


Fig. 228. Lateral view of thorax: *Oc. s. idahoensis*

55(53). Palpus and proboscis dark scaled (fig. 229); lower mesepimeral setae absent (fig. 230) .. ............................................................................ (in part) *Oc. ventrovittis* (plate 23D)

Palpus and proboscis with some pale scales (fig. 231); lower mesepimeral setae present (fig. 232) .......................................................................................................... 56

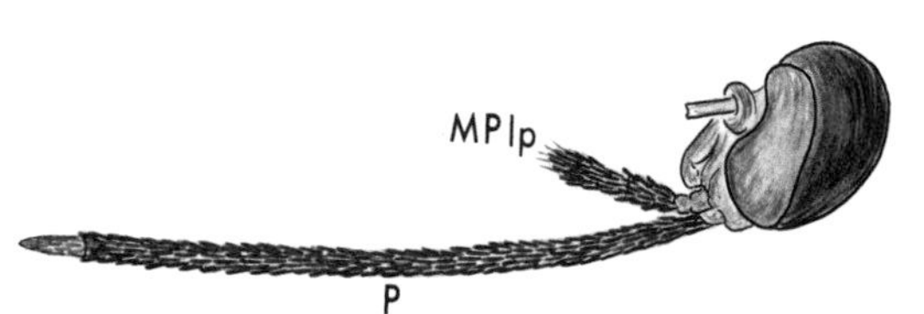


Fig. 229. Lateral view of head: *Oc. ventrovittis*

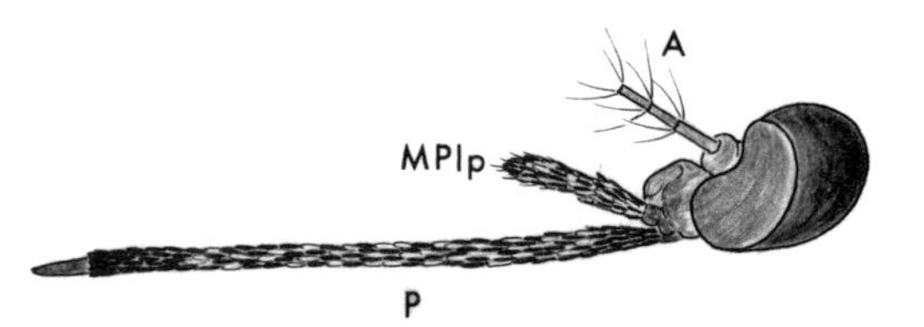


Fig. 231. Lateral view of head: *Oc. bicristatus*

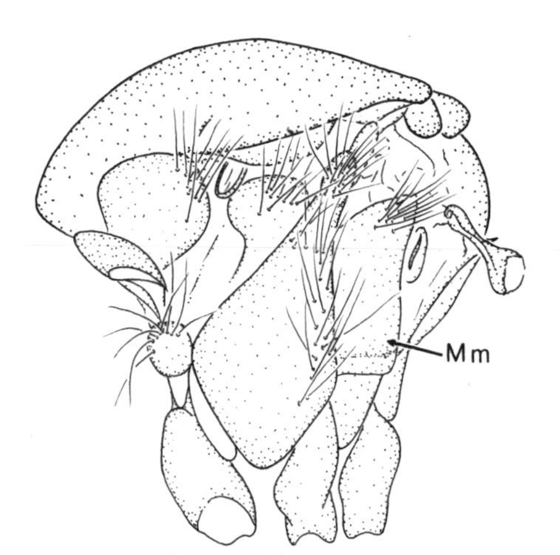


Fig. 230. Lateral view of thorax: *Oc. ventrovittis*

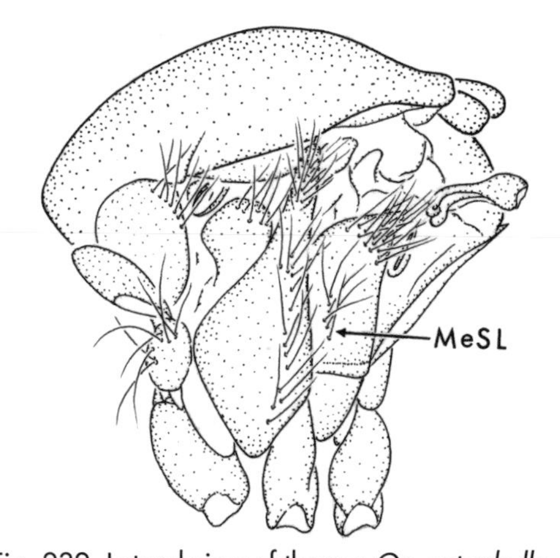


Fig. 232. Lateral view of thorax: *Oc. cataphylla*

56(55). Abdominal terga with broad basal pale bands and apical pale scales, often forming median longitudinal stripe (fig. 233); pale scales numerous on wing veins anterior to Cu (fig. 234) .................................................................................... *Oc. niphadopsis* (Plate 40A)

Abdominal terga with narrow basal pale bands without apical pale scales or longitudinal stripe (fig. 235); pale scales on wing confined to base of C and few scattered along C, Sc, and $R_1$ (fig. 236) .................................................................................................. 57

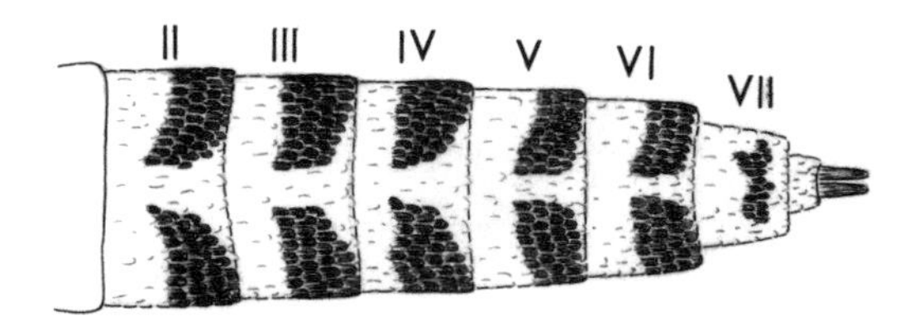


Fig. 233. Dorsal view of abdomen: *Oc. niphadopsis*

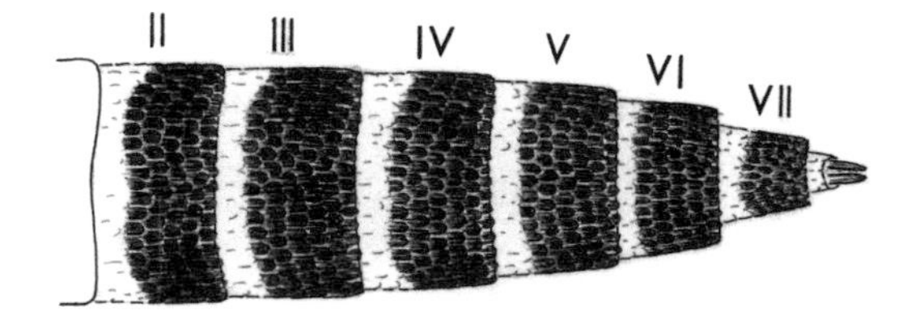


Fig. 235. Dorsal view of abdomen: *Oc. cataphylla*

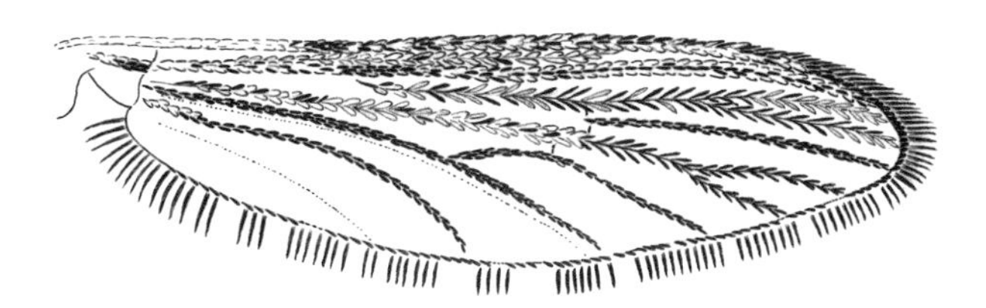

Fig. 234. Dorsal view of wing: *Oc. niphadopsis*

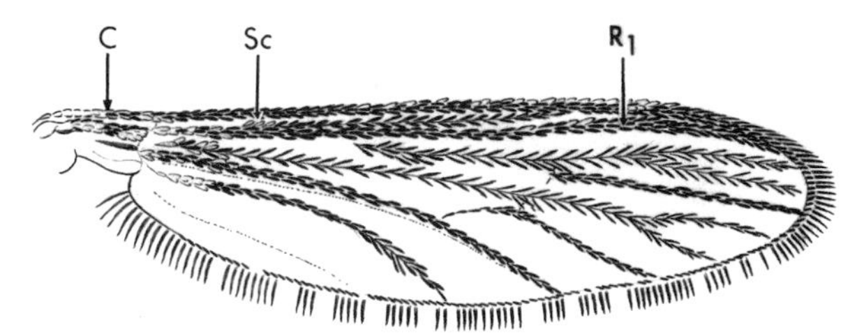


Fig. 236. Dorsal view of wing: *Oc. cataphylla*

57(56). Scutum with area of broad curved scales laterally at level of mesothoracic spiracle (fig. 237); palpi longer than basal 3 antennal flagellomeres (fig. 238) .................................................................................................................. *Oc. bicristatus* (plate 39A)

Scutum with only narrow scales laterally at level of mesothoracic spiracle (fig. 239); palpi shorter than basal 3 antennal flagellomeres (fig. 240) ........... *Oc. cataphylla* (plate 11D)

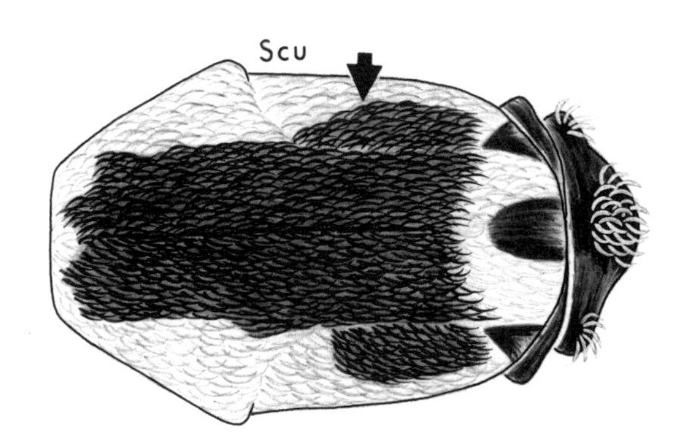


Fig. 237. Dorsal view of thorax: *Oc. bicristatus*

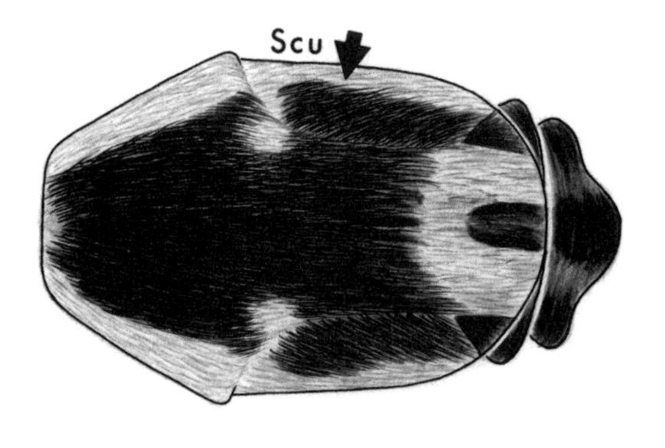


Fig. 239. Dorsal view of thorax: *Oc. cataphylla*

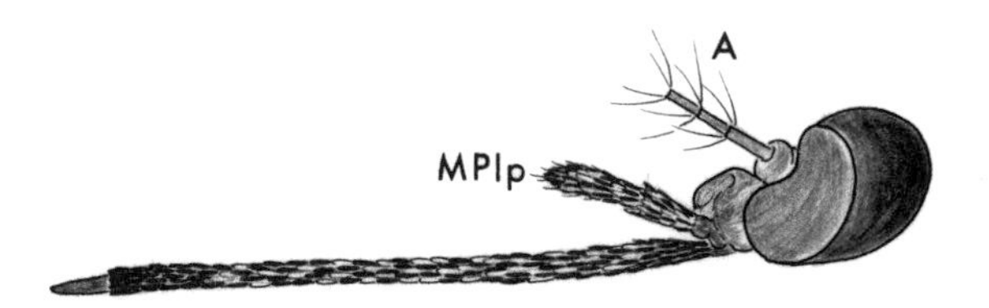


Fig. 238. Lateral view of head: *Oc.* bicristatus

Fig. 240. Lateral view of head: *Oc. cataphylla*

58(52). Hypostigmal area with scales (fig. 241) ........................................................................ 59

Hypostigmal area without scales (fig. 242) ................................................................ 62

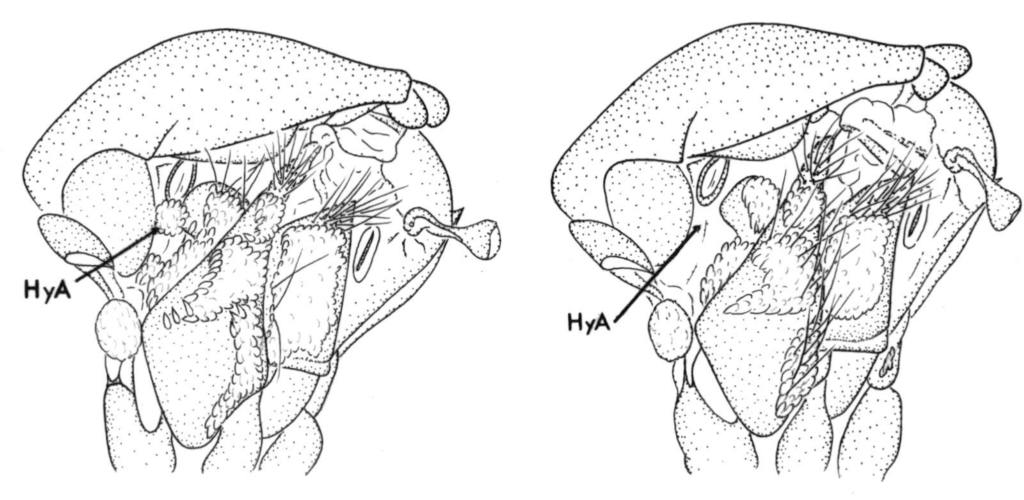


Fig. 241. Lateral view of thorax: *Oc. pullatus*

Fig. 242. Lateral view of thorax: *Oc. diantaeus*

59(58). Postprocoxal scale patch absent (fig. 243); palpi usually with some pale scales (fig. 244) ............................................................ 60

Postprocoxal scale patch present (fig. 245); palpi entirely dark scaled (fig. 246) ......... 61

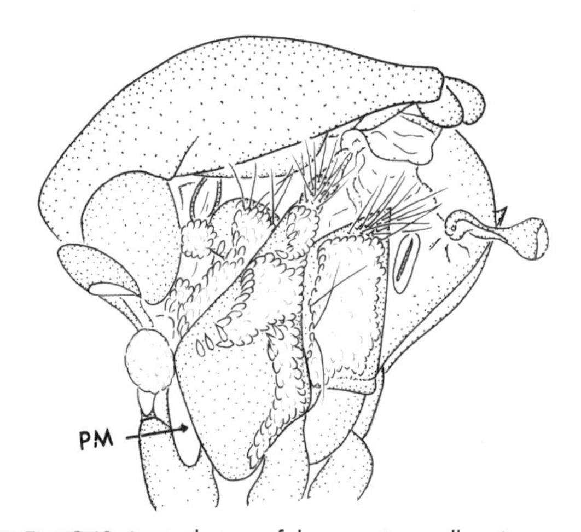


Fig. 243. Lateral view of thorax: *Oc. pullatus*)

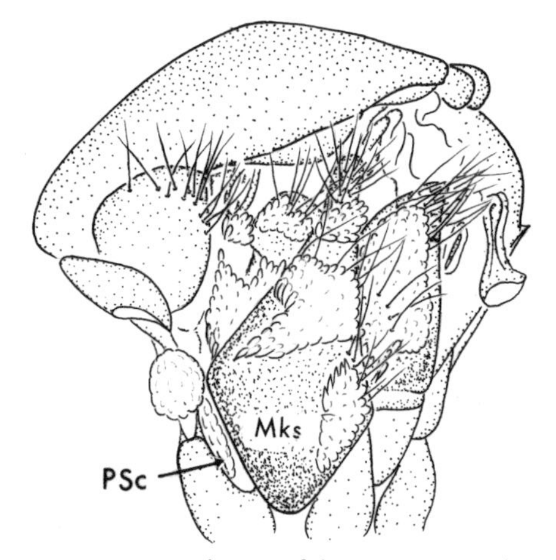


Fig. 245. Lateral view of thorax: *Oc. implicatus*

Fig. 244. Lateral view of head: *Oc. pullatus*

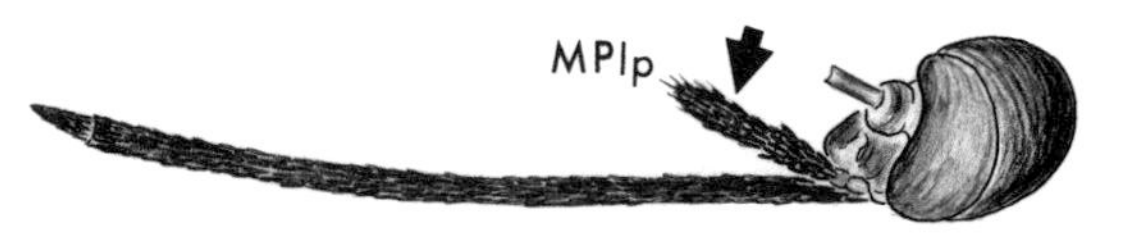


Fig. 246. Lateral view of head: *Oc. implicatus*

60(59). Scutum with scales nearly all unicolorous (fig. 247); mesepimeron sometimes without scales in ventral 0.25 (fig. 248) .................................. (in part) *Oc. intrudens* (plate 17A)

Scutum with pair of submedian stripes devoid of scales (fig. 249); mesepimeron usually with scales near to ventral margin (fig. 250) .............................. *Oc. pullatus* (plate 19B)

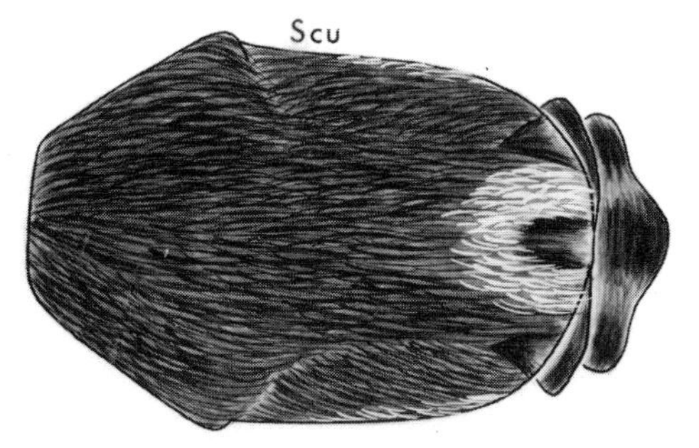


Fig. 247. Dorsal view of scutum: *Oc. intrudens*

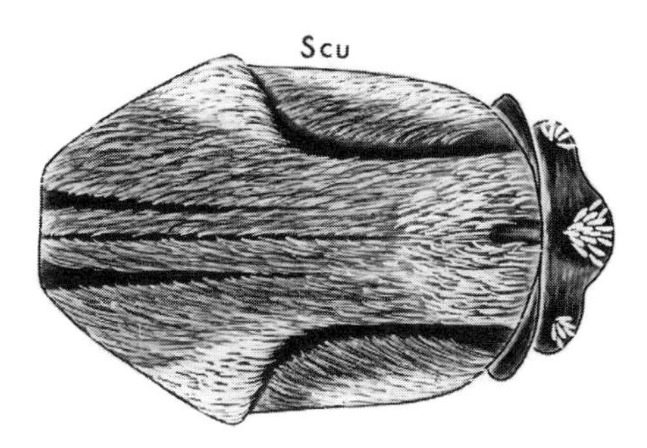


Fig. 249. Dorsal view of scutum: *Oc. pullatus*

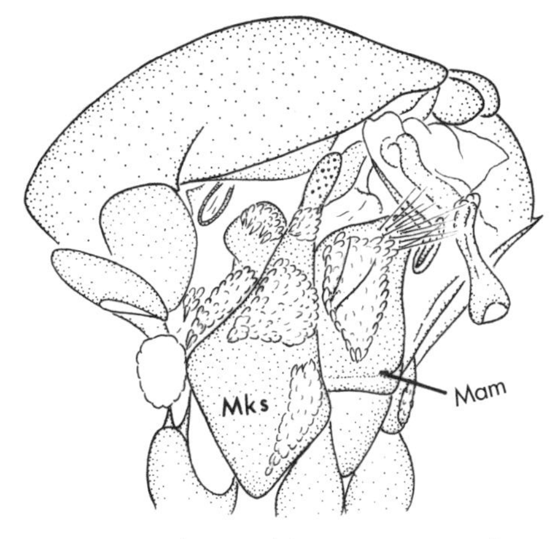


Fig. 248. Lateral view of thorax: *Oc. intrudens*

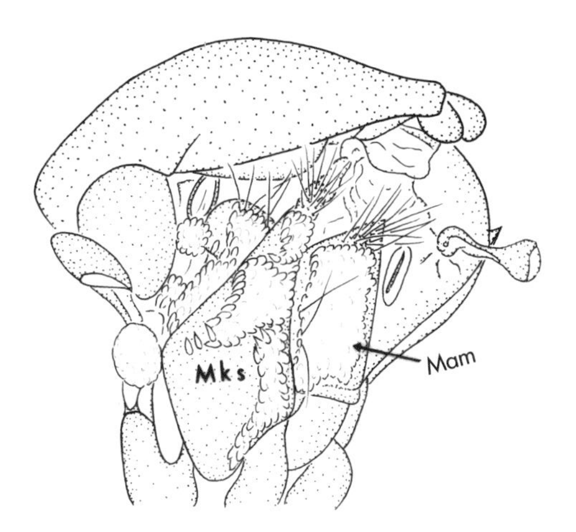


Fig. 250. Lateral view of thorax: *Oc. pullatus*

61(59). Mesokatepisternum with scales not extending to anterior angle, separated dorsally from posterior mesanepisternal scale patch (fig. 251); pale-scaled knee spots on all femora (fig. 252) .......................................................................(in part) *Oc. implicatus* (plate 16B)

Mesokatepisternum with scales extending to anterior angle, not separated from posterior mesanepisternal scale patch (fig. 253); femora without knee spots (fig. 254) .................. ....................................................................................... *Oc. provocans* (plate 19A)

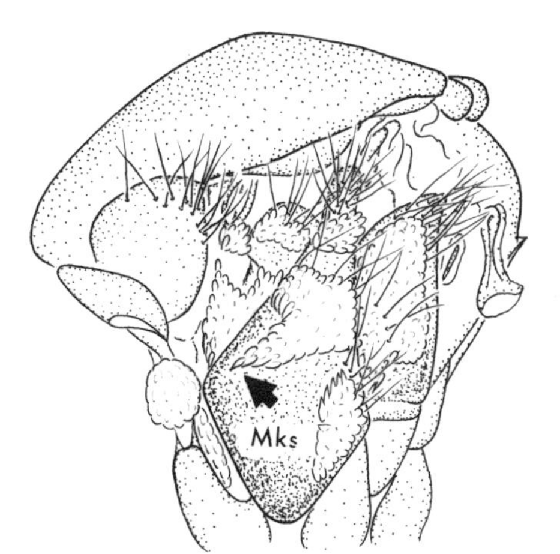


Fig. 251. Lateral view of thorax: *Oc. implicatus*

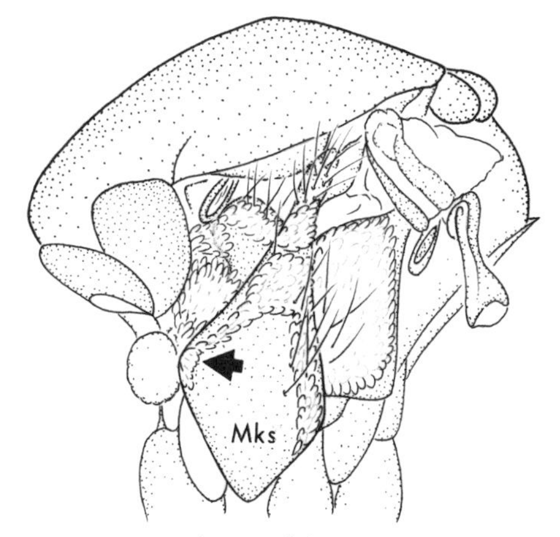


Fig. 253. Lateral view of thorax: *Oc. provocans*

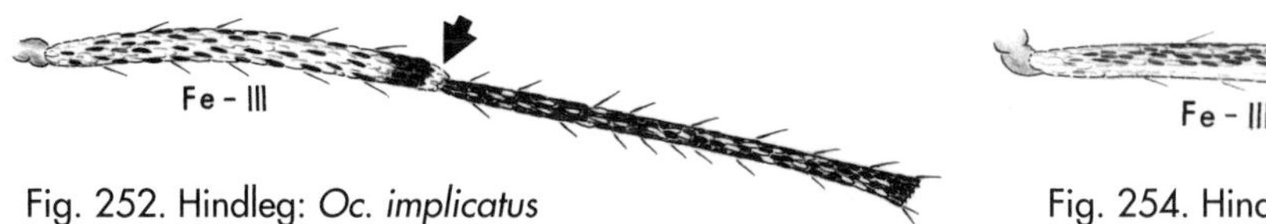


Fig. 252. Hindleg: *Oc. implicatus*

Fig. 254. Hindleg: *Oc. provocans*

62(58). Abdominal terga without basal pale bands, or if present, on fewer than 0.5 of segments (fig. 255) ........ 63

Abdominal terga usually with pale basal bands on segments I–VII, at least on more than 0.5, or if absent, then with stripe of pale scales laterally (fig. 256) ........ 66

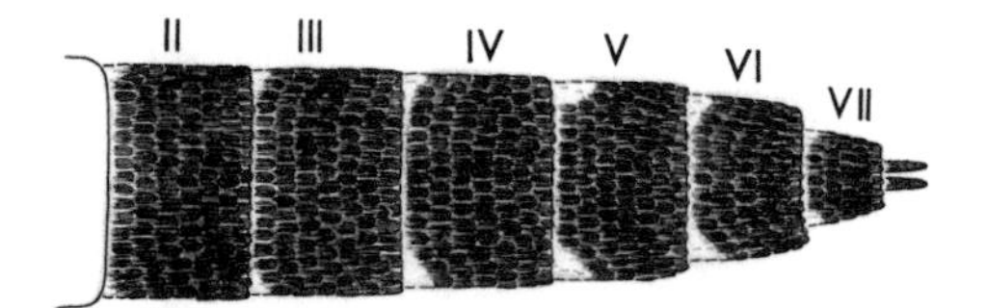


Fig. 255. Dorsal view of abdomen: *Oc. diantaeus*

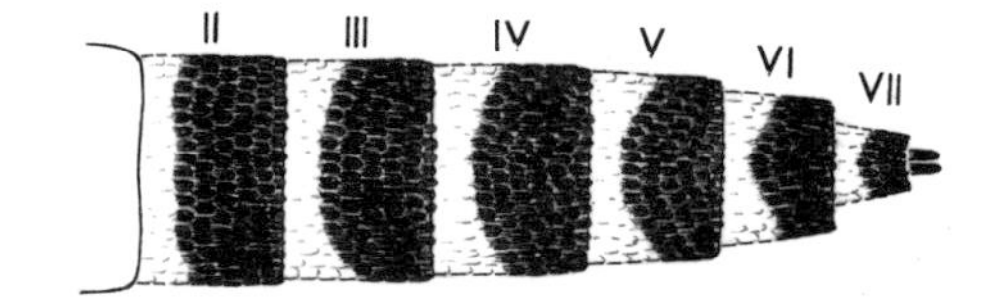


Fig. 256. Dorsal view of abdomen: *Oc. intrudens*

63(62). Abdominal sterna entirely pale scaled (fig. 257); forecoxa with at least some brown scales (fig. 258) ........ *Oc. aurifer* (plate 10D)

At least some abdominal sterna with dark scales apically (fig. 259); forecoxa with all pale scales (fig. 260) ........ 64

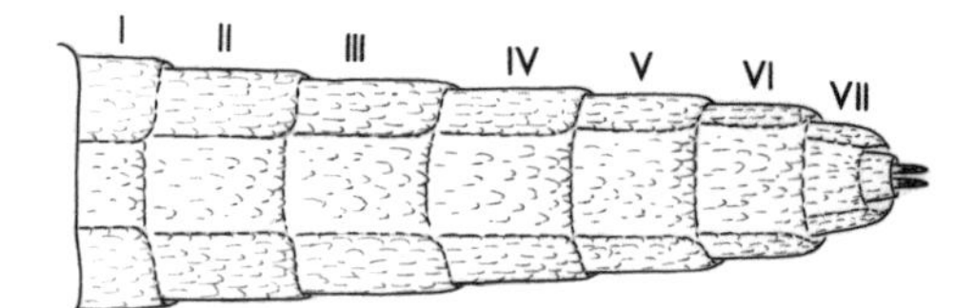


Fig. 257. Ventral view of abdomen: *Oc. aurifer*

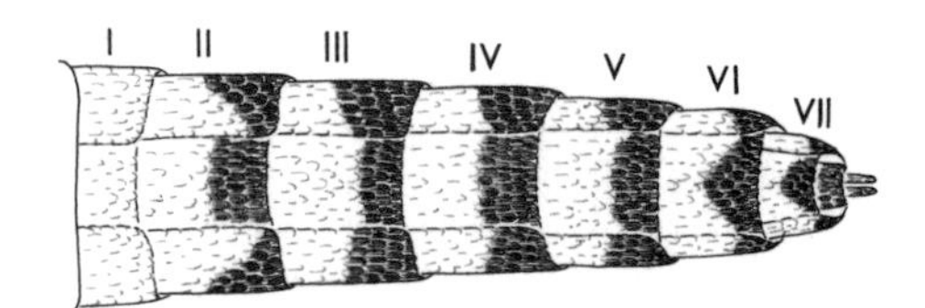


Fig. 259. Ventral view of abdomen: *Oc. thibaulti*

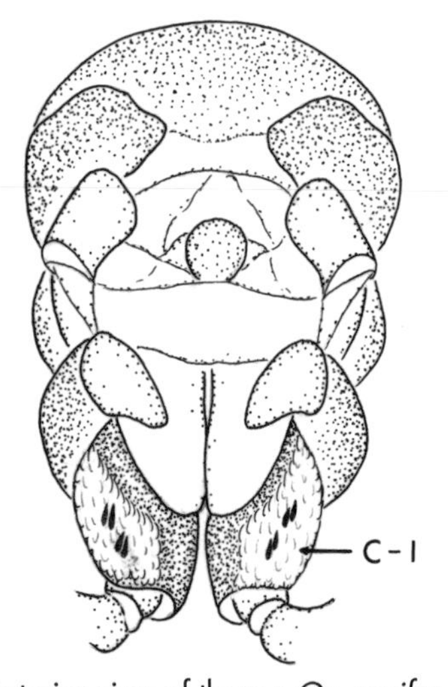


Fig. 258. Anterior view of thorax: *Oc. aurifer*

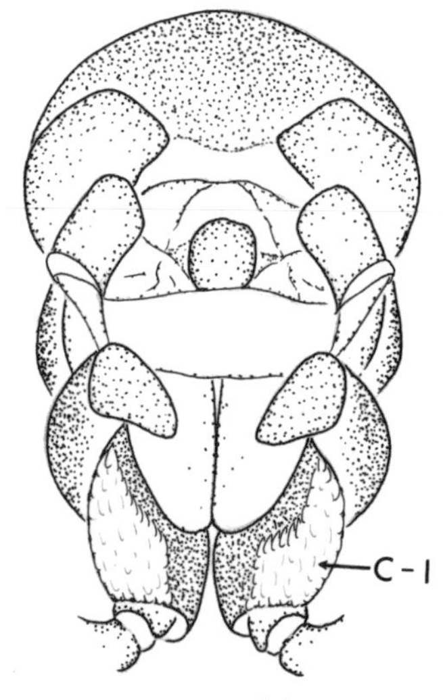


Fig. 260. Anterior view of thorax: *Oc. thibaulti*

64(63). Scutum with broad median longitudinal stripe of dark scales, broadening abruptly just posterior to scutal angle (fig. 261) ........ *Oc. thibaulti* (plate 22D)

Scutum with 2 narrower brown-scaled stripes, sometimes fused, if so, not distinctly broadening posteriorly (fig. 262) ........ 65

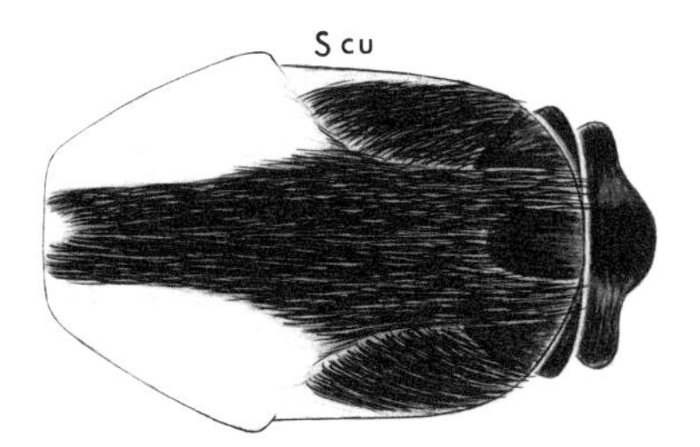


Fig. 261. Dorsal view of scutum: *Oc. thibaulti*

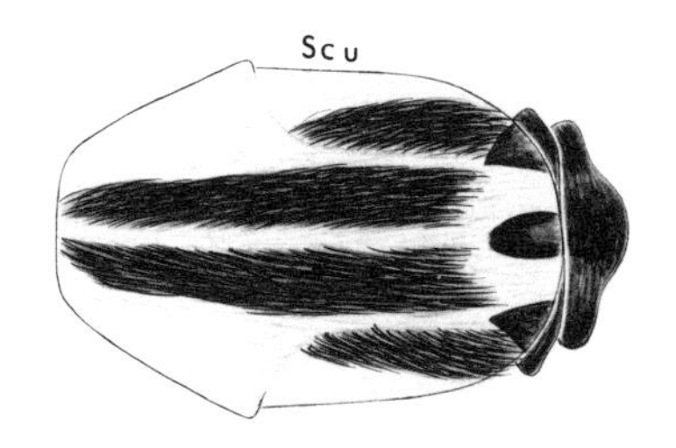


Fig. 262. Dorsal view of scutum: *Oc. decticus*

65(64). Mesokatepisternum with fewer than 10 setae, usually 5,6 (fig. 263); occiput with submedian spots of dark scales (fig. 264); metameron bare (fig. 263) .... *Oc. decticus* (plate 12D)

Mesokatepisternum with 10–20 setae (fig. 265); submedian spots on occiput lacking (fig. 266); metameron with small scale patch (fig. 265) ................. *Oc. diantaeus* (plate 13A)

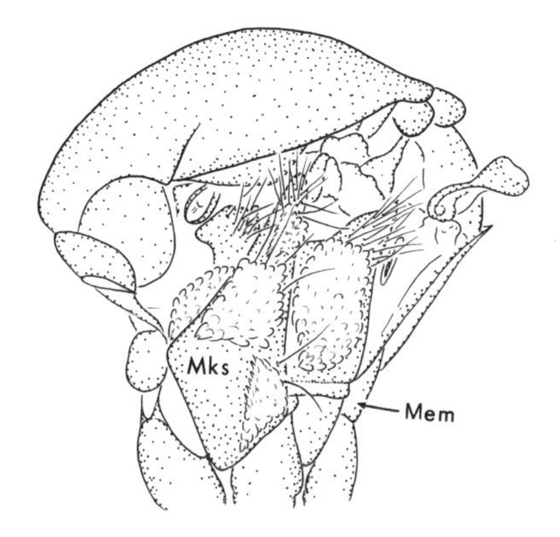


Fig. 263. Lateral view of thorax: *Oc. decticus*

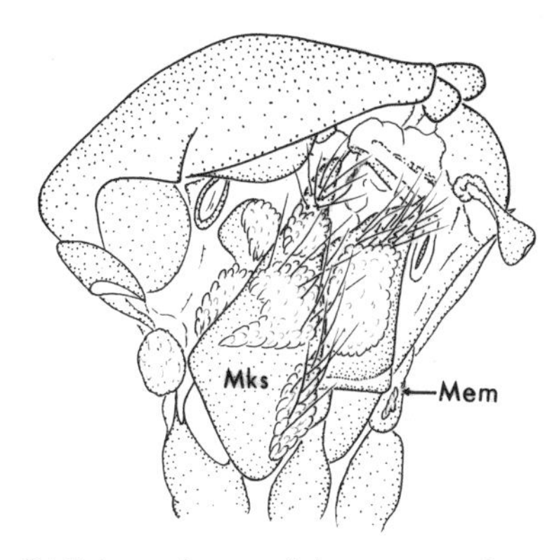


Fig. 265. Lateral view of thorax: *Oc. diantaeus*

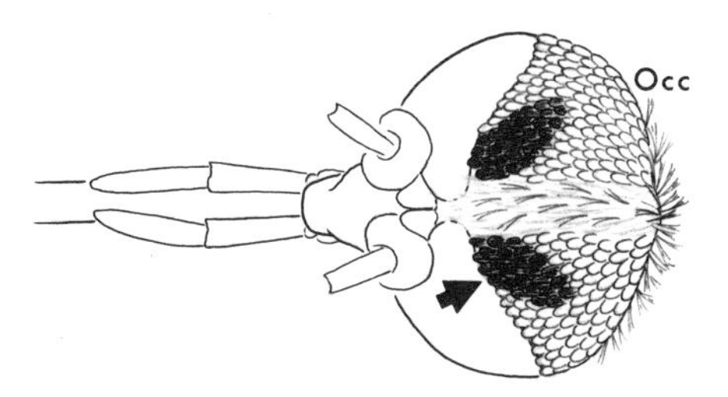


Fig. 264. Dorsal view of head: *Oc. decticus*

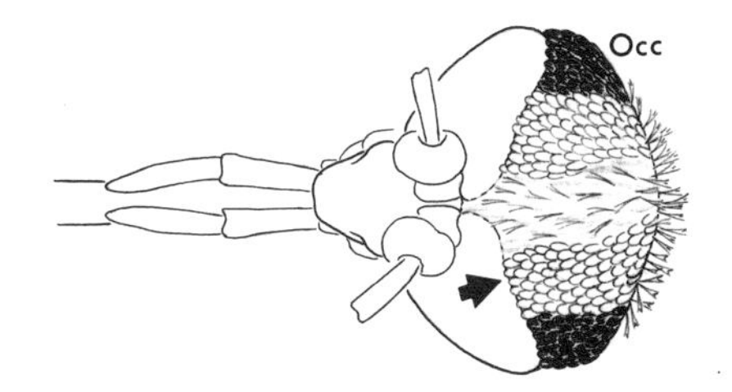


Fig. 266. Dorsal view of head: *Oc. diantaeus*

66(62). Postprocoxal scale patch absent (fig. 267) ................................................................. 67

Postprocoxal scale patch present (fig. 268) ................................................................. 75

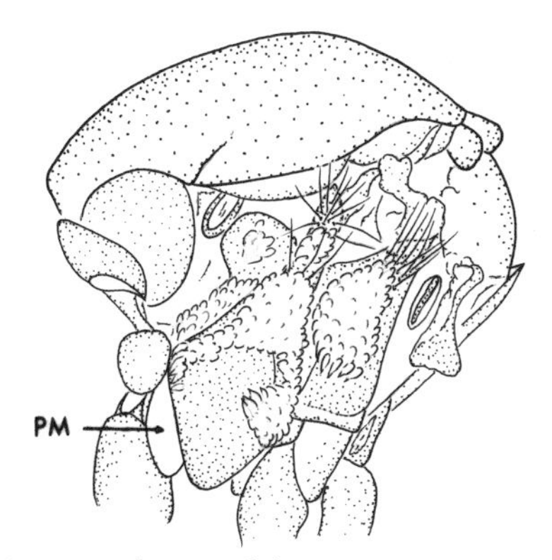


Fig. 267. Lateral view of thorax: *Oc. sticticus*

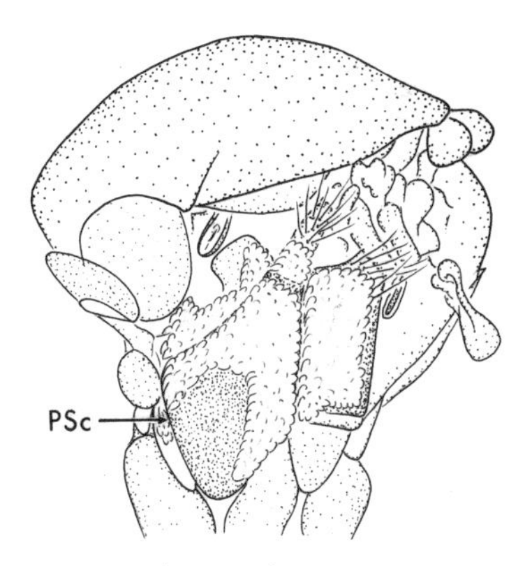


Fig. 268. Lateral view of thorax: *Oc. punctor*

67(66). Abdominal terga with median basal triangular patches of pale scales (fig. 269) .............. ................................................................................ *Oc. thelcter* (plate 22C)

Abdominal terga with basal pale scales with other pattern (fig. 270) .......................... 68

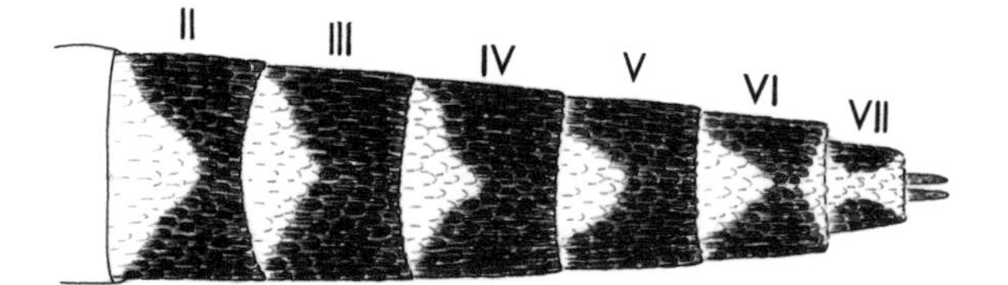


Fig. 269. Dorsal view of abdomen: *Oc. thelcter*

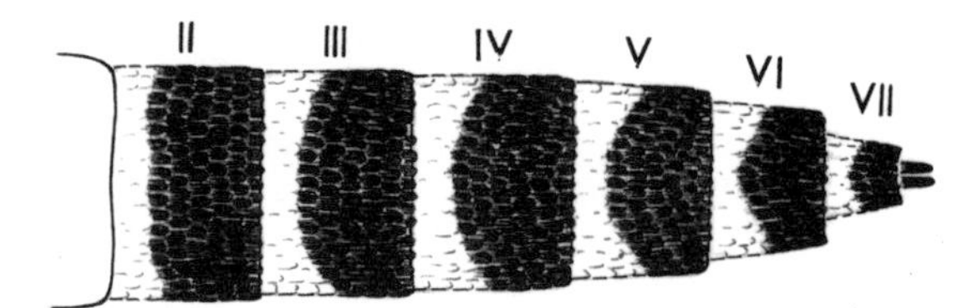


Fig. 270. Dorsal view of abdomen: *Oc. intrudens*

68(67). Scutum with unicolorous scales, or if median longitudinal stripe, its scales lighter than those laterally (fig. 271); mesokatepisternum with scales usually not extending to anterior angle (fig. 272) .................................................................................................. 69

Scutum with dark median longitudinal stripe (fig. 273); mesokatepisternum with scales extending to near anterior angle (fig. 274) ................................................................. 71

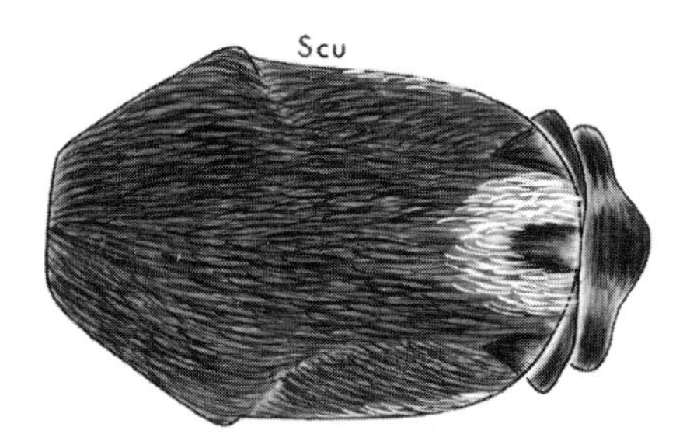


Fig. 271. Dorsal view of scutum: *Oc. intrudens*

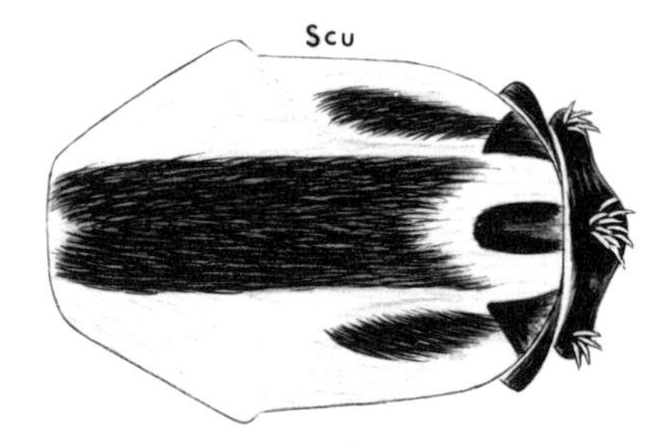


Fig. 273. Dorsal view of scutum: *Oc. sticticus*

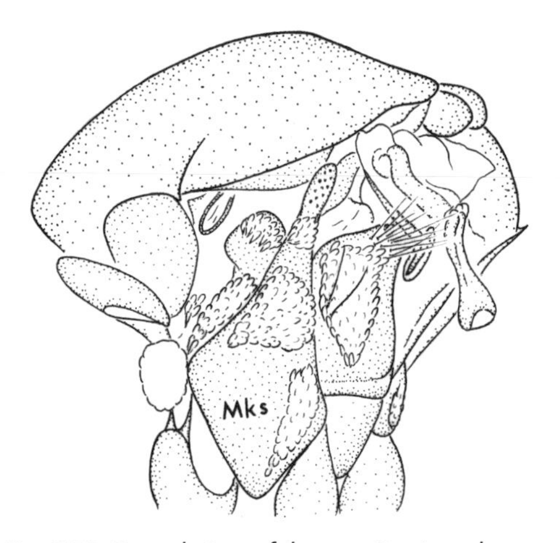


Fig. 272. Dorsal view of thorax: *Oc. intrudens*

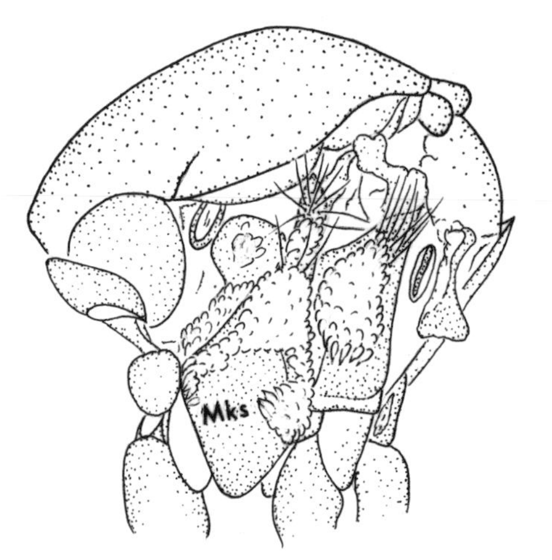


Fig. 274. Lateral view of thorax: *Oc. sticticus*

69(68). Forecoxa with patch of brown scales (fig. 275); subspiracular area bare (fig. 276) ......... ................................................................................ *Ae. cinereus* (plate 12B)

Forecoxa with pale scales, or with few dark scales only (fig. 277); subspiracular area with scales (fig. 278) .................................................................................................. 70

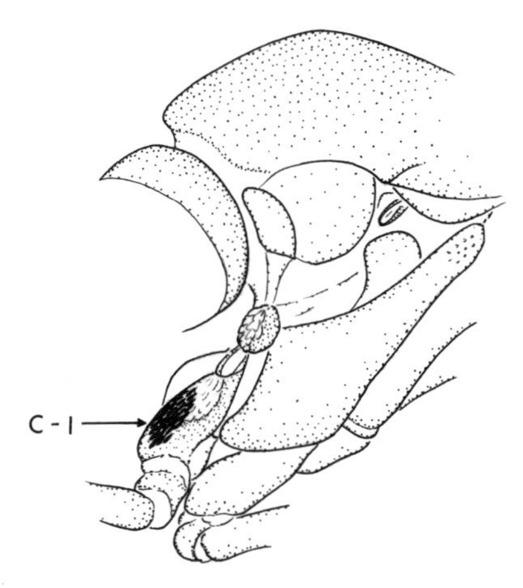


Fig. 275. Anterior view of thorax: *Ae. cinereus*

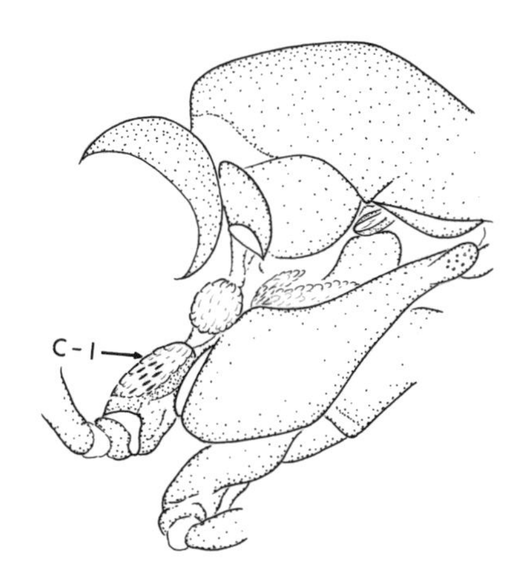


Fig. 277. Anterior view of thorax: *Oc. intrudens*

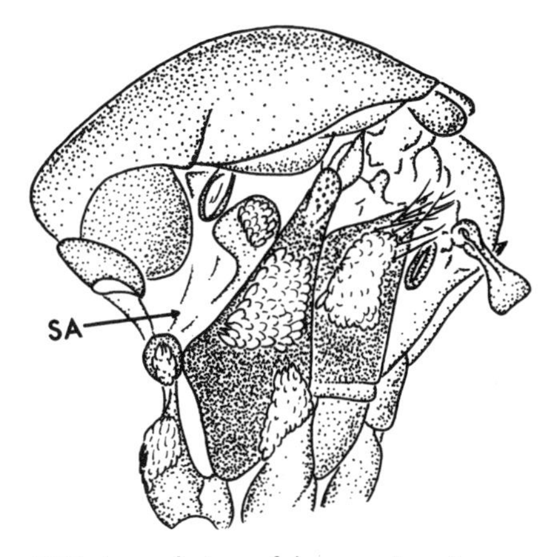


Fig. 276. Lateral view of thorax: *Ae. cinereus*

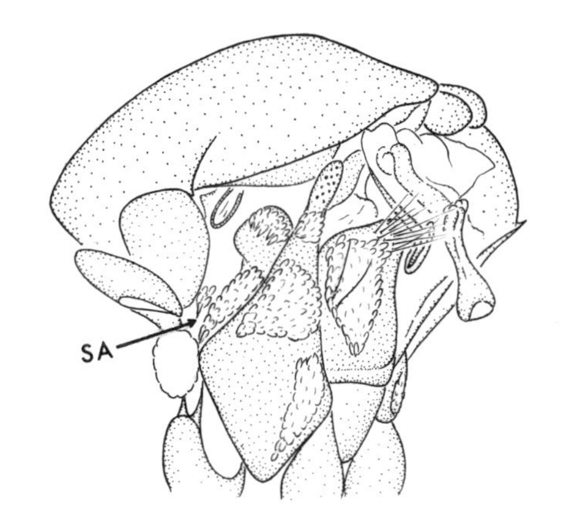


Fig. 278. Lateral view of thorax: *Oc. intrudens*

70(69). Lower mesepimeral setae present; metameron with scales (fig. 279) .............................. ........................................................................(in part) *Oc. intrudens* (plate 17A)

Lower mesepimeral setae absent; metameron bare (fig. 280) ....... *Oc. tortilis* (plate 39C)

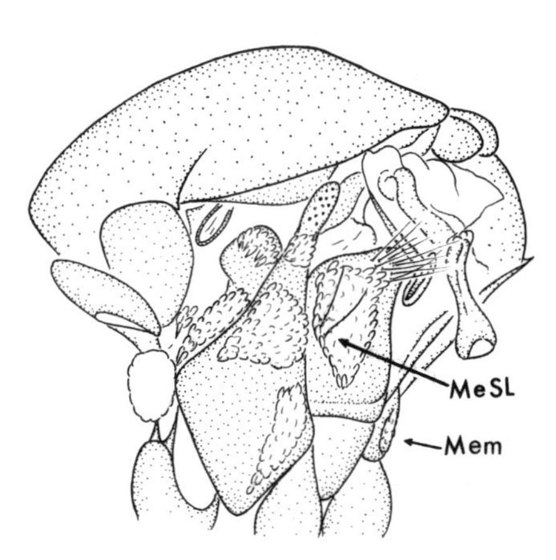


Fig. 279. Lateral view of thorax: *Oc. intrudens*

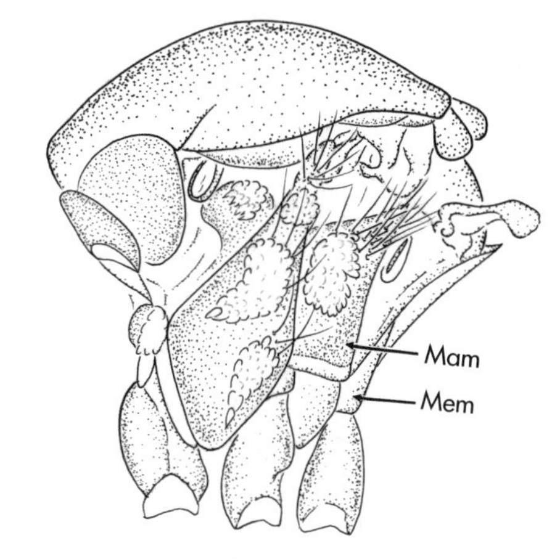


Fig. 280. Lateral view of thorax: *Oc. tortilis*

71(68). Scutum with submedian dark-scaled longitudinal band, varying in width, especially dark scales in fossa (fig. 281); foreclaw elongate (fig. 282) .................. *Oc. rempeli* (plate 20)

Median or submedian dark-scaled longitudinal band, when present on scutum, more uniform in width throughout, not covering fossa (fig. 283); foreclaw usually sharply curved distad to tooth (fig. 284) .................................................................................... 72

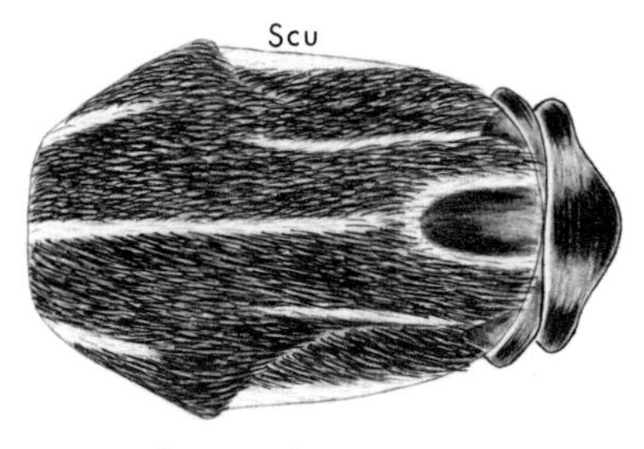


Fig. 281. Dorsal view of scutum: *Oc. rempeli*

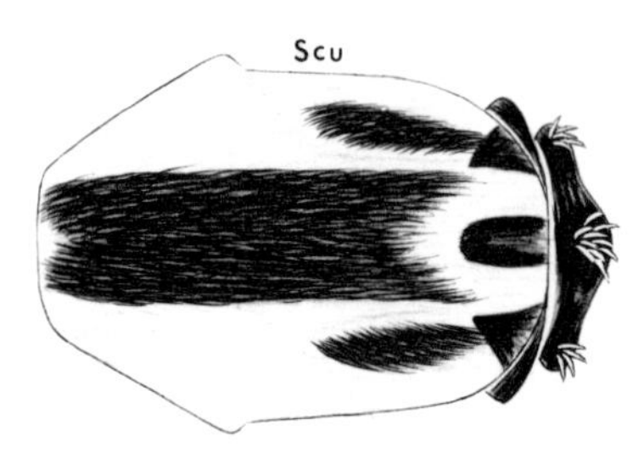


Fig. 283. Dorsal view of scutum: *Oc. sticticus*

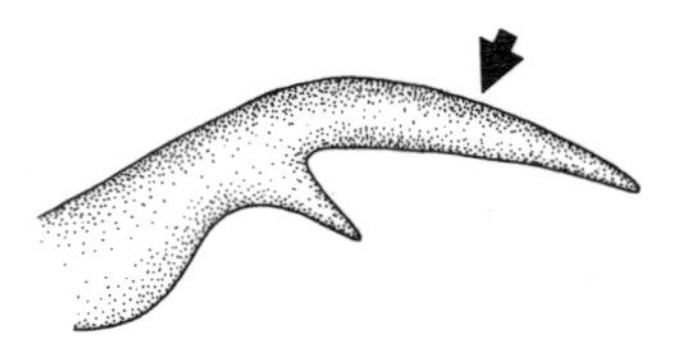

Fig. 282. Foreclaw: *Oc. rempeli*

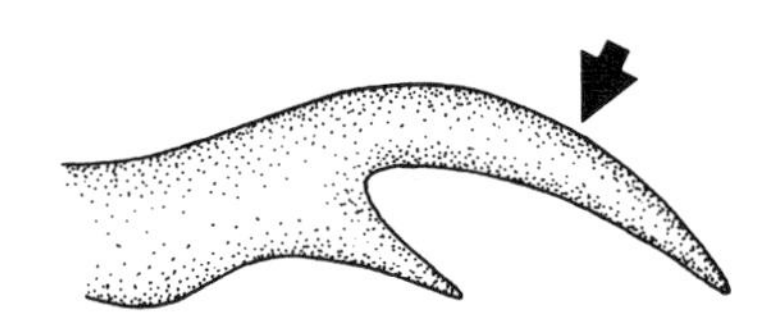

Fig. 284. Foreclaw: *Oc. sticticus*

72(71). Scutellar and supraalar setae yellowish (fig. 285); mesepimeron usually without lower setae, ventral 0.25 devoid of scales (fig. 286) ........................... *Oc. sticticus* (plate 21D)

.. Scutellar and supraalar setae brown or black (fig. 287); mesepimeron with lower seta, ventral 0.25 scaled (fig. 288) ........................................................................... 73

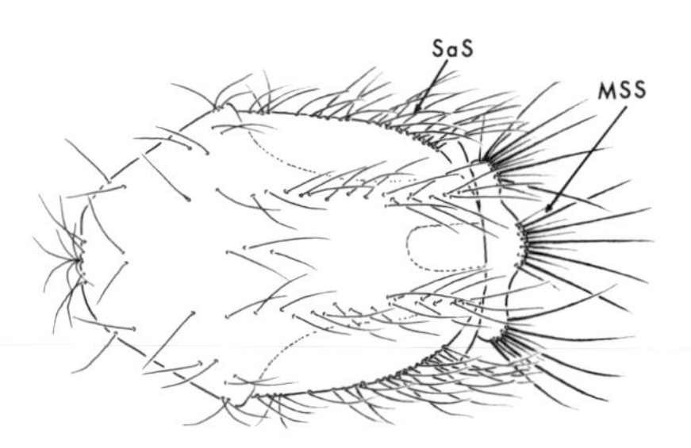


Fig. 285. Dorsal view of thorax: *Oc. sticticus*

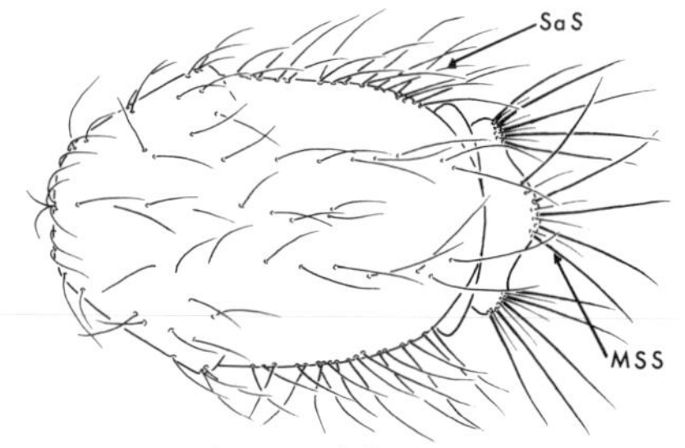


Fig. 287. Dorsal view of thorax: *Oc. communis*

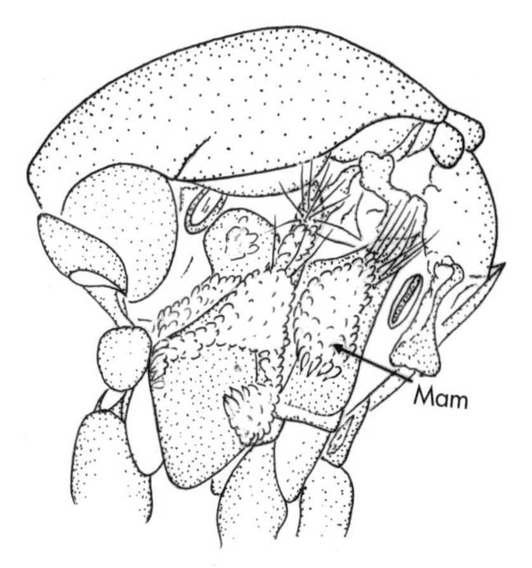


Fig. 286. Lateral view of thorax: *Oc. sticticus*

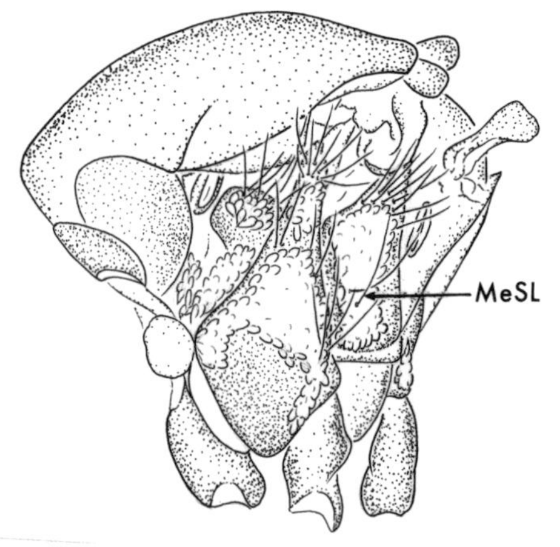


Fig. 288. Lateral view of thorax: *Oc. communis*

73(72). Tooth of hindclaw long, thin, claw usually curved abruptly distal to tooth (fig. 289) ..... ........................................................................................ *Oc. communis* (plate 12D)

Tooth of hindclaw short, claw usually curving more gradually distal to tooth (fig. 290). ........................................................................................................................ 74

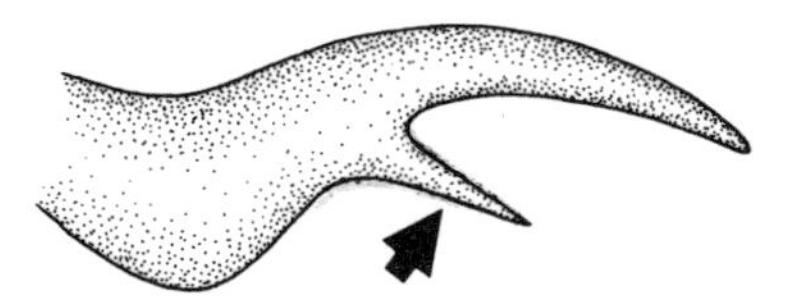

Fig. 289. Hindclaw: *Oc. communis*

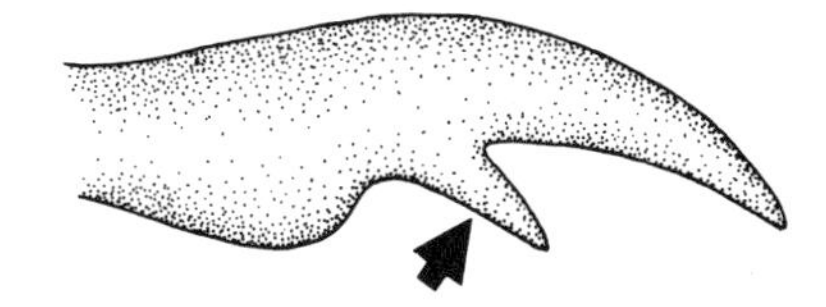

Fig. 290. Hindclaw: *Oc. nevadensis*

74(73). Upper mesepimeral setae usually 17–27, range 14–33; upper mesokatepisternal setae 5–8 (fig. 291) .................................................................................. *Oc. nevadensis* (plate 18A) *Oc. tahoensis* (plate 40D)

Upper mesepimeral setae 12–19, range 10–22; upper mesokatepiternal setae 4,5 (fig. 292) ........................................................................................ *Oc. churchillensis* (plate 12A)

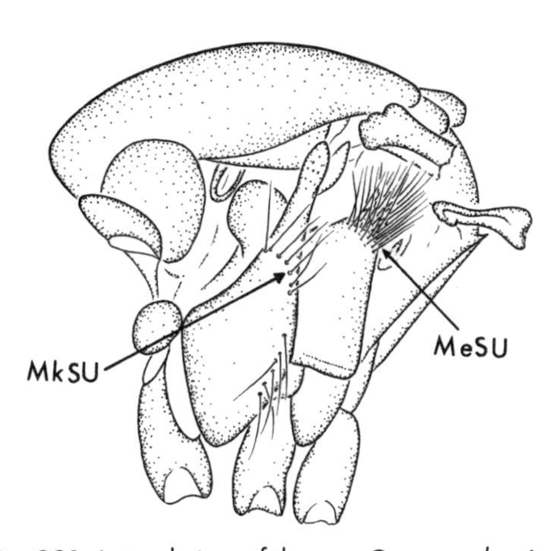


Fig. 291. Lateral view of thorax: *Oc. nevadensis*

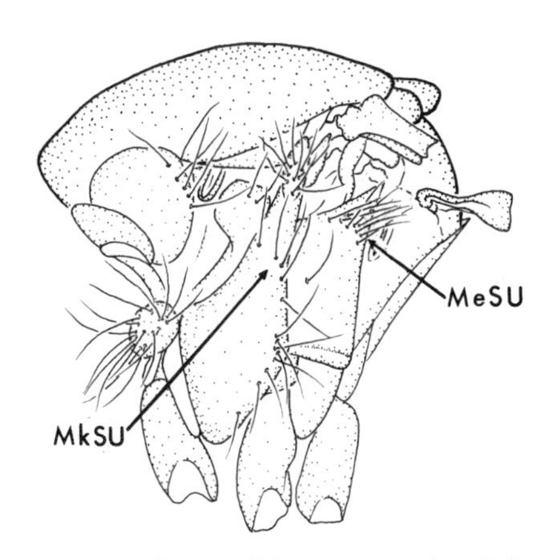


Fig. 292. Lateral view of thorax: *Oc. churchillensis*

75(66). Lower mesepimeral setae absent (fig. 293); pale basal band on abdominal tergum II narrowed, or completely interrupted, medially (fig. 294) ..................................................... ........................................................................... (in part) *Oc. ventrovittis* (plate 23D)

Lower mesepimeral setae present (fig. 295); pale basal band on abdominal tergum II scarcely narrower medially (fig. 296) .......................................................................... 76

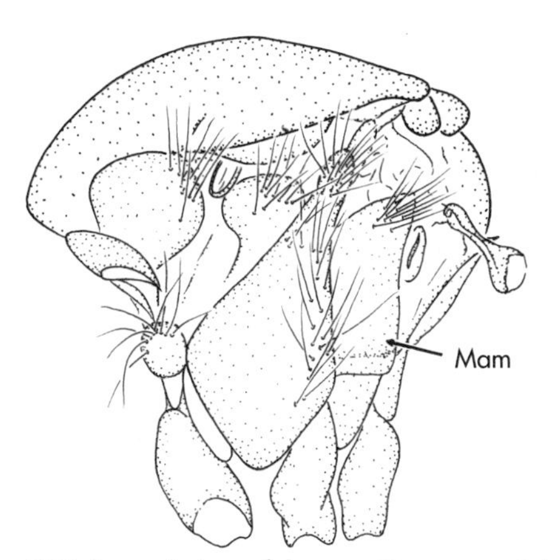


Fig. 293. Lateral view of thorax: *Oc. ventrovittis*

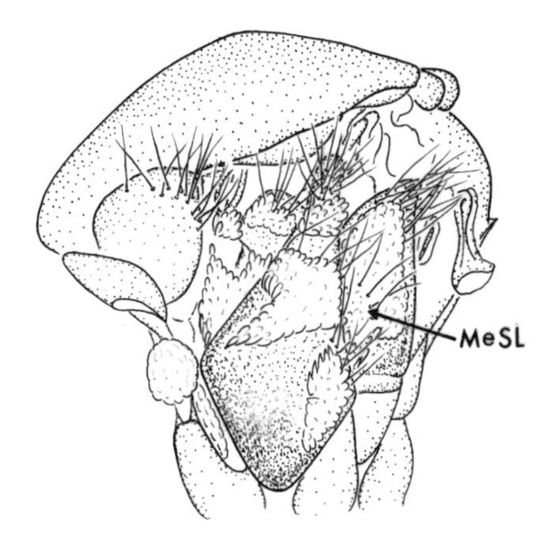


Fig. 295. Lateral view of thorax: *Oc. implicatus*

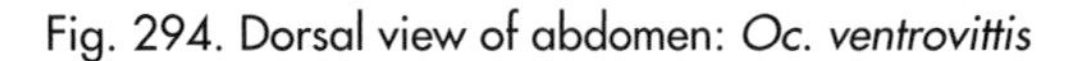

Fig. 294. Dorsal view of abdomen: *Oc. ventrovittis*

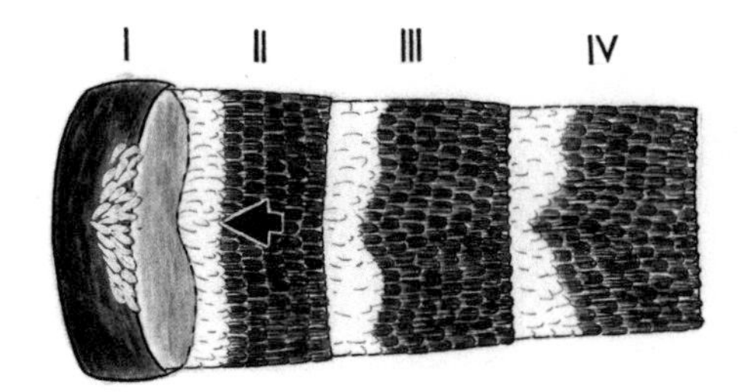


Fig. 296. Dorsal view of abdomen: *Oc. punctor*

76(75). Scutum with many long dark setae, hairy in appearance (fig. 297); postpronotum with setae scattered over posterior 0.5 (fig. 298) ........ 77

Scutum with few long setae, not hairy in appearance (fig. 299); postpronotum with setae in single or irregular double row along posterior border (fig. 300) ........ 78

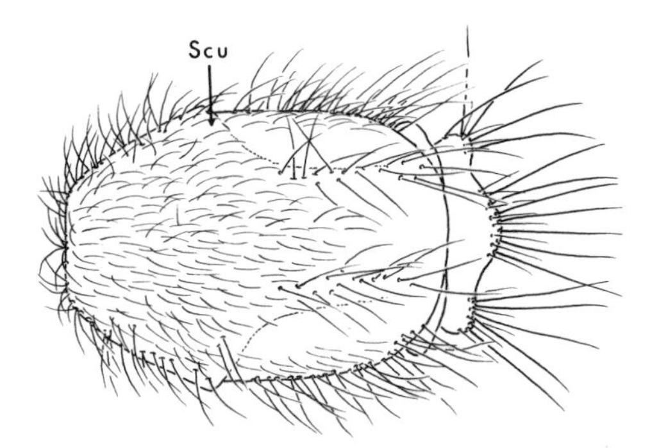


Fig. 297. Dorsal view of thorax: *Oc. impiger*

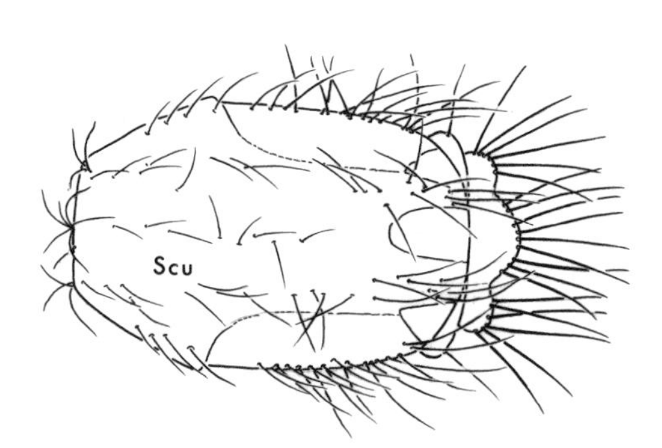


Fig. 299. Dorsal view of thorax: *Oc. pionips*

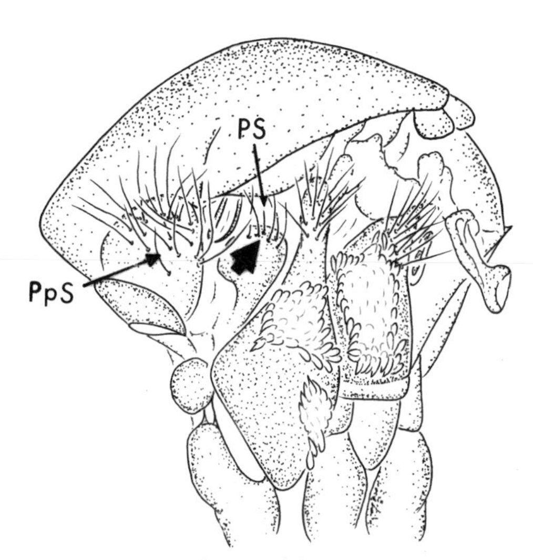


Fig. 298. Lateral view of thorax: *Oc. impiger*

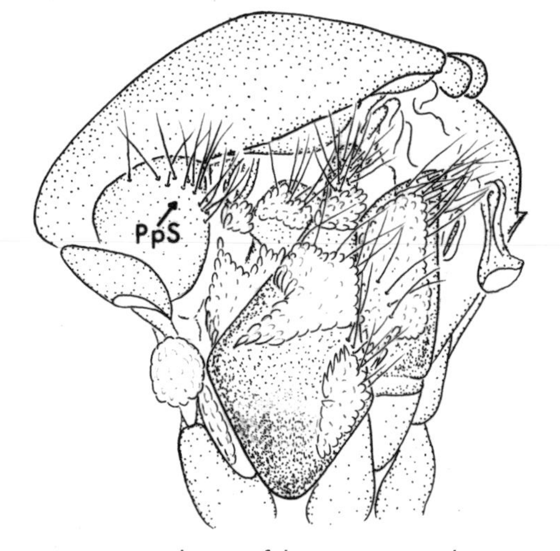


Fig. 300. Lateral view of thorax: *Oc. implicatus*

77(76). Foreclaw sharply bent apical to long tooth (fig. 301); postspiracular setae numbering 10 or fewer (fig. 302) ........ *Oc. impiger* (plate 16A)

Foreclaw elongate, very gradually curving distal to short tooth (fig. 303); postspiracular setae numbering 14 or more (fig. 304) ........ *Oc. nigripes* (plate 18B)

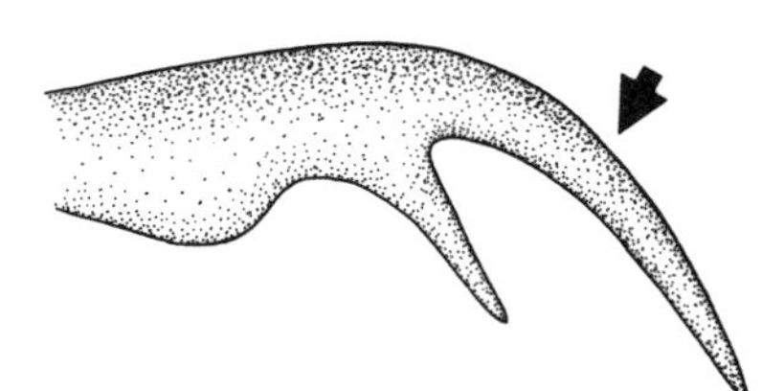

Fig. 301. Foreclaw: *Oc. impiger*

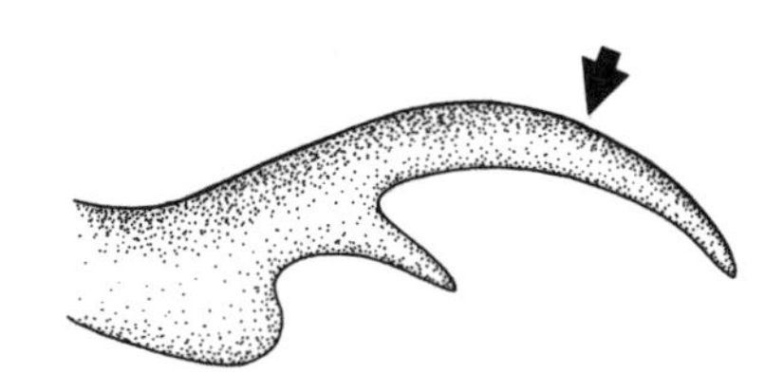

Fig. 303. Foreclaw: *Oc.* nigripes

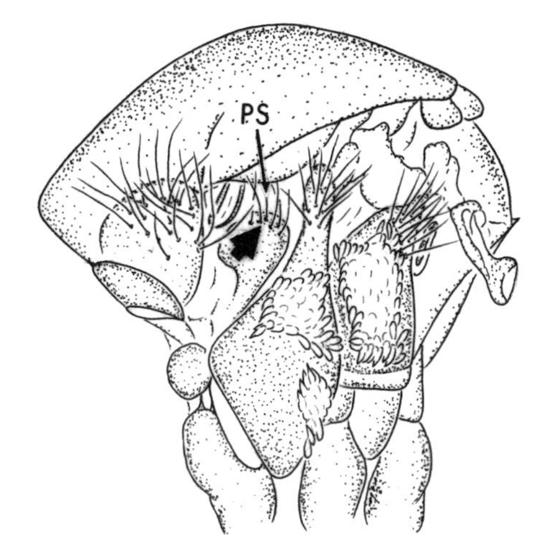


Fig. 302. Lateral view of thorax: *Oc. impiger*

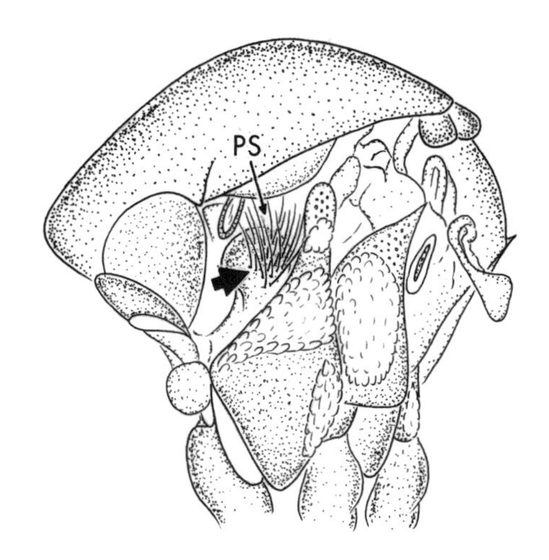


Fig. 304. Lateral view of thorax: *Oc. nigripes*

78(76). Proboscis with yellow gray scales ventrally; palpi with scattered pale scales (fig. 305); abdominal tergum VII nearly covered with pale scales (fig. 306) .................................. ............................................................................................ *Oc. schizopinax* (plate 20C)

Proboscis and palpi dark scaled (fig. 307); abdominal tergum VII with no more than 0.5 pale scaled (fig. 308) ........................................................................................... 79

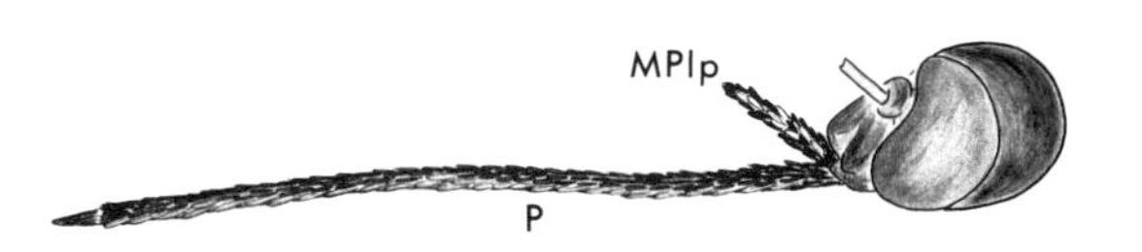


Fig. 305. Lateral view of head: *Oc. schizopinax*

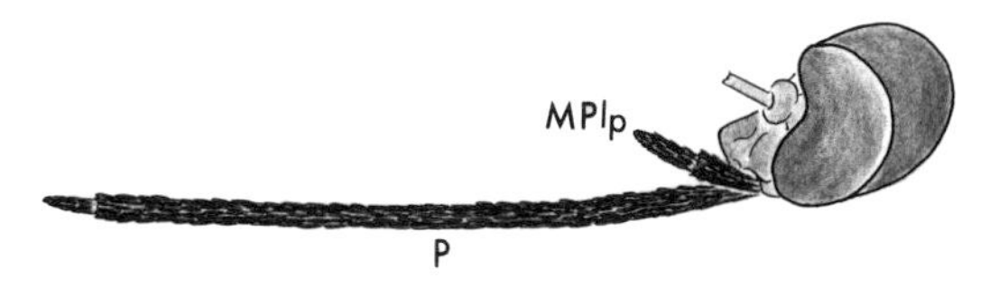


Fig. 307. Lateral view of head: *Oc. punctor*

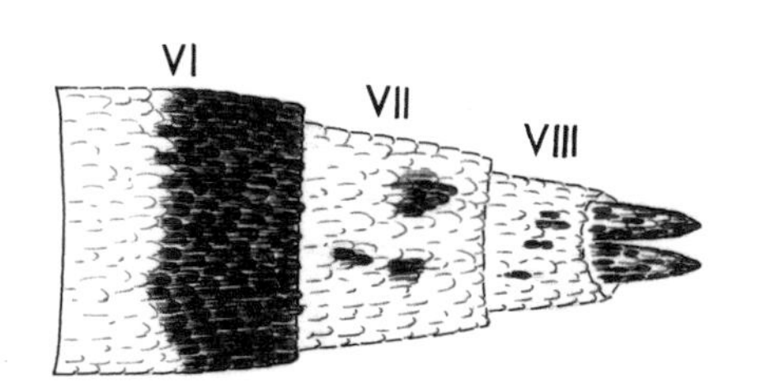


Fig. 306. Dorsal view of abdomen: *Oc. schizopinax*

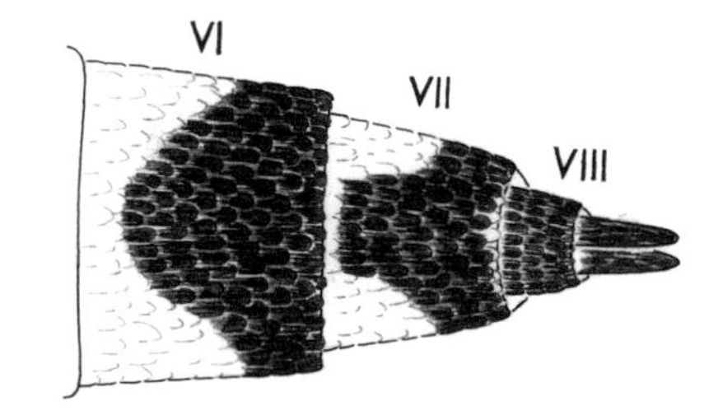


Fig. 308. Dorsal view of abdomen: *Oc. punctor*

79(78). Proepisternum without scales on anterior face, at least in ventral 0.5 (fig. 309) .......... 80

Proepisternum fully scaled on anterior face (fig. 310) ................................................ 81

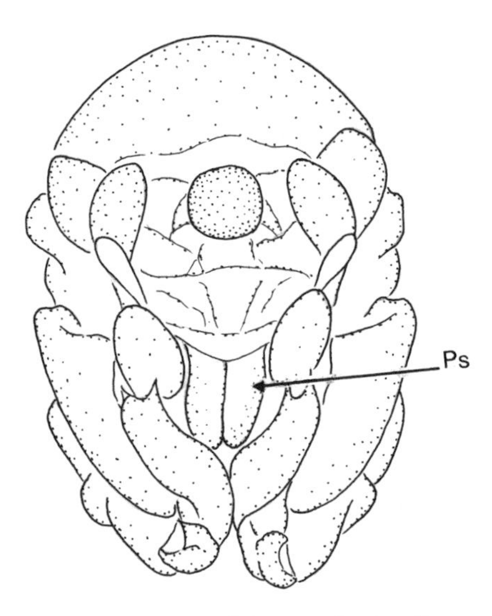


Fig. 309. Anterior view of thorax: *Oc. implicatus*

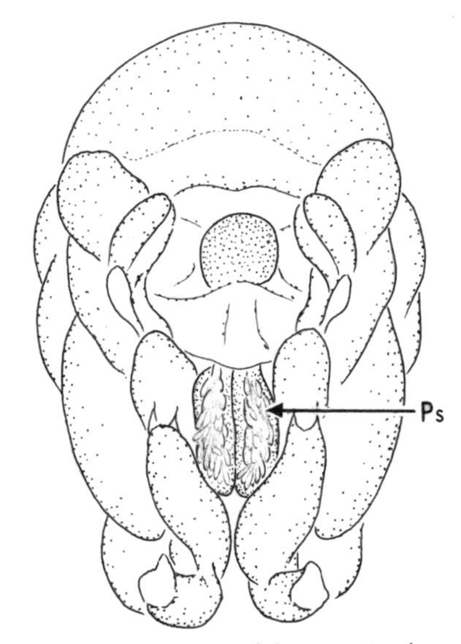


Fig. 310. Anterior view of thorax: *Oc. hexodontus*

80(79). Foreclaw sharply bent distal to tooth (fig. 311); mesokatepisternal scales not reaching anterior angle (fig. 312); wing with 7 or more pale scales at base of vein C (fig. 313) ..... ........................................................................ (in part) *Oc. implicatus* (plate 16B)

Foreclaw elongate, very gradually curving distal to tooth (fig. 314); mesokatepisternal scales reaching anterior angle (fig. 315); wing dark scaled or with fewer than 7 pale scales at base of vein C (fig. 316) .......................................... (in part) *Oc. punctor* (plate 19D) *Oc. aboriginis* (plate 9A)

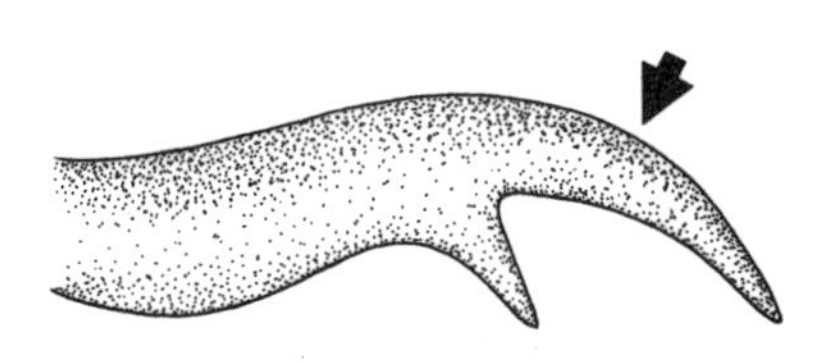

Fig. 311. Foreclaw: *Oc. implicatus*

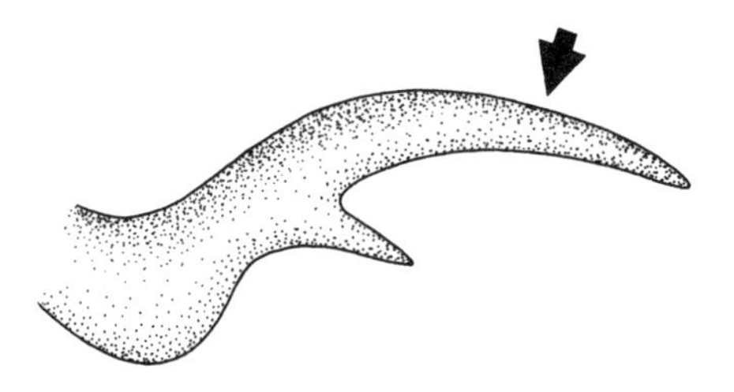

Fig. 314. Foreclaw: *Oc. punctor*

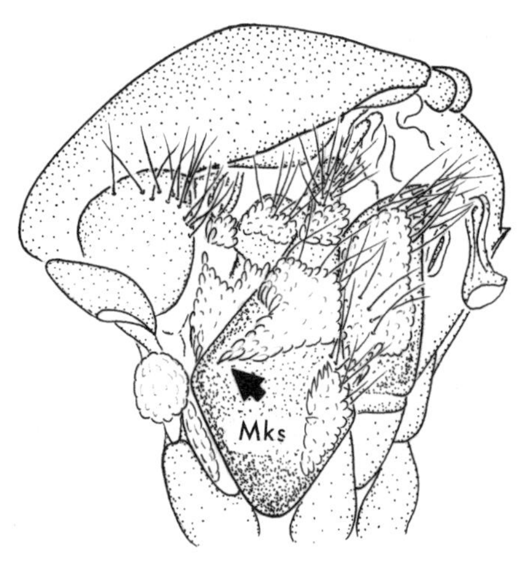


Fig. 312. Lateral view of thorax: *Oc. implicatus*

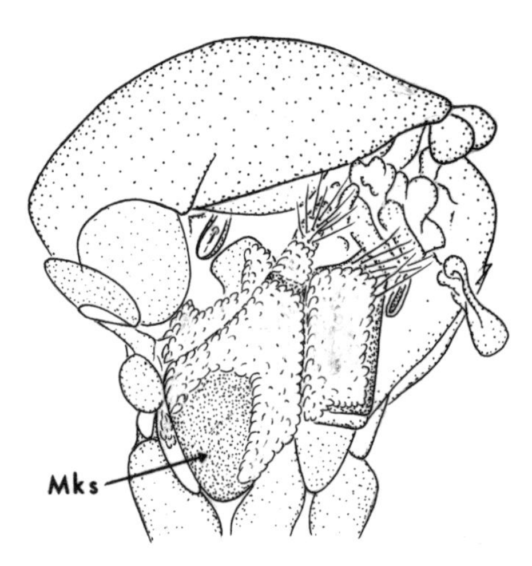


Fig. 315. Lateral view of thorax: *Oc. punctor*

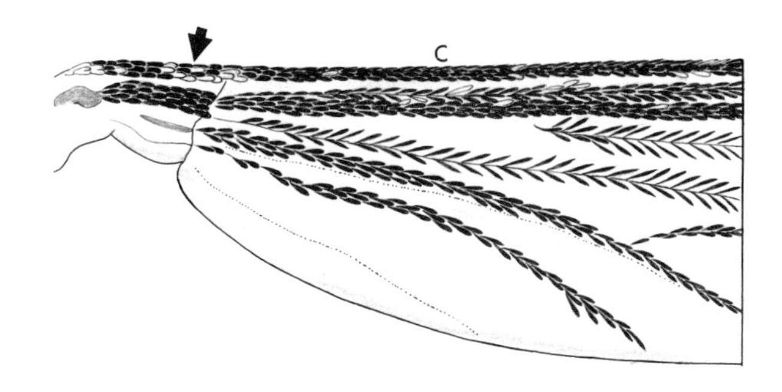


Fig. 313. Dorsal view of wing: *Oc. implicatus*

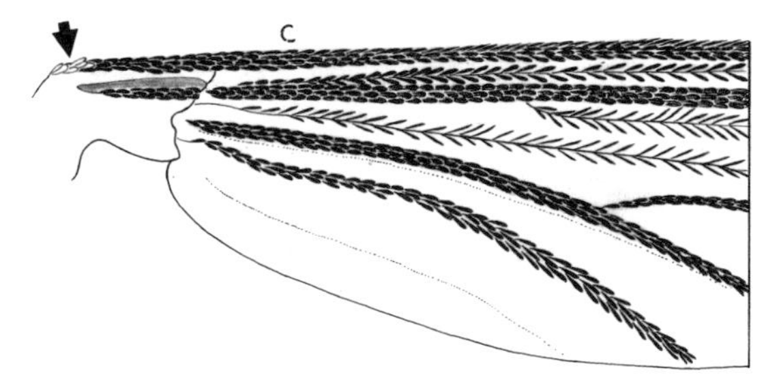


Fig. 316. Dorsal view of wing: *Oc. punctor*

81(79). Supraalar and scutellar setae dark brown or black (fig. 317); with 13 or more postmetasternal scales (fig. 318) ............................................................ *Oc. pionips* (plate 18D)

Supraalar and scutellar setae yellow brown (fig. 319); postmetasternal scales absent or with 2,3 scales only (fig. 320) ........................................................................ 82

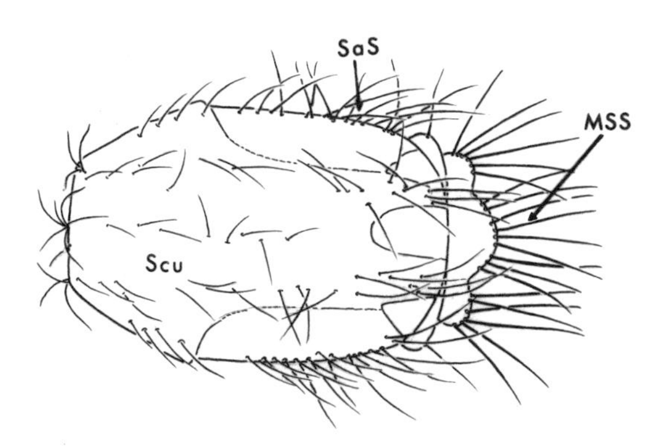


Fig. 317. Dorsal view of thorax: *Oc. pionips*

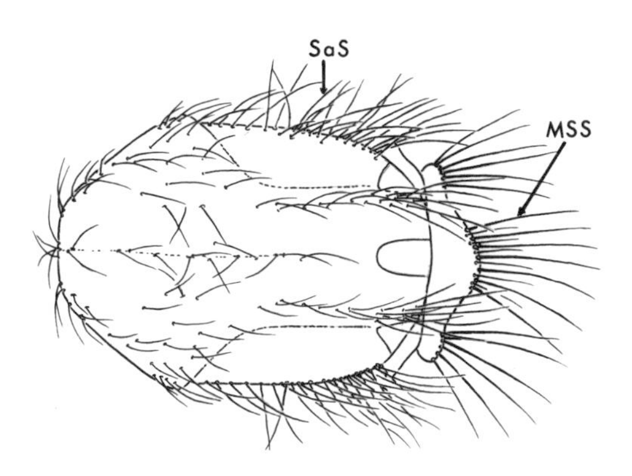


Fig. 319. Dorsal view of thorax: *Oc. hexodontus*

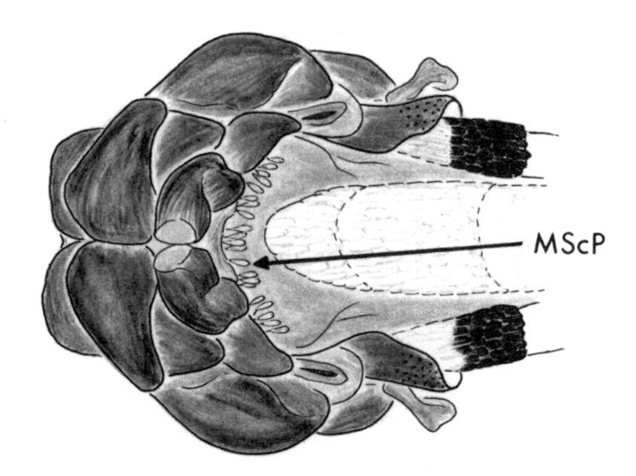


Fig. 318. Ventral view of abdomen: *Oc. pionips*

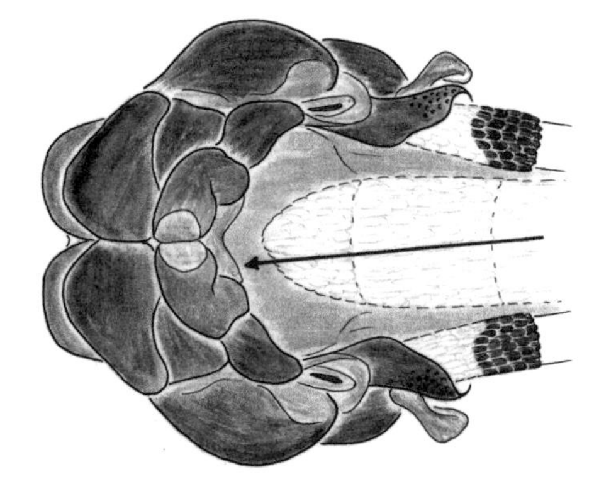

Fig. 320. Ventral view of abdomen: *Oc. hexodontus*

82(81). Large patch of pale scales at base of wing vein C (fig. 321); abdominal sterna III–VI pale scaled apically, or rarely with few dark scales (fig. 322) ....... *Oc. hexodontus* (plate 15D)

Wing dark scaled or with fewer than 8 pale scales at base of vein C (fig. 323); abdominal sterna III–VI with many dark scales apically (fig. 324) ... (in part) *Oc. punctor* (plate 19D)

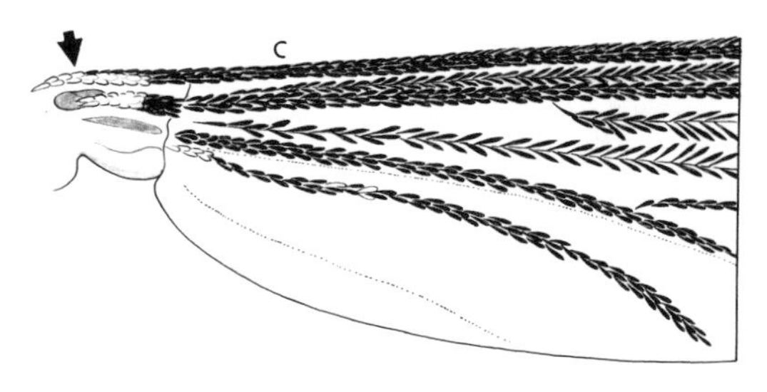


Fig. 321. Dorsal view of wing: *Oc. hexodontus*

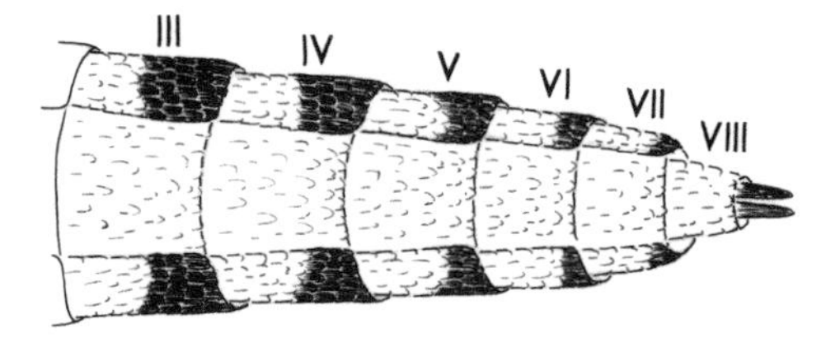


Fig. 322. Ventral view of abdomen: *Oc. hexodontus*

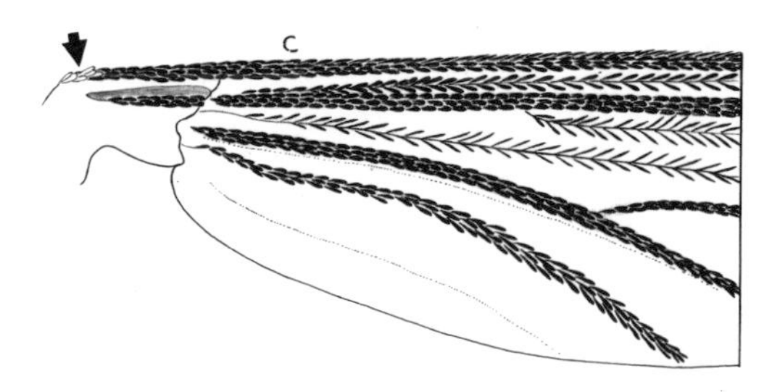


Fig. 323. Dorsal view of wing: *Oc. punctor*

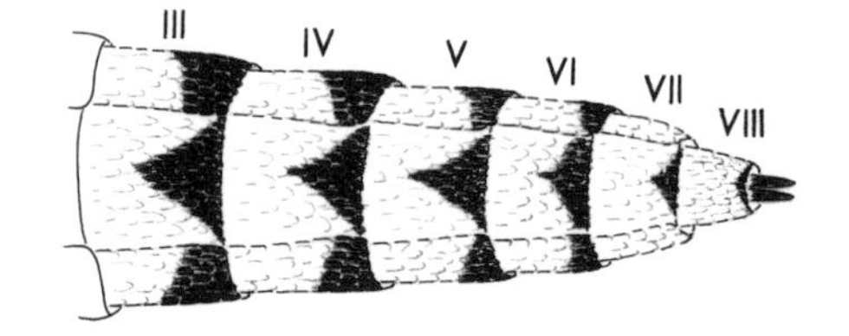


Fig. 324. Ventral view of abdomen: *Oc. punctor*

a. Found in eastern North America (fig. 325) ..................................... *Oc. abserratus* (plate 9C)
aa. Found in Alaska only (fig. 325) .................................................. *Oc. punctodes* (plate 19C)
aaa. Widely distributed in northern North America (fig. 325) ............. *Oc. punctor* (plate 19D)

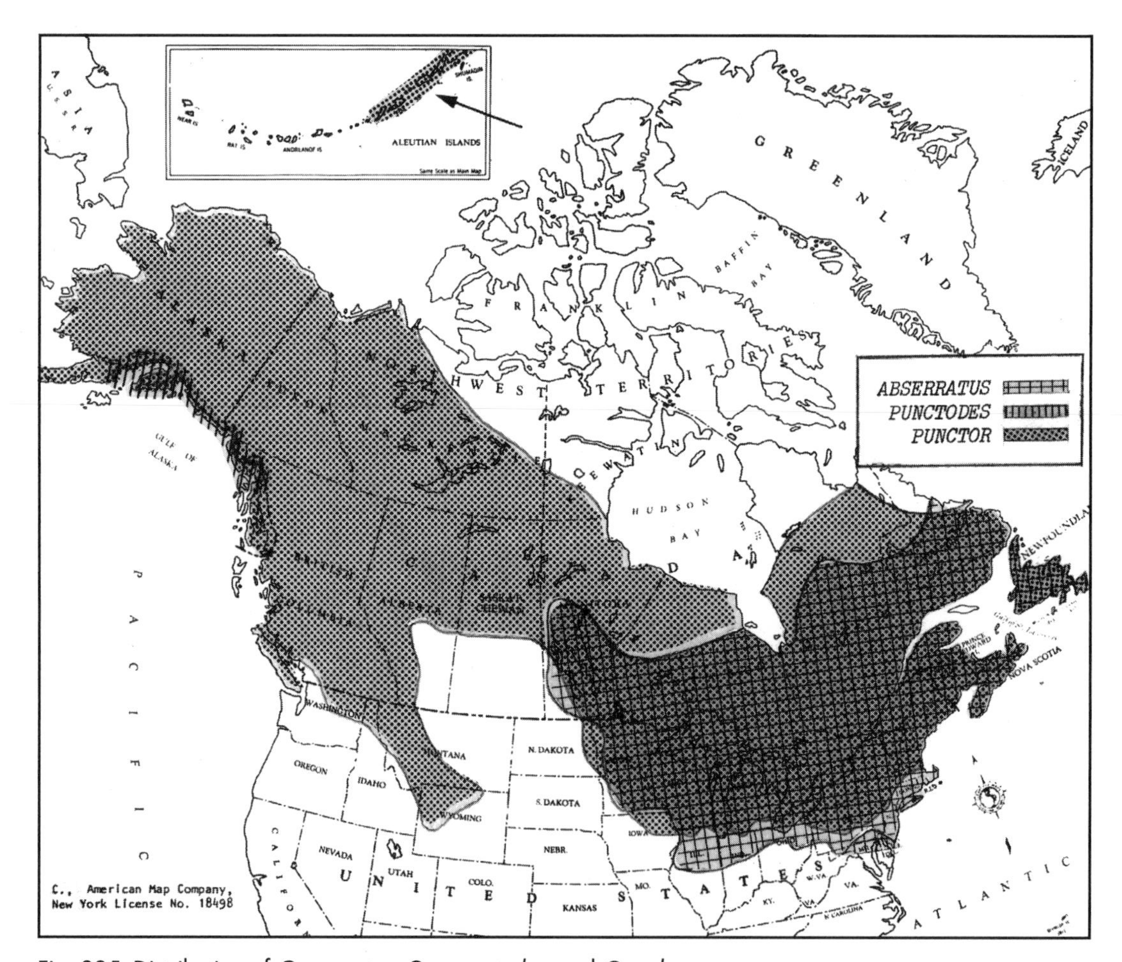


Fig. 325. Distribution of *Oc. punctor*, *Oc. punctodes*, and *Oc. abserratus*

## Key to Adult Females of Genus *Anopheles*

1. Wing with pale-scaled spots (fig. 326) .......... 2

   Wing entirely dark scaled or with silvery or coppery apical fringe spot (figs. 327, 328) ... 7

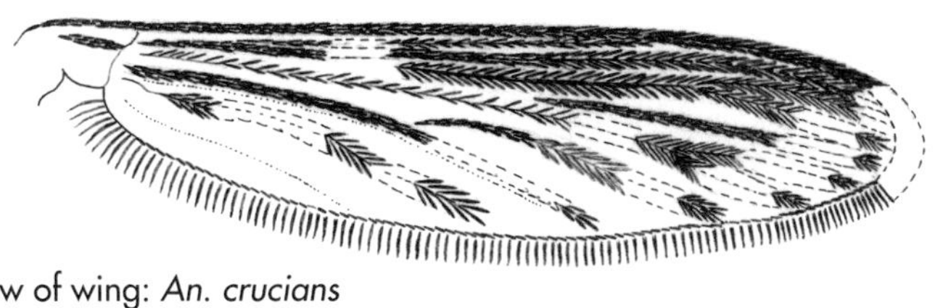

Fig. 326. Dorsal view of wing: *An. crucians*

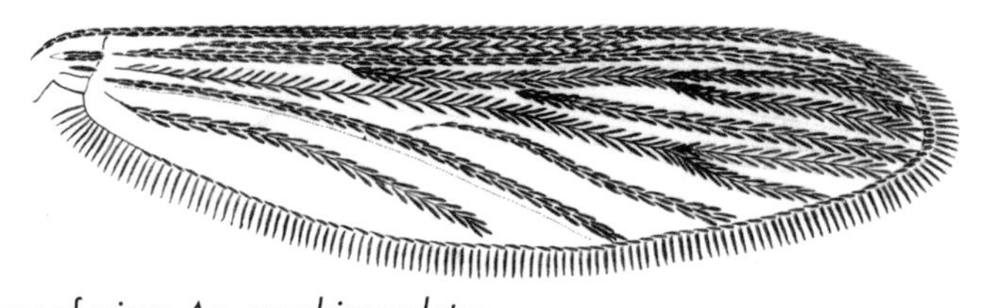

Fig. 327. Dorsal view of wing: *An. quadrimaculatus*

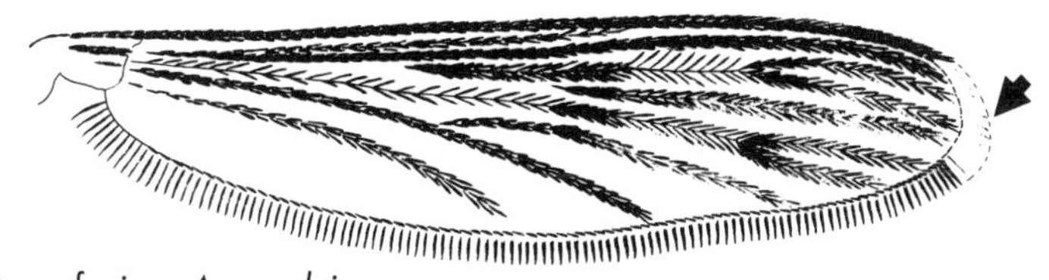

Fig. 328. Dorsal view of wing: *An. earlei*

2(1). Hindtarsomeres with apical 0.5 of 2 and all of 3,4, and 5 pale scaled, except for basal ring of dark scales on 5 (fig. 329) .......... *albimanus* (plate 40C)

   Hindtarsomeres dark scaled (fig. 330) .......... 3

Fig. 329. Hindleg: *An. albimanus*

Fig. 330. Hindleg: *An. punctipennis*

3(2). Wing vein C with apical pale spot, otherwise dark scaled; vein A with 3 dark spots (fig. 331) .......... *crucians* (plate 25B)
*bradleyi* (plate 26A)
*georgianus* (plate 26B)

   Wing vein C with apical and subcostal pale spots; vein A with 1 or 2 dark-scaled spots or lines (fig. 332) .......... 4

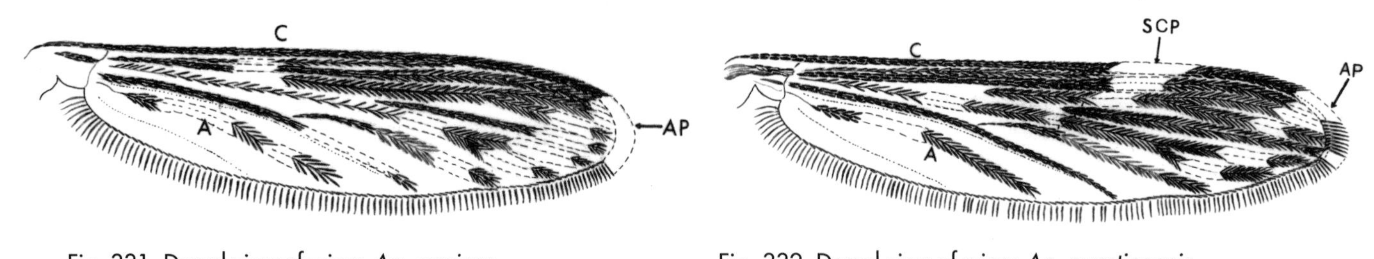


Fig. 331. Dorsal view of wing: *An. crucians*

Fig. 332. Dorsal view of wing: *An. punctipennis*

4(3). Palpi entirely dark scaled (fig. 333); wing vein $R_{4+5}$ and Cu with only dark scales (fig. 334) ........................................................................................................................... 5

Palpi with rings of pale scales (fig. 335); veins $R_{4+5}$ and Cu with long sections of pale scales centrally (fig. 336) ........................................................................................................ 6

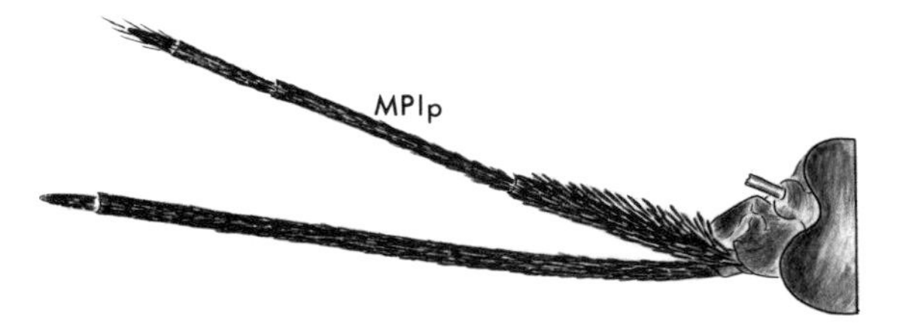


Fig. 333. Lateral view of head: *An. punctipennis*

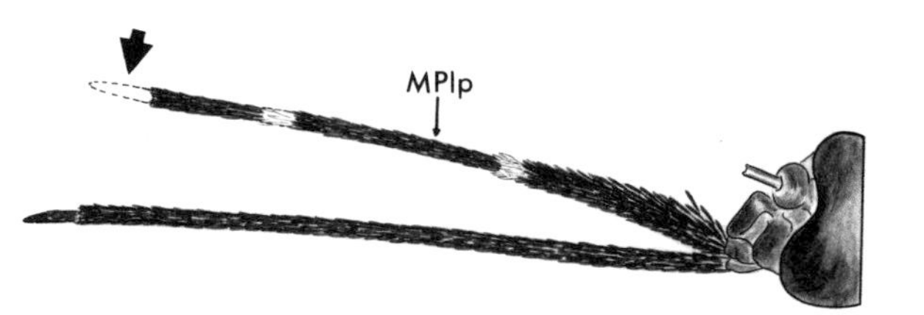


Fig. 335. Lateral view of head: *An. pseudopunctipennis*

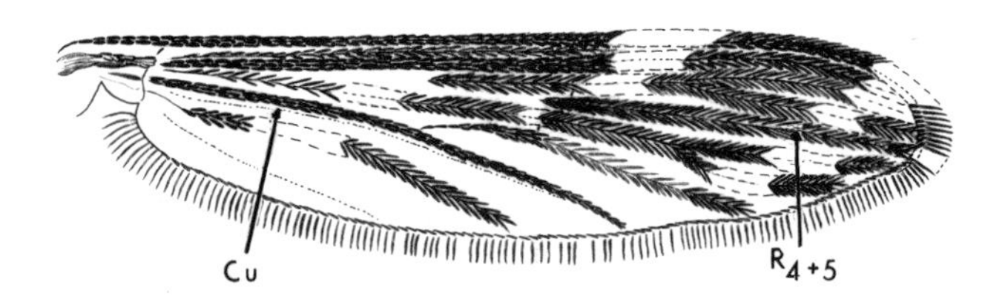


Fig. 334. Dorsal view of wing: *An. punctipennis*

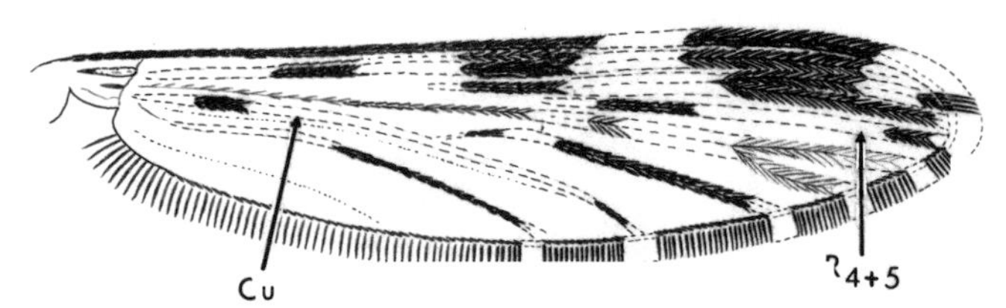


Fig. 336. Dorsal view of wing: *An. pseudopunctipennis*

5(4). Subcostal pale spot 0.5 or more length of preapical dark spot (fig. 337) ........................ .......................................................................................... *punctipennis* (plate 27A)

Subcostal pale spot much reduced, usually less than 0.3 length of preapical dark spot (fig. 338) ........................................................................................ *perplexens* (plate 32C)

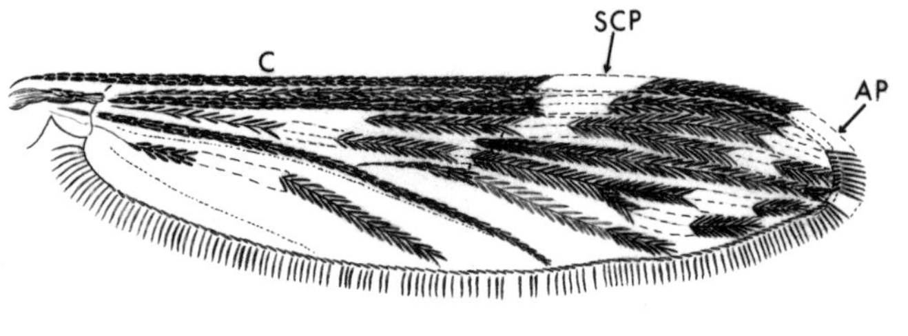


Fig. 337. Dorsal view of wing: *An. punctipennis*

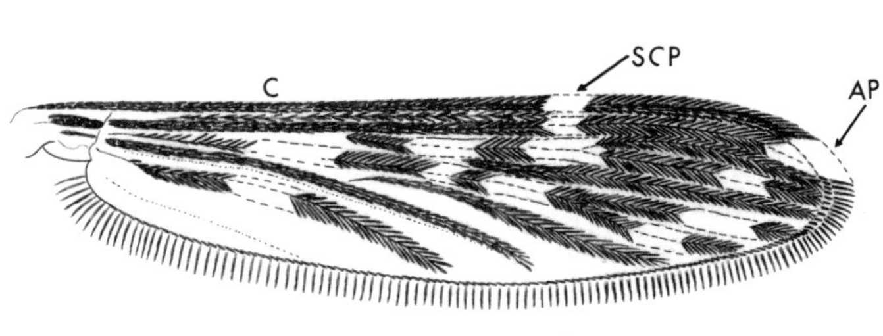


Fig. 338. Dorsal view of wing: *An. perplexens*

6(4).Wing vein M predominantly pale scaled (fig. 339); apical segment of palpus with pale scales (fig. 340) ........................................................................ *pseudopunctipennis* (plate 26D)

Vein M mostly dark scaled (fig. 341); apical segment of palpus with dark scales (fig. 342) ........................................................................................ *franciscanus* (plate 25D)

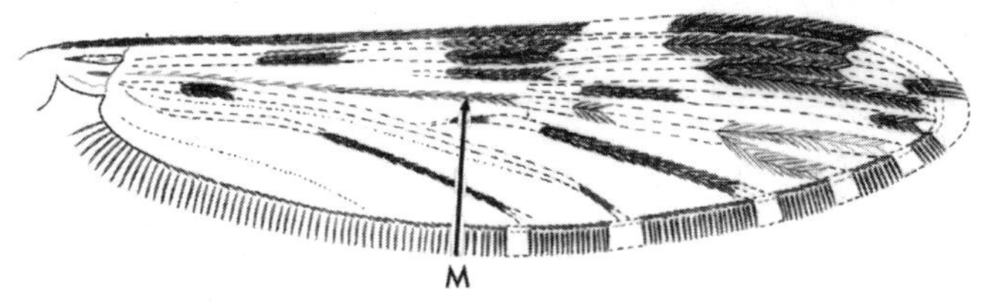


Fig. 339. Dorsal view of wing: *An. pseudopunctipennis*

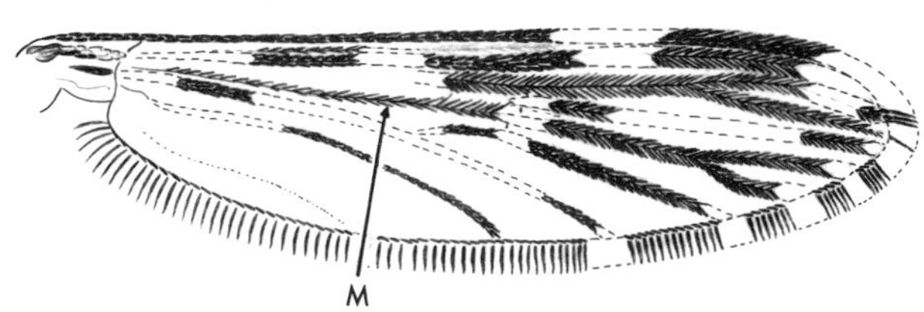


Fig. 341. Dorsal view of wing: *An. franciscanus*

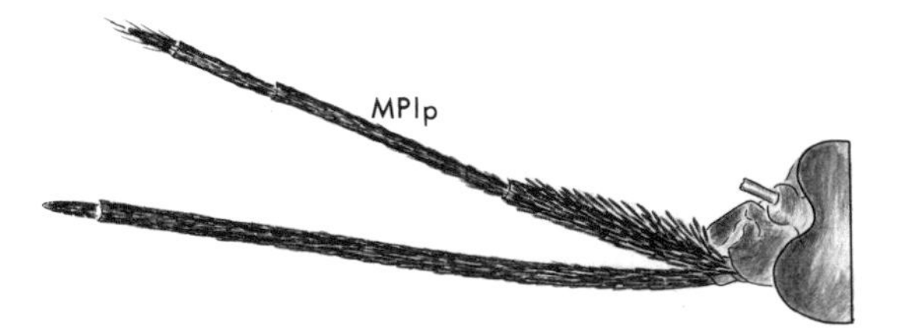


Fig. 340. Lateral view of head: *An. pseudopunctipennis*

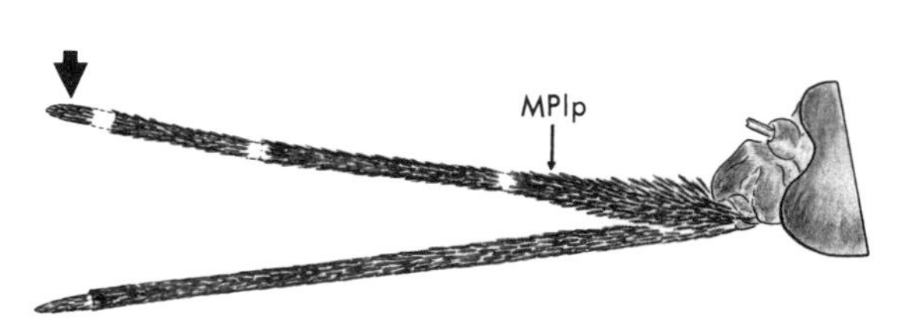


Fig. 342. Lateral view of head: *An. franciscanus*

7(1).Wing with silvery or coppery apical fringe spot (fig. 343) ................................................ 8

Wing entirely dark scaled (fig. 344) ................................................................................ 9

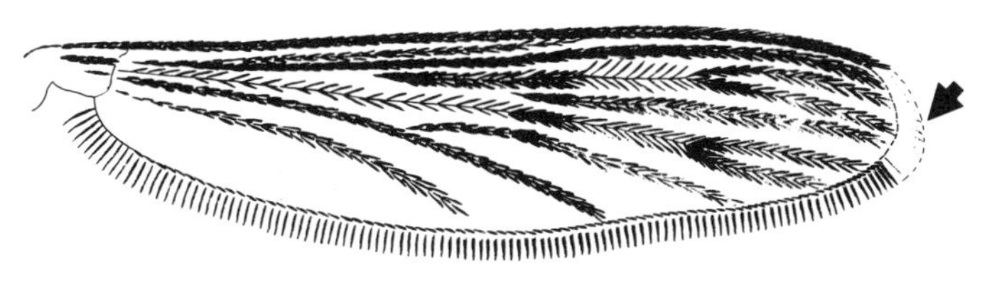

Fig. 343. Dorsal view of wing: *An. earlei*

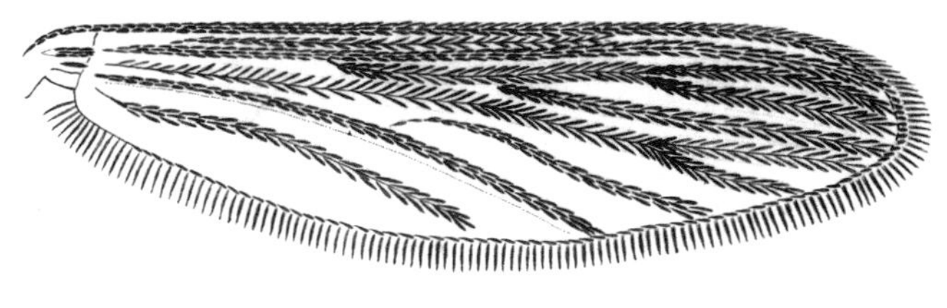

Fig. 344. Dorsal view of wing: *An. quadrimaculatus*

8(7).Numerous erect scales on dorsal surface of wing vein $R_{2+3}$ between its base dark spot and fork of veins $R_2$ and $R_3$, decumbent ventral scales of $R_{2+3}$ not visible from dorsal aspect (fig. 345) ........................................................................................................ *earlei* (plate 25C)

Vein $R_{2+3}$ bare or rarely with 1–2 erect scales on dorsal surface between its basal dark spot and fork of $R_2$ and $R_3$, decumbent ventral scales on $R_{2+3}$ visible from dorsal aspect (fig. 346) ........................................................................................... *occidentalis* (plate 42B)

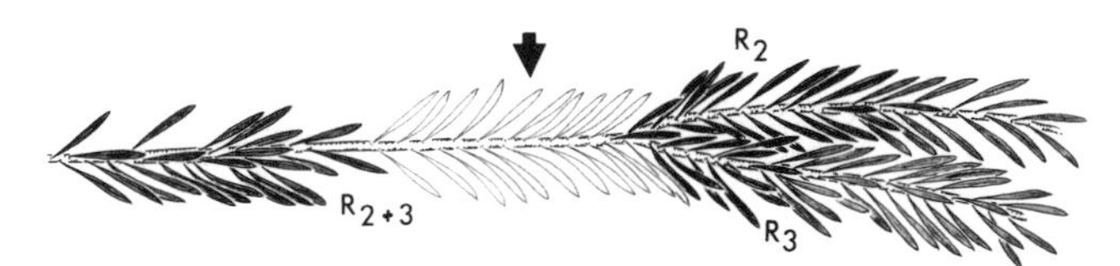


Fig. 345. Dorsal view of wing vein R: *An. earlei*

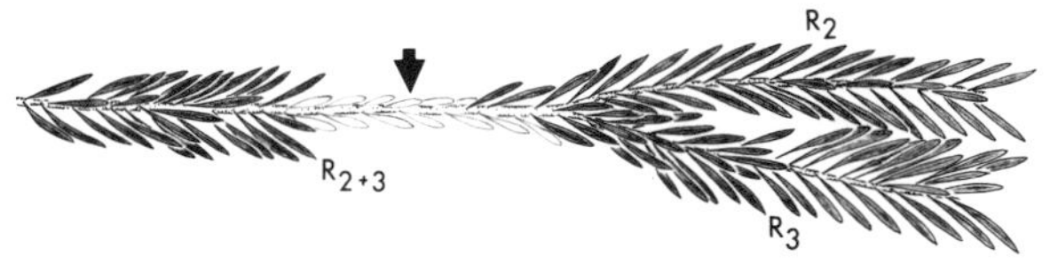


Fig. 346. Dorsal view of wing vein R: *An. occidentalis*

9(7). Wing unspotted (fig. 347); scutal setae about 0.5 width of scutum (fig. 348); small species, wing length about 3.0 mm ........................................................................................ 10

Wing spots of dark scales more or less distinct (fig. 349); scutal setae mostly shorter than 0.5 width of scutum (fig. 350); medium to large species, wing length 4.0 mm or more........ 11

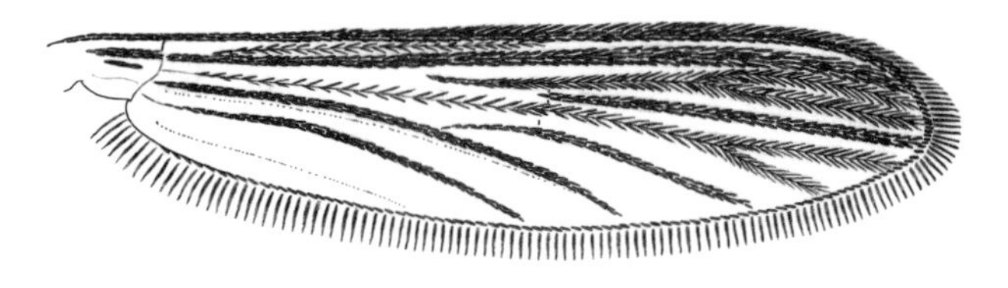

Fig. 347. Dorsal view of wing: *An. barberi*

Fig. 349. Dorsal view of wing: *An. quadrimaculatus*

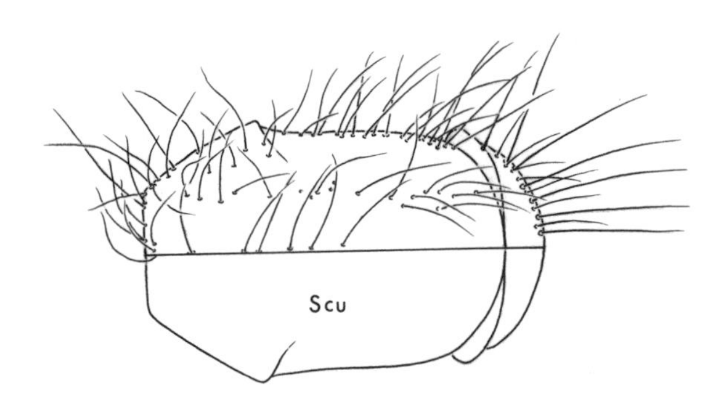


Fig. 348. Dorsal view of scutum: *An. barberi*

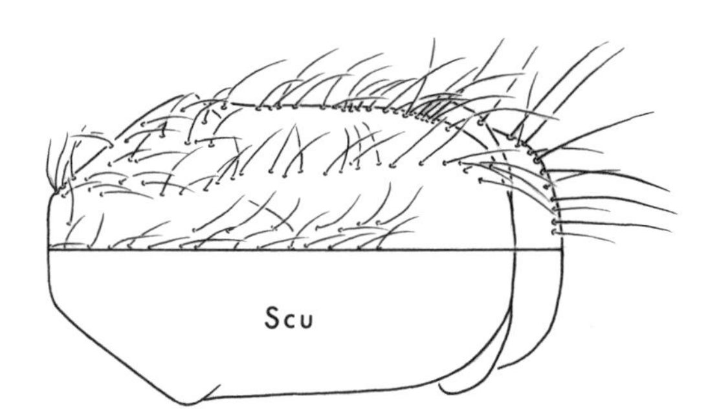


Fig. 350. Dorsal view of scutum: *An. freeborni*

10(9). Proepisternum with 6–11 setae; forecoxa with 19 or more setae; anterior acrostichal setae dark (fig. 351) .................................................................................... *barberi* (plate 24D)

Proepisternum with 2–5 setae; forecoxa with 18 or fewer setae; anterior acrostichal setae amber (fig. 352) ................................................................................ *judithae* (plate 41A)

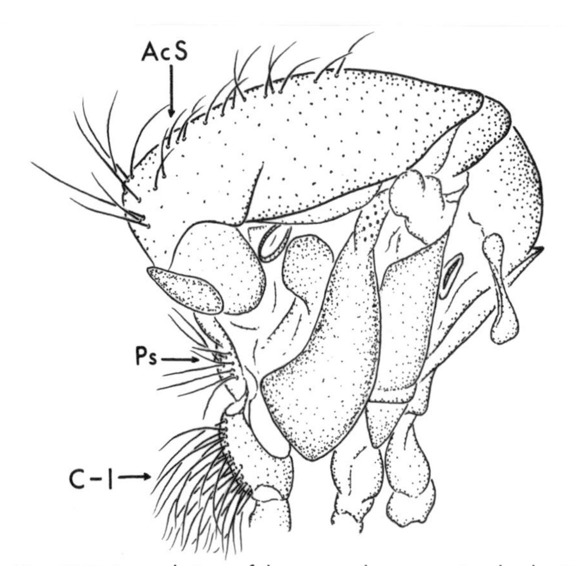


Fig. 351. Lateral view of thorax and scutum: *An. barberi*

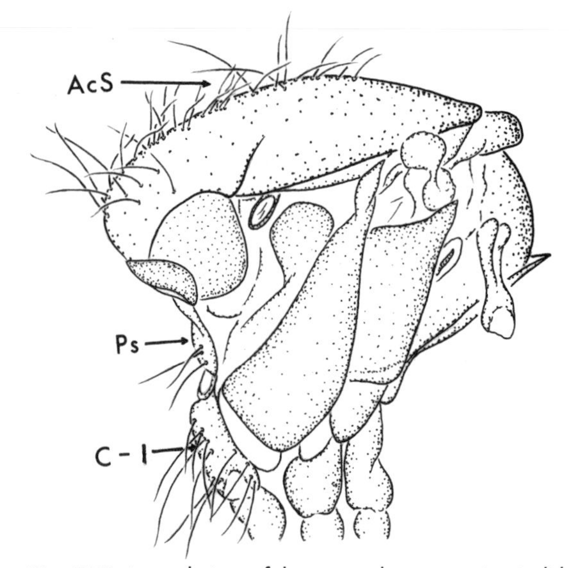


Fig. 352. Lateral view of thorax and scutum: *An. judithae*

11(9). Frontal tuft with some pale setae (fig. 353); wing with 4 distinct dark-scaled spots (fig. 354); palpi with dark scales (fig. 353) ........................................................................ 12

Frontal tuft entirely with dark setae (fig. 355); wing usually with dark-scaled spots indistinct (fig. 356); segments of palpi with or without apical pale rings (fig. 355)............ 17

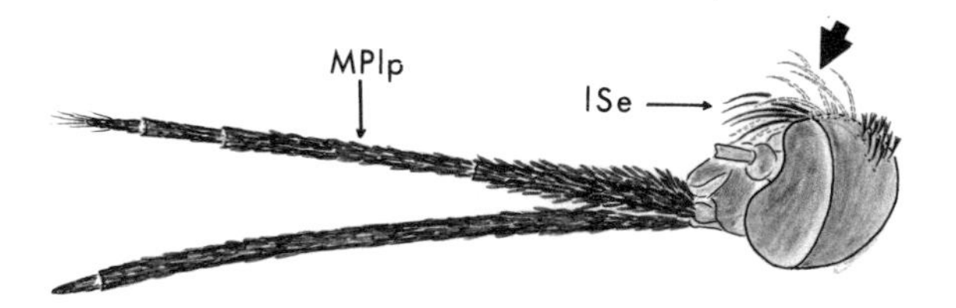


Fig. 353. Lateral view of head: *An. freeborni*

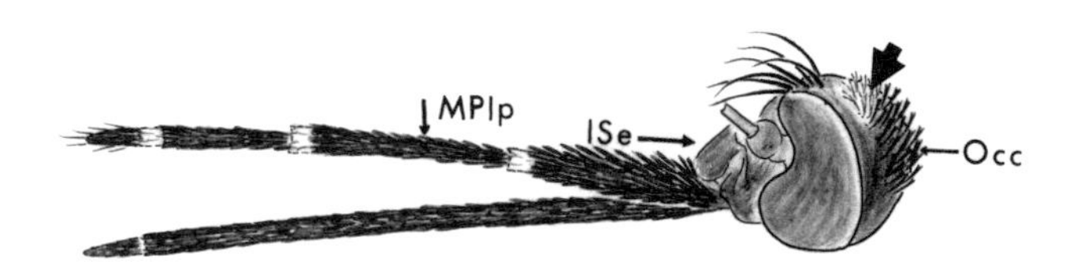


Fig. 355. Lateral view of head: *An. walkeri*

Fig. 354. Dorsal view of wing: *An. quadrimaculatus*

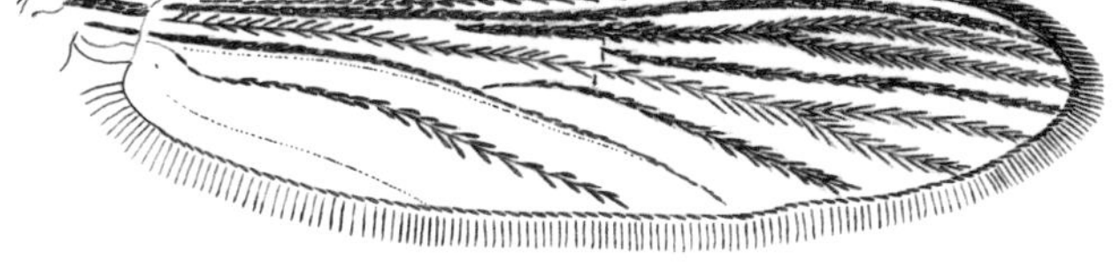

Fig. 356. Dorsal view of wing: *An. atropos*

12(11). Scales on basal part of wing vein Cu linear, with apices truncate (fig. 357) in western USA and Canada ....................................................................................*freeborni* (plate 26A)
........................................................................................................ *hermsi* (plate 42A)

Scales on base of vein Cu obovate with apices rounded (fig. 358): in eastern USA and Canada (*quadrimaculatus* group) ............................................................................. 13

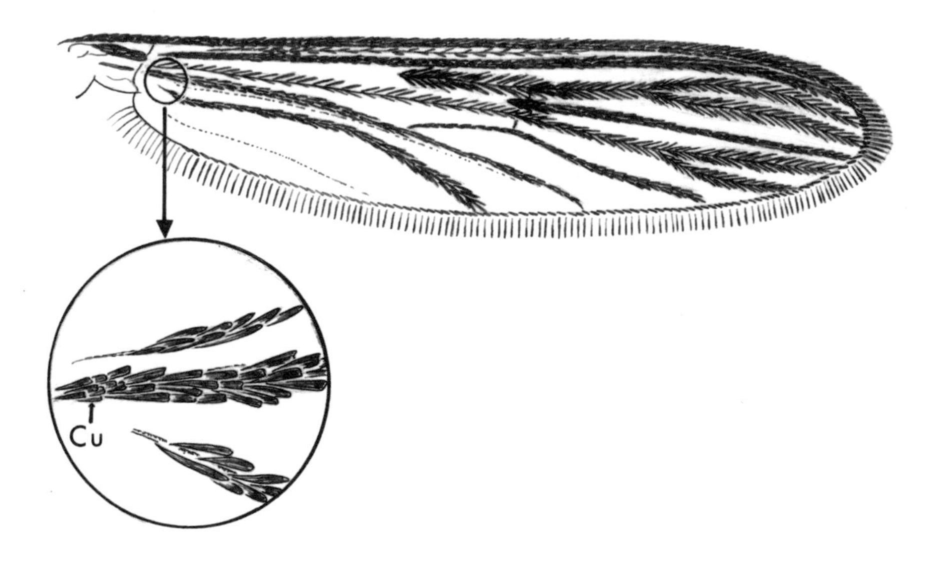


Fig. 357. Dorsal view of wing with basal cubital scales enlarged: *An. freeborni*

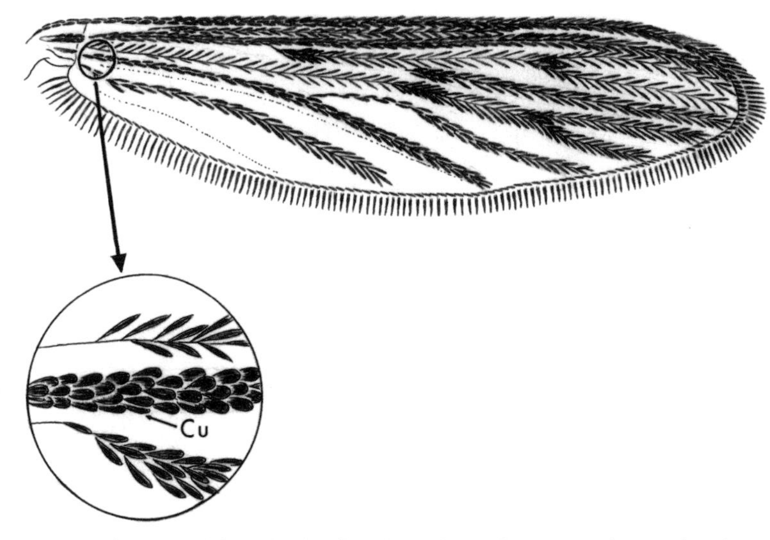


Fig. 358. Dorsal view of wing with basal cubital scales enlarged: *An. quadrimaculatus*

13(12). Upper proepisternum usually with 2–6 setae (fig. 359); mid- and usually foretibia with pale scales apically (fig. 360) .......... 14

Upper proepisternum usually with 7–26 setae (fig. 361); fore- and midtibiae dark scaled (fig. 362) .......... 15

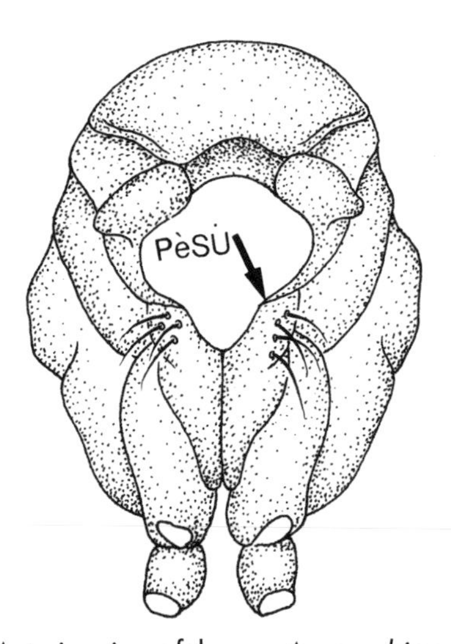


Fig. 359. Anterior view of thorax: *An. quadrimaculatus*

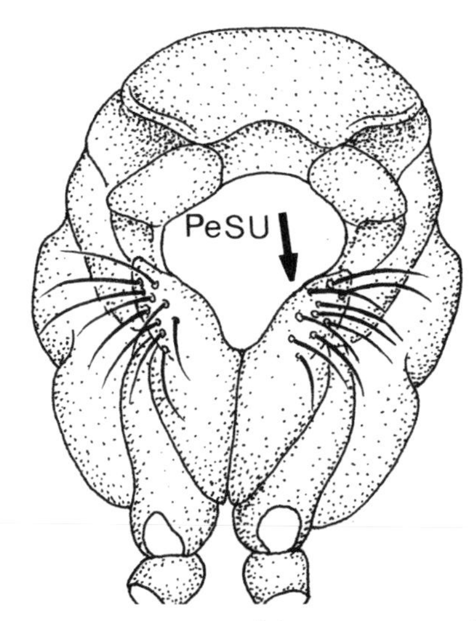


Fig. 361. Anterior view of thorax: *An. inundatus*

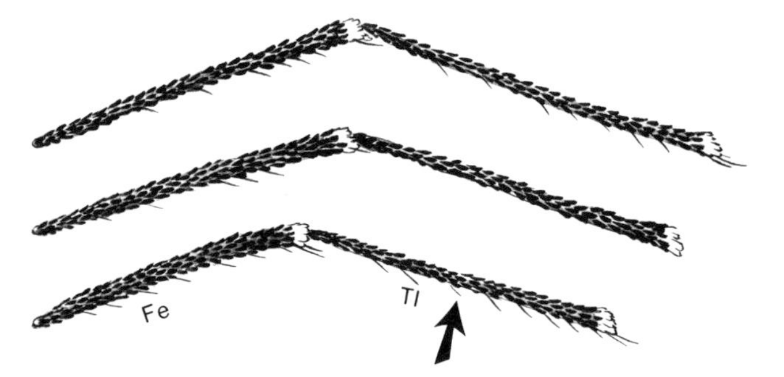


Fig. 360. Femora and tibiae: *An. quadrimaculatus*

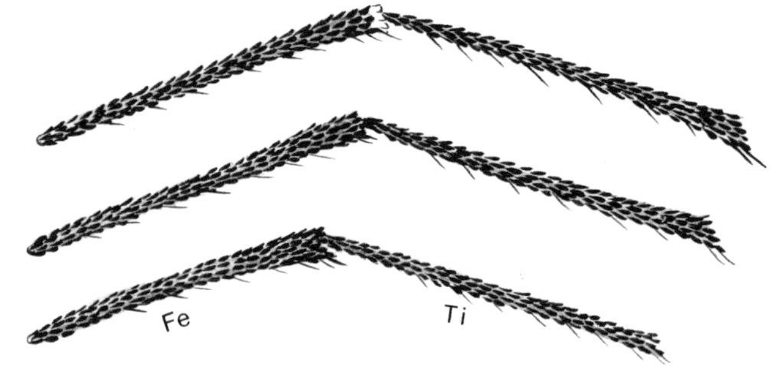


Fig. 362. Femora and tibiae: *An. inundatus*

14(13). Scutal fossa usually with 21–45 setae (fig. 363); prealar area usually with 6–12 setae (fig. 364); interocular area usually with 7–12 setae .................... *quadrimaculatus* (plate 27C)

Scutal fossa usually with 8–20 setae (fig. 365; prealar area usually 1–5 setae (fig. 366); interocular area usually with 4–6 setae .................................... *smaragdinus* (plate 27D)

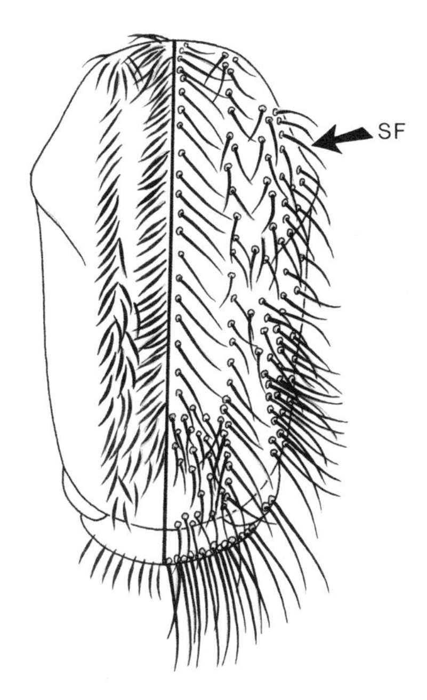


Fig. 363. Scutum: *An. quadrimaculatus*

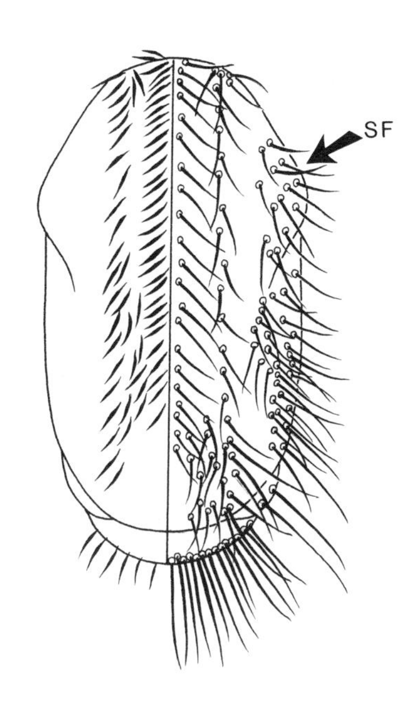


Fig. 365. Scutum: *An. smaragdinus*

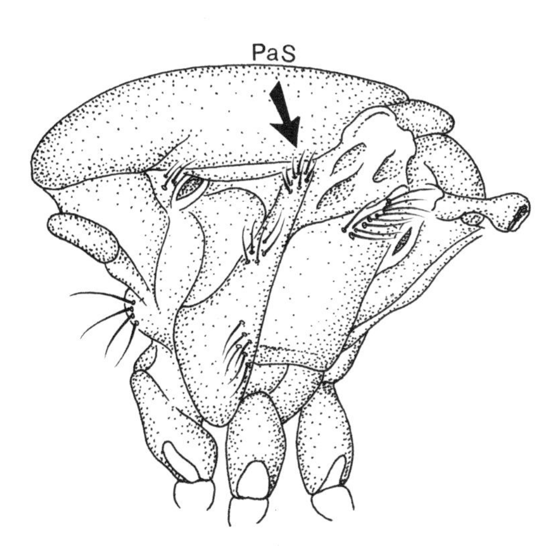


Fig. 364. Lateral view of thorax: *An. quadrimaculatus*

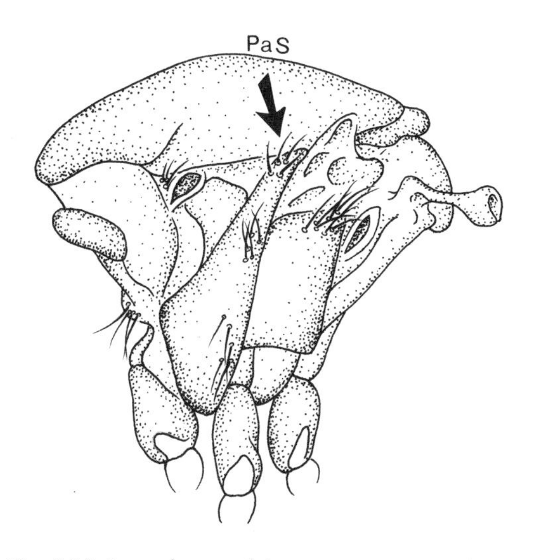


Fig. 366. Lateral view of thorax: *An. smaragdinus*

15(13). Dorsocentral area of scutum without piliform scales on anterior margin (fig. 367) ......... .......................................................................................... *diluvialis* (plate 40C)

Dorsocentral area of scutum with 2–10 golden piliform scales on anterior margin (fig. 368) ........................................................................................................................ 16

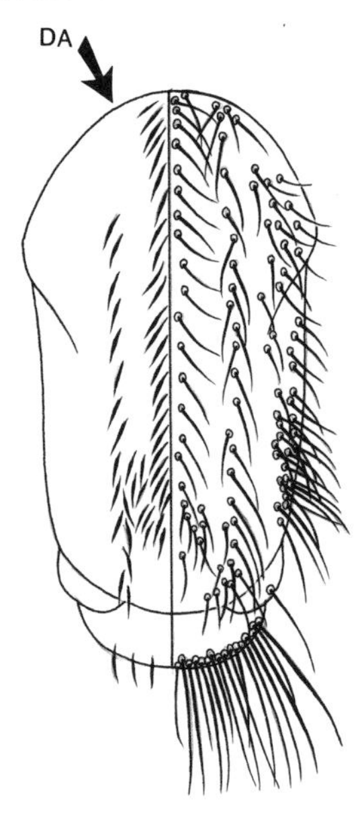

Fig. 367. Scutum: *An. diluvialis*

Fig. 368. Scutum: *An. inundatus*

16(15). Fore- and midfemora with knee spots (fig. 369); scutal fossa usually with 9–20 setae (fig. 370); palpi often shorter than proboscis (fig. 371); interocular area usually with 2–5 setae ........................................................................................................ *maverlius* (plate 26C)

Forefemora and usually midfemora dark scaled (fig. 372); scutal fossa usually with 21–32 setae (fig. 373); palpi longer than proboscis (fig. 374); interocular area usually with 6–9 setae ........................................................................................ *inundatus* (plate 40B)

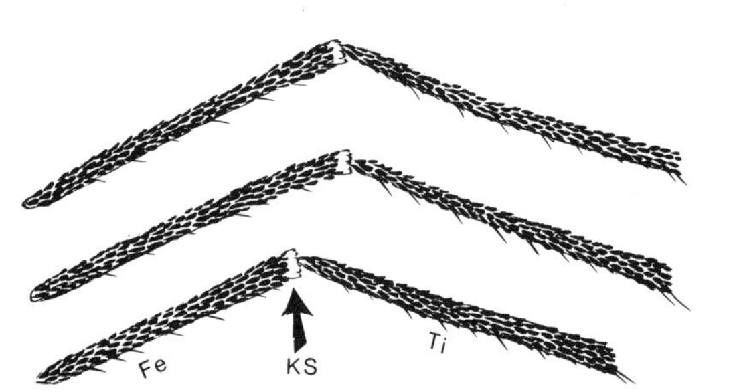


Fig. 369. Femora and tibiae: *An. maverlius*

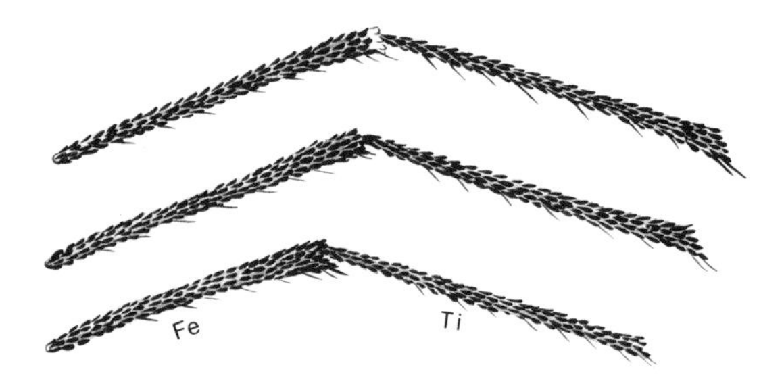


Fig. 372. Femora and tibiae: *An. inundatus*

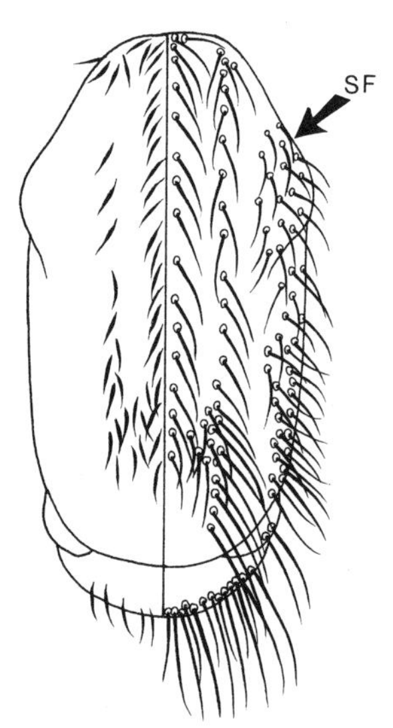


Fig. 370. Scutum: *An. maverlius*

Fig. 373. Scutum: *An. inundatus*

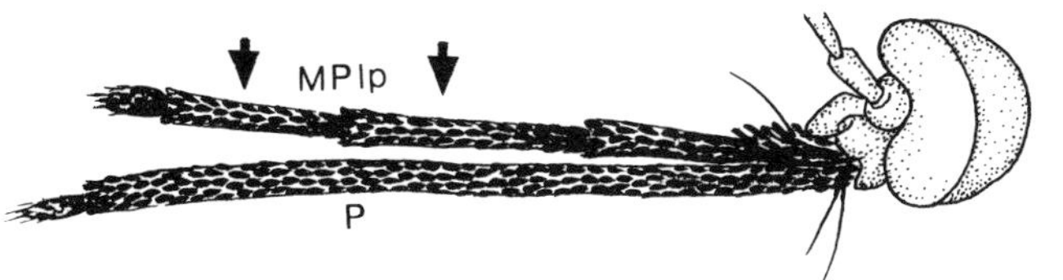

Fig. 371. Lateral view of head: *An. maverlius*

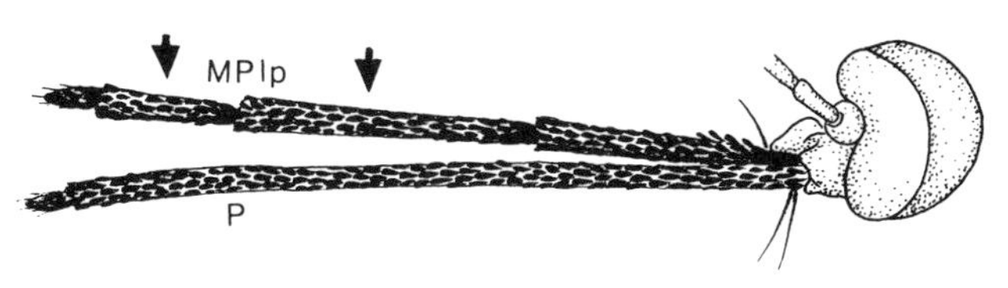


Fig. 374. Lateral view of head: *An. inundatus*

17(11). Capitellum of halter usually pale scaled (fig. 375); occiput with patch of pale scales medioanteriorly (fig. 376); femora with knee spots (fig. 377) ........... *walkeri* (plate 28A)

Capitellum of halter entirely dark scaled (fig. 378); occiput dark scaled (fig. 379); femora with few or no pale scales apically (fig. 380) .................................... *atropos* (plate 24C)

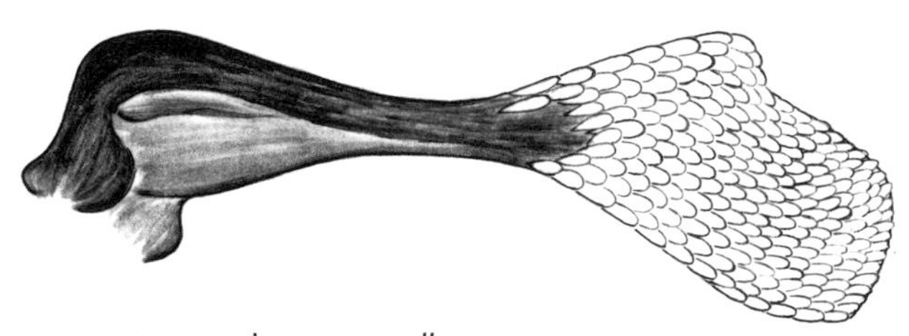
Fig. 375. Halter: *An. walkeri*

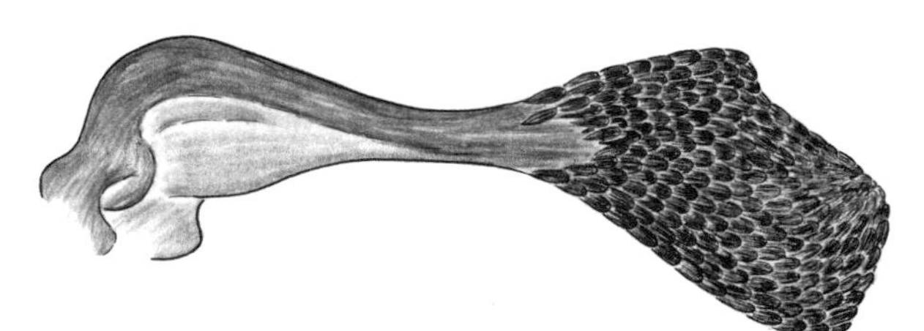
Fig. 378. Halter: *An. atropos*

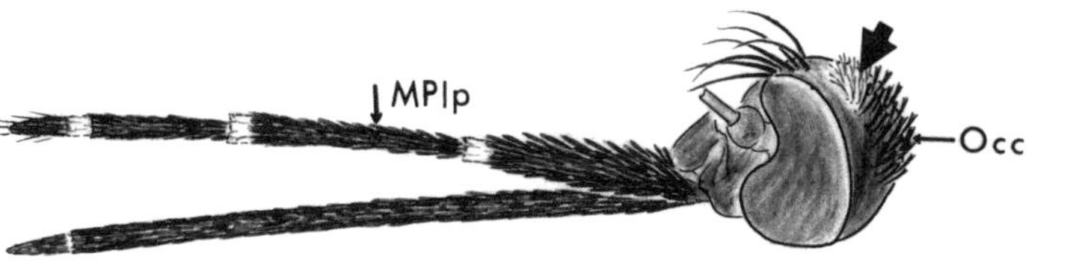


Fig. 376. Lateral view of head: *An. walkeri*

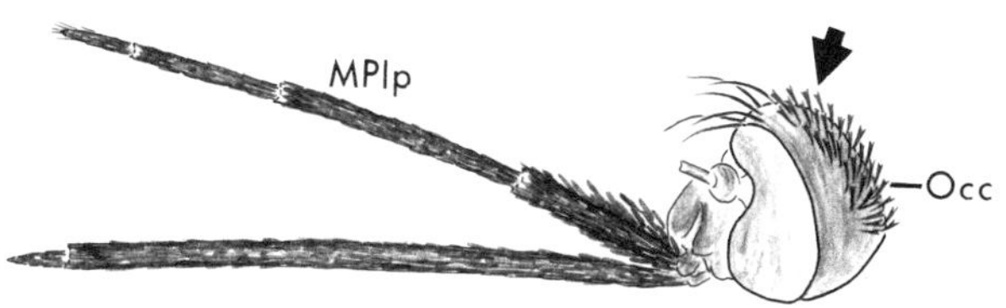


Fig. 379. Lateral view of head: *An. atropos*

Fig. 377. Hindleg: *An. walkeri*

Fig. 380. Hindleg: *An. atropos*

## Key to the Adult Females of the Genus *Culex*

1. Scutum with middorsal acrostichal setae (fig. 381); occiput with narrow appressed scales (fig. 382) .......... 2

   Scutum without middorsal acrostichal setae (fig. 383); occiput with broad appressed scales, at least on ocular line (fig. 384) (subgenus *Melanoconion*) .......... 20

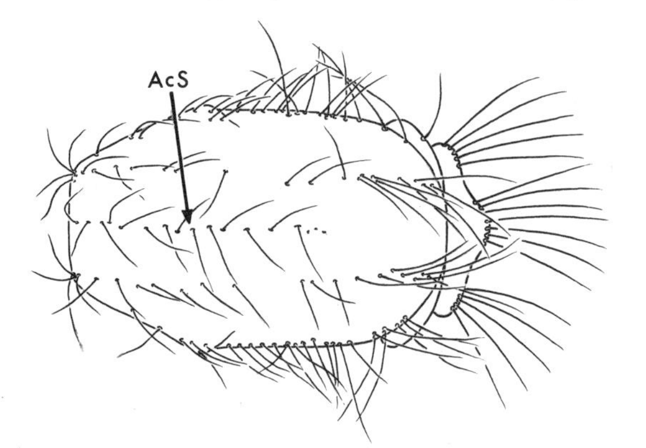


Fig. 381. Dorsal view of thorax: *Cx. pipiens*

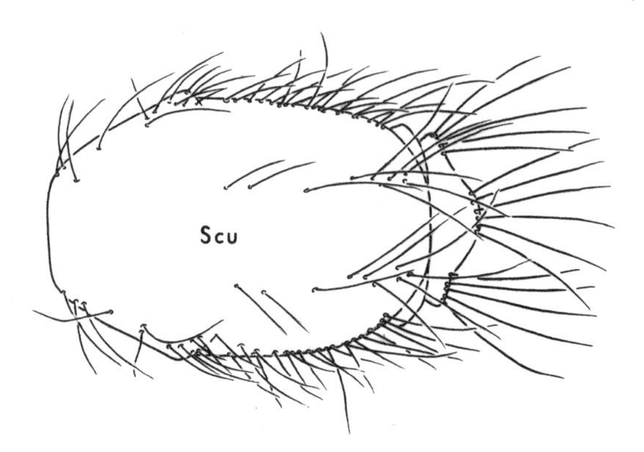


Fig. 383. Dorsal view of thorax: *Cx. erraticus*

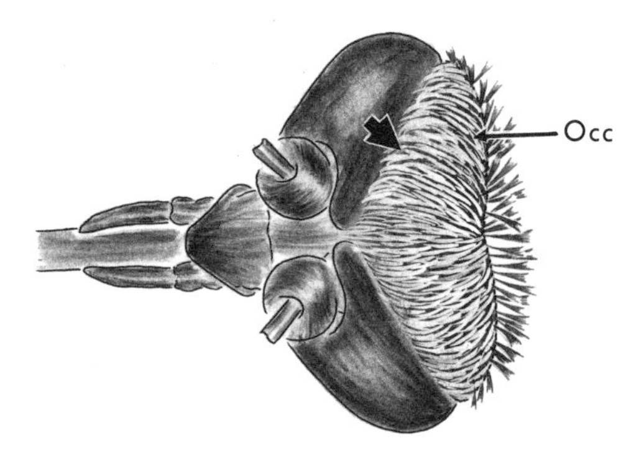


Fig. 382. Dorsal view of head: *Cx. pipiens*

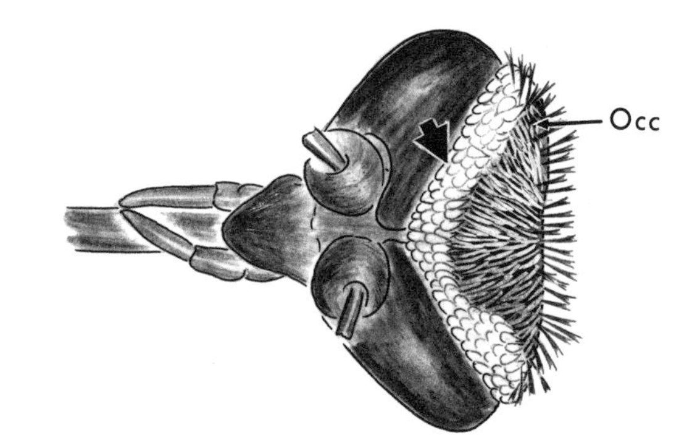


Fig. 384. Dorsal view of head: *Cx. erraticus*

2(1). Abdominal terga with bands or patches along basal border (fig. 385) (subgenus *Culex*) .. 3

   Abdominal terga with bands or lateral patches of pale scales along apical border, or sometimes all dark scaled (fig. 386) (subgenus *Neoculex*) .......... 16

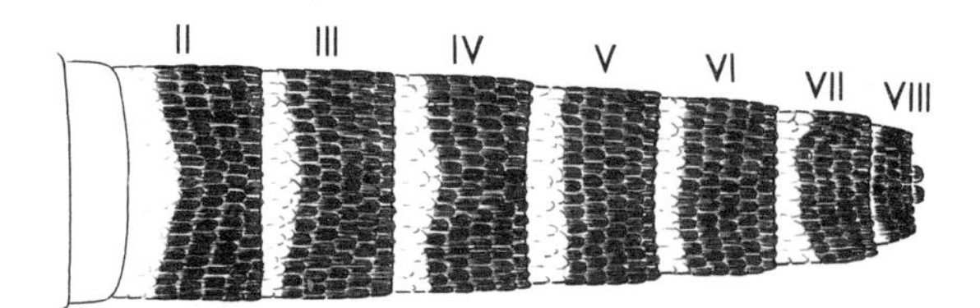


Fig. 385. Dorsal view of abdomen: *Cx. restuans*

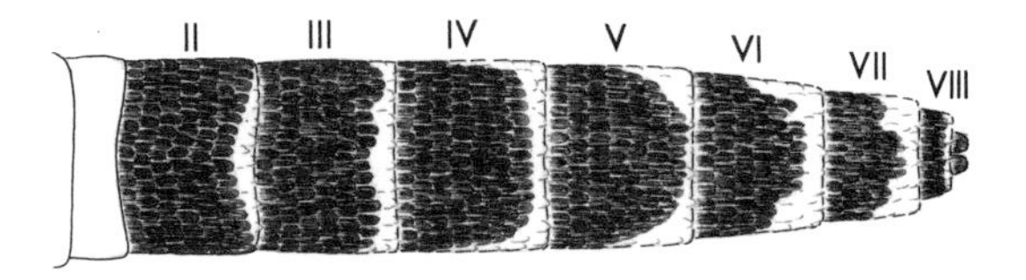


Fig. 386. Dorsal view of abdomen: *Cx. territans*

3(2). Hindtarsomeres with rather distinct basal and apical bands of pale scales (fig. 387)..... 4

   Hindtarsomeres dark scaled, or if with pale scales, then as very narrow basal bands (fig. 388) .......... 9

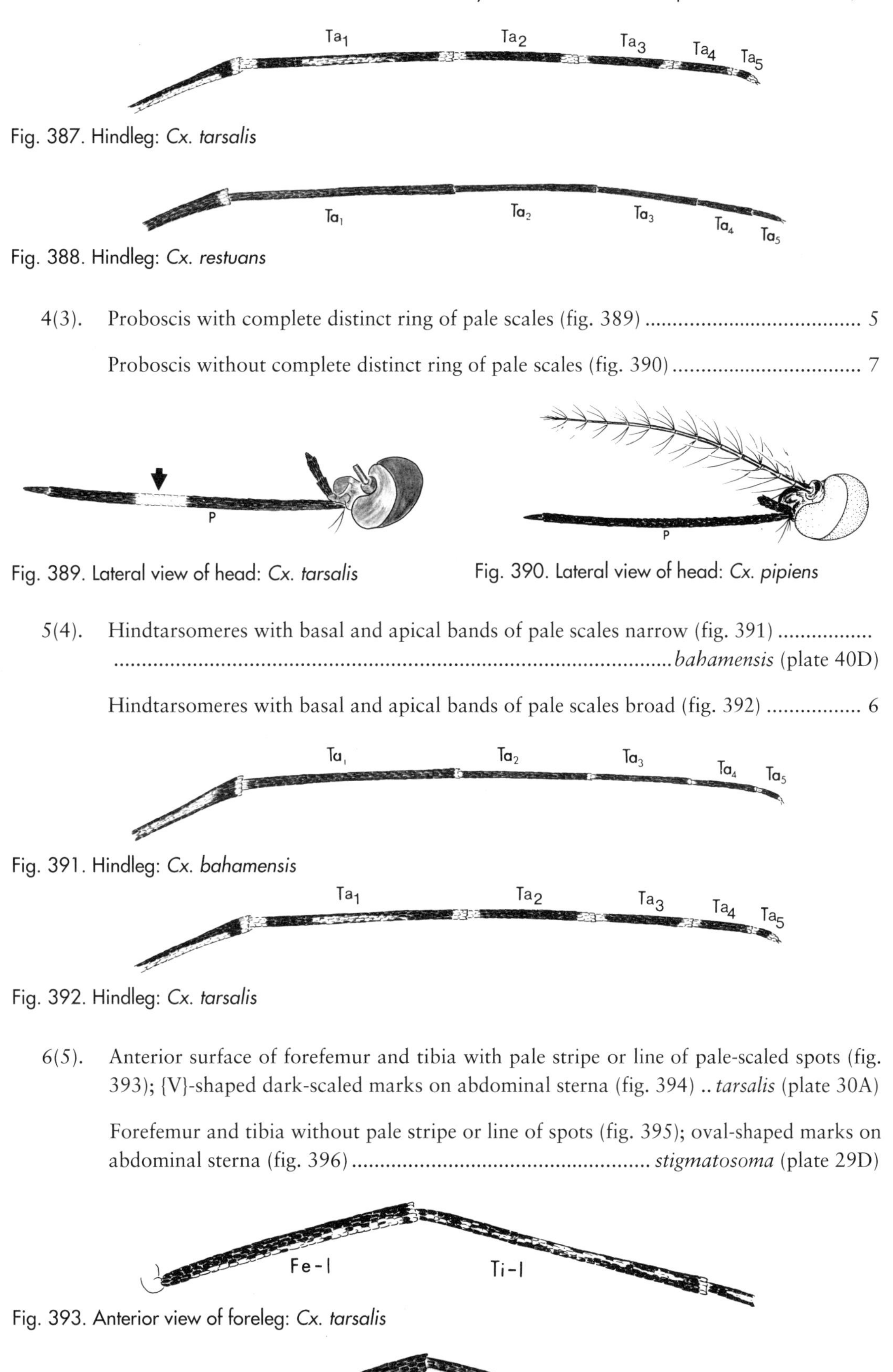


Fig. 387. Hindleg: *Cx. tarsalis*

Fig. 388. Hindleg: *Cx. restuans*

4(3). Proboscis with complete distinct ring of pale scales (fig. 389) ....................................... 5

Proboscis without complete distinct ring of pale scales (fig. 390) ................................. 7

Fig. 389. Lateral view of head: *Cx. tarsalis*

Fig. 390. Lateral view of head: *Cx. pipiens*

5(4). Hindtarsomeres with basal and apical bands of pale scales narrow (fig. 391) ................ ........................................................................................*bahamensis* (plate 40D)

Hindtarsomeres with basal and apical bands of pale scales broad (fig. 392) ................ 6

Fig. 391. Hindleg: *Cx. bahamensis*

Fig. 392. Hindleg: *Cx. tarsalis*

6(5). Anterior surface of forefemur and tibia with pale stripe or line of pale-scaled spots (fig. 393); {V}-shaped dark-scaled marks on abdominal sterna (fig. 394) .. *tarsalis* (plate 30A)

Forefemur and tibia without pale stripe or line of spots (fig. 395); oval-shaped marks on abdominal sterna (fig. 396) .................................................... *stigmatosoma* (plate 29D)

Fig. 393. Anterior view of foreleg: *Cx. tarsalis*

Fig. 395. Anterior view of foreleg: *Cx. stigmatosoma*

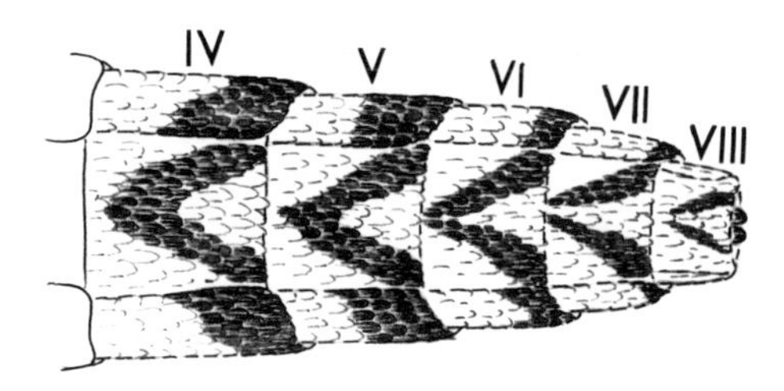


Fig. 394. Ventral view of abdomen: *Cx. tarsalis*

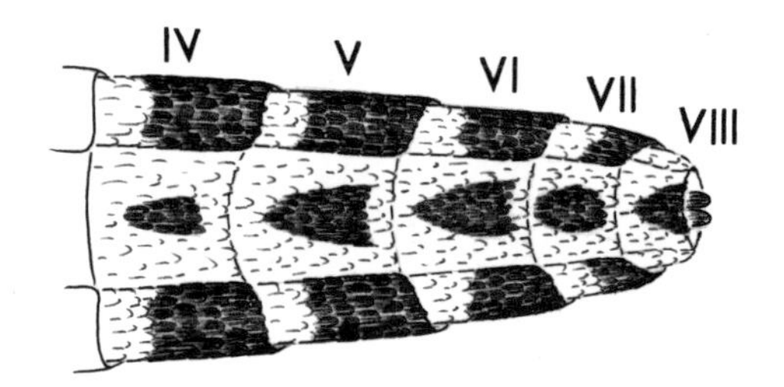


Fig. 396. Ventral view of abdomen: *Cx. stigmatosoma*

7(4). Abdominal sterna with median triangular areas of dark scales (fig. 397) ........................ ........................ *thriambus* (plate 30B)

Abdominal sterna without dark triangles mostly pale scaled (fig. 398) ........................ 8

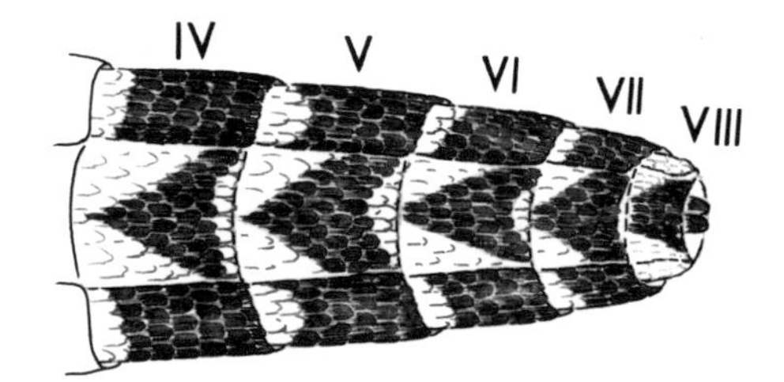


Fig. 397. Ventral view of abdomen: *Cx. thriambus*

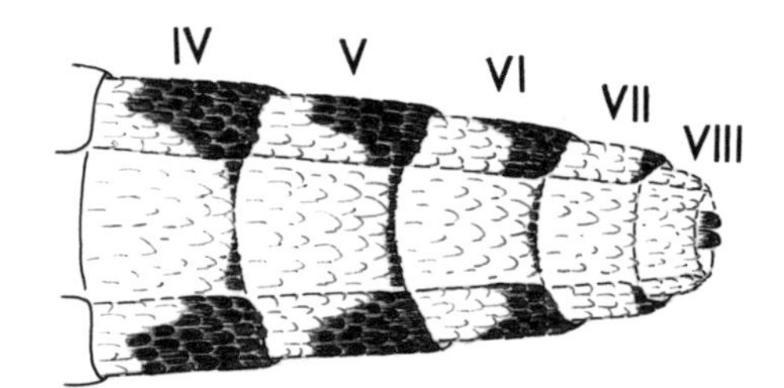


Fig. 398. Ventral view of abdomen: *Cx. coronator*

8(7). Hindtarsomere 5 with rings of pale scales basally and apically with dark scales medially (fig. 399) ........................ *coronator* (plate 42B)

Hindtarsomere 5 dark scaled with narrow ring of pale scales basally (fig. 400) ............. ........................ *declarator* (plate 41B)

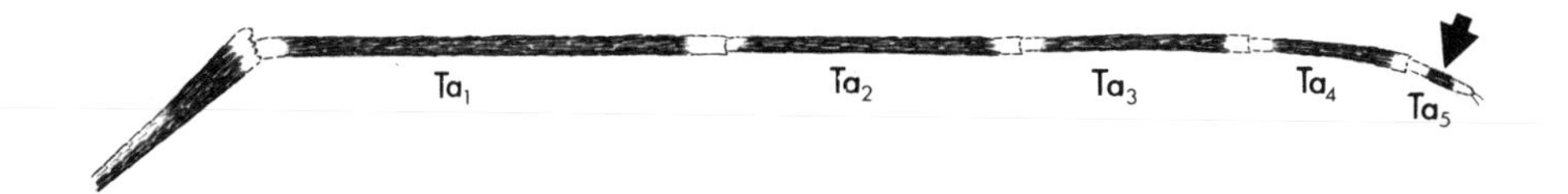


Fig. 399. Hindleg: *Cx. coronator*

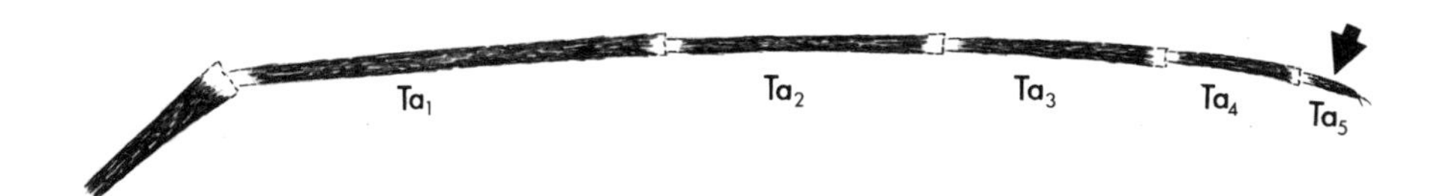


Fig. 400. Hindleg: *Cx. declarator*

9(3). Integument of scutum thoracic pleura and coxae reddish brown (fig. 401); scutum with hairlike golden brown scales (fig. 402) .................................. *erythrothorax* (plate 28B)

Integument of scutum thoracic pleura and coxae brown not reddish brown (fig. 403); scales of scutum narrow curved not hairlike (fig. 404) ................................................ 10

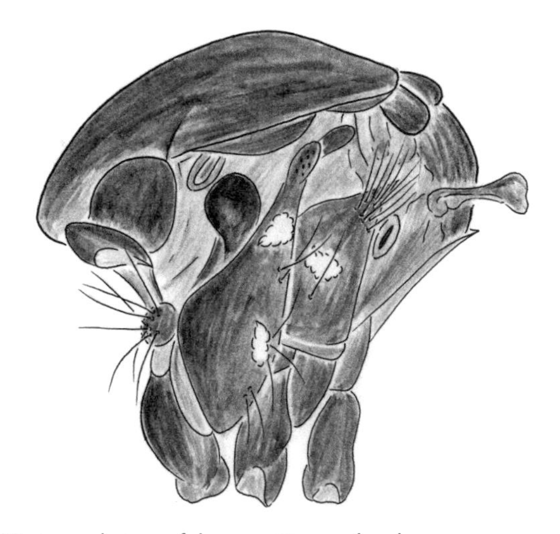

Fig. 401. Lateral view of thorax: *Cx. erythrothorax*

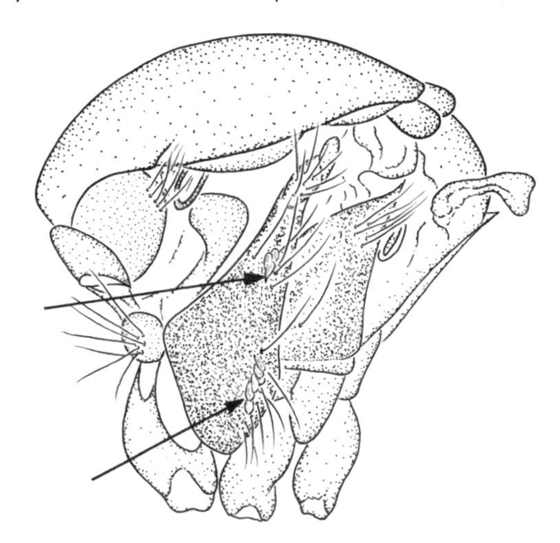

Fig. 403. Lateral view of thorax: *Cx. nigripalpus*

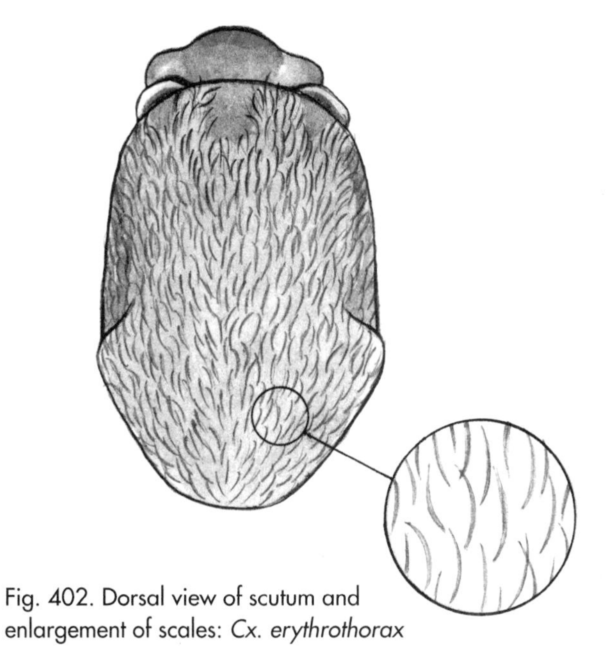

Fig. 402. Dorsal view of scutum and enlargement of scales: *Cx. erythrothorax*

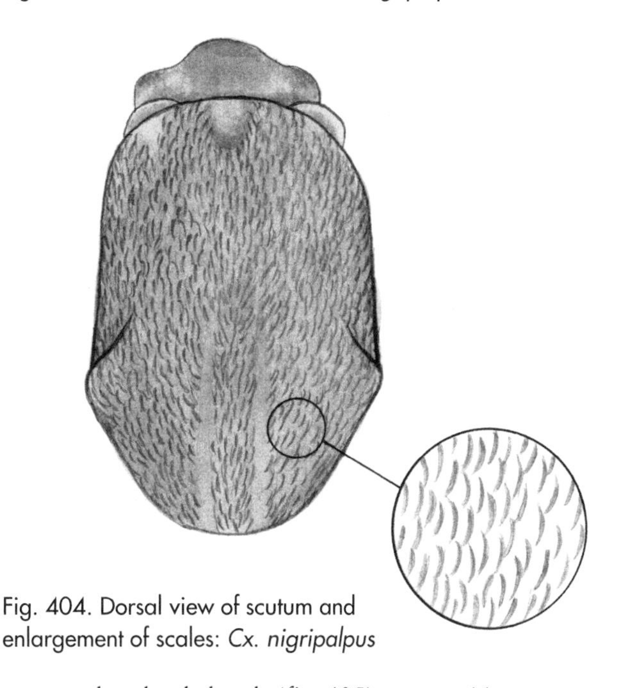

Fig. 404. Dorsal view of scutum and enlargement of scales: *Cx. nigripalpus*

10(9). Abdominal terga not banded or with only narrow basal pale bands (fig. 405) ............ 11

Abdominal terga with conspicuous basal pale bands (fig.406) ................................... 14

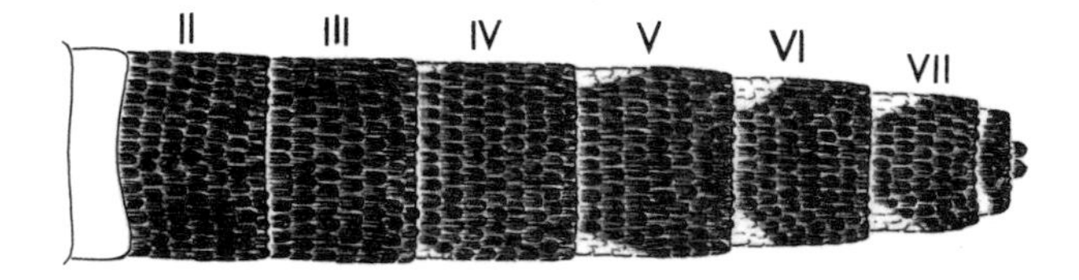


Fig. 405. Dorsal view of abdomen: *Cx. nigripalpus*

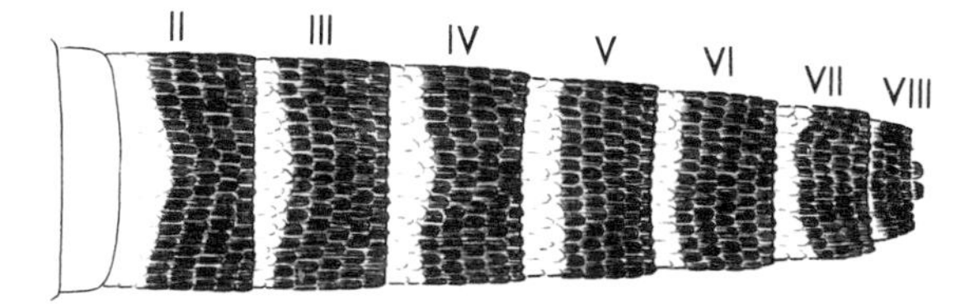


Fig. 406. Dorsal view of abdomen: *Cx. restuans*

11(10). Palpi long, 0.3 or more length of proboscis (fig. 407); lower mesepimeral seta absent (fig. 408) (subgenus *Micraedes*) ...................................................... *biscaynensis* (plate 41D)

Palpi short, no more than 0.25 length of proboscis (fig. 409); lower mesepimeral seta present (fig. 410) .......................................................................................... 12

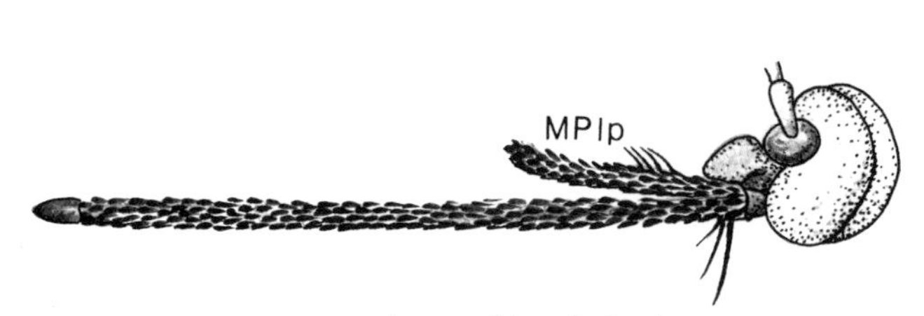


Fig. 407. Lateral view of head: *Cx. biscaynensis*

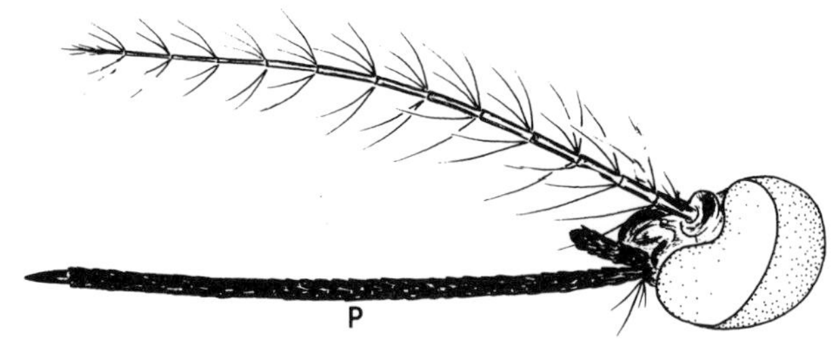


Fig. 409. Lateral view of head: *Cx. pipiens*

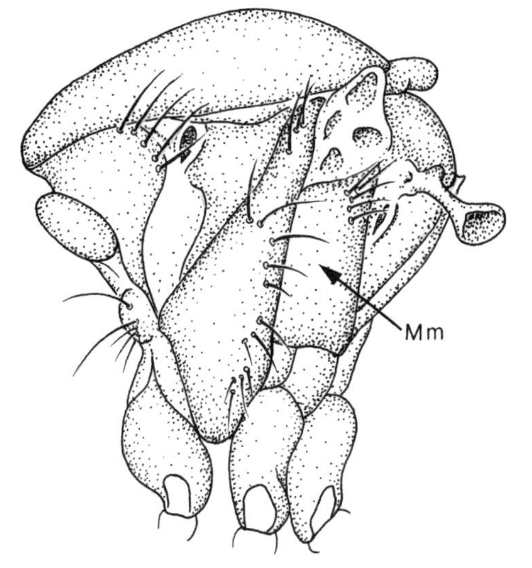


Fig. 408. Lateral view of thorax: *Cx. biscaynensis*

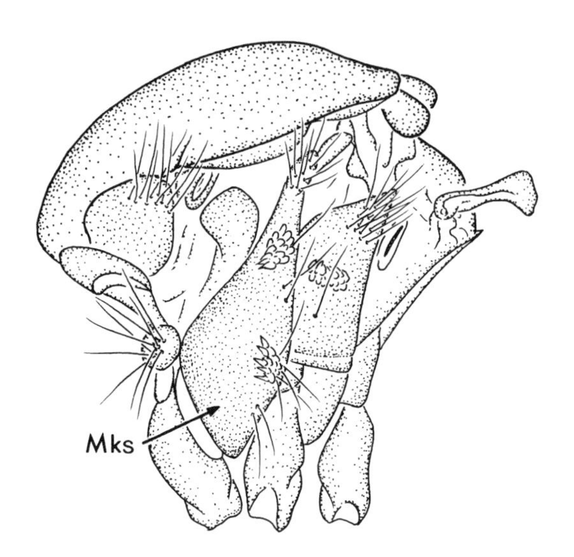


Fig. 410. Lateral view of thorax: *Cx. pipiens*

12(11). Scale patches on thoracic pleura in groups of fewer than 6 scales, mostly in lower meso-katepisternum (fig. 411); abdominal terga usually without basal pale bands, VII mostly dark scaled (fig. 412) ........ *nigripalpus* (plate 29A)

Thoracic pleura with several patches of pale scales, each with 6 or more scales (fig. 413); abdominal terga usually with narrow basal bands of white or yellow scales VII mostly dark or pale scaled (fig. 414) ........ 13

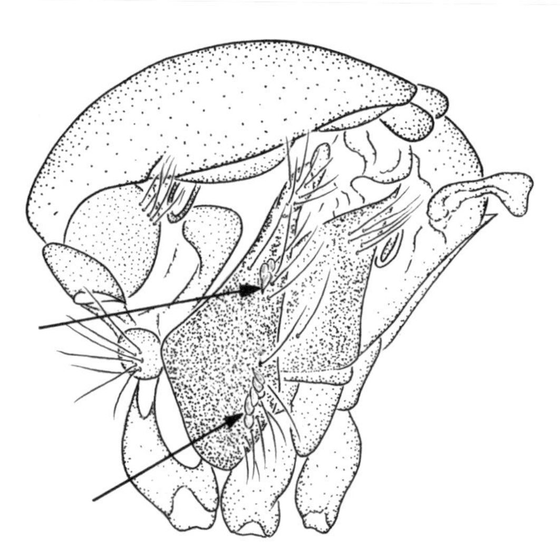

Fig. 411. Lateral view of thorax: *Cx. nigripalpus*

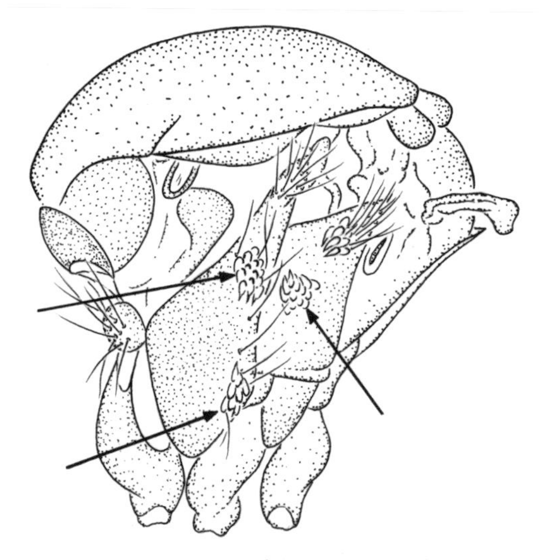

Fig. 413. Lateral view of thorax: *Cx. salinarius*

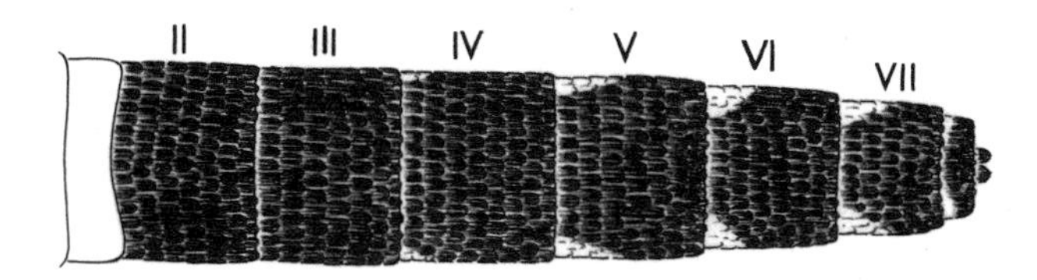


Fig. 412. Dorsal view of abdomen: *Cx. nigripalpus*

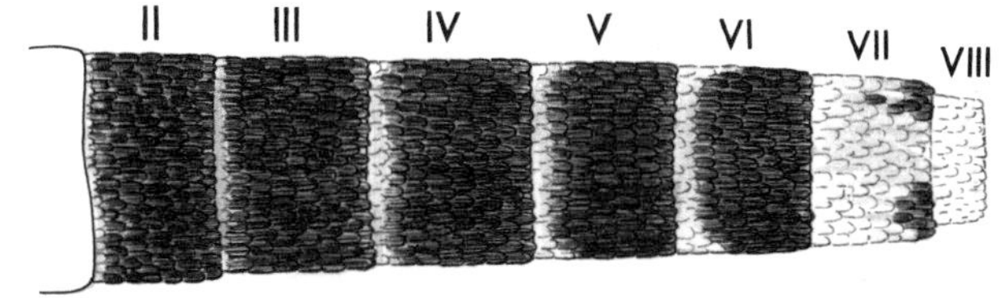


Fig. 414. Dorsal view of abdomen: *Cx. salinarius*

13(12). Abdominal tergum VII mostly with dingy yellow scales, terga II–VI with basolateral patches or narrow basal bands of dingy yellow scales sometimes blended with similar scales on apex of previous segment (fig. 415) ............................... *salinarius* (plate 29C)

Abdominal tergum VII mostly with dark scales terga II–VI with basolateral patches or narrow basal bands of whitish scales (fig. 416) ............................. *chidesteri* (plate 42A)

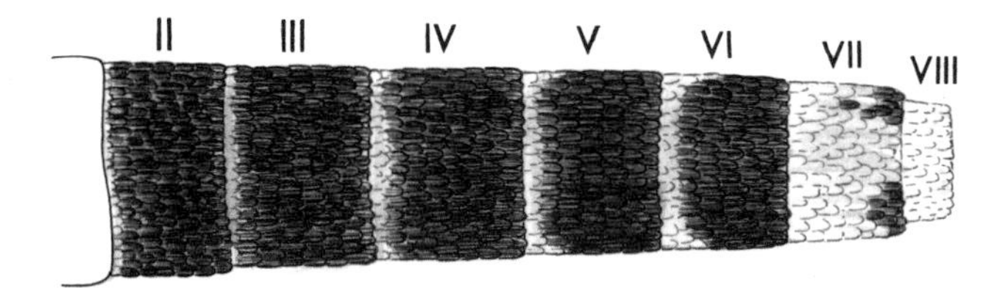


Fig. 415. Dorsal view of abdomen: *Cx. salinarius*

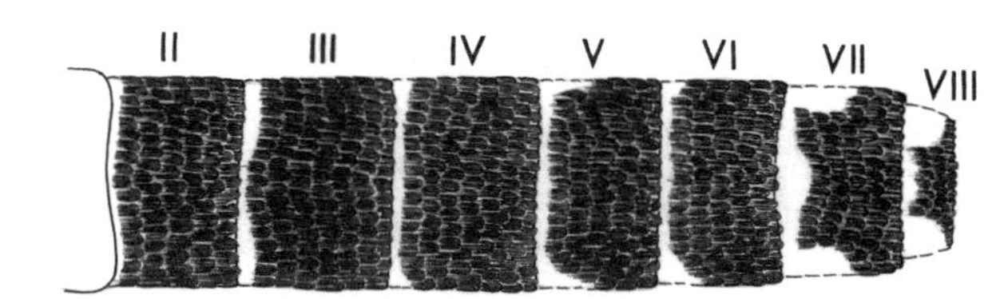


Fig. 416. Dorsal view of abdomen: *Cx. chidesteri*

14(10). Basal pale bands on abdominal terga rounded posteriorly, with marked sublateral constrictions, narrowly joined to large lateral pale patches (fig. 417); scutum without pale-scaled spots (fig. 418) ..................................................................................... *pipiens* (plate 28C)
*quinquefasciatus* (plate 28D)

Basal pale bands on abdominal terga more or less straight posteriorly, broadly joined to small lateral pale patches with only slight sublateral constriction, if at all (fig. 419); scutum with or without pale-scaled submedian spots near middle (fig. 420) .......................... 15

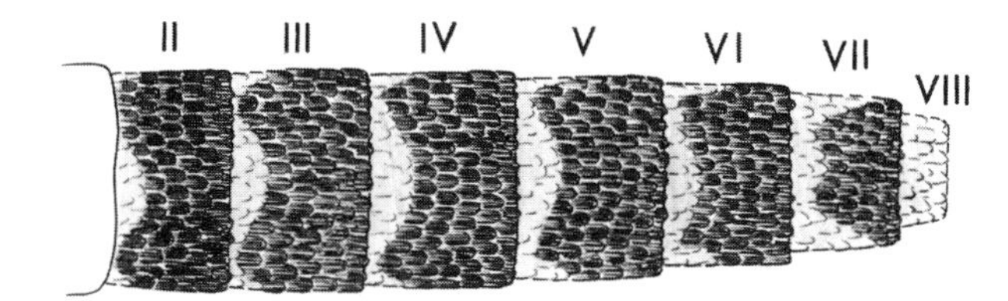


Fig. 417. Dorsal view of abdomen: *Cx. pipiens*

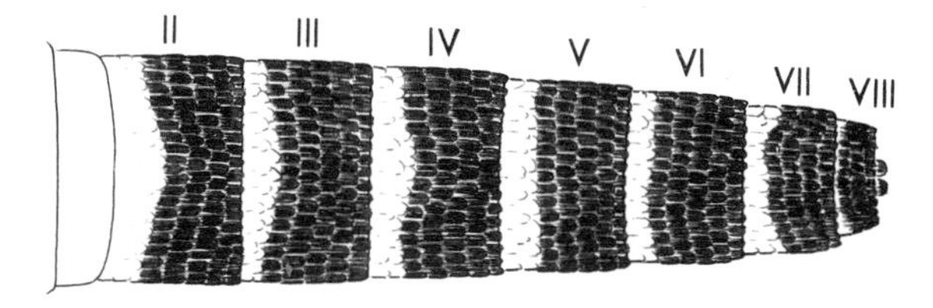


Fig. 419. Dorsal view of abdomen: *Cx. restuans*

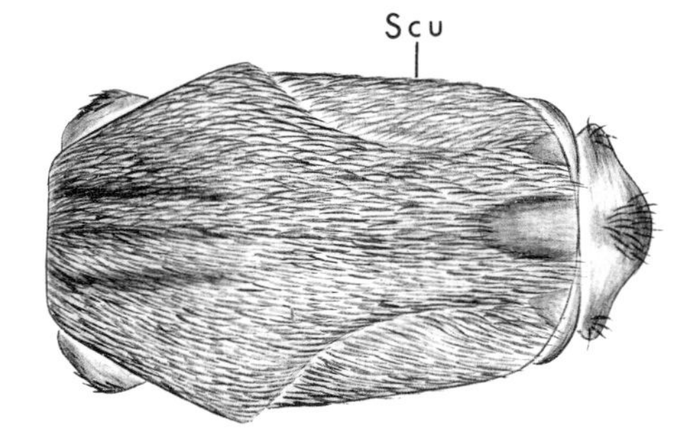


Fig. 418. Dorsal view of thorax: *Cx. pipiens*

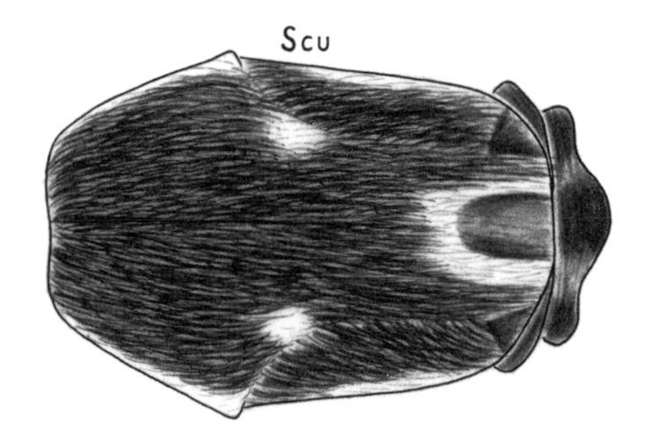


Fig. 420. Dorsal view of scutum: *Cx. restuans*

15(14). Wing cell $R_2$ 4.5 or more length of vein $R_{2+3}$ (fig. 421); scutum usually with pair of submedian pale-scaled spots near middle (fig. 422); medium sized species, wing length 4.0 mm or greater ........................................................................................ *restuans* (plate 29B)

Wing cell $R_2$ 3.0–4.0 length of vein $R_{2+3}$ (fig. 423); scutum without pale spots (fig. 424); small species, wing length 2.8 mm or less .................................. *interrogator* (plate 9A)

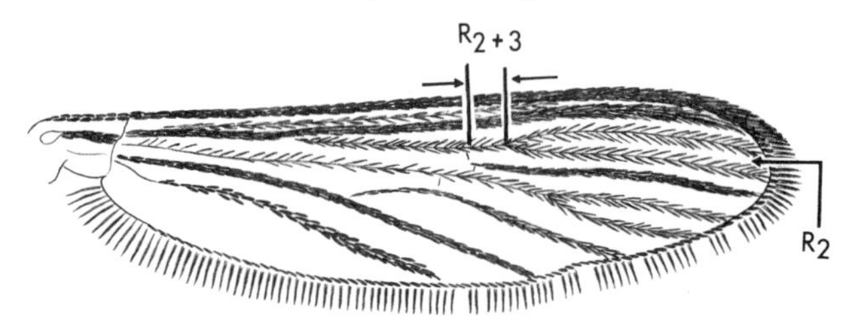


Fig. 421. Dorsal view of wing: *Cx. restuans*

Fig. 423. Dorsal view of wing: *Cx. interrogator*

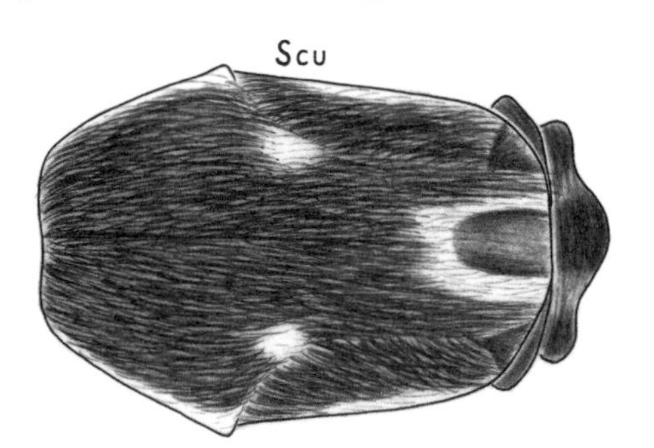


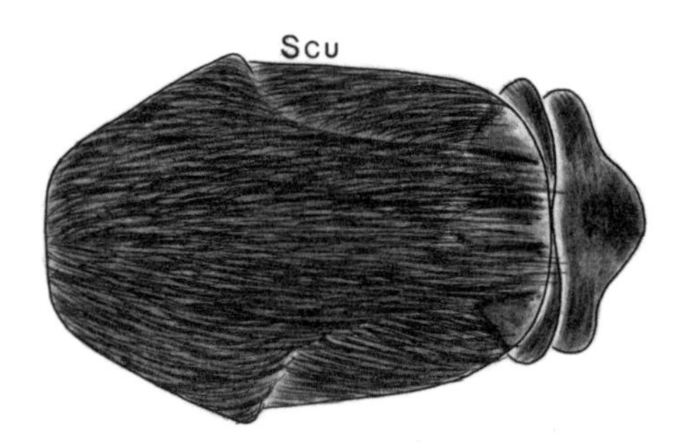


Fig. 422. Dorsal view of scutum: *Cx. restuans*

Fig. 424. Dorsal view of scutum: Cx. *interrogator*

16(2). Abdominal terga II,III with dorsum entirely dark scaled (fig. 425)..... *reevesi* (plate 41D)

Abdominal terga II,III with apical bands or apicolateral patches of pale scales extending onto dorsum (fig. 426) ........................................................................................ 17

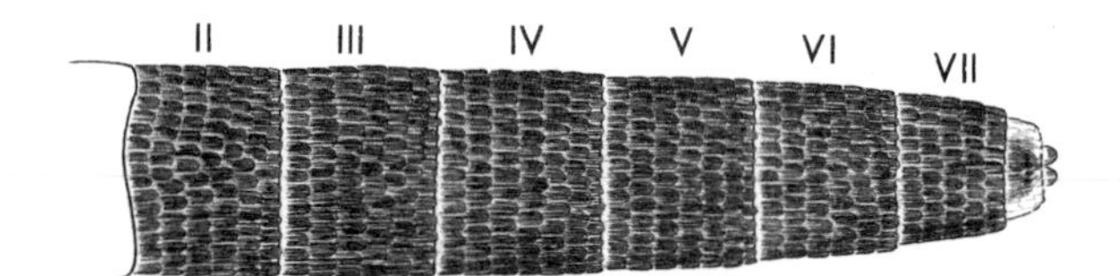


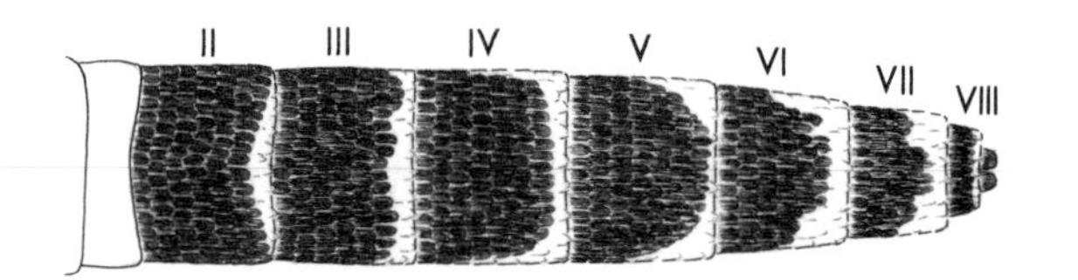


Fig. 425. Dorsal view of abdomen: *Cx. reevesi*

Fig. 426. Dorsal view of abdomen: *Cx. territans*

17(16). Apicolateral patches extending basally at least to 0.5 of tergum on IV–VI, usually connected to dorsoapical pale bands (fig. 427); palpi about 2.0 length of antennal flagellmere 4 (fig. 428) ........................................................................................ 18

Dorsoapical pale bands not markedly wider laterally, not extending basally more than 0.3 of terga IV–VI (fig. 429); palpi about 2.5–3.0 length of flagellomere 4 (fig. 430)........ 19

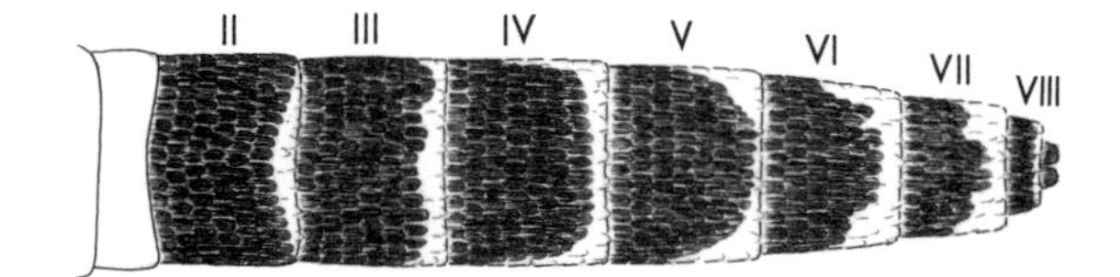


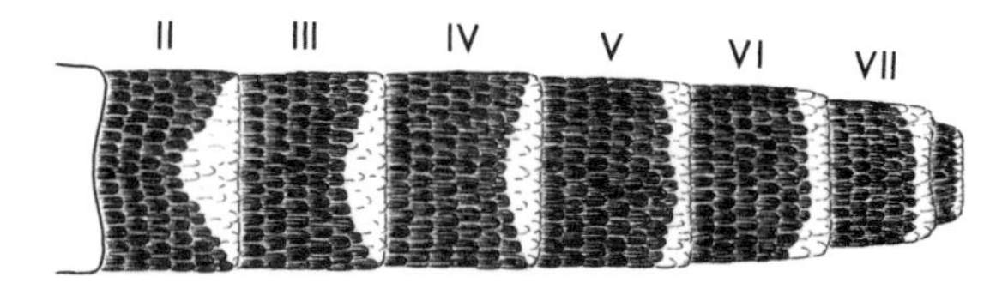


Fig. 427. Dorsal view of abdomen: *Cx. territans*

Fig. 429. Dorsal view of abdomen: *Cx. arizonensis*

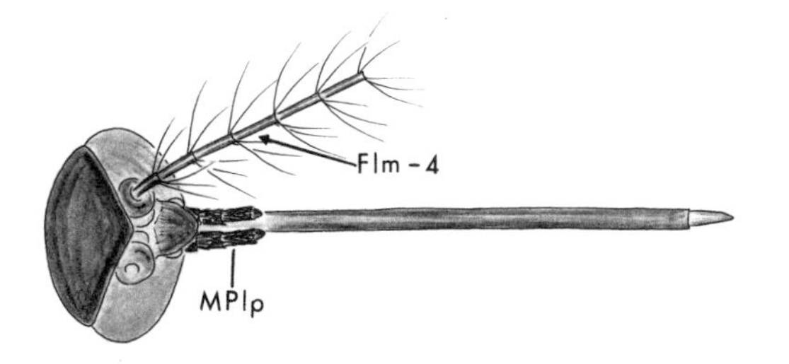


Fig. 428. Dorsal view of head: *Cx. territans*

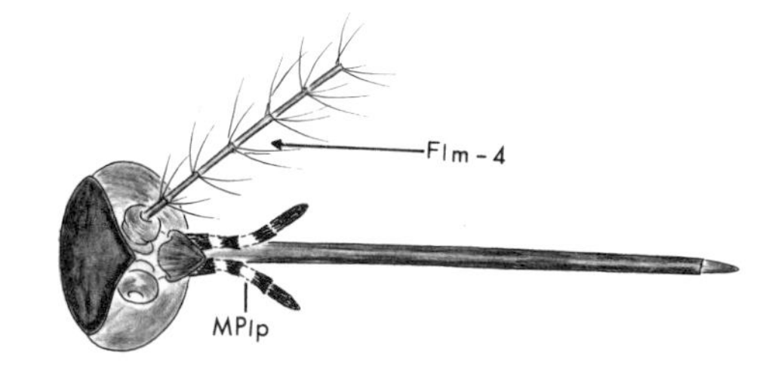


Fig. 430. Dorsal view of head: *Cx. apicalis*

18(17). Wing cell $R_2$ about 3.0 length of vein $R_{2+3}$ (fig. 431); apical and apicolateral scales of abdominal segments II–VII whitish (fig. 432) .................................. *territans* (plate 31D)

Wing cell $R_2$ 2.5 or less length of vein $R_{2+3}$ (fig. 433); apical and apicolateral scales on II–VII usually yellowish (fig. 434) ........................................................ *boharti* (plate 31C)

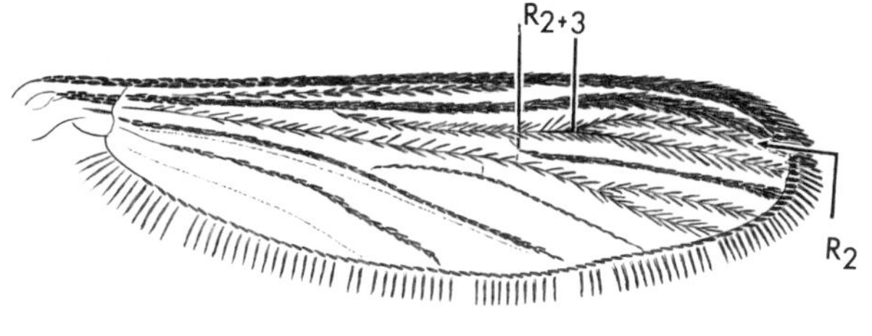


Fig. 431. Dorsal view of wing: *Cx. territans*

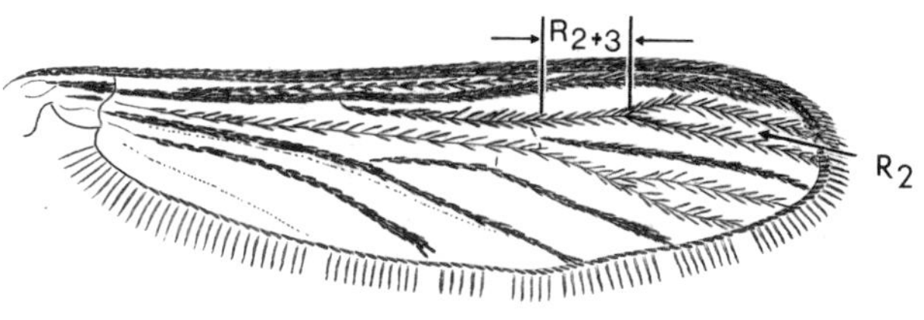


Fig. 433. Dorsal view of wing: *Cx. boharti*

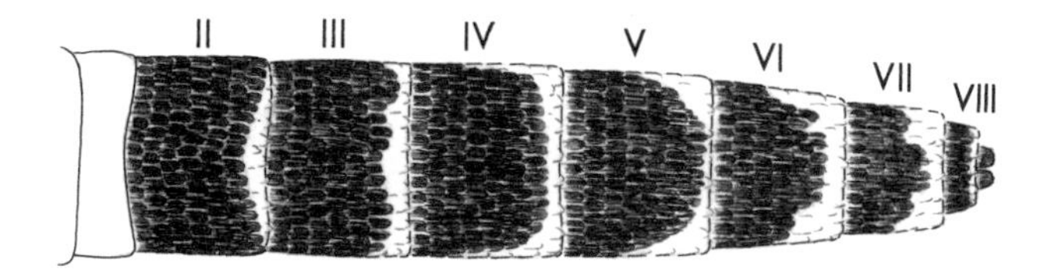


Fig. 432. Dorsal view of abdomen: *Cx. territans*

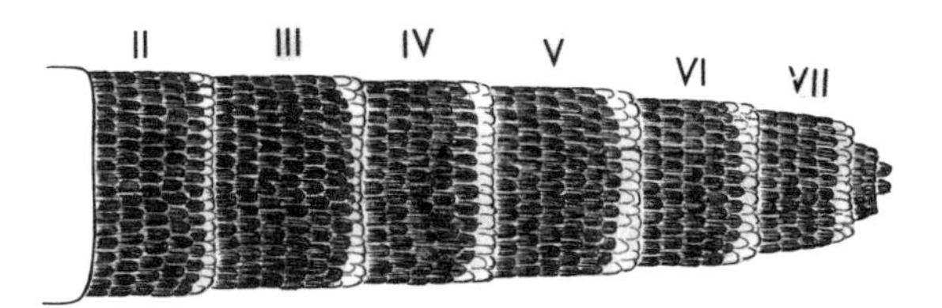


Fig. 434. Dorsal view of abdomen: *Cx. boharti*

19(17). Palpus with some pale scales (fig. 435) ............................................ *apicalis* (plate 31D)

Palpus entirely dark scaled (fig. 436) ......................................... *arizonensis* (plate 41D)

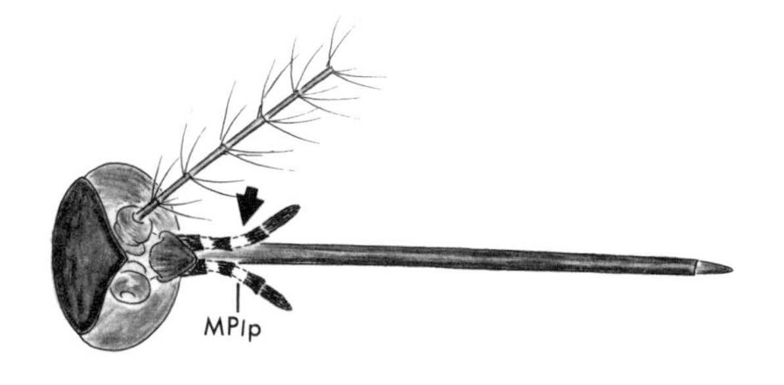


Fig. 435. Dorsal view of head: *Cx. apicalis*

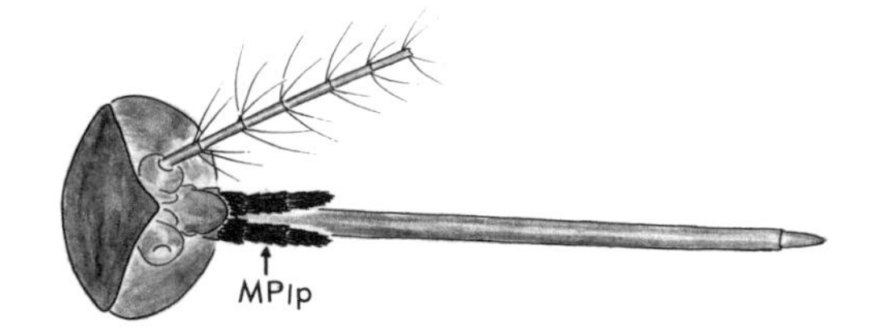


Fig. 436. Dorsal view of head: *Cx. arizonensis*

20(1). Mesepimeron with large patch of broad pale scales (fig. 437) ......... *erraticus* (plate 30C)

Mesepimeron usually unscaled, or with few narrow scales (fig. 438) ......................... 21

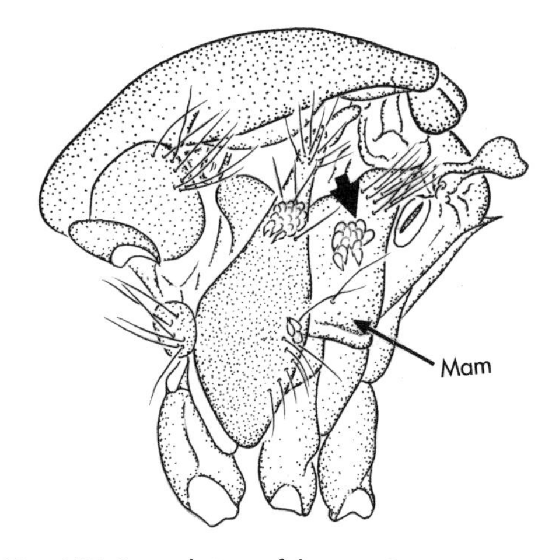

Fig. 437. Lateral view of thorax: *Cx. erraticus*

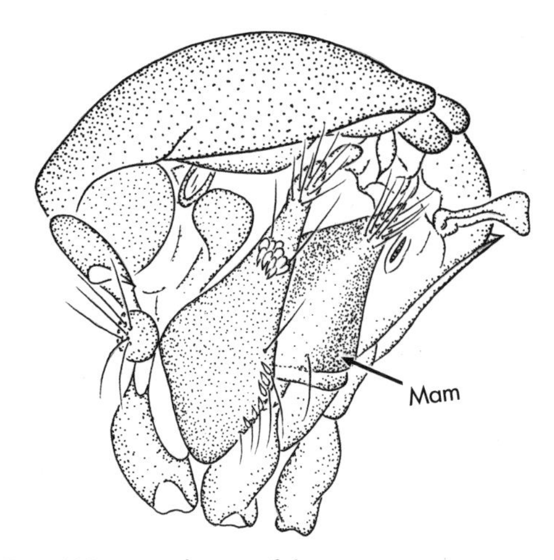


Fig. 438. Lateral view of thorax: *Cx. peccator*

21(20). Upper mesokatepisternum with patch of 6 or more pale scales; mesepimeron with light integumental area (fig. 439) .......... 22

Upper mesokatepisternum without scales or with fewer than 6; mesepimeron with or without light integumental area (fig. 440) .......... 24

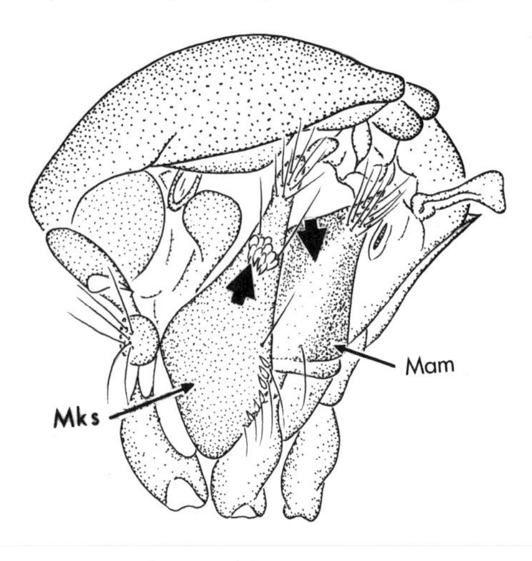


Fig. 439. Lateral view of thorax: *Cx. peccator*

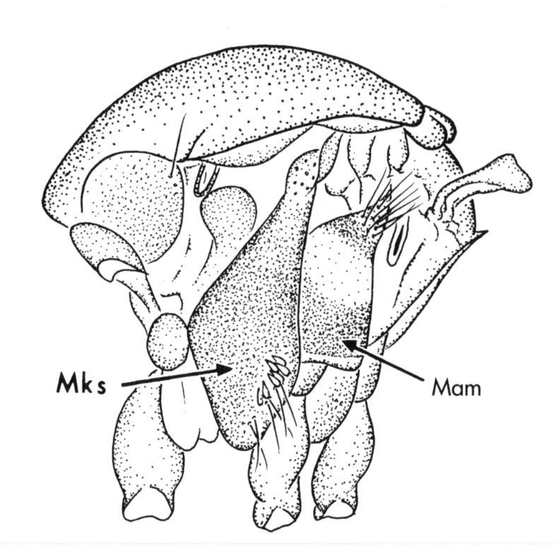


Fig. 440. Lateral view of thorax: *Cx. atratus*

22(21). Hindtarsomeres with pale bands on joints 1–4, tarsomere 5 entirely pale scaled (fig. 441) .......... *cedecei* (plate 42A)

Hindtarsomeres entirely dark scaled (fig. 442) .......... 23

Fig. 441. Hindleg: *Cx. cedecei*

Fig. 442. Hindleg: *Cx. peccator*

23(22). Occiput with broad dingy white scales anteromedially (fig. 443) .. *abominator* (plate 41C)

Occiput with broad dark brown scales anteromedially (fig. 444) .... *peccator* (plate 30D)
............................................................................................................ *anips* (plate 41B)

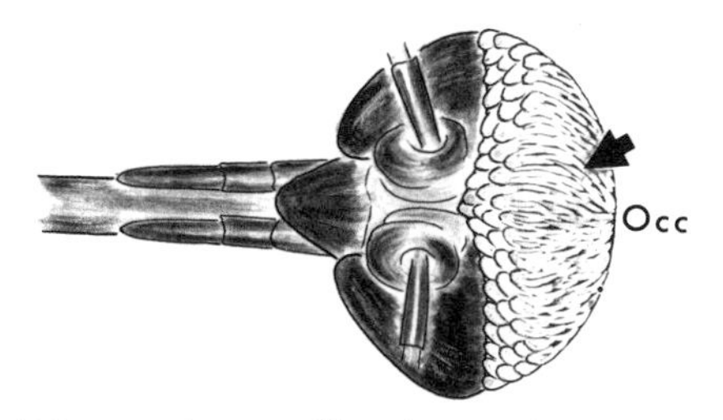


Fig. 443. Dorsal view of head: *Cx. abominator*

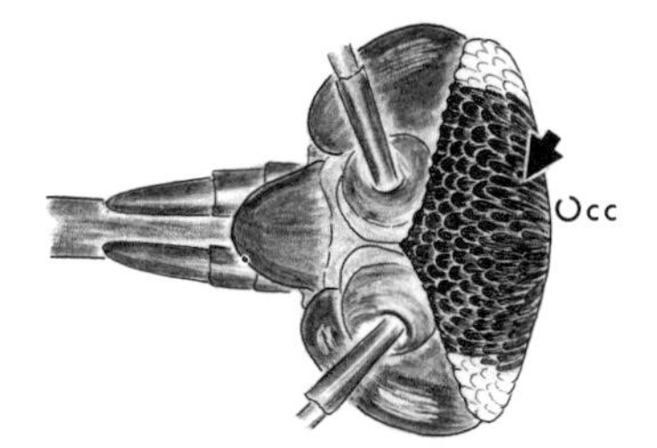


Fig. 444. Dorsal view of head: *Cx. peccator*

24(21). Mesepimeron with hairlike to ligulate scales, this sclerite and mesokatepisternum without pale spot or light integumental area (fig. 445)............................... *iolambdis* (plate 42B)

Mesepimeron without scales, this sclerite and mesokatepisternum with pale spot or light integumental area (fig. 446) ........................................................................................ 25

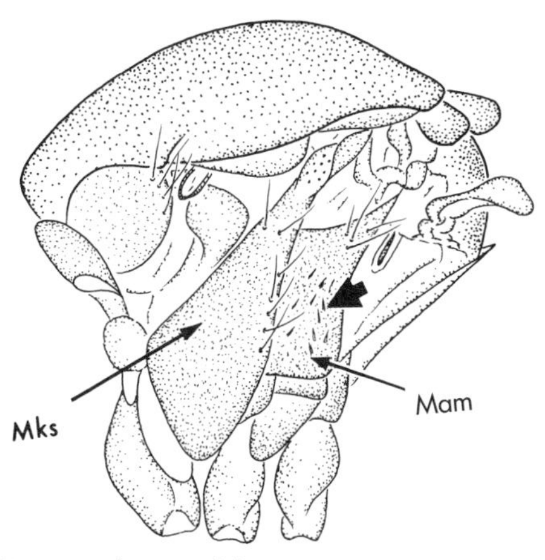


Fig. 445. Lateral view of thorax: *Cx. iolambdis*

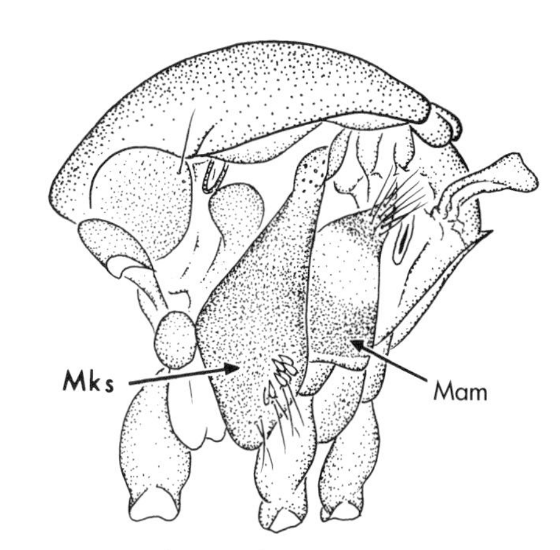


Fig. 446. Lateral view of thorax: *Cx. atratus*

25(24). Mesepimeron with distinct pale spot connected to anterior border, with dark area ventrally continuous with dark central area of mesokatepisternum (fig. 447) ... *atratus* (plate 41A)

Mesepimeron without distinct pale spot, with part of integument light in color (fig. 448)
............................................................................................................ 26

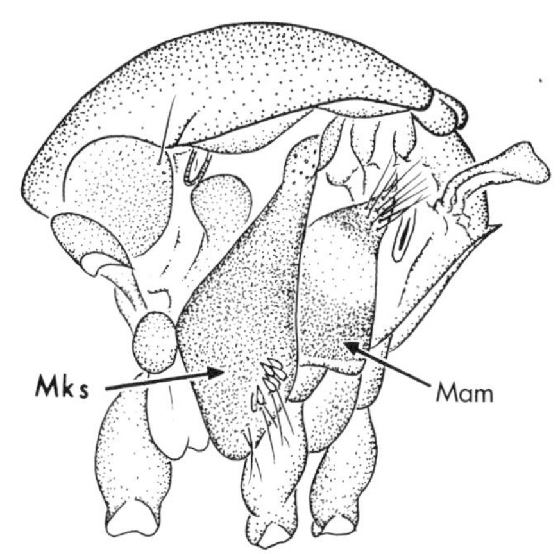


Fig. 447. Lateral view of thorax: *Cx. atratus*

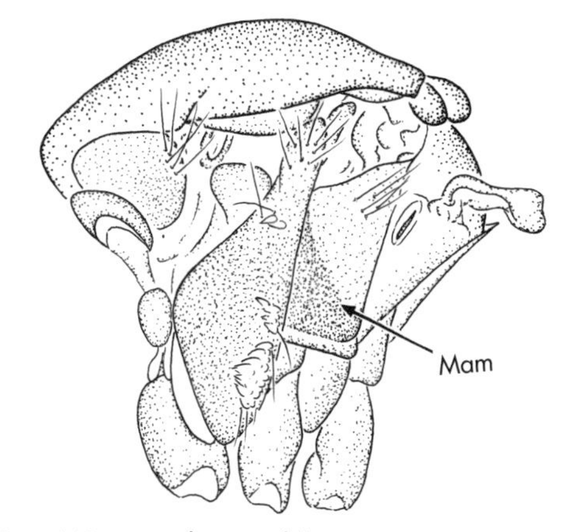


Fig. 448. Lateral view of thorax: *Cx. pilosus*

26(25). Mesepimeron with light integumental area covering dorsal 0.6; part of mesokatepisternum below ventral border of mesepimeron with width:length ratio of 1.2–1.3:1 (fig. 449) .... ........................................................ *pilosus* (plate 31A)

Mesepimeron with light integumental area confined to border; part of mesokatepisternum below ventral border of mesepimeron with width:length ratio of 1:1 (fig. 450) ............. ........................................................ *mulrennani* (plate 41B)

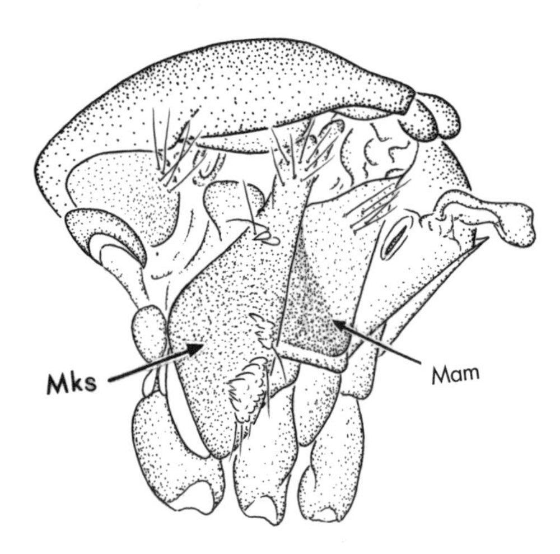


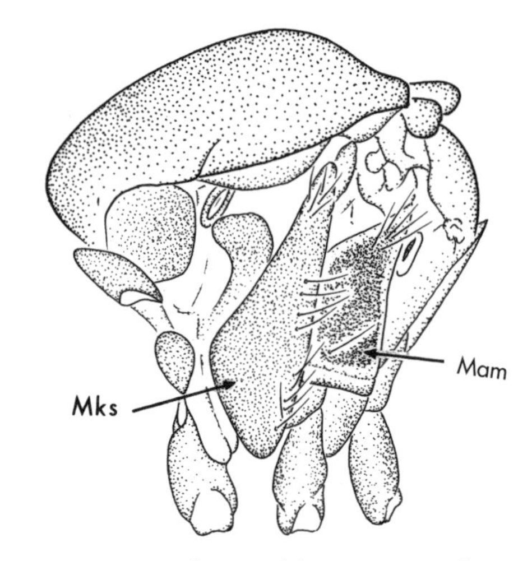

Fig. 449. Lateral view of thorax: *Cx. pilosus*

Fig. 450. Lateral view of thorax: *Cx. mulrennani*

## Key to Adult Females of Genus *Culiseta*

1. Abdominal terga without basal pale bands (fig. 451) .................... *melanura* (plate 33A)

   Abdominal terga with distinct basal pale bands (fig. 452) .......................................... 2

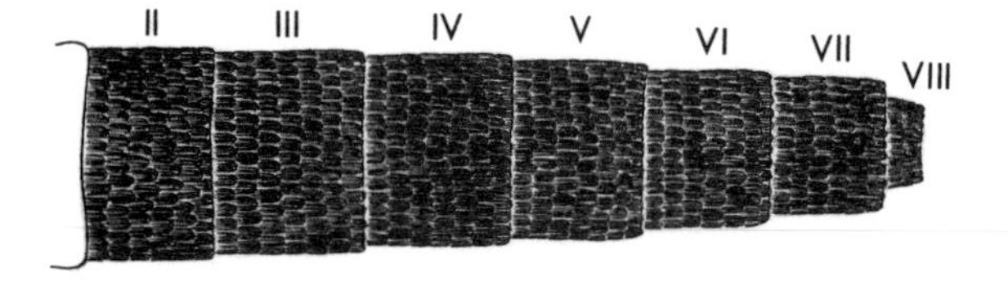


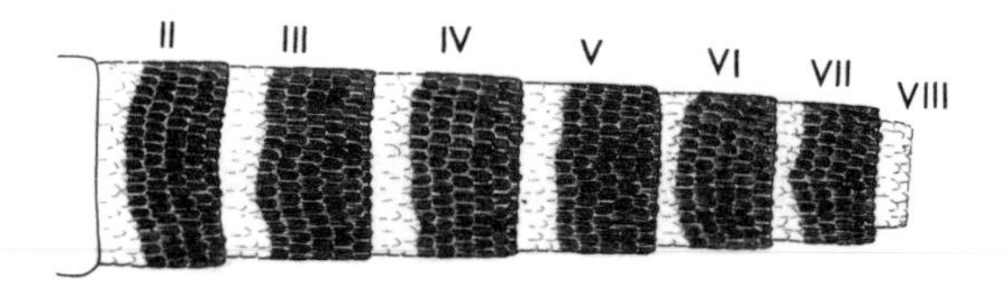


Fig. 451. Dorsal view of abdomen: *Cs. melanura*

Fig. 452. Dorsal view of abdomen: *Cs. morsitans*

2(1). Hindtarsomeres with pale-scaled bands on some segments (fig. 453) ........................... 3

Hindtarsomeres unbanded (fig. 454) .......................................................................... 7

Fig. 453. Hindleg: *Cs. morsitans*

Fig. 454. Hindleg: *Cs. impatiens*

3(2). Hindtarsomeres with broad pale bands, covering 0.25–0.33 of hindtarsomere 2 (fig. 455); crossveins of wing with scales (fig. 456) ........................................................................ 4

Hindtarsomeres with narrow pale bands, covering less than 0.1 of hindtarsomere 2 (fig. 457); crossveins without scales (fig. 458) ........................................................................ 5

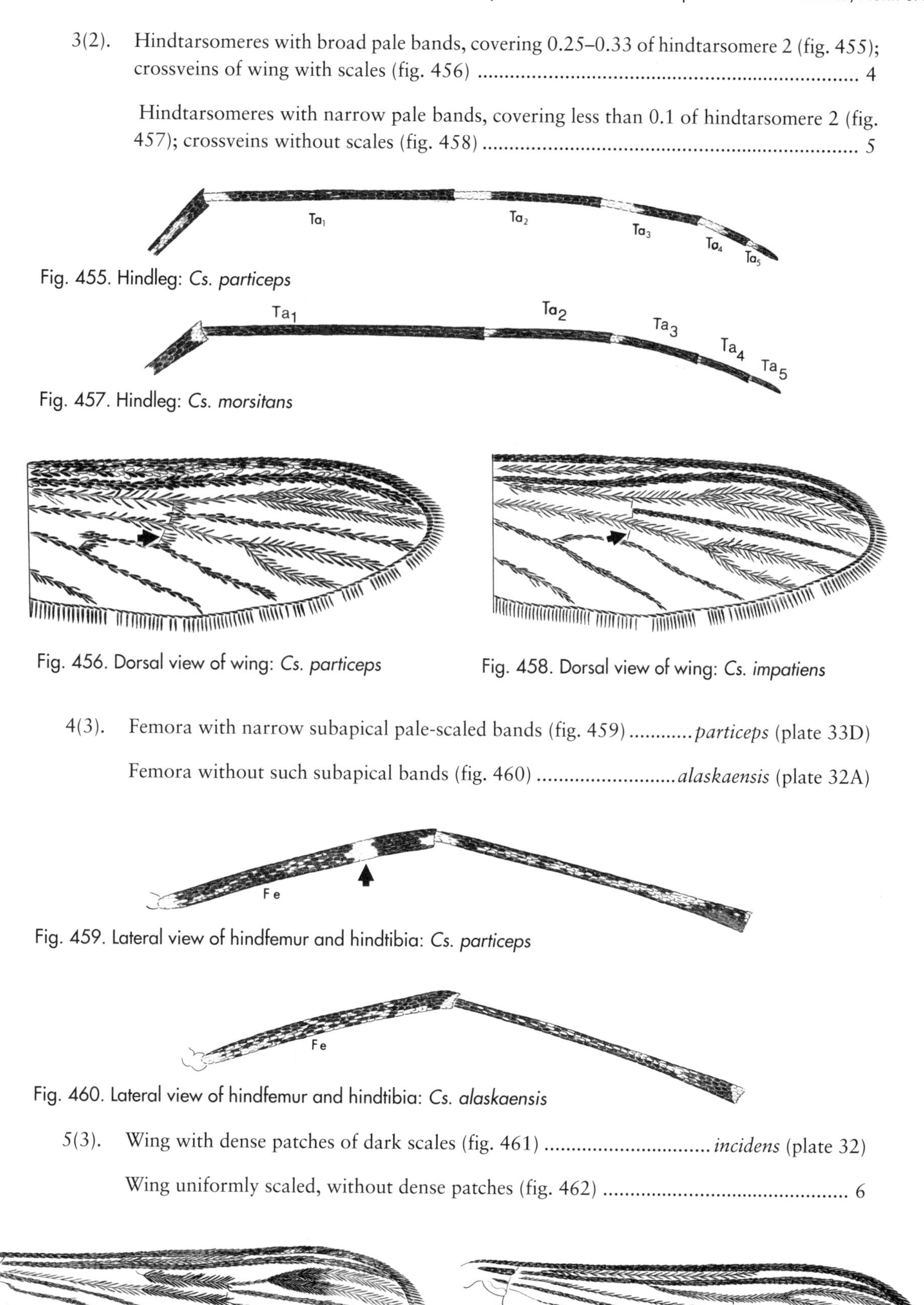


Fig. 455. Hindleg: *Cs. particeps*

Fig. 457. Hindleg: *Cs. morsitans*

Fig. 456. Dorsal view of wing: *Cs. particeps*

Fig. 458. Dorsal view of wing: *Cs. impatiens*

4(3). Femora with narrow subapical pale-scaled bands (fig. 459) ........... *particeps* (plate 33D)

Femora without such subapical bands (fig. 460) ......................... *alaskaensis* (plate 32A)

Fig. 459. Lateral view of hindfemur and hindtibia: *Cs. particeps*

Fig. 460. Lateral view of hindfemur and hindtibia: *Cs. alaskaensis*

5(3). Wing with dense patches of dark scales (fig. 461) .............................. *incidens* (plate 32)

Wing uniformly scaled, without dense patches (fig. 462) ............................................. 6

Fig. 461. Dorsal view of wing: *Cs. incidens*

Fig. 462. Dorsal view of wing: *Cs. impatiens*

6(5). Abdominal terga with pale bands on apices as well as bases, pale scales with brownish tinge, not white (fig. 463) ........................................................... *minnesotae* (plate 33B)

Abdominal terga with pale bands on bases only, pale scales whitish (fig. 464) ................ ................................................................................................ *morsitans* (plate 33C)

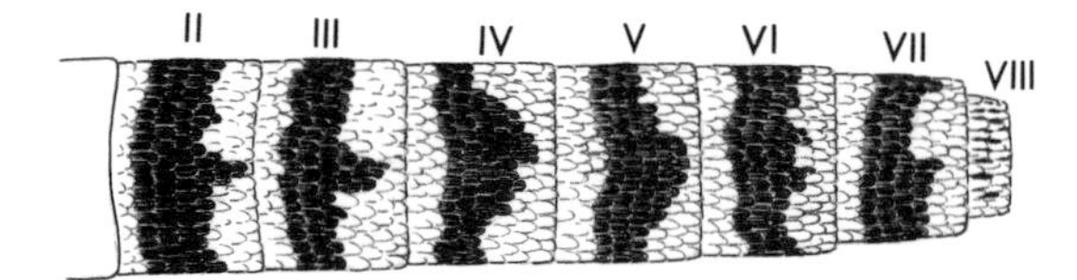


Fig. 463. Dorsal view of abdomen: *Cs. minnesotae*

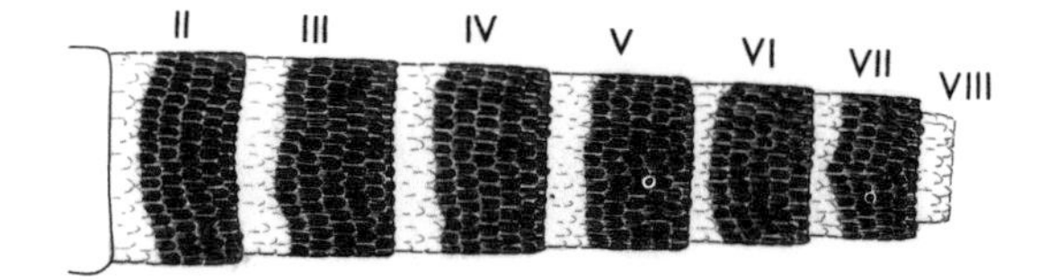


Fig. 464. Dorsal view of abdomen: *Cs. morsitans*

7(2). Wing with dark and pale scales intermixed on anterior veins (fig. 465); hindtarsomeres 1,2 with dark and pale scales (fig. 466) .............................................. *inornata* (plate 32D)

Wing and hindtarsomeres dark scaled (figs. 467, 468) .................. *impatiens* (plate 32B)

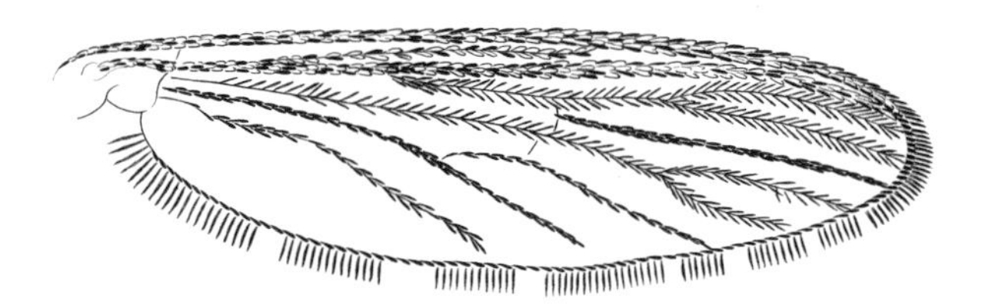

Fig. 465. Dorsal view of wing: *Cs. inornata*

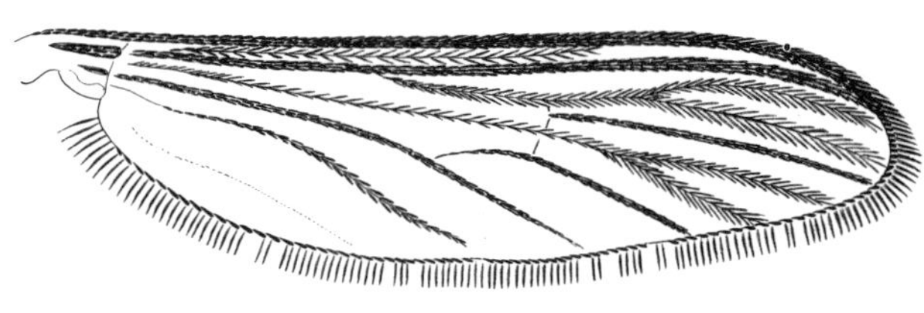

Fig. 467. Dorsal view of wing: *Cs. impatiens*

Fig. 466. Hindleg: *Cs. inornata*

Fig. 468. Hindleg: *Cs. impatiens*

### Key to Adult Females of Genus *Deinocerites*

1. Mesepimeron with patch of translucent scales (fig. 469) .................. *pseudes* (plate 39C)

Mesepimeron without scales (fig. 470) ........................................................................ 2

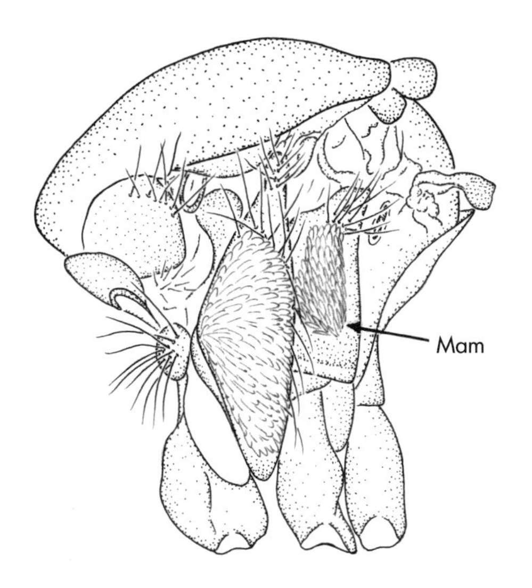


Fig. 469. Lateral view of thorax: *De. pseudes*

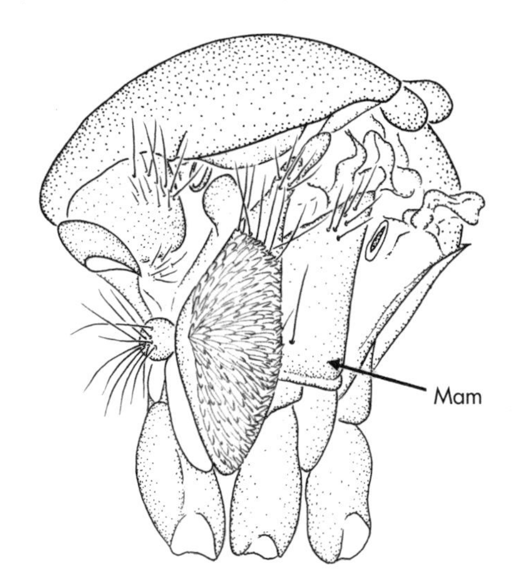


Fig. 470. Lateral view of thorax: *De. cancer*

2(1). Cerci with long spatulate apical or subapical setae (fig. 471); medium-sized species, wing length 2.9 mm ........ *cancer* (plate 39D)

Cerci without specialized setae (fig. 472); small species, wing length about 2.5 mm ........ *mathesoni* (plate 42C)

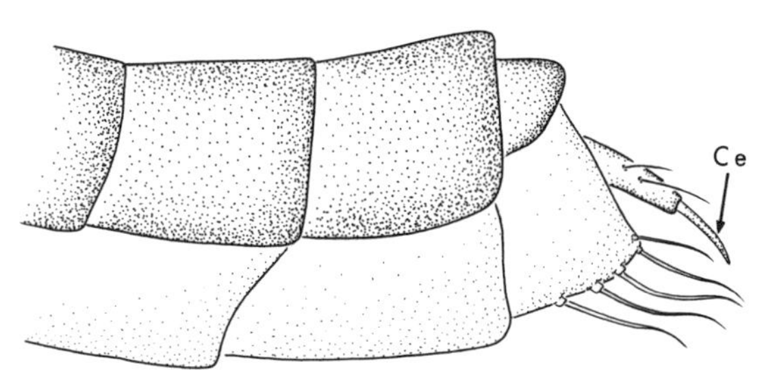


Fig. 471. Lateral view of abdominal segments VII–X: *De. cancer*

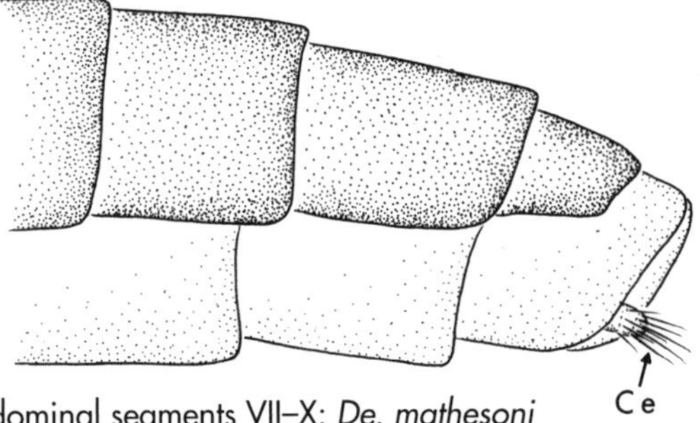


Fig. 472. Lateral view of abdominal segments VII–X: *De. mathesoni*

## Key to Adult Females of Genus *Mansonia*

Apex of abdominal tergum VII with row of short dark spiniforms (fig. 473); ventral surface of proboscis mostly dark scaled (fig. 474) ........ *titillans* (plate 33D)

Apex of tergum VII without spiniforms (fig. 475); ventral surface of proboscis with patch of pale scales (fig. 476) ........ *dyari* (plate 40A)

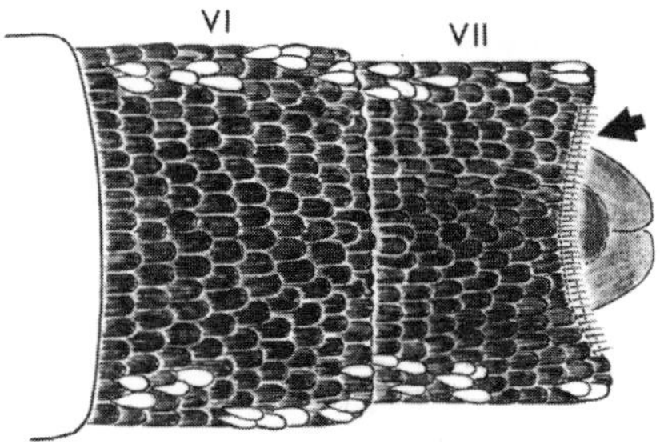

Fig. 473. Dorsal view of abdominal tergum VII: *Ma. titillans*

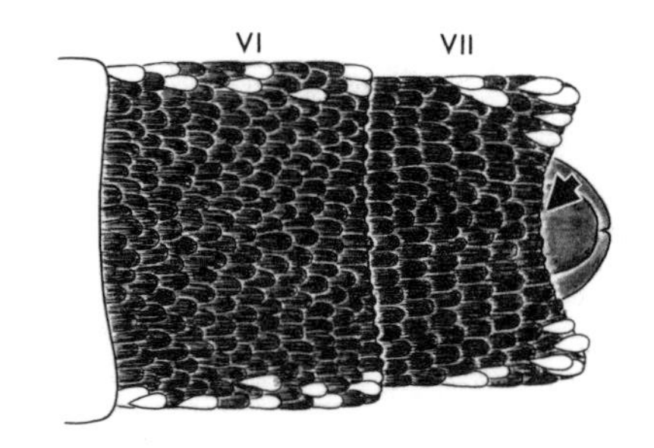


Fig. 475. Dorsal view of abdominal tergum VII: *Ma. dyari*

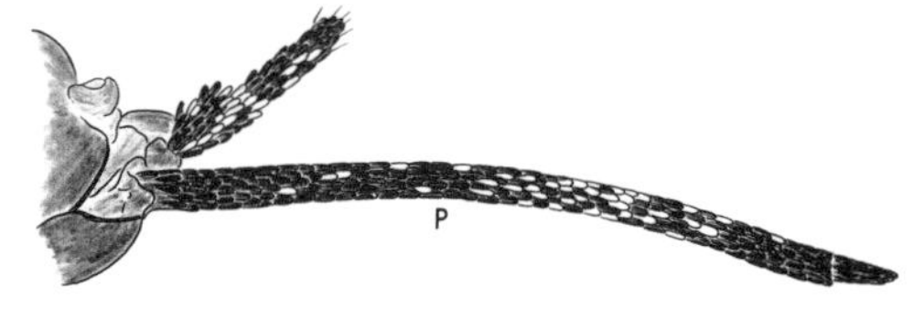


Fig. 474. Ventrolateral view of head: *Ma. titillans*

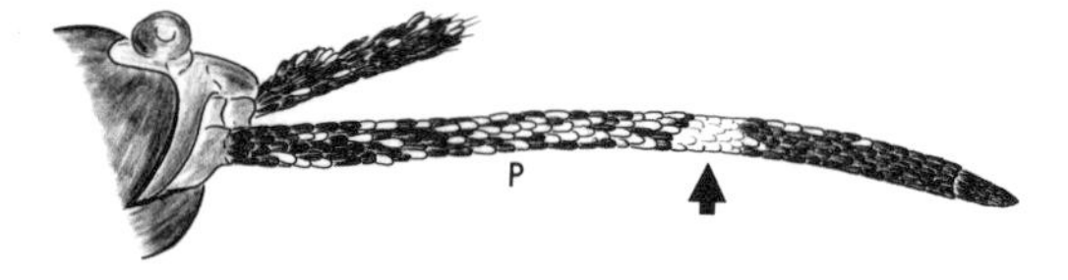


Fig. 476. Ventrolateral view of head: *Ma. dyari*

## Key to Adult Females of Genus *Orthopodomyia*

1. Proepisternum with transverse line of pale scales on anterior face (fig. 477); base of wing vein A dark scaled (fig. 478); lines of scales on mesokatepisternum very narrow (fig. 479) ........................................ *kummi* (plate 42D)

   Proepisternum with anterior face bare (fig. 480); base of vein A pale scaled (fig. 481); lines of scales on mesokatepisternum broad (fig. 482) ........................................ 2

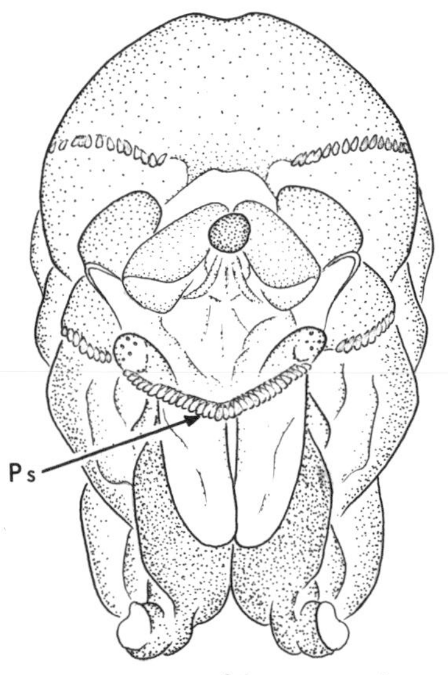


Fig. 477. Anterior view of thorax: *Or. kummi*

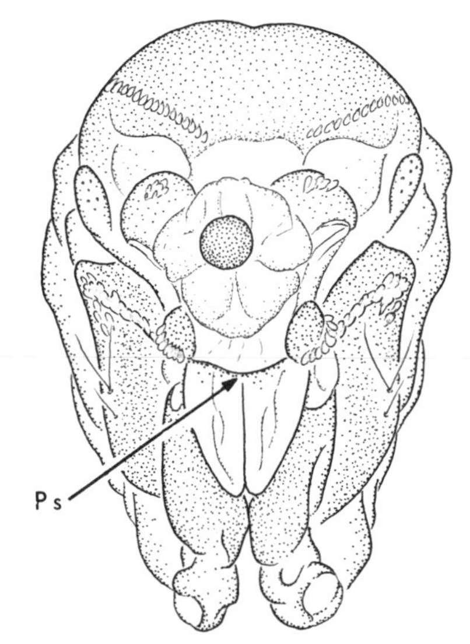


Fig. 480. Anterior view of thorax: *Or. alba*

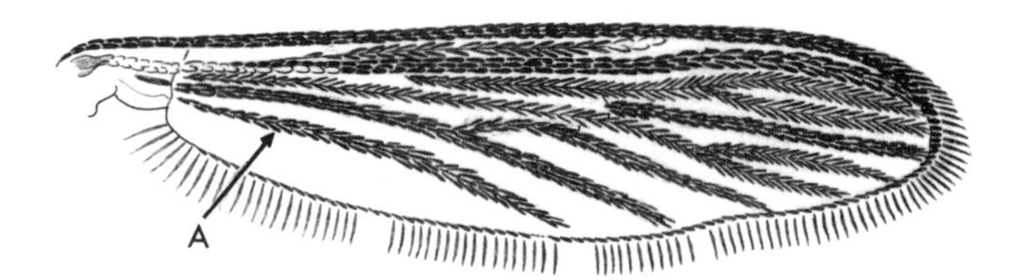


Fig. 478. Dorsal view of wing: *Or. kummi*

Fig. 481. Dorsal view of wing: *Or. signifera*

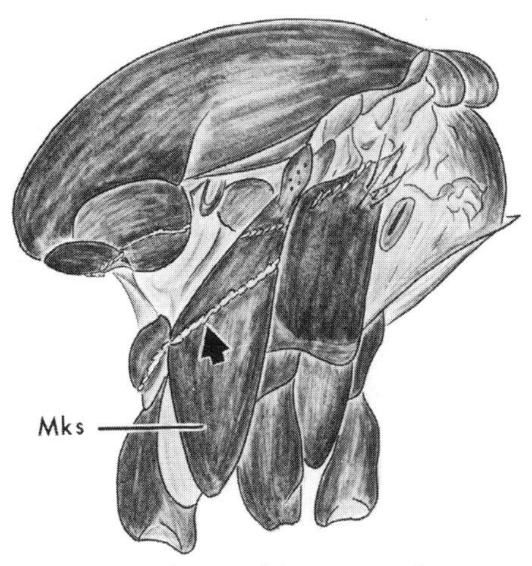

Fig. 479. Lateral view of thorax: *Or. kummi*

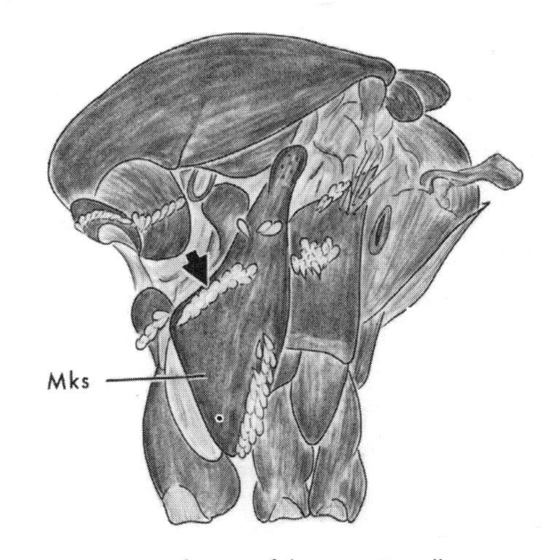


Fig. 482. Lateral view of thorax: *Or. alba*

2(1). Lower mesokatepisternal setae 4 or more (fig. 483); base of wing vein $R_{4+5}$ usually with patch of pale scales (fig. 484) ........................................................ *signifera* (plate 34C)

Lower mesokatepisternal setae 0–2 (fig. 485); base of vein $R_{4+5}$ usually dark scaled (fig. 486) .......................................................................................... *alba* (plate 34B)

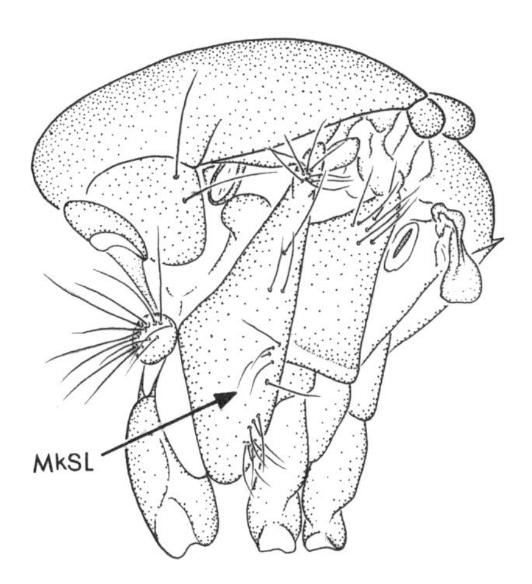


Fig. 483. Lateral view of thorax: *Or. signifera*

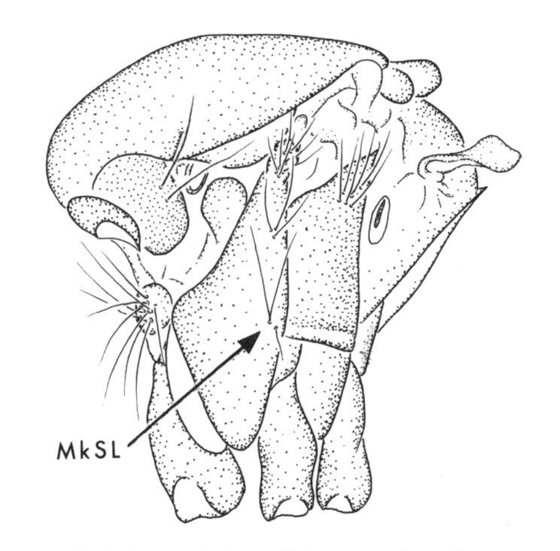


Fig. 485. Lateral view of thorax: *Or. alba*

Fig. 484. Dorsal view of wing: *Or. signifera*

Fig. 486. Dorsal view of wing: *Or. alba*

## Key to Adult Females of Genus *Psorophora*

1. Wing dark and pale scales on all veins (fig. 487); femora with more or less distinct narrow subapical band of pale scales (fig. 488) (subgenus *Grabhamia*) .................................. 2

Wing scales entirely dark or with only a few pale scales on vein C and Sc (fig. 489); femora without subapical pale bands (fig. 490) ...................................................................... 5

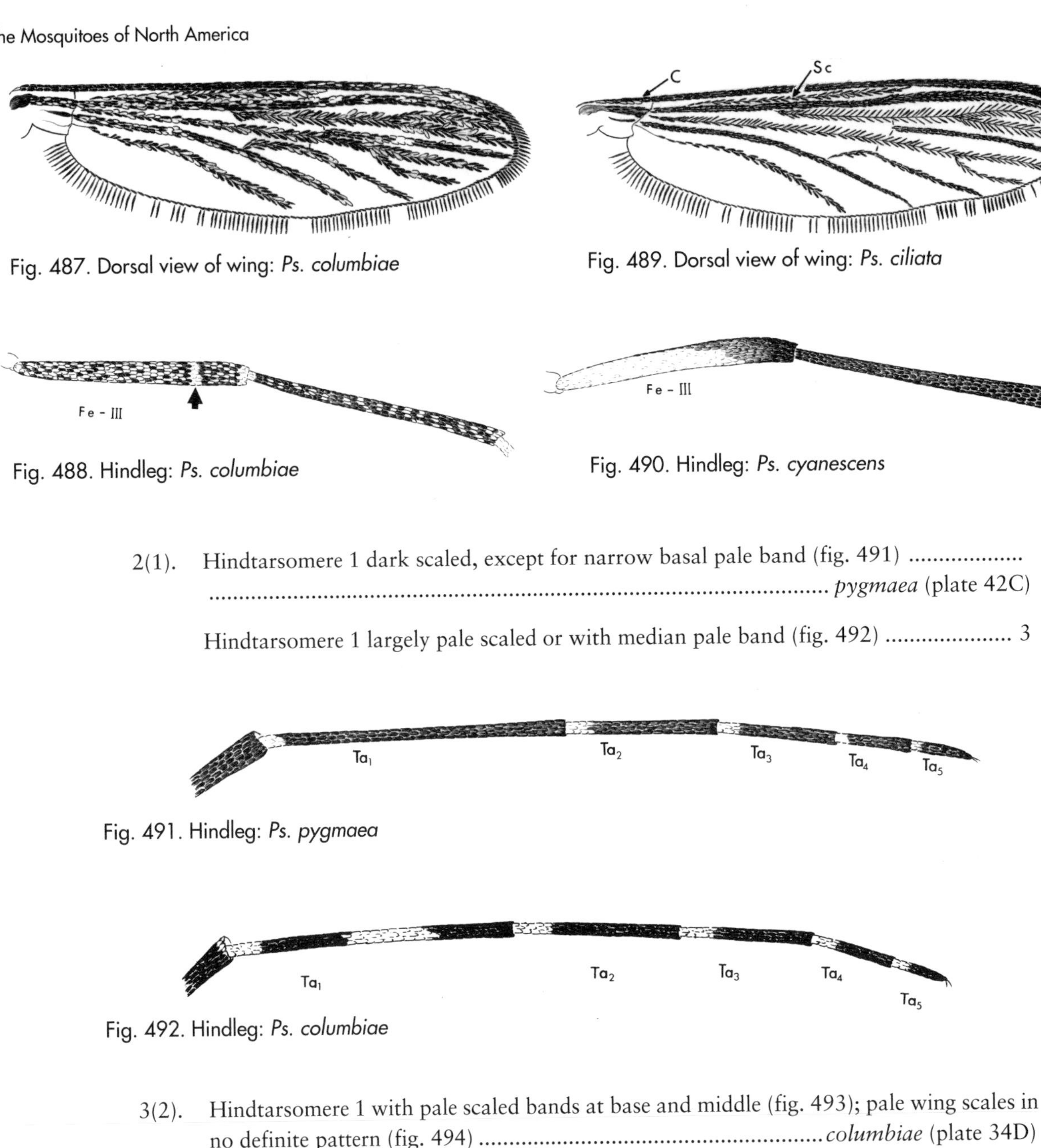


Fig. 487. Dorsal view of wing: *Ps. columbiae*

Fig. 489. Dorsal view of wing: *Ps. ciliata*

Fig. 488. Hindleg: *Ps. columbiae*

Fig. 490. Hindleg: *Ps. cyanescens*

2(1). Hindtarsomere 1 dark scaled, except for narrow basal pale band (fig. 491) ........................................................................................................ *pygmaea* (plate 42C)

Hindtarsomere 1 largely pale scaled or with median pale band (fig. 492) ..................... 3

Fig. 491. Hindleg: *Ps. pygmaea*

Fig. 492. Hindleg: *Ps. columbiae*

3(2). Hindtarsomere 1 with pale scaled bands at base and middle (fig. 493); pale wing scales in no definite pattern (fig. 494) ........................................................ *columbiae* (plate 34D)

Hindtarsomere 1 largely pale scaled (fig. 495); wing with definite areas of pale and dark scales (fig. 496) ........................................................................................................ 4

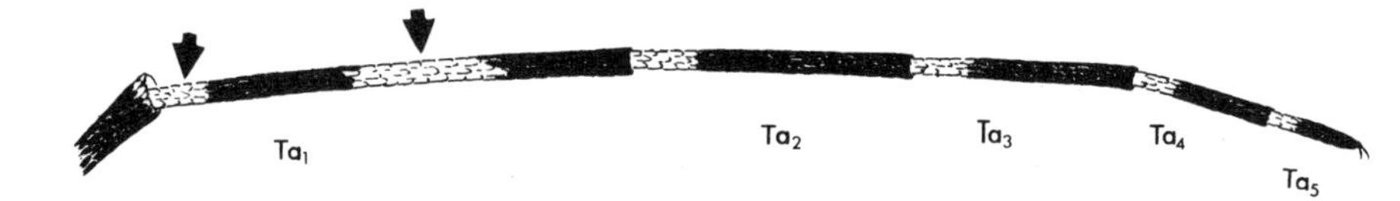


Fig. 493. Hindleg: *Ps. columbiae*

Fig. 495. Hindleg: *Ps. discolor*

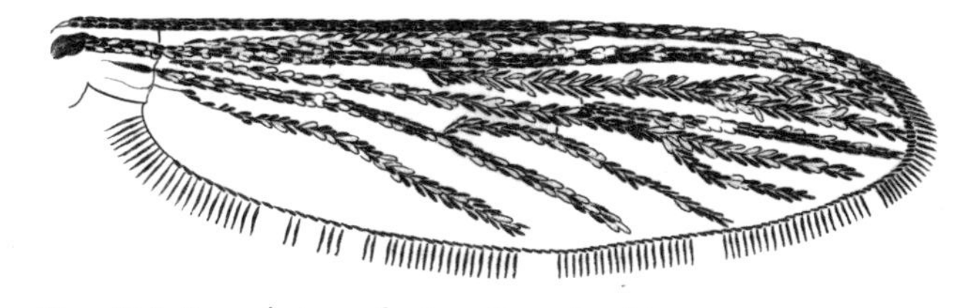

Fig. 494. Dorsal view of wing: *Ps. columbiae*

Fig. 496. Dorsal view of wing: *Ps. discolor*

4(3). Wing fringe with alternating spots of dark and pale scales, vein A pale scaled apically (fig. 497) ........................................................................................ *signipennis* (plate 37A)

Wing fringe uniformly dark scaled, vein A with dark scales apically (fig. 498) .............. ........................................................................................................ *discolor* (plate 35C)

Fig. 497. Dorsal view of wing: *Ps. signipennis*

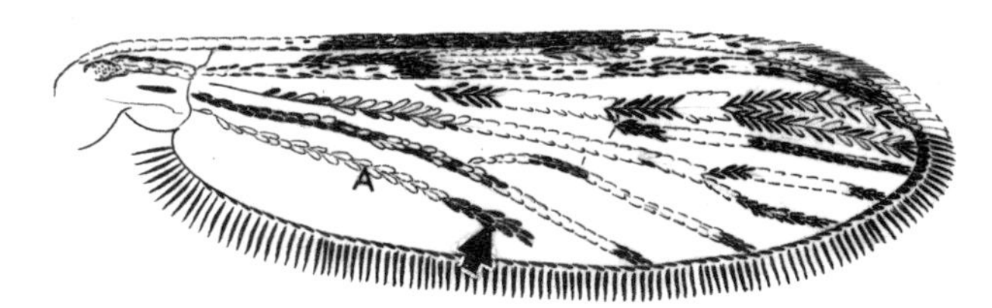


Fig. 498. Dorsal view of wing: *Ps. discolor*

5(1). Apices of hindfemur and tibia with long erect scales, shaggy, hindtarsomere 5 not entirely pale scaled (fig. 499) (subgenus *Psorophora*) ............................................................. 6

Apices of hindfemur and tibia usually without erect scales; if somewhat shaggy, then hindtarsomere 5 entirely pale scaled (figs. 500, 501) (subgenus *Janthinosoma*) .................. 7

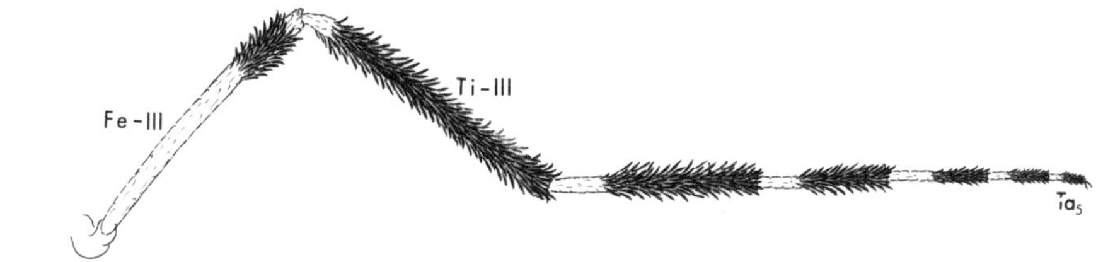


Fig. 499. Hindleg: *Ps. ciliata*

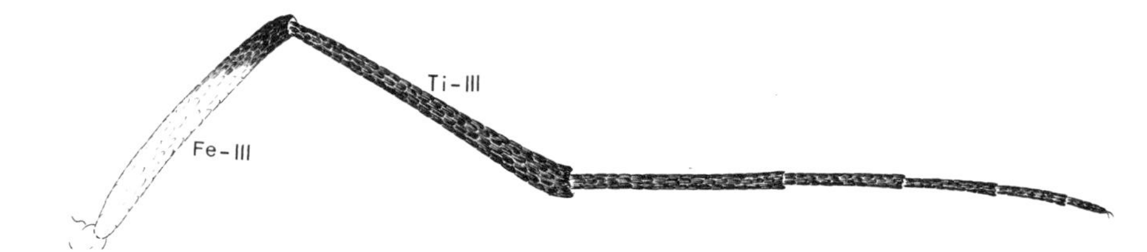


Fig. 500. Hindleg: *Ps. cyanescens*

Fig. 501. Hindleg: *Ps. ferox*

6(5). Scutum with narrow median longitudinal stripe of golden scales (fig. 502); proboscis yellow scaled in distal 0.5 except for labella (fig. 503) ........................... *ciliata* (plate 35A)

Scutum with median longitudinal stripe of dark brown scales (fig. 504); proboscis dark scaled (fig. 505) ........................................................................... *howardii* (plate 36B)

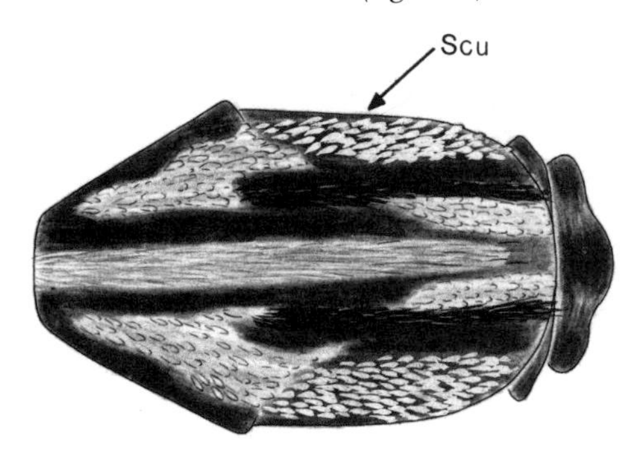


Fig. 502. Dorsal view of scutum: *Ps. ciliata*

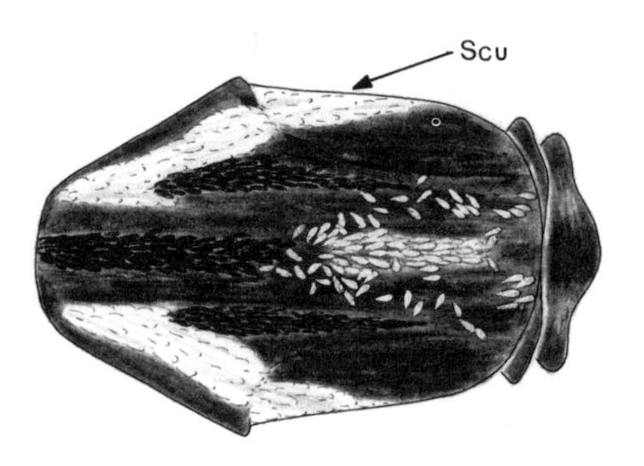


Fig. 504. Dorsal view of scutum: *Ps. howardii*

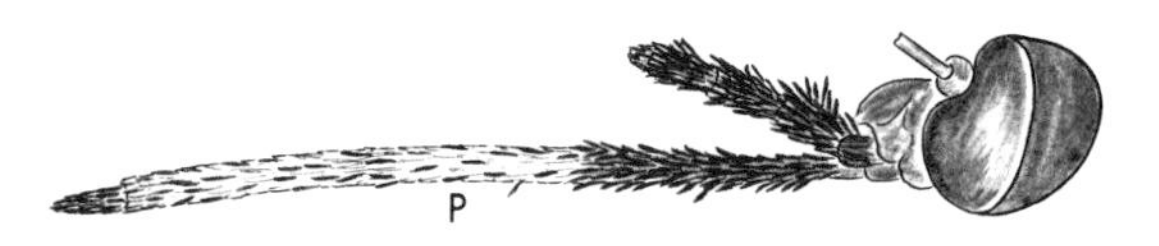


Fig. 503. Lateral view of head: *Ps. ciliata*

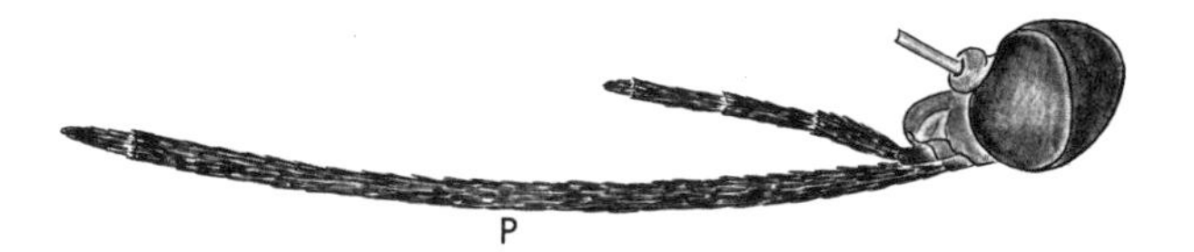


Fig. 505. Lateral view of head: *Ps. howardii*

7(5). Hindtarsomeres dark scaled (fig. 506); abdominal terga with dorsal patches of golden scales (fig. 507) ........................................................................... *cyanescens* (plate 35B)

Hindtarsomeres with at least some pale scaling (fig. 508); abdominal terga with apicolateral patches of pale white to yellow scales (fig. 509) ................................................ 8

Fig. 506. Hindleg: *Ps. cyanescens*

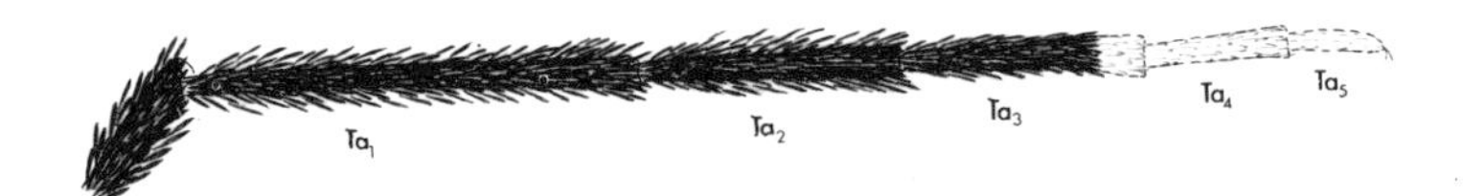


Fig. 508. Hindleg: *Ps. ferox*

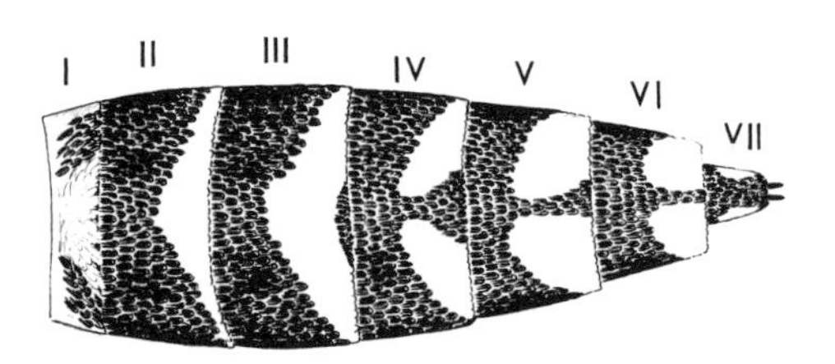


Fig. 507. Dorsal view of abdomen: *Ps. cyanescens*

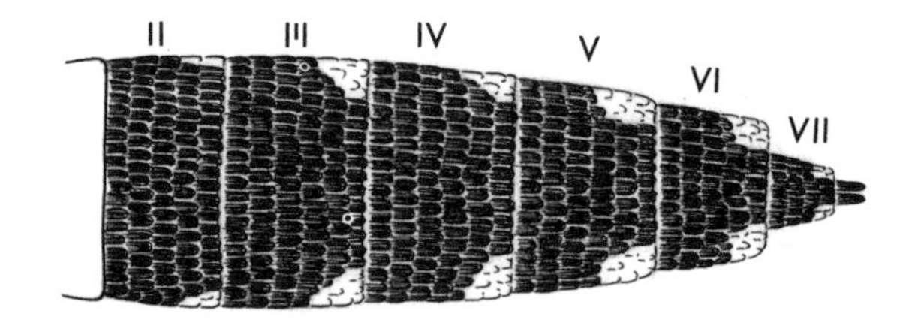


Fig. 509. Dorsal view of abdomen: *Ps. ferox*

8(7). Only hindtarsomere 4 pale scaled on at least one side, other hindtarsomeres dark scaled (fig. 510) .......... 9

Hindtarsomeres 4,5, and part of 3, or only hindtarsomere 5, pale scaled (fig. 511) .... 10

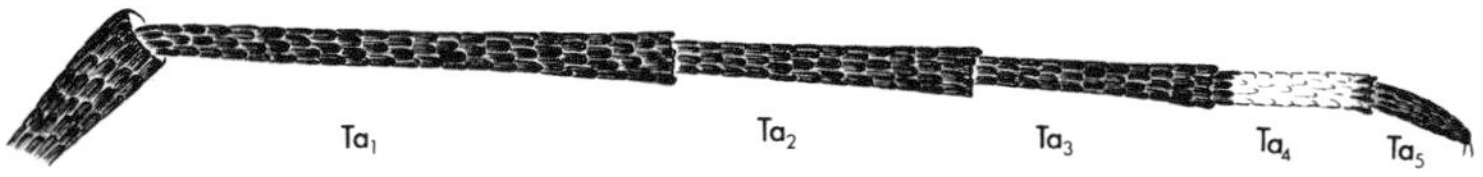

Fig. 510. Hindleg: *Ps. mathesoni*

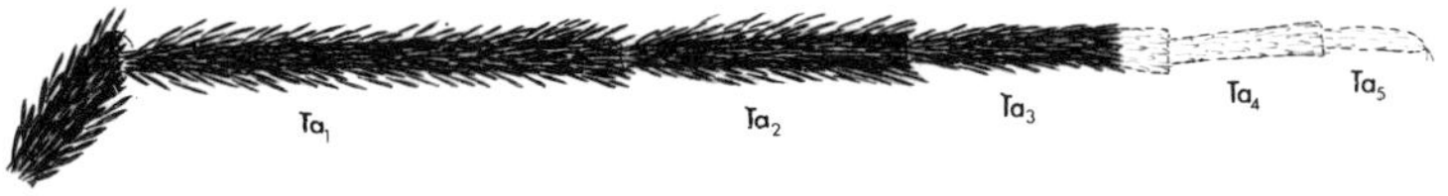

Fig. 511. Hindleg: *Ps. ferox*

9(8). Scutum covered with yellowish white scales (fig. 512) .......... *johnstonii* (plate 41C)

Scutum with broad longitudinal median stripe of dark scales, yellowish white scales laterally (fig. 513) .......... *mathesoni* (plate 36D)

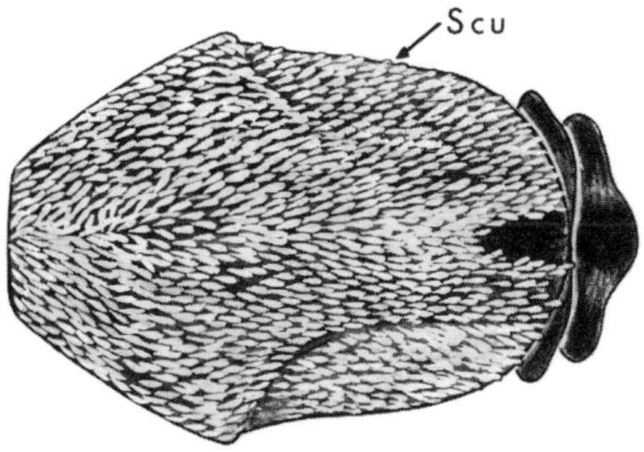

Fig. 512. Dorsal view of scutum: *Ps. johnstonii*

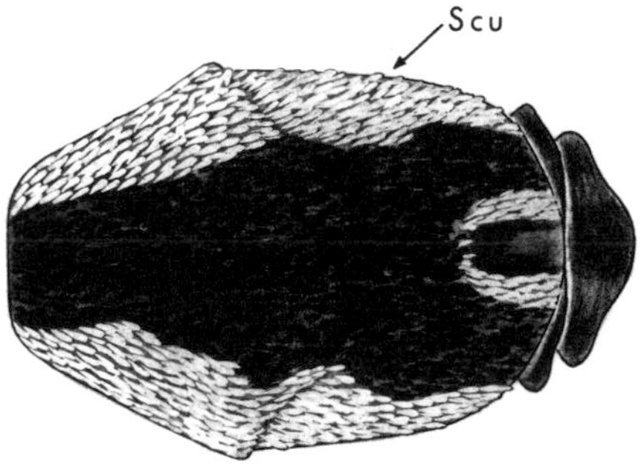

Fig. 513. Dorsal view of scutum: *Ps. mathesoni*

10(8). Hindtarsomere 5 pale scaled, others dark scaled (fig. 514) .......... *mexicana* (plate 42D)

Hindtarsomeres 4,5, and often part of 3 pale scaled (fig. 515) .......... 11

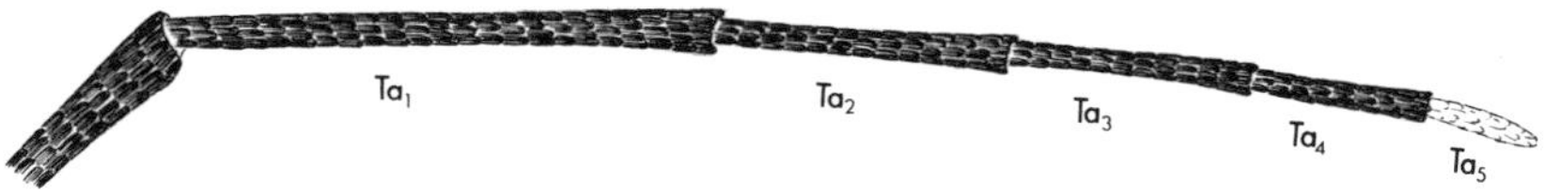

Fig. 514. Hindleg: *Ps. mexicana*

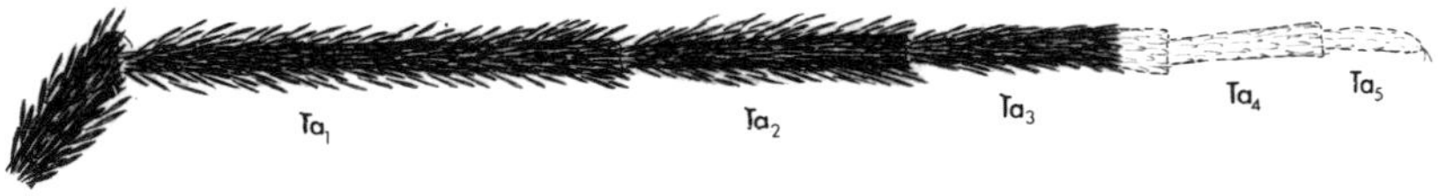

Fig. 515. Hindleg: *Ps. ferox*

11(10). Scutum with dark brown and golden yellow scales in no definite pattern (fig. 516); abdominal tergum I with purplish scales medially (fig. 517) .......... *ferox* (plate 35D)

Scutum with broad median longitudinal stripe of dark scales laterally (fig. 518); tergum I with pale scales medially (fig. 519) .......... 12

12(11). Femora with knee spots (fig. 520); palpi less than 0.25 length of proboscis, palpomere 4 subequal to palpomeres 1–3 (fig. 521) ............................................. *horrida* (plate 36A)

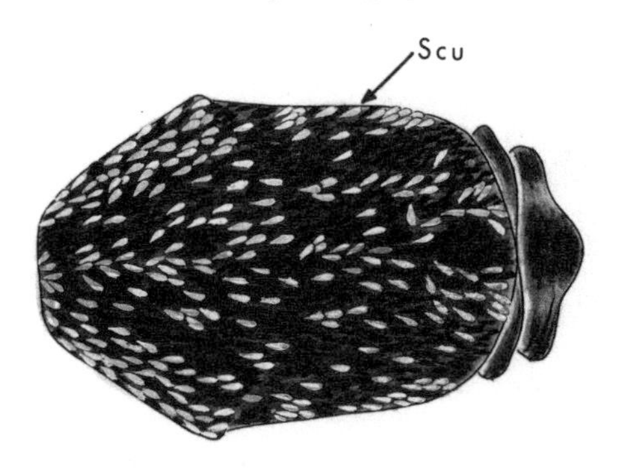


Fig. 516. Dorsal view of scutum: *Ps. ferox*

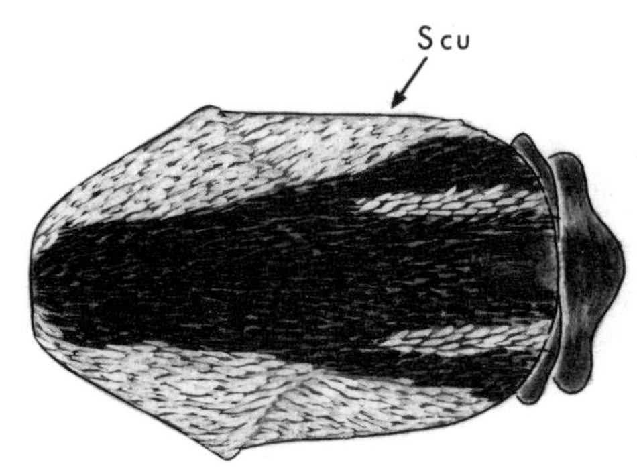


Fig. 518. Dorsal view of scutum: *Ps. horrida*

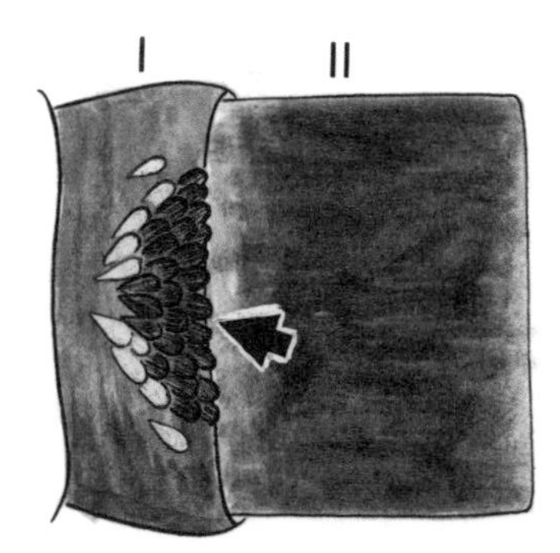


Fig. 517. Dorsal view of abdominal segments I,II: *Ps. ferox*

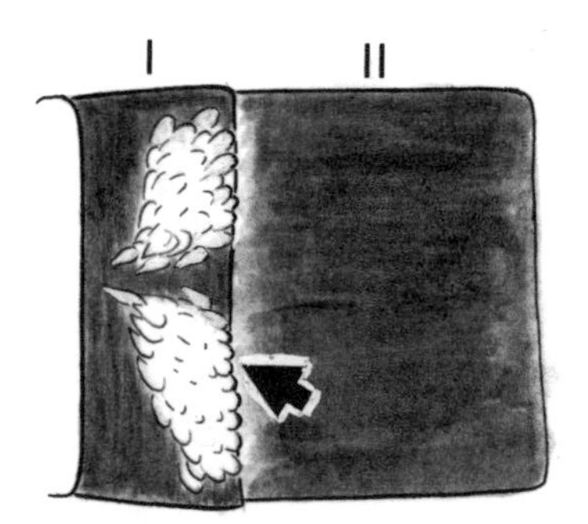


Fig. 519. Dorsal view of abdominal segments I,II: *Ps. horrida*

Femora without knee spots (fig. 522); palpi more than 0.3 length of proboscis, palpomere 4 1.5 length of palpomeres 1–3 (fig. 523) .................................. *longipalpus* (plate 36C)

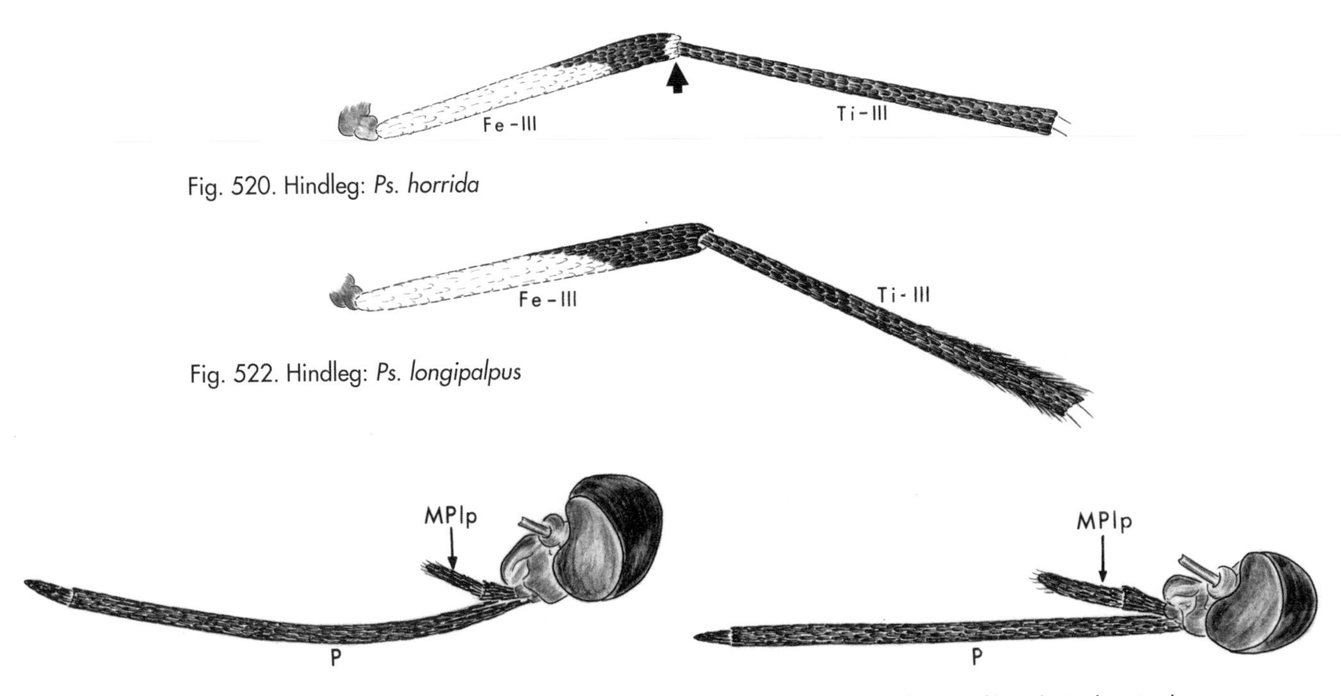


Fig. 520. Hindleg: *Ps. horrida*

Fig. 522. Hindleg: *Ps. longipalpus*

Fig. 521. Lateral view of head: *Ps. horrida*

Fig. 523. Lateral view of head: *Ps. longipalpus*

## Key to Adult Females of Genus *Uranotaenia*

1. Hindtarsomeres 4,5, and part of 3 pale scaled (fig. 524) ........................ *lowii* (plate 37D)

   Hindtarsomeres all dark scaled (fig. 525) ...................................................................... 2

Fig. 524. Hindleg: *Ur. lowii*

Fig. 525. Hindleg: *Ur. sapphirina*

2(1). Narrow median longitudinal stripe of iridescent blue scales on scutum and midlobe of scutellum (fig. 526) ..................................................................... *sapphirina* (plate 38A)

   Scutum and scutellum without longitudinal stripe of iridescent blue scales (fig. 527).... 3

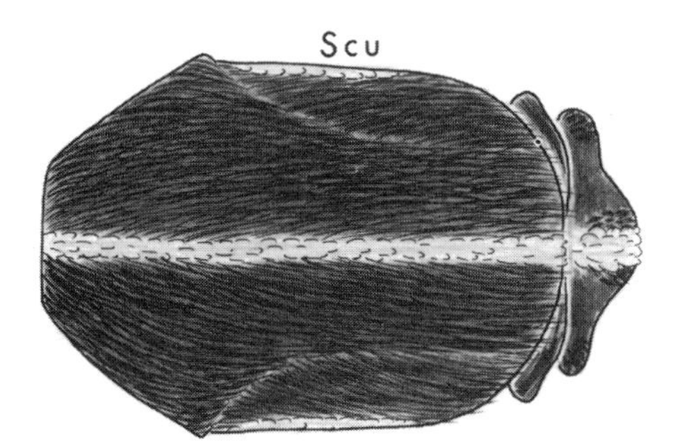

Fig. 526. Dorsal view of scutum: *Ur. sapphirina*

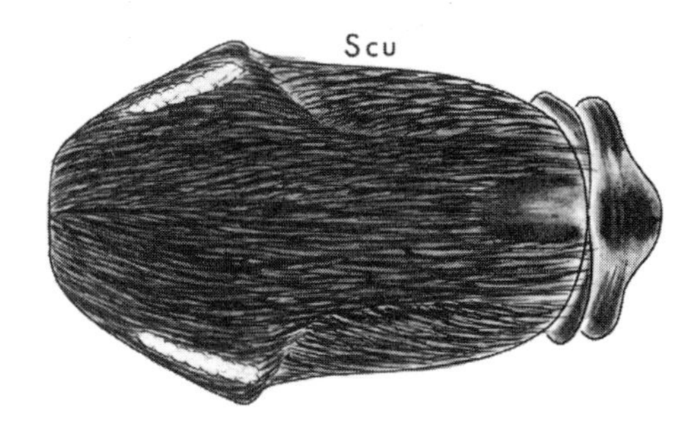


Fig. 527. Dorsal view of scutum: *Ur. a. anhydor*

3(2). Scutum with lateral line of iridescent blue scales incomplete, broken above mesothoracic spiracle (fig. 528) ........................................................................ *a. anhydor* (plate 41C)

   Scutum with continuous lateral line of iridescent blue scales from anterior promontory to wing base (fig. 529) .................................................................... *a. syntheta* (plate 37C)

Fig. 528. Dorsolateral view of thorax: *Ur. a. anhydor*

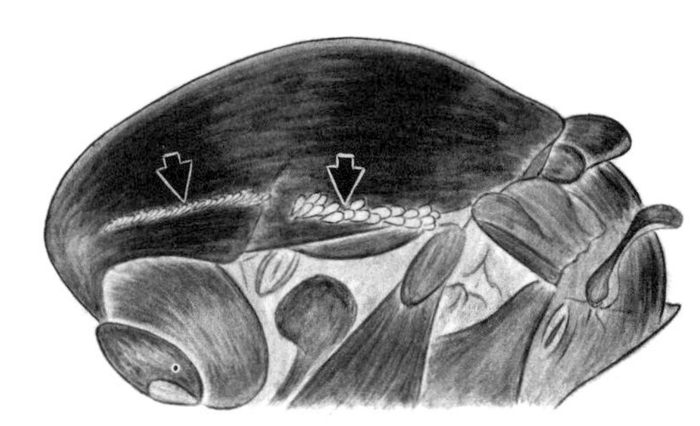

Fig. 529. Dorsolateral view of thorax: *Ur. a. syntheta*

## Key to Adult Females of Genus *Wyeomyia*

1. Antepronotum with silvery white scales (fig. 530); hindtarsomeres with basal patches of pale scales posteriorly (fig. 531) ........................................ *vanduzeei* (plate 40B)

   Antepronotum with mostly bluish to purple scales (fig. 532); hindtarsomeres with or without patches of pale scales (fig. 533) ........................................ 2

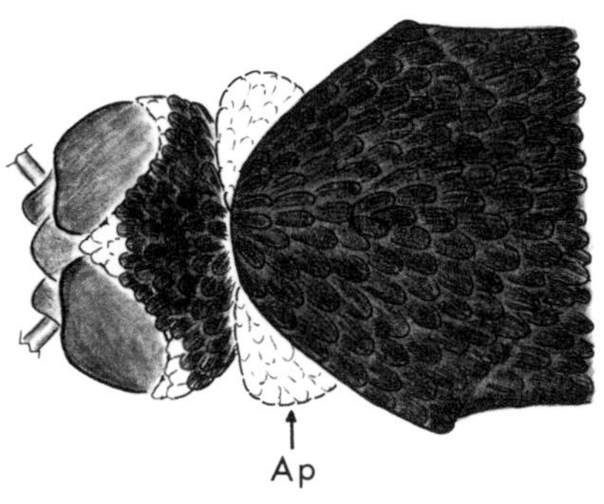


Fig. 530. Dorsal view of head and thorax: *Wy. vanduzeei*

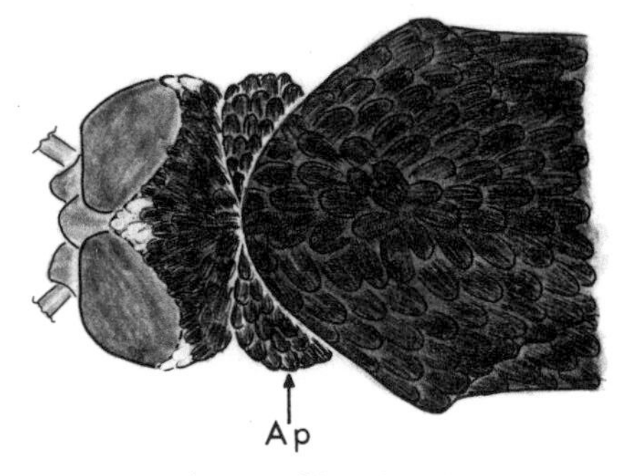


Fig. 532. Dorsal view of head and thorax: *Wy. smithii*

Fig. 531. Ventral view of hindleg: *Wy. vanduzeei*

Fig. 533. Ventral view of hindleg: *Wy. smithii*

2(1). Occiput with pale scales along ocular line (fig. 534); postpronotum with broad pale scales (fig. 535) ........................................ *mitchellii* (plate 37C)

   Occiput with dark scales along ocular line (fig. 536); postpronotum with overlapping dark scales (fig. 537) ........................................ *smithii* (plate 38B)

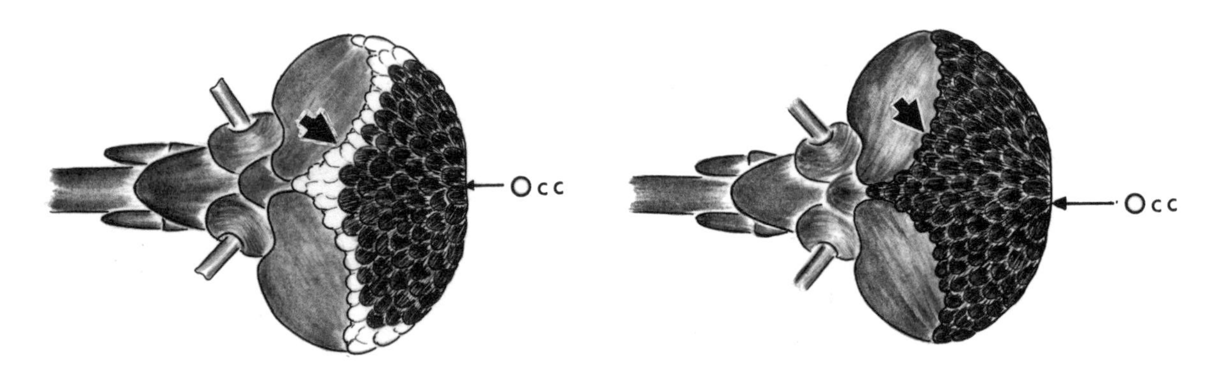


Fig. 534. Dorsal view of head: *Wy. mitchellii*

Fig. 536. Dorsal view of head: *Wy. smithii*

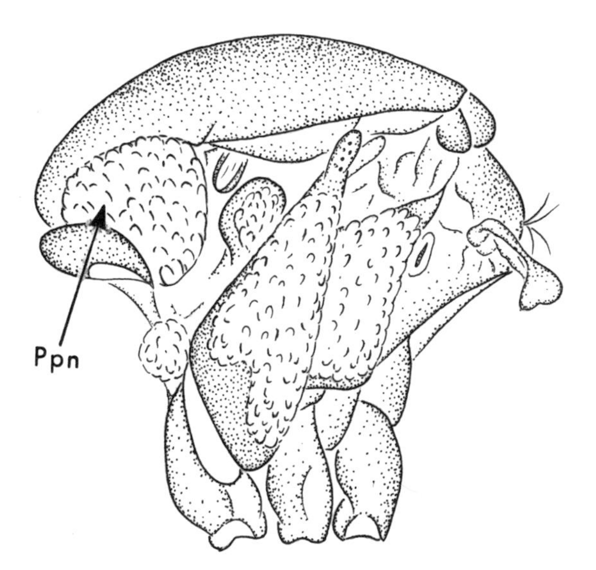

Fig. 535. Lateral view of thorax: *Wy. mitchellii*

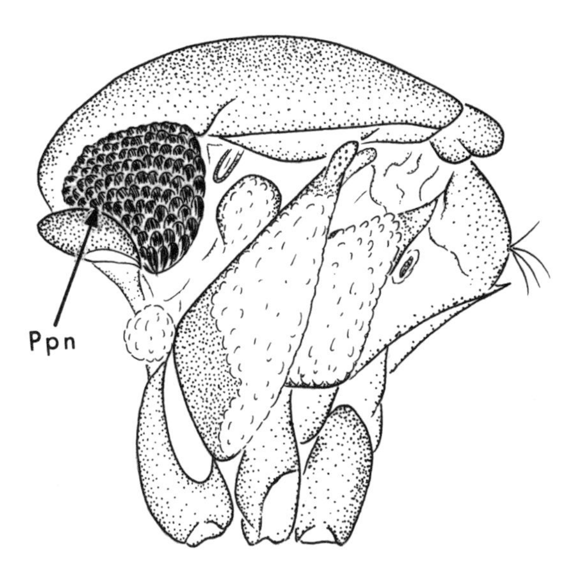


Fig. 537. Lateral view of thorax: *Wy. smithii*

# Morphology of Fourth Instar Mosquito Larvae

The fourth instar mosquito larval body, contrary to that of the adult, is composed largely of soft membranous tissue, but with some parts consisting of hardened sclerotized plates. This allows for the characteristic swimming movements and doubling of the body when the lateral palatal brushes are cleaned. The body is divided into the head, thorax, and abdomen. The head capsule is completely sclerotized, while the thorax and abdomen are largely membranous. The larval body is adorned with some 290 setae (plates 5, 6) and the study of their arrangement is called **chaetotaxy**. These, along with various kinds of spicules, are known collectively as the **vestiture**, i.e., protrusions from the cuticle of the integument or covering of the body (Harbach and Knight 1980), and are thus defined as cuticular projections. The organization and nomenclature of these structures is very important for larval identification. A complete treatment of the vestiture in general and of chaetotaxy in particular may be found in Harbach and Knight (1980); however, only those structures used in the present keys will be defined herein.

**Vestiture.** The two main components of the larval vestiture are spicules and setae (synonyms: hairs, hair tufts, bristles). In larvae whose thorax and/or abdomen are sparsely or densely covered with a pubescence, the spicules are called **aculeae**, and the cuticle is **aculeate** (fig. 714). Without this pile the surface would be smooth or glabrous. Where spicules are thornlike, varying from tiny to very coarse, they are termed **aciculae**, and the condition is known as **aciculate** (fig. 903). The lateral aspect of abdominal segment VIII, and also the siphon in many kinds of mosquito larvae, bear specialized projections (plates 6, 8). On abdominal segment VIII the structures are known as **comb scales** (CS) and they usually bear along their free posterior border a fringe of subequal spinules, or a median large spine and lateral smaller spinules. The **pecten** (PT) is a comblike row of spines borne on the **pecten plate** (PP) in anophelines and ventrolaterally on the **siphon** (S) in most culicine species. Each unit may have one or more lateral denticles on one or, less frequently, both margins (fig. 924). In the subgenus *Psorophora* the **pecten spines** (PS) are extended apically into long filaments (fig. 997).

Setae may be distinguished from spicules by the presence of an integumental **alveolus** from which the seta arises (see plate 7A–L). The alveolus is a cuplike socket, surrounded by a slightly swollen ring, and allows setal movement. Setae may be attached to the sclerotized structures, such as the head, siphon, and saddle, or directly to the membranous integument of the larval body. At times, the membrane may bear a special sclerite, to which one or more setae are attached, called the **setal support plate** (SSP). Setae can be simple, unbranched, or variously branched. Unbranched setae (plate 7A) are usually cylindrical and attenuated apically. They can be very thick and spinelike, in which case they are called **spiniform setae** (7B). **Branched setae** are composed of a main stem and ramifying members (7F). In some, the branches arise directly from the base and therefore have no stem or an extremely short one. Those that have only a few branches arising beyond the basal third of the main stem are termed **forked** (7E), while those with a very stout stem and many branches are called **fanlike setae** (7L). Setae with numerous regularly arranged branches arising on either side of the stem are **plumose** (7G). When setae have various stems with branches that are divided and subdivided so that they resemble branches of a tree, they are known as **dendritic** (7H). Specialized setae, characteristic of the genus *Anopheles*, have flattened moveable branches usually radiating horizontally from a short stout stem and are named **palmate** (7I,J). The branches are known as leaflets and can have smooth or serrate margins. The flat surface of the leaflet is the blade and it may have a terminal filament. The leaflets may be fully developed (7I) or partially formed (7J). The **lateral palatal brushes** (mouth brushes) are usually composed of unique specialized spicules, termed comb-tipped filaments (7K), bearing rigid processes apically on one side, like teeth of a comb.

Special mention needs to be made of seta 4 on abdominal segment X, a group of setal tufts known as the **ventral brush** (see figs. 701, 706). In most mosquito larvae, it is composed of a row of fanlike setae, some or all of which are usually attached to a heavily sclerotized network of bars, called the **grid** (G) (see plate 8A). It is composed of a number of **transverse grid bars** (TGB) connected to **lateral grid bars** (LGB). In some cases, the setae are joined to a setal support plate, known as a **boss**. Those setal tufts attached to the grid or boss are called **cratal setae** and those attached to the segment anterior to the cratal setae and grid are the **precratal setae** (fig. 969).

Single setae, or the components of branched setae, may be smooth or spiculate. Their parts may have thin needlelike processes that may vary in thickness and length. This condition is known as **aciculate** (7C), whereas if the processes are small and spinelike, the condition is **spinulate** (7D). If the setal parts have no processes, they are **smooth**.

The following abbreviations will be used in a discussion of the morphology of the larval body regions (plates 5, 6) and also in the larval keys.

| | | | |
|---|---|---|---|
| A | antenna | IV | abdominal segment IV |
| C | head | V | abdominal segment V |
| P | prothorax | VI | abdominal segment VI |
| M | mesothorax | VII | abdominal segment VII |
| T | metathorax | VIII | abdominal segment VIII |
| I | abdominal segment I | X | abdominal segment X |
| II | abdominal segment II | S | siphon |
| III | abdominal segment III | | |

## Head

The head is composed of a sclerotized capsule that bears the mouthparts and antennae anteriorly and the occipital foramen, the opening of the cranium to which the cervix (Cv) is attached, posteriorly. The shape of the head is distinctive in some mosquito larvae. Most have an ovate head, wider than long, with greatest width at the level of the eyes. In the genus *Deinocerites* the head is rather triangular, with the greatest width anteriorly at the level of the bases of the antennae. In the genus *Uranotaenia*, larval heads are longer than wide, while in the predatory larvae of the genera *Toxorhynchites* and *Psorophora* heads are quadrate.

For a few species the integument of the **dorsal apotome**, the large sclerite forming the dorsal aspect of the head, contains patterns of pigment that may be diagnostic. To evaluate this character correctly, examine the larval head under low magnification.

The mouthparts will not be discussed here. For their descriptions and understanding consult Gardner et al. (1973) and Harbach and Knight (1980). Dorsolateral to the mouthparts, of which the mandibles and maxillae are most obvious externally, is a lobe that bears a large brush formed of specialized spicules. The lobe is composed of the **lateral tormal process** (LTP) and the **lateral palatal plate** (LPP) and the brush, the **lateral palatal brush** (LPB) (= mouth brush). Usually the brush is made up of many comb-tipped filaments, but in the predatory larvae it consists of a few stout prehensile spicules (fig. 548).

**Setae of Head.** On the head are found 16 pairs of setae, of which setae 2-C to 9-C are used in identification. The letter "C" is used to indicate a seta that is located on the head. In the keys that follow only the numbers and letters or Roman numerals will be used in naming the larval setae.

The position of the setae in relation to one another is used in identification. In anophelines the two 2-C setae may be so close together that they are separated by less than the diameter of one of their alveoli (fig. 865), or they may be widely separated by more than the width of an alveolus (fig. 854). *Aedes cinereus* larva is distinguished by having 5-C, 6-C, and 7-C in a straight line (fig. 576), while others have 6-C anteriorly out of line.

In several species, the setae of the head are very coarse (e.g., *Oc. abserratus*, fig. 603), their diameters about equal almost to the apex, while in most larvae the setae of the head are attenuated, gradually tapering apically. Usually 4-C is a weak, small seta, but in some species of the subgenus *Protomacleaya* (e.g., *Oc. triseriatus*, fig. 696), it is well developed, multibranched.

In many instances, the size of a seta or relative size of one in comparison to another, the number of branches, the manner of branching, and the presence or absence of aciculae are all used as diagnostic characters. In some cases, the individual branches may be unequal, with some shorter than others (e.g., *Ps. longipalpus*, fig. 1031). Setae 5,6-C of larvae in the subgenus *Uranotaenia* are unique. They are very stout spiniform setae with spinulate surfaces (fig. 1033).

**Antennae.** The antenna is a cylindrical sensory appendage attached anterolaterally to the head. It bears six setae, 1-A to 6-A. In the genera *Coquillettidia* and *Mansonia* the antenna has an additional segment distal to the point of attachment of setae 2,3-A, called the **flagellum** (Fl) (see figs. 542, 544). Another unique variation of the antenna is its sinuate, inflated shape in *Ps. discolor* (fig. 998). In most species of the genus *Culex* (fig. 899), the antenna is markedly constricted in the distal 0.33, beyond the attachment of seta 1-A. The antennal length is significant; in most species it is shorter than the head, but in some it is as long as or much longer than the head (fig. 1004). In larvae of the subgenus *Psorophora*, the antenna is very short, hardly reaching the anterior margin of the head (fig. 996). The surface of the antenna is usually beset with spicules but may vary from none, as in *Oc. triseriatus* (fig. 696), to a few small spicules, as in *Oc. burgeri* (fig. 732), to many coarse spicules, as in *Oc. fitchii* (fig. 697). Some of the six antennal setae offer assistance in identification. The location of seta 1-A is diagnostic for some larvae. It may be near the middle, or may occur in the basal 0.33 or distal 0.33, depending on the species. The number and size of the branches of 1-A are also used. In several cases, comparing the size of setae 2,3-A between species is helpful (see *Ma. dyari* and *Cq. perturbans*, figs. 542, 544).

## Thorax

The thorax is an ovate unit of the body, somewhat wider than the head in well-nourished fourth instars. As in the adult, it consists of 3 segments, the pro-, meso-, and metathorax. They are distinguished by 3 distinct rows of setae, 0-P to 14-P on the prothorax, 1-M to 14-M on the mesothorax, and 1-T to 13-T on the metathorax. The integument of the thorax is sometimes aculeate. This is most easily detected under a compound microscope. Check the edges of the thorax on the vertical surface where debris, often found covering the body of mature larvae, does not seem to accumulate. The nonaculeate surface is called smooth or glabrous and is the more usual condition.

Of the 42 pairs of setae available on the thorax, only 10 are used in the larval keys. Setae 1-P, 3-P, and 7-P have diagnostic size and/or number of branches useful in separating species of several genera. In culicine larvae setae 1–3-P are in line, very close to one another, so it is hard to distinguish them. Likewise, often they are borne on a setal support plate (see fig. 1035). Seta 1-M is particularly useful in separating a number of *Ochlerotatus* larvae. In most species, it is a short seta, but in several it is long and rather stout. It is compared in the keys to the length of the antenna or to seta 2-M or 3-M. For the other thoracic setae, number of branches or size is used.

### Abdomen

The larval abdomen consists of 10 segments, each designated by the appropriate Roman numeral. The first 7 segments are very similar, segment I bearing 12 setae and II through VII, 15 setae. Segments VIII–X are functionally specialized and morphologically different from the others. Segment IX does not exist as a distinct morphological unit, but is incorporated into VIII and X and will not be used in the keys.

In anophelines, abdominal segments I–VII possess the **tergal plate** (TP) anteriorly and may also have one or more **accessory tergal plates** (ATP), as in figs. 884, 886. They do not ordinarily occur in culicine larvae, but some species of *Orthopodomyia* have well-developed tergal plates in VII,VIII. *Uranotaenia* and some *Psorophora* larvae have lateral sclerites on VIII, known as **comb plates** (CP), to which the comb scales are attached, and *Toxorhynchites* larvae have numerous small **setal support plates** on their thoracic and abdominal segments, a larger one laterally on VIII (fig. 549).

**Segments I–VII.** Although there are 86 pairs of setae on abdominal segments I–VII, only 24 are used as key characters. Seta 1 is developed as a palmate type in some or all abdominal segments I–VII of anopheline larvae. The fully developed palmate setae usually have 10 or more large leaflets, and when one is in its normal position, it is spread to at least 150 degrees. The number of segments with fully formed palmate setae varies with the species. Segments I–III and VII sometimes have palmate setae not fully developed, expressed as 0.5 or 0.7 as large (figs. 856, 860). Seta 6 (lateral abdominal hair by some authors) is used in a number of instances. It is usually a very prominent seta on each segment, especially on I,II. It is plumose on those 2 segments in anophelines and aciculate, commonly double or triple, in culicines. In 2 species of *Anopheles*, *barberi* and *judithae* (fig. 831), 6-I–VI are plumose. When seta 6 is more than single on segments III–VI, it is usually diagnostic for the species on which it occurs (e.g., *Oc. taeniorhynchus*, fig. 626). Its size may also be characteristic, as in *Ps. horrida* (fig. 1024). Seta 0 is usually a tiny, single seta in anophelines, but in *An. crucians* (fig. 850) it is well developed, with 4 or more branches. The other setae found on I–VII and employed in the keys are 2,3,7, and 13. Their size and number of branches are traits of certain species. In *Oc. monticola* (fig. 724) 1 and 13-IV,V are similar in size and number of branches, while in *Oc. varipalpus* (fig. 726) they are not. These 2 setae are located dorsoventrally opposite each other on the segment.

**Segment VIII.** Mosquito larvae are metapneustic, that is, the only functional external orifices of the respiratory system, the **spiracular openings** (SOp), are located on abdominal segment VIII (see plate 8). These openings are surrounded by the **spiracular apparatus** (Sap). In anophelines this structure is sessile, while in culicines it is borne on the apex of the sclerotized tube, the **siphon** (S). There are only 5 setae on the segment, 1-VIII to 5-VIII. Laterally, in all larvae, except those of the genus *Toxorhynchites*, there occur the **comb scales** (CS). They may be arranged in a single row, double row, or irregular patch. There may be as few as 4, as in *Oc. papago* (fig. 701), or as many as 70, as in *Oc. pionips* (fig. 783). The total number, within ranges, is diagnostic and used throughout the keys. Among those larvae of the subgenus *Melanoconion*, *Cx. abominator* (fig. 948) has short comb scales without a narrow elongation in the middle, while the others are rather slipper shaped, elongate and narrow in the middle, as in *Cx. iolambdis* (fig. 949). The character of the median spine and comparison of its size to that of the subapical spinules are extensively utilized. The size of the median spine ranges from only slightly larger than the subapical spinules, as in *Oc. melanimon* (fig. 811), to very long with tiny subapical and lateral spinules, as in *Oc. riparius* (fig. 687). Extreme development of the median spine occurs in some larvae. In the subgenus *Protomacleaya*, the whole posterior projection of the comb scale is a rather bluntly rounded spine, fringed all along the edges with tiny spinules (fig. 699). In *Oc. nevadensis*, larvae sometimes have 3 large median spines (286). In some larvae the subapical spinules are almost as stout as the median spine (e.g., *Ae. aegypti*, fig. 708).

**Spiracular apparatus.** The spiracular apparatus (Sap) is a 5-lobed valve that closes the **spiracular openings** (SOp) during submersion of the larva, protecting them. The 5 lobes are the **anterior spiracular lobe** (ASL), the two **anterolateral spiracular lobes** (LSL), and the two **posterolateral spiracular lobes** (PSL). They are moveable, flaplike projections and bear a total of 11 pairs of setae, 3-S to 13-S. Seta 6-S is unusually long in the larva of one species of *Psorophora* (fig. 1005). The posterolateral spiracular lobes are prolonged into taillike processes in *An. pseudopunctipennis* (fig. 844). In North America, the genera *Coquillettidia* and *Mansonia* have the spiracular apparatus highly modified for piercing the roots of certain aquatic plants, in which the larvae find a source of air. It is in the form of an attenuated tube, bearing hooklike teeth at the apex, inner and outer spiracular teeth (IST, OST), and a row of serrations on the anterior surface, known as the saw (SAW) (plate 8B). Such modified apparatus possesses 4 visible pairs of setae, 1, 2, 6, and 8-S.

**Siphon.** The siphon (S) in culicines is one of the most useful structures in identification. Its size and shape vary considerably. The length divided by basal width ratio is expressed as the **siphon index**. Harbach and Knight (1960) have defined it as the ratio of the length to the median width, but since so many descriptions of North American mosquito larvae have used the index as the ratio of the length to the basal width, the latter system is followed here. Actually, in most instances it makes very little difference, but for the larvae of the *Psorophora* subgenera *Janthinosoma* and *Grabhamia*, where the siphon is swollen medially, measurements would be dissimilar. In the species treated here the index varies from 1.4 (*Oc. togoi*, fig. 790) to 10.0 (*Cx. cedecei*, fig. 942). Attached to the base of the siphon is a small lateral sclerite, the **siphon acus** (SA). In some species, it is absent (*Oc. papago*, fig. 701) while in others it is detached from the siphon base and "floating" in its basal membrane (*Oc. hendersoni*, fig. 741).

**Pecten** (Pt). Five North American genera, *Coquillettidia*, *Mansonia*, *Orthopodomyia*, *Toxorhynchites*, and *Wyeomyia*, have no pecten spines on the siphon (see fig. 546). The pecten spines (PS) in the larvae of those genera bearing them are variable and offer good characters that are used in the keys. A common variant is the distalmost spines being widely spaced from the others. In the keys they are termed "detached apically" (e.g., *Oc. excrucians*, fig. 634). The pecten may be very short, with few spines, as in *Ps. columbiae* (fig. 999), or extend almost to the apex of the siphon, as in *Oc. cataphylla* (fig. 636). The number of spines and the proportion of the siphon to which it extends from the base are used in the keys. In some species several apical spines are quite large, and their length is compared to the apical diameter of the siphon, as in *Oc. fitchii* (fig. 746), or to the length of seta 2-S, as in *Oc. campestris* (fig. 802). The pecten spine usually has 1–4 lateral denticles on its ventral edge, or less frequently on the dorsal edge, too, but their number varies from 3 in *Cx. territans* (fig. 924), to about 20 in *Cx. anips* (fig. 952). The siphon may be adorned with other types of spicules. It may be covered with aciculae, as in *Cx. bahamensis* (fig. 890), or a set of spines near the apex, as in *Cx. coronator* (fig. 900).

**Siphon setae.** The siphon ordinarily bears two pairs of setae, 1-S and 2-S; however, when there are several present, the basalmost is named 1a-S, then in sequence 1b, 1c, 1d-S, etc., proceeding distally (Belkin 1950). Seta 2-S is small, preapical, and located anteriorly. The length, curvature, and presence or absence of a secondary tooth are all useful characters (see figs. 924, 925). The position of 1-S with respect to the pecten is useful for separating groups of species in *Ochlerotatus*. Normally 1-S is attached distal to the apicalmost pecten spine. At times it is attached basal to the distalmost spine and described as being "attached within the pecten" (see *Oc. tormentor*, fig. 594).

The number of setae and their positions on the siphon are diagnostic in many species. Several species of subgenus *Rusticoides* (e.g., *provocans*, fig. 567), have at least 1a-S to 1c-S. A trait of *Culex* larvae is the presence of 3 or more pairs of setae on the siphon. The total number is often characteristic, and in many instances the penultimate seta is dorsally out of line (fig. 892). They are also frequently in a straight line and larvae in the subgenus *Melanoconion* have an additional one or more subdorsal small setae (fig. 925). The genus *Culiseta* has as its principal recognizing feature a pair of basal ventrolateral setae, 1-S (see plate 8A). Furthermore, some species of subgenus *Culiseta* bear a row of short spicules just distal to the pecten (fig. 957). In some larvae, the siphonal setae are irregularly placed (e.g., *Cx. restuans*, fig. 896, and *Wy smithii*, fig. 1044). The length of seta 1-S is compared to many other structural dimensions, e.g., basal or apical diameter (fig. 676), total length (fig. 1003), and distance from its alveolus to the apex of the siphon (fig. 992). Likewise, its location at or distal to the middle of the siphon is peculiar to some larvae (see *Oc. melanimon*, fig. 635). Of course, the number of branches of 1-S vary and are employed in the keys.

**Segment X.** This highly modified abdominal segment, commonly called the anal segment, is the most posterior. It possesses a large sclerite, the **saddle** (Sa), which partially or entirely encircles the segment, usually with 2 pairs of anal papillae, the homeostatic cylindrical organs attached terminally to the segment, and four pairs of setae, 1-X to 4-X.

**Saddle.** In most larvae there is a single saddle sclerite, but those of the genus *Deinocerites* bear small ones dorsally and ventrally. Of the remaining culicine genera, larvae of *Haemagogus*, *Wyeomyia*, *Aedes*, some *Ochlerotatus*, and *Cx. bahamensis* possess saddles that do not completely encircle segment X. It is often necessary to determine the extent to which the saddle encircles the segment. Some are small and do not exceed even 0.5 the distance to the midventral line (e.g., *Oc. atropalpus*, fig. 643), in which case seta 1-X is attached ventral to the saddle. On the other hand, some larvae have very long, though incomplete saddles, almost reaching to the midventral line (e.g., *Oc. punctodes*, fig. 750). At times it is extremely difficult to determine the exact size of the saddle sclerite of larvae that have been mounted in Canada balsam for many years because of clearing action of the mountant. Very fine focusing by a compound microscope with 200–400X magnification will help to locate its ventral edge. Some saddles are deeply incised along the ventral margin, as in *Oc. euedes* (fig. 677); in a number of larvae of the genera *Ochlerotatus*, *Haemagogus*, and *Culiseta*, the saddles have prominent aciculae along the posterior border that vary in size with the species (see figs. 824, 827).

**Anal papillae** (APP). Of those species treated here two have larvae with only one pair of anal papillae: *Cx. bahamensis* (fig. 891) and northern *Wy. smithii* (fig. 1040). *Oc. dupreei* larvae are unique for having very long anal papillae, about 8.0 the length of the saddle, tracheated and darkly pigmented (fig. 606). At the other extreme, those species with larvae that breed in brackish water have very small anal papillae (see *Oc. taeniorhynchus*, fig. 604). It is customary to express the length of the anal papillae as a ratio to the length of the saddle, known as the anal papilla–saddle index and computed by dividing the length of the papilla by the dorsal length of the saddle.

**Setae of segment X.** Setae of segment X provide differentiating characters. The length of seta 1-X, the saddle seta, is used in the *Ochlerotatus* key (figs. 796, 797). It is commonly compared with the saddle length. Setae 2,3-X are known collectively as the dorsal brush; 2-X is ordinarily multibranched and 3-X long and single. *Oc. abserratus* larvae are unusual in that both these setae are long and single (fig. 602). Seta 4-X is composed of a variable number of paired and unpaired setae. The most anterior seta is designated as 4a-X; proceeding posteriorly the setae are then 4b, 4c, 4d-X, etc. This group of setae acts as a rudder during swimming. It is particularly well developed in the larvae of genus *Psorophora*, in which the numerous precratal, fanlike setae usually extend anteriorly more than 0.5 the length of the segment (fig. 556). Contrarily, it is poorly developed in treehole inhabiting larvae belonging to subgenera *Abraedes*, *Kompia*, *Protomacleaya*, and the *varipalpus* group of *Ochlerotatus*, as well as in those larvae of the genera *Coquillettidia* and *Mansonia*, which attach themselves to roots of plants. They have no more than 3–7 pairs of setae in the brush (see figs. 740, 743). In some of these larvae a boss is present for attachment of the setae instead of a grid (see fig. 701). The number of branches in the two caudalmost setae (*Oc. sierrensis*, fig. 718), or in the two anteriormost setae (*Oc. brelandi*, fig. 743), is diagnostic. The position of the ventral brush is important in distinguishing those *Ochlerotatus* larvae possessing a completely circular saddle. In them the setae are confined to that part of the segment posterior to the saddle. The total number of fanlike setae is distinctive for a number of species (e.g., *Oc. zoosophus*, fig. 719, and *Cs. minnesotae*, fig. 963). In *Wyeomyia* larvae no regular rudder-like ventral brush (seta 4-X) is present. Seta 4 is nothing more than a pair of long or short ventrolateral setae posteriorly on the segment (see figs. 1038, 1040).

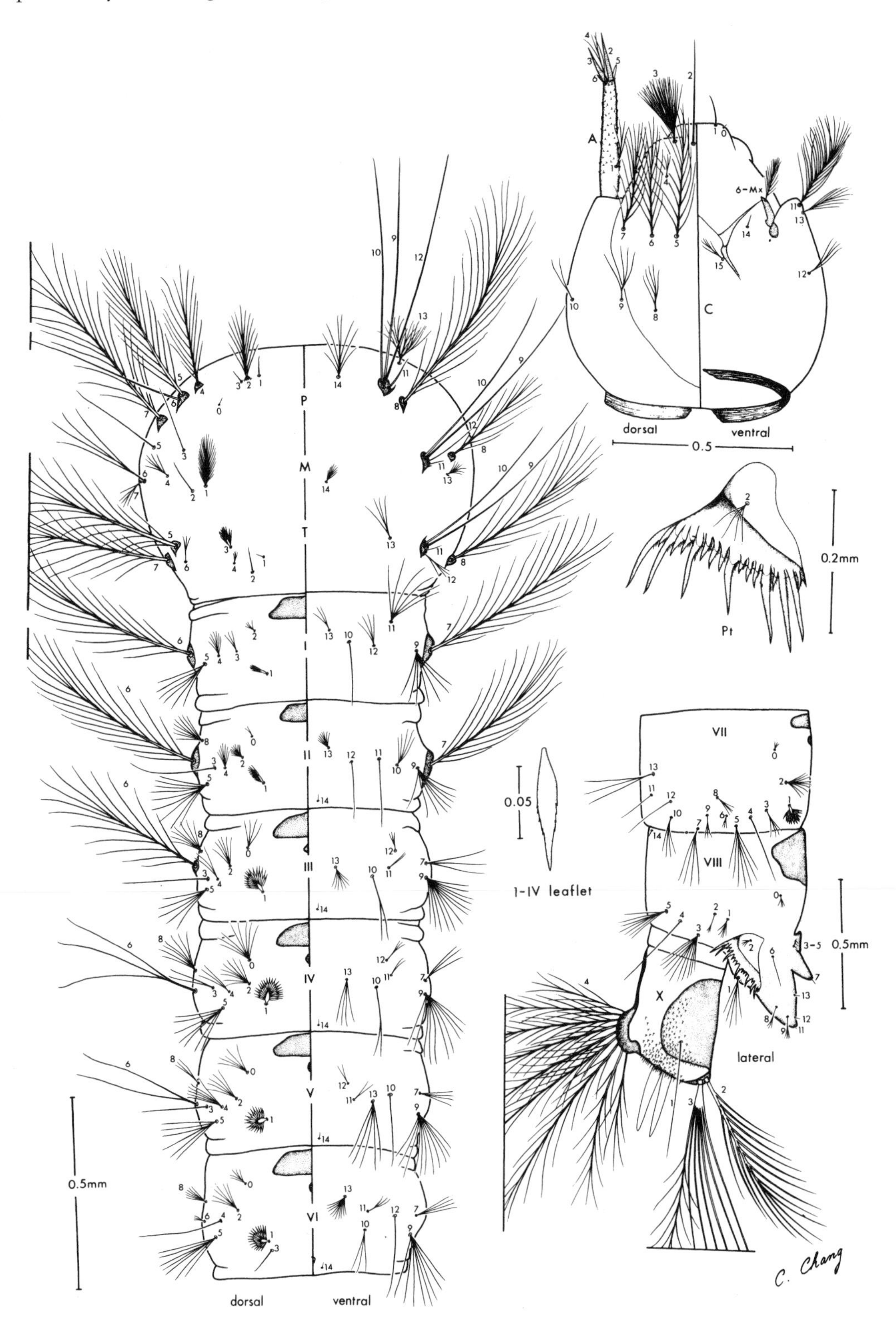


Plate 5. Fourth stage anopheline larva; dorsal left, ventral right.

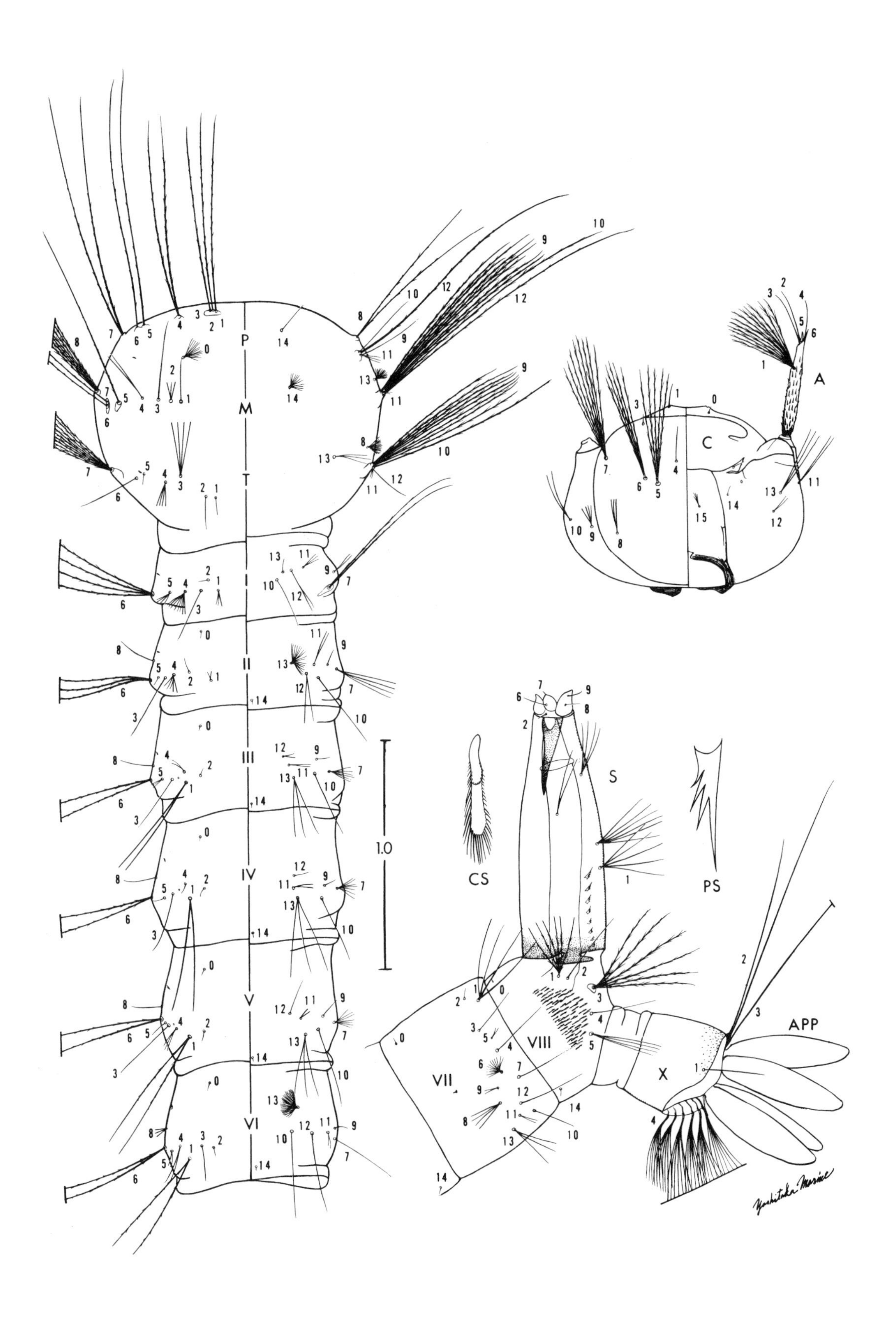


Plate 6. Fourth stage culicine larva; dorsal left, ventral right.

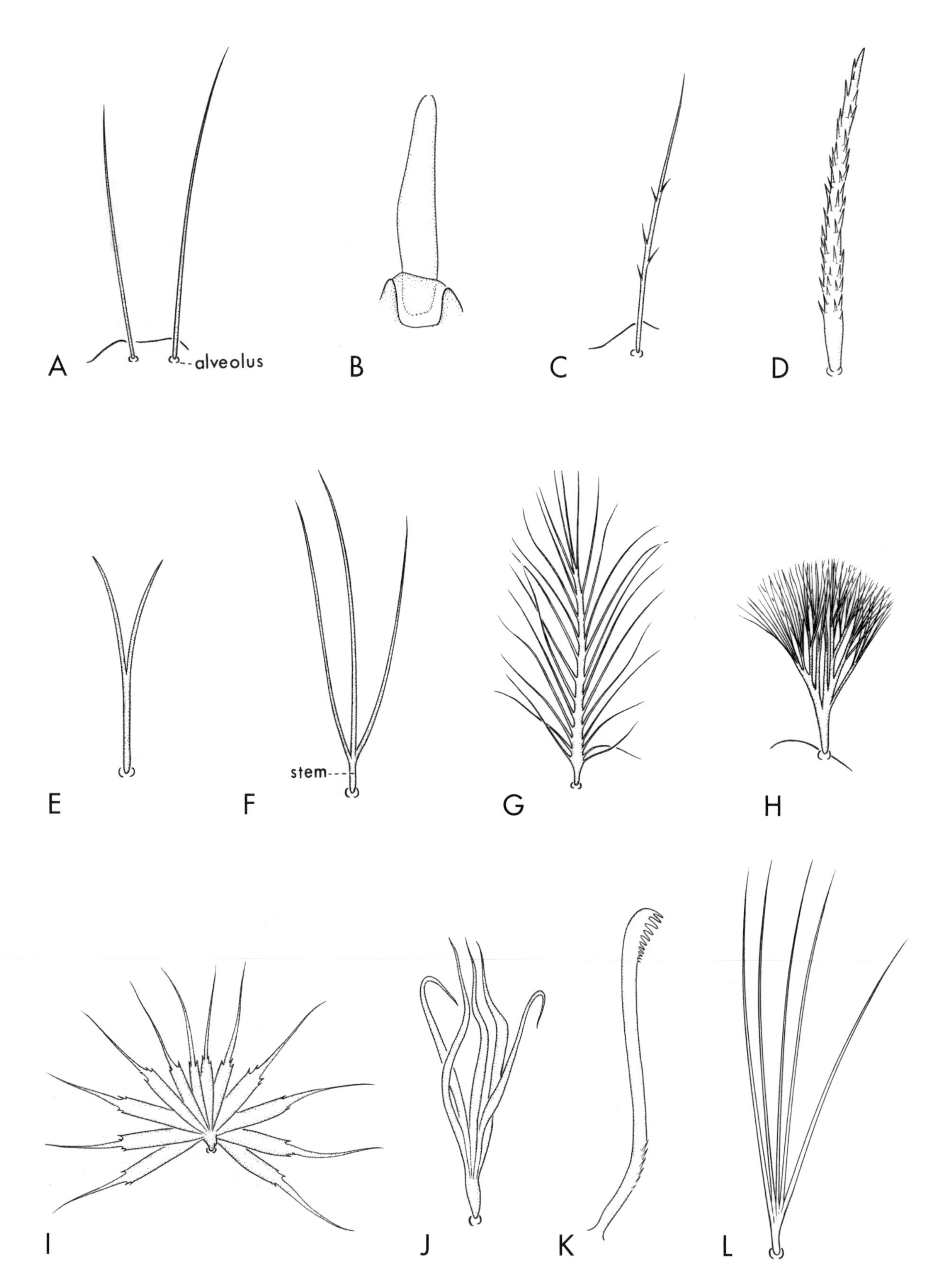


Plate 7. Examples of kinds of setae found in mosquito larvae. A. Unbranched smooth setae; B. Spiniform seta; C. Unbranched aciculate seta; D. Spinulate spiniform seta; E. Forked seta; F. Branched seta; G. Plumose seta; H. Dendritic seta; I. Palmate seta, fully developed; J. Palmate seta, 0.5 developed; K. Comb-tipped filament; L. Fanlike seta of ventral brush.

## Abbreviations in Plate 8

APP anal papilla
ASL anterior spiracular lobe
ASLP anterior spiracular lobe plate
C comb
CS comb scales
G grid
IST inner spiracular teeth
LGB lateral grid bar
LSL anterolateral spiracular lobe
LSLP anterolateral spiracular lobe plate
MdP median plate
OST outer spiracular teeth
PP pecten plate
PS pecten spines
PSL posterolateral spiracular lobe
PSLP posterolateral spiracular lobe plate
PSP posterior spiracular plate
Pt pecten
S siphon
Sa saddle
SA siphon acus
SAd spiracular apodeme
SAp spiracular apparatus
SAW saw
SOp spiracular opening
TGB transverse grid bar
VII abdominal segment VII
VIII abdominal segment VIII
X abdominal segment X (anal segment)
2-S seta 2 of siphon

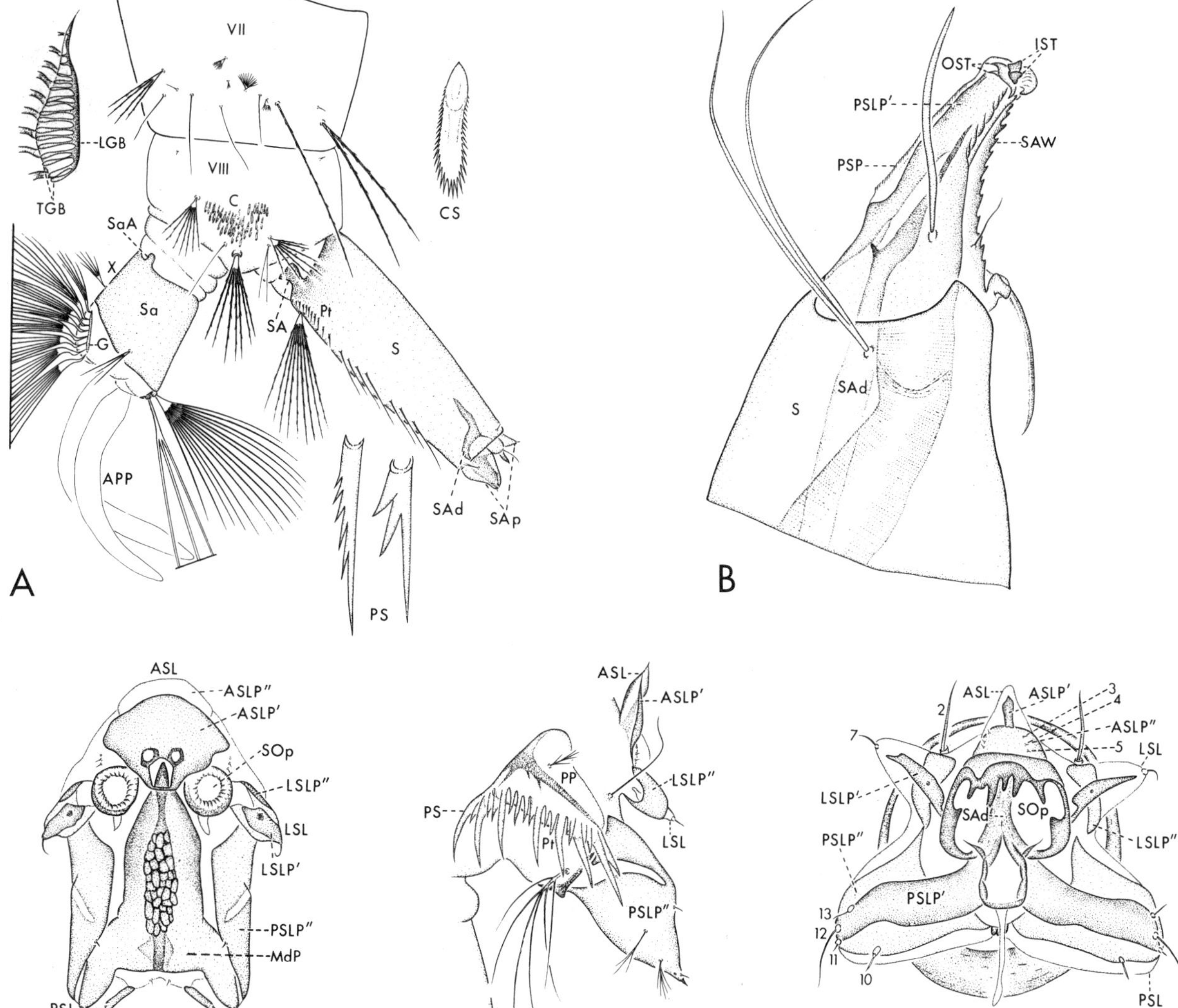


Plate 8. Morphology of terminal abdominal segments of mosquito larvae. *A*. Segments VII–X of *Culiseta*; *B*. Siphon and spiracular apparatus of *Mansonia*; *C,D*. Spiracular apparatus of *Anopheles*; *C*. dorsal view, *D*. lateral view; *E*. Dorsal view of spiracular apparatus of *Culex*.

# Keys to the Fourth Instar Larvae of the Mosquitoes of North America, North of Mexico

## Key to Genera of Fourth Instar Mosquito Larvae

1. Respiratory siphon absent; at least some abdominal terga with seta 1 palmate (fig. 538) ... ........................................................................................................... *Anopheles*

   Respiratory siphon present; seta 1 on abdominal terga never palmate (fig. 539) ............. 2

Fig. 538. Lateral view of abdominal segments IV–X: *An. quadrimaculatus*

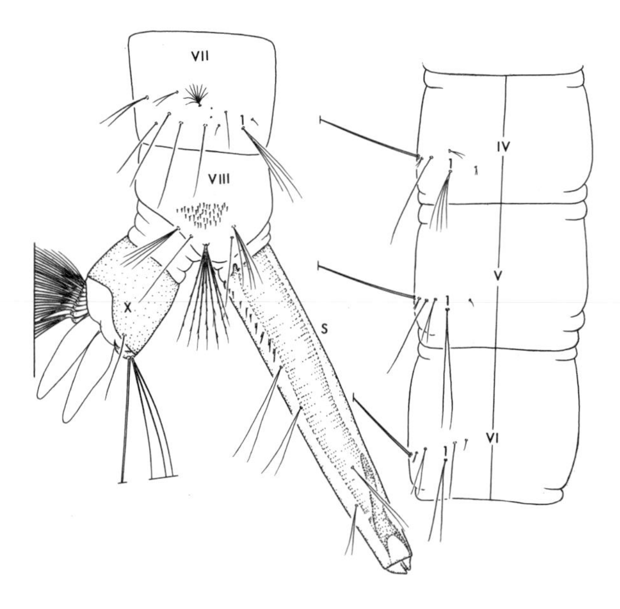


Fig. 539. Dorsal and lateral view of abdominal segments IV–X: *Cx. pipiens*

2(1). Siphon attenuated apically, with dorsal saw, adapted for piercing plant tissue (fig. 540) 3

Siphon not attenuated apically, not adapted for piercing plant tissue (fig. 541) ........... 4

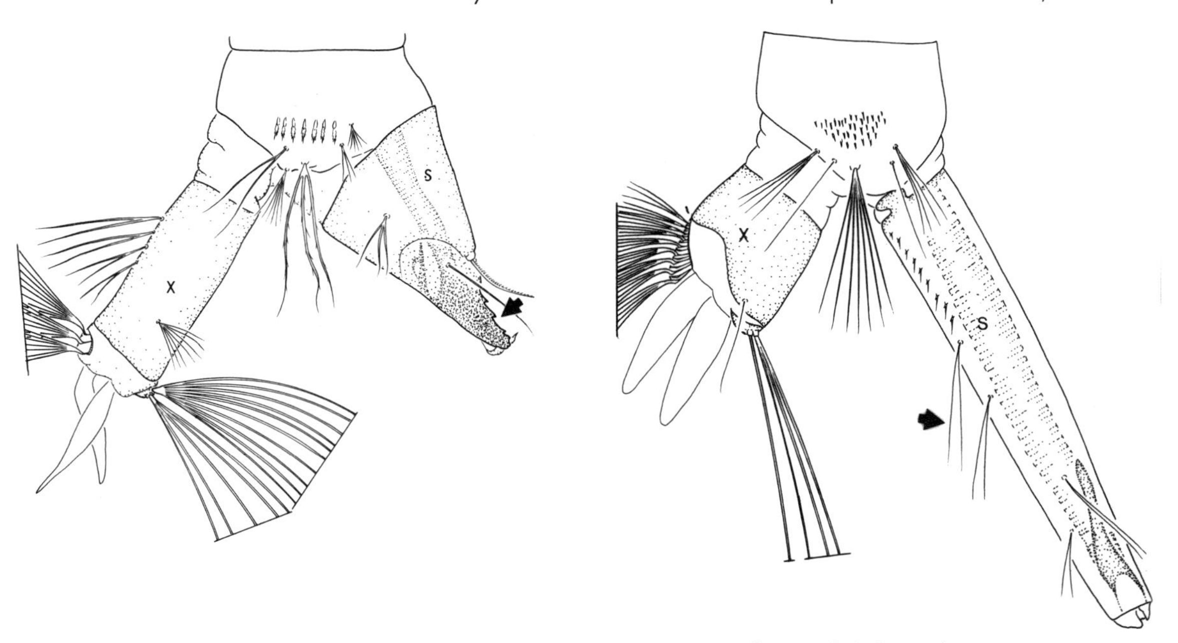


Fig. 540. Lateral view of abdominal segments VIII–X: *Ma. dyari*

Fig. 541. Lateral view of abdominal segments VIII–X: *Cx. pipiens*

3(2). Setae 2,3-A about length of antennal flagellum, or longer (fig. 542); saddle bearing 3,4 robust precratal setae (fig. 543) .......................................................................... *Mansonia*

Setae 2,3-A much shorter than antennal flagellum (fig. 544); saddle without precratal setae, or if present, no more than 2 thin setae posteriorly (fig. 545) ........................................... ................................................................................ *Coquillettidia perturbans* (plate 34A)

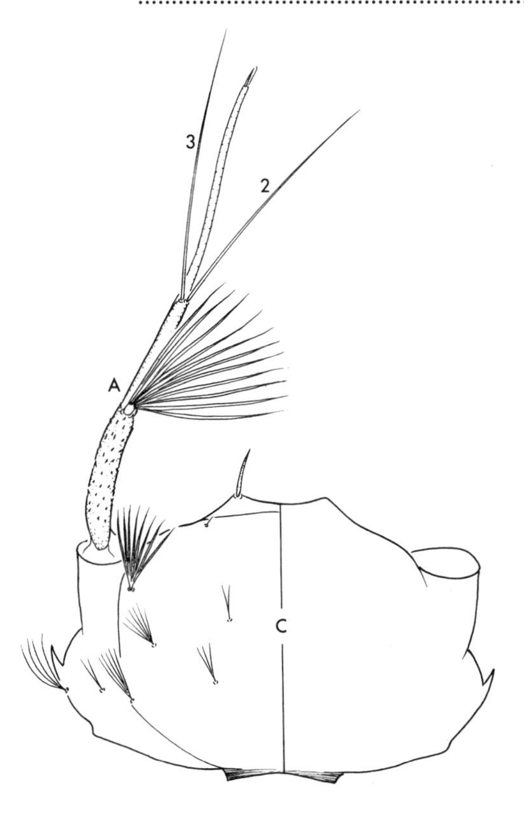


Fig. 542. Dorsal view of head: *Ma. dyari*

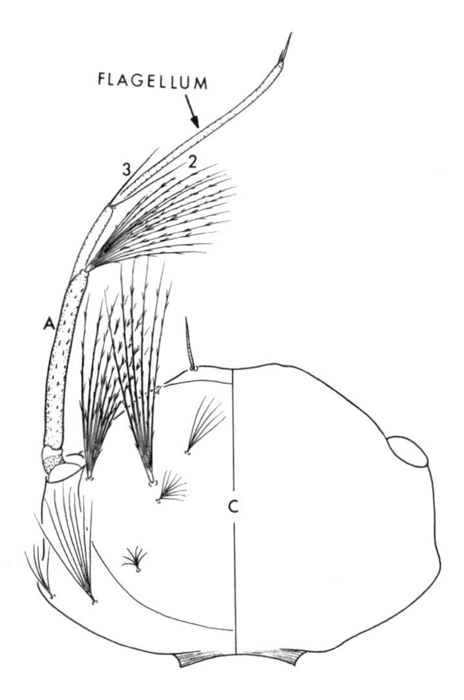


Fig. 544. Dorsal view of head: *Cq. perturbans*

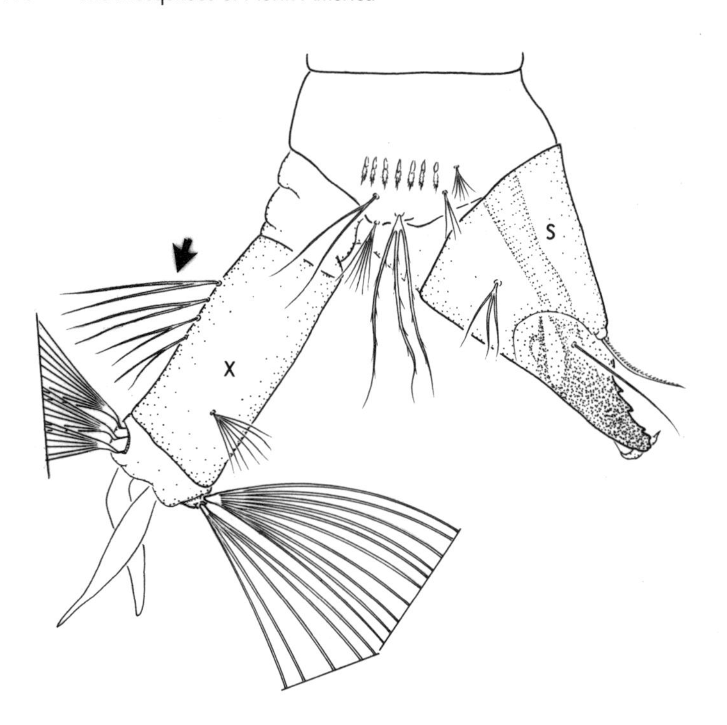


Fig. 543. Lateral view of abdominal segments VIII–X: *Ma. dyari*

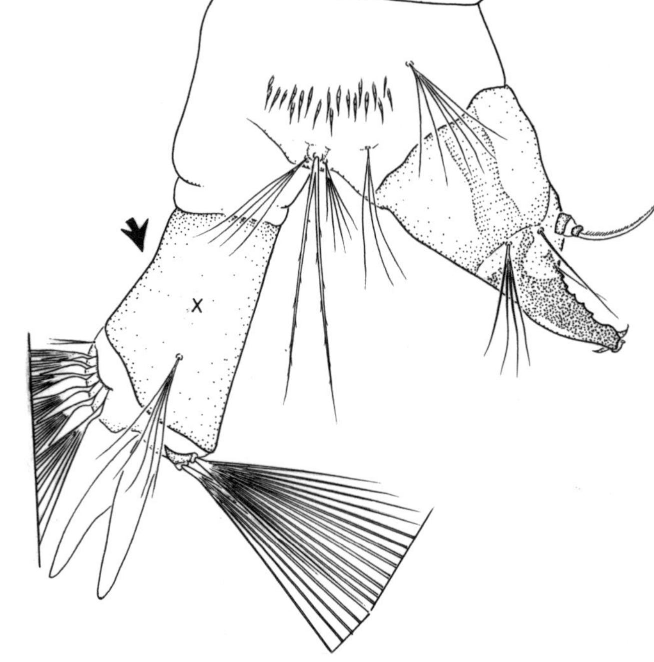


Fig. 545. Lateral view of abdominal segments VIII–X: *Cq. perturbans*

4(2). Siphon without pecten spines (fig. 546) .......... 5

Siphon with pecten spines (fig. 547) .......... 7

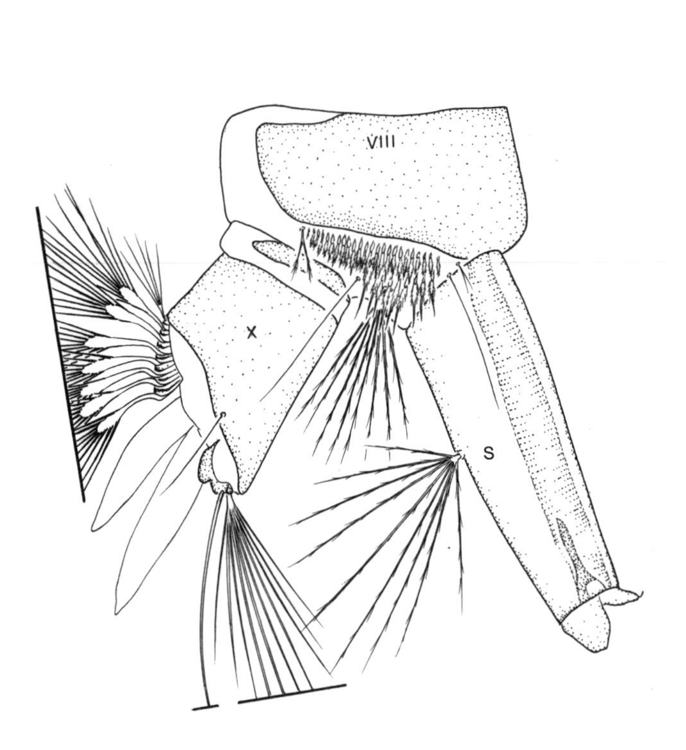


Fig. 546. Lateral view of abdominal segments VIII–X: *Or. signifera*

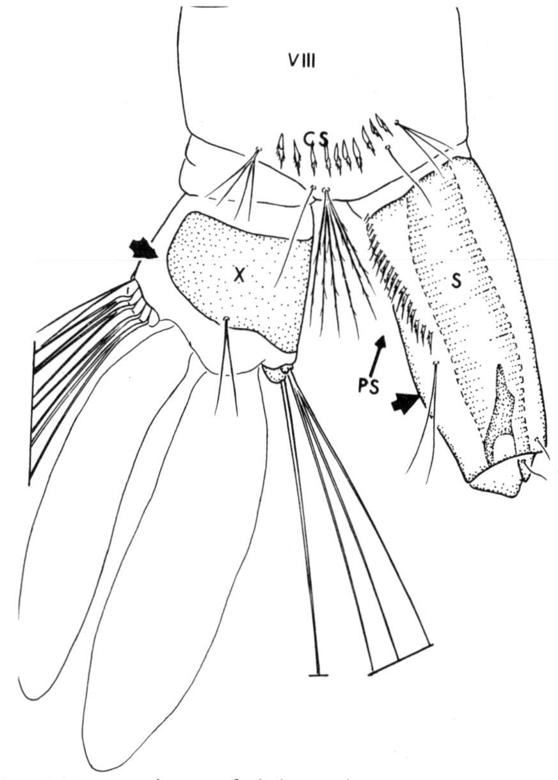


Fig. 547. Lateral view of abdominal segments VIII–X: *Ae. aegypti*

5(4). Lateral palatal brush composed of few stout curved rods (fig. 548); comb scales absent (fig. 549) ........ *Toxorhynchites r. rutilus* (plate 42D)
*Toxorhynchites r. septentrionalis* (plate 37B)
*Toxorhynchites moctezuma* (plate 42C)

Lateral palatal brush composed of numerous thin, sometimes pectinate, filaments (fig. 550); with comb scales(fig. 551) ........ 6

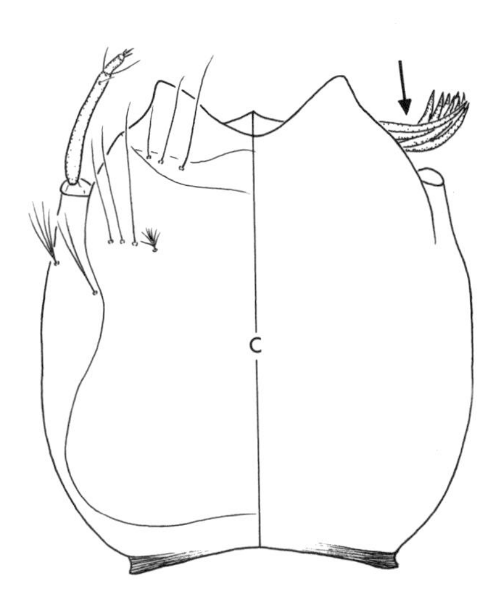


Fig. 548. Dorsal view of head: *Tx. r. septentrionalis*

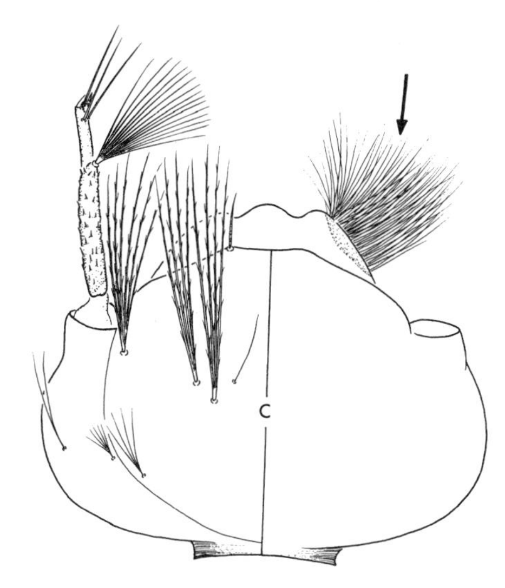


Fig. 550. Dorsal view of head: *Cx. pipiens*

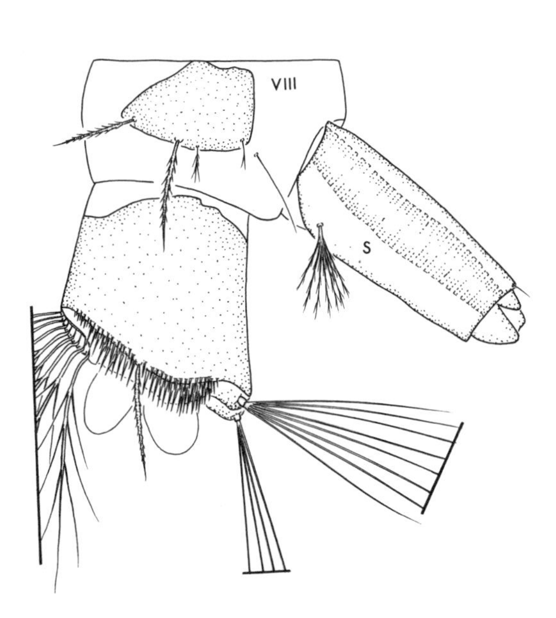


Fig. 549. Lateral view of abdominal segments VIII–X: *Tx. r. septentrionalis*

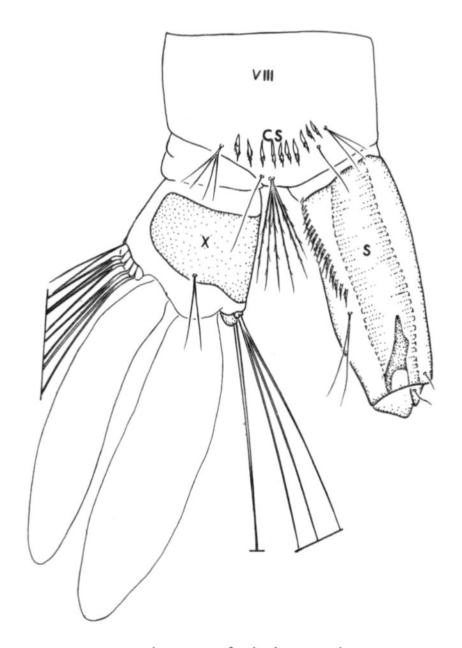


Fig. 551. Lateral view of abdominal segments VIII–X: *Ae. aegypti*

6(5). Segment X without median ventral brush, seta 4-X a pair of ventroposterolateral setae; comb scales in single row (fig. 552) ........ *Wyeomyia*

Segment X with seta 4-X a well-developed median ventral brush; comb scales in 2 rows (fig. 553) ........ *Orthopodomyia*

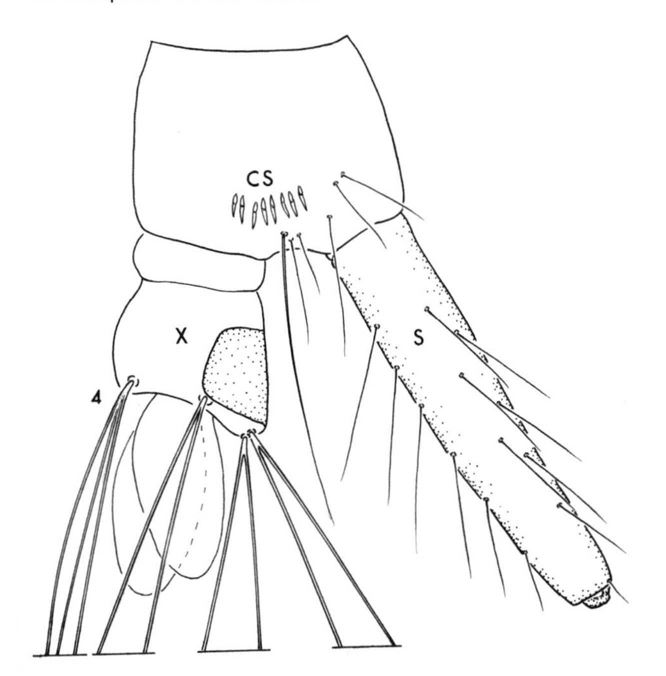


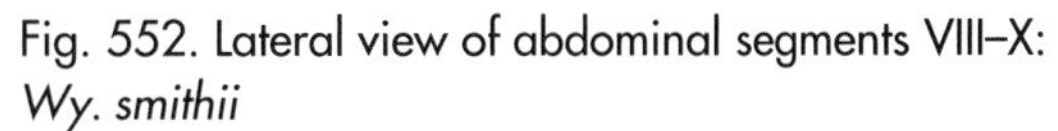

Fig. 552. Lateral view of abdominal segments VIII–X: *Wy. smithii*

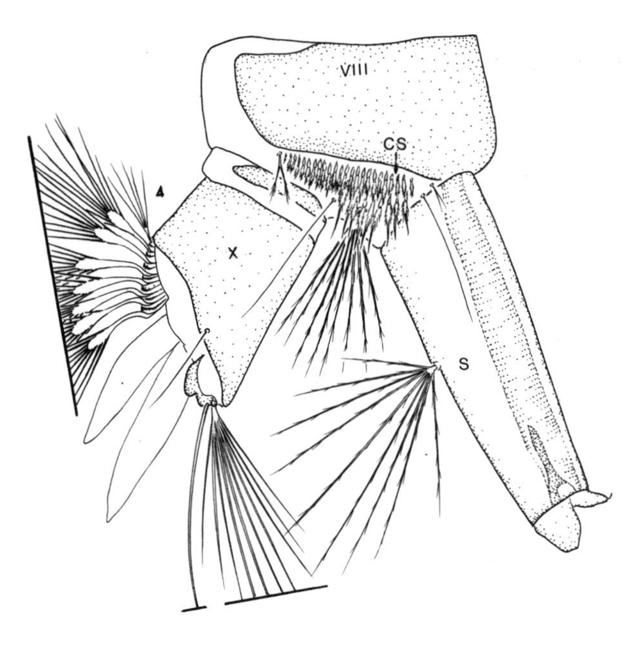


Fig. 553. Lateral view of abdominal segments VIII–X: *Or. signifera*

7(4). Segment VIII with large lateral comb plate bearing comb scales (fig. 554); head longer than wide (fig. 555) .......... *Uranotaenia*

Segment VIII without comb plate or, if present, small (fig. 556); head wider than long (fig. 557) .......... 8

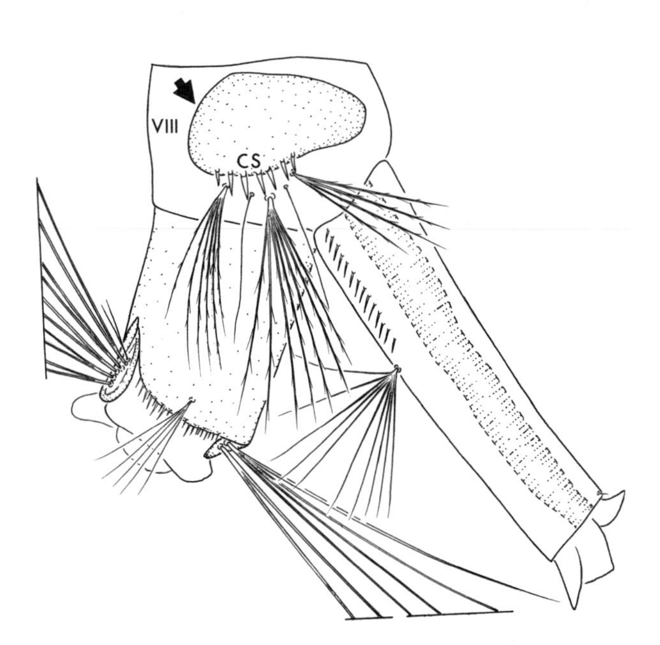


Fig. 554. Lateral view of abdominal segments VIII–X: *Ur. sapphirina*

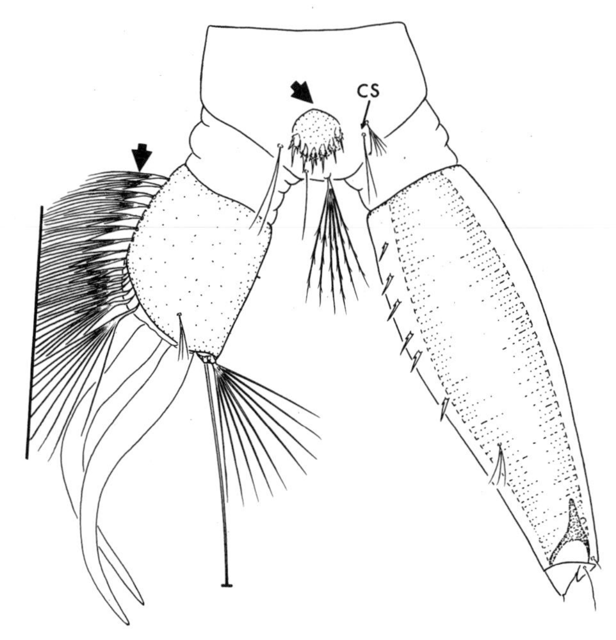


Fig. 556. Lateral view of abdominal segments VIII–X: *Ps. columbiae*

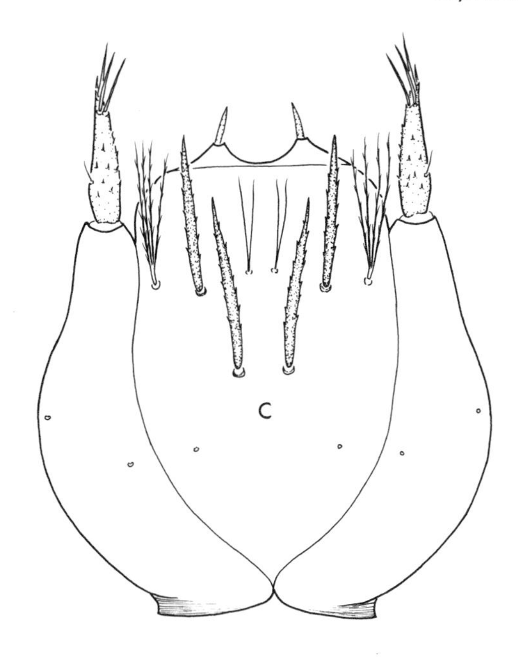


Fig. 555. Dorsal view of head: *Ur. sapphirina*

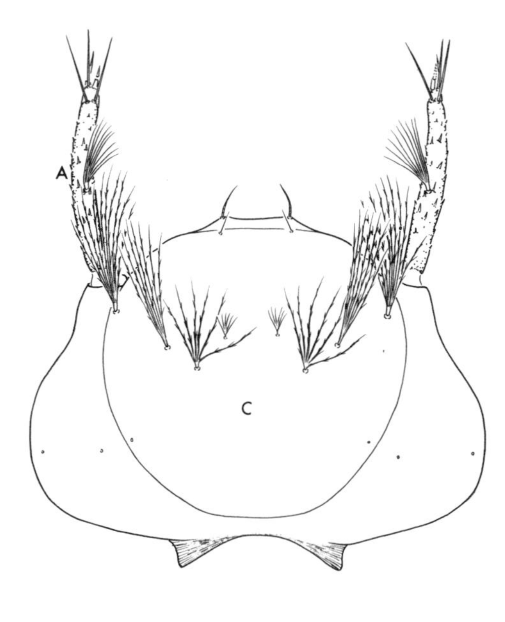


Fig. 557. Dorsal view of head: *Ps. columbiae*

8(7). Head capsule widest near level of bases of antennae (fig. 558); segment X with dorsal and ventral sclerotized plates (fig. 559) .................................................................. *Deinocerites*

Head capsule widest in posterior 0.5 (fig. 560); segment X with single sclerotized saddle (fig. 561) ........................................................................................................................ 9

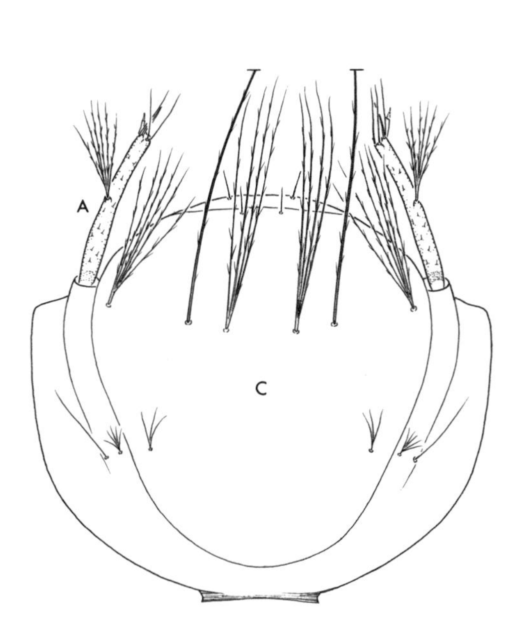

Fig. 558. Dorsal view of head: *De. pseudes*

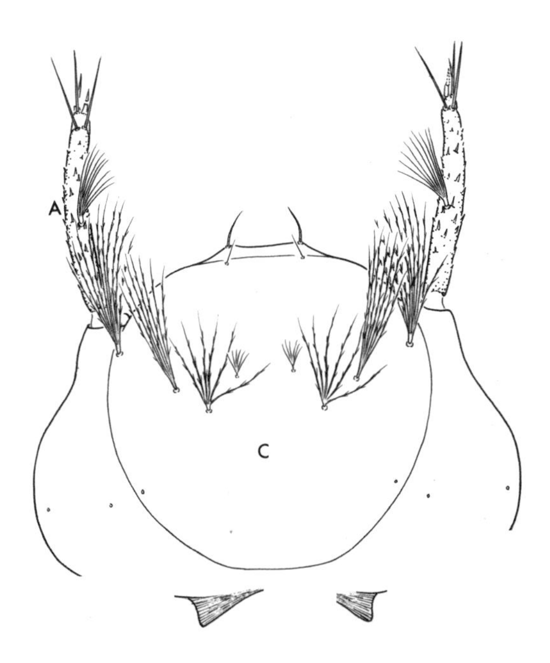


Fig. 560. Dorsal view of head: *Ps. columbiae*

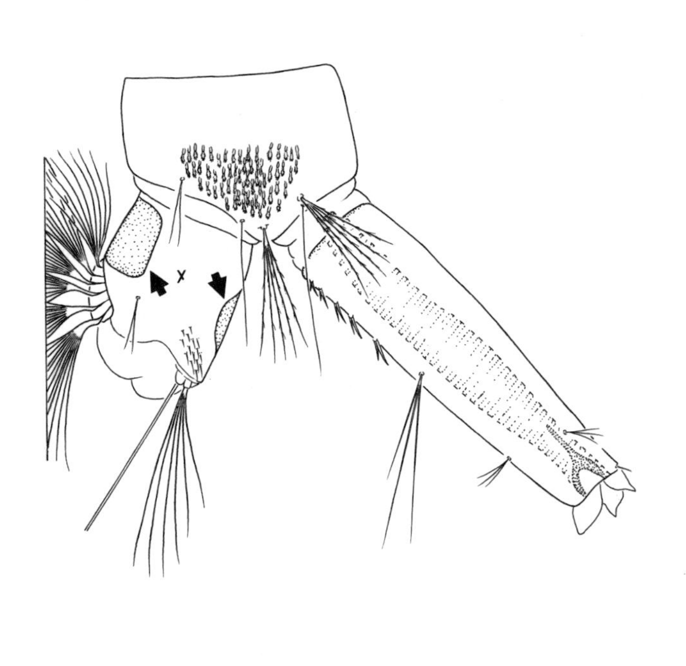


Fig. 559. Lateral view of abdominal segments VIII–X: *De. pseudes*

Fig. 561. Lateral view of abdominal segments VIII–X: *Ae. aegypti*

9(8). Siphon with at least a basal pair of ventral setae (fig. 562) ................................ *Culiseta*

Siphon with setae elsewhere, not ventrally near base (fig. 563) ................................ 10

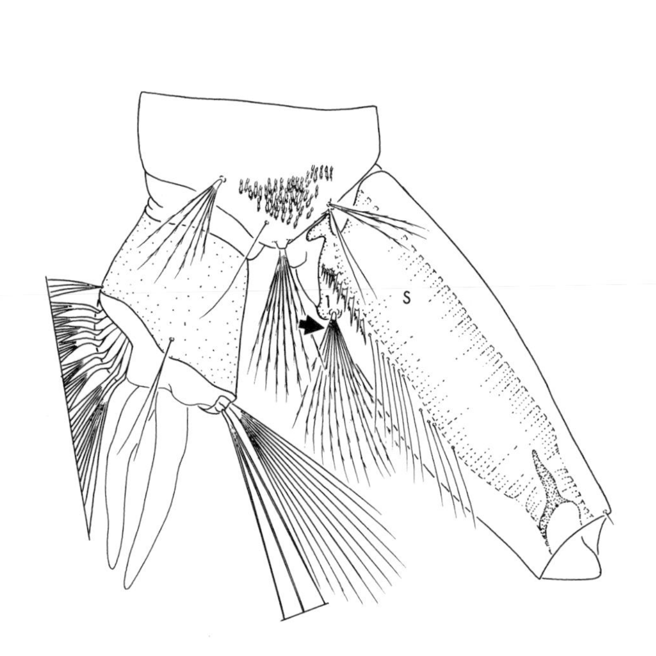


Fig. 562. Lateral view of abdominal segments VIII–X: *Cs. inornata*

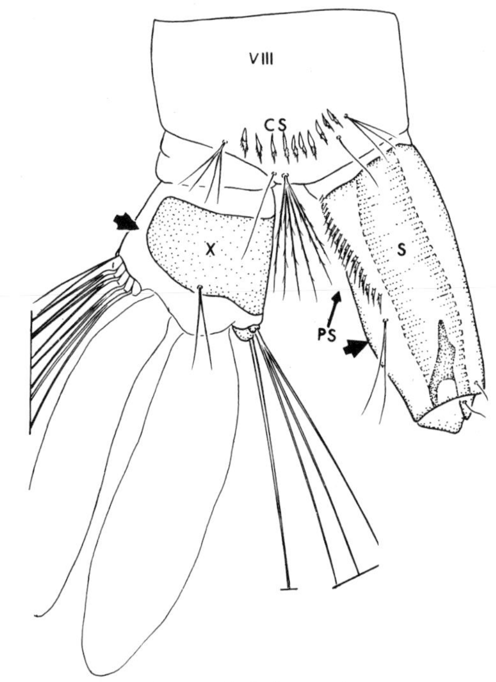


Fig. 563. Lateral view of abdominal segments VIII–X: *Ae. aegypti*

10(9). Siphon with 3 or more pairs of setae (fig. 564) ................................................ 11

Siphon with 1 pair of setae (fig. 565) ................................................................ 12

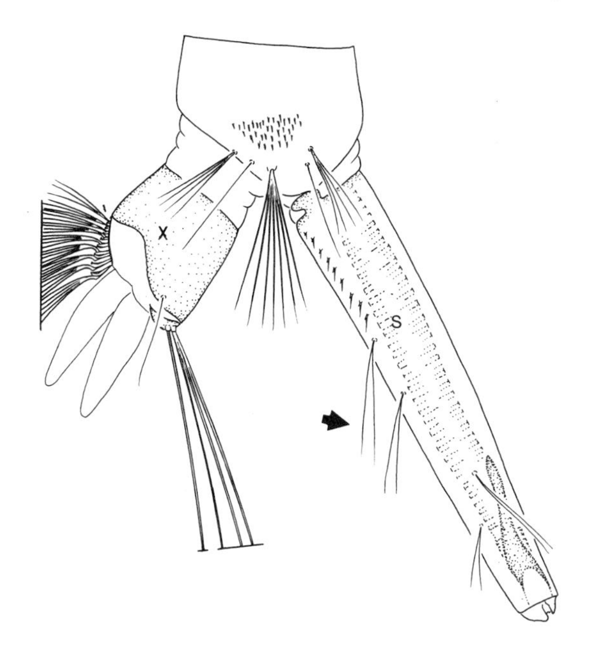


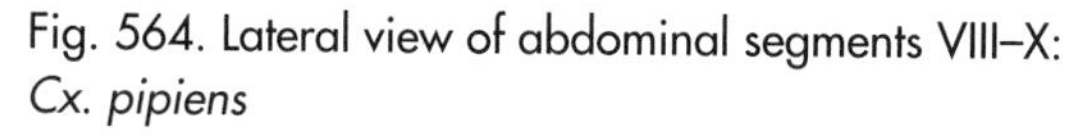

Fig. 564. Lateral view of abdominal segments VIII–X: *Cx. pipiens*

Fig. 565. Lateral view of abdominal segments VIII–X: *Ae. aegypti*

11(10). Saddle completely encircling segment X (fig. 566) ................................................ *Culex*

Saddle not completely encircling segment X (fig. 567) ............................ (in part) *Aedes*
(in part) *Ochlerotatus*

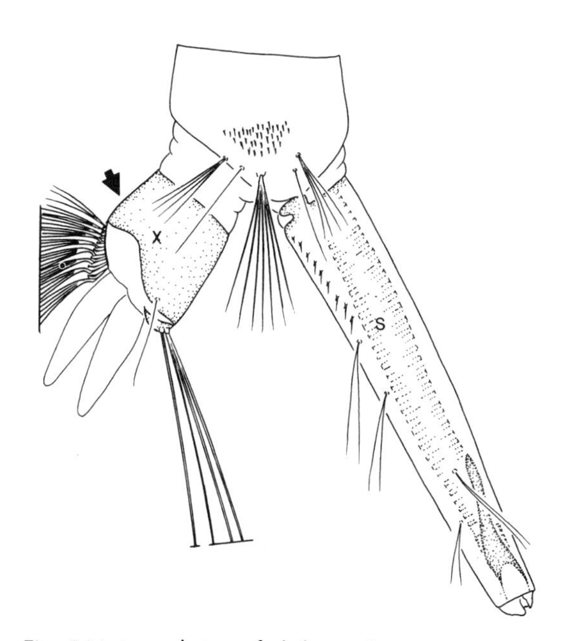


Fig. 566. Lateral view of abdominal segments VIII–X: *Cx. pipiens*

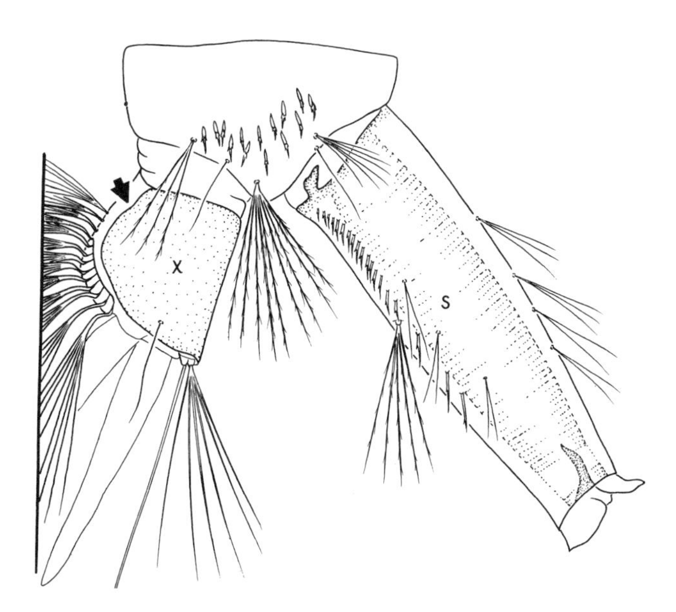


Fig. 567. Lateral view of abdominal segments VIII–X: *Oc. provocans*

12(10). Saddle completely encircling segment X, pierced along midventral line by row of precratal setal tufts (fig. 568) ........................................................................................ *Psorophora*

Saddle usually not encircling segment X, but if so, then setal tufts of seta 4-X confined posterior to it (fig. 569) ........................................................................................ 13

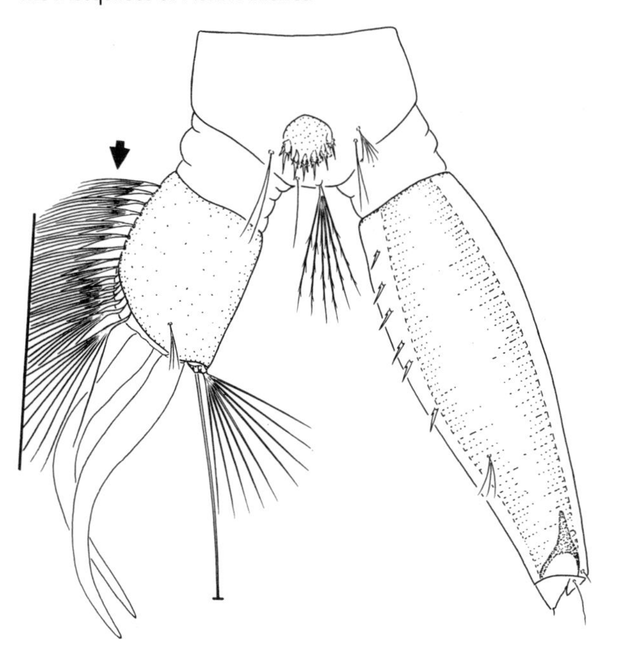

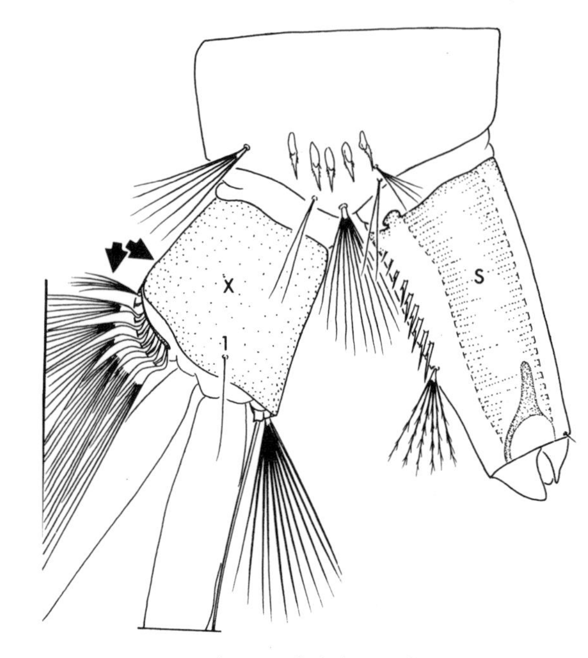


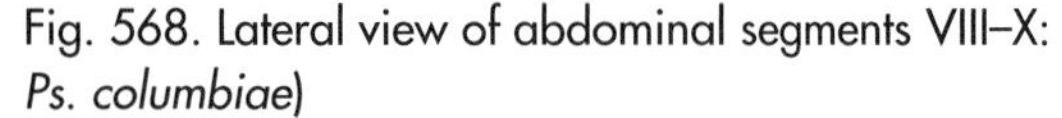

Fig. 568. Lateral view of abdominal segments VIII–X: *Ps. columbiae*)

Fig. 569. Lateral view of abdominal segments VIII–X: *Oc. atlanticus*

13(12). Saddle completely encircling segment X (fig. 570) ........................ (in part) *Ochlerotatus*

Saddle not completely encircling segment X (fig. 571) ................................................ 14

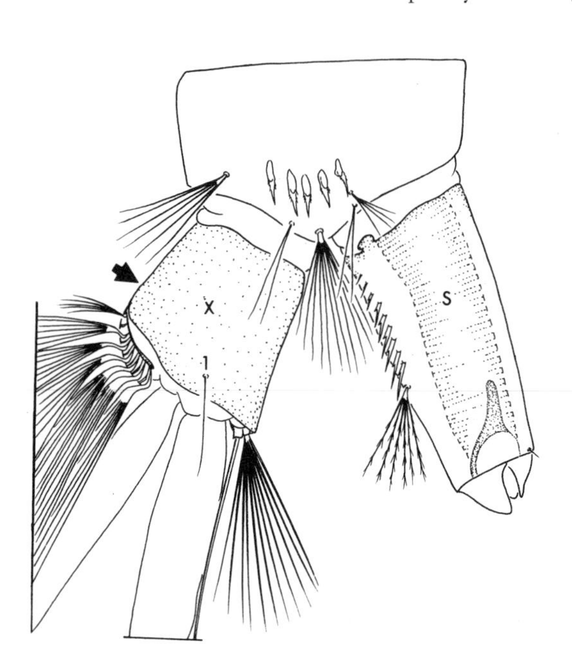

Fig. 570. Lateral view of abdominal segments VIII–X: *Oc. atlanticus*

Fig. 571. Lateral view of abdominal segments VIII–X: *Ae. aegypti*)

14(13). Saddle bearing prominent aciculae on posterior border; seta 3-VII well developed, longer than tergum, single (fig. 572) ..................................... *Haemagogus equinus* (plate 40D)

Saddle with at most small aciculae; seta 3-VII weak, shorter than its tergum, single or multibranched (fig. 573) ......................................................................... (in part) *Aedes*
(in part) *Ochlerotatus*

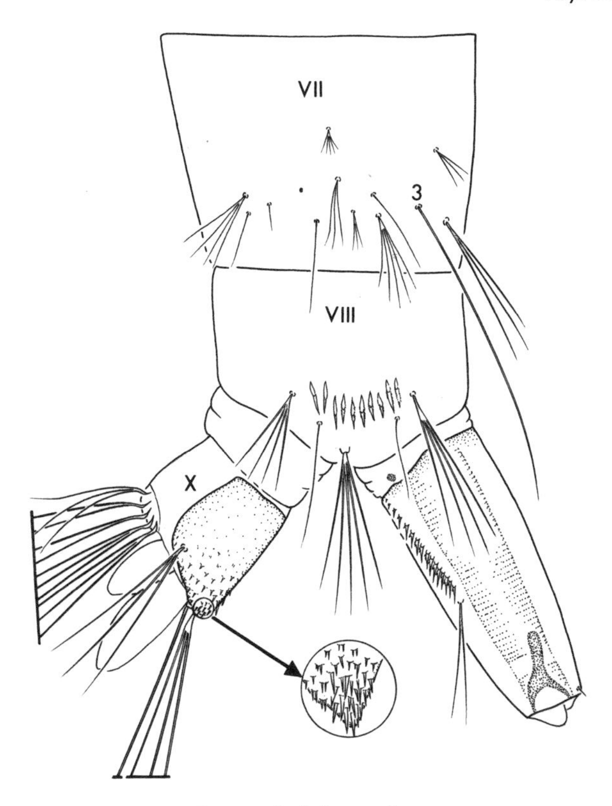


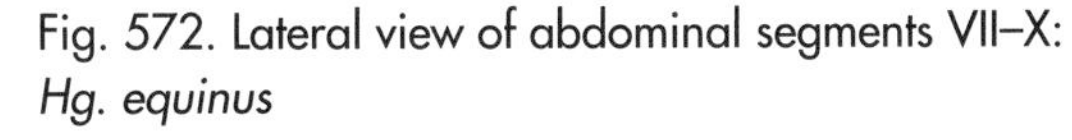
Fig. 572. Lateral view of abdominal segments VII–X: *Hg. equinus*

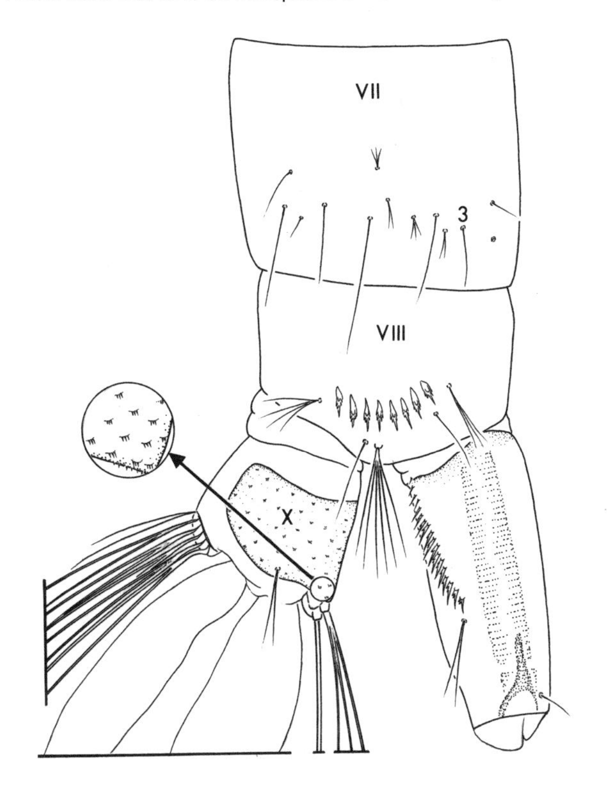


Fig. 573. Lateral view of abdominal segments VII–X: *Ae. aegypti*

## Key to Fourth Instar Larvae of Genera *Aedes* (*Ae.*) and *Ochlerotatus* (*Oc.*)

1. Siphon with more than 1 pair of setae, excluding seta 2-S (fig. 574) ............................ 2

   Siphon with 1 pair of setae, excluding seta 2-S (fig. 575) ............................................ 4

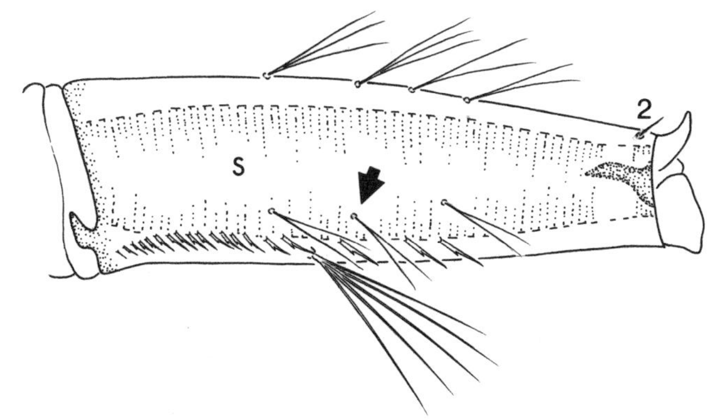


Fig. 574. Lateral view of siphon: *Oc. provocans*

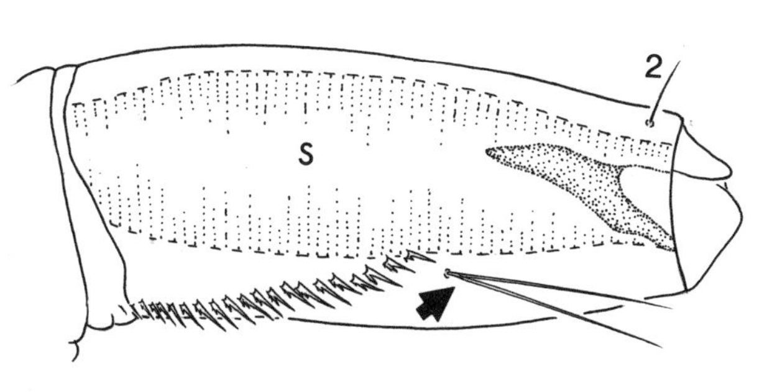


Fig. 575. Lateral view of siphon: *Ae.aegypti*

2(1). Bases of setae 5,6, and 7-C nearly in straight line (fig. 576) .......... Ae. *cinereus* (plate 12B)

Base of seta 6-C distinctly anterior to setae 5-C and 7-C (fig. 577) ................................ 3

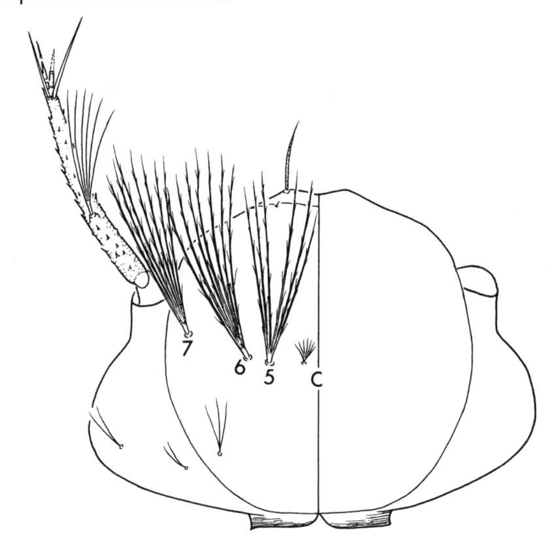


Fig. 576. Dorsal view of head: *Ae. cinereus*

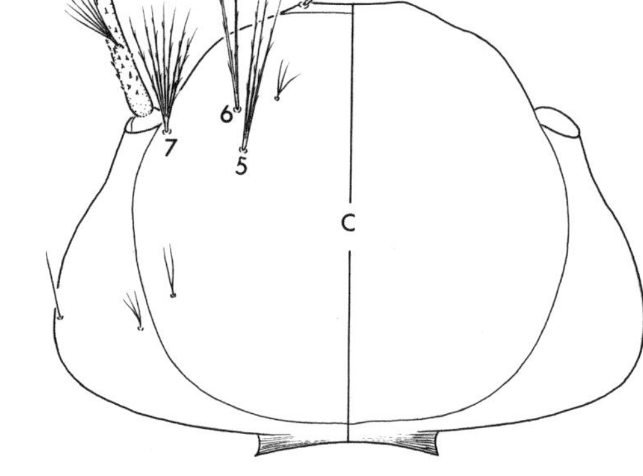


Fig. 577. Dorsal view of head: *Oc. provocans*

3(2). Siphon with 4,5 pairs of subdorsal setae; segment VIII with 14–16 comb scales (fig. 578) .. ........................................................................................ *Oc. provocans* (plate 19A)

Siphon with 1 pair of subdorsal setae; segment VIII with 4–6 comb scales (fig. 579) ......... ........................................................................................ *Oc. bicristatus* (plate 39A)

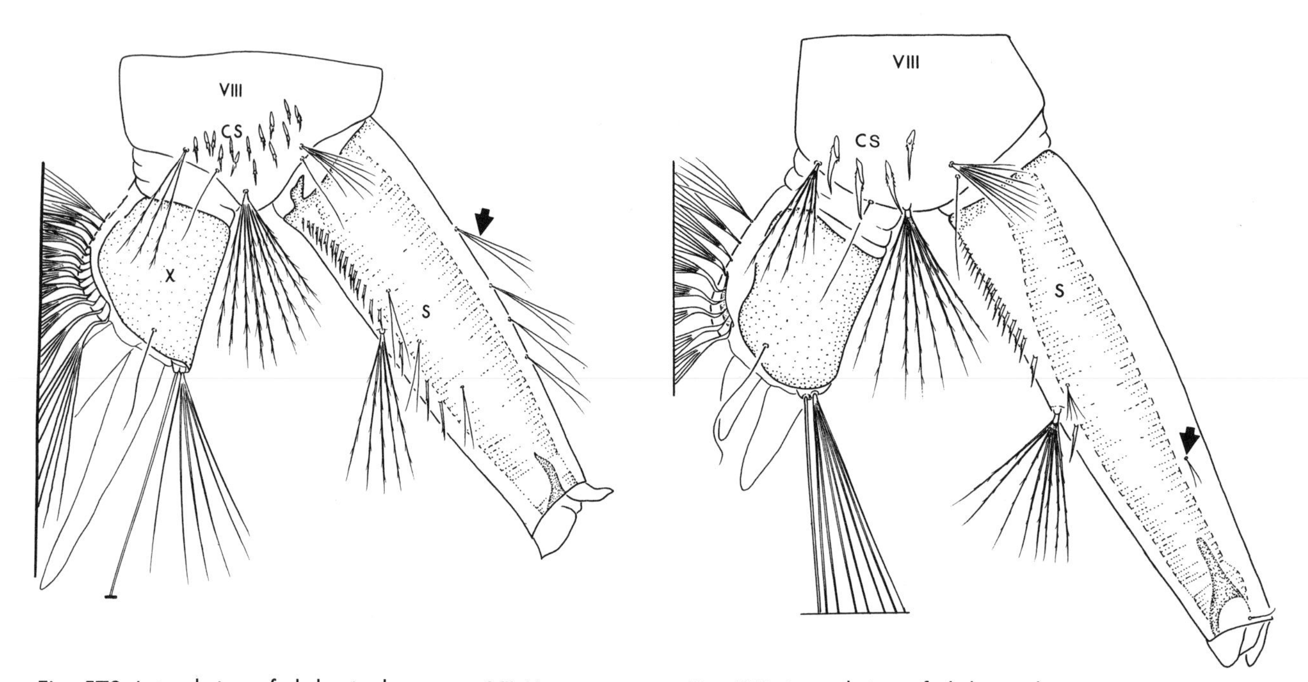


Fig. 578. Lateral view of abdominal segments VII–X: *Oc.provocans*

Fig. 579. Lateral view of abdominal segments VIII–X: *Oc. bicristatus*

4(1). Saddle completely encircling segment X (fig. 580) ........................................................... 5

Saddle not completely encircling segment X (fig. 581) ............................................... 23

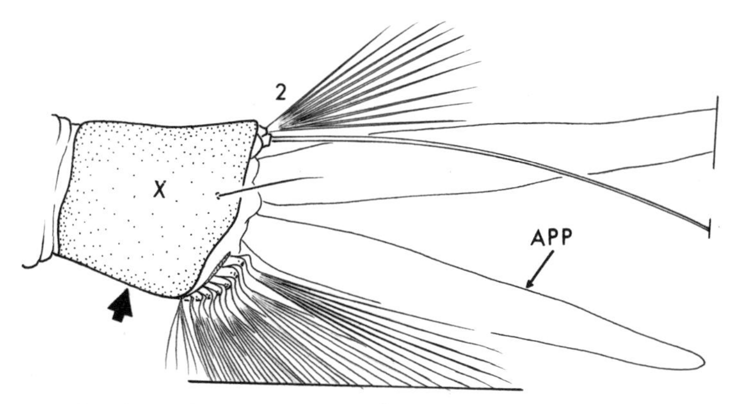


Fig. 580. Lateral view of abdominal segment X: *Oc. atlanticus*

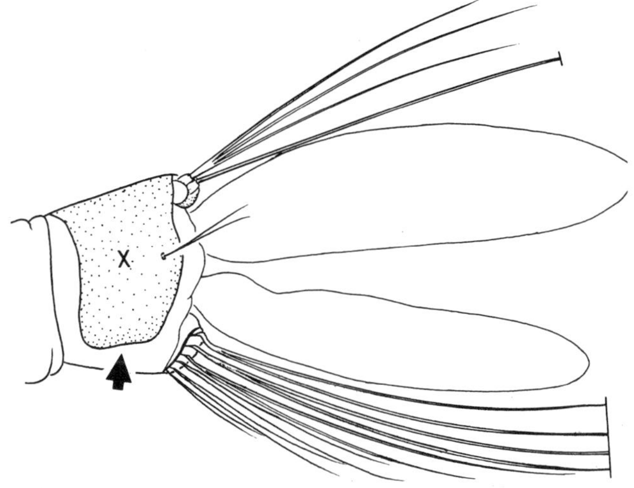


Fig. 581. Lateral view of abdominal segment X: *Ae. aegypti*

5(4). Pecten on siphon with 1 or more distal spines detached apically (fig. 582) .................. 6

Pecten spines more or less evenly spaced (fig. 583) ...................................................... 9

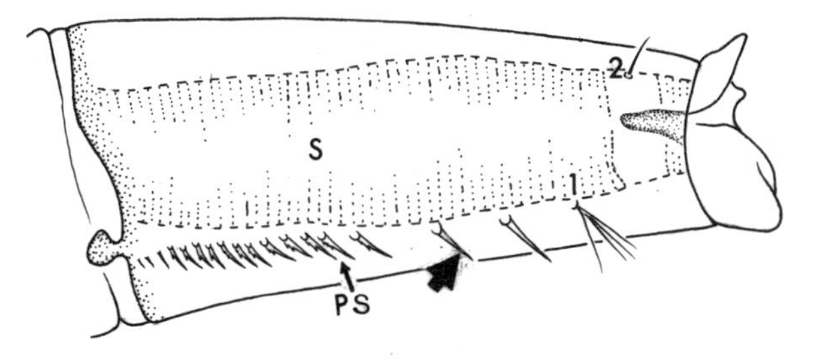


Fig. 582. Lateral view of siphon: *Oc. nigromaculis*

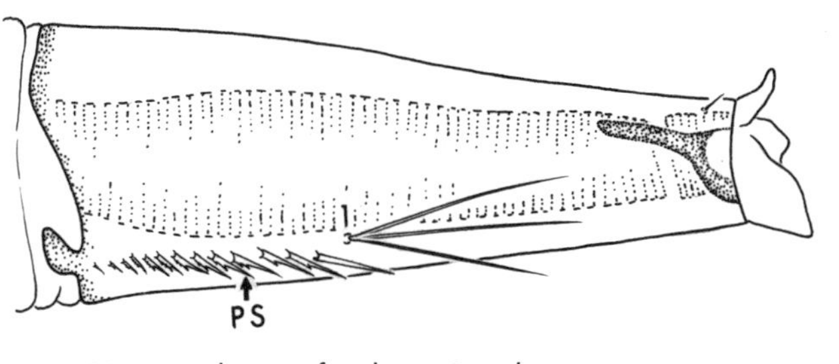


Fig. 583. Lateral view of siphon: *Oc. abserratus*

6(5). Comb scales with median spine at least 4.0 length of minute basal spinules (fig. 584); seta 1-S attached distal to pecten, or sometimes within it (fig. 585) .......................................... 7

Comb scales with median spine no more than 2.0 length of subapical spinules, or fringed with subequal spinules (fig. 586); seta 1-S within pecten (fig. 587).......................................... 8

Fig. 584. Comb scale: *Oc. nigromaculis*

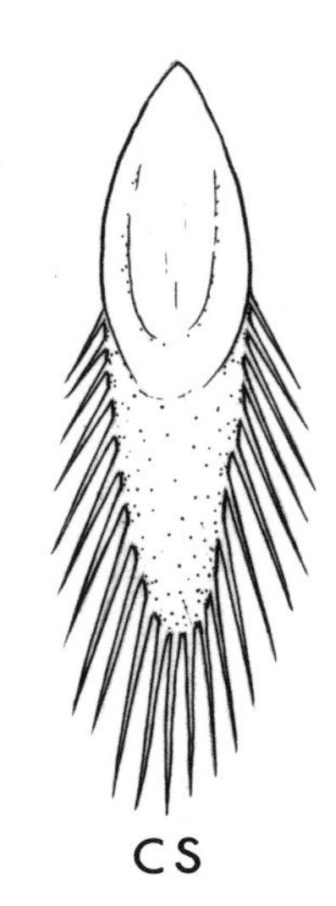


Fig. 586. Comb scale: *Oc. fulvus pallens*

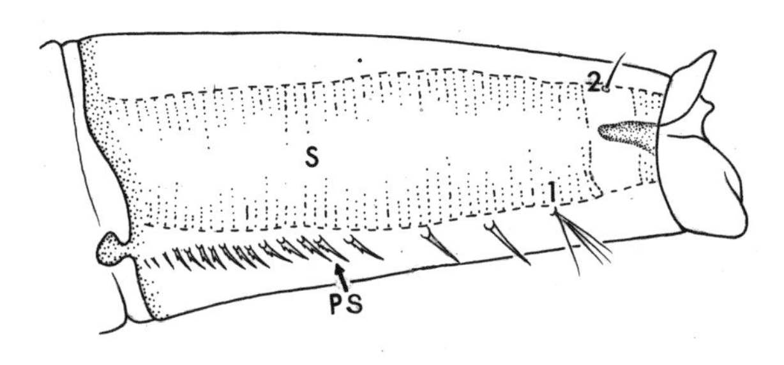


Fig. 585. Lateral view of siphon: *Oc. nigromaculis*

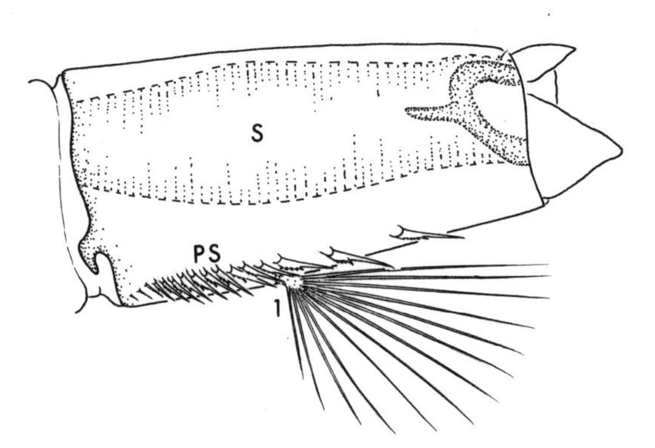


Fig. 587. Lateral view of siphon: *Oc. fulvus pallens*

7(6).Seta 1-S with branches less than 0.5 length of basal diameter of siphon; seta 2-S nearly equal to length of apical pecten spine (fig. 588) .......................... *Oc. nigromaculis* (plate 18C)

Seta 1-S with branches at least equal to basal diameter of siphon; seta 2-S less than 0.5 length of apical pecten spine (fig. 589) .................................................... *Oc. nigripes* (plate 18B)

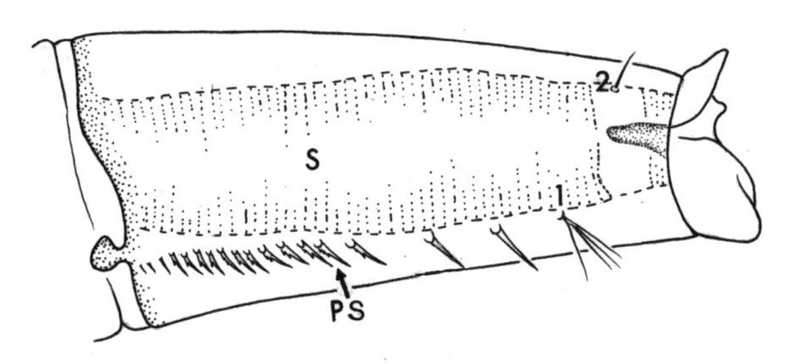


Fig. 588. Lateral view of siphon: *Oc. nigromaculis*

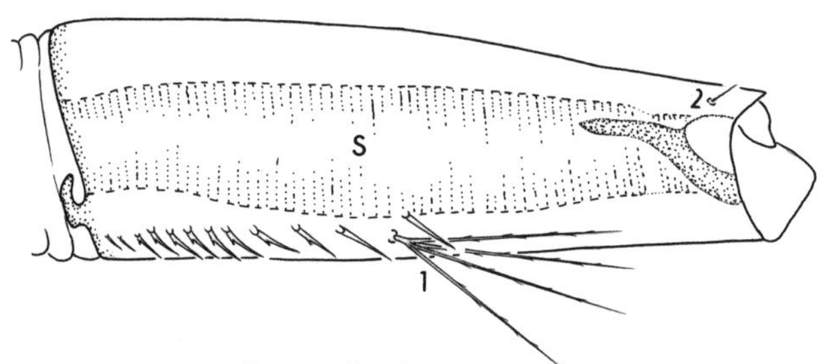


Fig. 589. Lateral view of siphon: *Oc. nigripes*

8(6). Comb scale fringed with subequal spinules (fig. 590); seta 6-C usually double or triple (fig. 591) ........................................................................... Oc. *fulvus pallens* (plate 15A)

Comb scale with median spine markedly longer than subapical spinules (fig. 592); seta 6-C single (fig. 593) ........................................................................... *Oc. thelcter* (plate 22C)

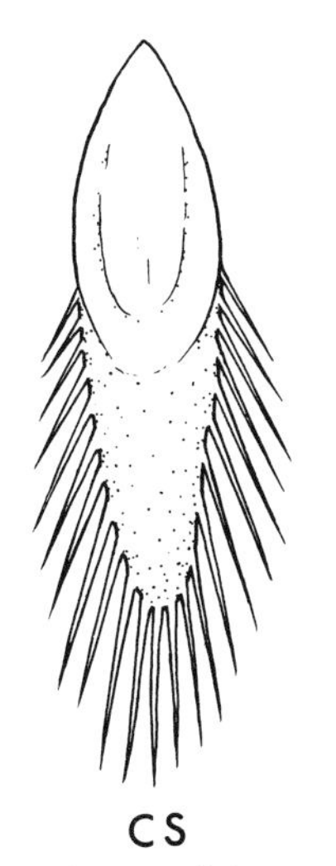


Fig. 590. Comb scale: *Oc. fulvus pallens*

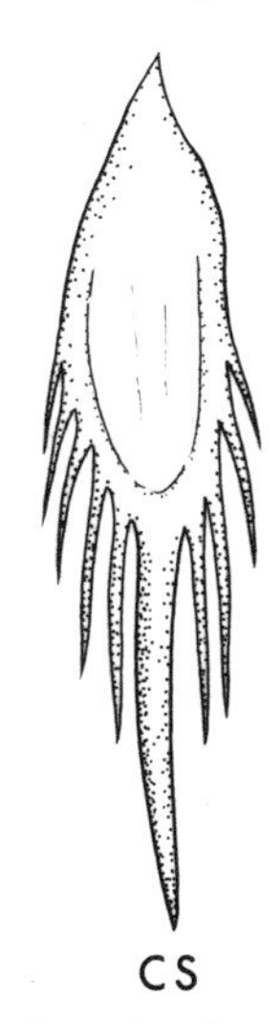


Fig. 592. Comb scale: *Oc. thelcter*

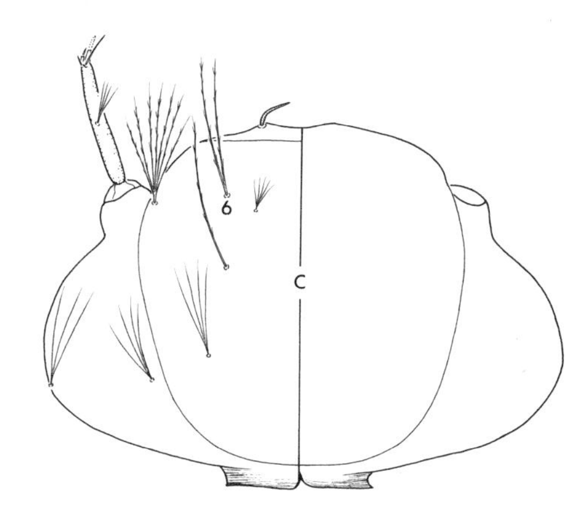


Fig. 591. Dorsal view of head: *Oc. fulvus pallens*

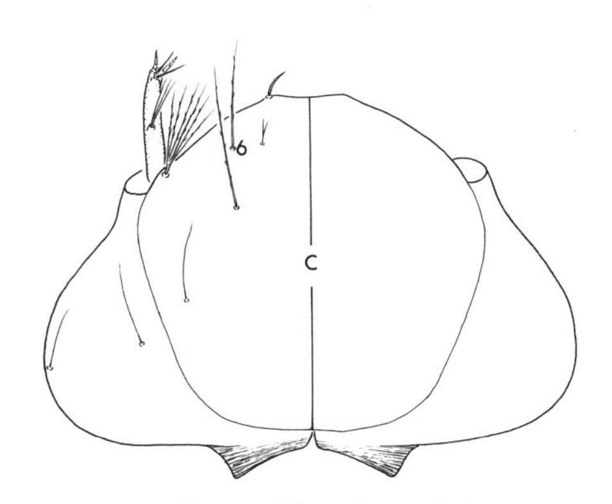


Fig. 593. Dorsal view of head: *Oc. thelcter*

9(5). Seta 1-S attached within pecten (fig. 594) .......................................................................... 10

Seta 1-S attached distal to pecten (fig. 595) ....................................................................11

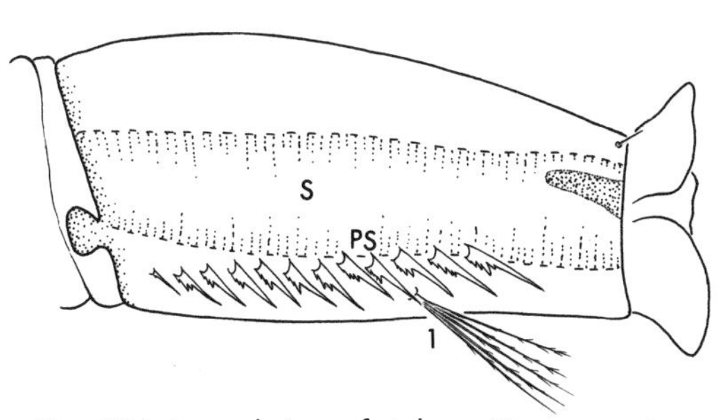


Fig. 594. Lateral view of siphon: *Oc. tormentor*

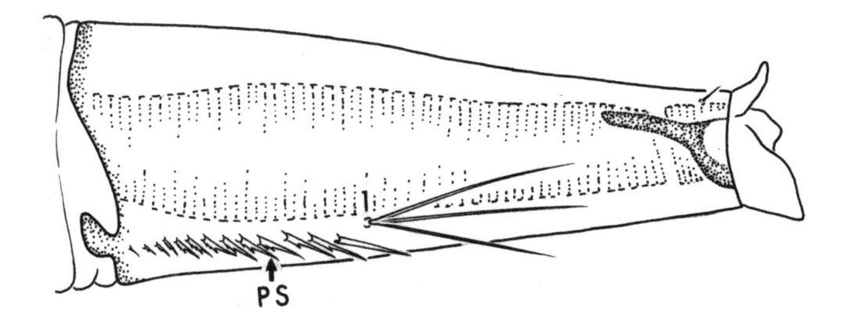


Fig. 595. Lateral view of siphon: *Oc. abserratus*

10(9).Comb scales 30–40, evenly fringed with subequal spinules (fig. 596) ................................
............................................................................................ *Oc. bimaculatus* (plate 39A)

Comb scales 9–12, with large median spine and minute basal spinules (fig. 597)............ ........................................................................................ *Oc. tormentor* (plate 23A)

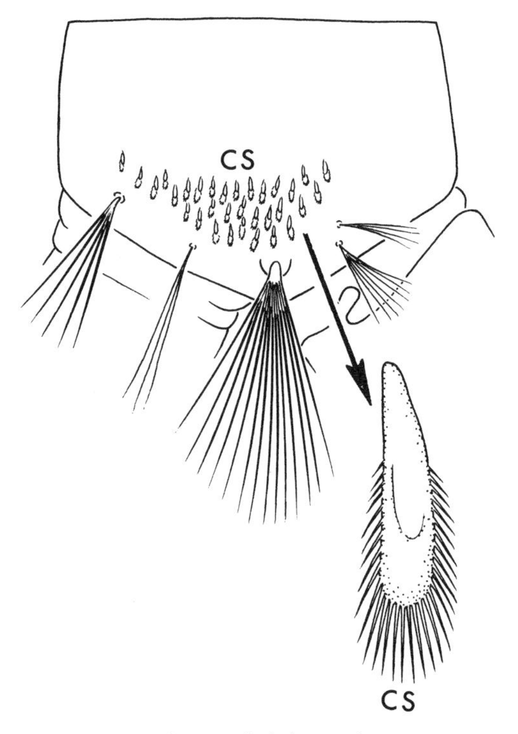


Fig. 596. Lateral view of abdominal segment VIII: *Oc. bimaculatus*

Fig. 597. Lateral view of abdominal segment VIII: *Oc. tormentor*

11(9). Comb scale with apical spine at least 4.0 length of subapical spinules (fig. 598); thoracic integument smooth (fig. 599) ........................................................................................ 12

Comb scale with apical spine not more than 3.0 length of subapical spinules or fringed by subequal spinules (fig. 600); thoracic integument aculeate (fig. 601) ........................... 18

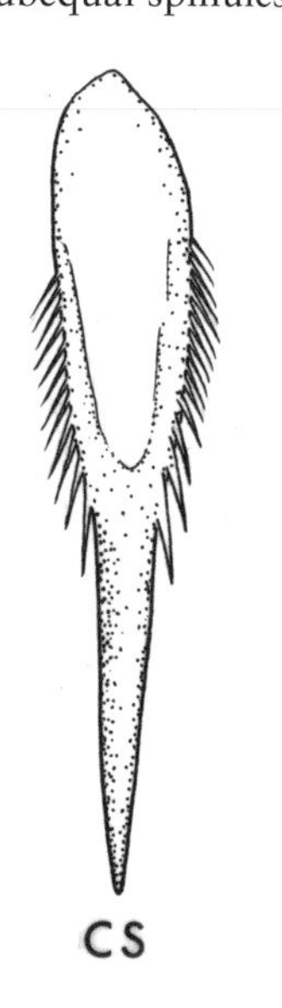

Fig. 598. Comb scale: *Oc. atlanticus*

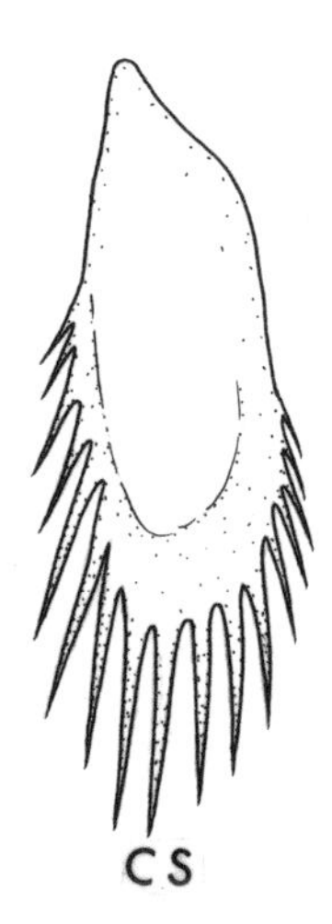


Fig. 600. Comb scale: *Oc. taeniorhynchus*

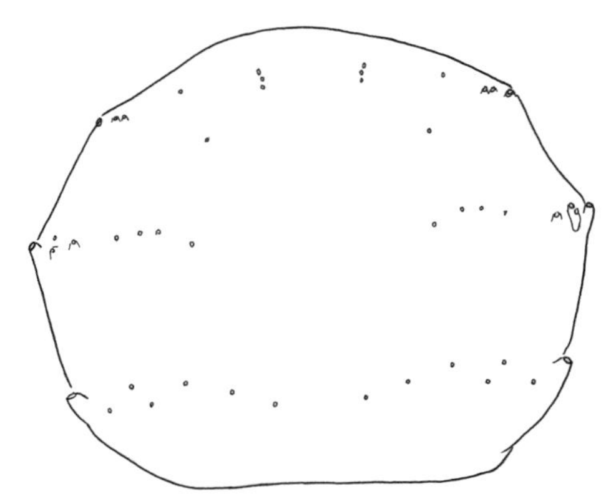

Fig. 599. Dorsal view of thorax: *Oc. sollicitans*

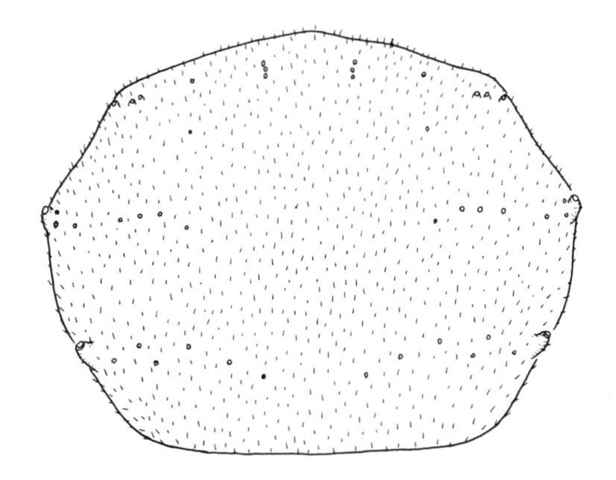

Fig. 601. Dorsal view of thorax: *Oc. taeniorhynchus*

12(11). Setae 2,3-X both single (fig. 602); most setae on head and body coarse, about equal in diameter throughout (fig. 603) ............................................ *Oc. abserratus* (plate 9C)

Seta 2-X multibranched, 3-X single (fig. 604); head and body setae finely attenuated apically (fig. 605) ............................................................................................ 13

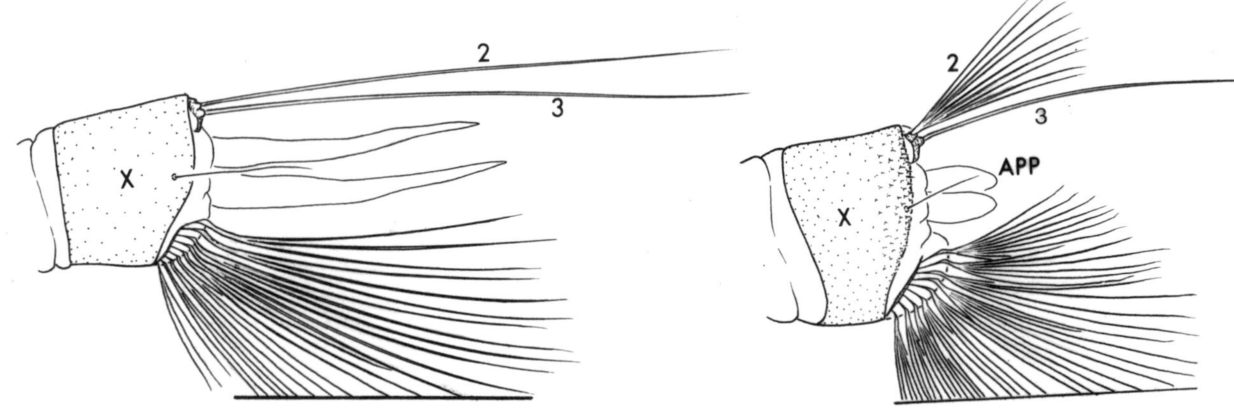


Fig. 602. Lateral view of abdominal segment X: *Oc. abserratus*

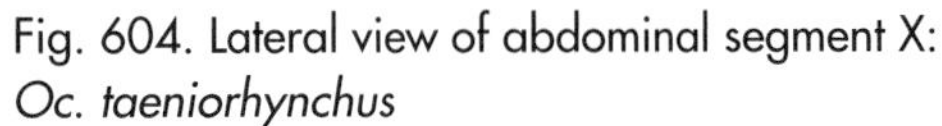

Fig. 604. Lateral view of abdominal segment X: *Oc. taeniorhynchus*

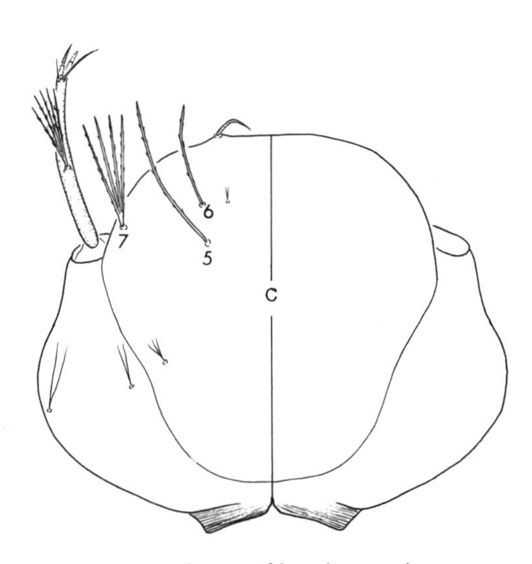


Fig. 603. Dorsal view of head: *Oc. abserratus*

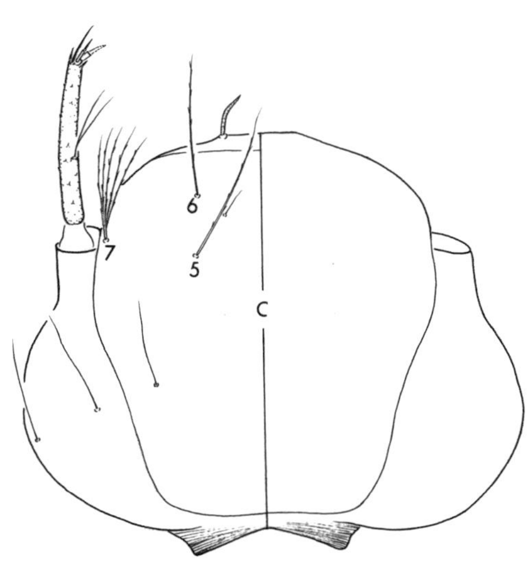


Fig. 605. Dorsal view of head: *Oc. taeniorhynchus*

13(12). Anal papilla–saddle index at least 8.0, papilla with darkly pigmented tracheae; seta 2-X with 2,3 branches (fig. 606) ........................................................ *Oc. dupreei* (plate 13C)

Anal papilla–saddle index at most 5.0, usually much less, papilla lacking dark tracheae; seta 2-X with 4 or more branches (fig. 607) ............................................................... 14

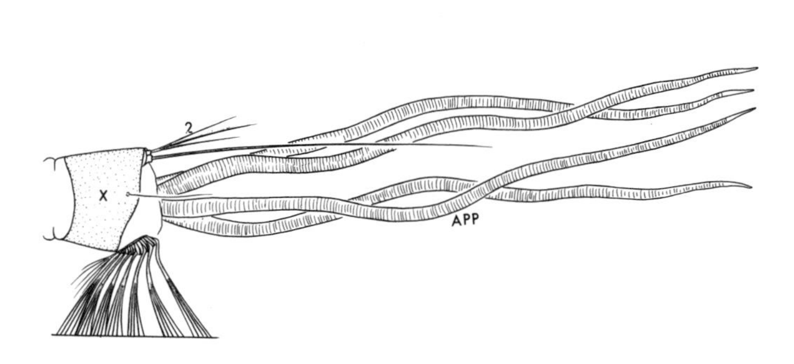


Fig. 606. Lateral view of abdominal segment X: *Oc. dupreei*

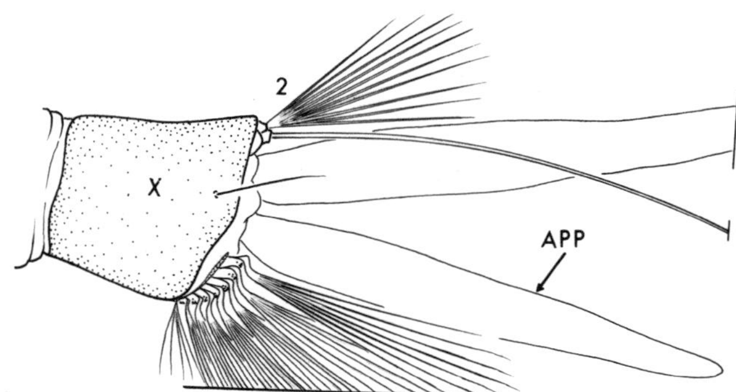


Fig. 607. Lateral view of abdominal segment X: *Oc. atlanticus*

14(13). Comb scales 4–9, large (fig. 608) ..................................................................................... 15

Comb scales 10–30, small (fig. 609) ............................................................................ 16

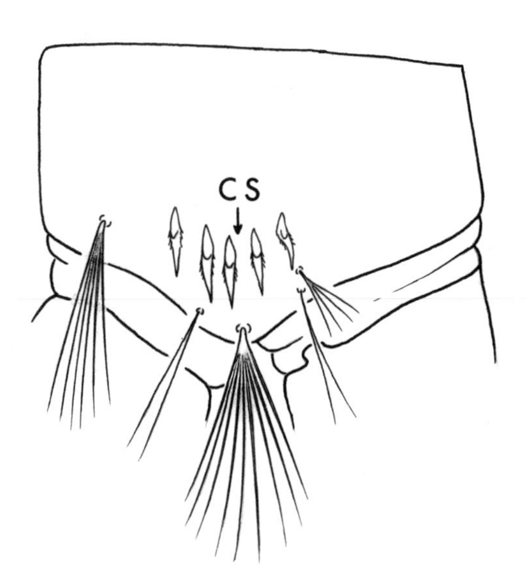


Fig. 608. Lateral view of abdominal segment VIII: *Oc. atlanticus*

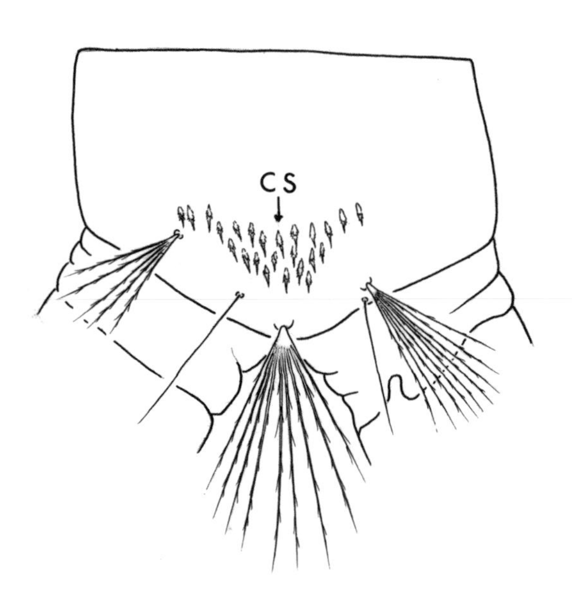


Fig. 609. Lateral view of abdominal segment VIII: *Oc. sollicitans*

15(14). Siphon index about 2.0; seta 1-X shorter than saddle (fig. 610) . *Oc. atlanticus* (plate 10B)

Siphon index about 3.0; seta 1-X equal to length of saddle, or longer (fig. 611) ............ .................................................................................................. *Oc. hexodontus* (plate 15D)

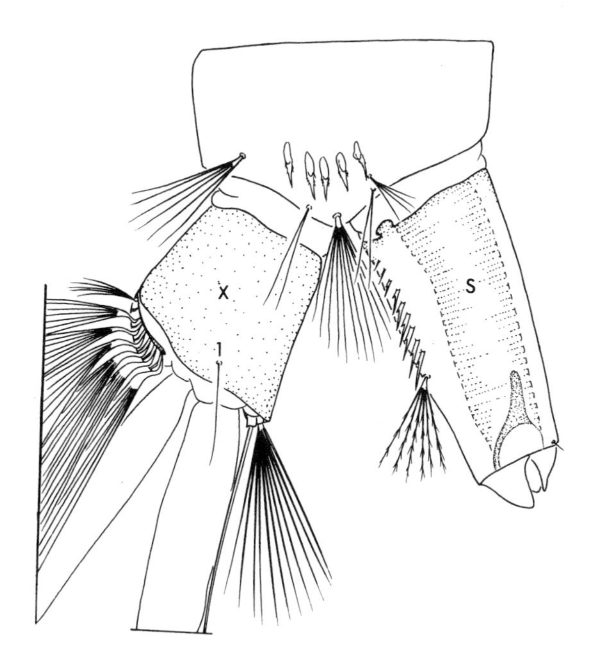


Fig. 610. Lateral view of abdominal segment X: *Oc. atlanticus*

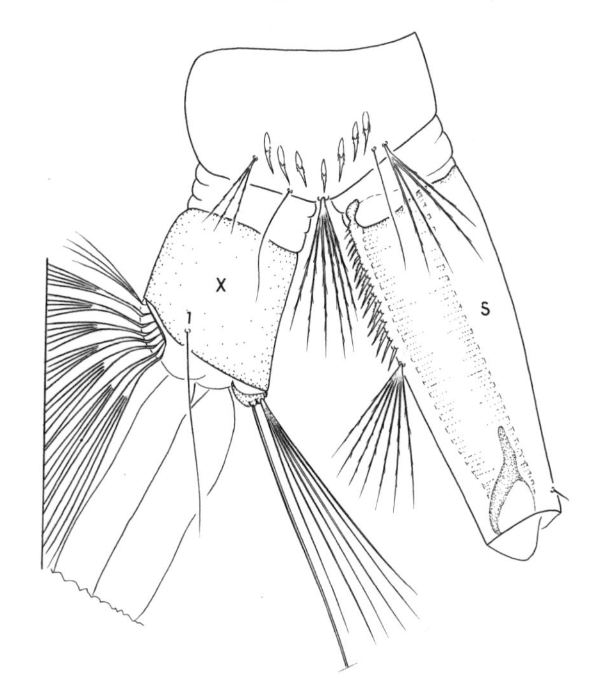


Fig. 611. Lateral view of abdominal segments VIII–X: *Oc. hexodontus*

16(14). Seta 2-S much shorter than apical pecten spine; seta 1-X subequal to saddle (fig. 612) ... ........................................................................................ *Oc. punctor* (plate 19D)

Seta 2-S equal to length of apical pecten spine, seta 1-X shorter than saddle (fig. 613) .. 17

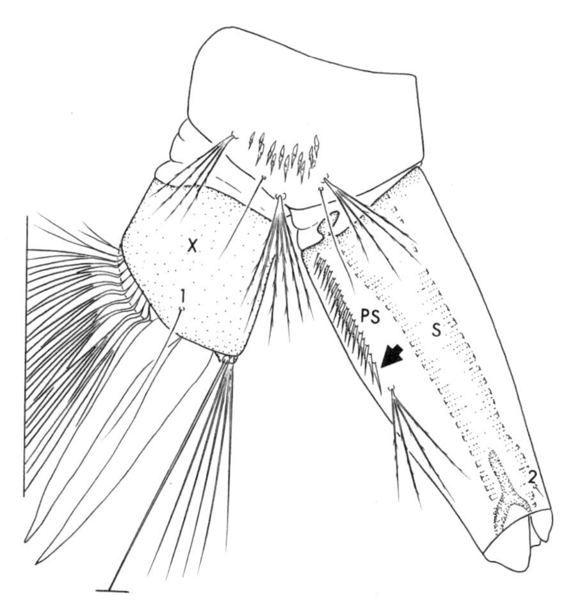


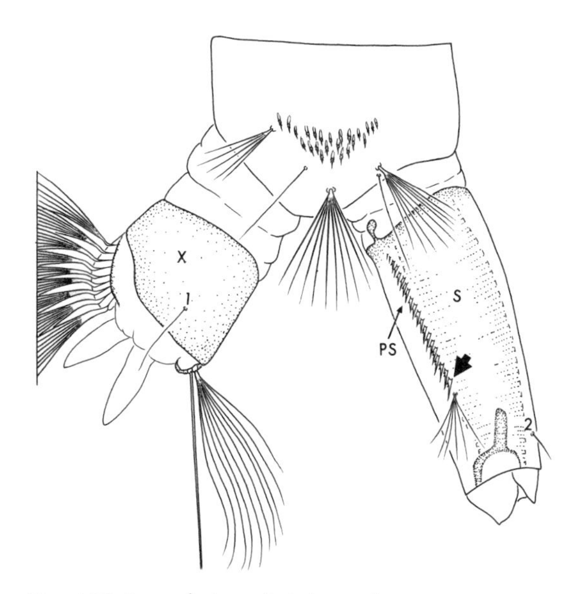


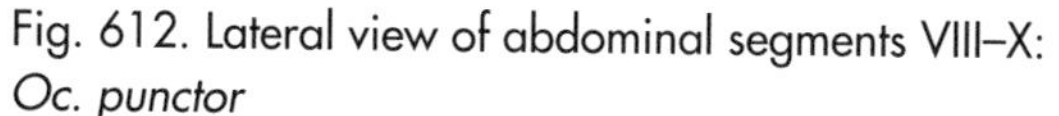

Fig. 612. Lateral view of abdominal segments VIII–X: *Oc. punctor*

Fig. 613. Lateral view of abdominal segments VIII–X: *Oc. sollicitans*

17(16). Siphon index 3.0–3.5; pecten not reaching middle of siphon (fig. 614); setae 5,6-C coarse, about equal in diameter near apex (fig. 615) ........................ *Oc. mitchellae* (plate 17D)

Siphon index 2.0–2.5; pecten reaching to middle of siphon or more distally (fig. 616); setae 5,6-C attenuated apically (fig. 617) ...................................... *Oc. sollicitans* (plate 21A)

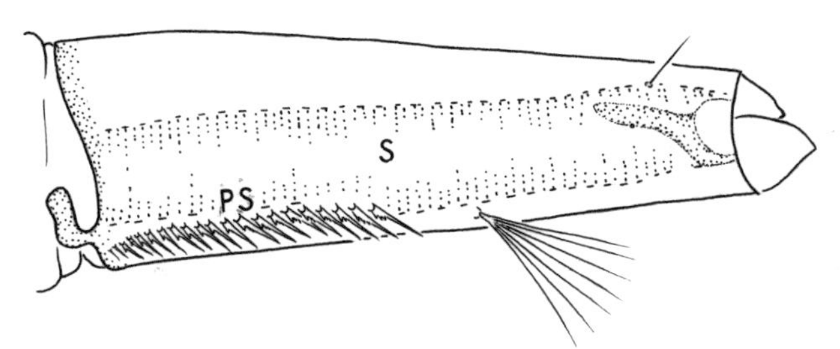


Fig. 614. Lateral view of siphon: *Oc. mitchellae*

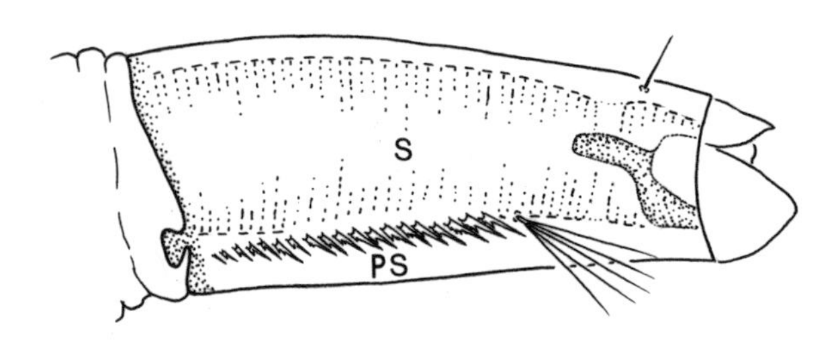


Fig. 616. Lateral view of siphon: *Oc. sollicitans*

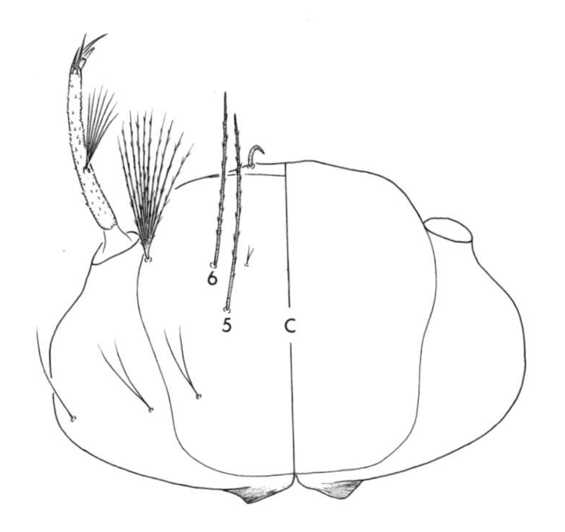


Fig. 615. Dorsal view of head: *Oc. mitchellae*

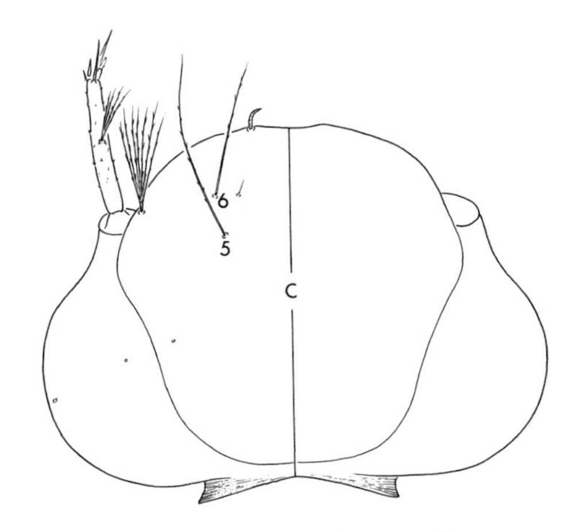


Fig. 617. Dorsal view of head: *Oc. sollicitans*

18(11). Comb scale with apical spine about 2.0–3.0 length of subapical spinules (fig. 618) .... 19

Comb scale with apical spine subequal to subapical spinules, or only slightly stouter and longer (fig. 619) .......... 20

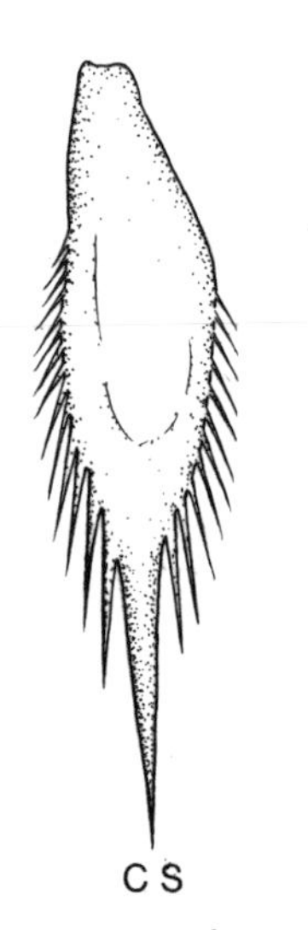


Fig. 618. Comb scale: *Oc. infirmatus*

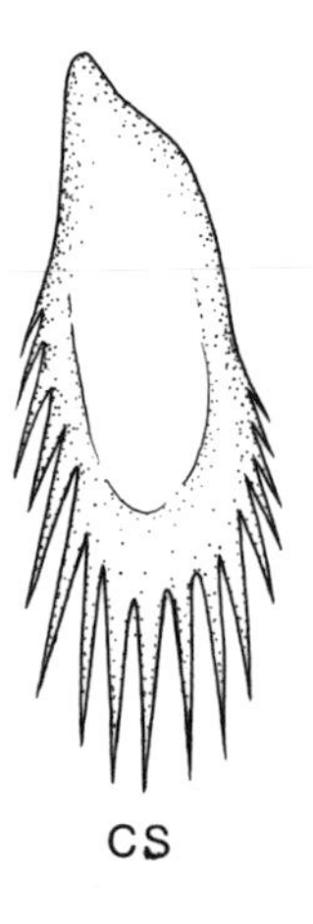


Fig. 619. Comb scale: *Oc. taeniorhynchus*

19(18). Median spine of comb scale 6.0 broader at base, or more, and 2.0–3.0 longer than subapical spinules (fig. 620) .......... *Oc. infirmatus* (plate 16D)

Median spine of comb scale no more than 2.0 broader at base and less than 2.0 longer than subapical spinules (fig. 621) .......... *Oc. trivittatus* (plate 23C)

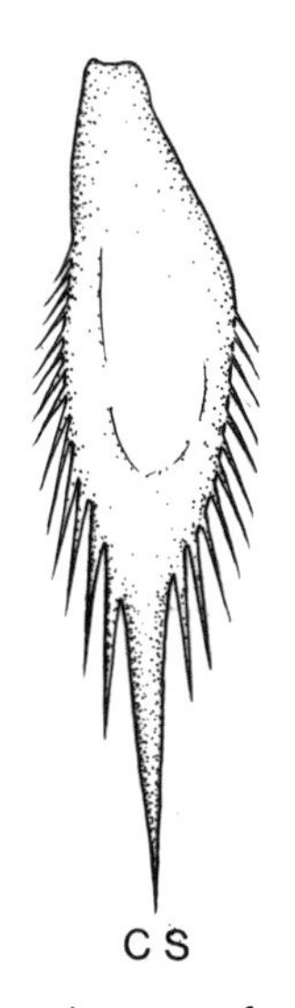

Fig. 620. Comb scale: *Oc. infirmatus*

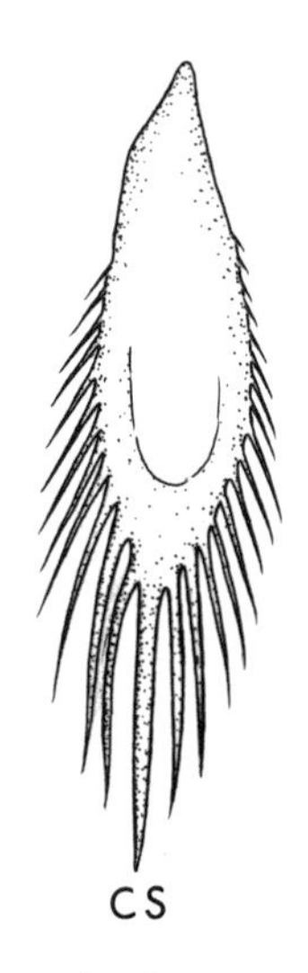


Fig. 621. Comb scale: *Oc. trivittatus*

20(18). Siphon index about 3.5 (fig. 622); thoracic integument glabrous (fig. 623) .......................................................................................................... *Oc. rempeli* (plate 20A)

Siphon index no more than 3.0 (fig. 624); thoracic integument aculeate (fig. 625) ..... 21

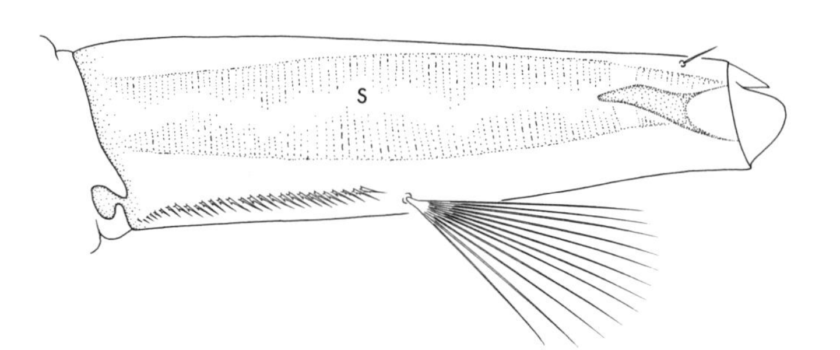


Fig. 622. Lateral view of siphon: *Oc. rempeli*

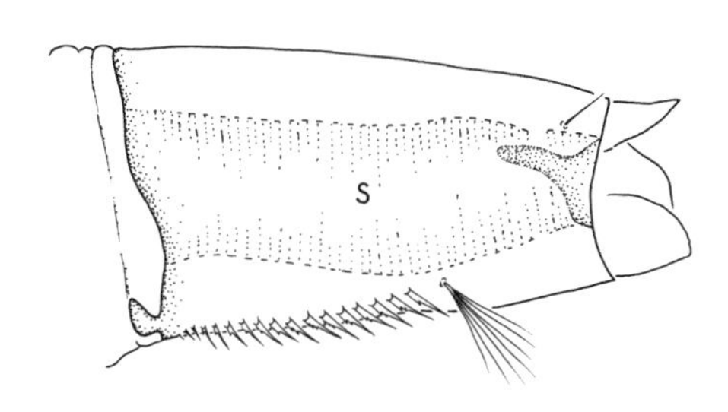


Fig. 624. Lateral view of siphon: *Oc. taeniorhynchus*

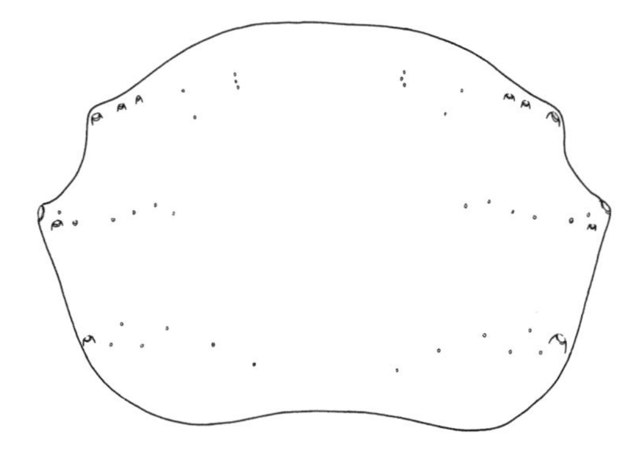

Fig. 623. Dorsal view of thorax: *Oc. rempeli*

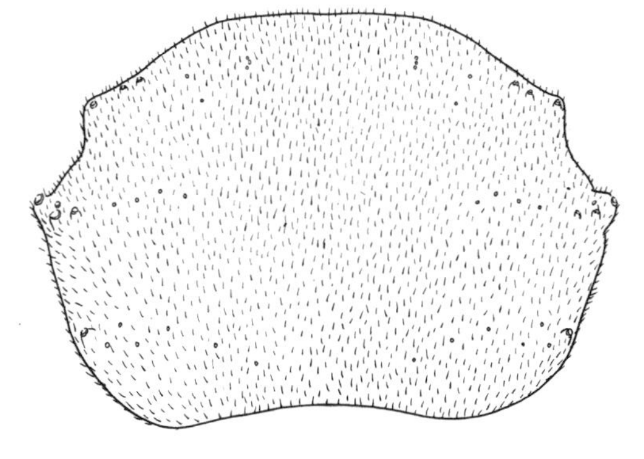

Fig. 625. Dorsal view of thorax: *Oc. scapularis*

21(20). Setae 6-III–V with 2–5 branches (fig. 626); anal papilla–saddle index 0.5 or less (fig. 627) .......................................................................................... *Oc. taeniorhynchus* (plate 22B)

Setae 6-III–V single (fig. 628); anal papilla–saddle index 1.0 or more (fig. 629).......... 22

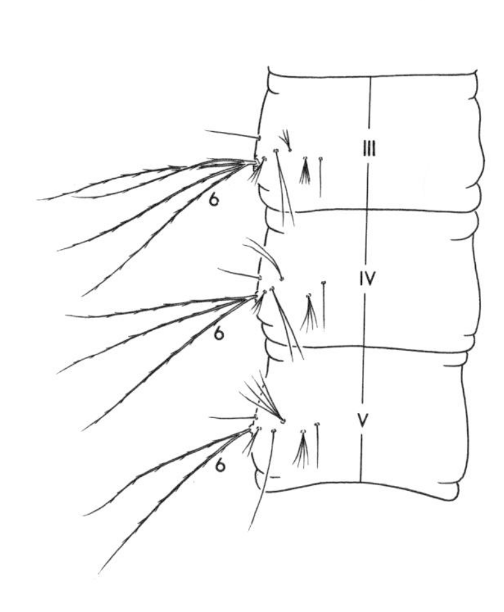


Fig. 626. Dorsal view of abdominal segments III–V: *Oc. taeniorhynchus*

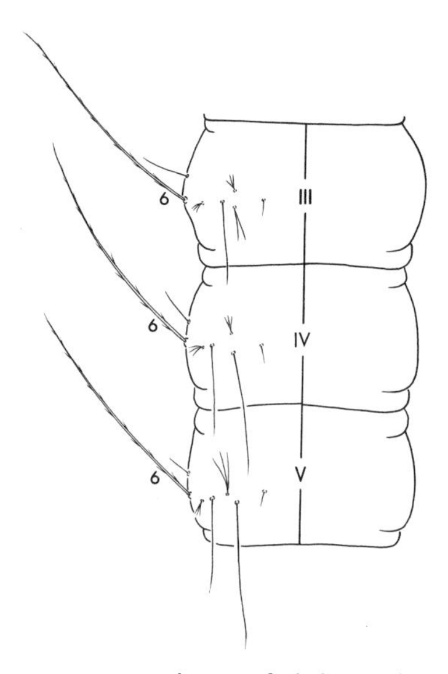


Fig. 628. Dorsal view of abdominal segments III–V: *Oc. scapularis*

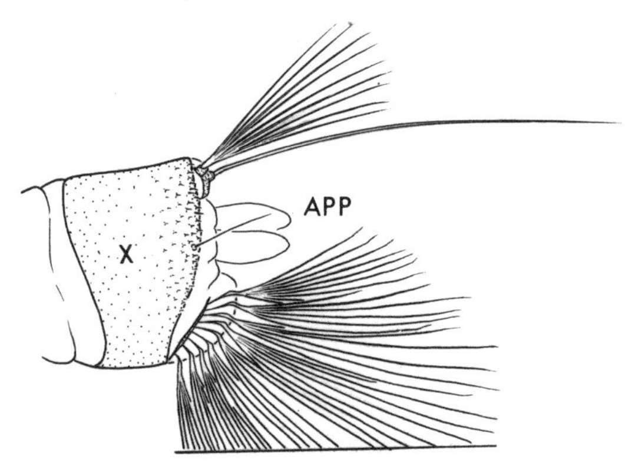


Fig. 627. Lateral view of abdominal segment X: *Oc. taeniorhynchus*

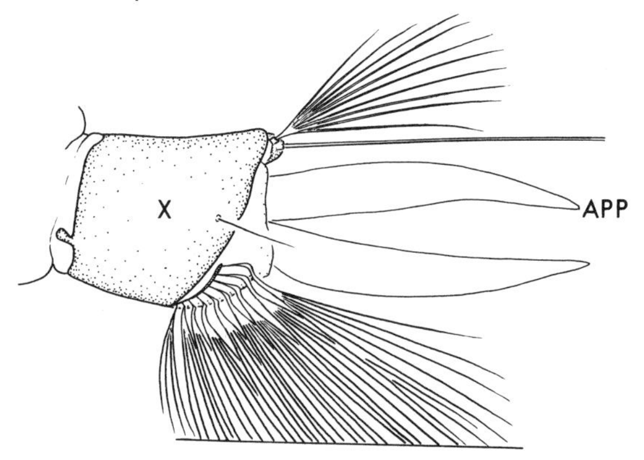


Fig. 629. Lateral view of abdominal segment X: *Oc. scapularis*

22(21). Seta 13-III long, single (fig. 630); thoracic integument densely aculeate (fig. 631) .......... ........................................................................................ *Oc. scapularis* (plate 23D)

Seta 13-III short, multibranched (fig. 632); thoracic integument sparsely aculeate (fig. 633) ........................................................................................ *Oc. tortilis* (plate 39C)

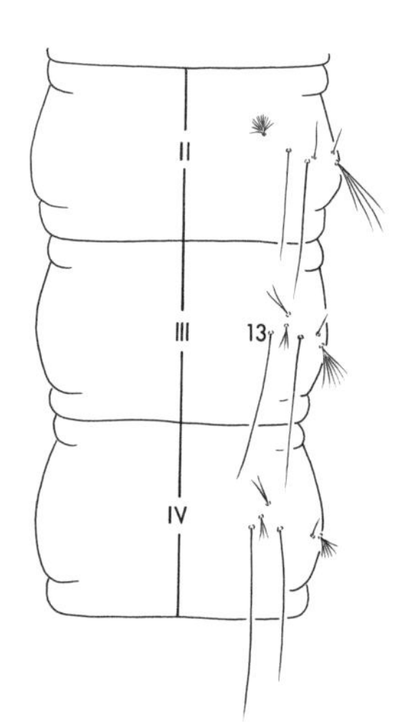


Fig. 630. Ventral view of abdominal sterna II–IV: *Oc. scapularis*

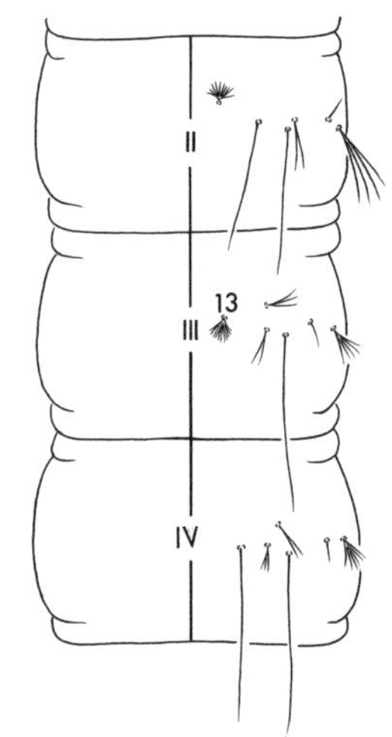


Fig. 632. Ventral view of sterna II–IV: *Oc. tortilis*

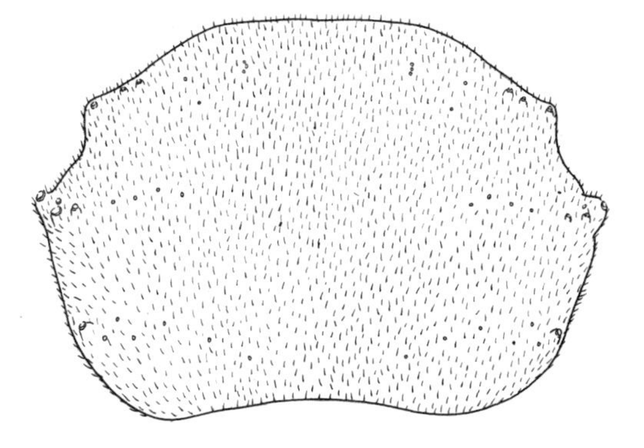

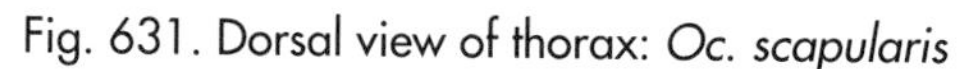

Fig. 631. Dorsal view of thorax: *Oc. scapularis*

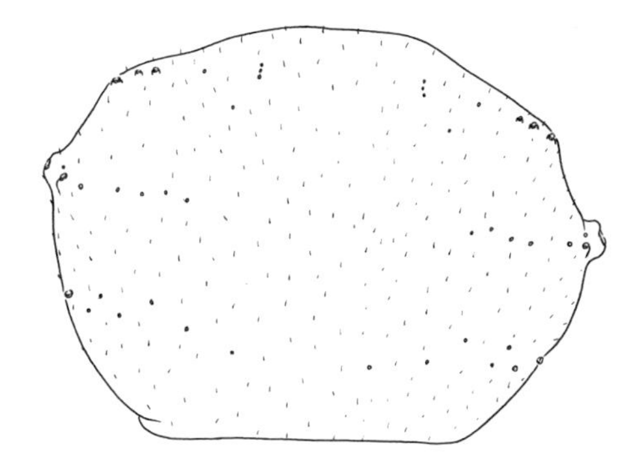

Fig. 633. Dorsal view of thorax: *Oc. tortilis*

23(4). Pecten on siphon with 1 or more spines detached distally (fig. 634) .......................... 24

Pecten with spines more or less evenly spaced (fig. 635) .......................................... 43

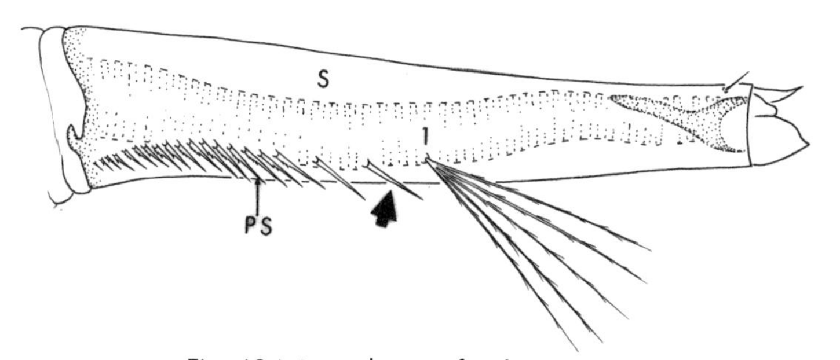


Fig. 634. Lateral view of siphon: *Oc. excrucians*

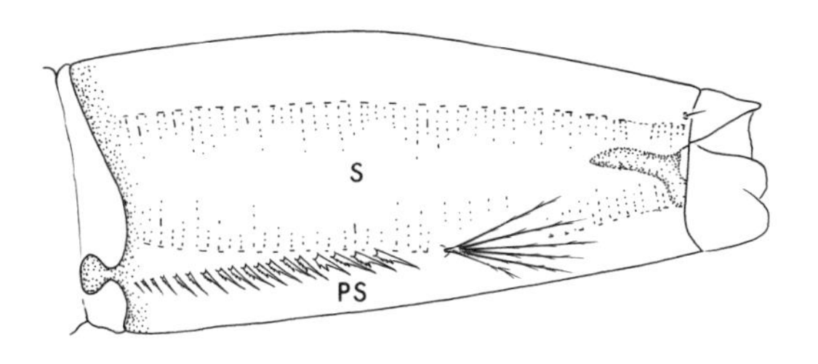


Fig. 635. Lateral view of siphon: *Oc. melanimon*

24(23). Seta 1-S attached within pecten (fig. 636) ................................................................ 25

Seta 1-S attached distal to pecten (fig. 637) ............................................................... 28

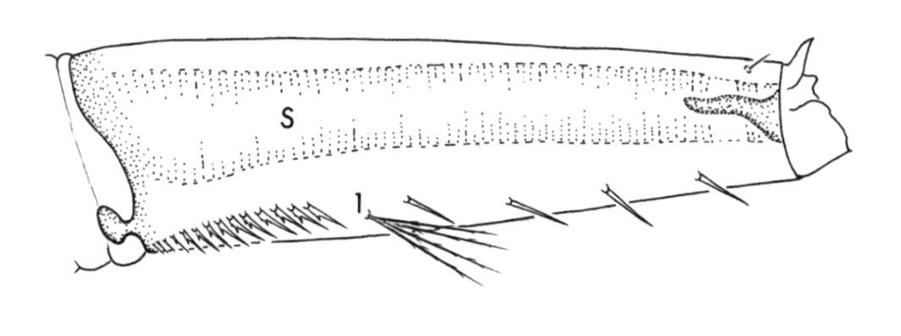


Fig. 636. Lateral view of siphon: *Oc. cataphylla*

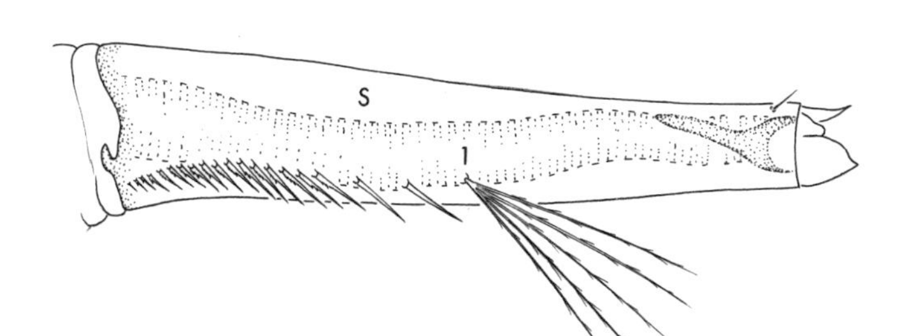


Fig. 637. Lateral view of siphon: *Oc. excrucians*

25(24). Setae 5,6-C with 3–6 branches, placed far forward, anterior to 7-C (fig. 638) ................
.......................................................................... *Oc. japonicus japonicus* (plate 39D)

Setae 5,6-C single or double, at least 5-C placed posterior to seta 7-C (fig. 639) ........ 26

Fig. 638. Dorsal view of head: *Oc. j. japonicus*

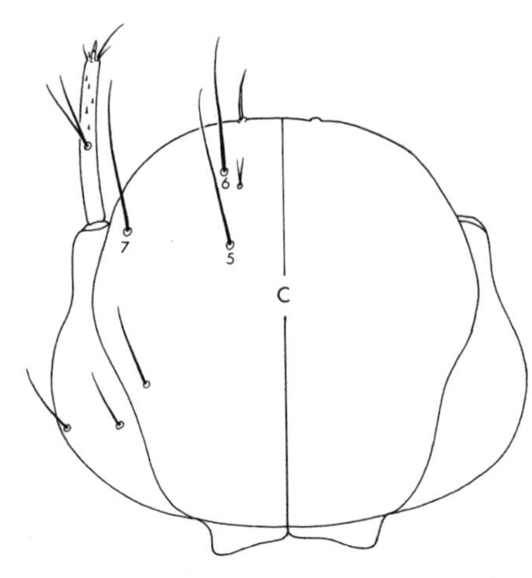


Fig. 639. Dorsal view of head: *Oc. atropalpus*

26(25). Comb scale with large apical spine and short lateral spinules (fig. 640); seta 1-X attached to saddle (fig. 641) .................................................. *Oc. cataphylla* (plate 11D)

Comb scales fringed with subequal spinules (fig. 642); seta 1-X attached ventral to saddle (fig. 643) .................................................. 27

Fig. 640. Comb scale: *Oc. cataphylla*

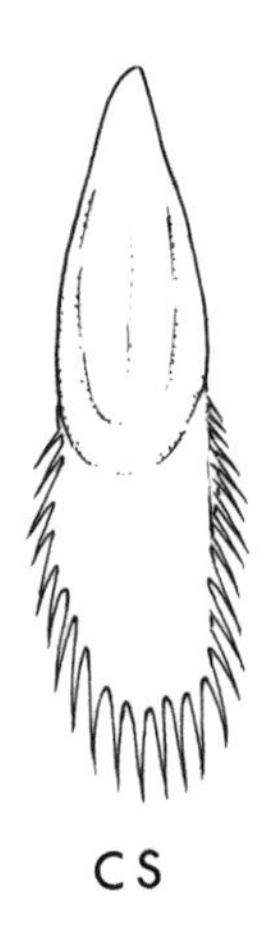


Fig. 642. Comb scale: *Oc. atropalpus*

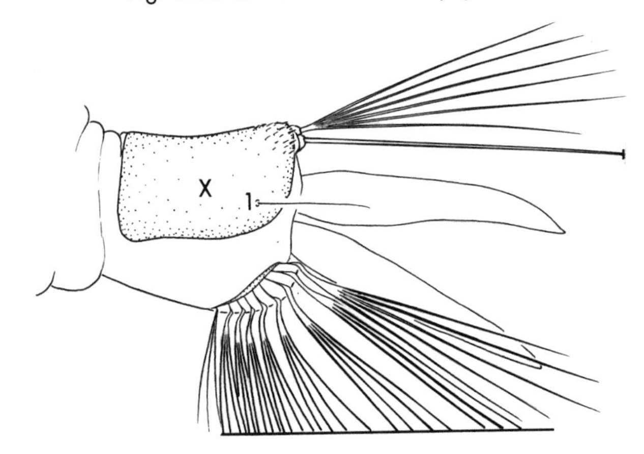


Fig. 641. Lateral view of abdominal segment X: *Oc. cataphylla*

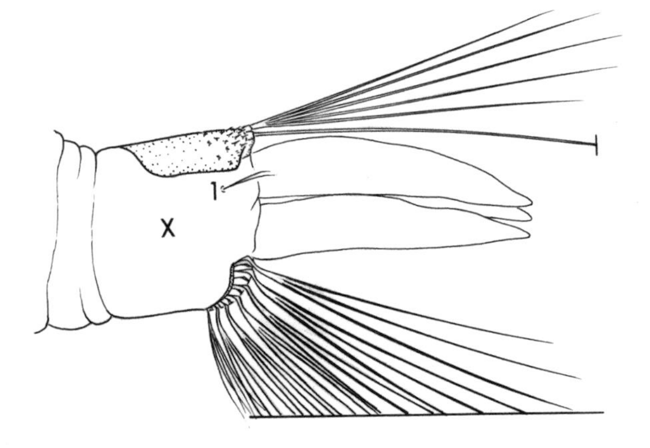


Fig. 643. Lateral view of abdominal segment X: *Oc. atropalpus*

27(26). Seta 1-M long, reaching anterior to level of seta 0-P (fig. 644); with 34 or more comb scales (usually 34–62) (fig. 645) ..................................................... *Oc. atropalpus* (plate 10C)

Seta 1-M short, reaching only near level of seta 0-P (fig. 646); usually with fewer than 34 comb scales (18–34) (fig. 647) .................................................. *Oc. epactius* (plate 13D)

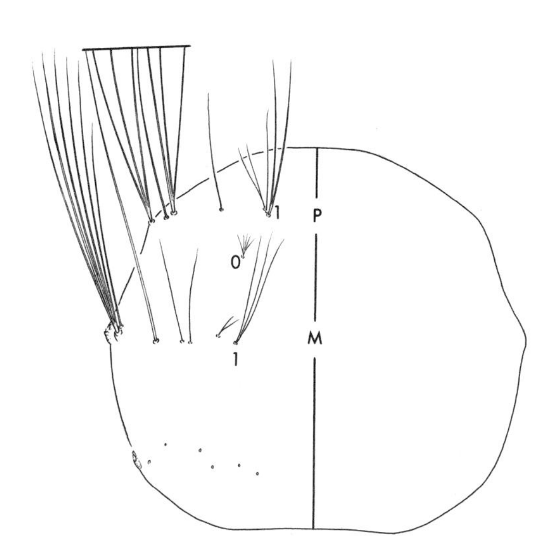


Fig. 644. Dorsal view of thorax: *Oc. atropalpus*

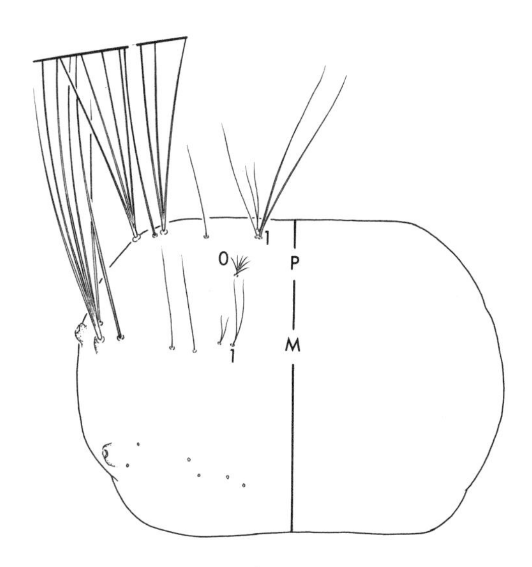


Fig. 646. Dorsal view of thorax: *Oc. epactius*

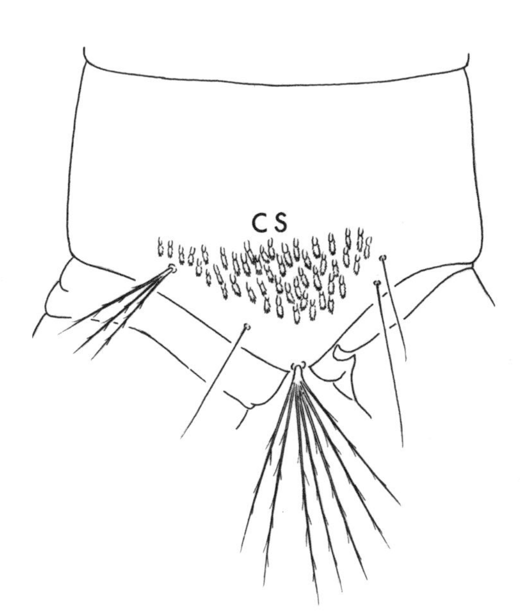


Fig. 645. Lateral view of abdominal segment VIII: *Oc. atropalpus*

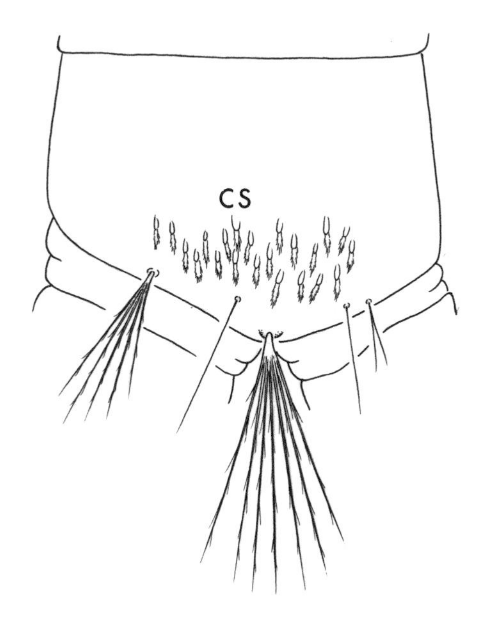


Fig. 647. Lateral view of abdominal segment VIII: *Oc. epactius*

28(24). Antenna equal to length of head capsule, or longer (fig. 648) ..................................... 29

Antenna shorter than head capsule (fig. 649) ............................................................ 30

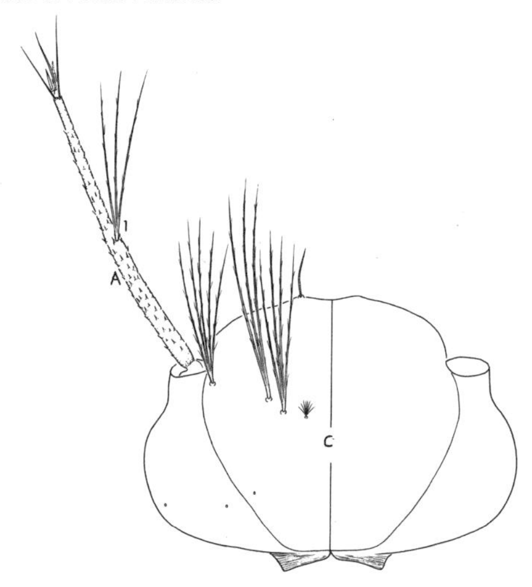


Fig. 648. Dorsal view of head: *Oc. diantaeus*

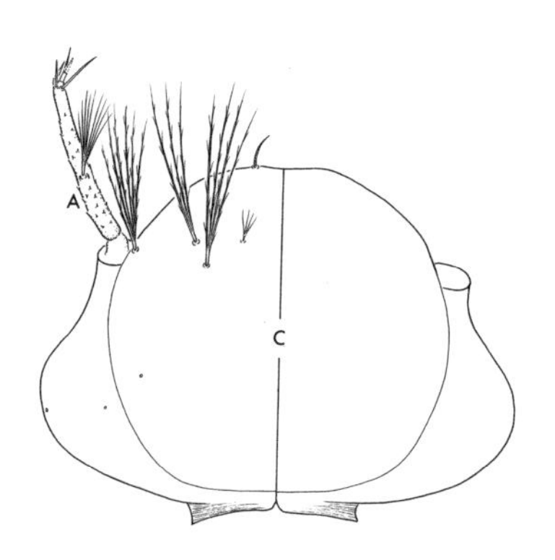


Fig. 649. Dorsal view of head: *Ae. vexans*

29(28). Seta 1-A attached near middle of antenna (fig. 650); with 15 or fewer comb scales in irregular row (fig. 651) ........................................................ *Oc. diantaeus* (plate 13A)

Seta 1-A attached to distal 0.4 of antenna (fig. 652); with 20 or more comb scales in patch (fig. 653) ........................................................................... *Oc. aurifer* (plate 10D)

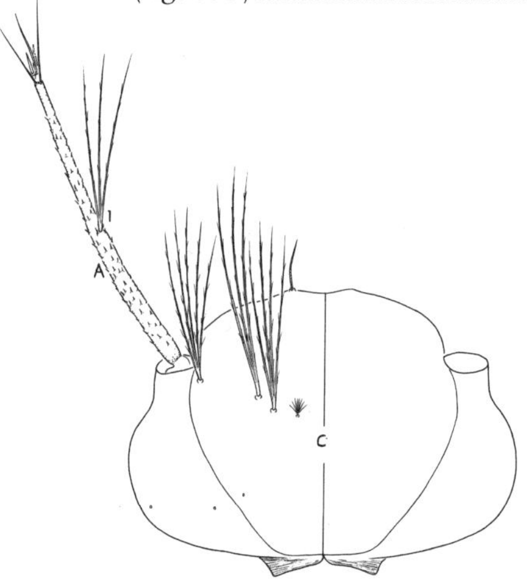


Fig. 650. Dorsal view of head: *Oc. diantaeus*

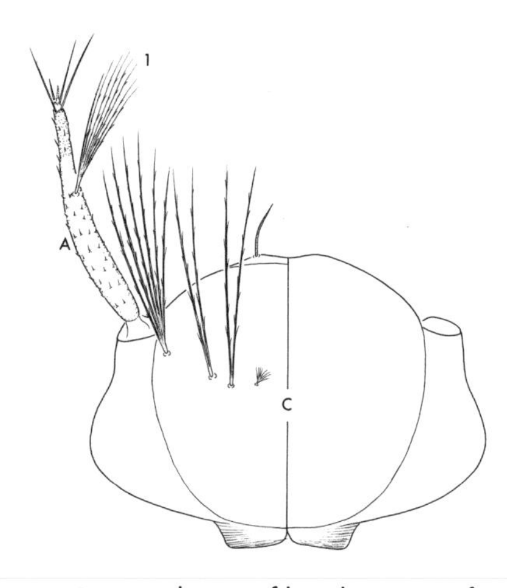


Fig. 652. Dorsal view of head: *Oc. aurifer*

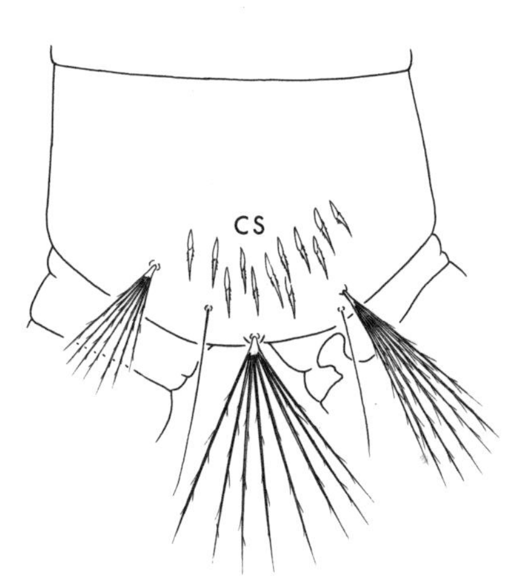


Fig. 651. Lateral view of abdominal segment VIII: *Oc. diantaeus*

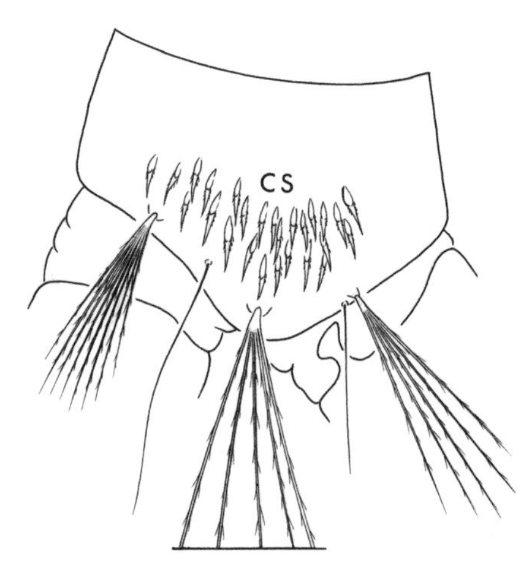


Fig. 653. Lateral view of abdominal segment VIII: *Oc. aurifer*

30(28). Thorax and abdomen with integument aculeate (fig. 654) .......................................... 31

Thorax and abdomen with integument glabrous (fig. 655) .......................................... 32

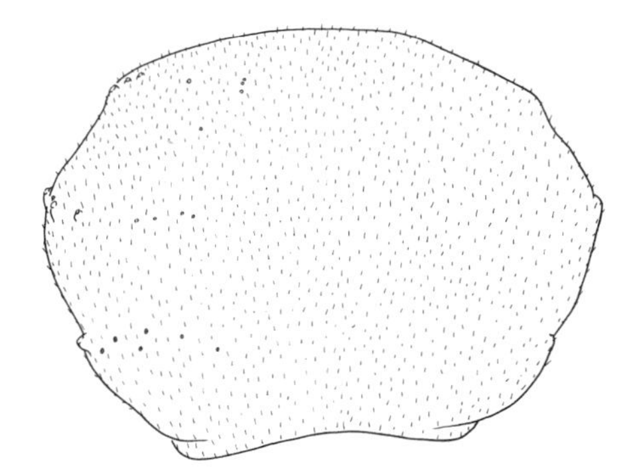

Fig. 654. Dorsal view of thorax: *Oc. s. spencerii*

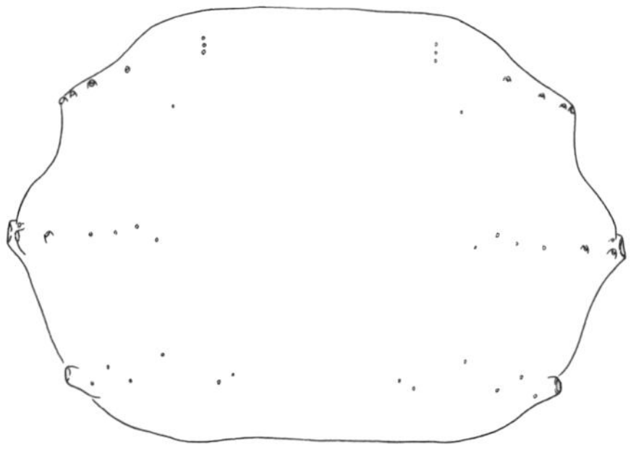

Fig. 655. Dorsal view of thorax: *Oc. campestris*

31(30). Comb scales 13 or fewer (fig. 656); median spine of comb scale broad at base (fig. 657) .. ........................................................................................ *Oc. s. spencerii* (plate 21C)

Comb scales 14 or more (fig. 658); median spine of comb scale narrow at base (fig. 659) ........................................................................................ *Oc. s. idahoensis* (plate 21B)

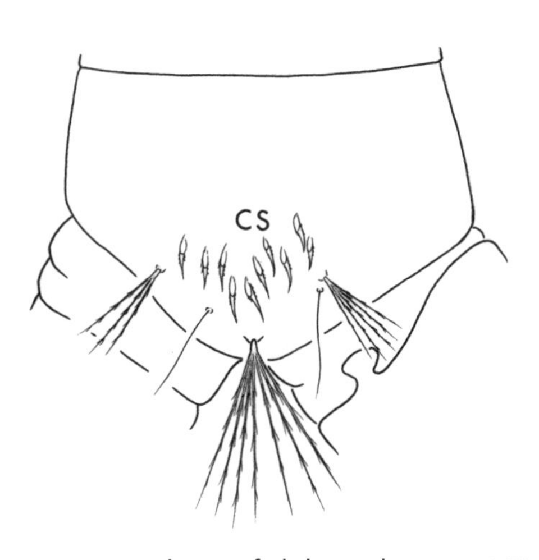


Fig. 656. Lateral view of abdominal segment VIII: *Oc. s. spencerii*

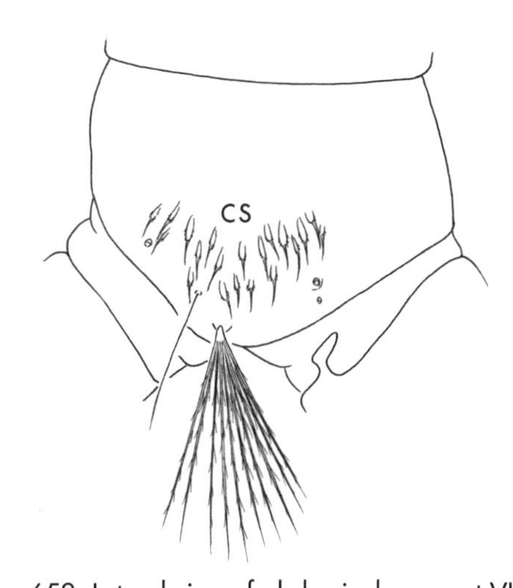


Fig. 658. Lateral view of abdominal segment VIII: *Oc. s. idahoensis*

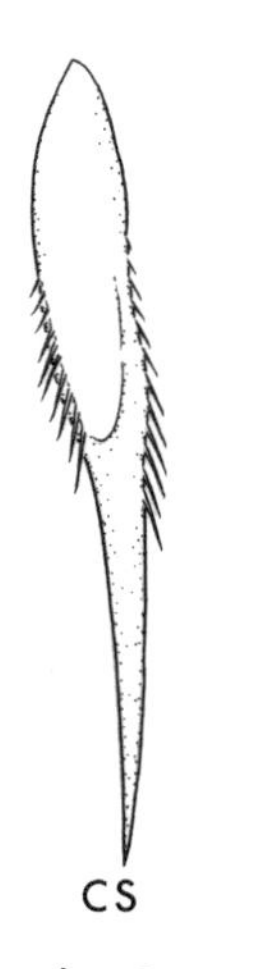


Fig. 657. Comb scale: *Oc. s. spencerii*

Fig. 659. Comb scale: *Oc. s. idahoensis*

32(30). Comb scales in patch of 18 or more (fig. 660) .......................................................... 33

Comb scales in single or irregular double row, usually 17 or fewer (fig. 661) ............ 36

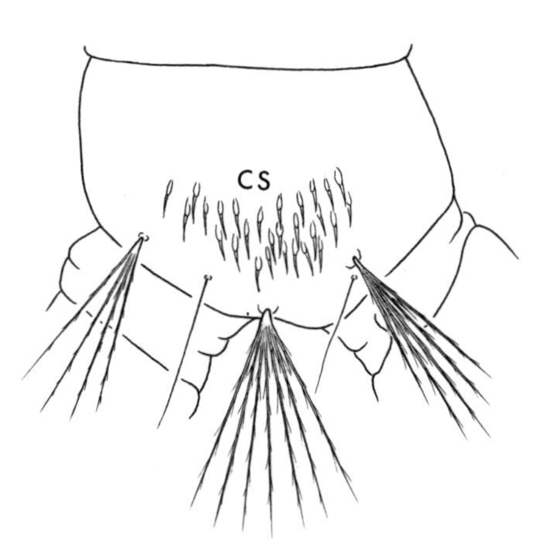


Fig. 660. Lateral view of abdominal segment VIII: *Oc. excrucians*

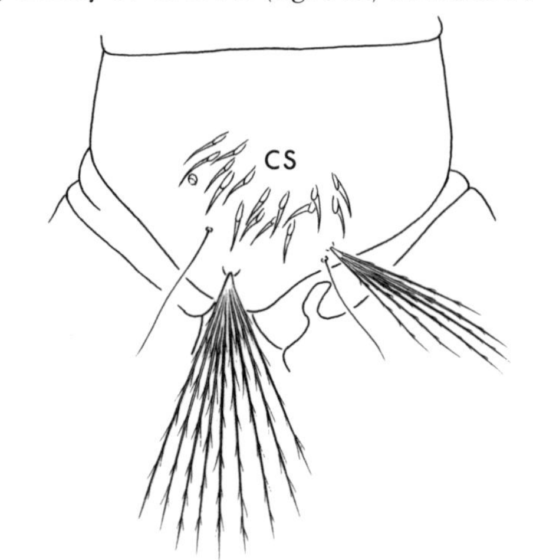


Fig. 661. Lateral view of abdominal segment VIII: *Oc. intrudens*

33(32). Siphon slender, index about 5.0 (fig. 662); seta 6 usually single on III–VI (fig. 663) ....... ................................................................................... *Oc. excrucians* (plate 14B)

Siphon stouter, index not more than 4.0 (fig. 664); seta 6 double on III–VI (fig. 665).... 34

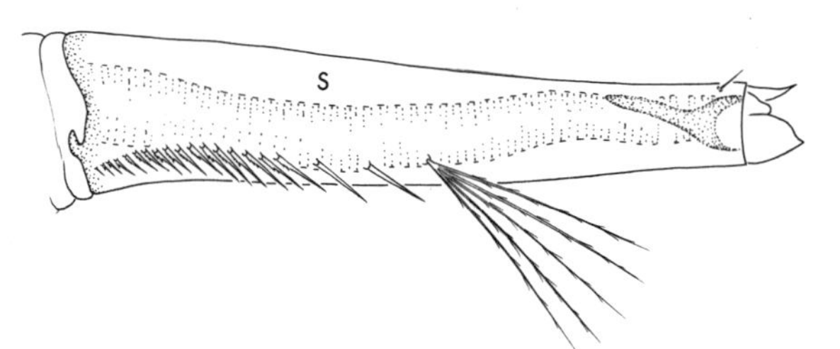


Fig. 662. Lateral view of siphon: *Oc. excrucians*

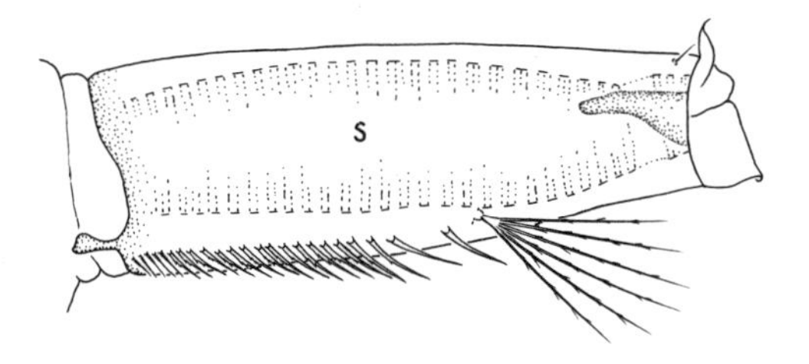


Fig. 664. Lateral view of siphon: *Oc. campestris*

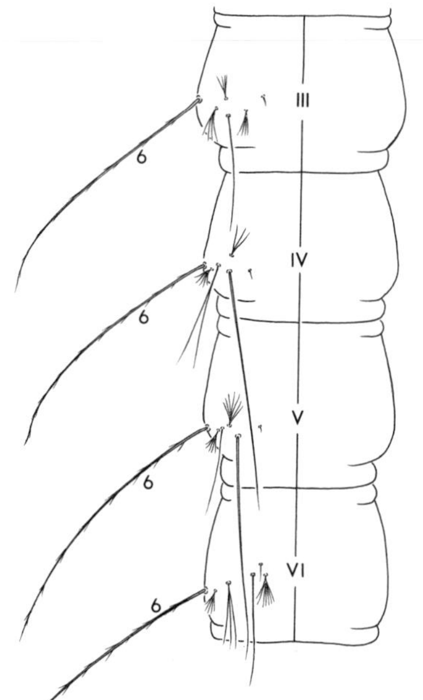


Fig. 663. Dorsal view of abdominal segments III–VI: *Oc. excrucians*

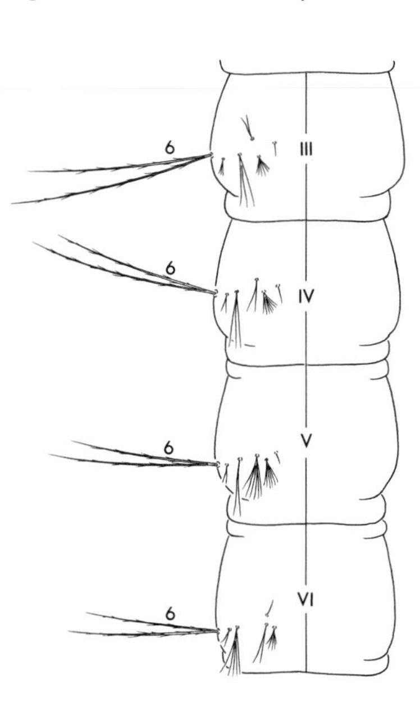


Fig. 665. Dorsal view of abdominal segments III–VI: *Oc. campestris*

34(33). Pecten reaching distal to middle of siphon (fig. 666); seta 1-M longer than antenna (fig. 667) .......................................................................... (in part) *Oc. campestris* (plate 11A)

Pecten not reaching middle of siphon (fig. 668); seta 1-M shorter than antenna (fig. 669) .......................................................................................................................... 35

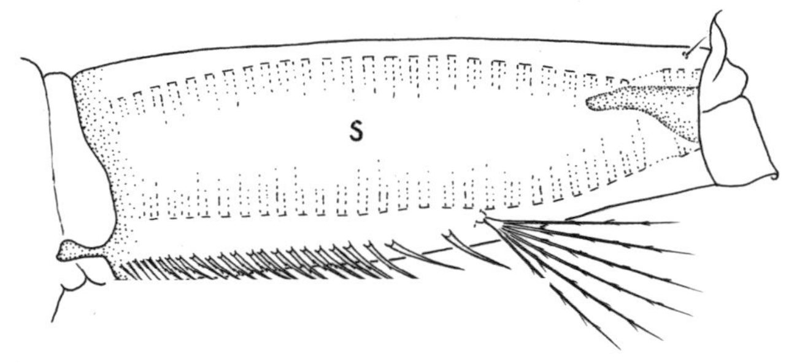


Fig. 666. Lateral view of siphon: *Oc. campestris*

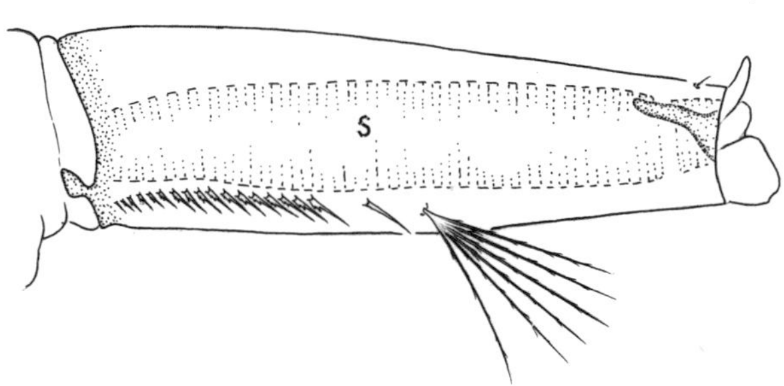


Fig. 668. Lateral view of siphon: *Oc. flavescens*

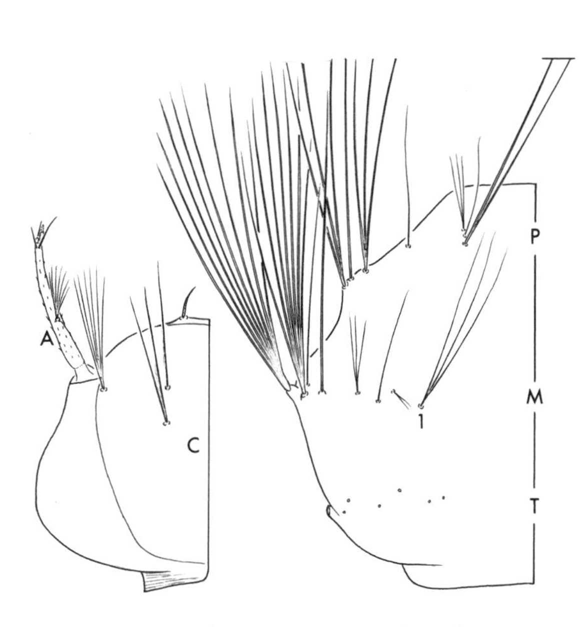


Fig. 667. Dorsal view of thorax and head: *Oc. campestris*

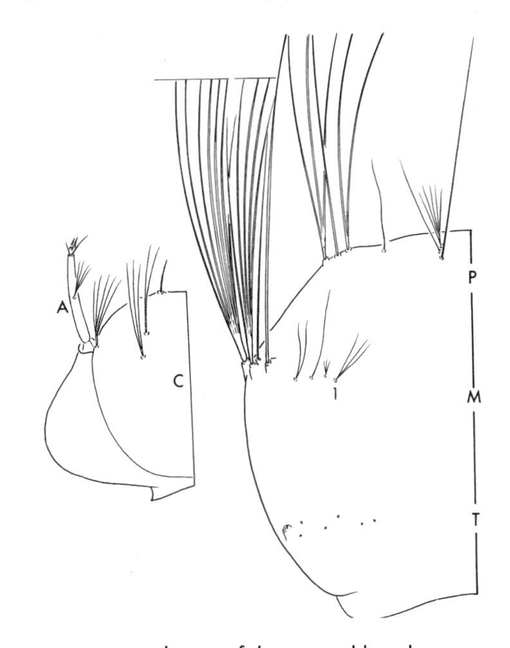


Fig. 669. Dorsal view of thorax and head: *Oc. flavescens*

35(34). Siphon index 3.5–4.0 (fig. 670); body integument glabrous (fig. 671) ........................... .......................................................................... (in part) *Oc. flavescens* (plate 14D)

Siphon index 4.5–5.0 (fig. 672); body integument aculeate (fig. 673) ............................ .................................................................................. *Oc. aloponotum* (plate 10A)

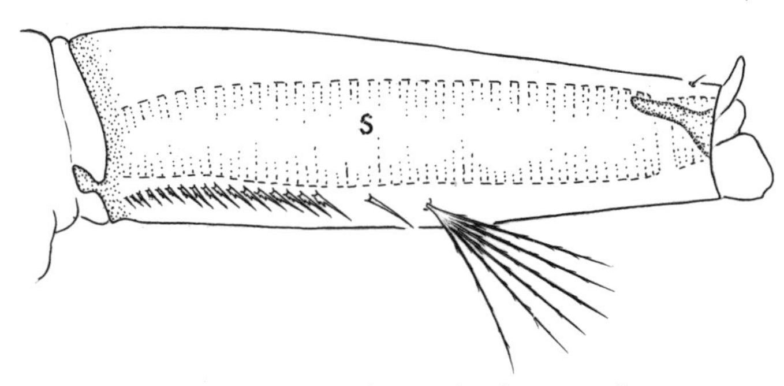


Fig. 670. Lateral view of siphon: *Oc. flavescens*

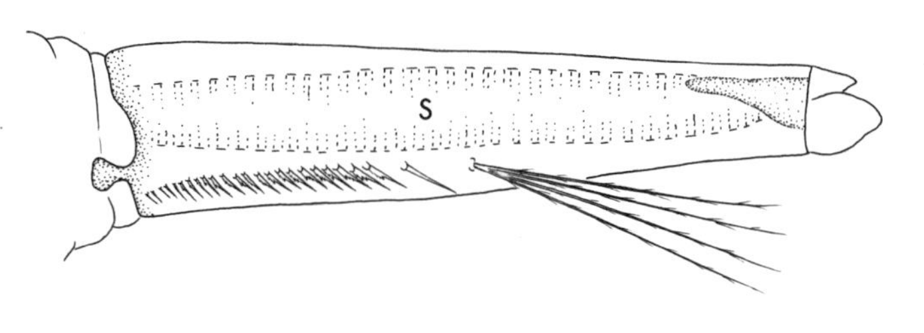


Fig. 672. Lateral view of siphon: *Oc. aloponotum*

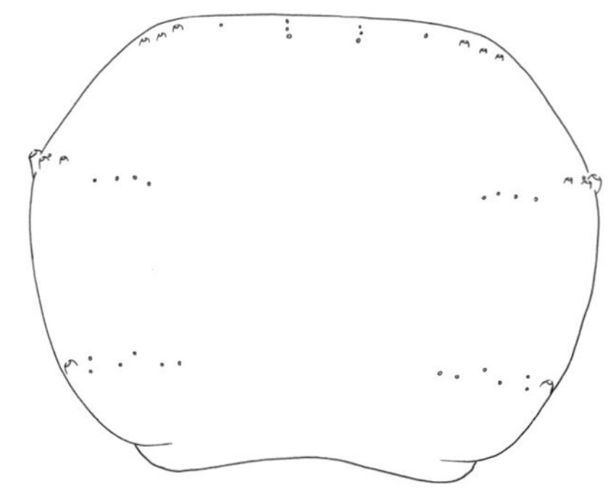

Fig. 671. Dorsal view of thorax: *Oc. flavescens*

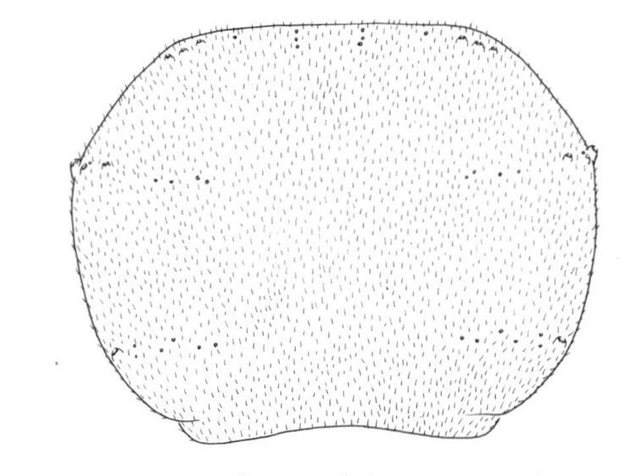

Fig. 673. Dorsal view of thorax: *Oc. aloponotum*

36(32). Seta 5-C with 3 or more branches (fig. 674) .......... 37

Seta 5-C single or double, rarely triple on both sides (fig. 675) .......... 39

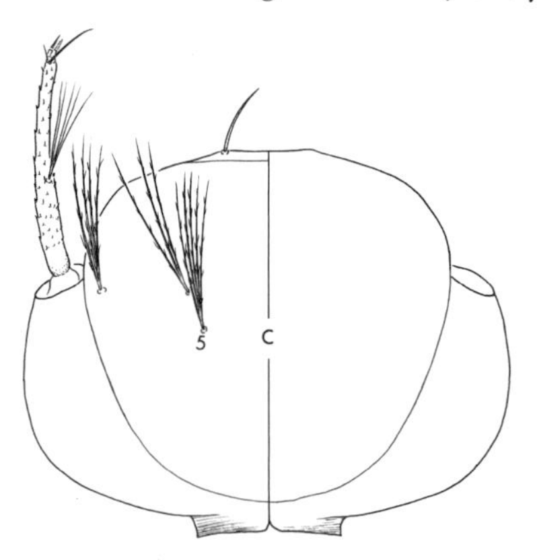


Fig. 674. Dorsal view of head: *Oc. intrudens*

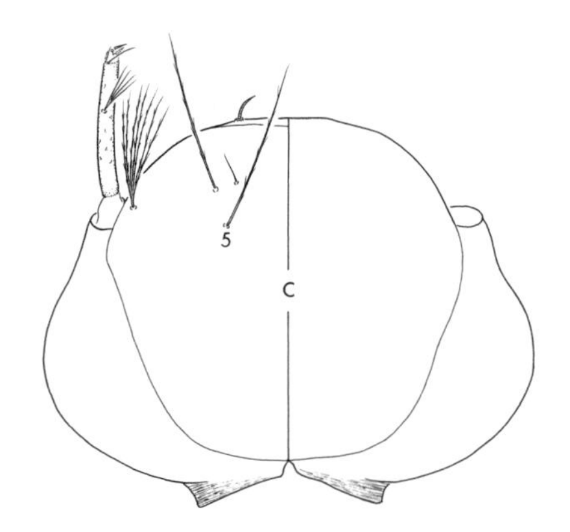


Fig. 675. Dorsal view of head: *Oc. niphadopsis*

37(36). Branches of seta 1-S rarely more than 0.5 length of basal diameter of siphon; saddle not incised on ventral margin (fig. 676) .......... *Ae. vexans* (plate 24A)

Branches of seta 1-S equal to length of basal diameter of siphon; saddle deeply incised on ventral margin (fig. 677) .......... 38

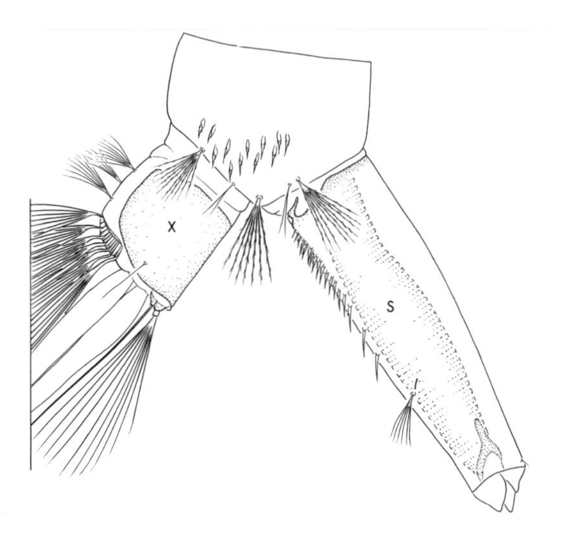


Fig. 676. Lateral view of abdominal segments VIII–X: *Ae. vexans*

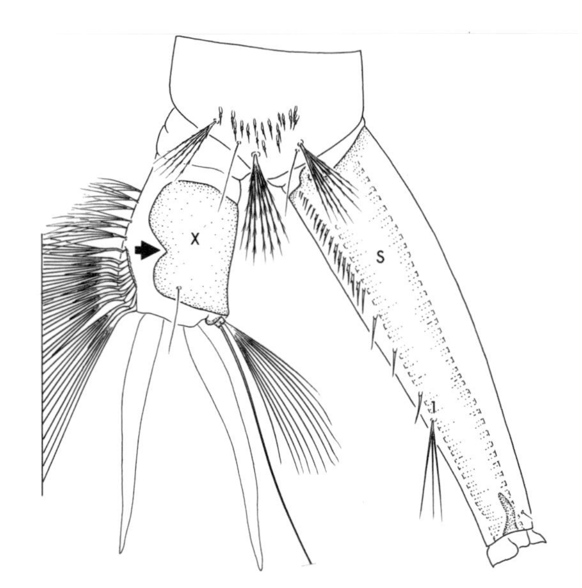


Fig. 677. Lateral view of abdominal segments VIII–X: *Oc. euedes*

38(37). Seta 6 usually single in III–VI (fig. 678); seta 1-S with 4 or more branches (fig. 679) ...... ........................................................................................ *Oc. intrudens* (plate 17A)

Seta 6 usually double on III–VI (fig. 680); seta 1-S double or triple (fig. 681) ................ ........................................................................................ (in part) *Oc. euedes* (plate 14A)

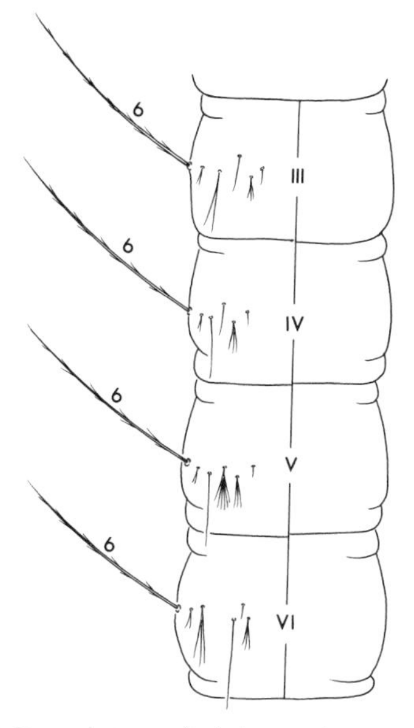


Fig. 678. Dorsal view of abdominal segments III–VI: *Oc. intrudens*

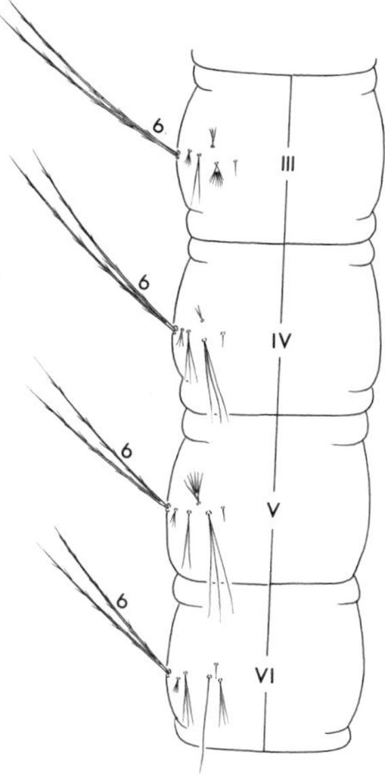


Fig. 680. Dorsal view of abdominal segments III–VI: *Oc. euedes*

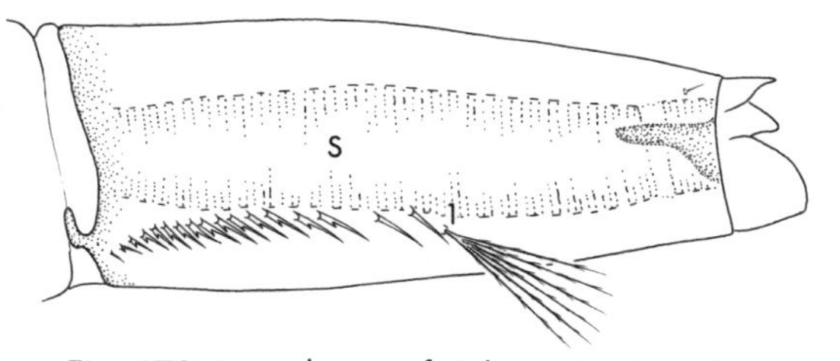


Fig. 679. Lateral view of siphon: *Oc. intrudens*

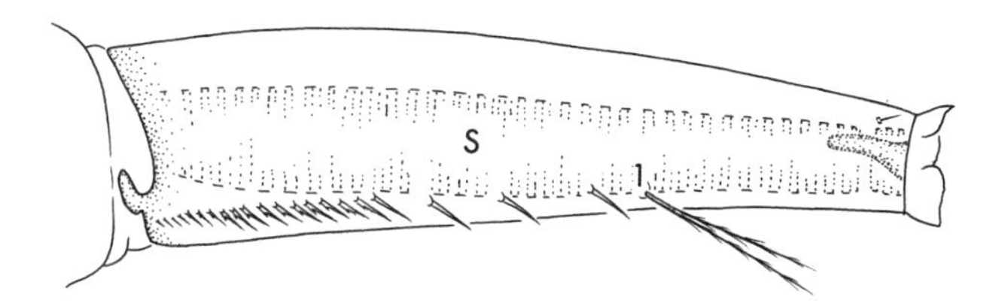


Fig. 681. Lateral view of siphon: *Oc. euedes*

39(36). Antenna at least 0.6 length of head capsule; setae 5–7-C coarse, of about equal diameter throughout (fig. 682) ............................................................ *Oc. decticus* (plate 12D)

Antenna not more than 0.5 length of head capsule; setae 5–7-C gradually tapering apically (fig. 683) ........................................................................................................ 40

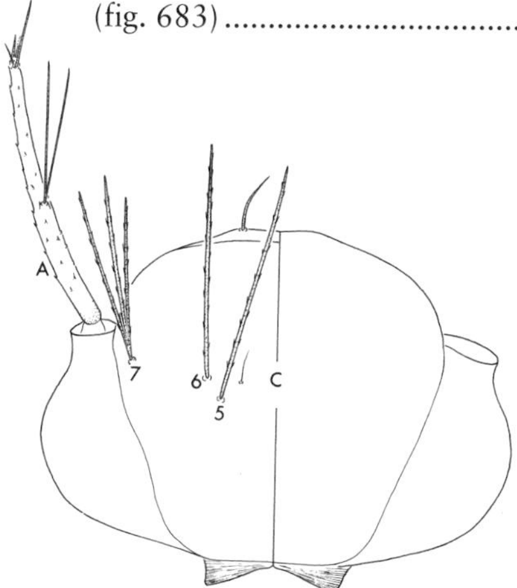


Fig. 682. Dorsal view of head: *Oc. decticus*

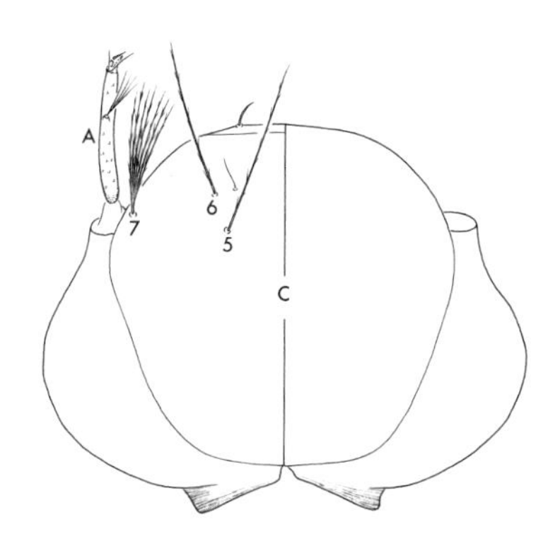


Fig. 683. Dorsal view of head: *Oc. niphadopsis*

40(39). Median spine of comb scale 2.0 length of subapical spinules (fig. 684); seta 1-M equal to 3-M, or longer (fig. 685); pecten confined to basal 0.3 of siphon (fig. 686) .................... ........................................................................................ *Oc. niphadopsis* (plate 40A)

Median spine of comb scale 4.0 length of subapical spinules, or more (fig. 687); seta 1-M shorter than 3-M (fig. 688); pecten on basal 0.5 of siphon, or more (fig. 689) ........... 41

Fig. 684. Comb scale: *Oc. niphadopsis*

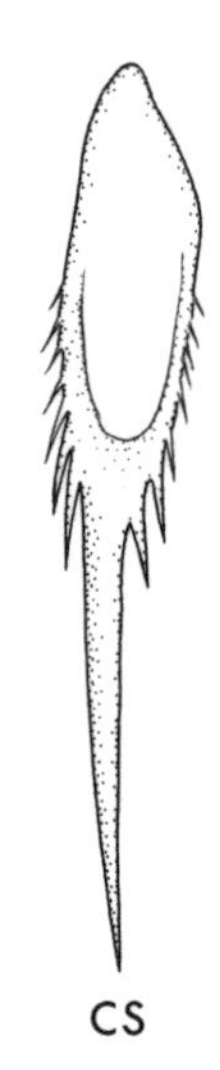


Fig. 687. Comb scale: *Oc. riparius*

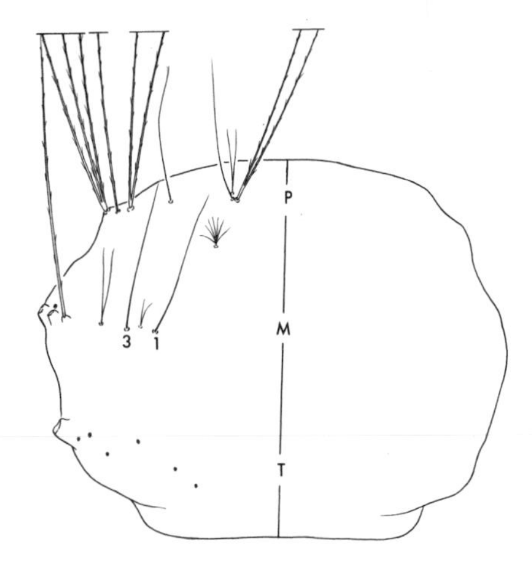


Fig. 685. Dorsal view of thorax: *Oc. niphadopsis*

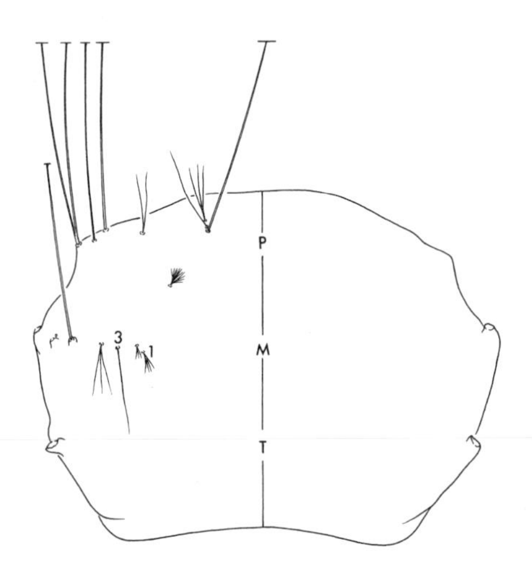


Fig. 688. Dorsal view of thorax: *Oc. riparius*

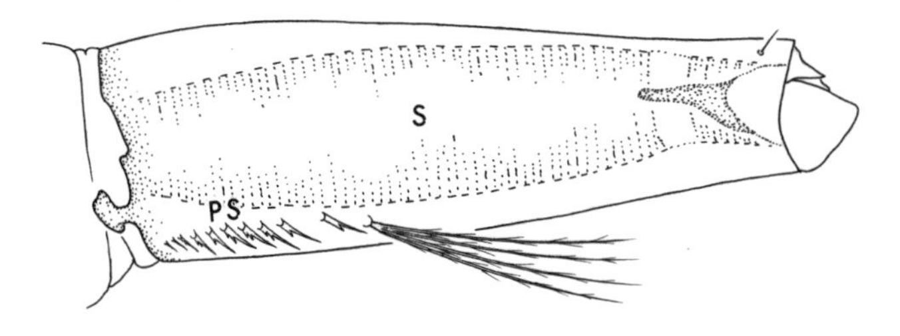


Fig. 686. Lateral view of siphon: *Oc. niphadopsis*

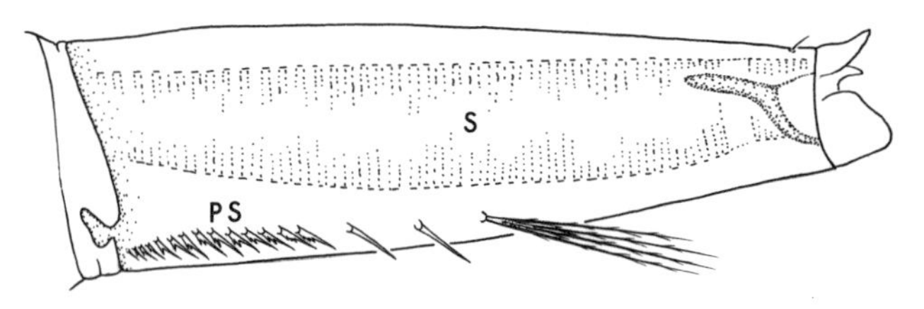


Fig. 689. Lateral view of siphon: *Oc. riparius*

41(40). Comb with 12 or more scales; pecten on siphon with 18 or more spines (fig. 690) ......... ......................................................................................... (in part) *Oc. euedes* (plate 14A)

Comb with 11 or fewer scales; pecten with 17 or fewer spines (fig. 691) .................... 42

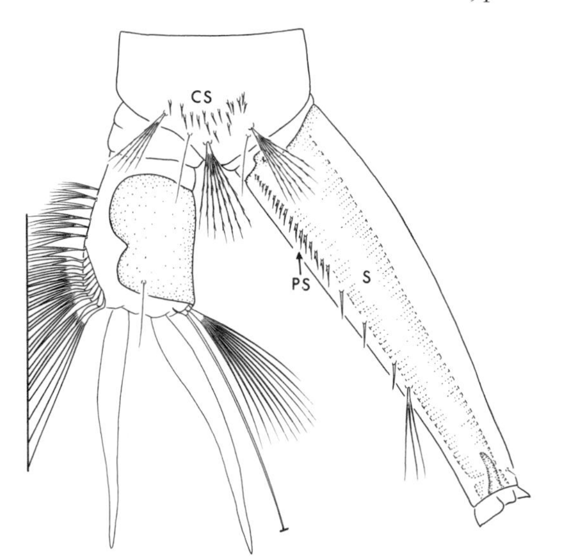


Fig. 690. Lateral view of abdominal segments VIII–X: *Oc. euedes*

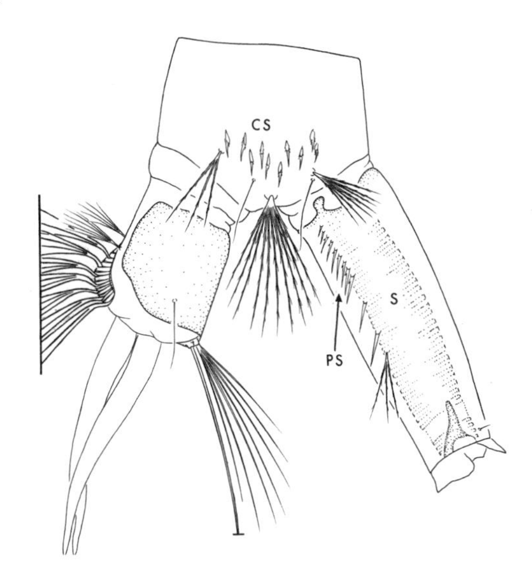


Fig. 691. Lateral view of abdominal segments VIII–X: *Oc. ventrovittis*

42(41). Setae 5,6-C double (fig. 692); saddle incised along ventral margin (fig. 693) ................. ......................................................................................... *Oc. riparius* (plate 20B)

Setae 5,6-C single (fig. 694); saddle not incised on ventral margin (fig. 695) ................. ......................................................................................... *Oc. ventrovittis* (plate 23D)

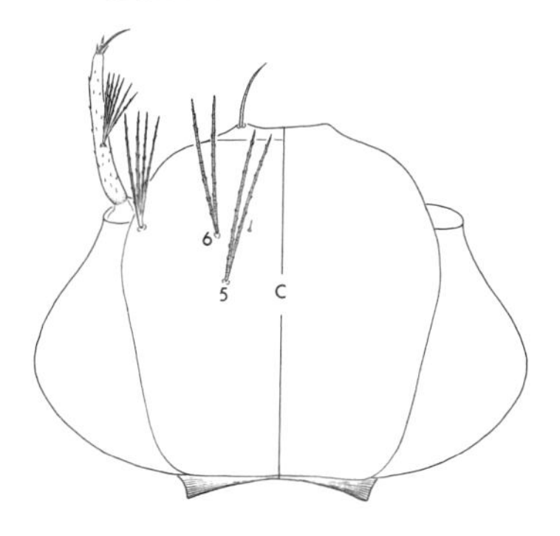


Fig. 692. Dorsal view of head: *Oc. riparius*

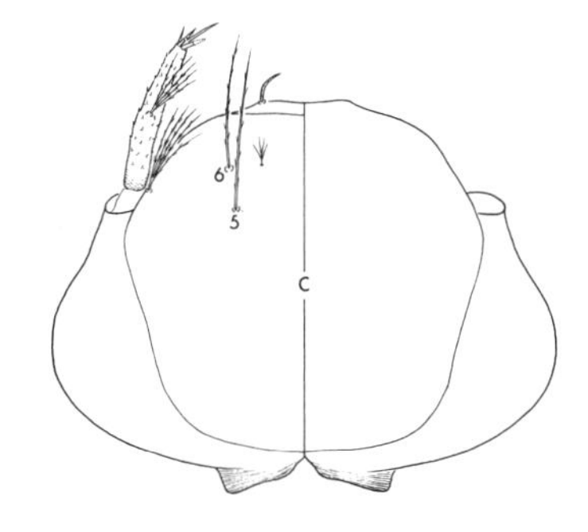


Fig. 694. Dorsal view of head: *Oc. ventrovittis*

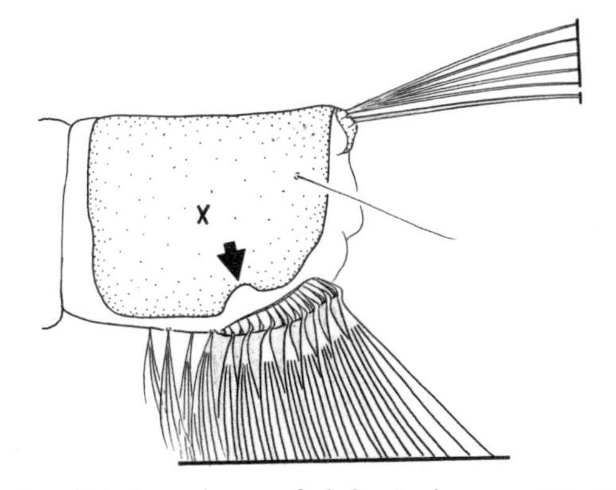


Fig. 693. Lateral view of abdominal segment X: *Oc. riparius*

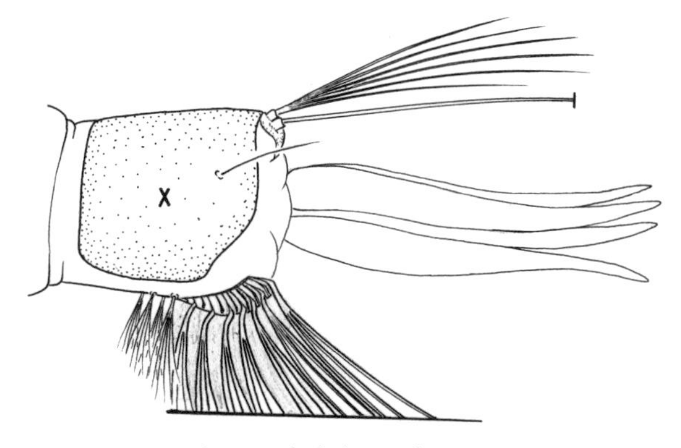


Fig. 695. Lateral view of abdominal segment X: *Oc. ventrovittis*

43(23). Seta 1-A single or double, antenna usually smooth, or with tiny spinules (fig. 696) .... 44

Seta 1-A with more than 3 branches, antenna with prominent coarse spinules (fig. 697) ... ............................................................................................................ 58

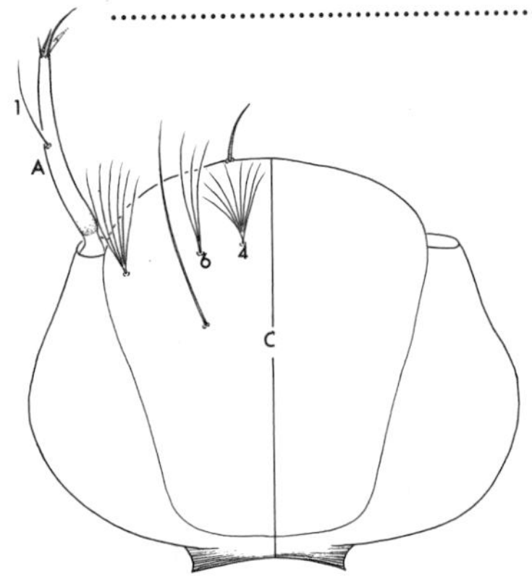

Fig. 696. Dorsal view of head: *Oc. triseriatus*

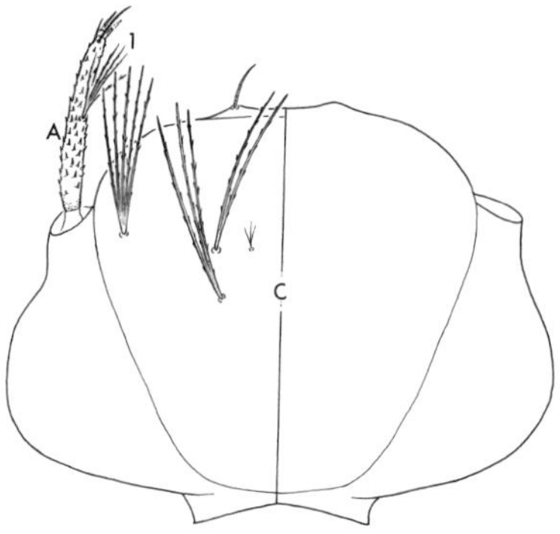

Fig. 697. Dorsal view of head: *Oc. fitchii*

44(43). Comb with pointed unfringed median spine (fig. 698) ............................................ 45

Comb scale rather blunt apically, evenly fringed with short spinules (fig. 699) .......... 49

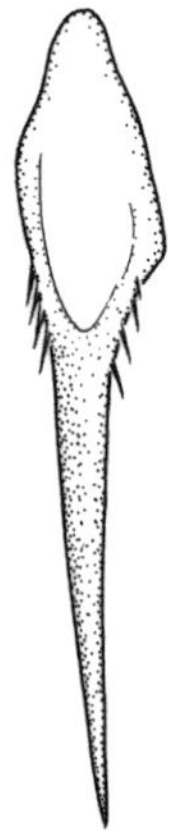

Fig. 698. Comb scale: *Oc. purpureipes*

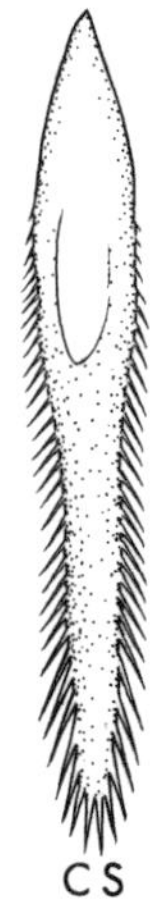

Fig. 699. Comb scale: *Oc. triseriatus*

45(44). Seta 1-A short, not reaching more than 0.75 of distance to apex of antenna (fig. 700); siphon without acus (fig. 701) .......................................................................... 46

Seta 1-A long, at least reaching near apex of antenna (fig. 702); siphon with large acus (fig. 703) ........................................................................................................ 48

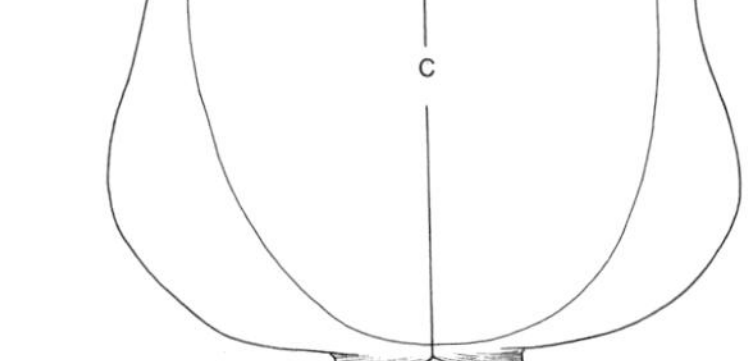


Fig. 700. Dorsal view of head: *Ae. aegypti*

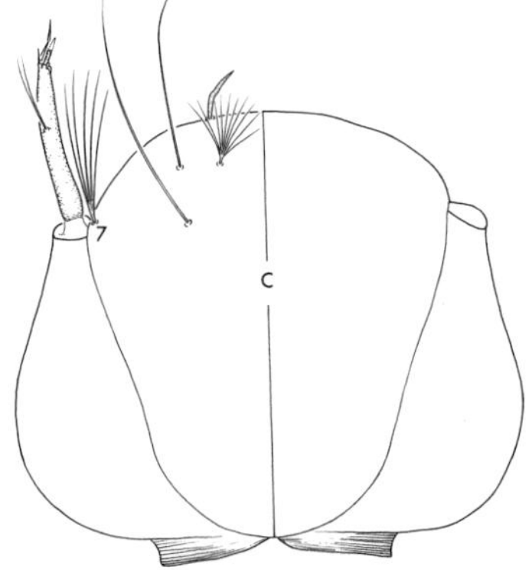


Fig. 702. Dorsal view of head: *Oc. muelleri*

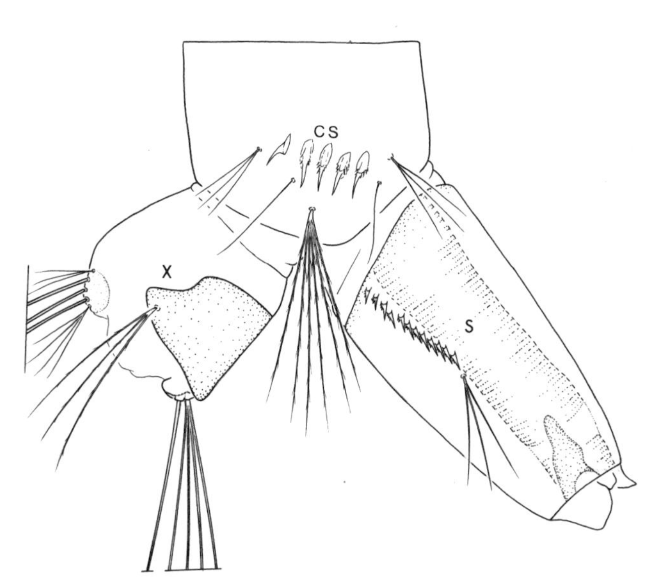


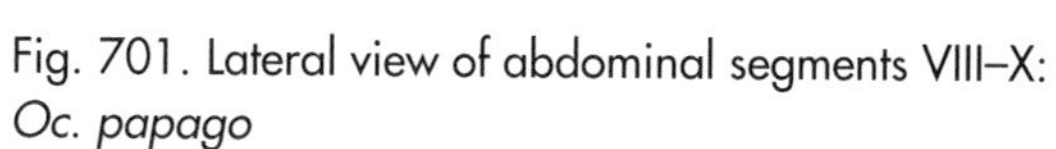
Fig. 701. Lateral view of abdominal segments VIII–X: *Oc. papago*

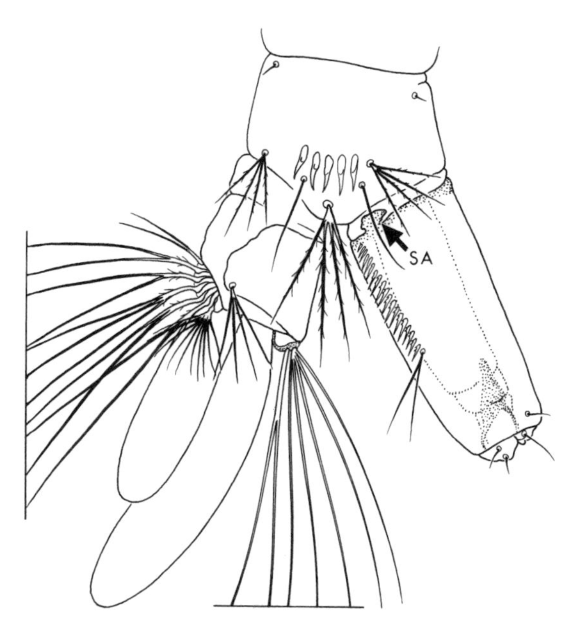


Fig. 703. Lateral view of abdominal segments VIII–X: *Oc. purpureipes*

46(45). Abdominal segment VIII with 3–5 comb scales (fig. 704); seta 1-C stout, broad, short (fig. 705) .......................................................................................... *Oc. papago* (plate 40B)

Abdominal segment VIII with 6–12 comb scales (fig. 706); seta 1-C long, thin (fig. 707) .. .......................................................................................................................... 47

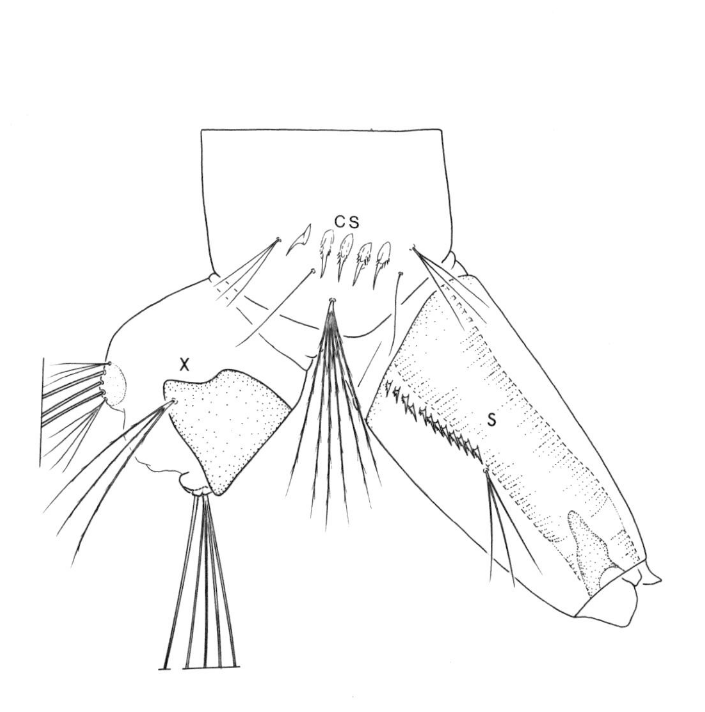


Fig. 704. Lateral view of abdominal segments VIII–X: *Oc. papago*)

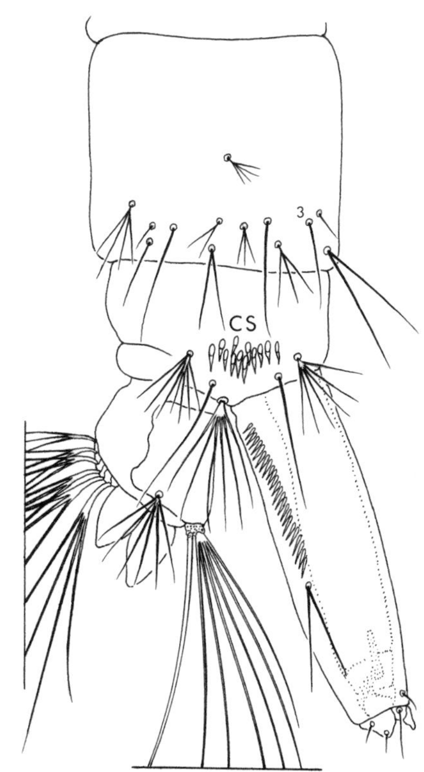


Fig. 706. Lateral view of abdominal segments VIII–X: *Oc. triseriatus*

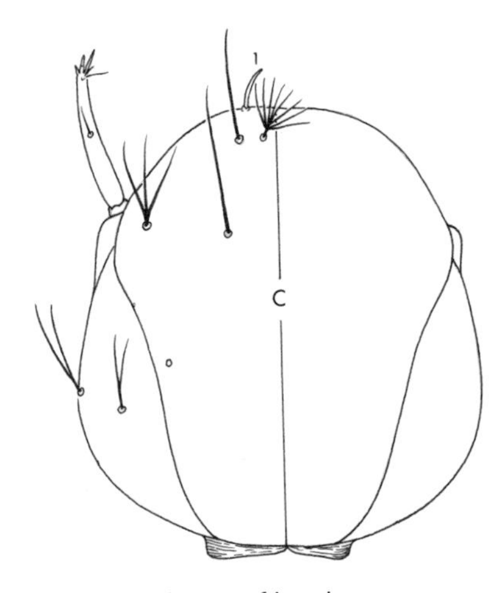


Fig. 705. Dorsal view of head: *Oc. papago*

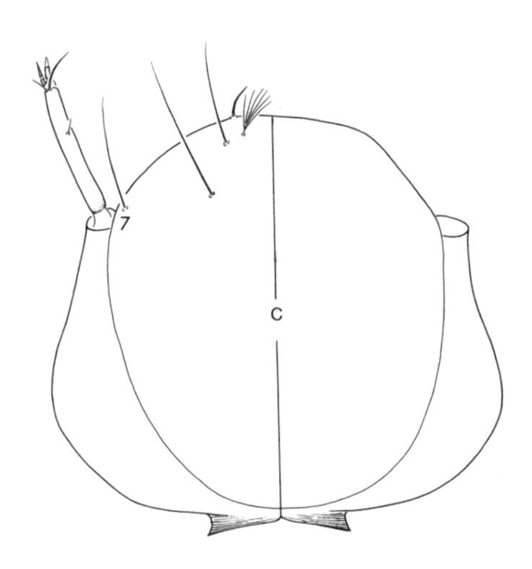


Fig. 707. Dorsal view of head: *Ae. aegypti*

47(46). Comb scales with strong subapical spines (fig. 708); setal support plate of setae 9–12-M,T with prominent spine (fig. 709); seta 7-C single (fig. 710) ............. *Ae. aegypti* (plate 9B)

Comb scales with laterobasal fringe of fine spicules (fig. 711); setal support plate of setae 9–12-M,T with short thin spine (fig. 712); seta 7-C double (fig. 713) ............................ ........................................................................................ *Ae. albopictus* (plate 9D)

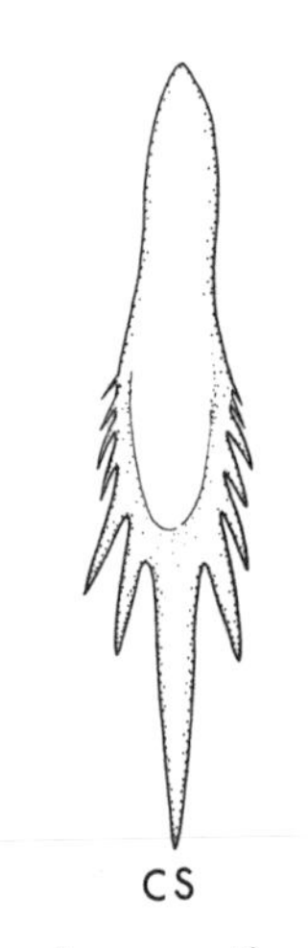


Fig. 708. Comb scale: *Ae. aegypti*

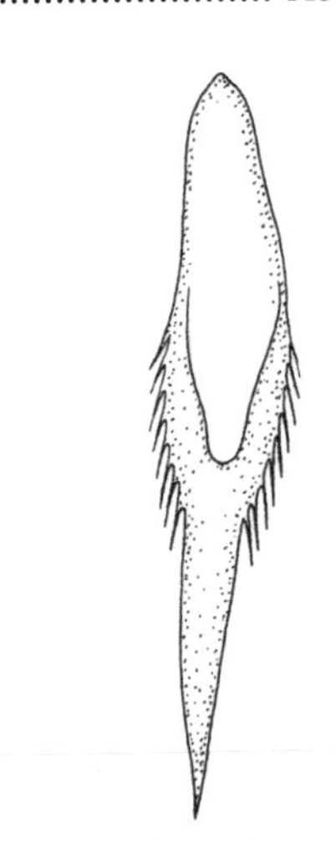

Fig. 711. Comb scale: *Ae. albopictus*

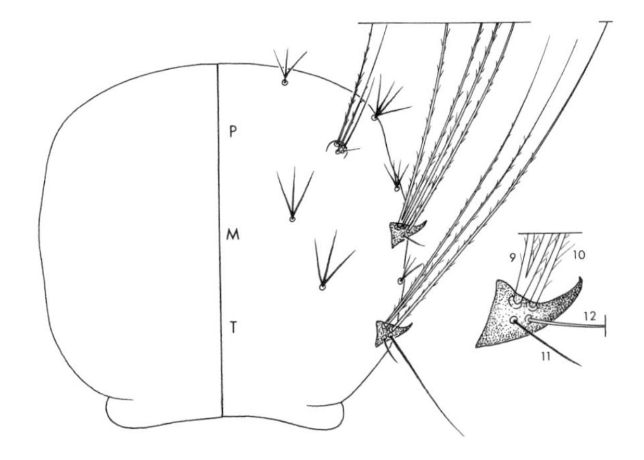


Fig. 709. Ventral view of thorax: *Ae. aegypti*

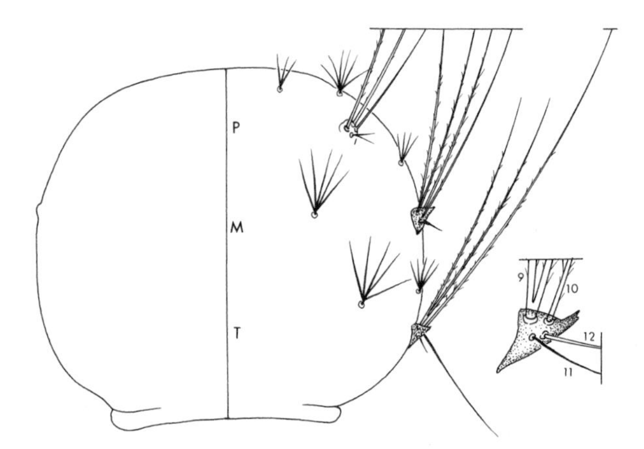


Fig. 712. Ventral view of thorax: *Ae. albopictus*

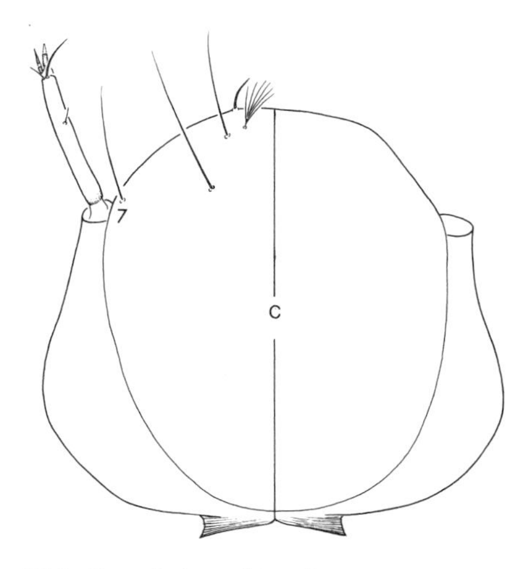

Fig. 710. Dorsal view of head: *Ae. aegypti*

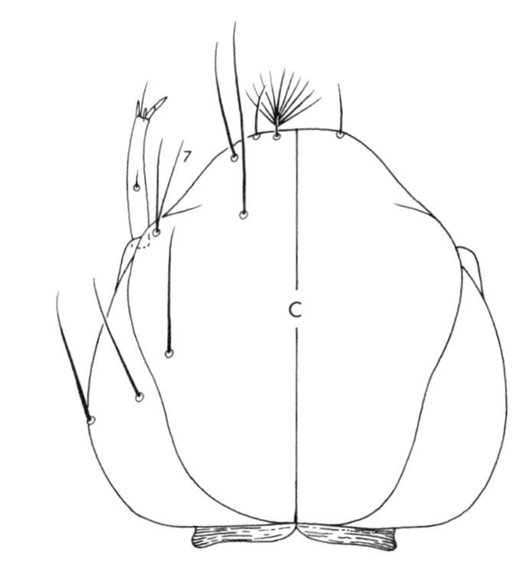


Fig. 713. Dorsal view of head: *Ae. albopictus*

48(45). Integument of thorax and abdomen aculeate (fig. 714); with 3–7 comb scales (fig. 715) ... .................................................................................... *Oc. purpureipes* (plate 40C)

Integument of thorax and abdomen glabrous (fig. 716); with 8–12 comb scales (fig. 717) .................................................................................... *Oc. muelleri* (plate 39C)

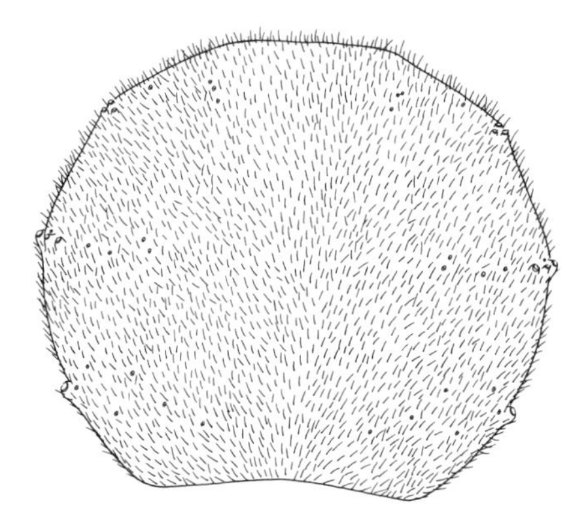
Fig. 714. Dorsal view of thorax: *Oc. purpureipes*

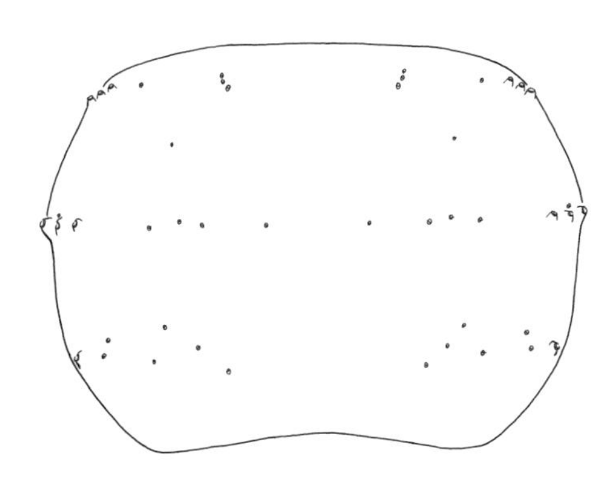
Fig. 716. Dorsal view of thorax: *Ae. aegypti*

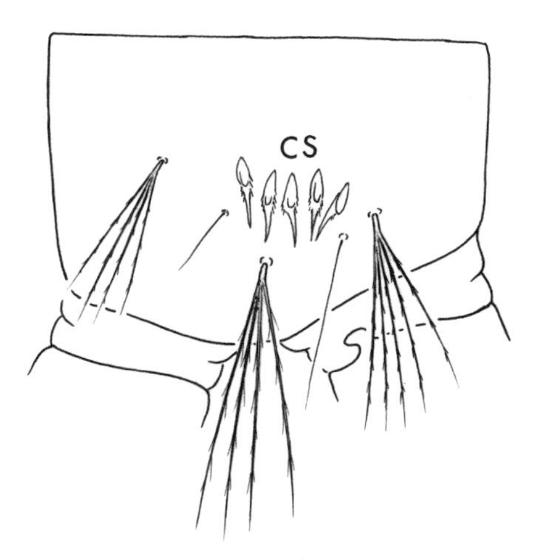


Fig. 715. Lateral view of abdominal segment VIII: *Oc. purpureipes*

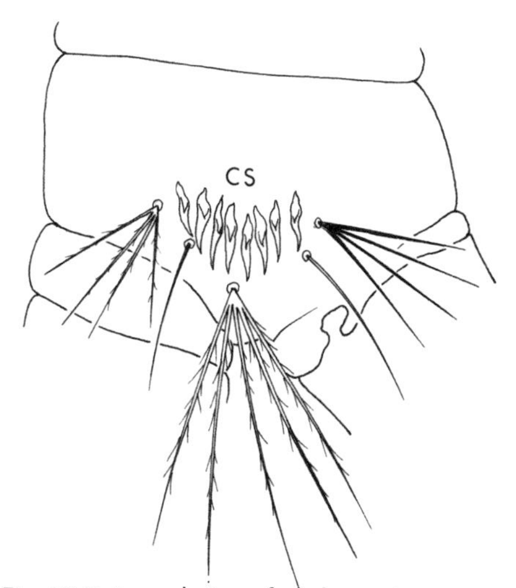


Fig. 717. Lateral view of abdominal segment VIII: *Oc. muelleri*

49(44). Seta 4-X (ventral brush) with 2 most caudal setae single or double, usually with total of 6 fanlike setae (fig. 718) ........................................................ 50

Seta 4-X with at least one of the 2 most caudal setae 3-branched or more, usually with a total of either 5 or 7 fanlike setae (fig. 719) ........................................................ 53

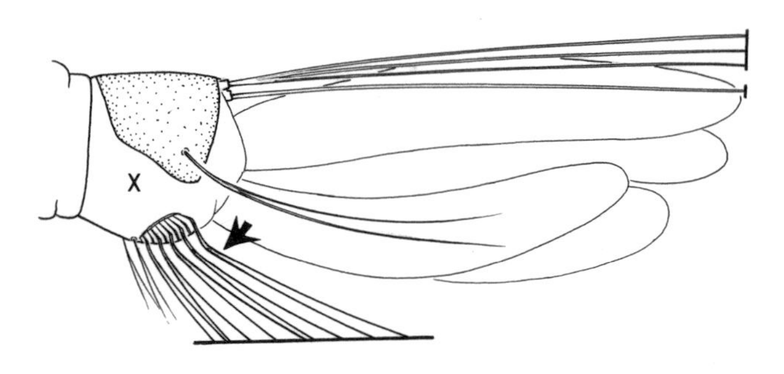


Fig. 718. Lateral view of abdominal segment X: *Oc. sierrensis*

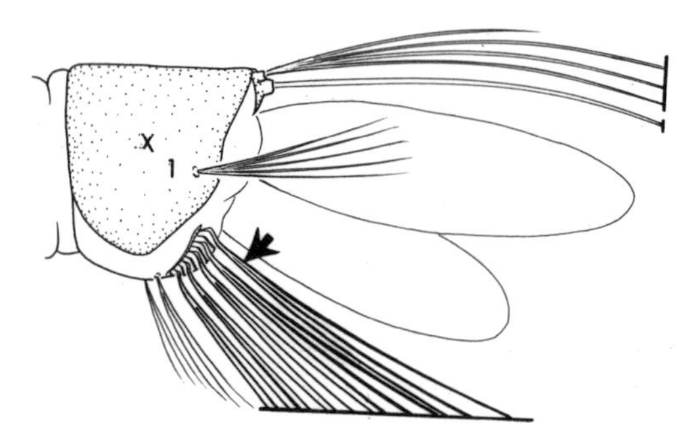


Fig. 719. Lateral view of abdominal segment X: *Oc. zoosophus*

50(49). Siphon index 2.7–3.0, not inflated at middle, diameter at apex more than 0.5 width at widest point; comb scales usually more than 15 (fig. 720) ....... *Oc. sierrensis* (plate 20D)

Siphon index 2.5 or less, inflated at middle and sharply reduced at apex, less than 0.5 width at widest point: comb scales usually fewer than 15 (fig. 721) ........................................................ 51

Fig. 720. Lateral view of abdominal segments VIII–X: *Oc. sierrensis*

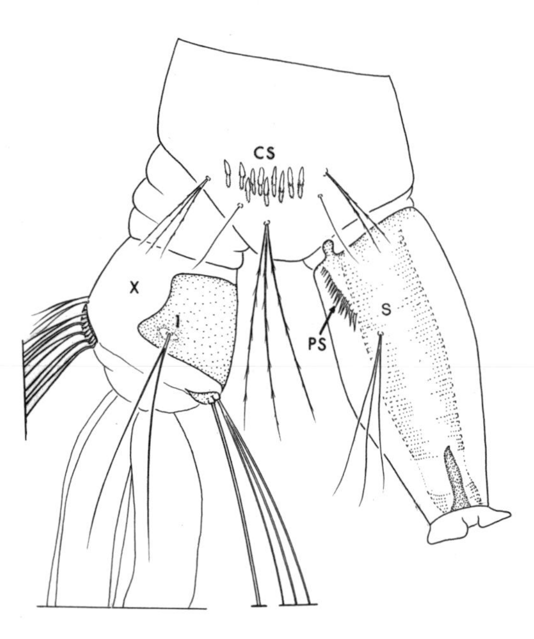


Fig. 721. Lateral view of abdominal segments VIII–X: *Oc. monticola*

51(50). Seta 1-X usually single, rarely double; pecten with 7–11 spines, restricted to basal 0.2 of siphon (fig. 722) ........................................................ *Oc. deserticola* (plate 39D)

Seta 1-X double, rarely single; pecten with 10–15 spines, on basal 0.25 of siphon (fig. 723) ........................................................ 52

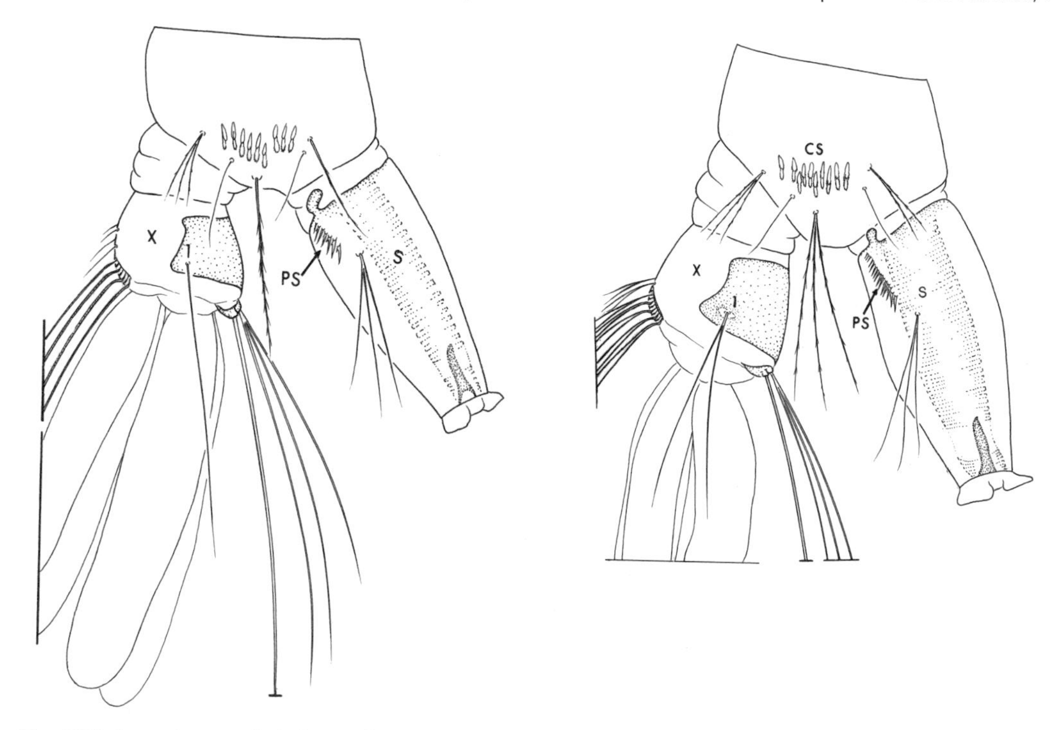


Fig. 722. Lateral view of abdominal segments VIII–X: *Oc. deserticola*

Fig. 723. Lateral view of abdominal segments VIII–X: *Oc. monticola*

52(51). Setae 1,13-IV,V similar in size and number of branches (fig. 724); seta 7-P mostly triple (fig. 725) ........................................................................ *Oc. monticola* (plate 40A)

Setae 1-IV,V with more branches and usually weaker than 13-IV,V (fig. 726); seta 7-P usually double (fig. 727) ........................................................ *Oc. varipalpus* (plate 40D)

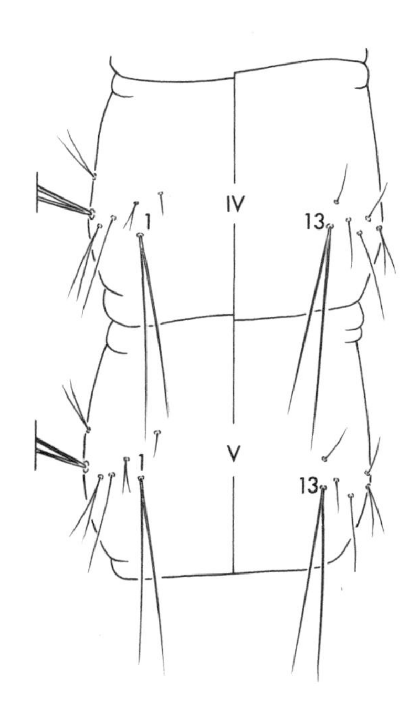


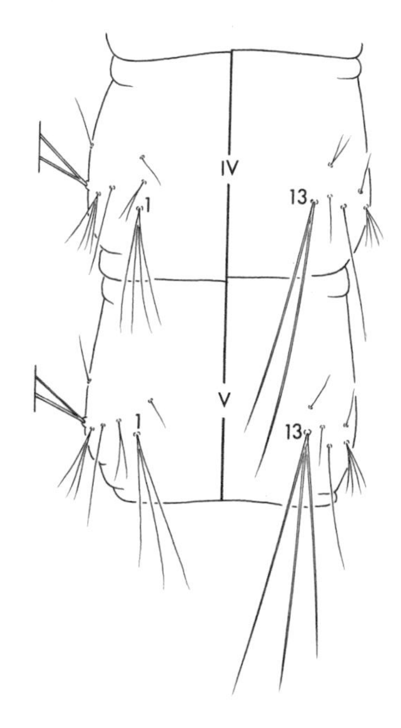


Fig. 724. Dorsal/ventral view of abdominal segments IV,V: *Oc. monticola*

Fig. 726. Dorsal/ventral view of abdominal segments IV,V: *Oc. varipalpus*

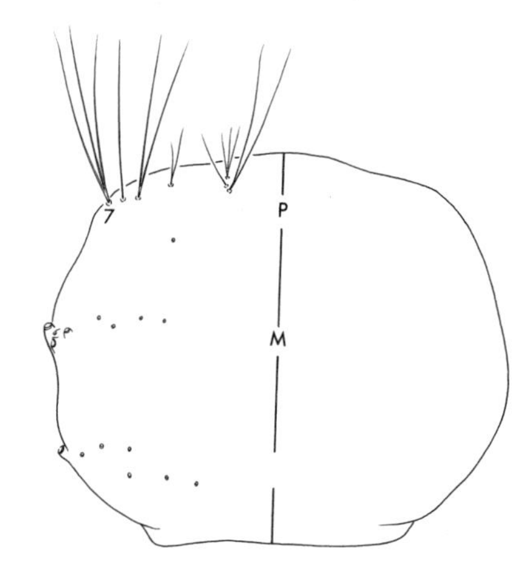


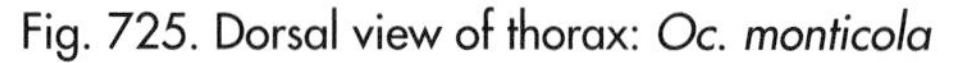
Fig. 725. Dorsal view of thorax: *Oc. monticola*

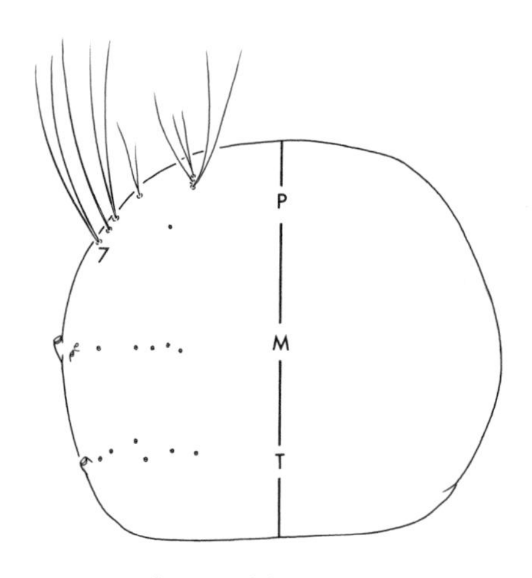


Fig. 727. Dorsal view of thorax: *Oc. varipalpus*

53(49). Setae 2-III–VI with at least 5 branches; seta 9-III–V longer and stronger than 7-III–V (fig. 728); seta 3-VII long, reaching almost to base of siphon, single; comb scales in single row in form of chevron (fig. 729) ............................................ *Oc. bahamensis* (plate 39A)

Setae 2-III–VI single or double; setae 9-III–V shorter and weaker than setae 7-III–V (fig. 730); seta 3-VII short, not reaching middle of segment VIII, with 2 or more branches; comb scales in single row or triangular patch (fig. 731) .................................................... 54

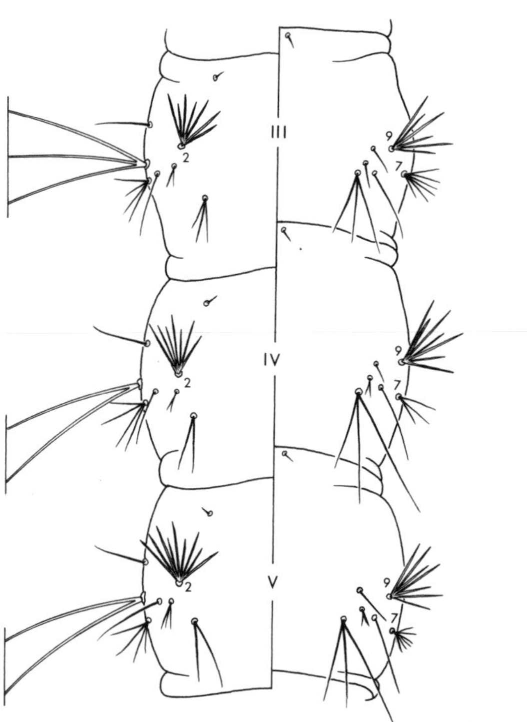


Fig. 728. Dorsal/ventral view of abdominal segments III–V: *Oc. bahamensis*

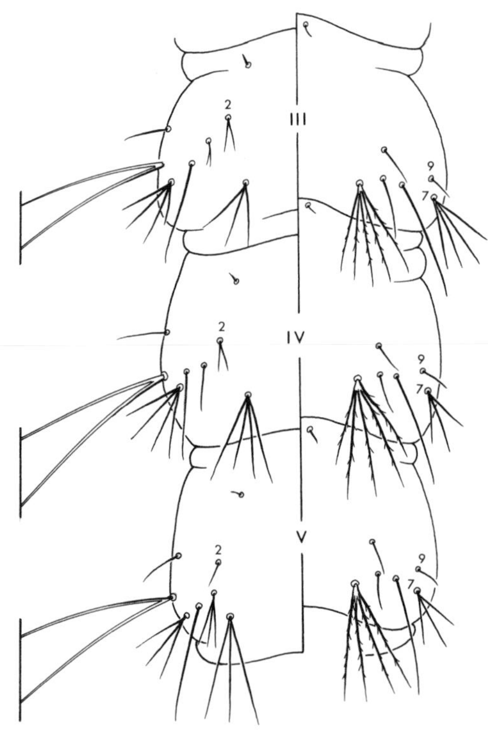


Fig. 730. Dorsal/ventral view of abdominal segments III–V: *Oc. triseriatus*

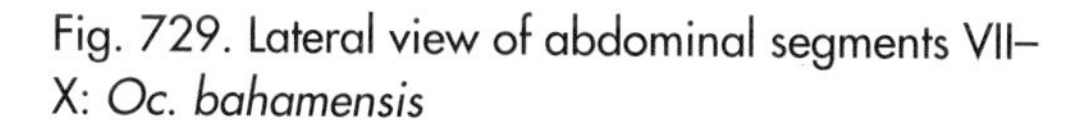

Fig. 729. Lateral view of abdominal segments VII–X: *Oc. bahamensis*

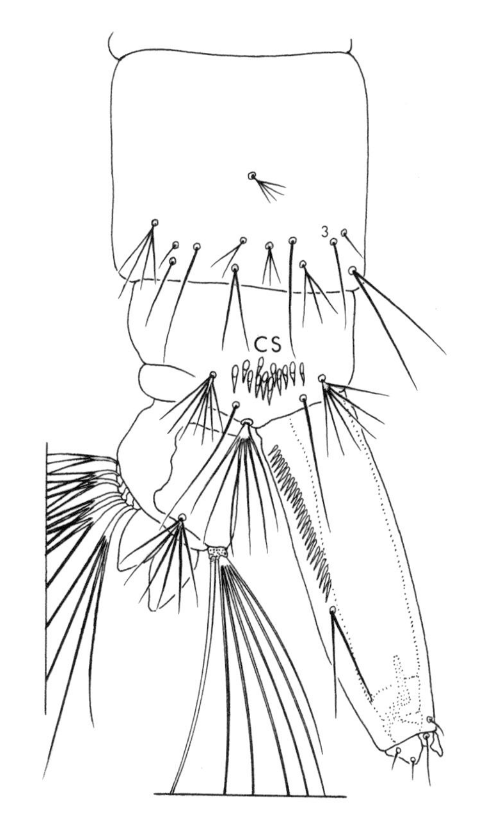


Fig. 731. Lateral view of abdominal segments VIII–X: *Oc. triseriatus*

54(53). Seta 4-C weak, usually with 7 or fewer branches, nearer to seta 6-C than to middorsal line (fig. 732); comb with 20 or more scales (fig. 733) ...................... *Oc. burgeri* (plate 39B)

Seta 4-C strong, with 8 or more branches, nearer to middorsal line than to seta 6-C (fig. 734); comb with 15 or fewer scales (fig. 735) ............................................................ 55

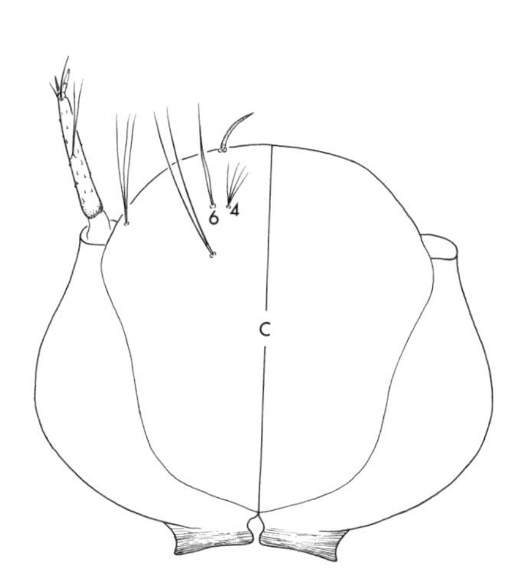


Fig. 732. Dorsal view of head: *Oc. burgeri*

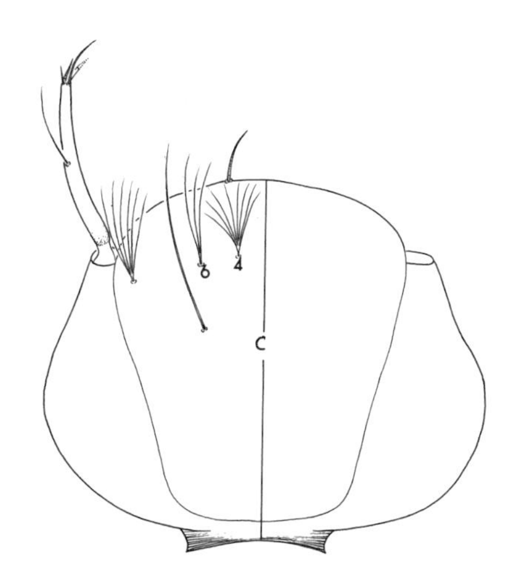


Fig. 734. Dorsal view of head: *Oc. triseriatus*

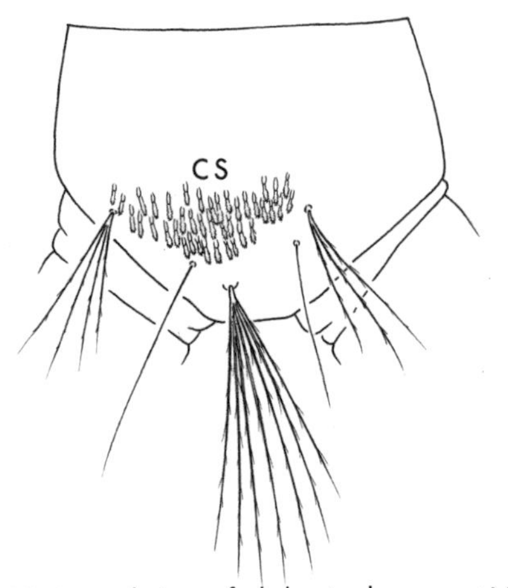


Fig. 733. Lateral view of abdominal segment VIII: *Oc. burgeri*

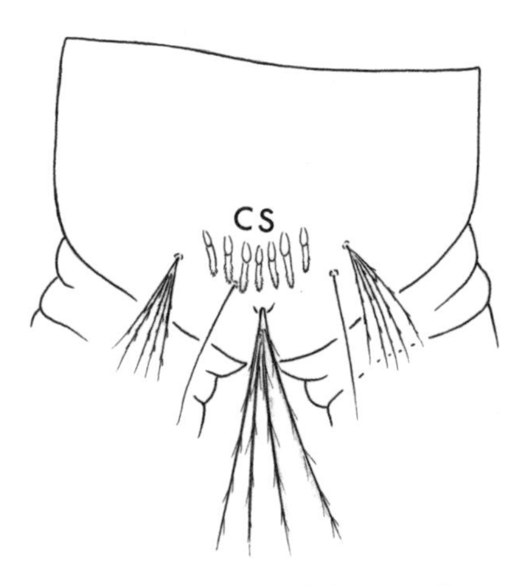


Fig. 735. Lateral view of abdominal segment VIII: *Oc. triseriatus*

55(54). Saddle extending more than 0.6 distance to midventral line, seta 1-X attached considerably dorsad of ventral border of saddle; seta 4-X with 7 pairs of fanlike setae (fig. 736) ........ ........................................................................................ *Oc. zoosophus* (plate 24B)

Saddle not extending more than 0.6 distance to midventral line, seta 1-X attached near ventral border of saddle; seta 4-X with 5 or 6 pairs of fanlike setae (fig. 737) ........... 56

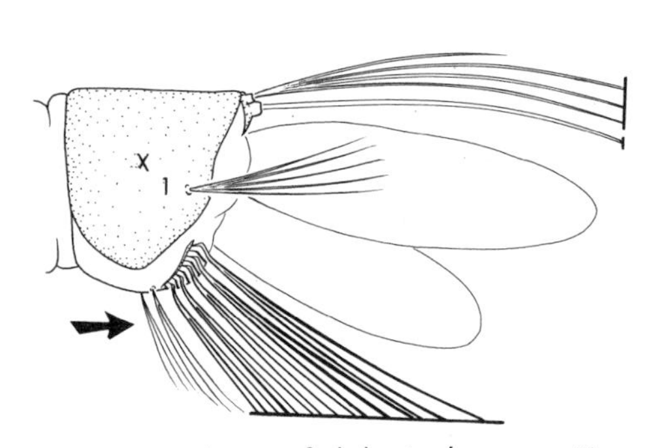


Fig. 736. Lateral view of abdominal segment X: *Oc. zoosophus*

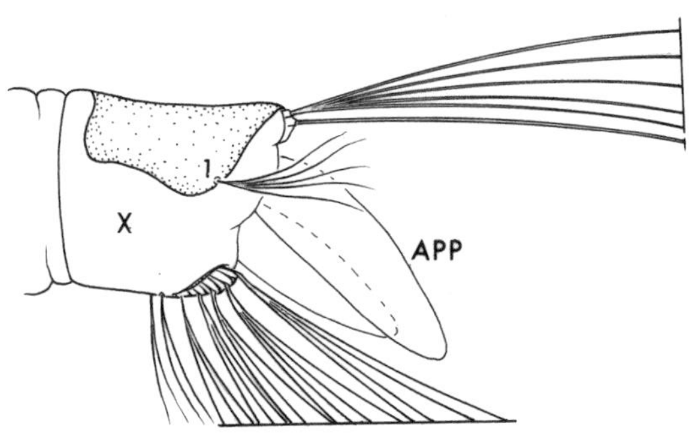


Fig. 737. Lateral view of abdominal segment X: *Oc. triseriatus*

56(55). Seta 4-X with 6 pairs of fanlike setae (fig. 738); acus usually attached to siphon or detached and situated close to its base (fig. 739); anal papillae not bulbous, dorsal pair longer than ventral pair (fig. 738) ............................................................ *Oc. triseratus* (plate 23B)

Seta 4-X with 5 pairs of fanlike setae (fig. 740); acus detached and removed from base of siphon (fig. 741); both pairs of anal papillae about same length, bulbous (fig. 740) ... 57

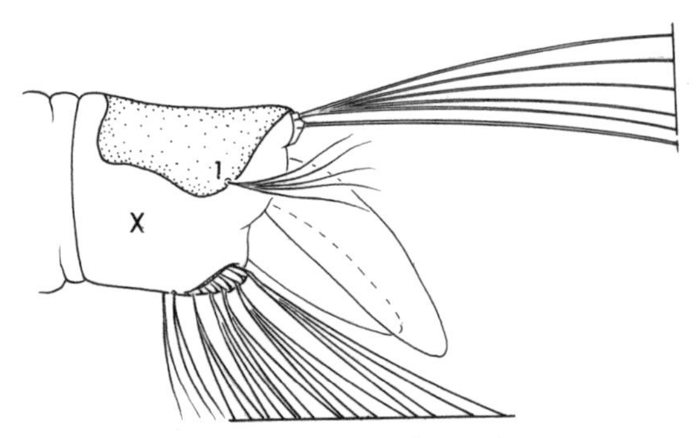


Fig. 738. Lateral view of abdominal segment X: *Oc. triseriatus*)

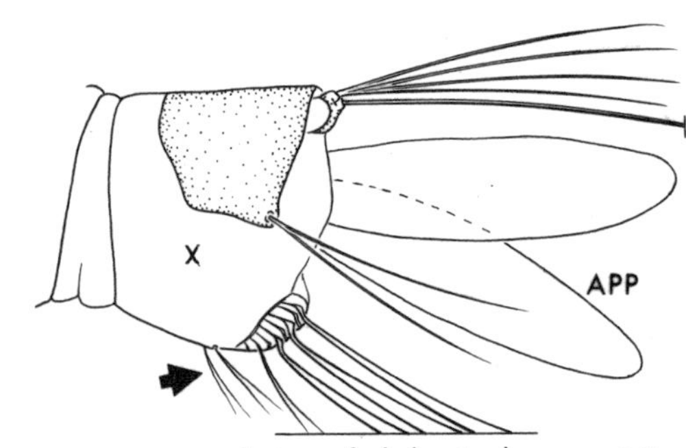


Fig. 740. Lateral view of abdominal segment X: *Oc. hendersoni*

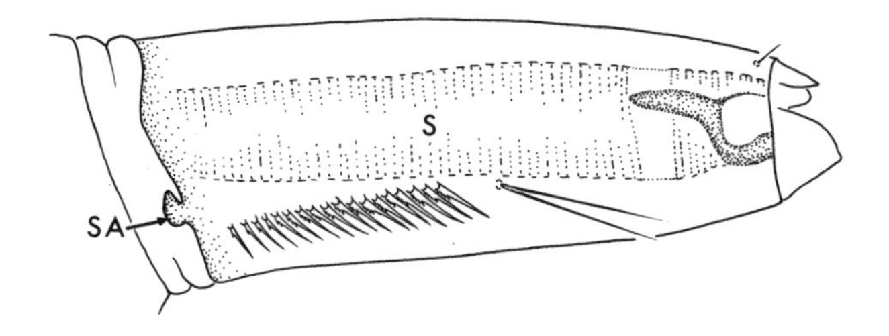


Fig. 739. Lateral view of siphon: *Oc. triseriatus*

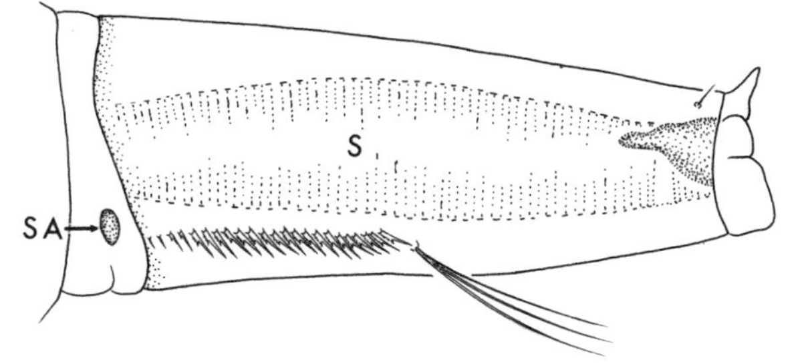


Fig. 741. Lateral view of siphon: *Oc. hendersoni*

57(56). Seta 4-X with 2 most anterior setae double (fig. 742) .......... *Oc. hendersoni* (plate 15C)

Seta 4-X with 2 most anterior setae with 3 or 4 branches (fig. 743) ............................... .......................................................................................... *Oc. brelandi* (plate 39B)

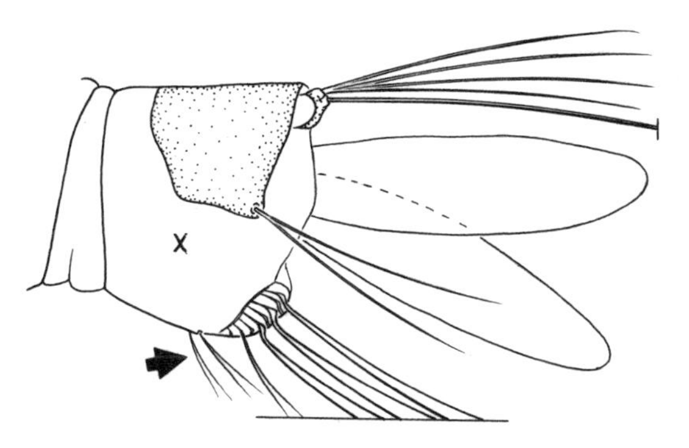


Fig. 742. Lateral view of abdominal segment X: *Oc. hendersoni*)

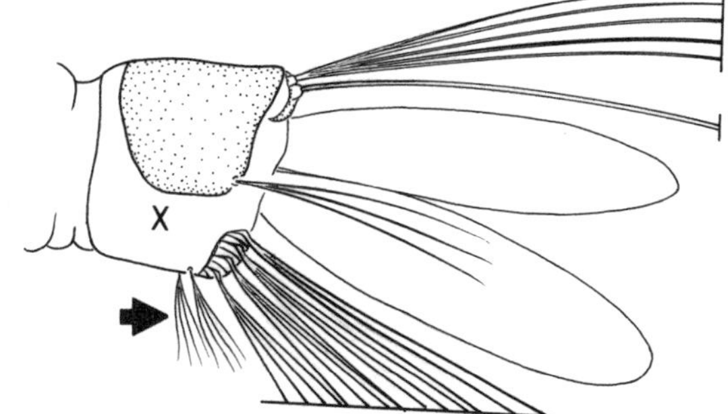


Fig. 743. Lateral view of abdominal segment X: *Oc. brelandi*

58(43). Individual comb scale with median spine 1.5 or more length of subapical spinules (fig. 744) ............................................................................................................................ 59

Individual comb scale fringed with subequal spinules or with median spine less than 1.5 length of subapical spinules (fig. 745) ........................................................................ 69

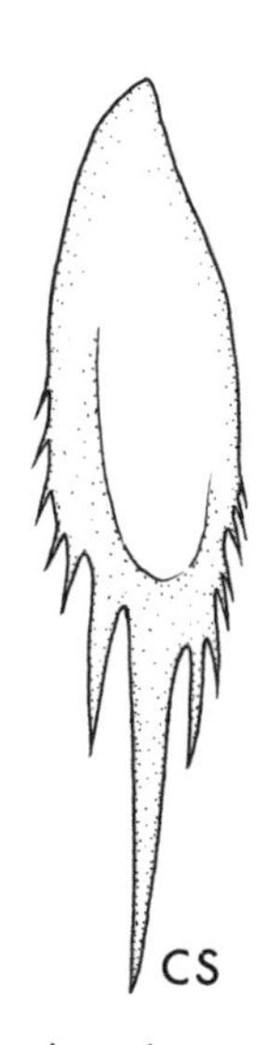

Fig. 744. Comb scale: *Oc. impiger*

Fig. 745. Comb scale: *Oc. cantator*

59(58). Siphon index 4.0–5.0; apical pecten spine nearly equal to apical diameter of siphon (fig. 746) .......................................................................................... *Oc. fitchii* (plate 14C)

Siphon index usually less than 4.0; apical pecten spine not more than 0.5 apical diameter of siphon (fig. 747) .......................................................................................... 60

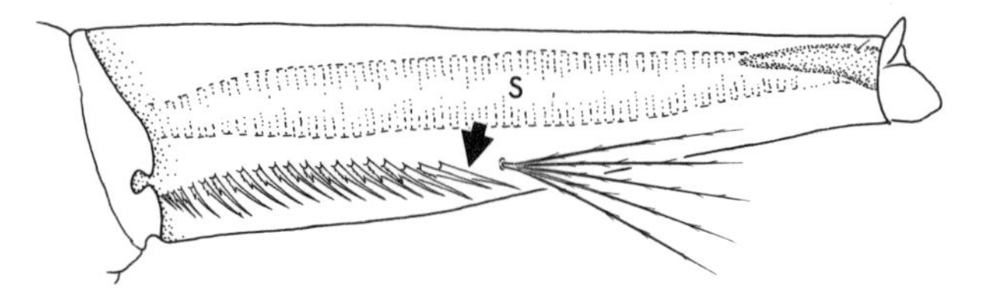


Fig. 746. Lateral view of siphon: *Oc. fitchii*

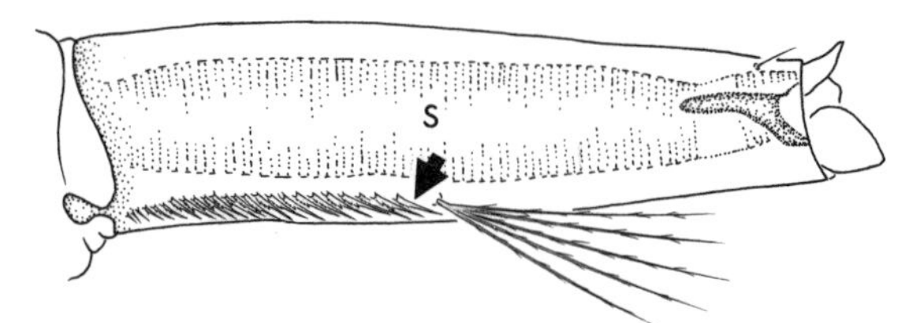


Fig. 747. Lateral view of siphon: *Oc. c. canadensis*

60(59). Comb with 8–16 scales (fig. 748) .......................................................................................... 61

Comb with 18 or more scales (fig. 749) .......................................................................................... 62

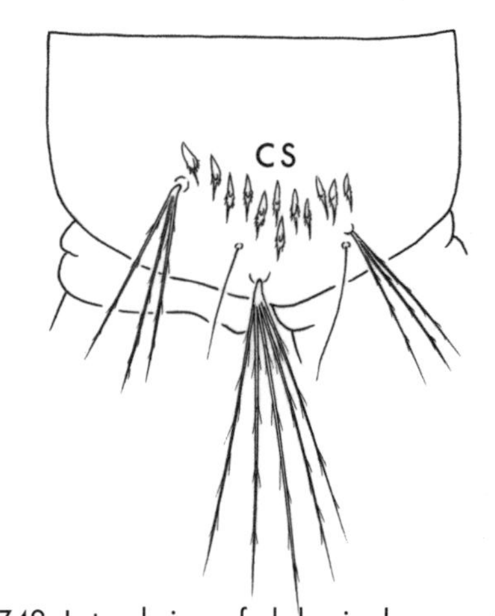


Fig. 748. Lateral view of abdominal segment VIII: *Oc. impiger*

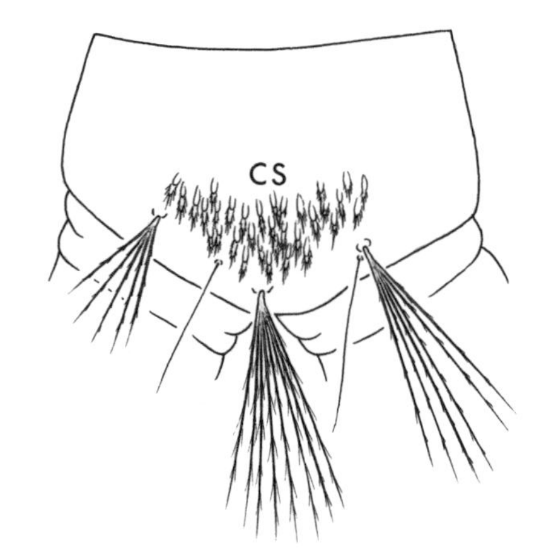


Fig. 749. Lateral view of abdominal segment VIII: *Oc. stimulans*

61(60). Saddle extending to near midventral line; anal papilla–saddle index less than 1.5 (fig. 750) .......................................................................................... *Oc. punctodes* (plate 19C)

Saddle extending only about 0.5 to midventral line; anal papilla–saddle index 2.0 or more (fig. 751) .......................................................................................... *Oc. impiger* (plate 16A)

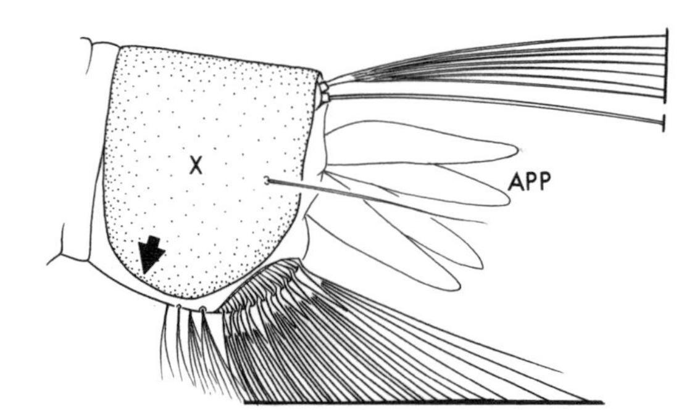


Fig. 750. Lateral view of abdominal segment X: *Oc. punctodes*

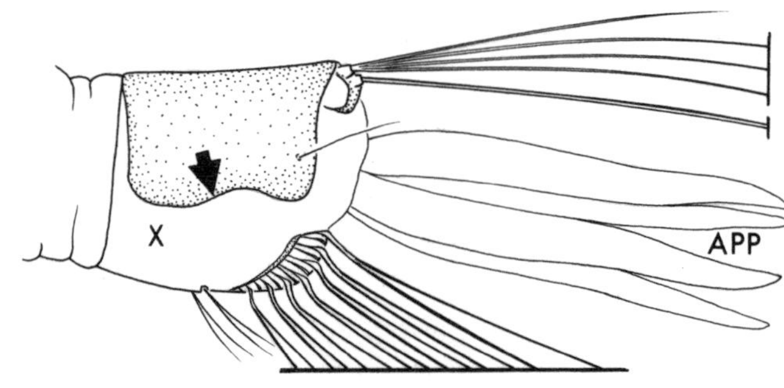


Fig. 751. Lateral view of abdominal segment X: *Oc. impiger*

62(60). Seta 1-X shorter than saddle (fig. 752) .................... 63

Seta 1-X longer than saddle (fig. 753) .................... 68

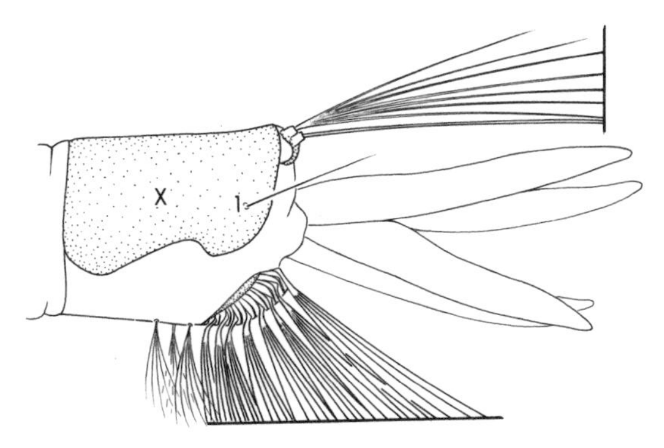


Fig. 752. Lateral view of abdominal segment X: *Oc. stimulans*

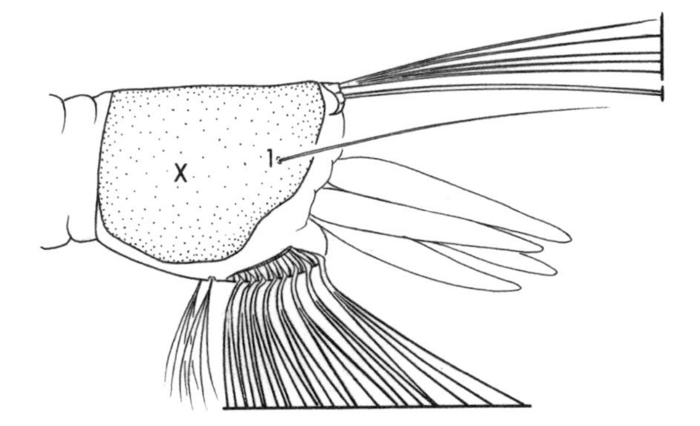


Fig. 753. Lateral view of abdominal segment X: *Oc. aboriginis*

63(62). Setae 5,6-C single, rarely double (fig. 754) .................... 64

Setae 5-C with 2–4 branches, 6-C usually double (fig. 755) .................... 66

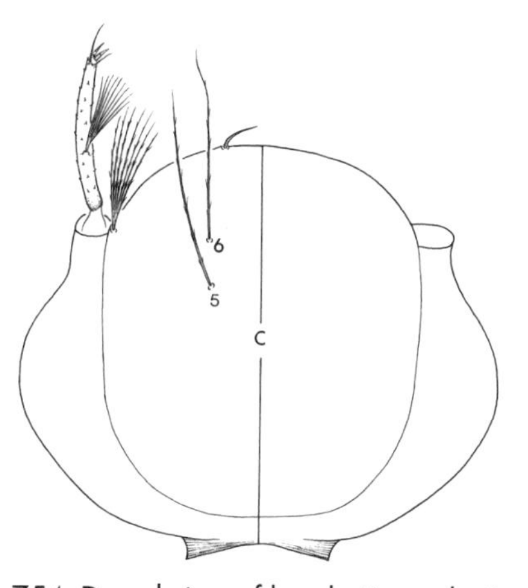


Fig. 754. Dorsal view of head: *Oc. melanimon*

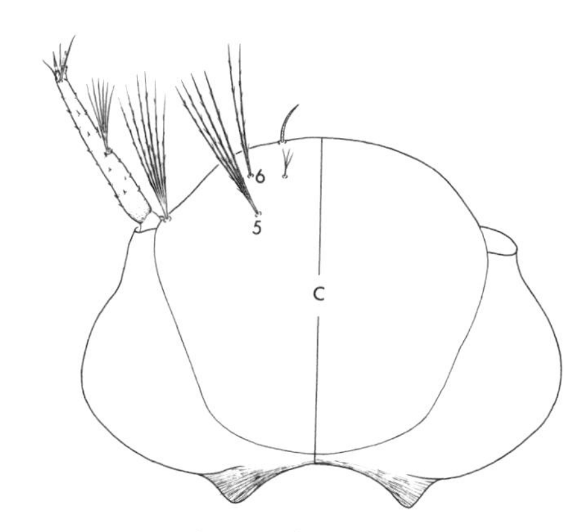


Fig. 755. Dorsal view of head: *Oc. sticticus*

64(63). Seta 1 attached distal to middle of siphon (fig. 756); seta 1-M about equal to 2-M in length (fig. 757) .................... (in part) *Oc. melanimon* (plate 17B)

Seta 1 attached about at middle of siphon (fig. 758); seta 1-M longer than 2-M (fig. 759) .................... 65

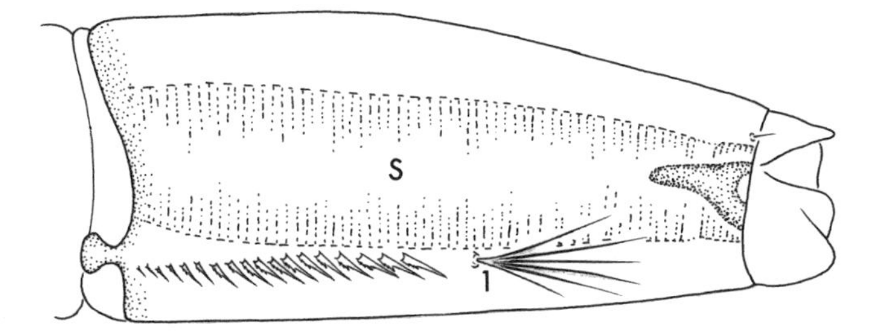


Fig. 756. Lateral view of siphon: *Oc. melanimon*

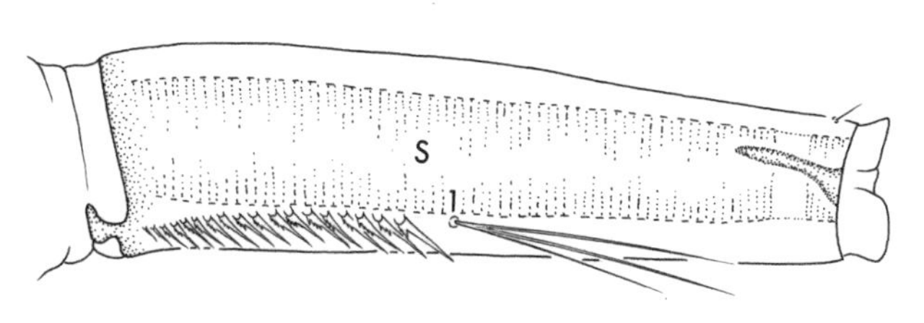


Fig. 758. Lateral view of siphon: *Oc. stimulans*

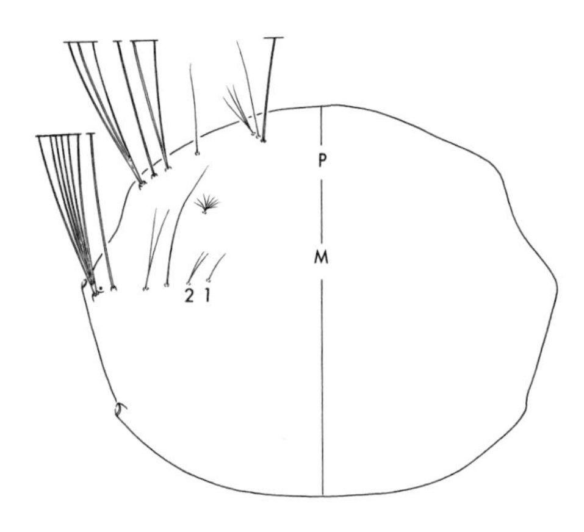


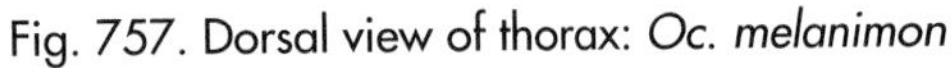
Fig. 757. Dorsal view of thorax: *Oc. melanimon*

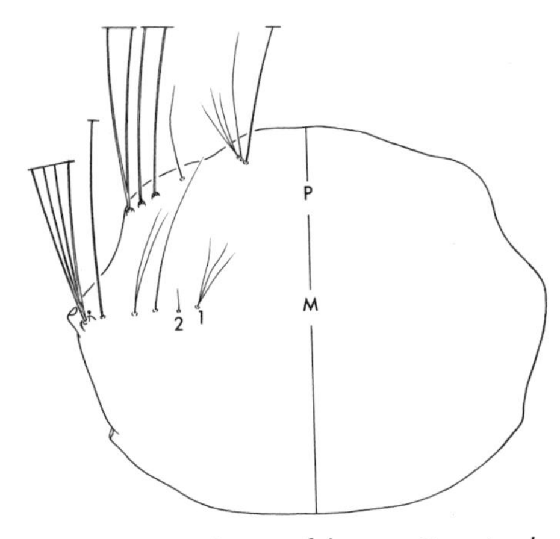


Fig. 759. Dorsal view of thorax: *Oc. stimulans*

65(64). Comb scale with 1–3 median spines 2.0 or more length of subapical spinules (fig. 760); usually with more than 35 comb scales (fig. 761) ................ *Oc. nevadensis* (plate 18A)

Comb scale with median spine about 1.5 length of subapical spinules (fig. 762); usually fewer than 35 comb scales (fig. 763) ....................................... *Oc. stimulans* (plate 22A)

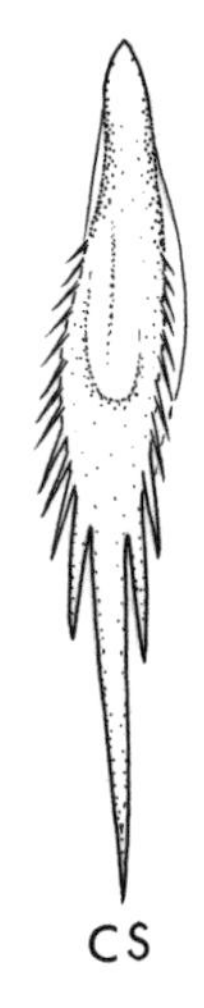


Fig. 760. Comb scale: *Oc. nevadensis*

Fig. 762. Comb scale: *Oc. stimulans*

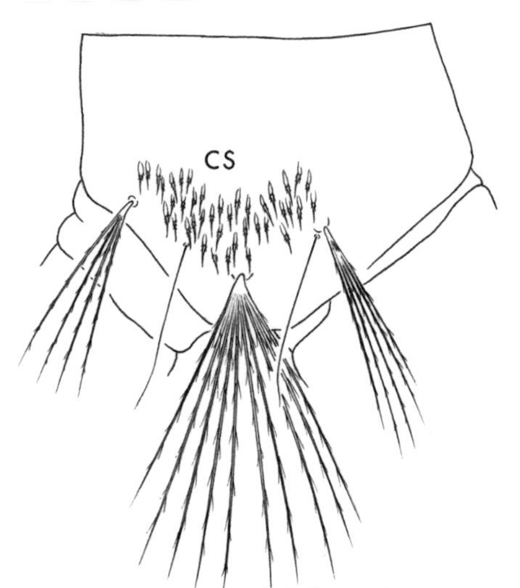


Fig. 761. Lateral view of abdominal segment VIII: *Oc. nevadensis*

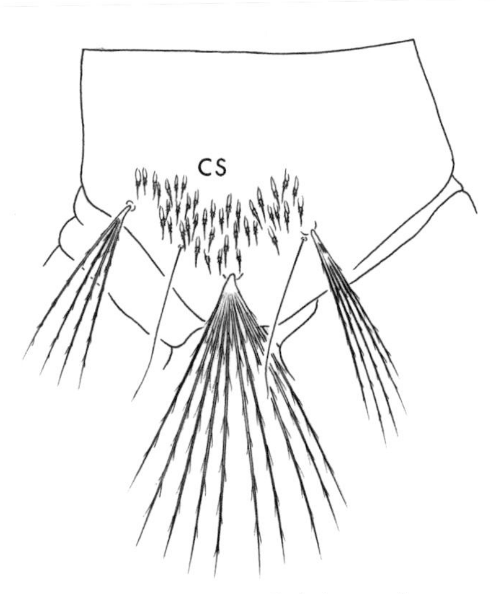


Fig. 763. Lateral view of abdominal segment VIII: *Oc. stimulans*

66(63). Seta 1-M longer than setae 3-M and 5-C (fig. 764, 765) ....... *Oc. mercurator* (plate 17C)

Seta 1-M shorter than setae 3-M and 5-C (fig. 766, 767) .......................................... 67

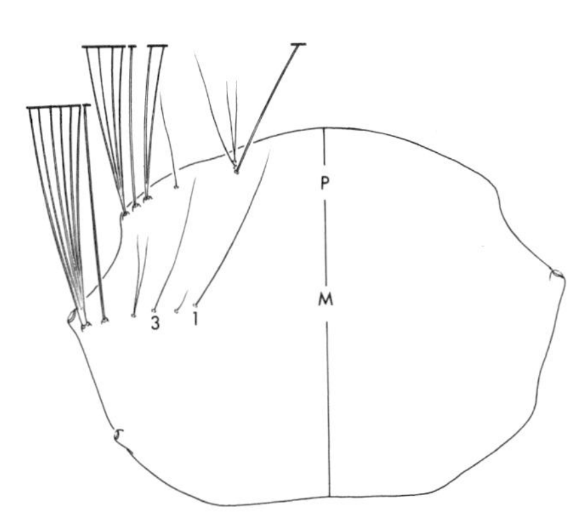


Fig. 764. Dorsal view of thorax: *Oc. mercurator*

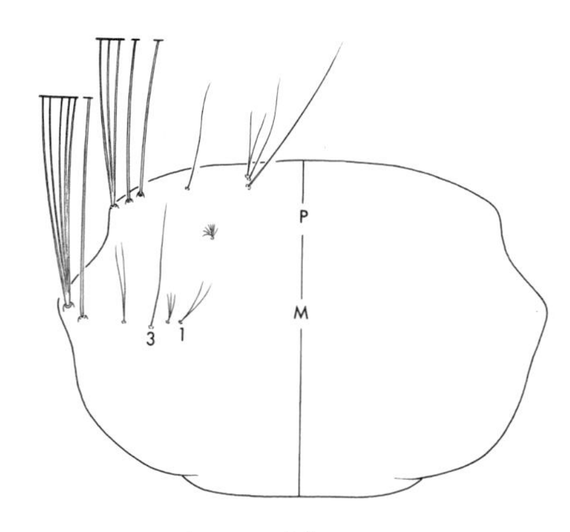


Fig. 766. Dorsal view of thorax: *Oc. sticticus*

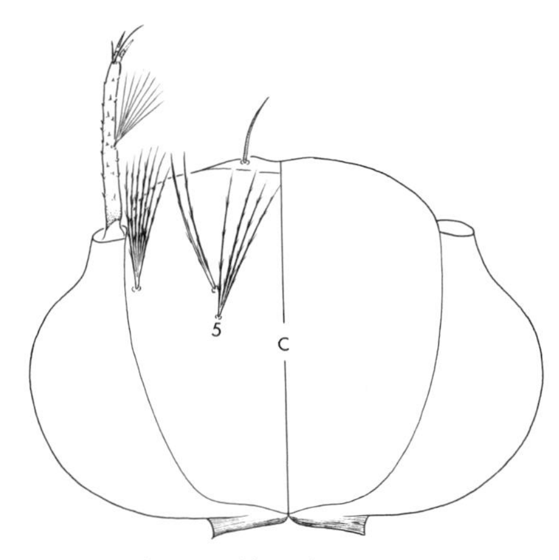


Fig. 765. Dorsal view of head: *Oc. mercurator*

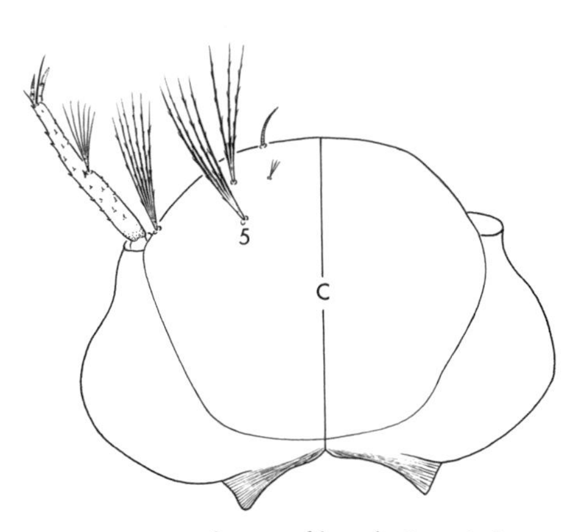


Fig. 767. Dorsal view of head: *Oc. sticticus*

67(66). Siphon index 3.2–4.0 (fig. 768); comb scale with stout subapical spinules (fig. 769) ...... ........................................................................ (in part) *Oc. flavescens* (plate 14D)

Siphon index 2.5–3.0 (fig. 770); comb scale with weak subapical spinules (fig. 771) ...... .......................................................................................... *Oc. sticticus* (plate 21D)

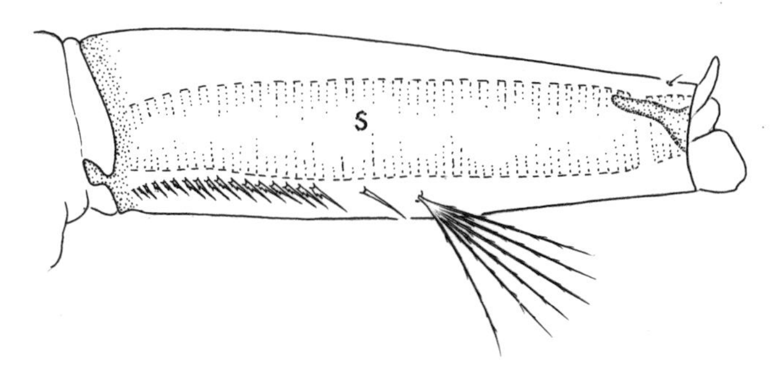


Fig. 768. Lateral view of siphon: *Oc. flavescens*

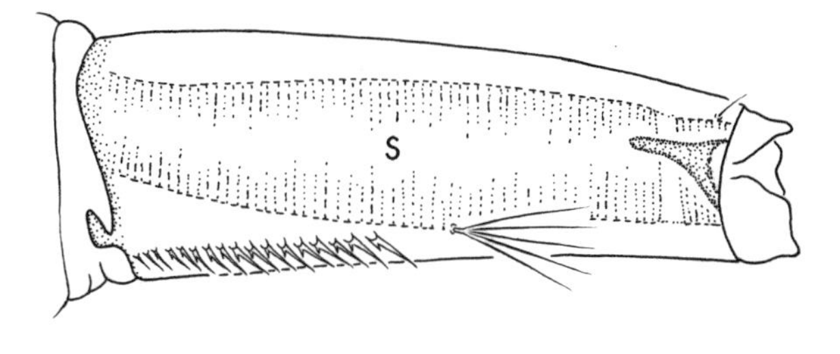


Fig. 770. Lateral view of siphon: *Oc. sticticus*

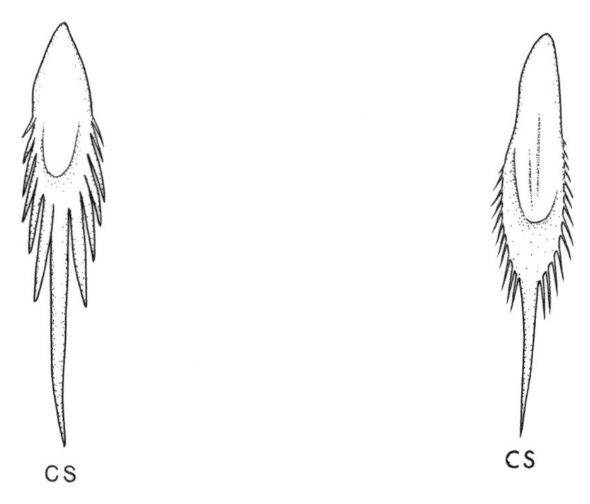


Fig. 769. Comb scale: *Oc. flavescens*

Fig. 771. Comb scale: *Oc. sticticus*

68(62). Posterior border of saddle aciculate (fig. 772); seta 1-M with 3–6 branches (fig. 773) .... ........................................................................................ *Oc. schizopinax* (plate 20C)

Posterior border of saddle without aciculae (fig. 774); seta 1-M single (fig. 775) ............ ........................................................................................ *Oc. aboriginis* (plate 9A)

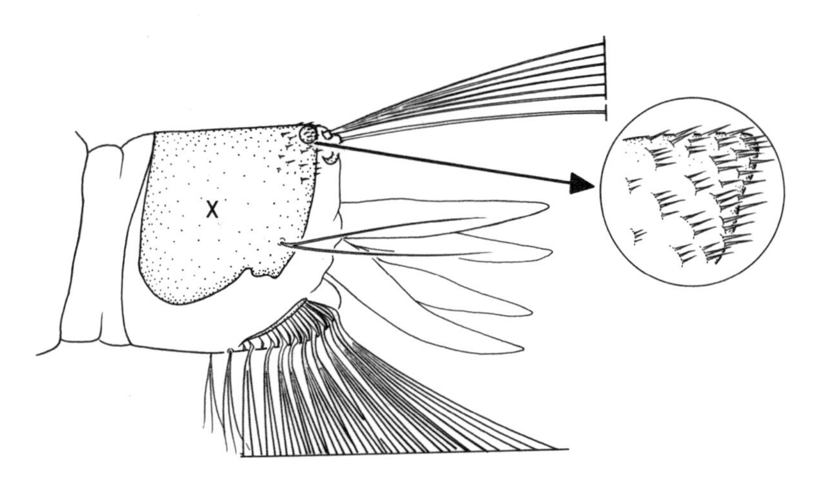


Fig. 772. Lateral view of abdominal segment X: *Oc. schizopinax*

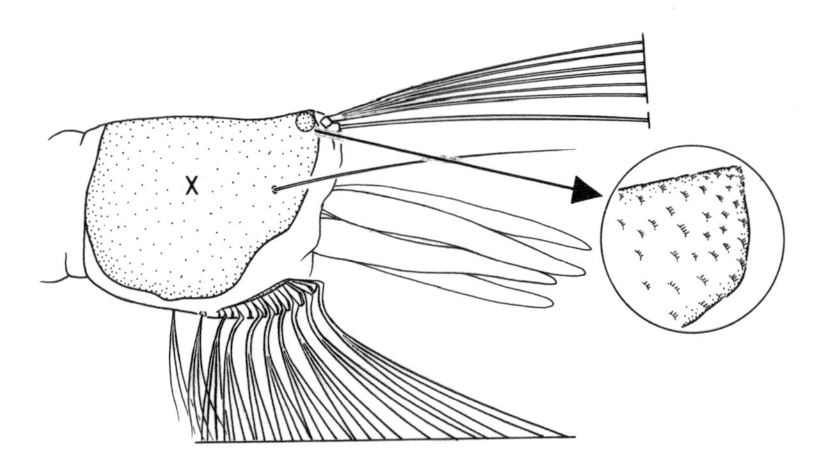


Fig. 774. Lateral view of abdominal segment X: *Oc. aboriginis*

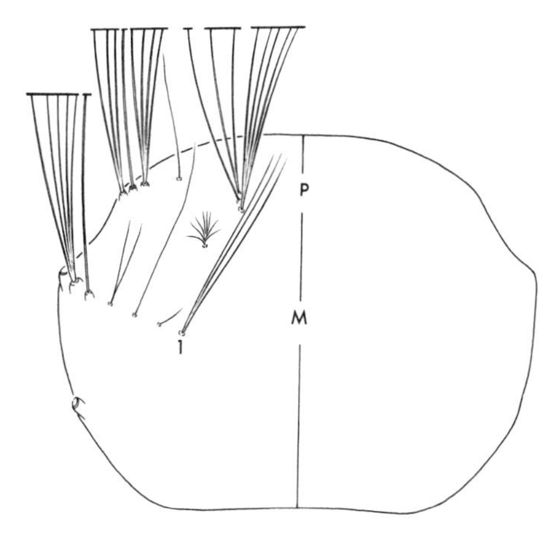


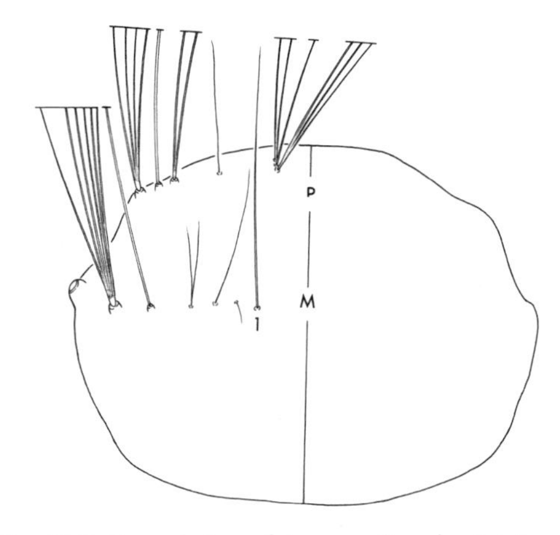


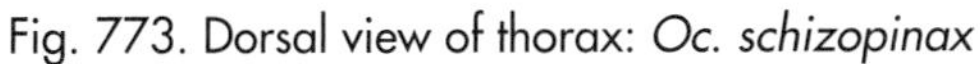

Fig. 773. Dorsal view of thorax: *Oc. schizopinax*

Fig. 775. Dorsal view of thorax: *Oc. aboriginis*

69(58). Seta 5-C with 4 or more and seta 6-C with 3 or more branches (fig. 776) .................. 70

Seta 5-C with 1–3 branches, rarely 4; seta 6-C single or double, rarely triple (fig. 777) . 75

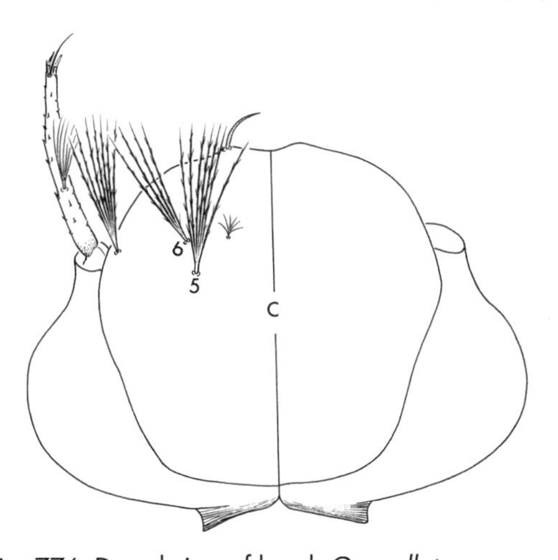


Fig. 776. Dorsal view of head: *Oc. pullatus*

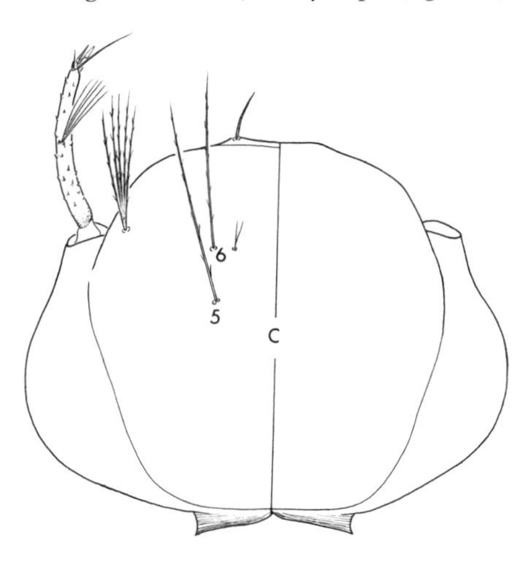


Fig. 777. Dorsal view of head: *Oc. dorsalis*

70(69). Seta 1-M about length of antenna or longer (figs. 778, 779) ....................................... 71

Seta 1-M much shorter than antenna (figs. 780, 781) ................................................. 73

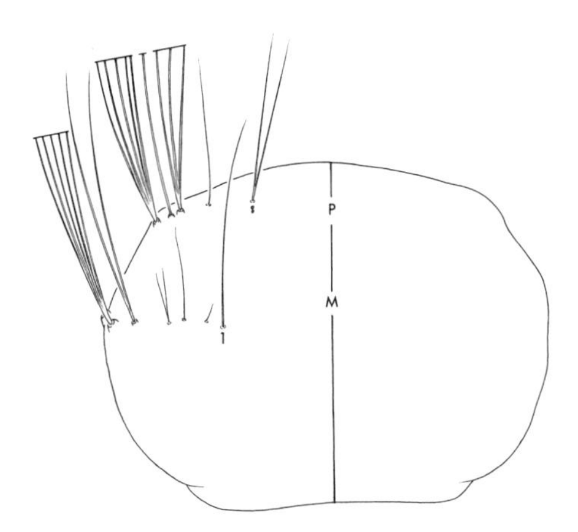


Fig. 778. Dorsal view of thorax: *Oc. pullatus*

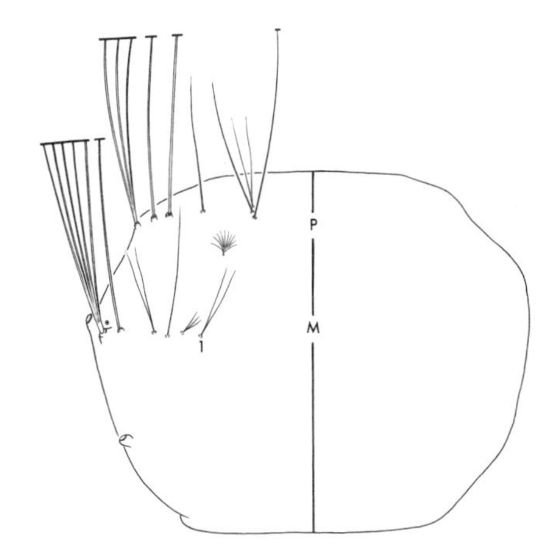


Fig. 780. Dorsal view of thorax: *Oc. c. canadensis*

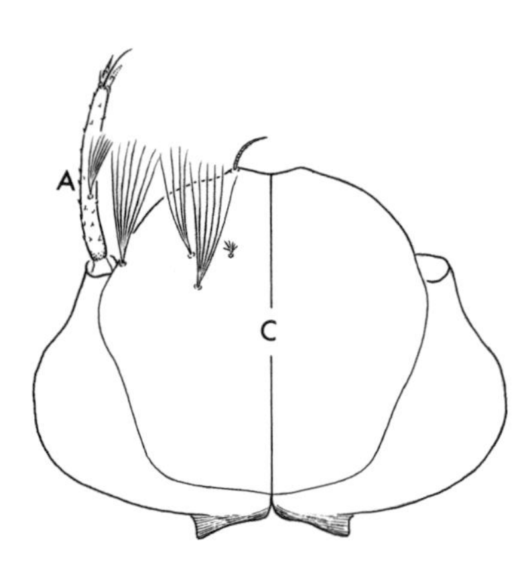


Fig. 779. Dorsal view of head: *Oc. pullatus*

Fig. 781. Dorsal view of head: *Oc. c. canadensis*

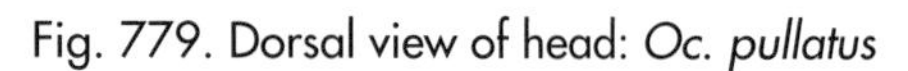

71(70). Seta 3-P single (fig. 782); with 70 or more comb scales (fig. 783) . *Oc. pionips* (plate 18D)

Seta 3-P double or triple (fig. 784); with 60 or fewer comb scales (fig. 785) .............. 72

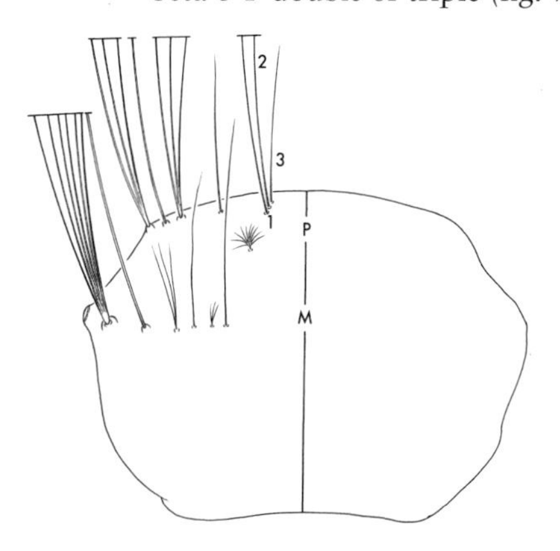


Fig. 782. Dorsal view of thorax: *Oc. pionips*

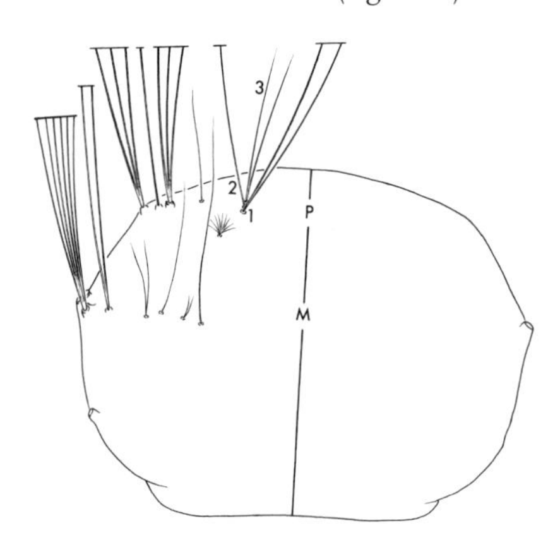


Fig. 784. Dorsal view of thorax: *Oc. pullatus*

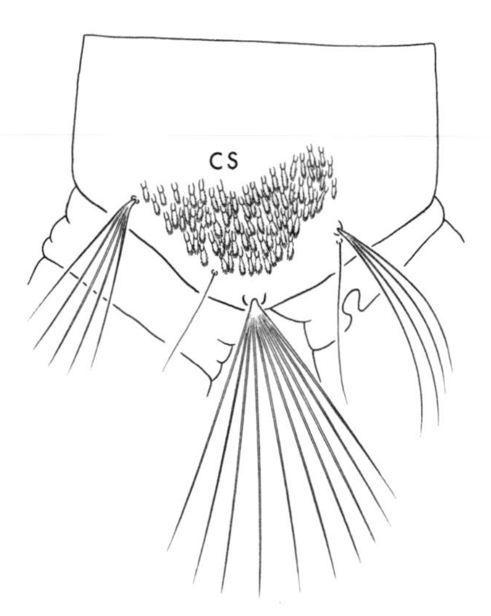


Fig. 783. Lateral view of abdominal segment VIII: *Oc. pionips*

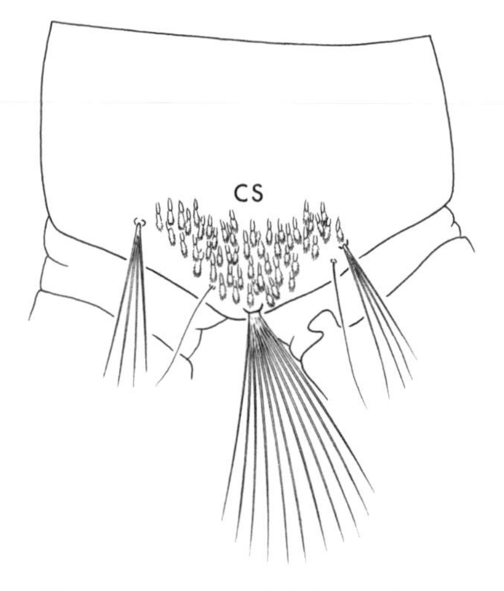


Fig. 785. Lateral view of abdominal segment VIII: *Oc. pullatus*

72(71). Seta 5-M branched (fig. 786); seta 1-X about 0.5 length of saddle (fig. 787) ................. ........................................................................................ *Oc. pullatus* (plate 19B)

Seta 5-M single (fig. 788); seta 1-X about equal to length of saddle (fig. 789) ................ ..................................................................................................... *Oc. cantator* (plate 11C)

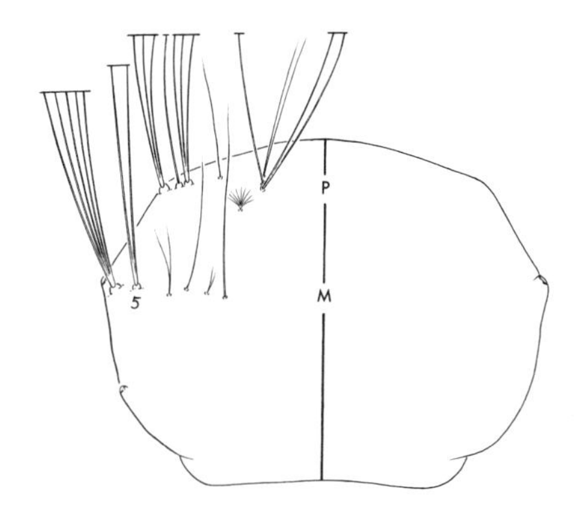


Fig. 786. Dorsal view of thorax: *Oc. pullatus*

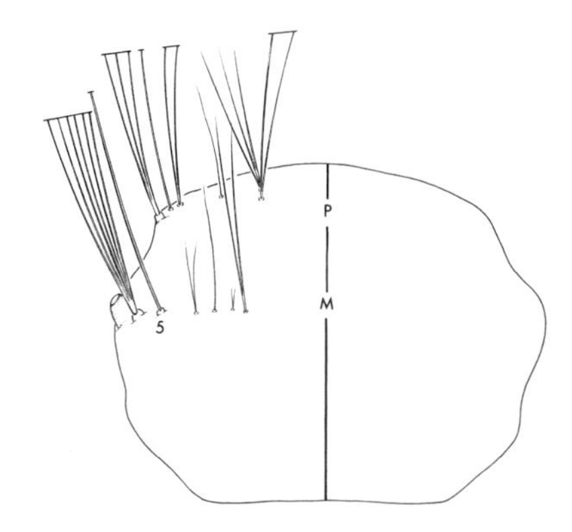


Fig. 788. Dorsal view of thorax: *Oc. cantator*

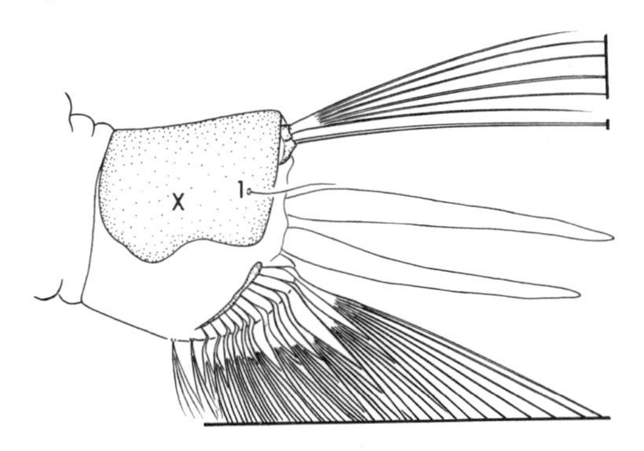


Fig. 787. Lateral view of abdominal segment X: *Oc. pullatus*

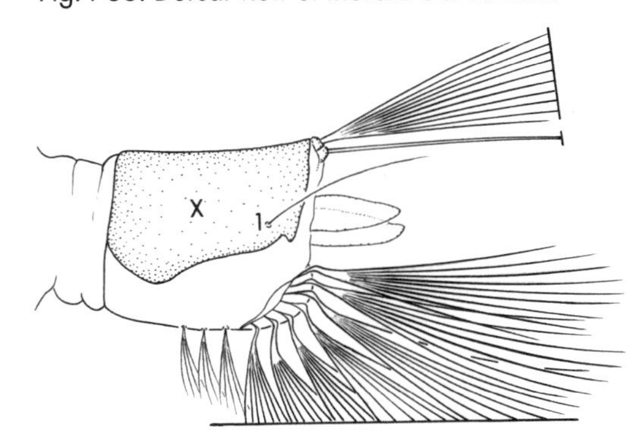


Fig. 789. Lateral view of abdominal segment X: *Oc. cantator*

73(70).Seta 1-X not attached to saddle; siphon index less than 2.5 (fig. 790) .......................... ..................................................................................................... *Oc. togoi* (plate 10B)

Seta 1-X attached to saddle; siphon index 3.0 or more (fig. 791) ............................... 74

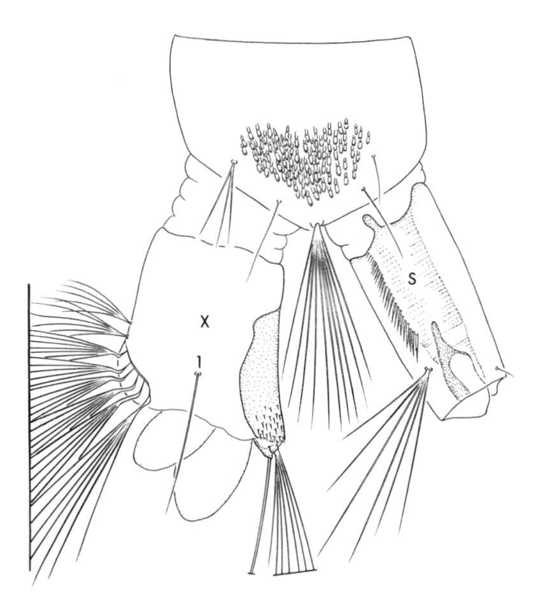


Fig. 790. Lateral view of abdominal segments VIII–X: *Oc. togoi*

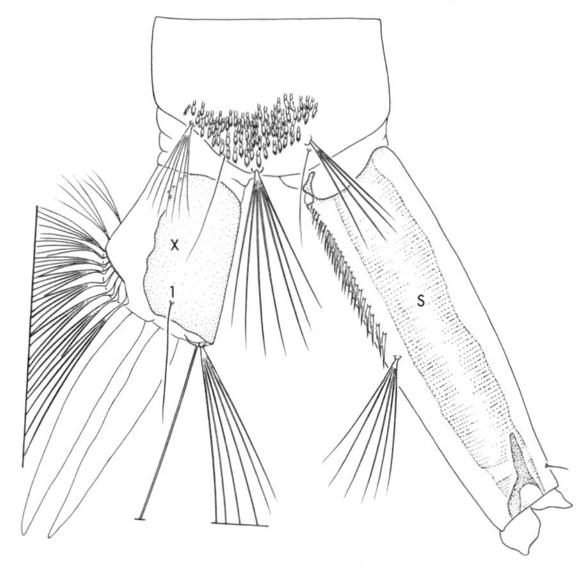


Fig. 791. Lateral view of abdominal segments VIII–X: *Oc. c. canadensis*

74(73). Comb scale with apical and subapical spines much stouter than lateral spinules (fig. 792); setae 6-I,II with 3,4 branches (fig. 793) .................................... *Oc. thibaulti* (plate 22D)

Comb scale fringed with subequal spinules (fig. 794); setae 6-I,II double (fig. 795) ........ .................................................................................. *Oc. c. canadensis* (plate 11B)
*Oc. c. mathesoni* (plate 39B)

Fig. 792. Comb scale: *Oc. thibaulti*

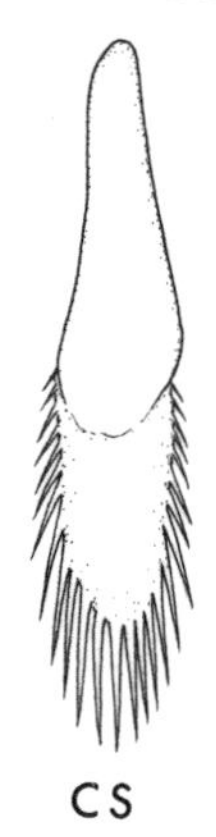


Fig. 794. Comb scale: *Oc. c. canadensis*

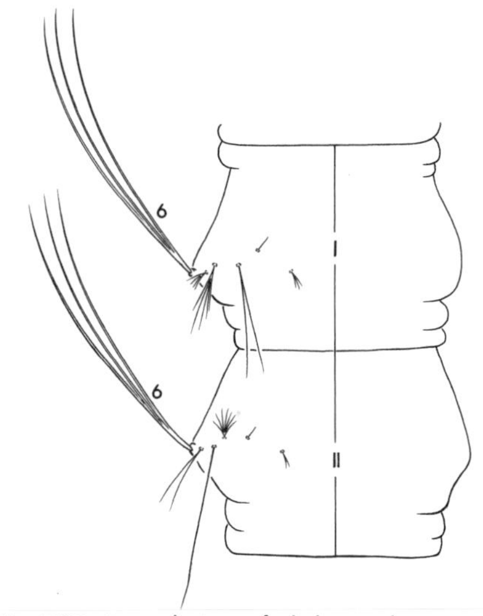


Fig. 793. Dorsal view of abdominal segments I,II: *Oc. thibaulti*

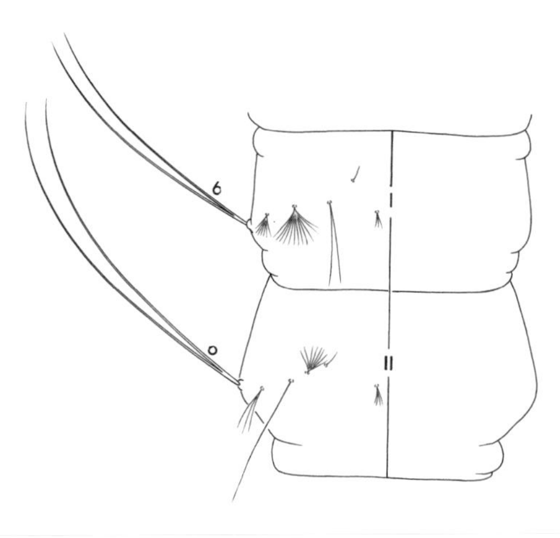


Fig. 795. Dorsal view of abdominal segments I,II: *Oc. c. canadensis*

75(69). Seta 1-X equal to length of saddle or longer (fig. 796) ........... *Oc. squamiger* (plate 40B)

Seta 1-X shorter than saddle (fig. 797) ....................................................................... 76

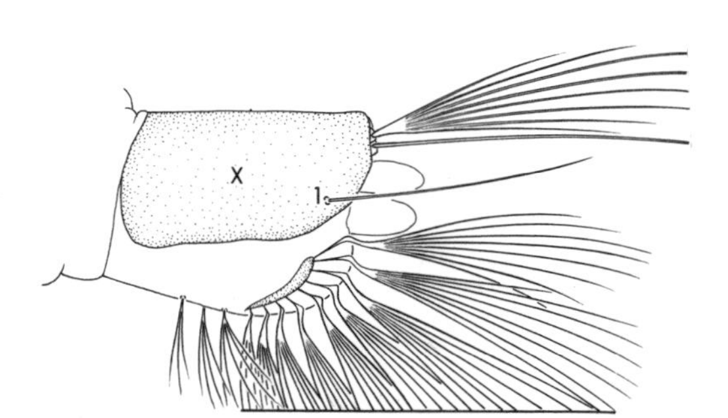


Fig. 796. Lateral view of abdominal segment X: *Oc. squamiger*

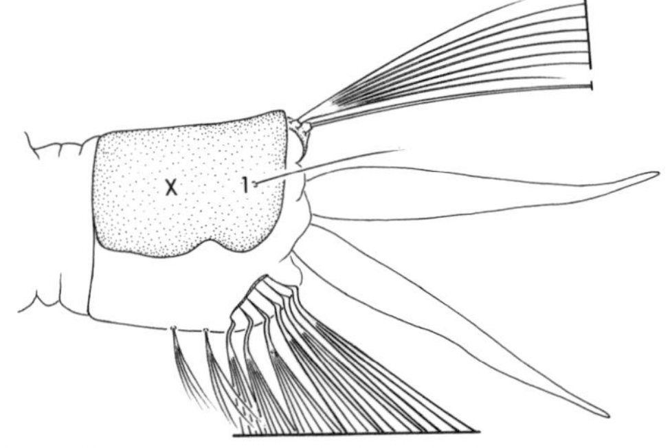


Fig. 797. Lateral view of abdominal segment X: *Oc. communis*

76(75). Seta 1-M about equal to length of antenna or longer (fig. 798, 799) ......................... 77

Seta 1-M shorter than antenna (fig. 800, 801) ......................................................... 79

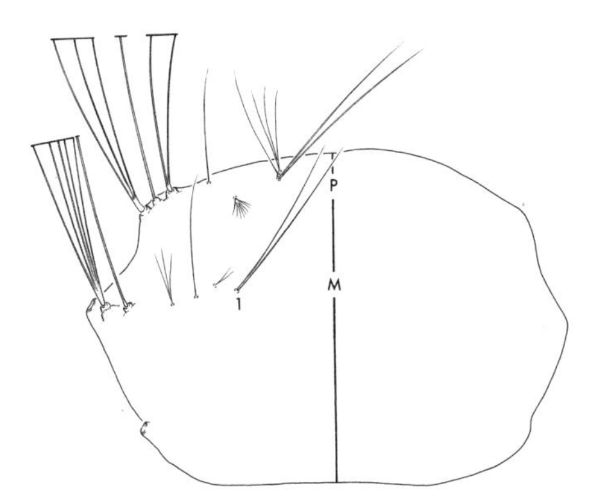


Fig. 798. Dorsal view of thorax: *Oc. dorsalis*

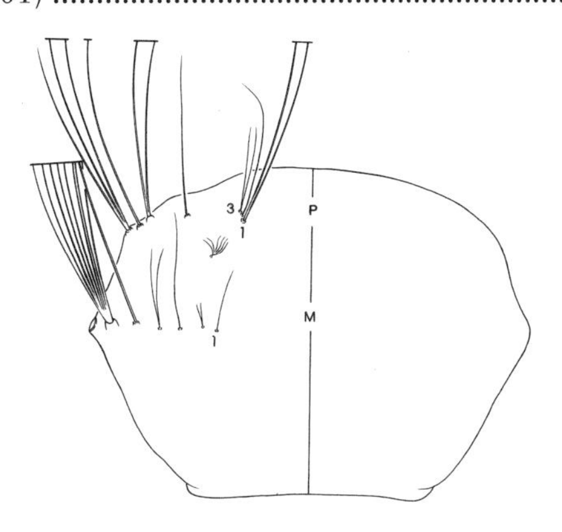


Fig. 800. Dorsal view of thorax: *Oc. increpitus*

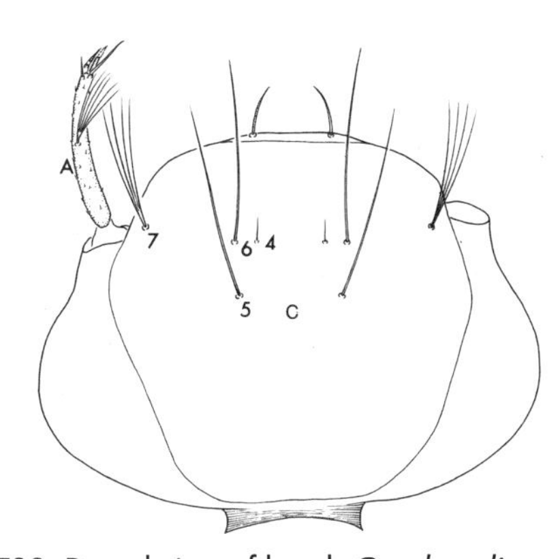


Fig. 799. Dorsal view of head: *Oc. dorsalis*

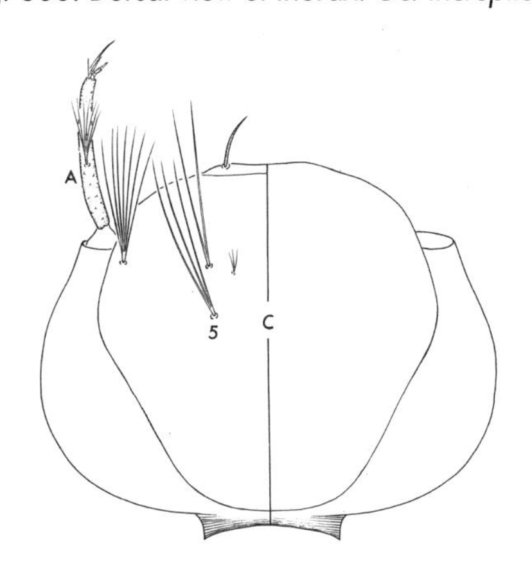


Fig. 801. Dorsal view of head: *Oc. increpitus*

77(76). Pecten extending to distal 0.5 of siphon, most apical spines 1,2 stouter than preceding two and about 2.0 length of seta 2-S (fig. 802) .............. (in part) *Oc. campestris* (plate 11A)

Pecten in basal 0.5 of siphon, the 2 most apical spines not much stouter than preceding two and about equal to seta 2-S (fig. 803) ..................................................................... 78

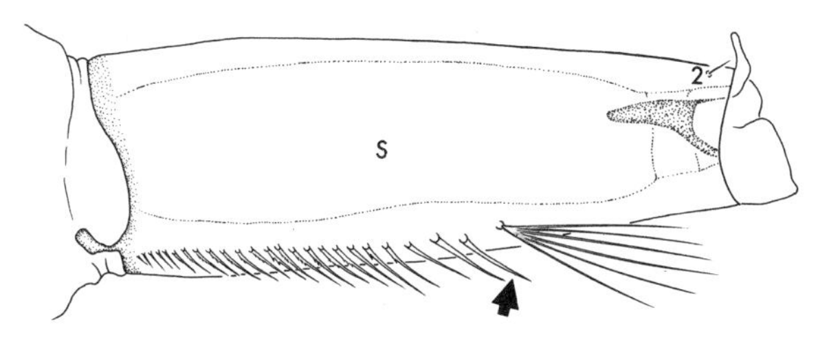


Fig. 802. Lateral view of siphon: *Oc. campestris*

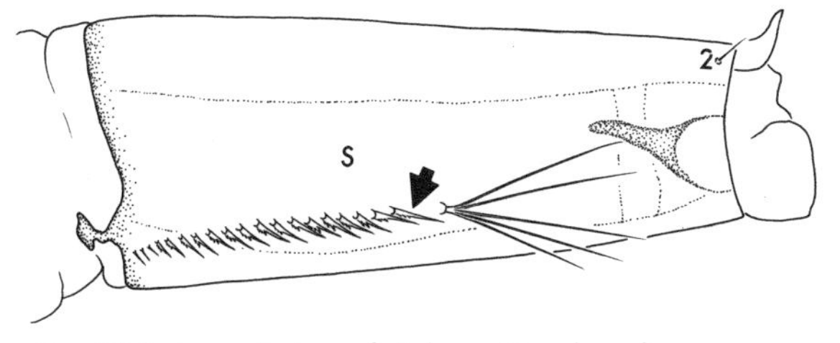


Fig. 803. Lateral view of siphon: *Oc. dorsalis*

78(77). Seta 1-X about 0.5 length of saddle (fig. 804); the 4 setae 5,6-C usually single, or total of single and branches of branched setae rarely more than 7 (fig. 805) .............................
.......................................................................................... *Oc. dorsalis* (plate 13B)

Seta 1-X almost equal to length of saddle (fig. 806); the 4 setae 5,6-C usually branched, the total of single and branches of branched setae usually 10, not fewer than 8 (fig. 807) ....
...................................................................................... *Oc. grossbecki* (plate 15B)

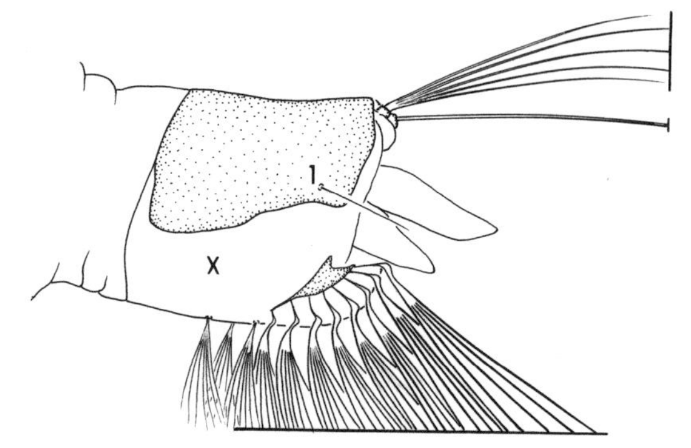


Fig. 804. Lateral view of abdominal segment X: *Oc. dorsalis*

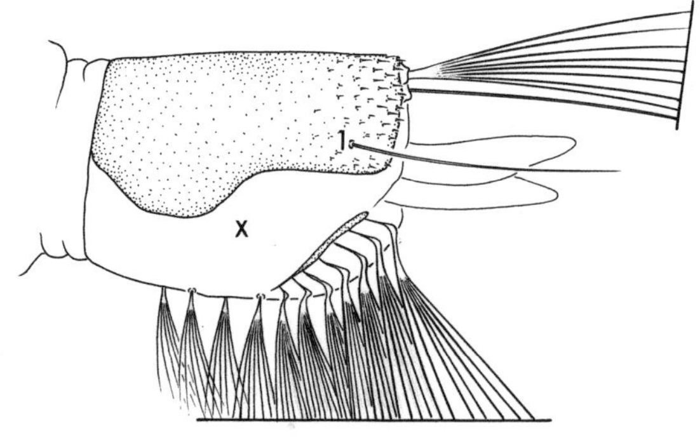


Fig. 806. Lateral view of abdominal segment X: *Oc. grossbecki*

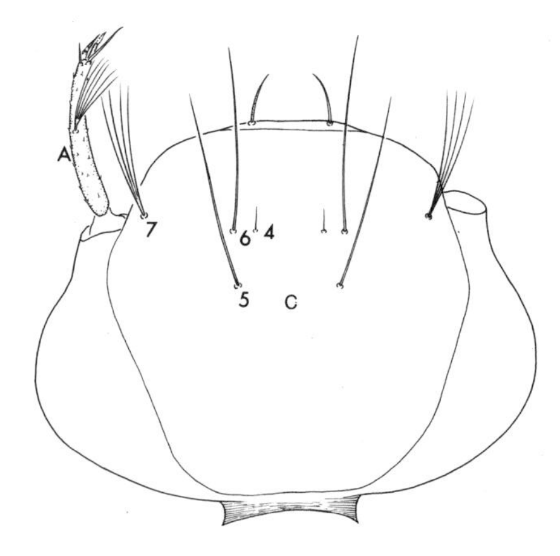


Fig. 805. Dorsal view of head: *Oc. dorsalis*

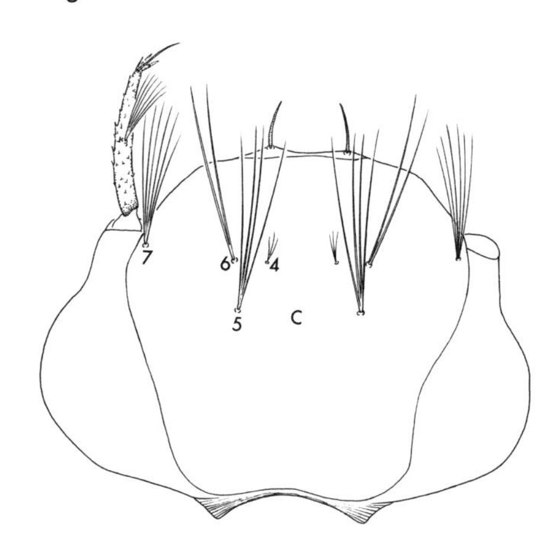


Fig. 807. Dorsal view of head: *Oc. grossbecki*

79(76). Comb scales 36 or more, with median spine no stouter than subapical spinules (figs. 808, 809) ........ 80

Comb scales fewer than 35, with median spine stouter than subapical spinules on at least some scales (figs. 810, 811) ........ 82

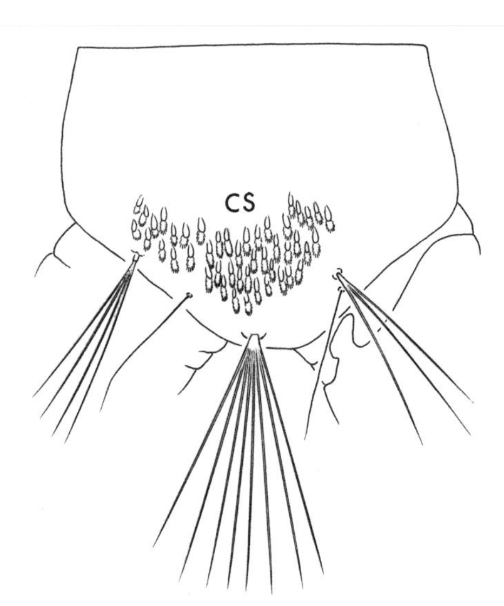


Fig. 808. Lateral view of abdominal segment VIII: *Oc. communis*

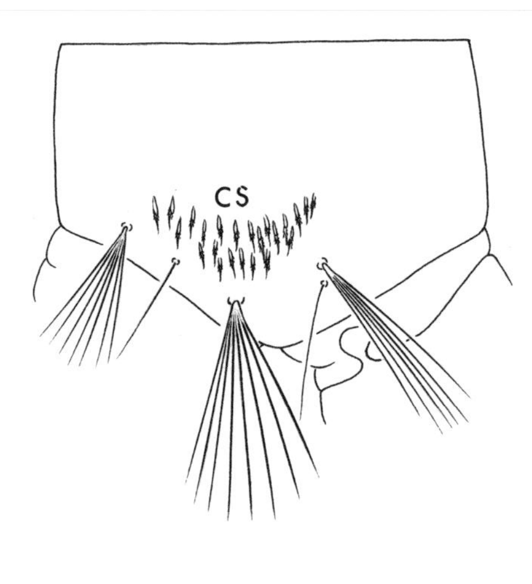


Fig. 810. Lateral view of abdominal segment VIII: *Oc. melanimon*

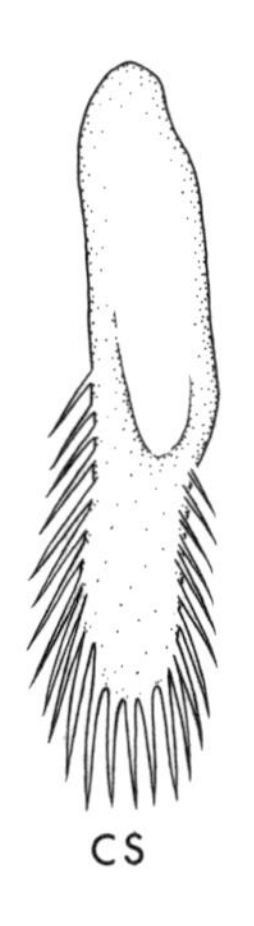


Fig. 809. Comb scale: *Oc. communis*

Fig. 811. Comb scale: *Oc. melanimon*

80(79). Average number of comb scales 36 (17–66); pecten usually with fewer than 21 spines (fig. 812) ........................................................................................ *Oc. tahoensis* (plate 40D)

Average number of comb scales 55 (25–75); pecten usually with 21 or more spines (fig. 813) ........................................................................................................................ 81

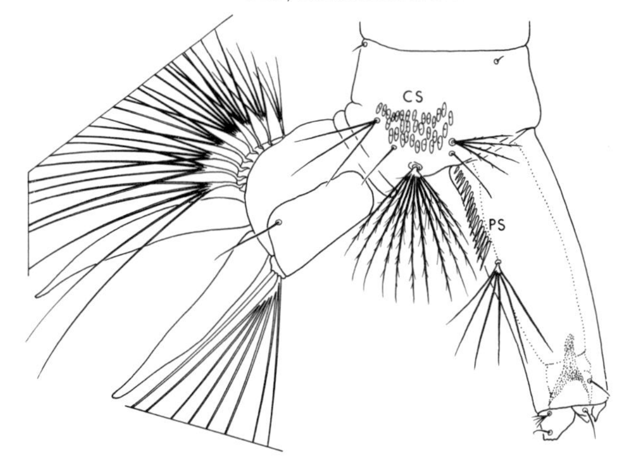


Fig. 812. Lateral view of abdominal segments VIII–X: *Oc. tahoensis*

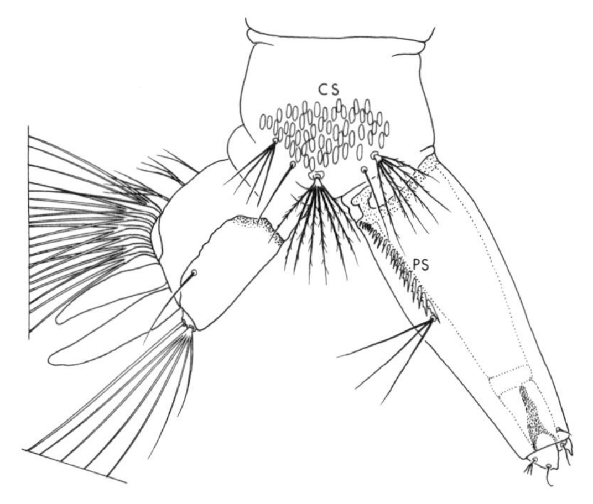


Fig. 813. Lateral view of abdominal segments VIII–X: *Oc. communis*

81(80). Head narrow, usually 1.05–1.17 mm wide (fig. 814); siphon short, usually 0.89–1.01 mm long (fig. 815) ...................................................................... *Oc. churchillensis* (plate 12A)

Head broad, usually 1.2–1.33 mm wide (fig. 816); siphon long, usually 1.0–1.14 mm long (fig. 817) .................................................................................. *Oc. communis* (plate 12C)

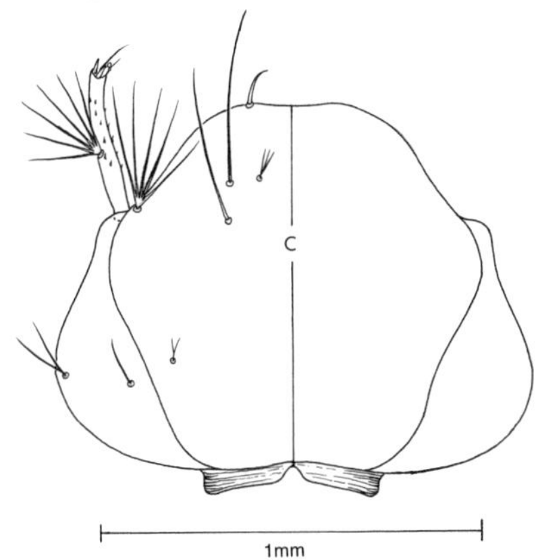


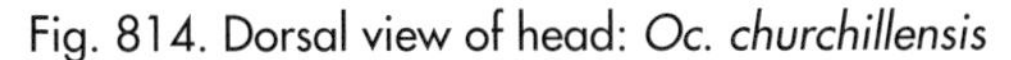


Fig. 814. Dorsal view of head: *Oc. churchillensis*

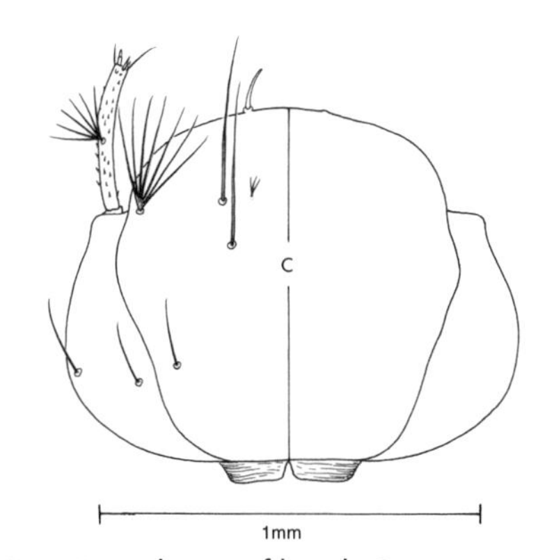


Fig. 816. Dorsal view of head: *Oc. communis*

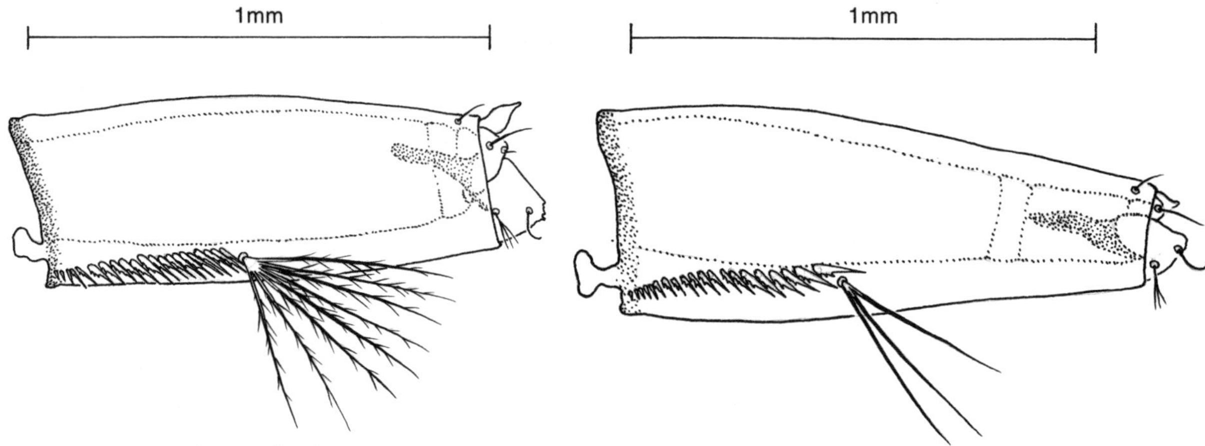


Fig. 815. Lateral view of siphon: *Oc. churchillensis*

Fig. 817. Lateral view of siphon: *Oc. communis*

82(79). Pecten extending distal to middle of siphon (fig. 818); setae 1-IV,V short, multibranched (fig. 819) .............................................................. (in part) *Oc. melanimon* (plate 17B)

Pecten confined to basal 0.5 of siphon (fig. 820); setae 1-IV,V long, single to triple (fig. 821) .............................................................. 83

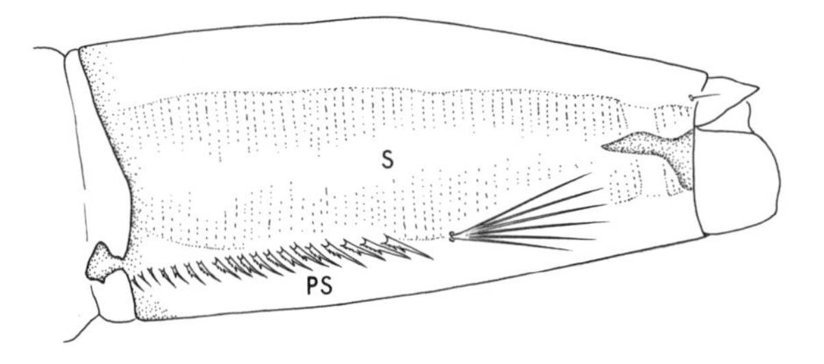


Fig. 818. Lateral view of siphon: *Oc. melanimon*

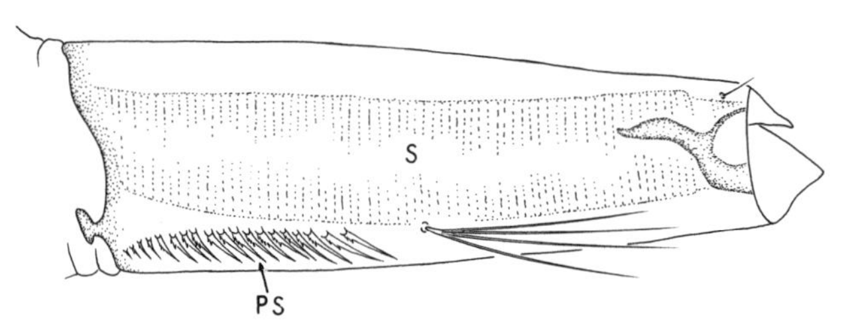


Fig. 820. Lateral view of siphon: *Oc. increpitus*

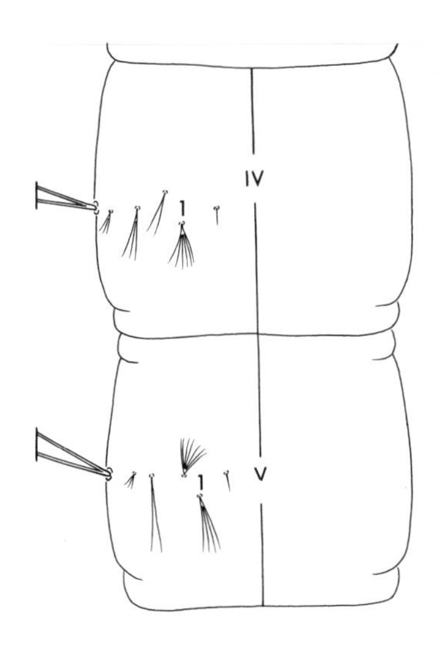


Fig. 819. Dorsal view of abdominal segments IV,V: *Oc. melanimon*

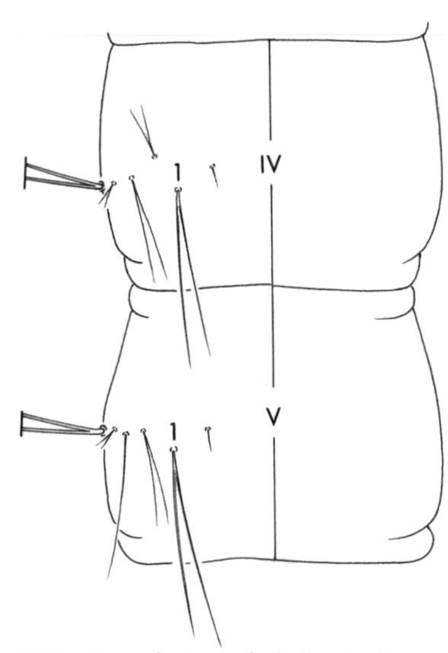


Fig. 821. Dorsal view of abdominal segments IV–V: *Oc. increpitus*

83(82). Comb scales with apical spine longer than subapical spinules (fig. 822); seta 7-P usually double (fig. 823); posterior border of saddle finely aciculate (fig. 824) .......................... .......................................................................................... *Oc. implicatus* (plate 16B)

Comb scales with apical spine subequal to subapical spinules (fig. 825); seta 7-P usually with 3 or 4 branches (fig. 826); posterior border of saddle coarsely aciculate (fig. 827) ........ 84

Fig. 822. Comb scale: *Oc. implicatus*

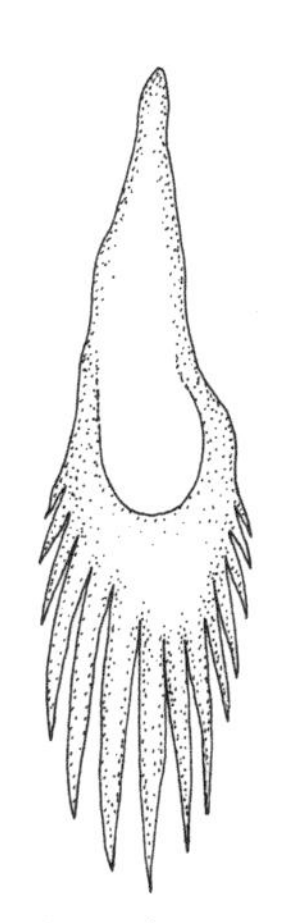

Fig. 825. Comb scale: *Oc. increpitus*

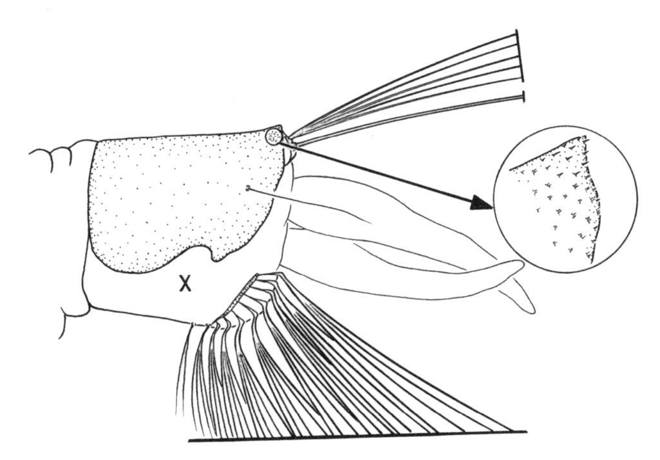


Fig. 824. Lateral view of abdominal segment X: *Oc. implicatus*

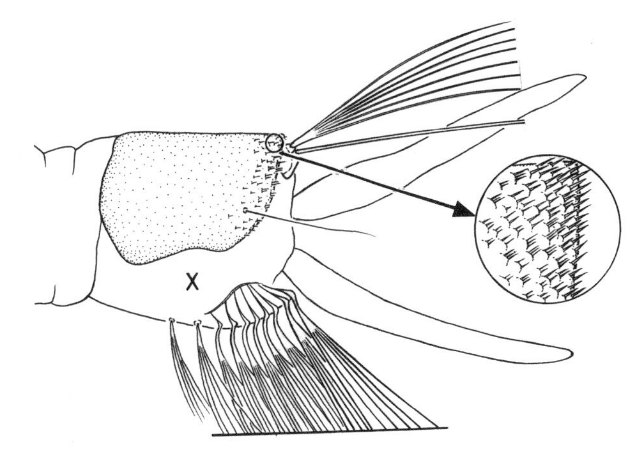


Fig. 827. Lateral view of abdominal segment X: *Oc. increpitus*

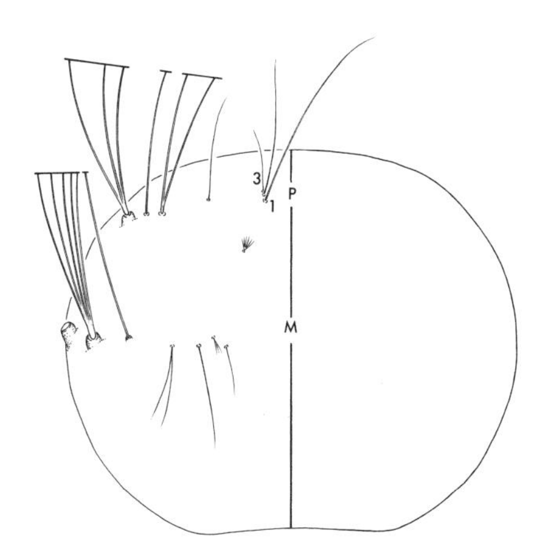


Fig. 823. Dorsal view of thorax: *Oc. implicatus*

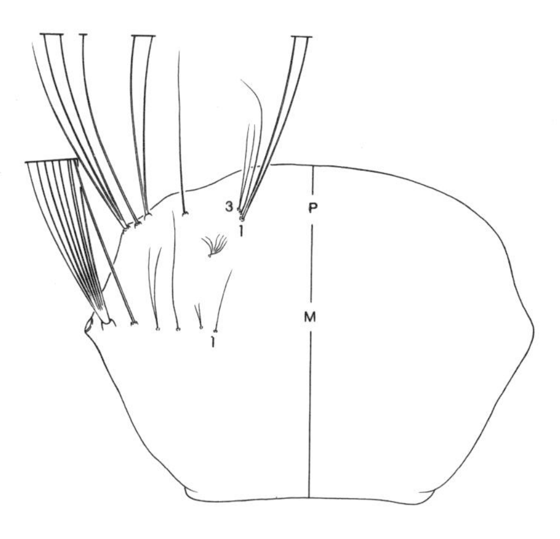


Fig. 826. Dorsal view of thorax: *Oc. increpitus*

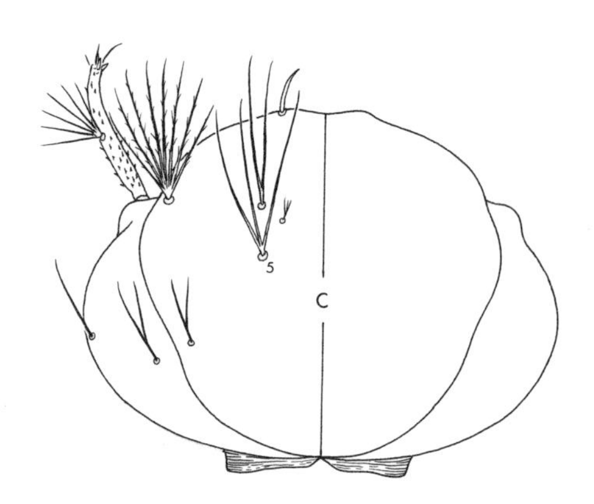

Fig. 828. Dorsal view of head: *Oc. clivis*

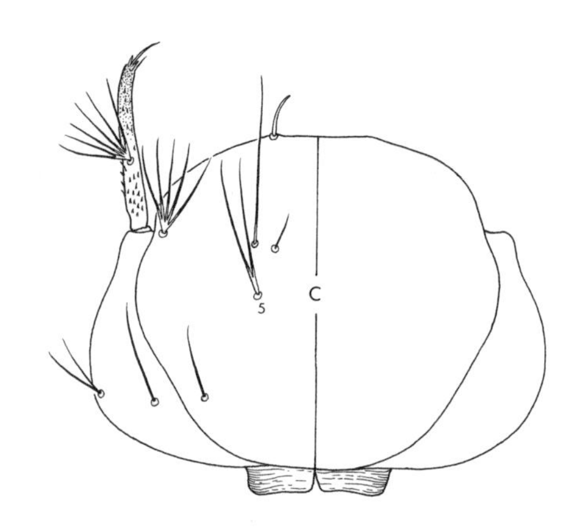


Fig. 829. Dorsal view of head: *Oc. washinoi*

84(83). Seta 5-C usually triple (fig. 626) .................................................... *Oc. clivis* (plate 39C)

Seta 5-C usually single or double (fig. 829) ............................ *Oc. increpitus* (plate 16C)
*Oc. washino* (plate 41A)

## Key to Fourth Instar Larvae of Genus *Anopheles*

1. Setae 5–7-C small, single or double (fig. 830); setae 6-I–VI plumose (fig. 831) ............. 2

   Seta 5–7-C large, multibranched (fig. 832); setae 6-IV–VI not plumose (fig. 833) ......... 3

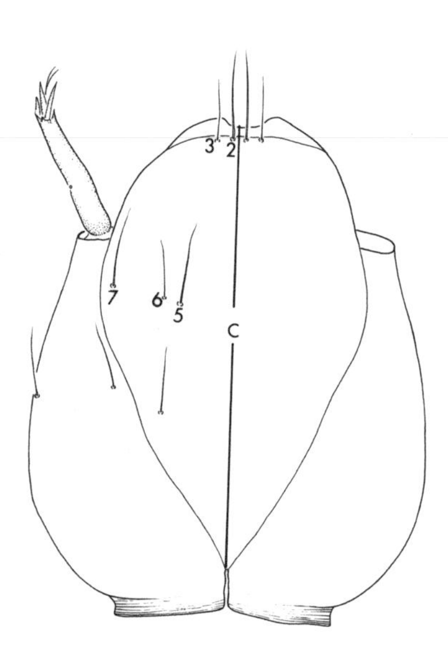


Fig. 830. Dorsal view of head: *An. judithae*

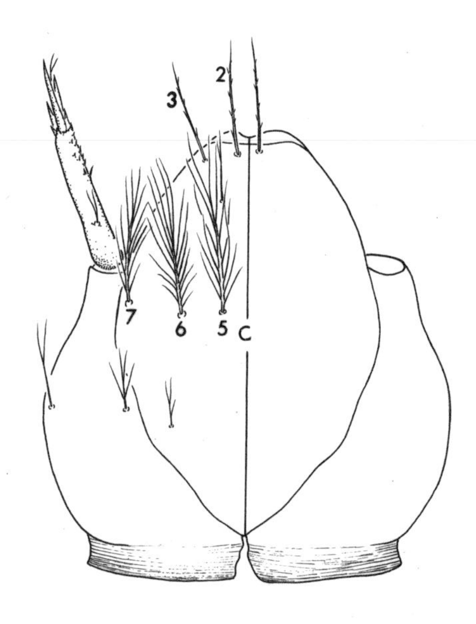


Fig. 832. Dorsal view of head: *An. albimanus*

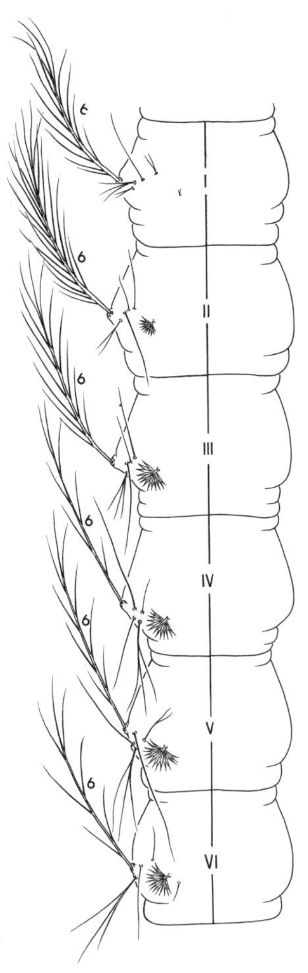


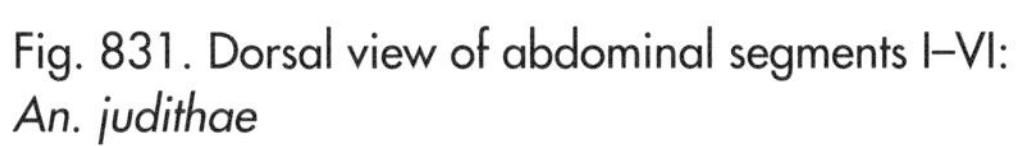

Fig. 831. Dorsal view of abdominal segments I–VI: *An. judithae*

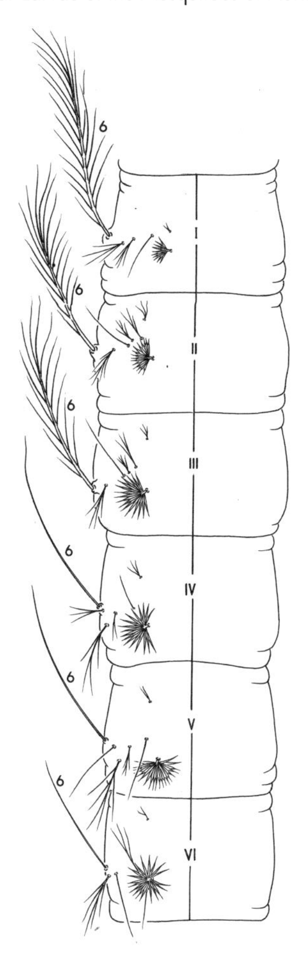


Fig. 833. Dorsal view of abdominal segments I–VI: *An. albimanus*

2(1). Setae 2-C widely separated, closer to setae 3-C than to each other (fig. 834); setae 13-II–V and VII usually with 3 branches (fig. 835) ....................................... *barberi* (plate 24D)

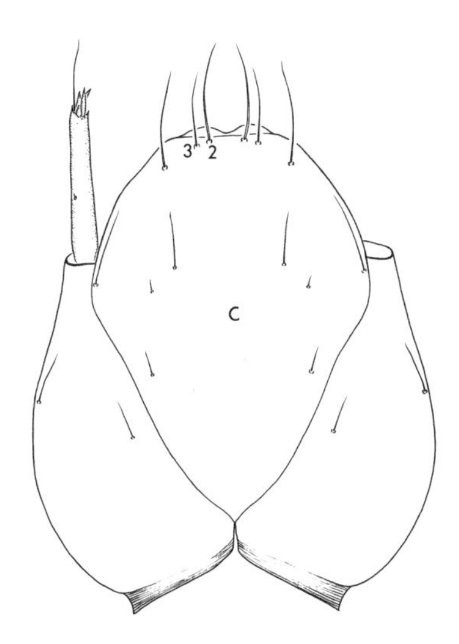


Fig. 834. Dorsal view of head: *An. barberi*

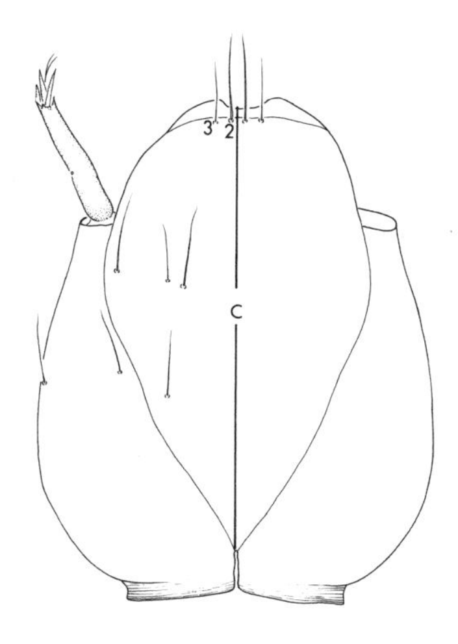


Fig. 836. Dorsal view of head: *An. judithae*

Setae 2-C close together, closer to each other than to setae 3-C (fig. 836); setae 13-II–V and VII usually single (fig. 837) ............................................................ *judithae* (plate 41A)

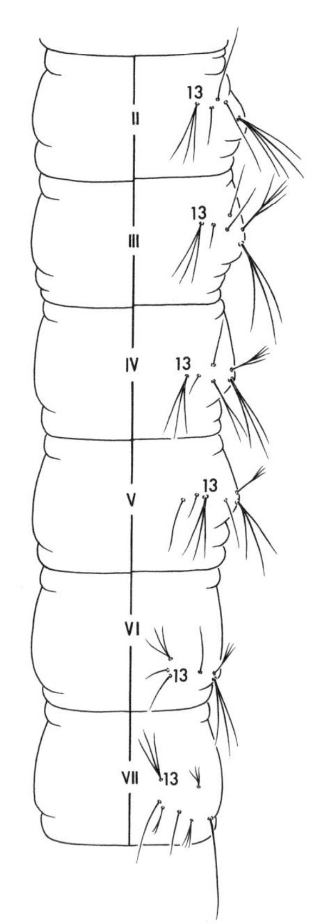


Fig. 835. Ventral view of abdominal segments II–VII: *An. barberi*

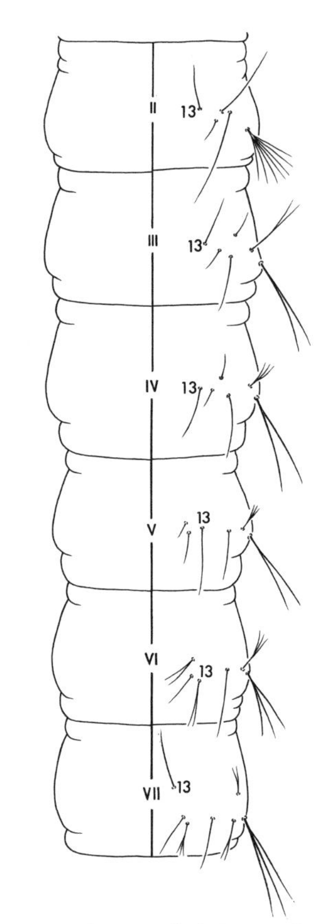


Fig. 837. Ventral view of abdominal segments II–VII: *An. judithae*

3(1).Seta 3-C unbranched (fig. 838) ......................................................................................... 4

Seta 3-C with 5 or more branches (fig. 839) ................................................................ 6

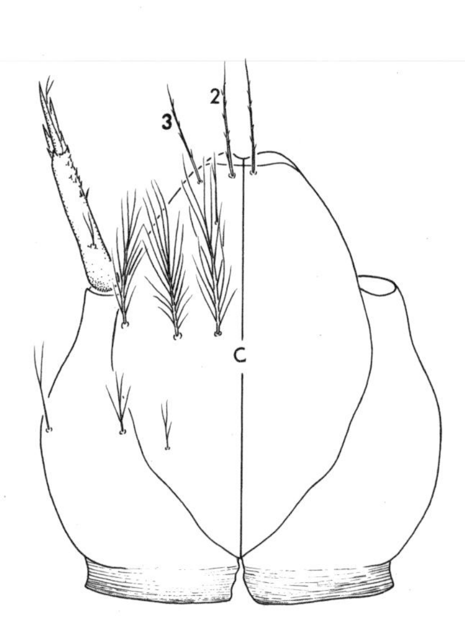


Fig. 838. Dorsal view of head: *An. albimanus*

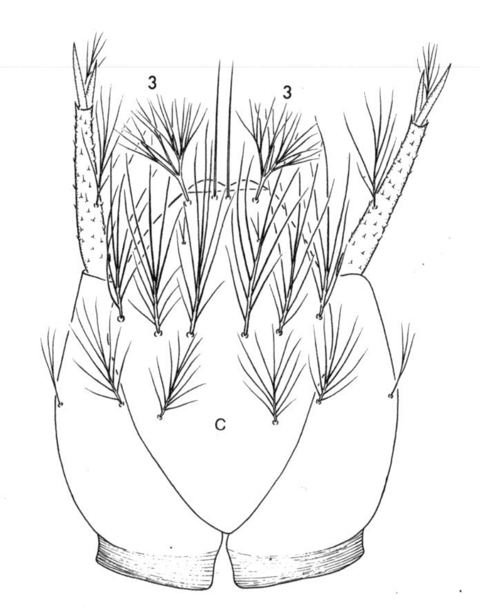


Fig. 839. Dorsal view of head: *An. quadrimaculatus*

4(3). Setae 1-I–VII palmate, leaflets with margins smooth (fig. 840); setae 2,3-C aciculate (fig. 841) .......... *albimanus* (plate 40C)

Seta 1 palmate in III–VII, leaflets with serrate margins (fig. 842) setae 2,3-C smooth (fig. 843) .......... 5

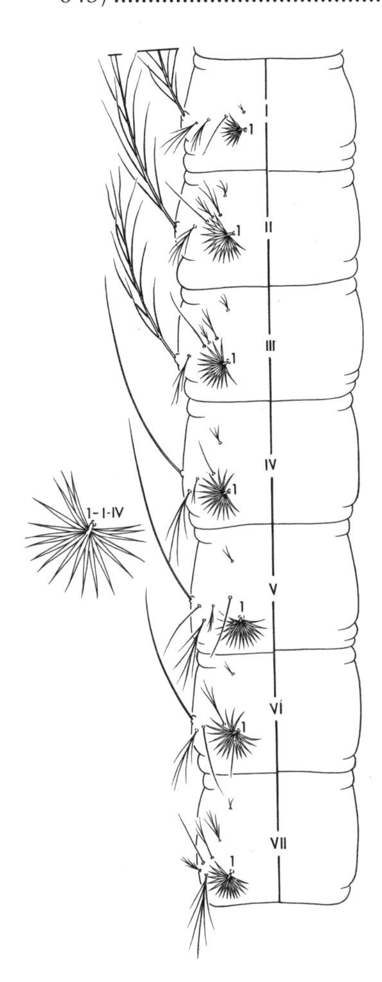


Fig. 840. Dorsal view of abdominal segments I–VII: *An. albimanus*

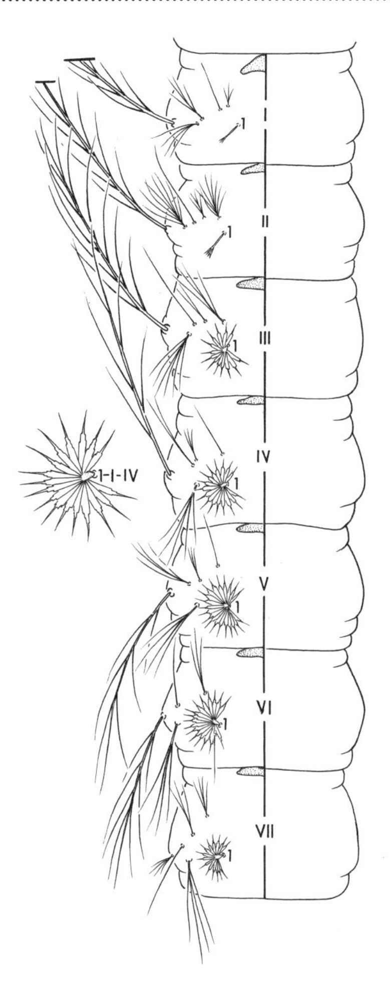


Fig. 842. Dorsal view of abdominal segments I–VII: *An. pseudopunctipennis*

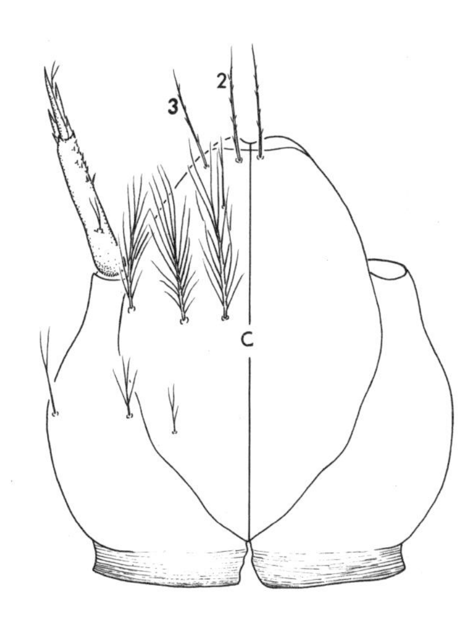


Fig. 841. Dorsal view of head: *An. albimanus*

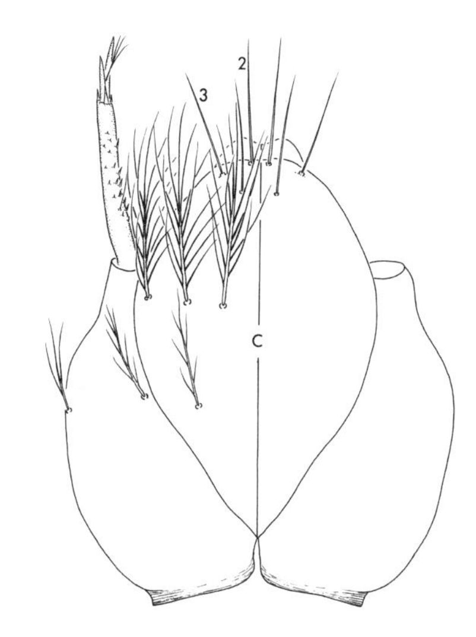


Fig. 843. Dorsal view of head: *An. pseudopunctipennis*

5(4).Spiracular apparatus with caudal margin of posterolateral spiracular lobes produced into elongated dark processes (fig. 844); seta 2-IV single (fig. 845) ........................................ ........................................................................................ *pseudopunctipennis* (plate 26D)

Spiracular apparatus without elongated process on caudal margin of posterolateral spiracular lobe (fig. 846); seta 2-IV usually double or triple (fig. 847)..... *franciscanus* (plate 25D)

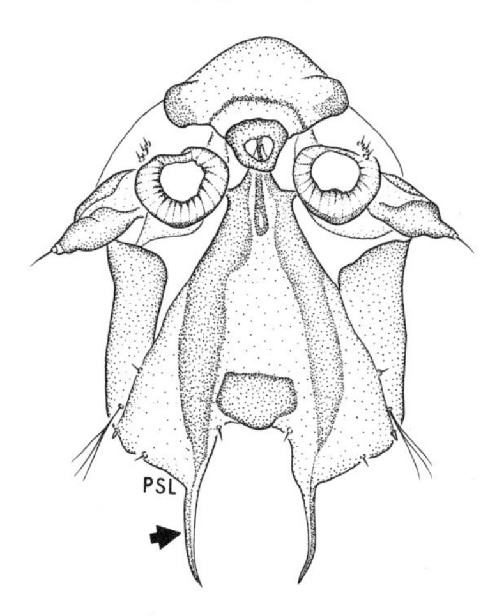


Fig. 844. Spiracular apparatus: *An. pseudopunctipennis*

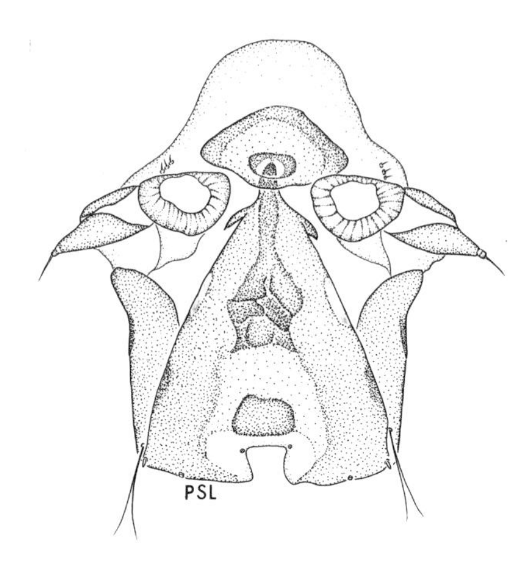


Fig. 846. Spiracular apparatus: *An. franciscanus*

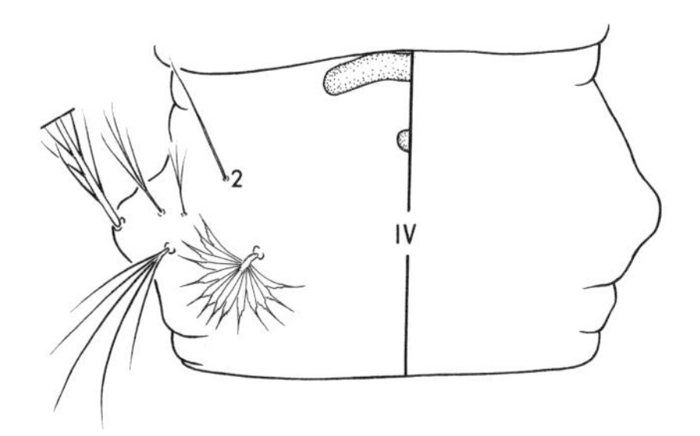


Fig 845. Dorsal view of abdominal segment IV: *An. pseudopunctipennis*

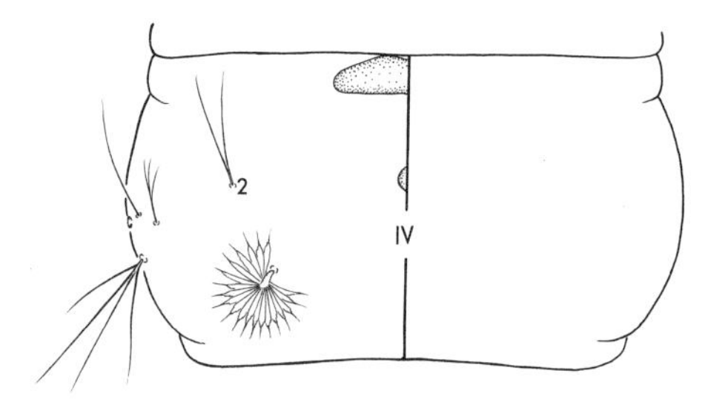


Fig. 847. Dorsal view of abdominal segment IV: *An. franciscanus*

6(3). Seta 3-C with fewer than 11 branches (fig. 848) .............................. *atropos* (plate 24C)

Seta 3-C dendritic, densely branched (fig. 849) ............................................................ 7

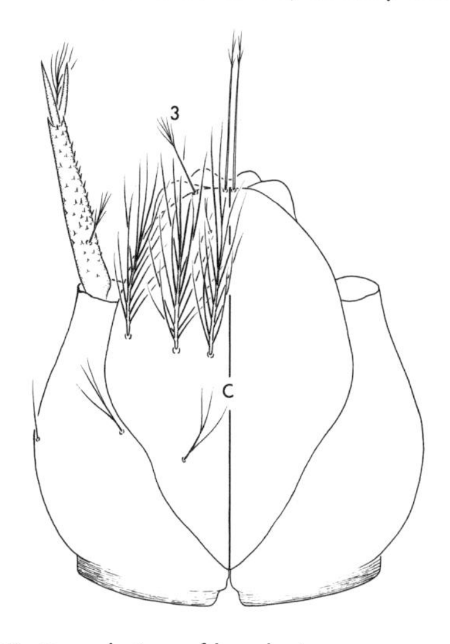


Fig. 848. Dorsal view of head: *An. atropos*

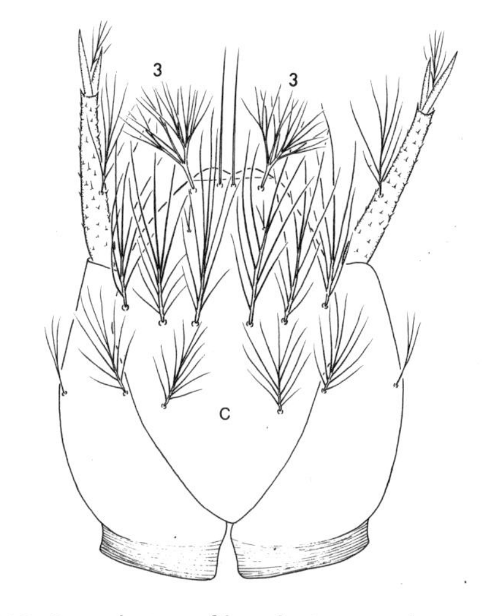


Fig. 849. Dorsal view of head: *An. quadrimaculatus*

7(6). Seta 0 well developed on IV,V, with 4 or more branches, about equal to 2-IV,V (fig. 850) ........ *crucians* (plate 25B)

Seta 0 minute on IV,V, single to triple, much smaller than 2-IV,V (fig. 851) ........ 8

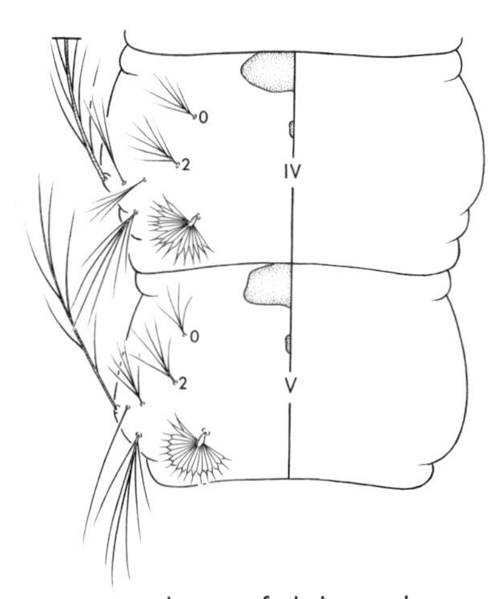


Fig. 850. Dorsal view of abdominal segments IV,V: *An. crucians*

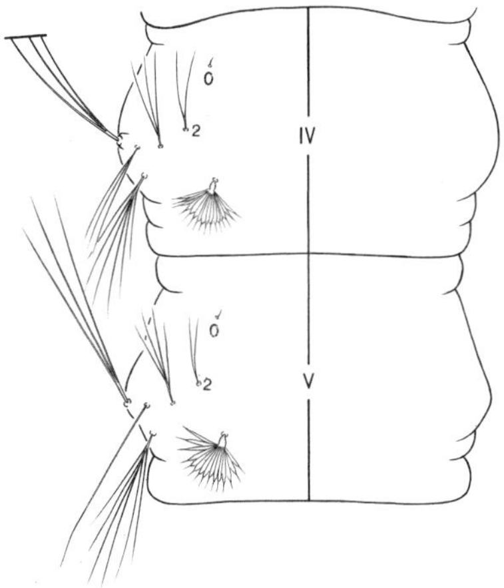

Fig. 851. Dorsal view of abdominal segments IV,V: *An. punctipennis*

8(7). Seta 2-C simple, sparsely aciculate toward apex (fig. 852); seta 1-P with 3–5 strong branches from near base (fig. 853) ........ *walkeri* (plate 28A)

Seta 2-C simple or forked in outer 0.5, without aciculae (fig. 854); seta 1-P weak, single or branched in outer 0.5 (fig. 855) ........ 9

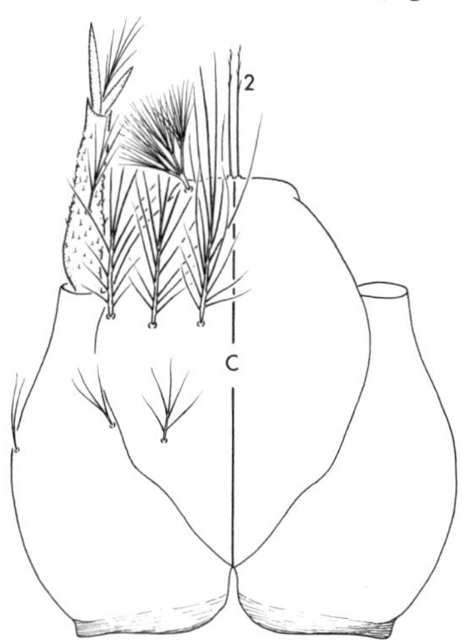


Fig. 852. Dorsal view of head: *An. walkeri*

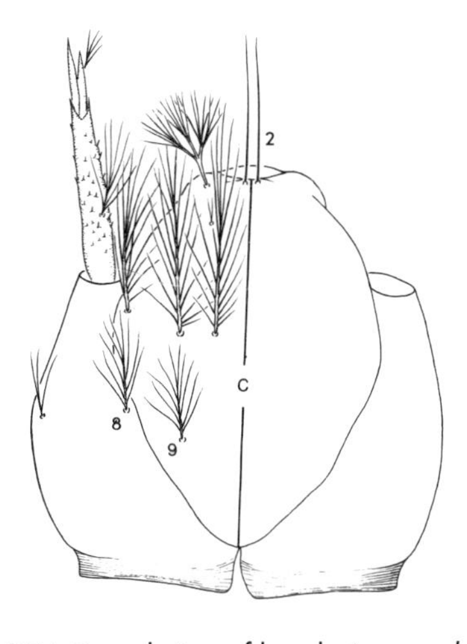


Fig. 854. Dorsal view of head: *An. quadrimaculatus*

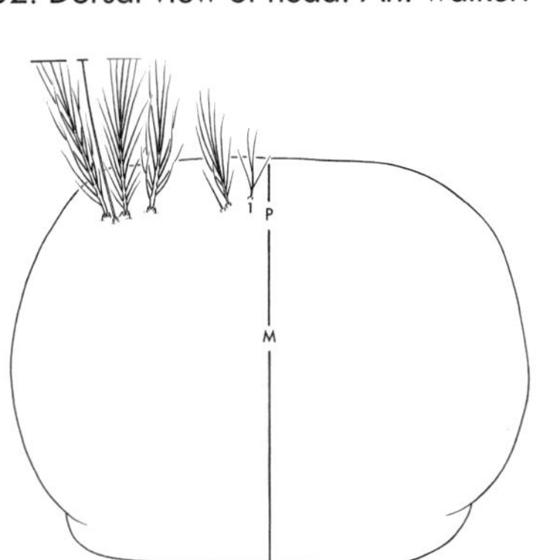


Fig. 853. Dorsal view of thorax: *An. walkeri*

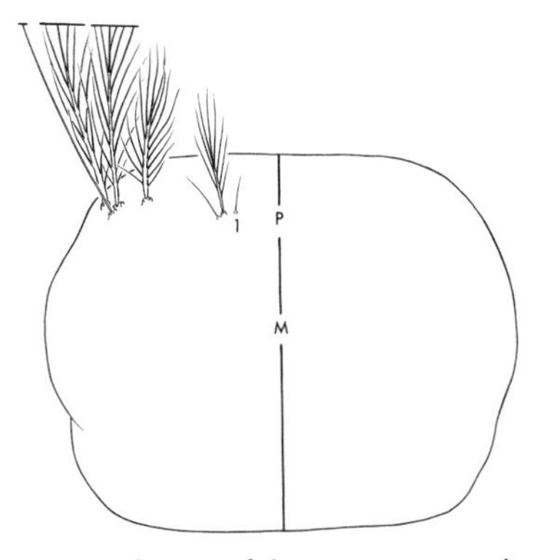


Fig. 855. Dorsal view of thorax: *An. quadrimaculatus*

9(8). Setae 1-IV–VI fully palmate, 1-III and VII not more than 0.7 as large, leaflets usually with fine marginal serrations (fig. 856) ........ 10

Setae 1-III–VII fully palmate, apical 0.5 of leaflets with coarse marginal serrations (fig. 857) ........ 11

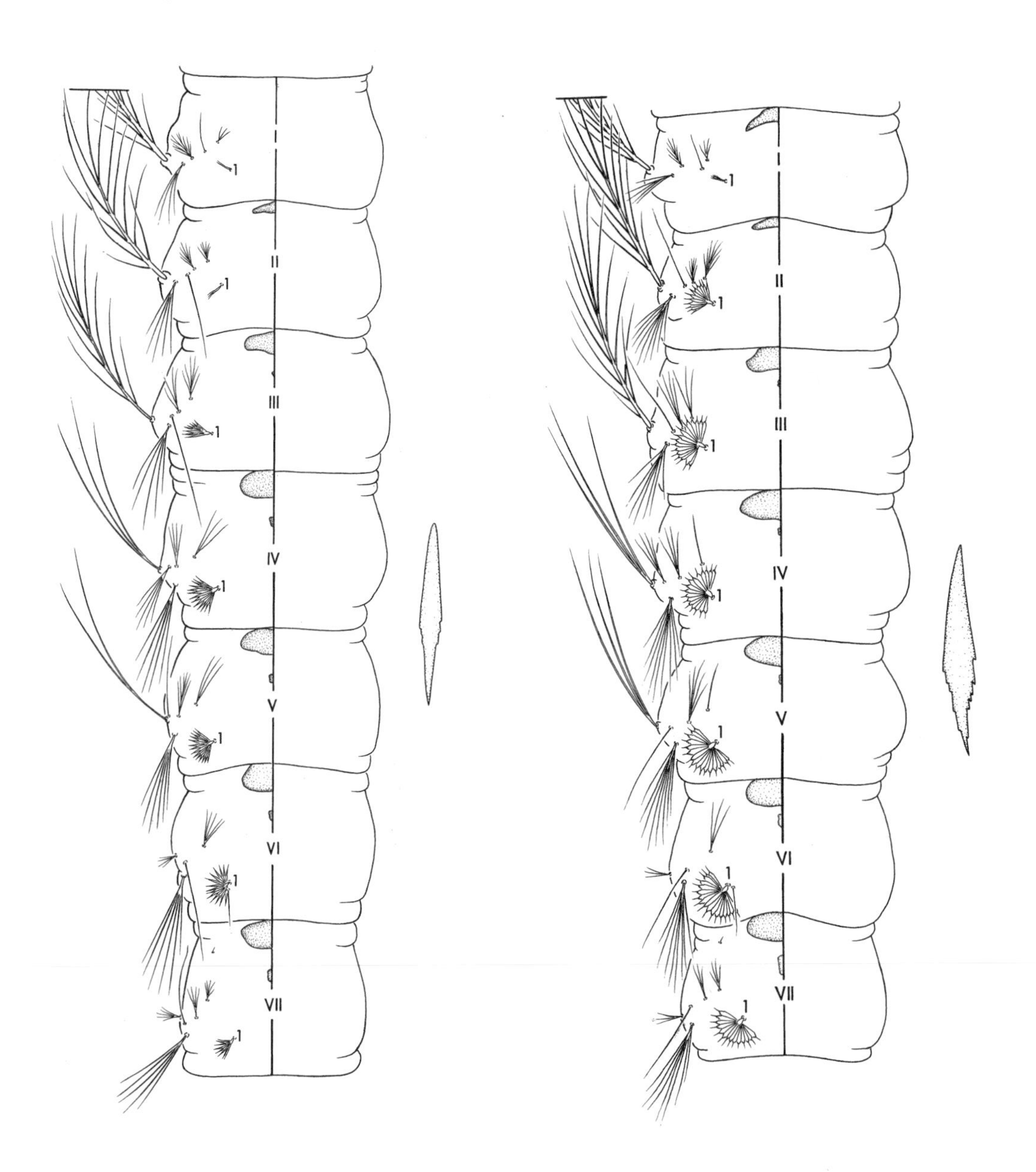


Fig. 856. Dorsal view of abdominal segments I–VII: *An. bradleyi*

Fig. 857. Dorsal view of abdominal segments I–VII: *An. quadrimaculatus*

10(9). Seta 1-III better developed palmate seta than 1-I (fig. 858); seta 5-II usually with fewer than 9 branches (fig. 859) ........ *bradleyi* (plate 25A)

Seta 1-III not much better developed palmate seta than 1-I (fig. 860); seta 5-II with 9 or more branches (fig. 861) ........ *georgianus* (plate 26B)

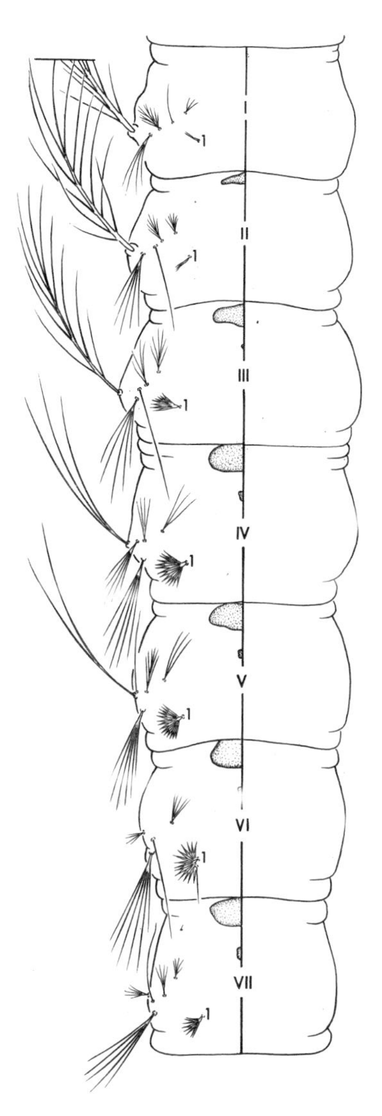


Fig. 858. Dorsal view of abdominal segments I–VII: *An. bradleyi*

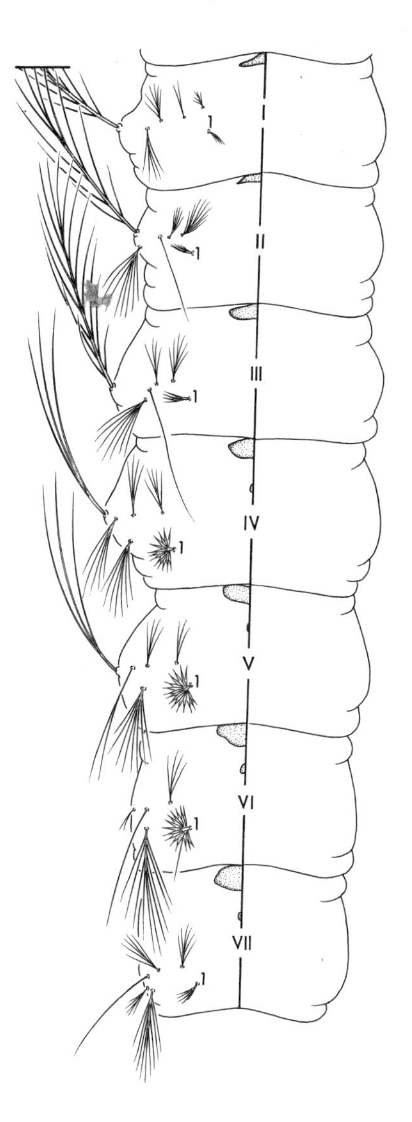


Fig. 860. Dorsal view of abdominal segments I–VII: *An. georgianus*

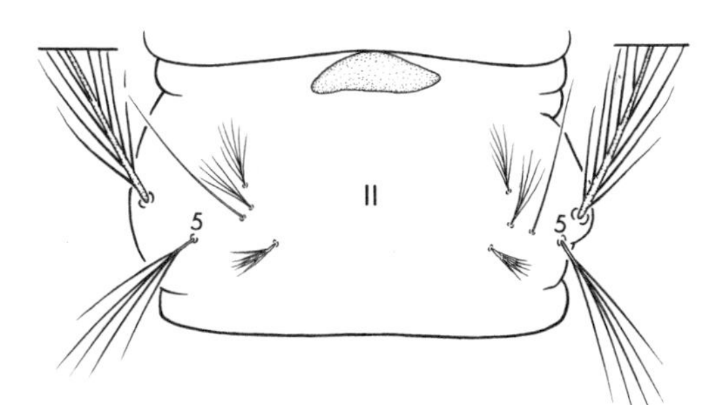


Fig. 859. Dorsal view of abdominal segment II: *An. bradleyi*

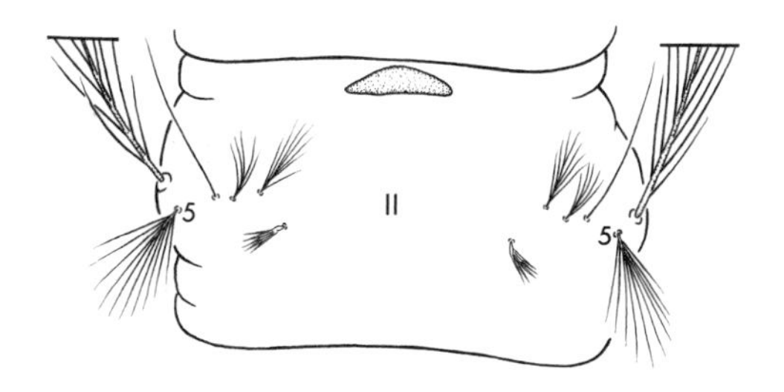


Fig. 861. Dorsal view of abdominal segment II: *An. georgianus*

11(9). Seta 2-C usually with 2–5 branches in outer 0.5 (fig. 862) .................. *earlei* (plate 25C)

Seta 2-C single (fig. 863) .................................................................................. 12

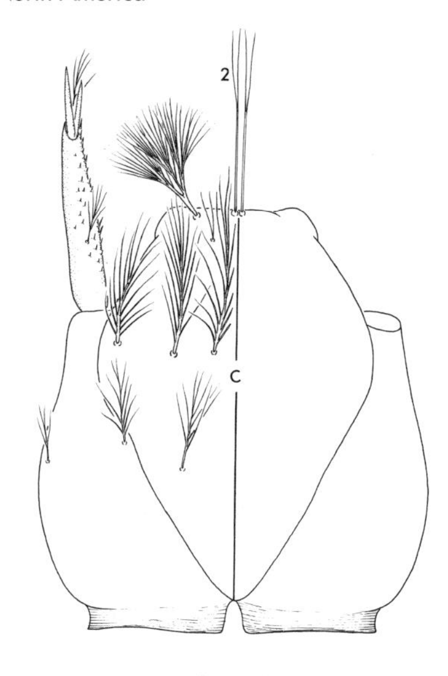


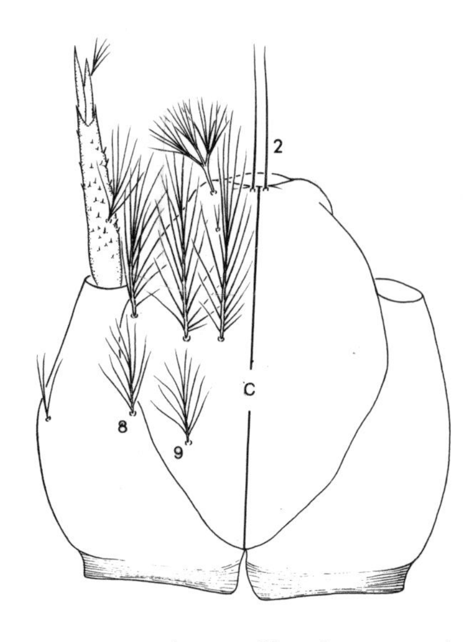

Fig. 862. Dorsal view of head: *An. earlei*

Fig. 863. Dorsal view of head: *An. quadrimaculatus*

12(11). Alveoli of setae 2-C separated by more than diameter of one alveolus; setae 8,9-C large, usually with 8–10 branches (fig. 864) ............................ (*quadrimaculatus* complex) 13

Alveoli of setae 2-C closer together than diameter of one alveolus; setae 8,9-C smaller, usually with 5–7 branches (fig. 865) .......................................................................... 17

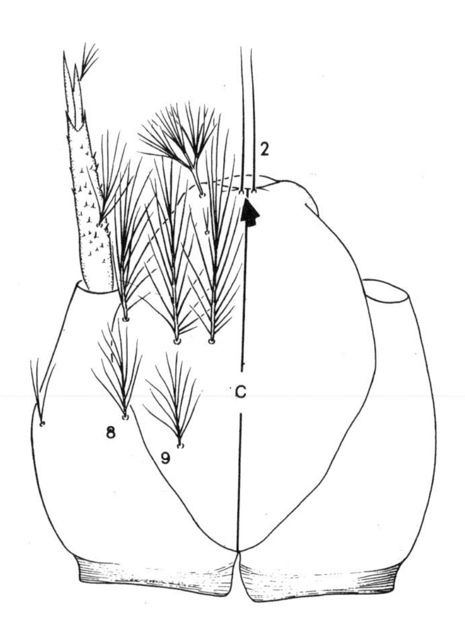


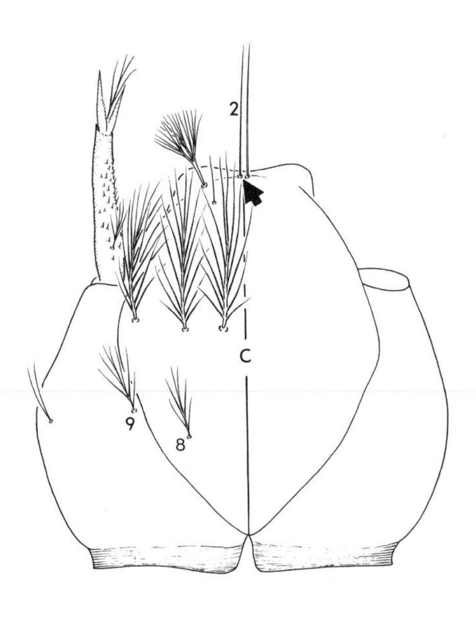

Fig. 864. Dorsal view of head: *An. quadrimaculatus*

Fig. 865. Dorsal view of head: *An. punctipennis*

13(12). Sum of both setae 1-A usually 18 or more branches (fig. 866); sum of both setae 8-III plus 8-VI usually 19 or more branches (fig. 867); seta 2-C 1.29 or less length of seta 3-C (fig. 866) .......................................................................................... *maverlius* (plate 26C)

Sum of both setae 1-A usually 17 or fewer branches (fig. 868); sum of setae 8-III plus 8-VI usually 18 or fewer branches (fig. 869); seta 2-C 1.30 or greater length of 3-C (fig. 868) .......................................................................................................................... 14

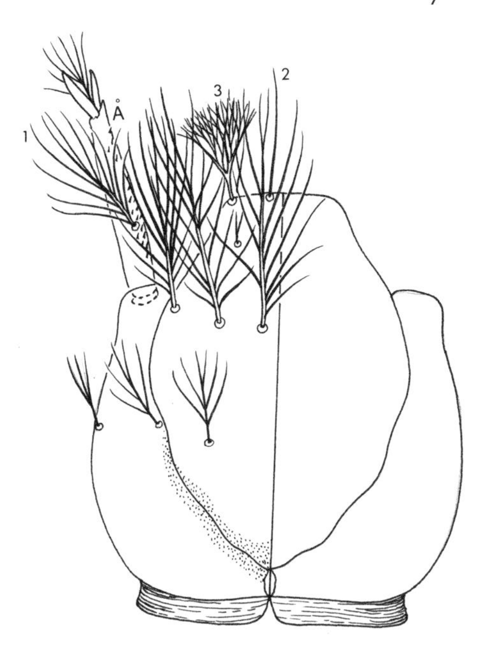


Fig. 866. Dorsal view of head: *An. maverlius*

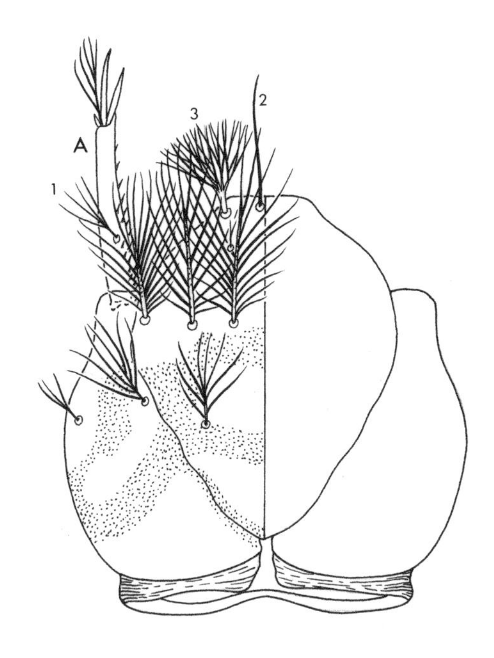


Fig. 868. Dorsal view of head: *An. quadrimaculatus*

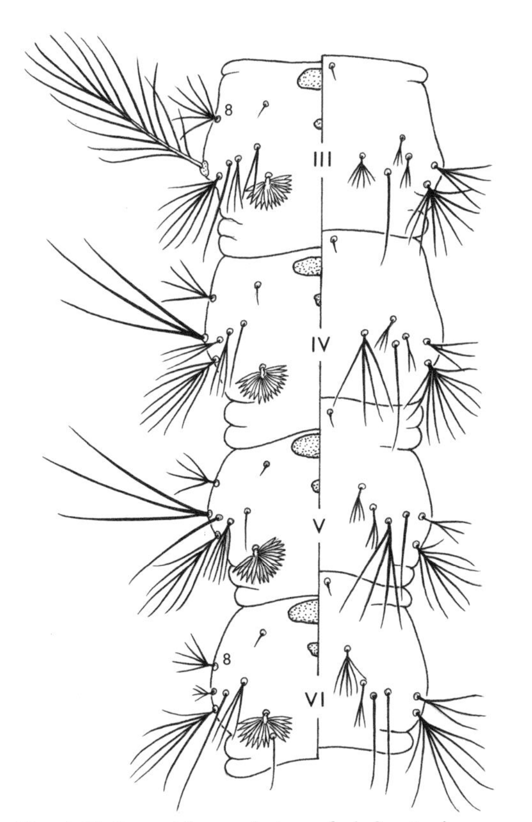


Fig. 867. Dorsal/ventral view of abdominal segments III–VI: *An. maverlius*

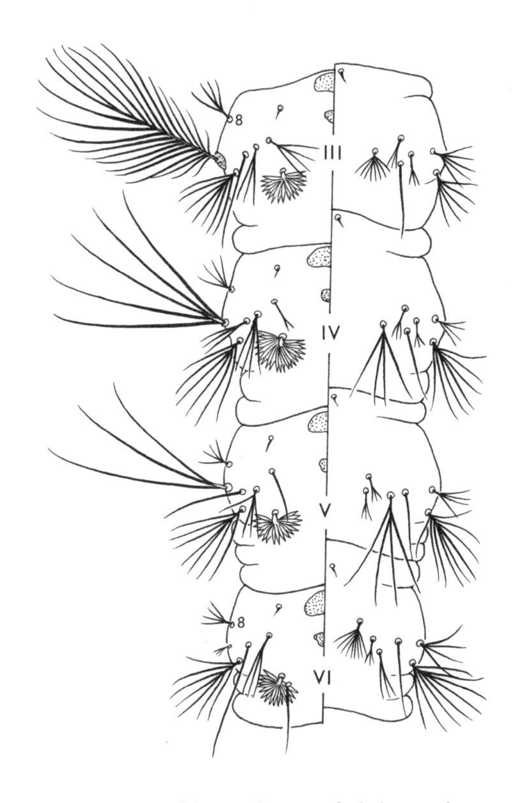


Fig. 869. Dorsal/ventral view of abdominal segments III–VI: *An. quadrimaculatus*

14(13). Pecten plates each with 6–8 long spines (fig. 870); seta 1-A usually inserted 0.31 or more from base of antenna, whose length is 0.31 mm or longer (fig. 871) .......................... 15

One or both pecten plates each with 9–11 long spines (fig. 872); seta 1-A usually inserted 0.30 or less from base of antenna, whose length is 0.30 mm or shorter (fig. 873) ....... 16

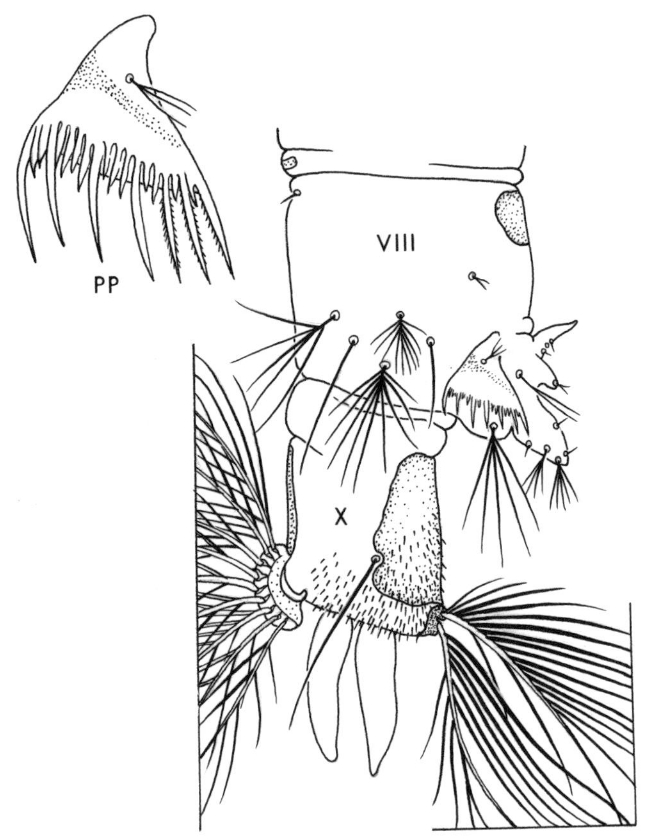


Fig. 870. Lateral view of abdominal segments VIII–X: *An. quadrimaculatus*

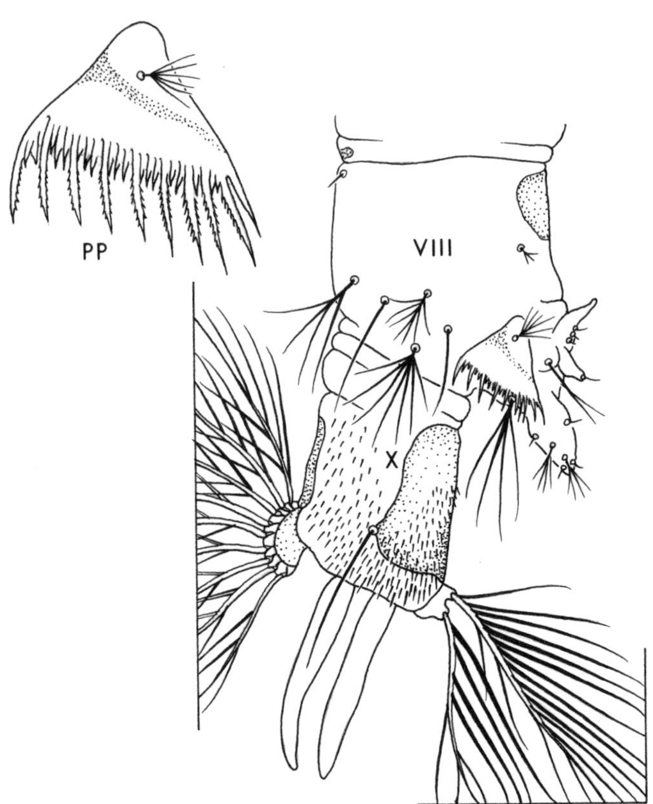


Fig. 872. Lateral view of abdominal segments VIII–X: *An. inundatus*

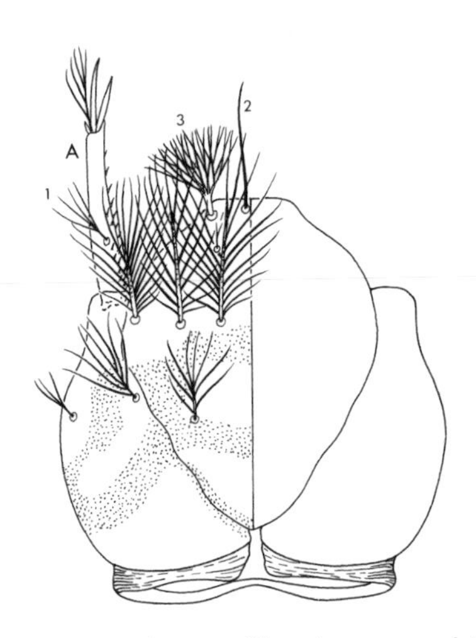


Fig. 871. Dorsal view of head: *An. quadrimaculatus*

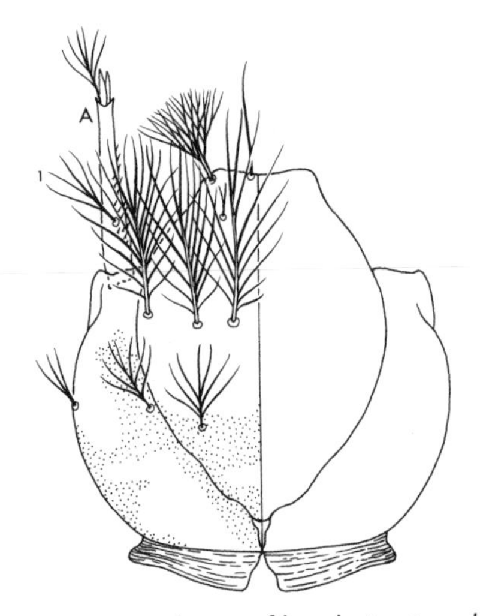


Fig. 873. Dorsal view of head: *An. inundatus*

15(14). Sum of both setae 3-C, 25–63 branches, more or less widely separated distally (fig. 874); sum of both setae 8-V usually 7–10 branches (fig. 875); segment VIII usually with anteromedian sternal plate (fig. 876) .......................................... *quadrimaculatus* (plate 27B)

Sum of both setae 3-C usually 64 or more branches, bunched together distally (fig. 877); sum of both setae 8-V usually 6 or fewer branches (fig. 878); segment VIII usually without anteromedian sternal plate (fig. 879) ...................................... *smaragdinus* (plate 27D)

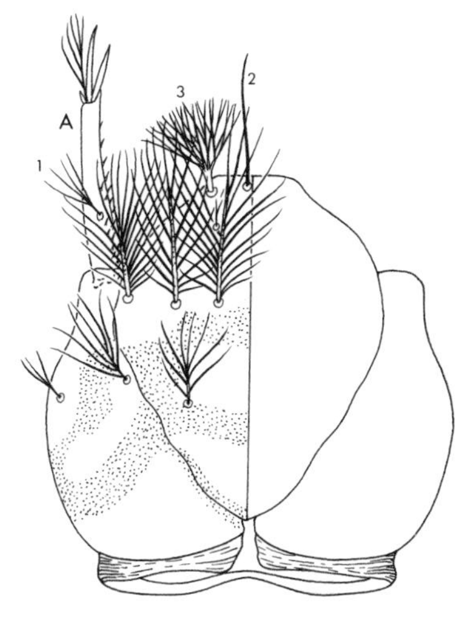


Fig. 874. Dorsal view of head: *An. quadrimaculatus*

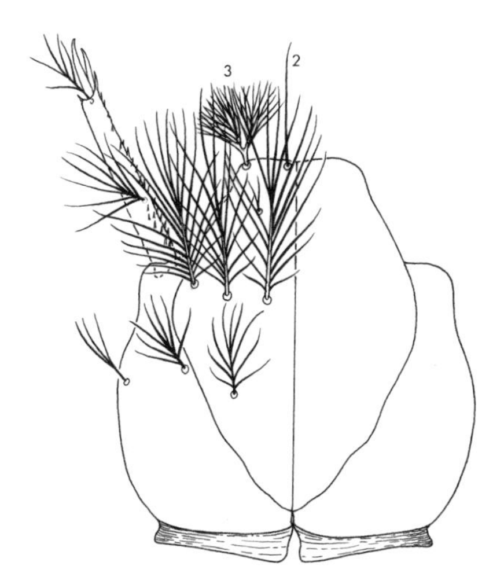


Fig. 877. Dorsal view of head: *An. smaragdinus*

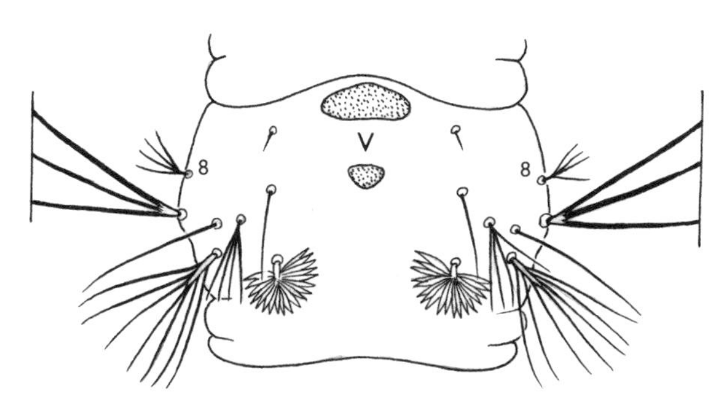


Fig. 875. Dorsal view of abdominal segment V: *An. quadrimaculatus*

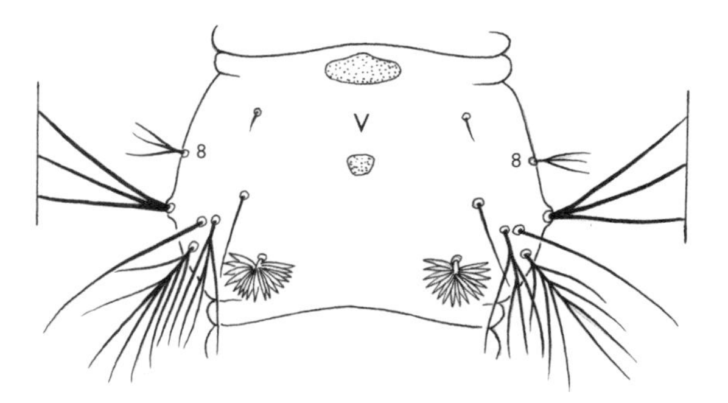


Fig. 878. Dorsal view of abdominal V: *An. smaragdinus*

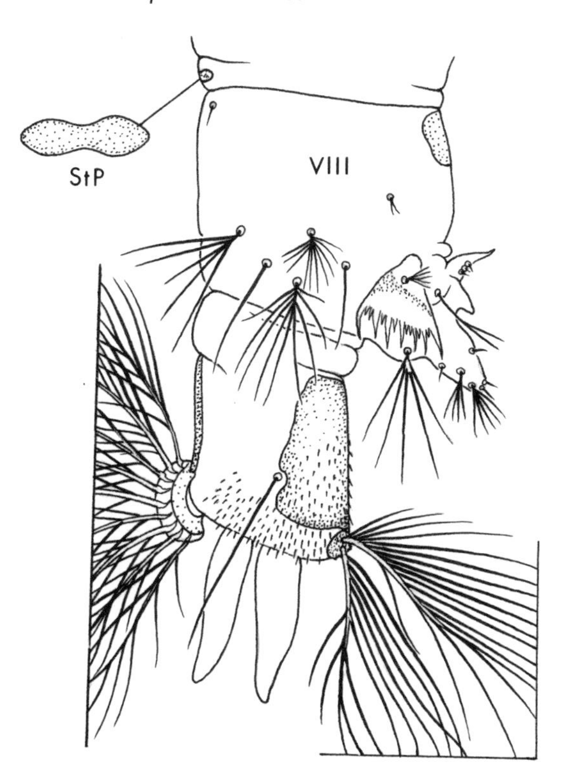


Fig. 876. Lateral view of abdominal segments VII–X: *An. quadrimaculatus*

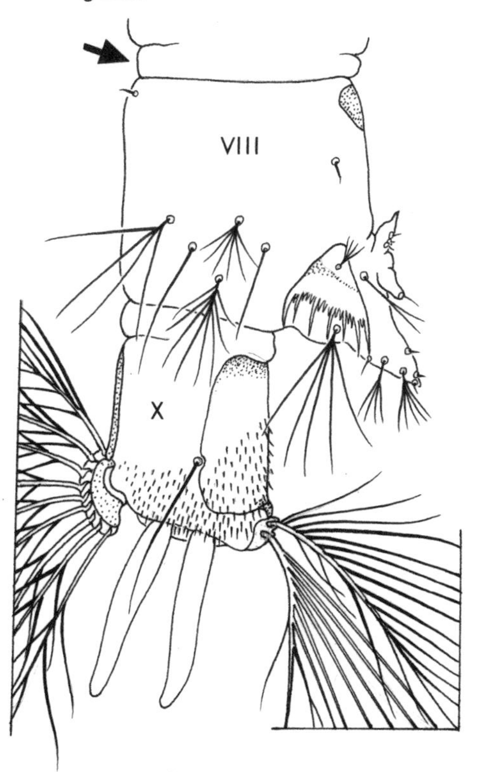


Fig. 879. Lateral view of abdominal segments VII–X: *An. smaragdinus*

16(14). Sum of both setae 8-II and 9-II usually 25 or fewer branches; sum of both setae 2-I and 9-I, 24 or fewer branches (fig. 880); sum of both setae 14-P often 17 or more branches .... ................................................................................................ *diluvialis* (plate 40C)

Sum of both setae 8-II and 9-II usually 26 or more branches; sum of both setae 2-I and 9-I usually 25 or more branches (fig. 881); sum of both 14-P usually 16 or fewer branches ................................................................................................ *inundatus* (plate 40B)

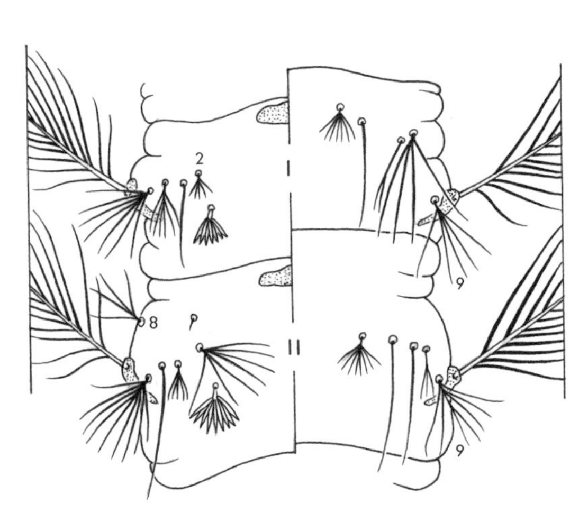


Fig. 880. Dorsal/ventral view of thorax and abdominal segments I,II: *An. diluvialis*

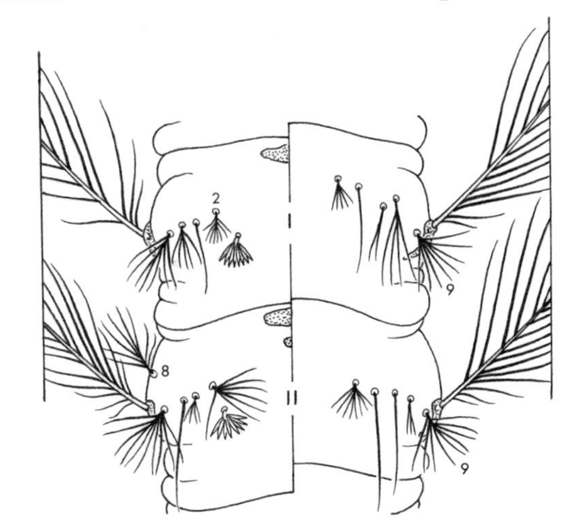


Fig. 881. Dorsal/ventral view of thorax and abdominal segments I,II: *An. inundatus*

17(12). Setae 2-IV,V usually single (fig. 882) .......................................... *occidentalis* (plate 42B)
(in part) *perplexens* (plate 32C)

Setae 2-IV–V usually double or triple (fig. 883) ........................................................ 18

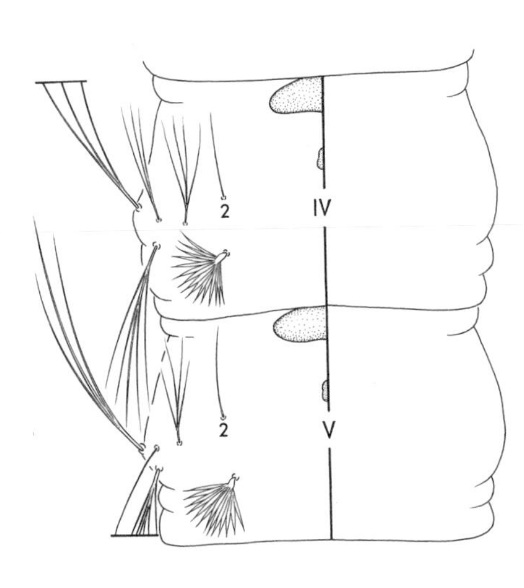


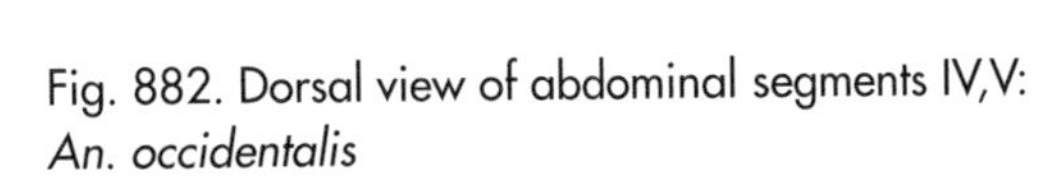

Fig. 882. Dorsal view of abdominal segments IV,V: *An. occidentalis*

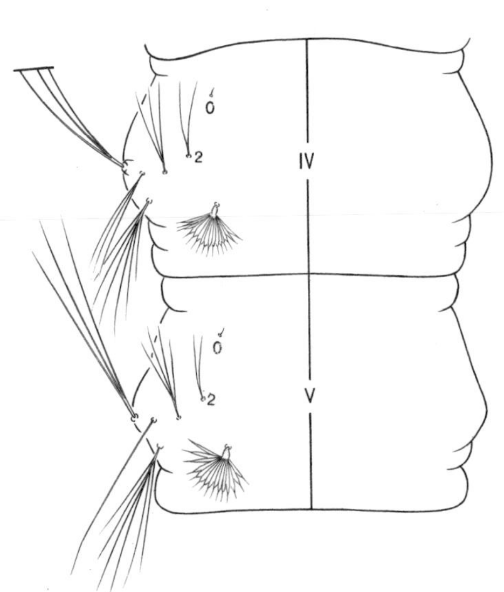


Fig. 883. Dorsal view of abdominal segments IV,V: *An. quadrimaculatus*

18(17). Segments IV,V with 3 small accessory tergal plates (fig. 884); seta 1-A attached at or distal to basal 0.33 of antenna; dorsal apotome with integument spotted (fig. 885) ................ ................................................................................................ *freeborni* (plate 26A)
*hermsi* (plate 42A)

Segments IV,V with only 1 accessory tergal plate on IV,V (fig. 886); seta 1-A attached within basal 0.33 of antenna; dorsal apotome with integument irregularly banded (fig. 887) ........................................................................................ *punctipennis* (plate 27A)
(in part) *perplexens*(plate 32C)

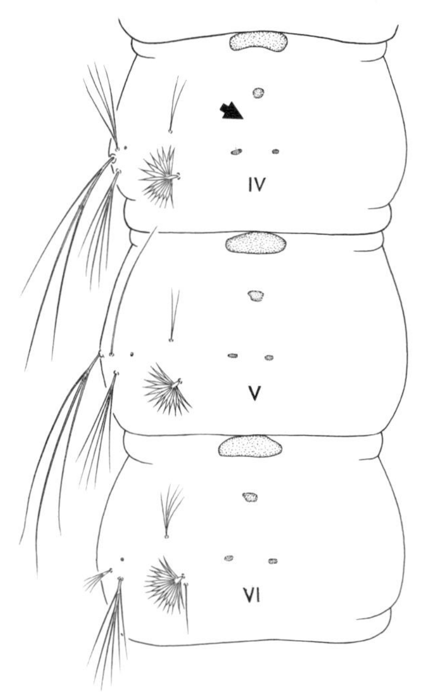


Fig. 884. Dorsal view of abdominal segments IV–VI: *An. freeborni*

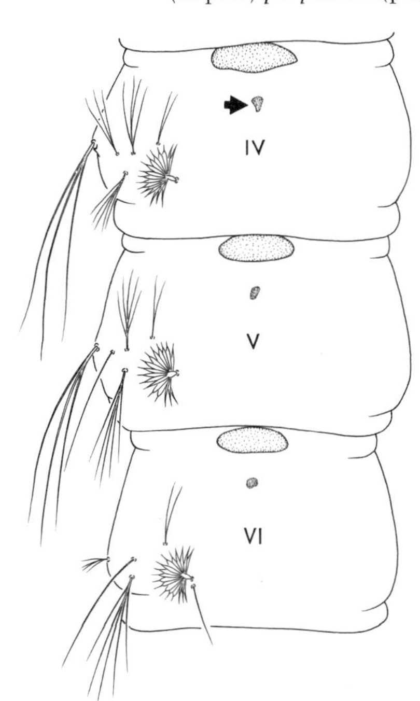


Fig. 886. Dorsal view of abdominal segments IV–VI: *An. punctipennis*

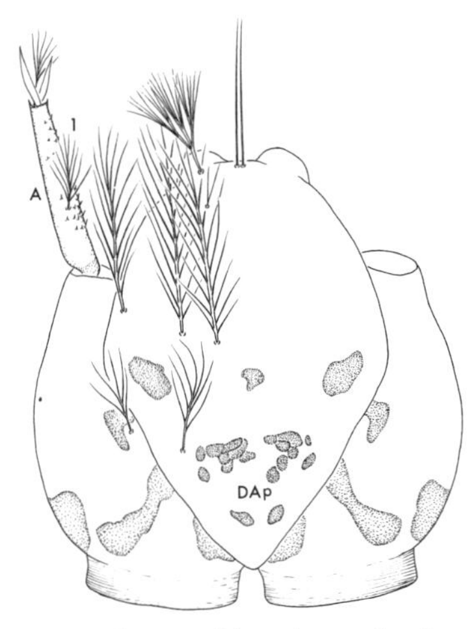


Fig. 885. Dorsal view of head: *An. freeborni*

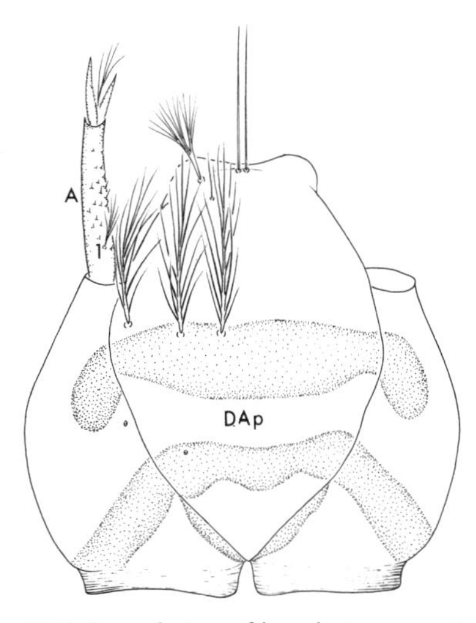


Fig. 887. Dorsal view of head: *An. punctipennis*

*Alternate key:*

a. Remainder of total branches of setae 3-VII,VIII less sum of branches of 5-I and 2-II greater than 2 ........................................................................................ *punctipennis*

Remainder of above combination 2 or less ........................................................ aa

aa. Sum of branches of setae 13-III, 2-IV, 2-V, and 2-VI, 28 or more ..................... *freeborni*

Sum of above less than 28 ........................................................................ aaa

aaa. Sum of setae in aa above, 22 or less .................................................... *occidentalis*

Sum of setae in aa above greater than 22, less than 28 ....................................... *hermsi*

## Key to Fourth Instar Larvae of Genus *Culex*

1. Seta 6-C with 3 or more branches (fig. 888) (subgenus *Culex*) .................................... 2

   Seta 6-C single or double (fig. 889) .......................................................................... 15

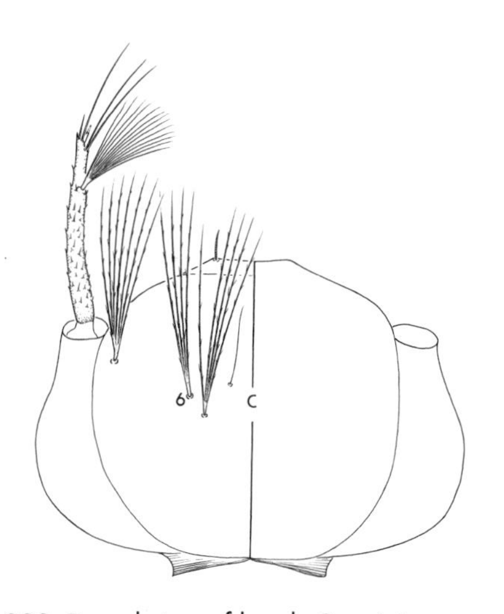


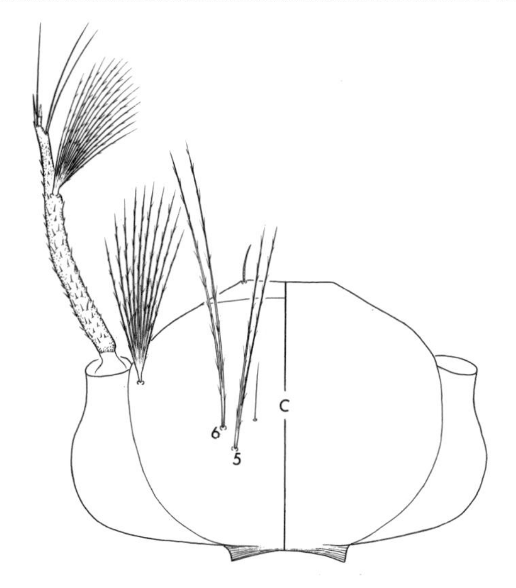

Fig. 888. Dorsal view of head: *Cx. pipiens*

Fig. 889. Dorsal view of head: *Cx. territans*

2(1). Siphon with large aciculae apically (fig. 890); segment X with 2 anal papillae (fig. 891) ............................................................................................. *bahamensis* (plate 40D)

   Siphon not aciculate (fig. 892); segment X with 4 anal papillae (fig. 893) .................... 3

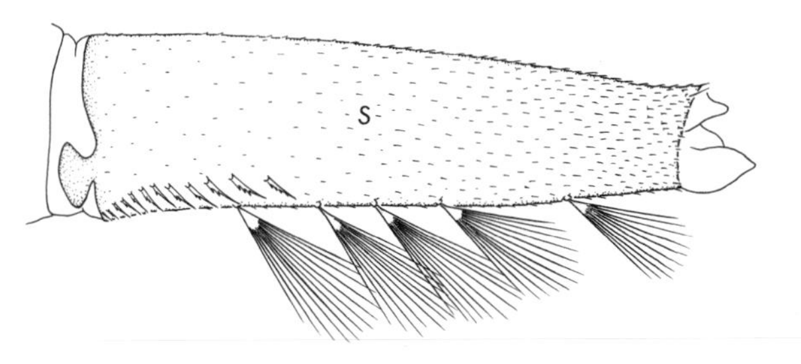


Fig. 890. Lateral view of siphon: *Cx. bahamensis*

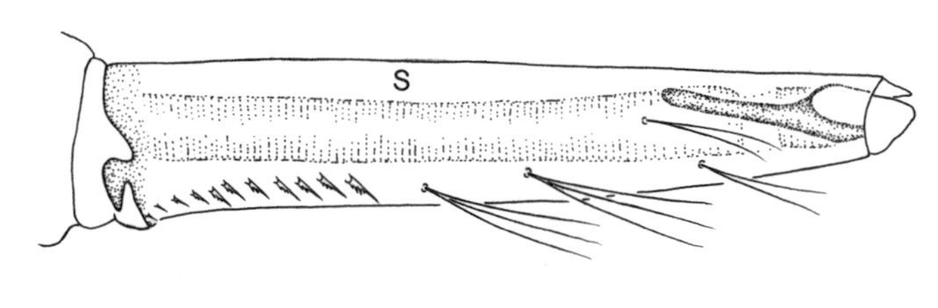


Fig. 892. Lateral view of siphon: *Cx. pipiens*

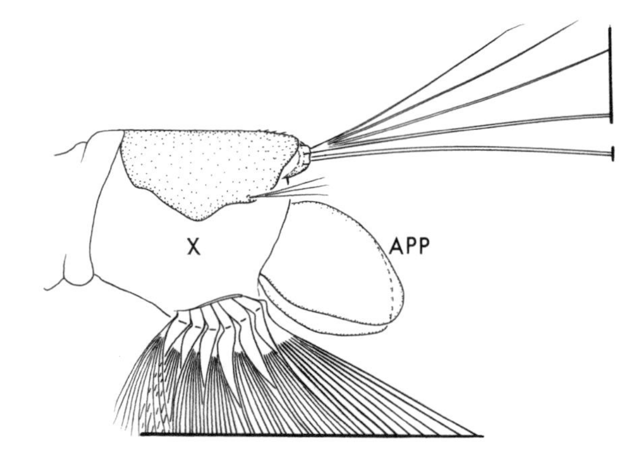


Fig. 891. Lateral view of abdominal segment X: *Cx, bahamensis*

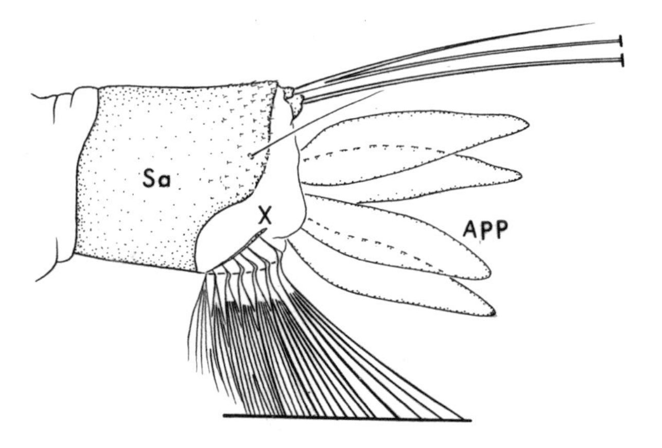


Fig. 893. Lateral view of abdominal segment X: *Cx. pipiens*

3(2). Pecten reaching distal 0.75 of siphon, apical 4,5 spines large (fig. 894) ........................ ........................ *interrogator* (plate 9A)

Pecten confined to basal 0.33 of siphon, without large apical spines (fig. 895) ............ 4

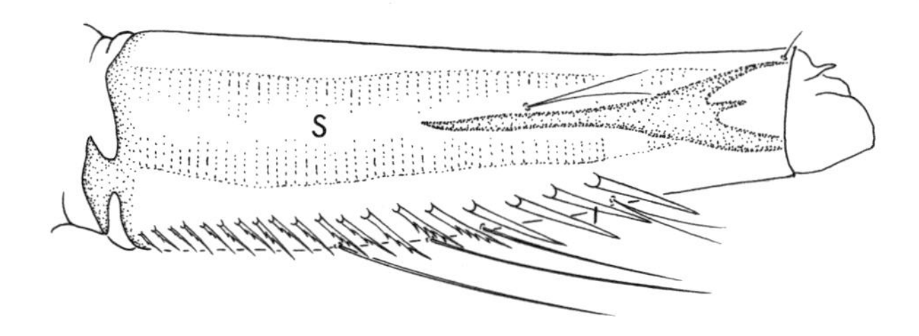


Fig. 894. Lateral view of siphon: *Cx. interrogator*

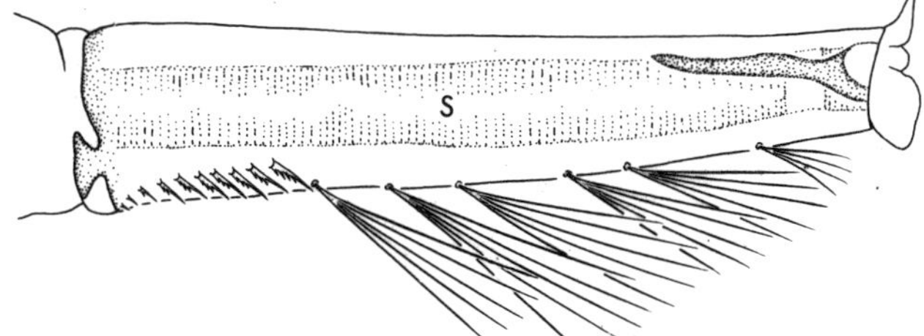


Fig. 895. Lateral view of siphon: *Cx. tarsalis*

4(3). Siphon setae long, irregularly placed, mostly single (fig. 896) ........................ 5

Siphon setae linear, sometimes with 1,2 pairs dorsally out of line, mostly branched (fig. 897) ........................ 6

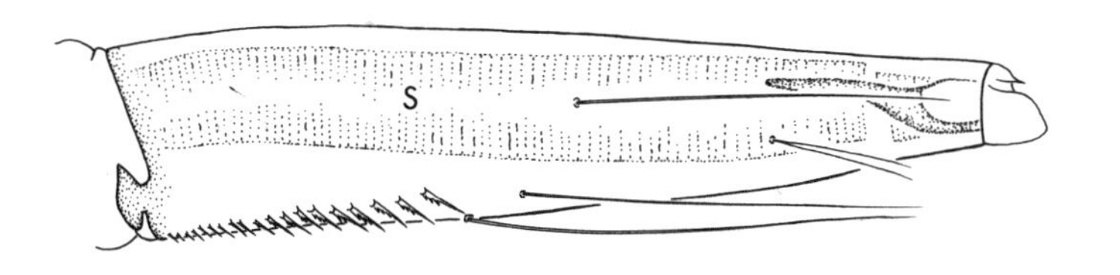


Fig. 896. Lateral view of siphon: *Cx. restuans*

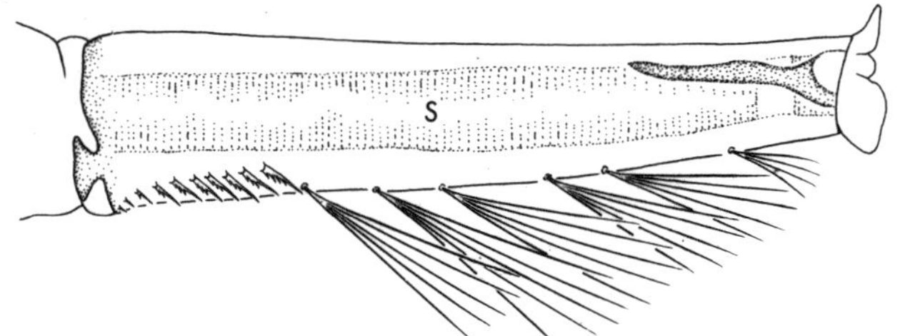


Fig. 897. Lateral view of siphon: *Cx. tarsalis*

5(4). Antenna not markedly constricted distally, seta 1-A attached near middle (fig. 898) ...... ........................ *restuans* (plate 29B)

Seta 1-A attached at constriction in outer 0.33 of antenna, distal part more slender (fig. 899) ........................ *thriambus* (plate 30B)

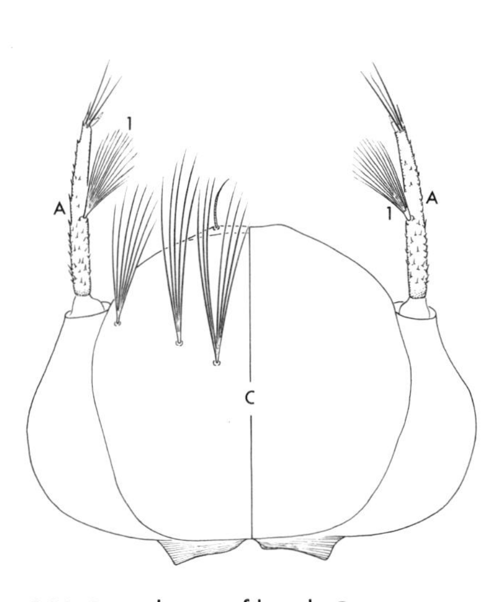


Fig. 898. Dorsal view of head: *Cx. restuans*

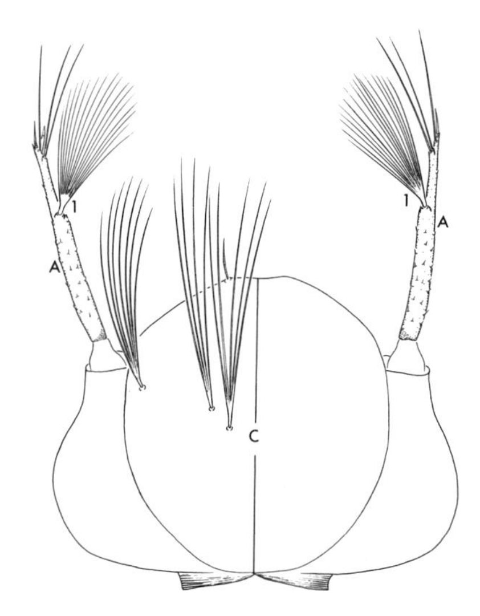


Fig. 899. Dorsal view of head: *Cx. thriambus*

6(4). Siphon with several spines near apex (fig. 900) ............................ *coronator* (plate 42B)

Siphon without spines near apex (fig. 901) ................................................................ 7

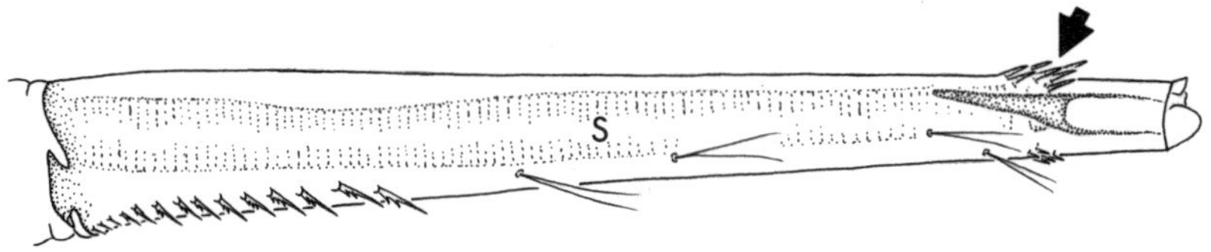

Fig. 900. Lateral view of siphon: *Cx. coronator*

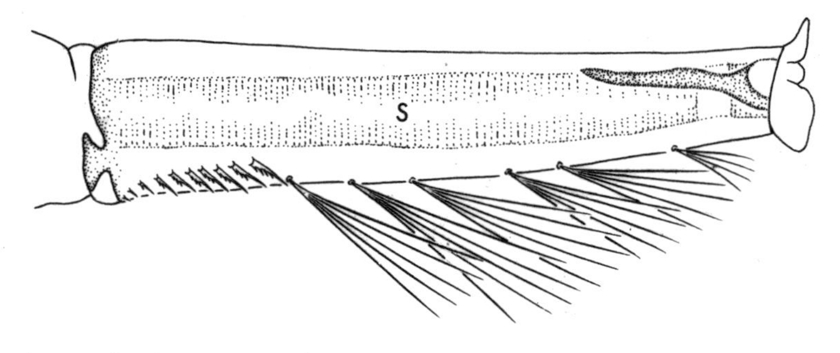


Fig. 901. Lateral view of siphon: *Cx. tarsalis*

7(6). Siphon with setae in straight line, usually with 5–9 pairs (fig. 902) ............................. 8

Siphon with 3–5 pairs of setae not all in straight line, 1,2 pairs dorsally out of line (fig. 903) ........................................................................................................................ 9

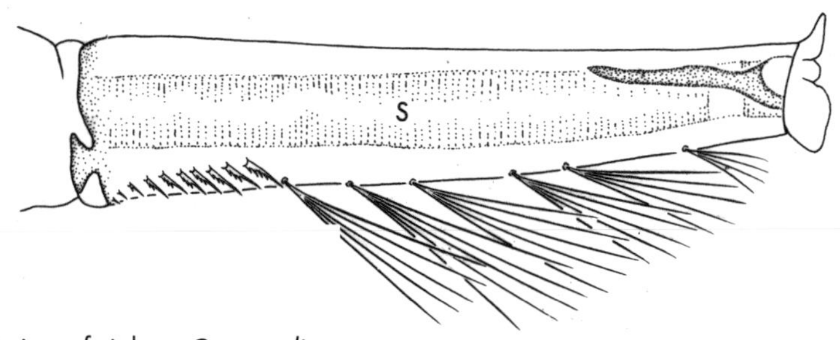


Fig. 902. Lateral view of siphon: *Cx. tarsalis*

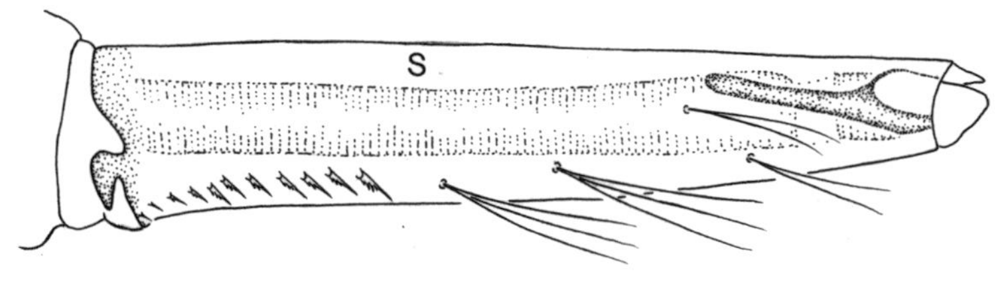


Fig. 903. Lateral view of siphon: *Cx. pipiens*

8(7). Siphon usually with 5 pairs of setae, index 4.5– 5.5 (fig. 904) .............. *tarsalis* (plate 30A)

Siphon with 6–9 pairs of setae, index 8.0 or more (fig. 905) ........... *chidesteri* (plate 42A)

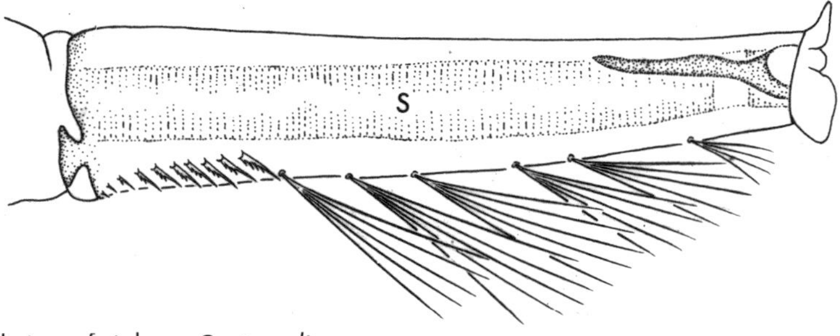


Fig. 904. Lateral view of siphon: *Cx. tarsalis*

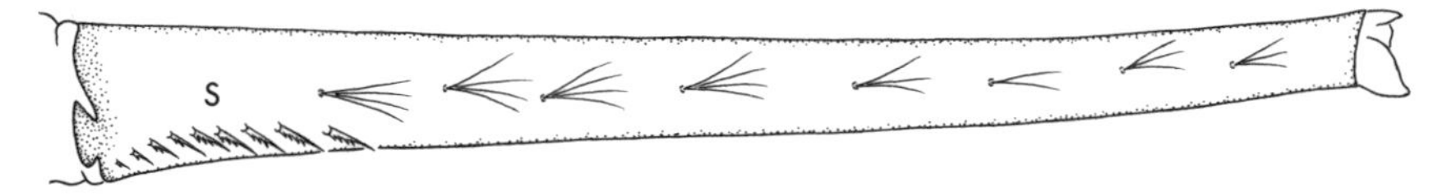


Fig. 905. Lateral view of siphon: *Cx. chidesteri*

9(7). Siphon with 3 pairs of setae (fig. 906) ............................................ *declarator* (plate 41)

Siphon with 4,5 pairs of setae (fig. 907) .................................................................. 10

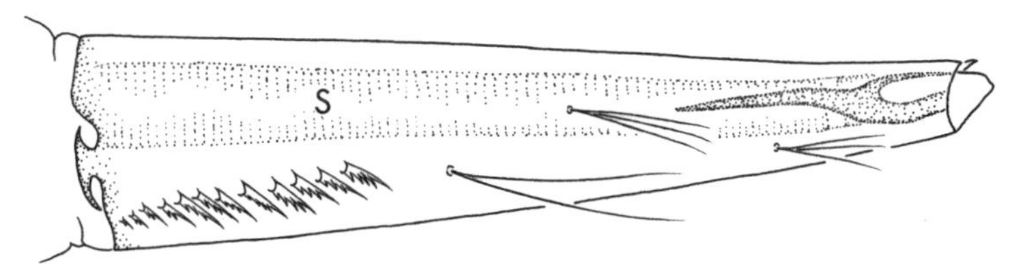


Fig. 906. Lateral view of siphon: *Cx. declarator*

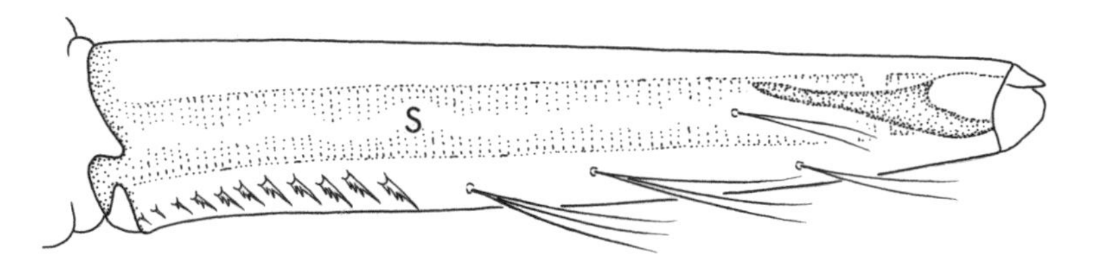


Fig. 907. Lateral view of siphon: *Cx. pipiens*

10(9). Siphon index 4–0–5.0 (fig. 908) ........................................................................ 11

Siphon index 6.0–8.0 (fig. 909) ......................................................................... 12

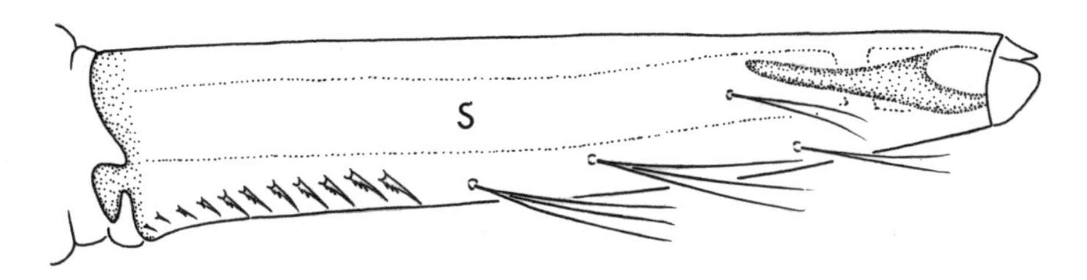


Fig. 908. Lateral view of siphon: *Cx. pipiens*

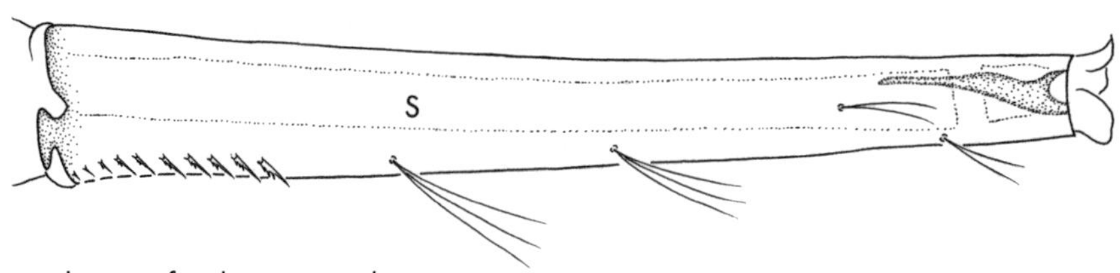


Fig. 909. Lateral view of siphon: *Cx. salinarius*

11(10). Aciculae on dorsoposterior border of saddle much larger than those at dorsal middle (fig. 910); setae 6-III,IV triple (fig. 911) ............................................ *stigmatosoma* (plate 29D)

Aciculae on dorsoposterior border of saddle not much larger than those at dorsal middle (fig. 912); setae 6-III,IV usually single or double (fig. 913) ......................... *pipiens* (plate 28C)
*quinquefasciatus* (plate 28D)

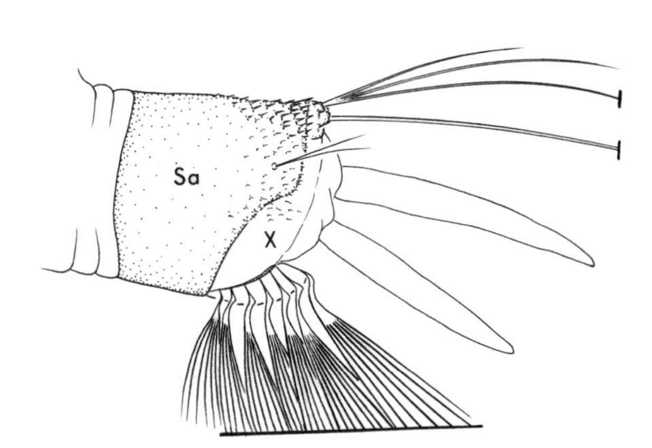


Fig. 910. Lateral view of abdominal segment X: *Cx. stigmatosoma*

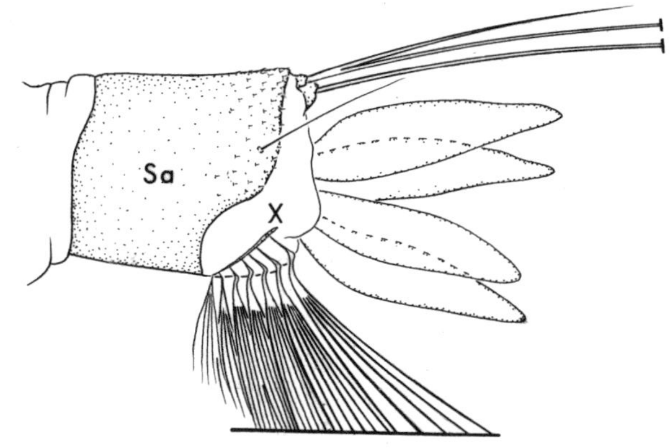


Fig. 912. Lateral view of abdominal segment X: *Cx. pipiens*

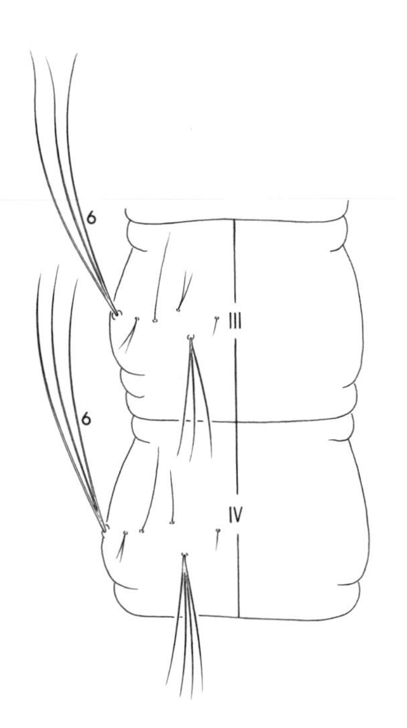


Fig. 911. Dorsal view of abdominal segments III,IV: *Cx. stigmatosoma*

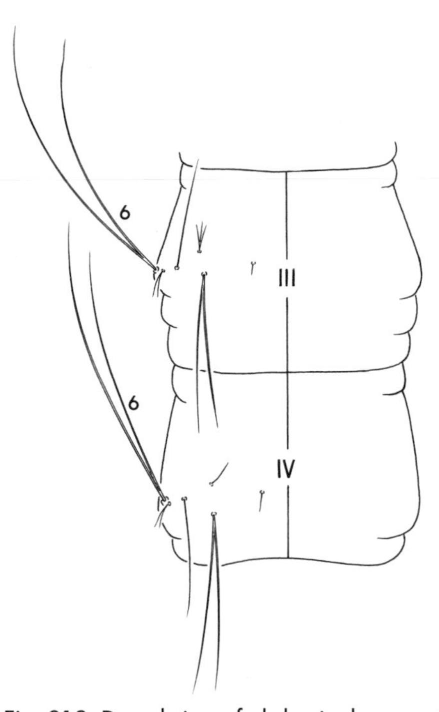


Fig. 913. Dorsal view of abdominal segments III,IV: *Cx. pipiens*

12(10). Thoracic integument with fine aculeae; seta 1-M subequal to 2-M (fig. 914); seta 1-X single (fig. 915) ........................................ *nigripalpus* (plate 29A)

Thoracic integument glabrous; seta 1-M much longer than 2-M (fig. 916); seta 1-X usually double (fig. 917) ........................................ 13

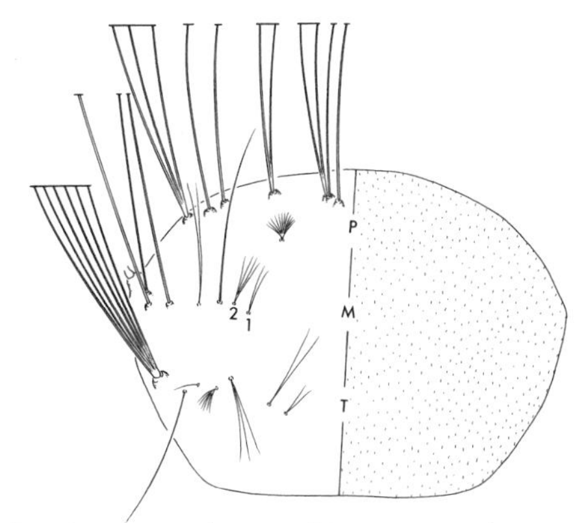

Fig. 914. Dorsal view of thorax: *Cx. nigripalpus*

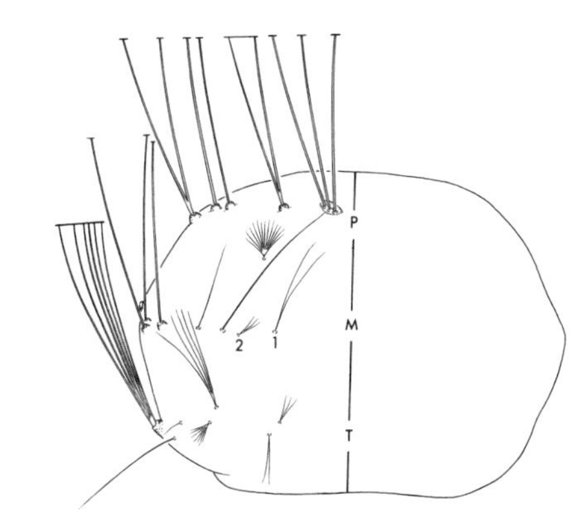

Fig. 916. Dorsal view of thorax: *Cx. salinarius*

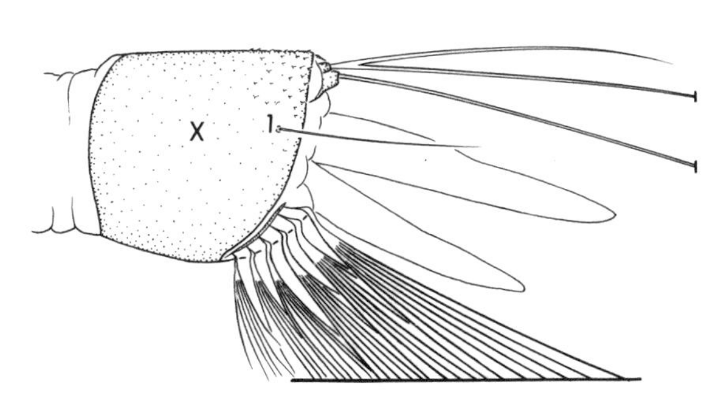

Fig. 915. Lateral view of abdominal segment X: *Cx. nigripalpus*

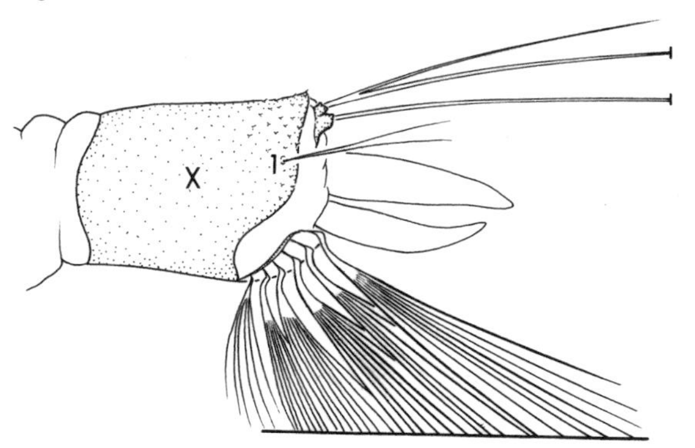

Fig. 917. Lateral view of abdominal segment X: *Cx. salinarius*

13(12). Thorax and abdomen with stellate setae (fig. 918); posterior margin of saddle with long spines (fig. 919) (subg. *Micraedes*) ........................................ *biscaynensis* (plate 41D)

Thorax and abdomen without stellate setae (fig. 920); posterior margin of saddle without large spines (fig. 921) ........................................ 14

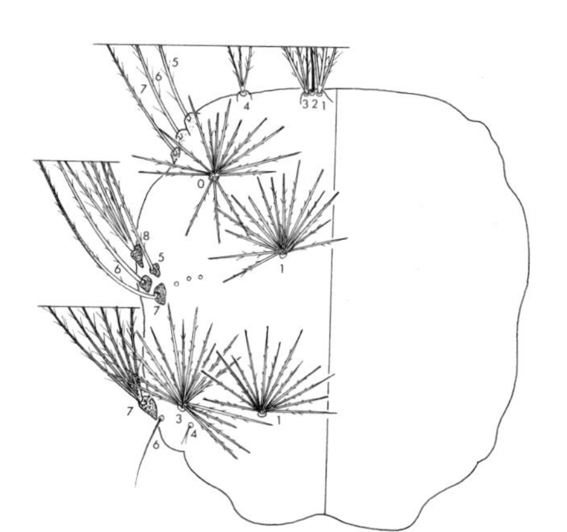
Fig. 918. Dorsal view of thorax: *Cx. biscaynensis*

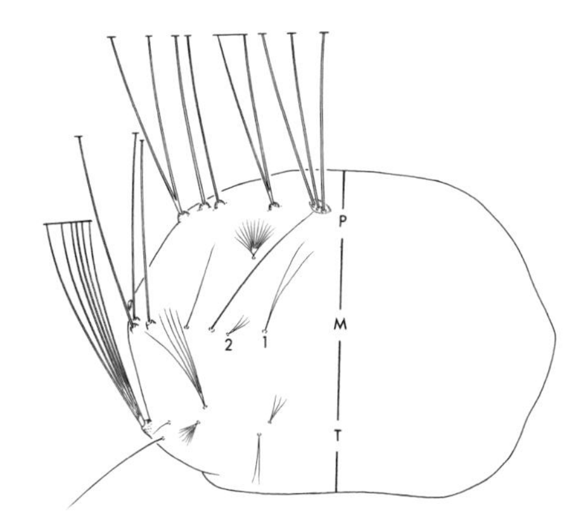

Fig. 920. Dorsal view of thorax: *Cx. salinarius*

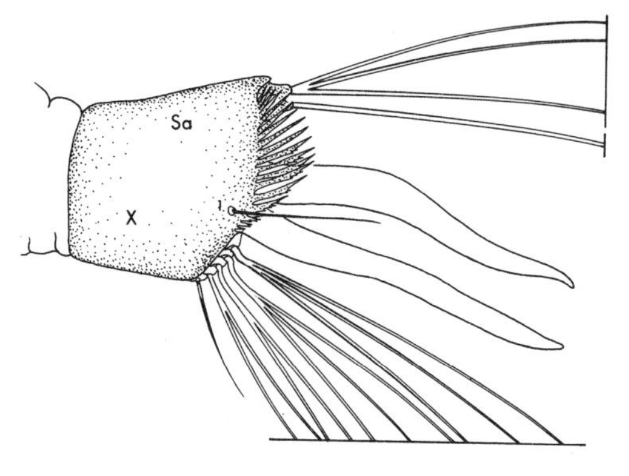


Fig. 919. Lateral view of abdominal segment X: *Cx. biscaynensis*

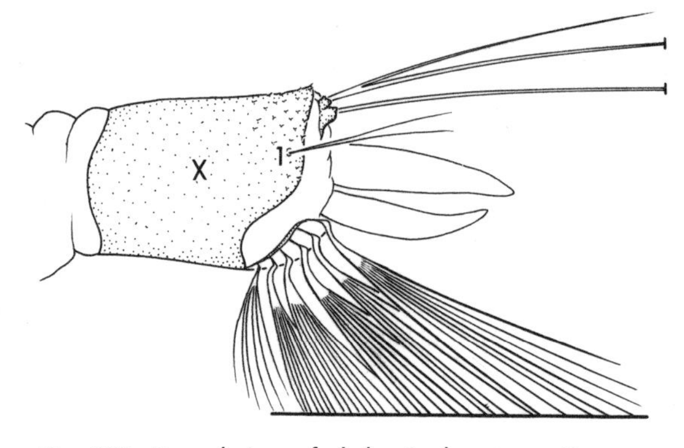


Fig. 921. Dorsal view of abdominal segment X: *Cx. salinarius*)

14(13). Siphon usually with 5 pairs of setae, most often 2 pairs dorsally out of line (fig. 922) .... ........................................................................................... *erythrothorax* (plate 28B)

Siphon usually with 4 pairs of setae, only 1 pair dorsally out of line (fig. 923) ............... ........................................................................................... *salinarius* (plate 29C)

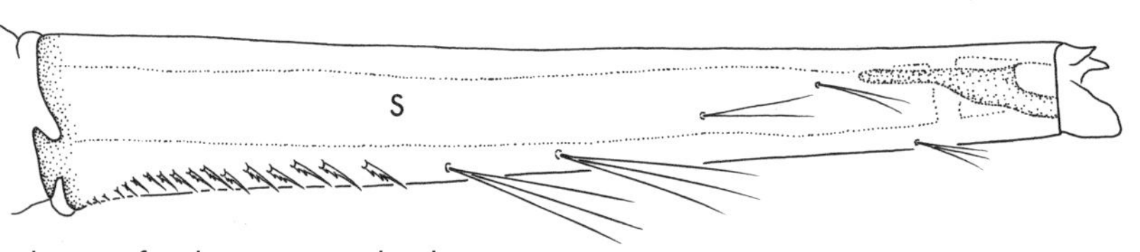


Fig. 922. Lateral view of siphon: *Cx. erythrothorax*

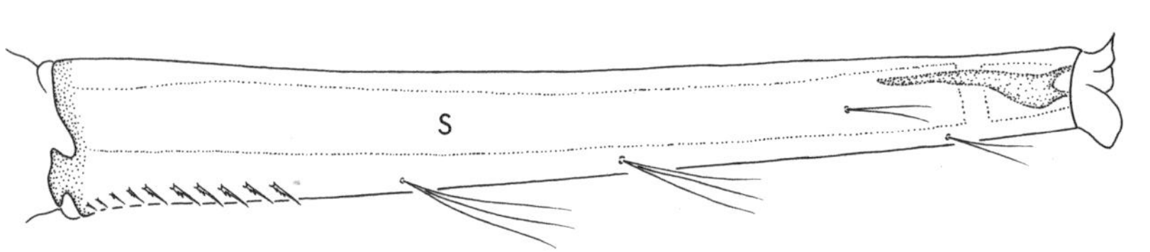


Fig. 923. Lateral view of siphon: *Cx. salinarius*

15(1). Pecten spines with 1–4 lateral denticles; seta 2-S straight; siphon without subdorsal setae (fig. 924) (subg. *Neoculex*) ........................................................................................... 16

Pecten spines with 10 or more denticles; seta 2-S strongly curved; siphon with 1 or more pairs of subdorsal setae (fig. 925) (subg. *Melanoconion*) ........................................... 20

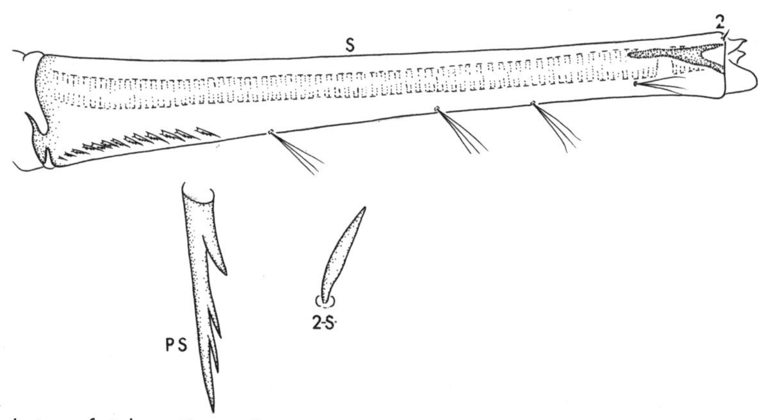


Fig. 924. Lateral view of siphon: *Cx. territans*

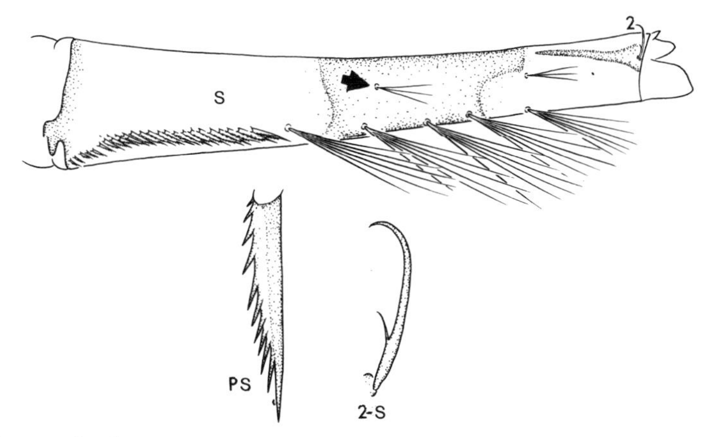


Fig. 925. Lateral view of siphon: *Cx. peccator*

16(15). Setae 5,6-C about equal in length, double (fig. 926); seta 1a-S about 1.5 longer than distance from its alveolus to base of siphon (fig. 927) .................... *arizonensis* (plate 41D)

Seta 6-C longer than 5-C, usually not both double (fig. 928); seta 1a-S not more than 1.2 longer than distance from alveolus to base of siphon (fig. 929) ................................. 17

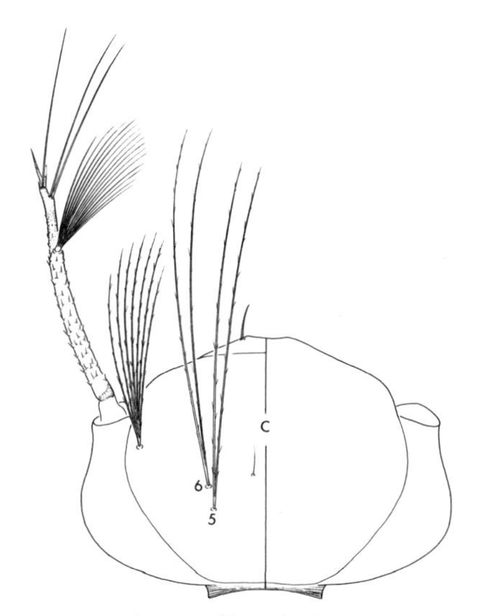


Fig. 926. Dorsal view of head: *Cx. arizonensis*

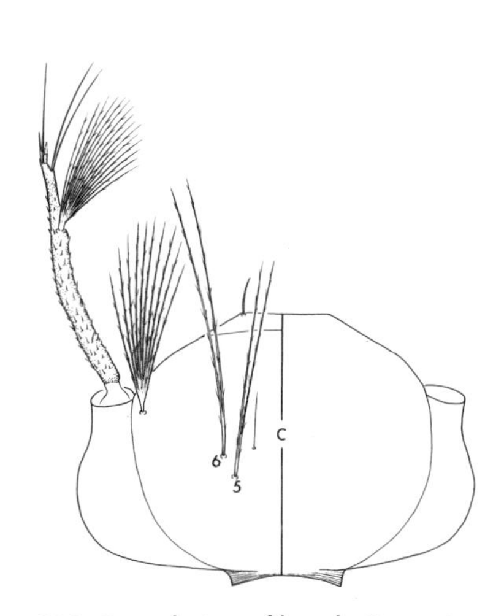


Fig. 928. Dorsal view of head: *Cx. territans*

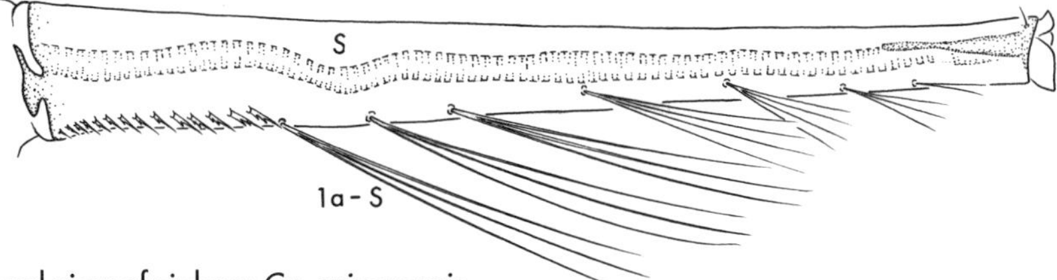


Fig. 927. Lateral view of siphon: *Cx. arizonensis*

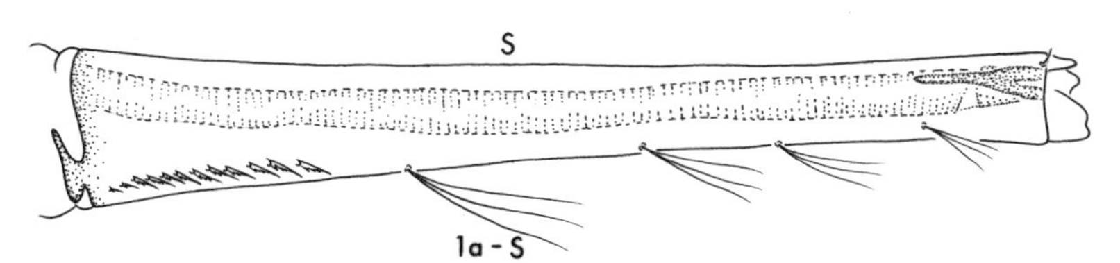


Fig. 929. Lateral view of siphon: *Cx. territans*

17(16). Siphon 6.0–8.0 longer than most basal seta, index 7.0–9.0 (fig. 930) ... *apicalis* (plate 31B)

Siphon less than 6.0 longer than most basal seta, index usually less than 7.0 (fig. 931) ........................................ 18

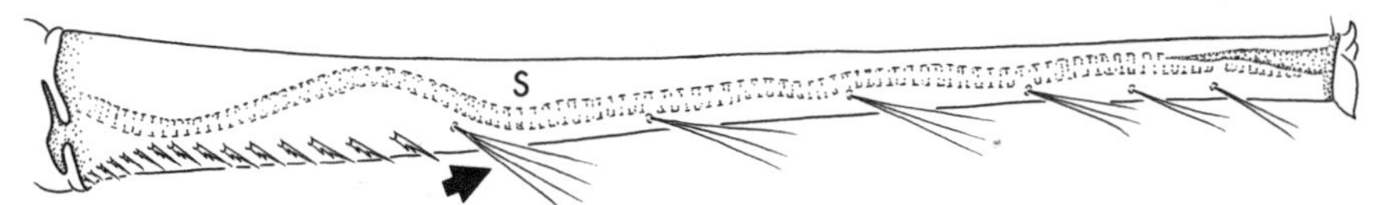


Fig. 930. Lateral view of siphon: *Cx. apicalis*

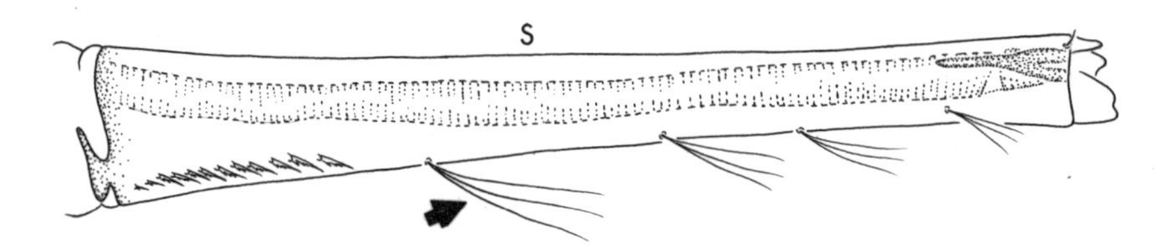


Fig. 931. Lateral view of siphon: *Cx. territans*

18(17). Seta 5-C with 3 branches, 6-C double (fig. 932) ............................... *reevesi* (plate 41D)

Seta 5-C single or double, 6-C usually single (fig. 933) ............................................. 19

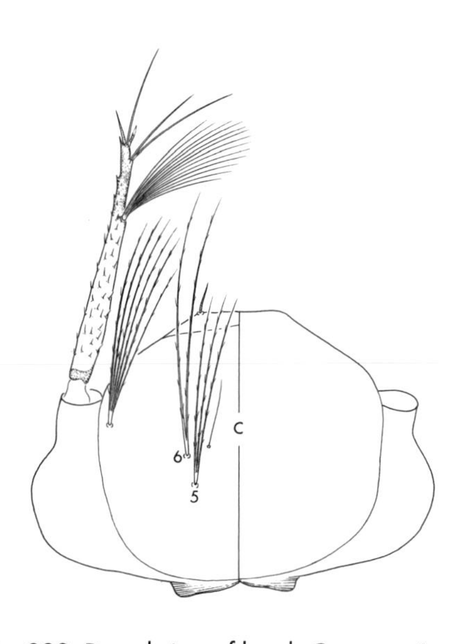


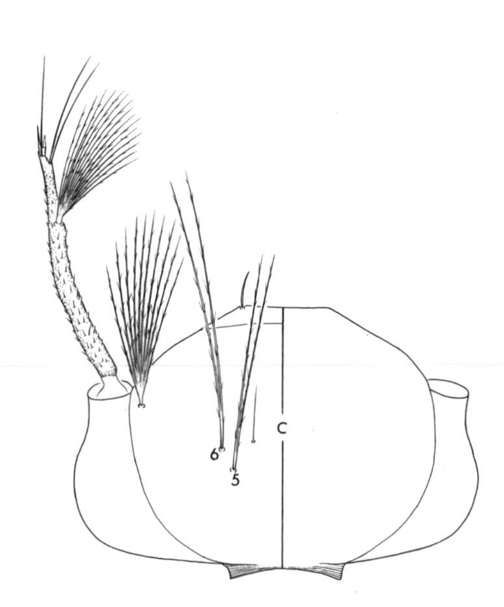

Fig. 932. Dorsal view of head: *Cx. reevesi*

Fig. 933. Dorsal view of head: *Cx. territans*

19(18). Seta 5-C single, occasionally double or triple (fig. 934); abdominal segments III–V evenly pigmented (fig. 935) ........................................................................ *territans* (plate 31D)

Seta 5-C double, rarely triple (fig. 936); abdominal segment III and V more darkly pigmented than IV (fig. 937) ................................................................ *boharti* (plate 31C)

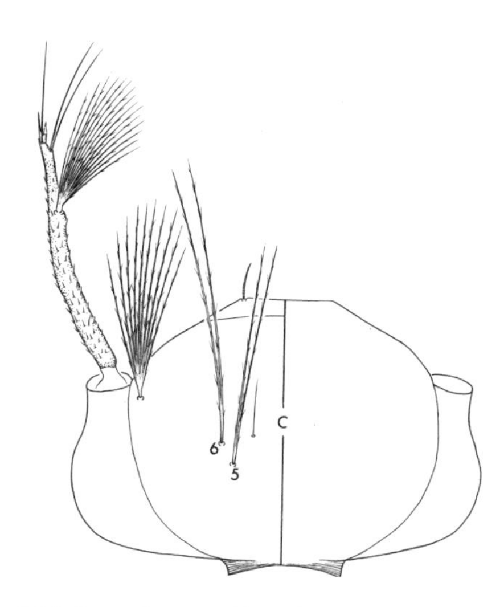


Fig. 934. Dorsal view of head: *Cx. territans*

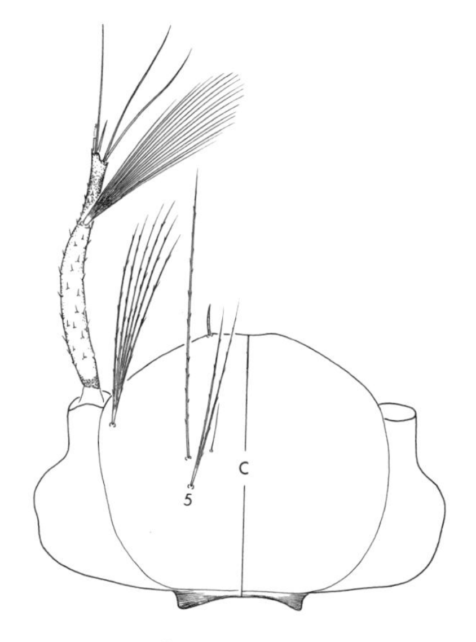


Fig. 936. Dorsal view of head: *Cx. boharti*

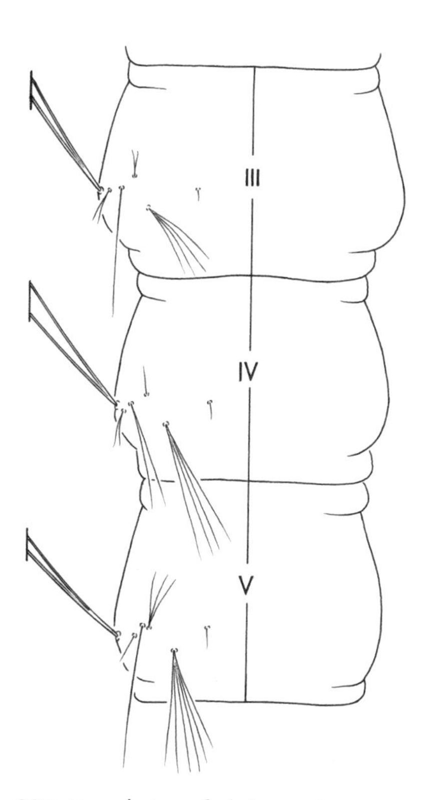


Fig. 935. Dorsal view of abdominal segments III–V: *Cx. territans*

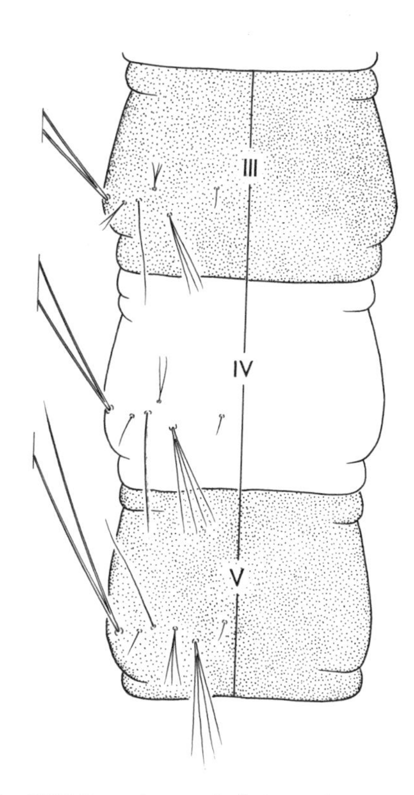


Fig. 937. Dorsal view of abdominal segments III–V: *Cx. boharti*

20(15). At least some comb scales with large median spine (fig. 938) .................................... 21

All comb scales evenly fringed with subequal spinules (fig. 939) ................................ 22

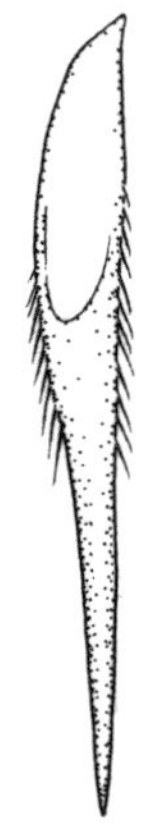

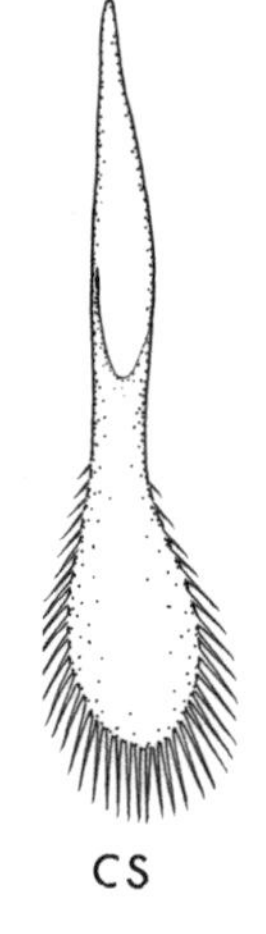


CS

Fig. 938. Comb scale: *Cx. pilosus*

Fig. 939. Comb scale: *Cx. atratus*

21(20). Siphon distinctly curved, index 4.5 or less, most distal seta very near apex (fig. 940) ..... ........................................................................................................ *pilosus* (plate 31A)

Siphon only slightly curved, if at all, index 6.0 or more, most distal seta not near apex (fig. 941) ........................................................................................ *erraticus* (plate 30C)

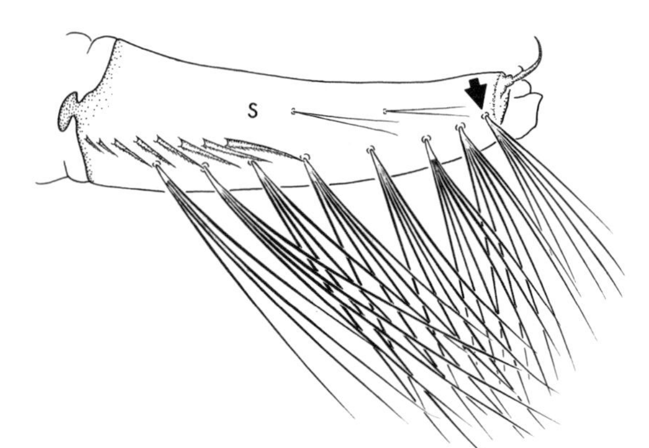


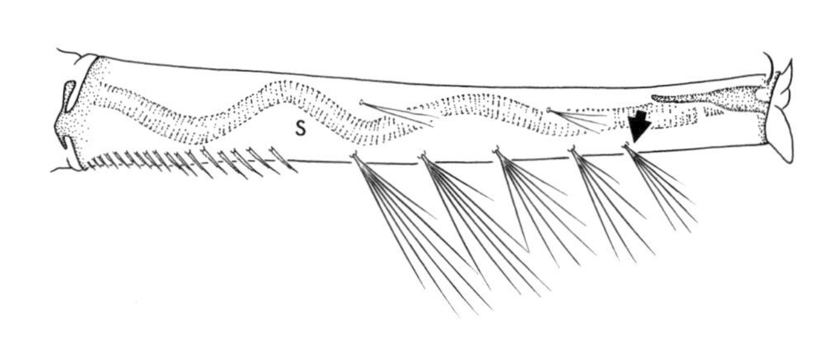


Fig. 940. Lateral view of siphon: *Cx. pilosus*

Fig. 941. Lateral view of siphon: *Cx. erraticus*

22(20). Siphon index more than 7.0 (fig. 942) ........................................................................ 23

Siphon index 7.0 or less (fig. 943) ............................................................................ 24

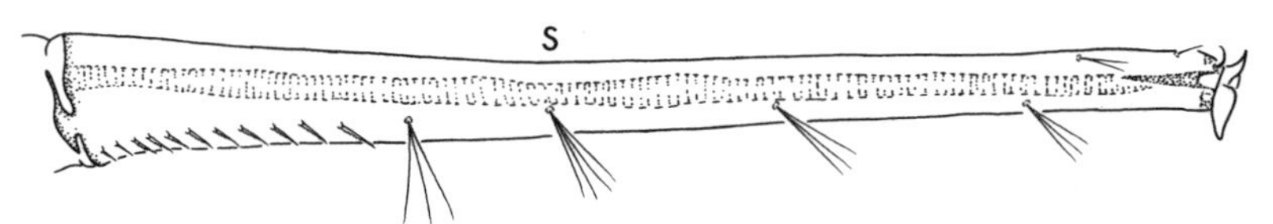


Fig. 942. Lateral view of siphon: *Cx. cedecei*

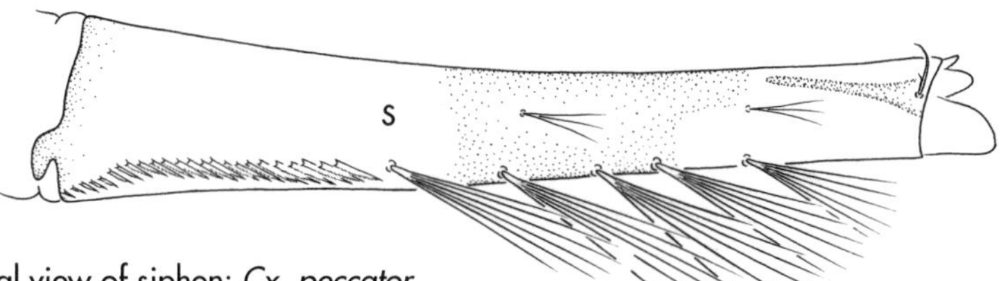


Fig. 943. Lateral view of siphon: *Cx. peccator*

23(22). Seta 7-I double (fig. 944); saddle not aciculate dorsoposteriorly (fig. 945) ........................................................................ *cedecei* (plate 42A)

Seta 7-I single (fig. 946); saddle with rather large aciculae dorsoposteriorly (fig. 947) ........................................................................ *atratus* (plate 41A)

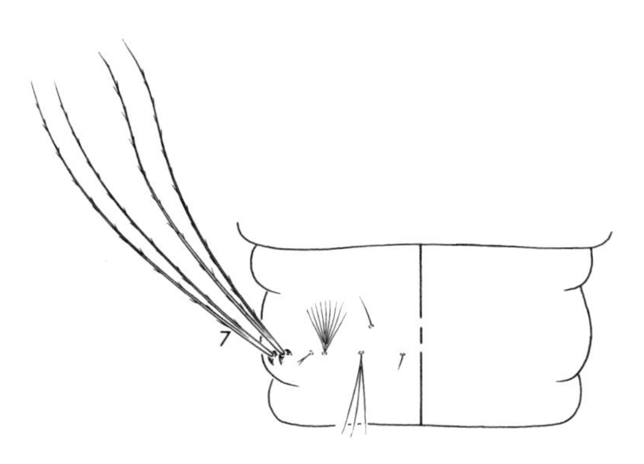


Fig. 944. Ventral view of abdominal segment I: *Cx. cedecei*

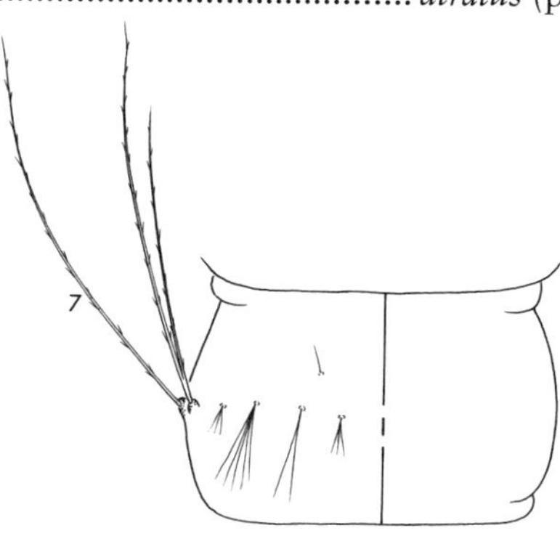


Fig. 946. Ventral view of abdominal segment I: *Cx. atratus*

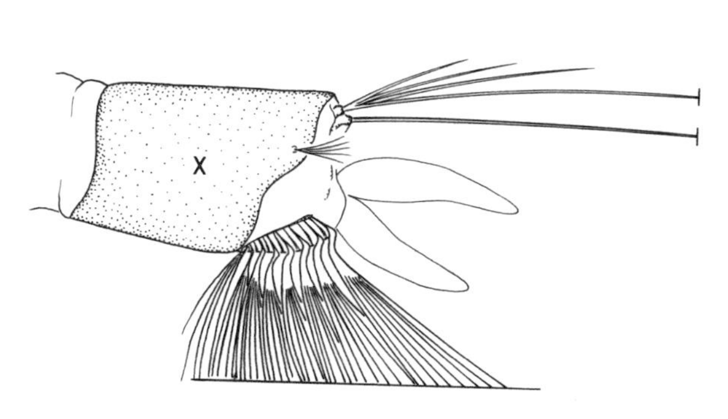


Fig. 945. Lateral view of abdominal segment X: *Cx. cedecei*

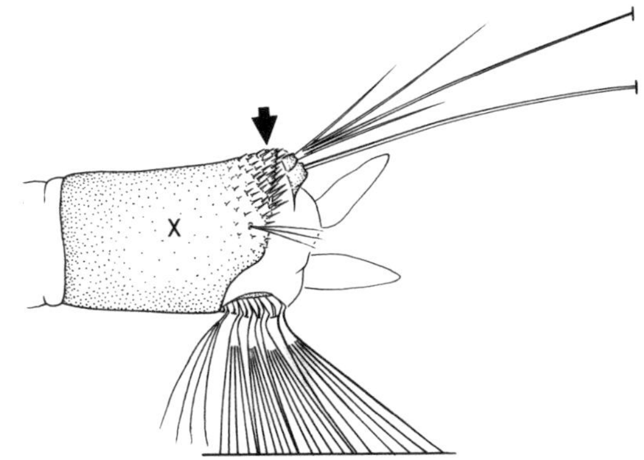


Fig. 947. Lateral view of abdominal segment X: *Cx. atratus*

24(22). Comb scale short, the fringed apical portion about length of basal portion, no elongation in middle (fig. 948) ........................................................................ *abominator* (plate 41C)

Comb scale long, with narrow elongation in middle between base and apical fringed portion (fig. 949) ........................................................................ 25

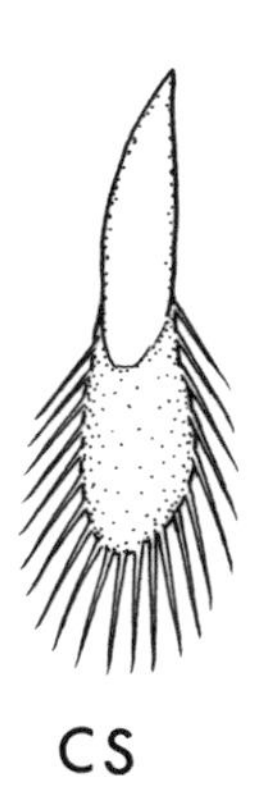


Fig. 948. Comb scale: *Cx. abominator*

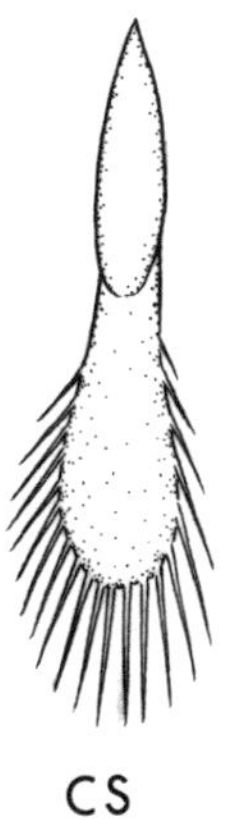


Fig. 949. Comb scale: *Cx. iolambdis*

25(24). Seta 5-C thin, much thinner than and 0.5 length of seta 6-C, without aciculae (fig. 950)

........................................................................ 26

Seta 5-C stout, about 0.75 length of 6-C, lightly aciculate (fig. 951) .......................... 27

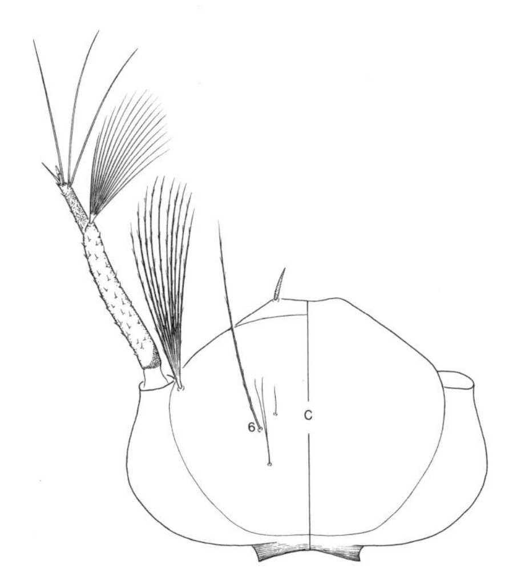


Fig. 950. Dorsal view of head: *Cx. peccator*

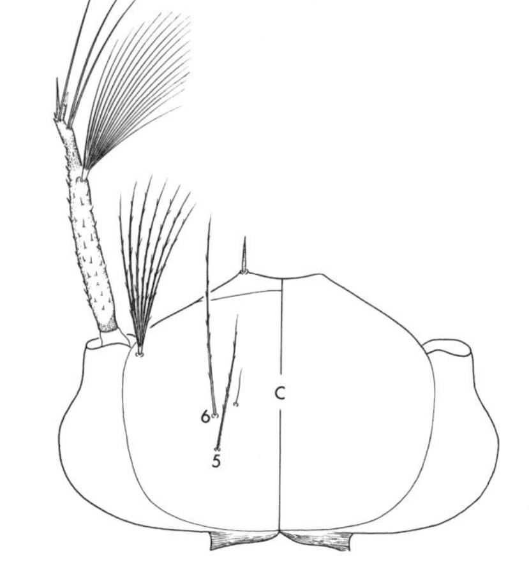


Fig. 951. Dorsal view of head: *Cx. iolambdis*

26(25). Seta 2-S without secondary tooth; pecten spine with 15 or more fine lateral denticles (fig. 952) ........................................................................ *anips* (plate 41B)

Seta 2-S with secondary tooth; pecten with fewer than 12 coarse lateral denticles (fig. 953) ........................................................................ *peccator* (plate 30D)

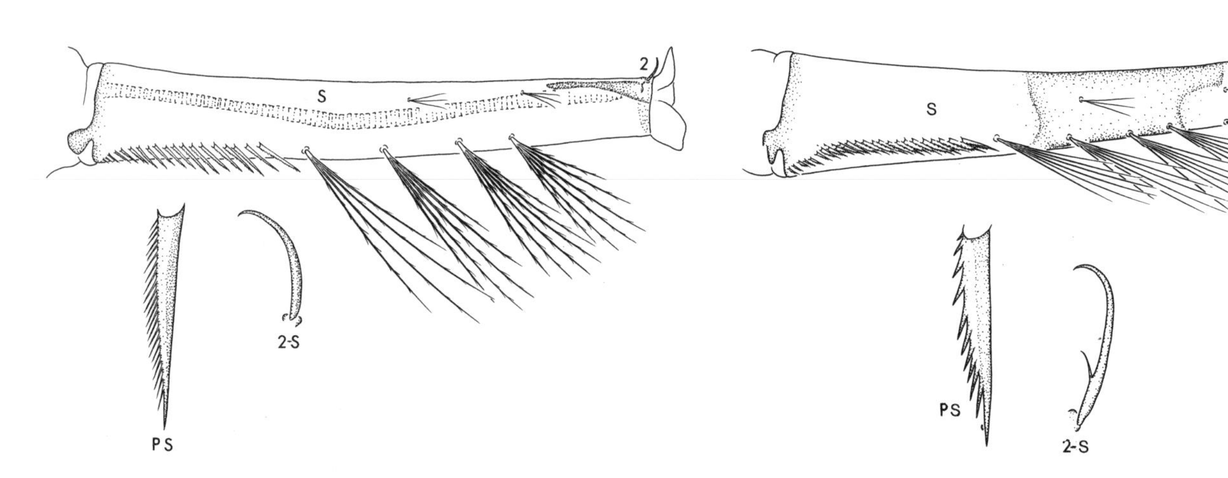


Fig. 952. Lateral view of siphon: *Cx. anips*

Fig. 953. Lateral view of siphon: *Cx. peccator*

27(25). Seta 5-C usually double; length of setae 1-C mostly no more than 0.6 the distance between their bases (fig. 954) ........................................................................ *iolambdis* (plate 42B)

Seta 5-C usually triple; length of setae 1-C mostly at least 0.7 the distance between their bases (fig. 955) ........................................................................ *mulrennani* (plate 41B)

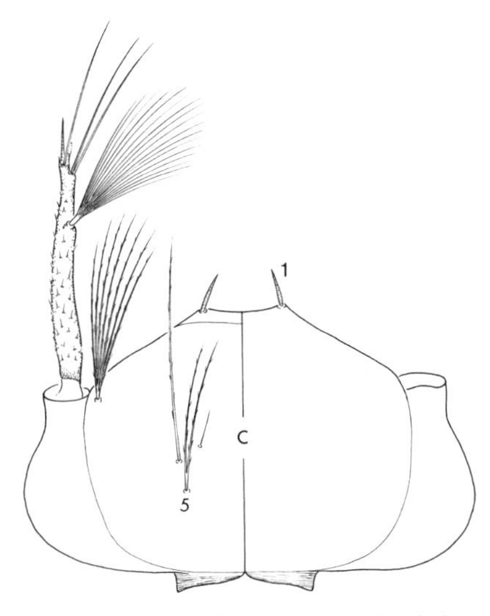


Fig. 954. Dorsal view of head: *Cx. iolambdis*

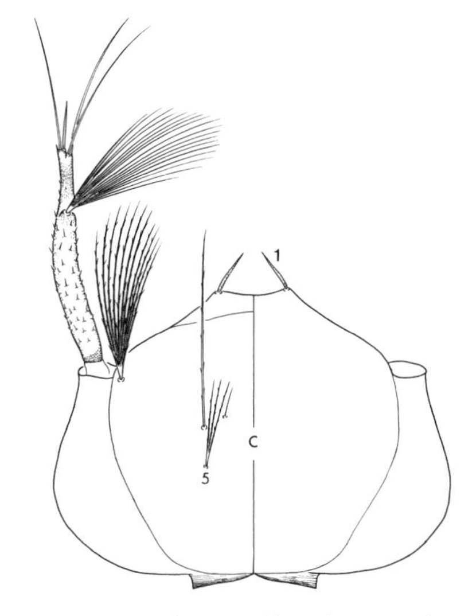


Fig. 955. Dorsal view of head: *Cx, mulrennani*

## Key to Fourth Instar Larvae of Genus *Culiseta*

1. Siphon with row of 8–14 setae along midventral line (fig. 956) (subgenus *Climacura*) ... ........................................................................................................ *melanura* (plate 33A)

   Siphon with setae otherwise distributed, no midventral row (fig. 957) .............................2

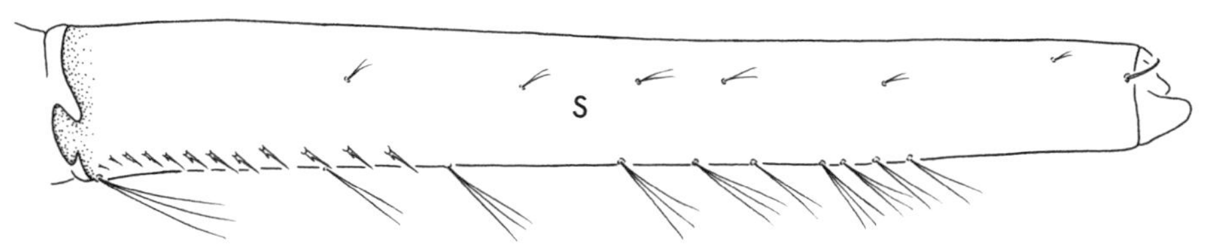


Fig. 956. Lateral view of siphon: *Cs. melanura*

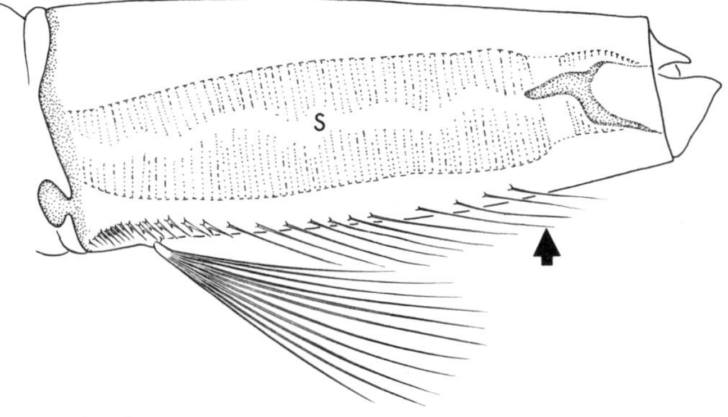


Fig. 957. Lateral view of siphon: *Cs. inornata*

2(1). Antenna longer than head, seta 1-A attached to distal 0.33 (fig. 958); siphon without row of single spicules distal to pecten (fig. 959) (subgenus *Culicella*) .................................. 3

Antenna shorter than head, seta 1-A attached near middle (fig. 960); siphon with row of single spicules distal to pecten (fig. 961) (subgenus *Culiseta*) ....................................... 4

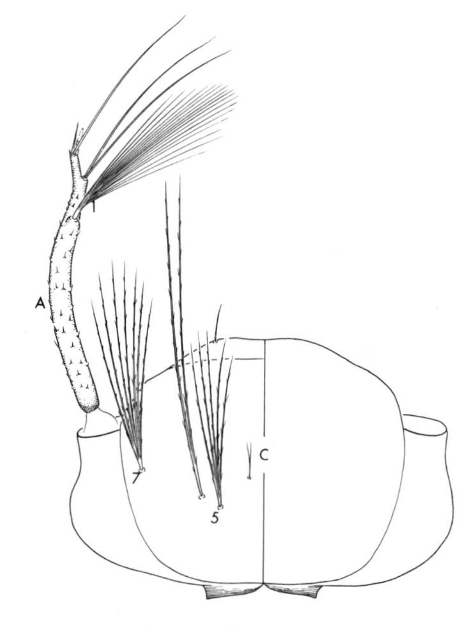

Fig. 958. Dorsal view of head: *Cs. morsitans*

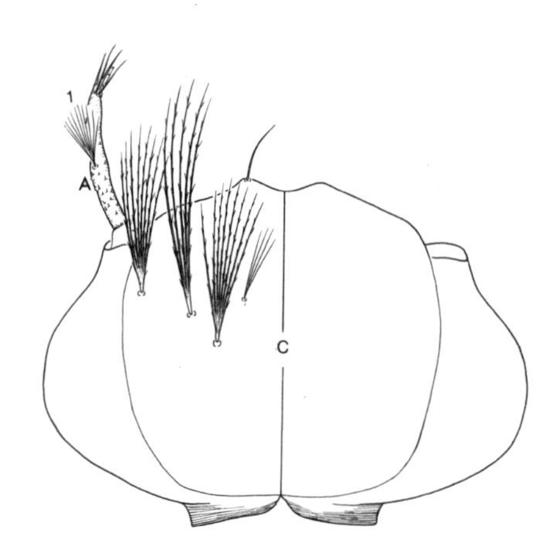

Fig. 960. Dorsal view of head: *Cs. inornata*

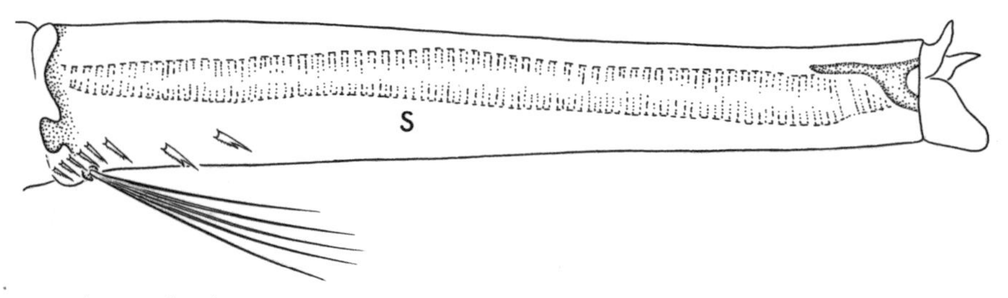

Fig. 959. Lateral view of siphon: *Cs. morsitans*

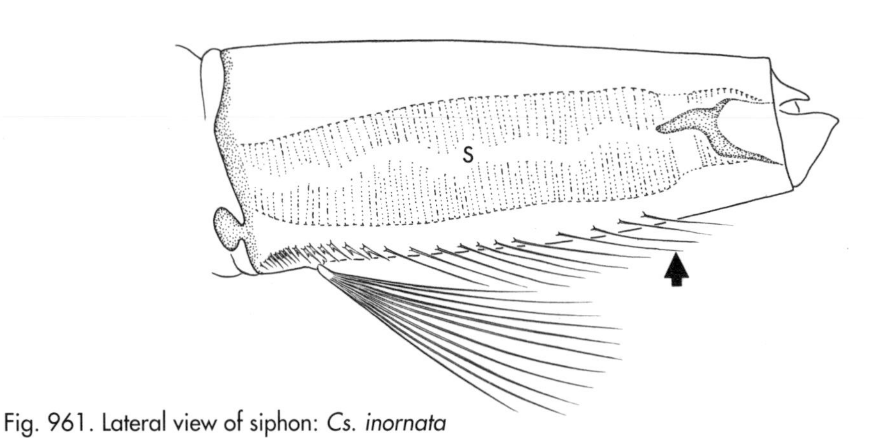

Fig. 961. Lateral view of siphon: *Cs. inornata*

3(2). Seta 5-C usually with 7 or more branches (fig. 962); seta 4-X with 16–18 fanlike setae (fig. 963); seta 7-C mostly with 9 or more branches (fig. 963) ........... *minnesotae* (plate 33B)

Seta 5-C usually with 5 or fewer branches (fig. 964); seta 4-X with 19–22 fanlike setae (fig. 965); seta 7-C mostly with 8 or fewer branches (fig. 965) ............. *morsitans* (plate 33C)

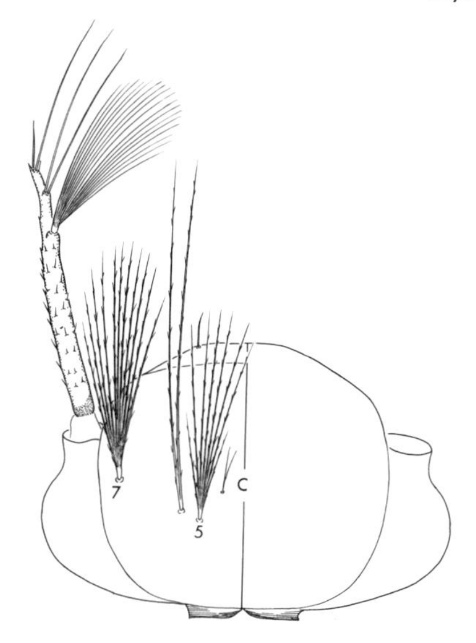

Fig. 962. Dorsal view of head: *Cs. minnesotae*

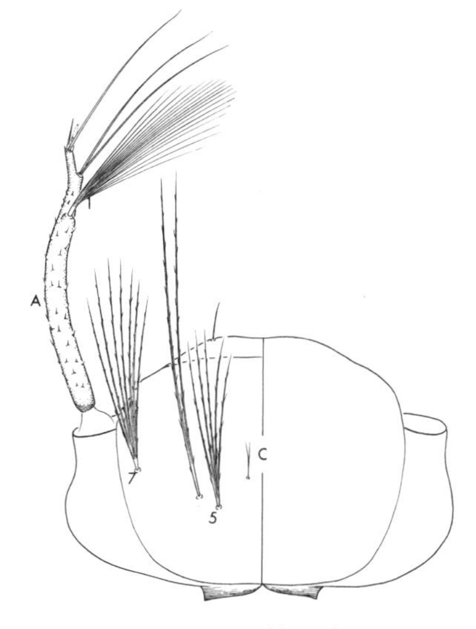

Fig. 964. Dorsal view of head: *Cs. morsitans*

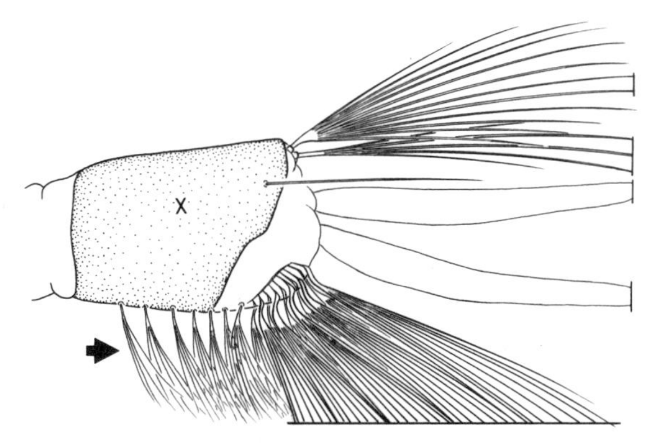

Fig. 963. Lateral view of abdominal segment X: *Cs. minnesotae*

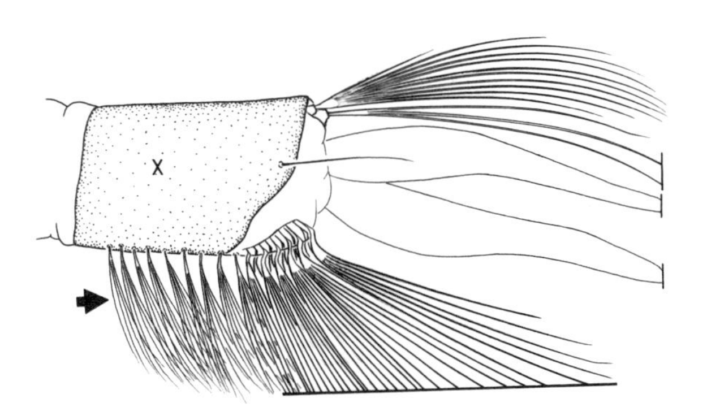

Fig. 965. Lateral view of abdominal segment X: *Cs. morsitans*

4(2). Setae 5,6-C similar in size and number of branches (fig. 966) ........ *impatiens* (plate 32B)

Seta 6-C with fewer branches and usually somewhat longer than 5-C (fig. 967) ........... 5

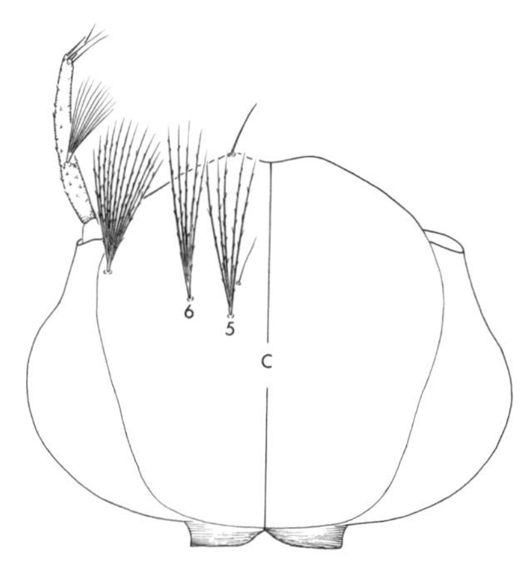

Fig. 966. Dorsal view of head: *Cs. impatiens*

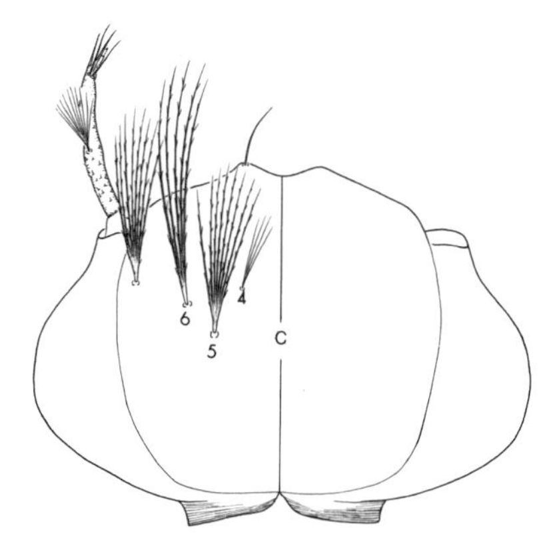

Fig. 967. Dorsal view of head: *Cs. inornata*

5(4). Seta 4-C nearly equal in size to seta 5,6-C (fig. 968); saddle with coarse aciculae dorso-posteriorly (fig. 969) ..................................................................... *particeps* (plate 33D)

Seta 4-C much shorter and with branches thinner than setae 5,6-C (fig. 970); saddle not aciculate dorsoposteriorly (fig. 971) .......................................................................... 6

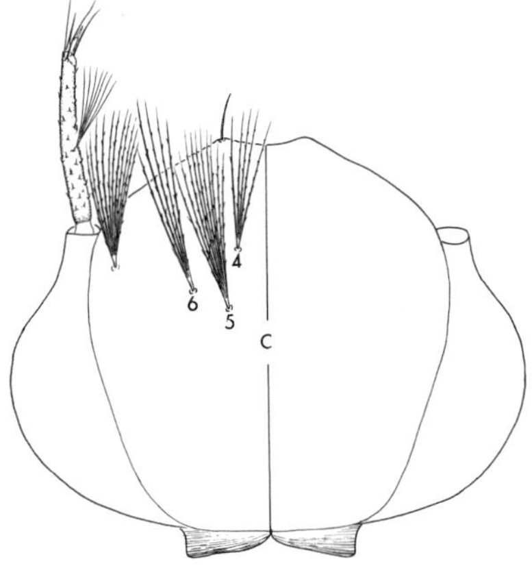


Fig. 968. Dorsal view of head: *Cs. particeps*

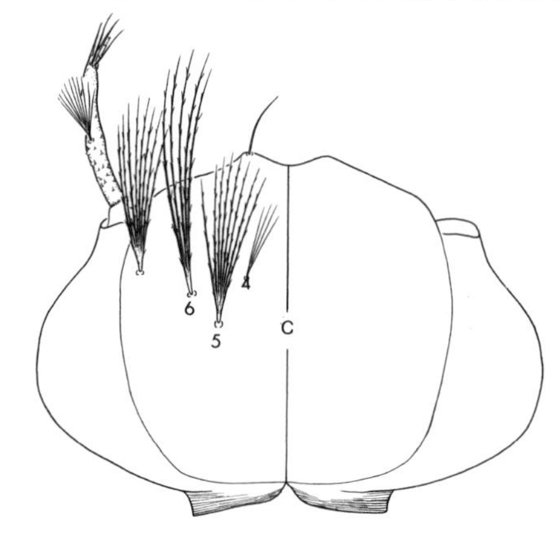


Fig. 970. Dorsal view of head: *Cs. inornata*

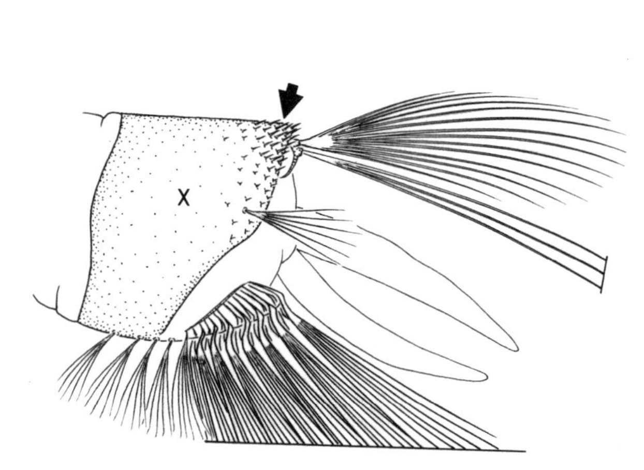


Fig. 969. Lateral view of abdominal segment X: *Cs. particeps*

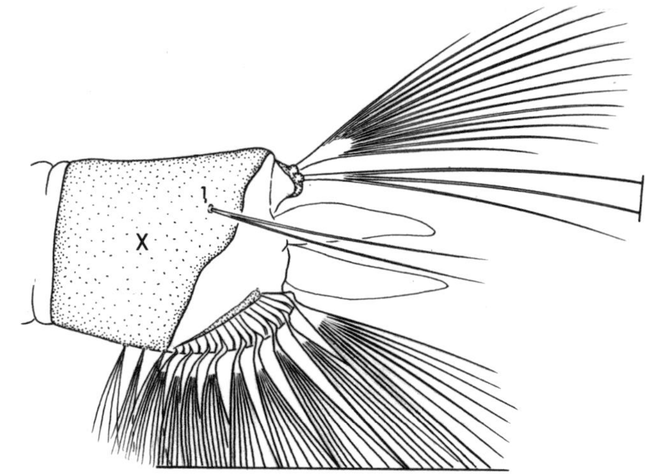


Fig. 971. Lateral view of abdominal segment X: *Cs. inornata*

6(5). Seta 1-X with rather strong branches equal to length of saddle or longer (fig. 972) ........ .......................................................................................... *inornata* (plate 32)

Seta 1-X with fine branches, shorter than saddle (fig. 973) .......................................... 7

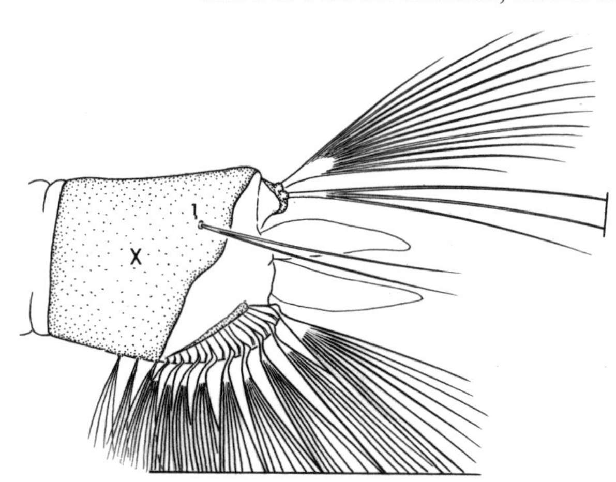


Fig. 972. Lateral view of abdominal segment X: *Cs. inornata)*

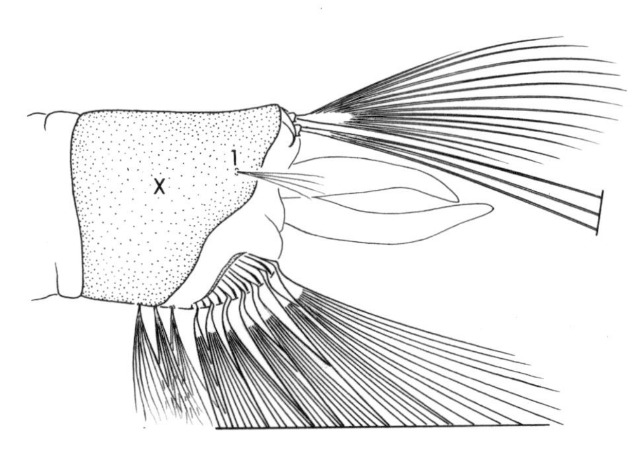


Fig. 973. Lateral view of abdominal segment X: *Cs. incidens*

7(6). Antenna robust, no more than 8.0 length of basal diameter, with many coarse spicules (fig. 974); setae 1,2-M both short, multibranched (fig. 975) .............. *alaskaensis* (plate 32A)

Antenna slender, 9.0 or more length of basal diameter, with fine spicules (fig. 976); seta 1-M single, much longer than multibranched seta 2-M (fig. 977) ....... *incidens* (plate 32C)

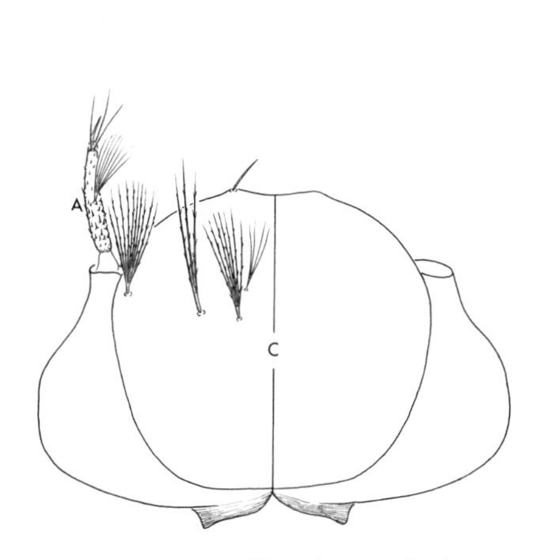


Fig. 974. Dorsal view of head: *Cs. alaskaensis*

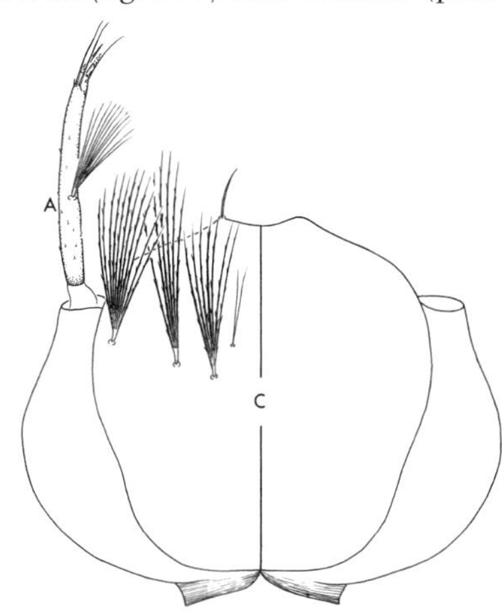


Fig. 976. Dorsal view of head: *Cs. incidens*

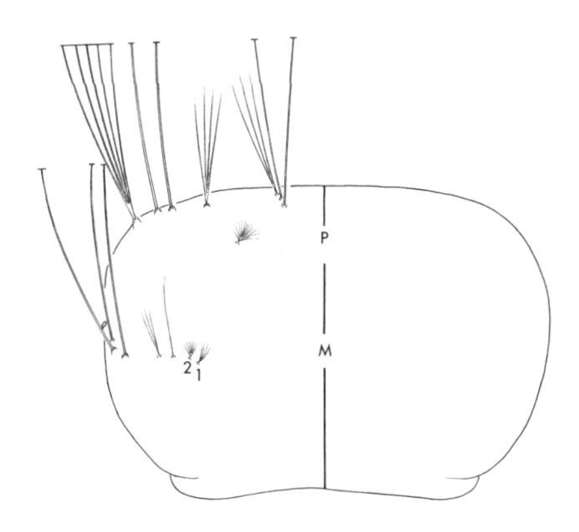


Fig. 975. Dorsal view of thorax: *Cs. alaskaensis*

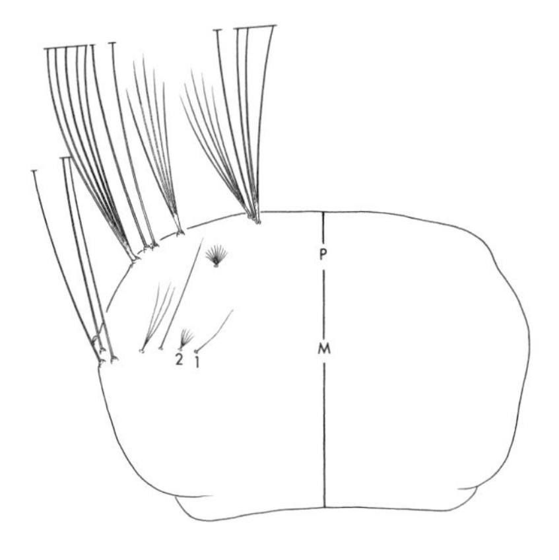


Fig. 977. Dorsal view of thorax: *Cs. incidens*

## Key to Fourth Instar Larvae of Genus *Deinocerites*

1. Seta 6-II single (fig. 978); seta 1-S with 4–6 branches (fig. 979) .... *mathesoni* (plate 42C)

   Seta 6-II double (fig. 980); seta 1-S double or triple (fig. 981) ....................................... 2

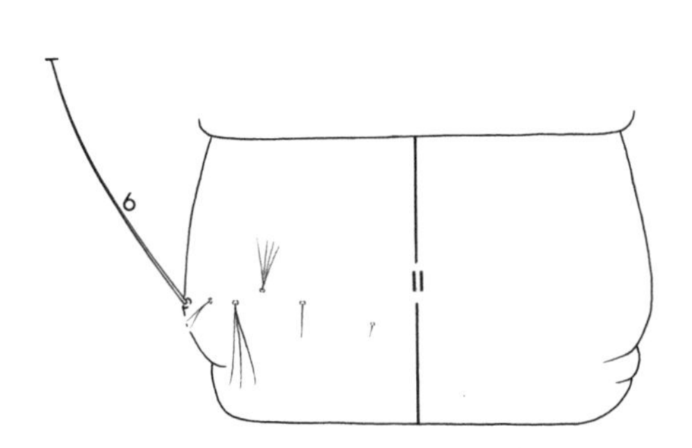


Fig. 978. Dorsal view of abdominal segment II: *De. mathesoni*

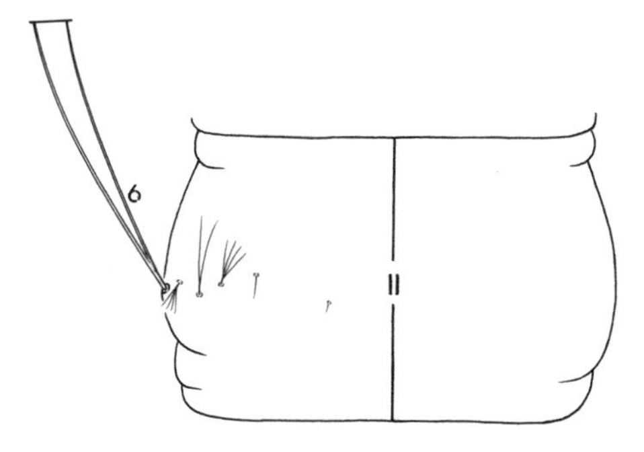


Fig. 980. Dorsal view of abdominal segment II: *De. pseudes*

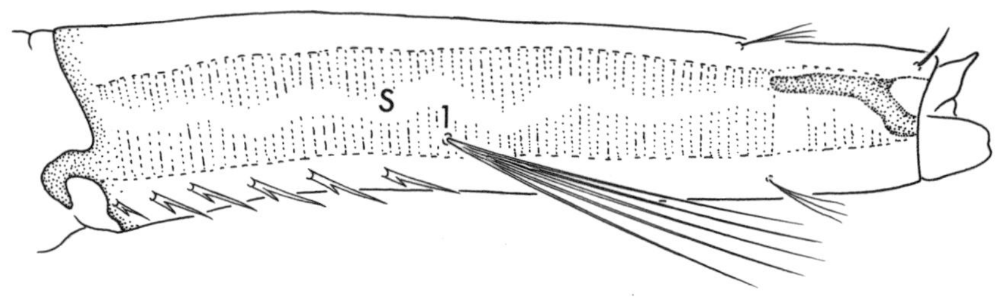


Fig. 979. Lateral view of siphon: *De. mathesoni*

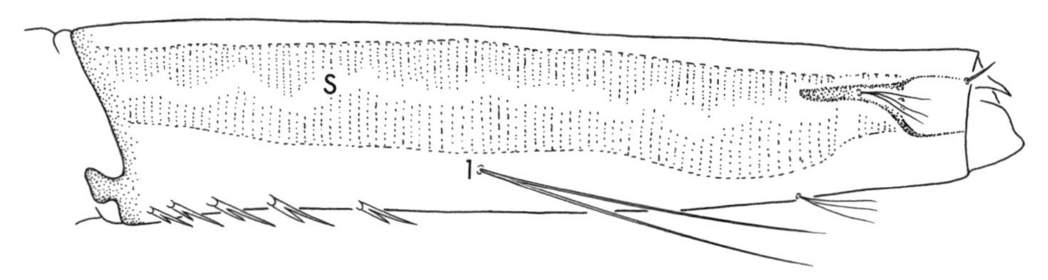


Fig. 981. Lateral view of siphon: *De. pseudes*

2(1). Seta 6-C double or triple (fig. 982) .................................... (in part) *pseudes* (plate 39C)

Seta 6-C single (fig. 983) .................................................................................. 3

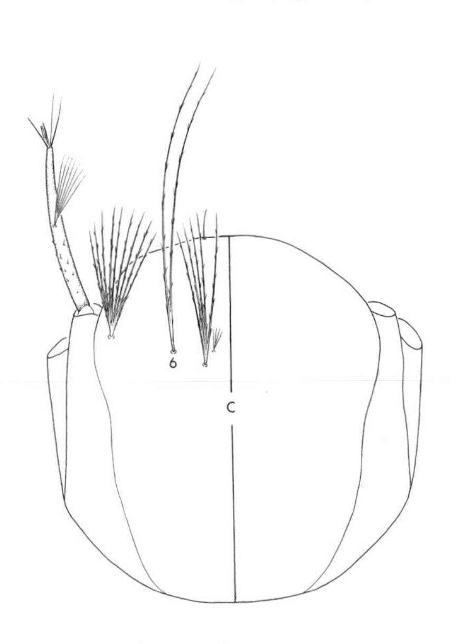


Fig. 982. Dorsal view of head: *De. pseudes*

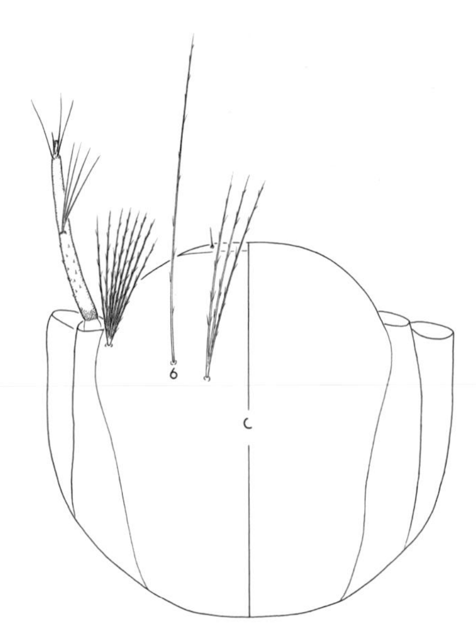


Fig. 983. Dorsal view of head: *De. cancer*

3(2). Seta 1-VIII usually with 5–7 branches; seta 1-VII long, frequently reaching base of siphon (fig. 984) ........................................................................ (in part) *pseudes* (plate 39C)

Seta 1-VIII usually with 3,4 branches; seta 1-VII shorter, not reaching base of siphon (fig. 985) ........................................................................................ *cancer* (plate 39D)

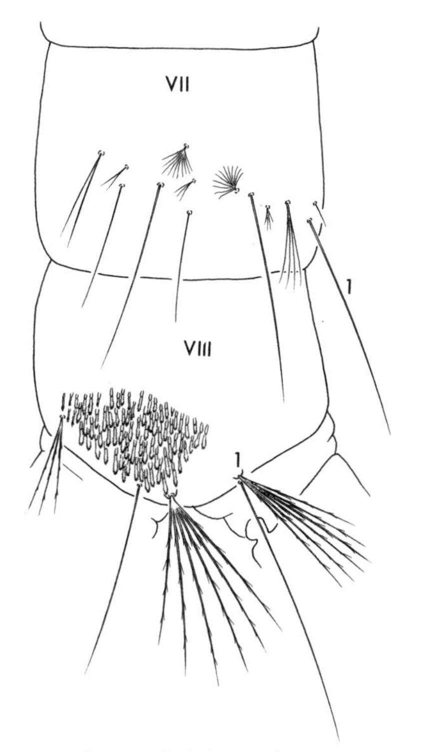


Fig. 984. Lateral view of abdominal segments VII–VIII: *De. pseudes*

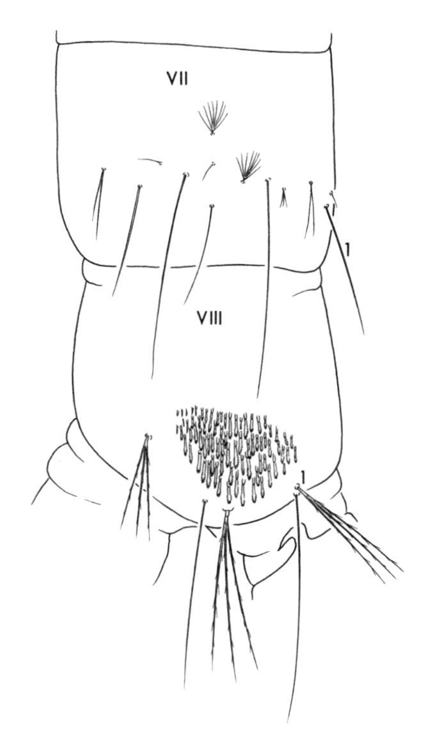


Fig. 985. Lateral view of abdominal segments VII–VIII: *De. cancer*

## Key to Fourth Instar Larvae of Genus *Mansonia*

Seta 4-X with 4 pairs of fanlike setae attached to grid (fig. 986); comb scale slender, with single spine apically (fig. 987) .......................................................... *titillans* (plate 33D)

Seta 4-X with 3 pairs of fanlike setae attached to grid (fig. 988); comb scale broader, with several stout subequal spinules (fig. 989) ............................................. *dyari* (plate 40A)

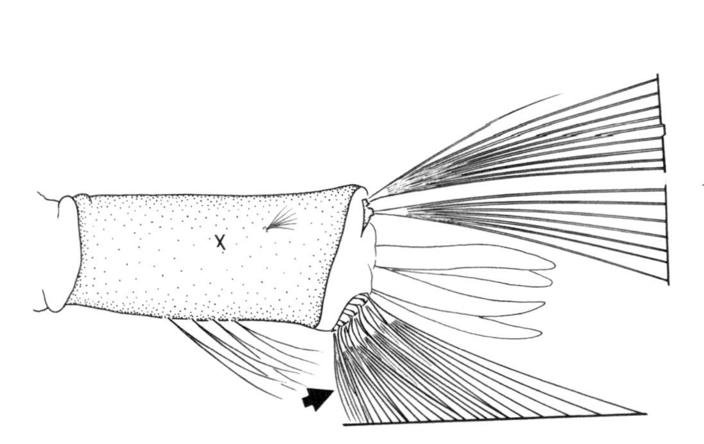


Fig. 986. Lateral view of abdominal segment X: *Ma. titillans*

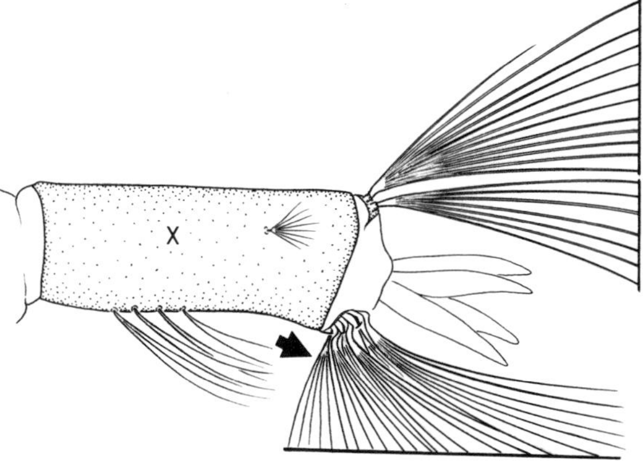


Fig. 988. Lateral view of abdominal segment X: *Ma. dyari*

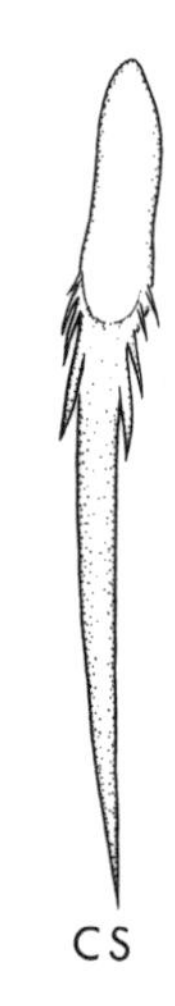


Fig. 987. Comb scale: *Ma. titillans*

Fig. 989. Comb scale: *Ma. dyari*

## Key to Fourth Instar Larvae of Genus *Orthopodomyia*

1. Seta 1-S usually with 3,4 branches, subequal in length to diameter of siphon at level of attachment; without large tergal plate on VIII (fig. 990) ......................... *alba* (plate 34B)

   Seta 1-S usually with 6 or more branches, much longer than diameter of siphon at point of attachment; with large tergal plate on VIII (fig. 991) .................................................... 2

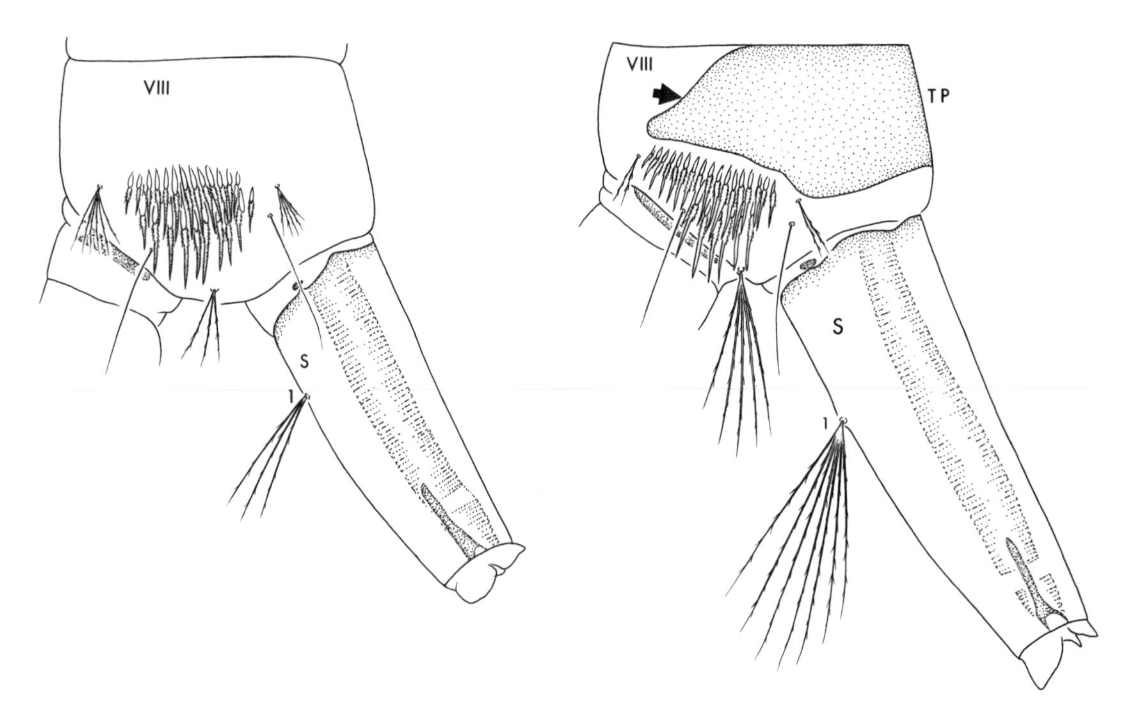


Fig. 990. Lateral view of abdominal segment VIII: *Or. alba*

Fig. 991. Lateral view of abdominal segment VIII: *Or. signifera*

2(1). Seta 1-S with branches longer than distance from its alveolus to apex of siphon (fig. 992); dorsal pair of anal papillae much longer than saddle (fig. 993) .......... *kummi* (plate 42D)

   Seta 1-S with branches no longer than distance from its alveolus to apex of siphon (fig. 994); dorsal pair of anal papillae no longer than saddle (fig. 995) ........... *signifera* (plate 34C)

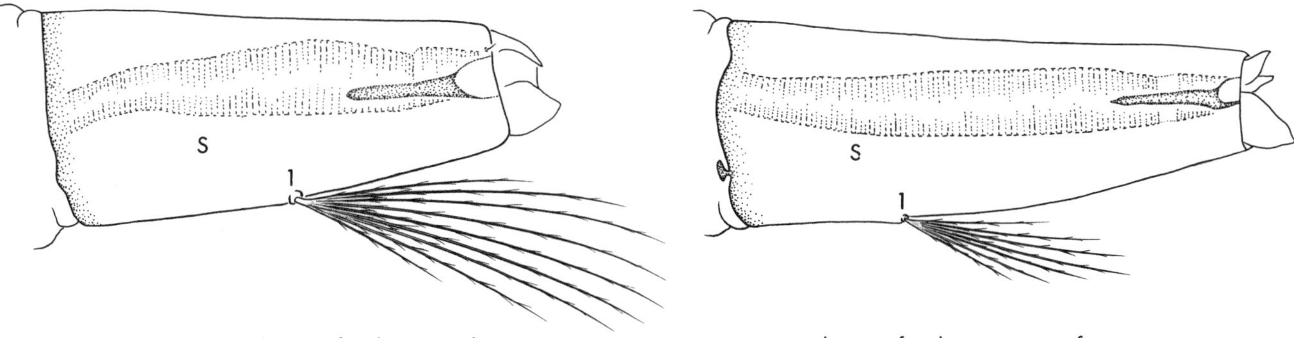


Fig. 992. Lateral view of siphon: *Or. kummi*

Fig. 994. Lateral view of siphon: *Or. signifera*

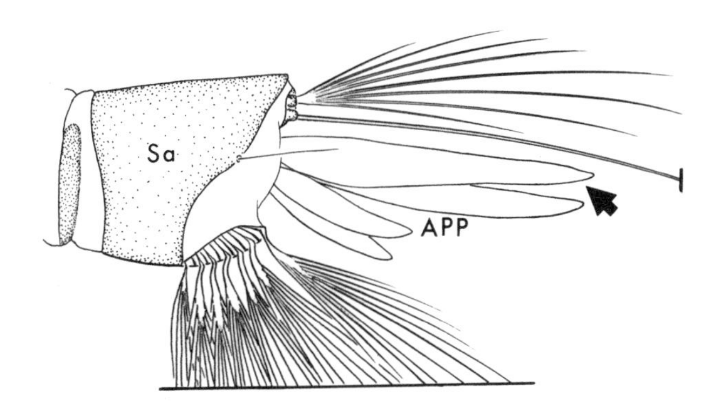


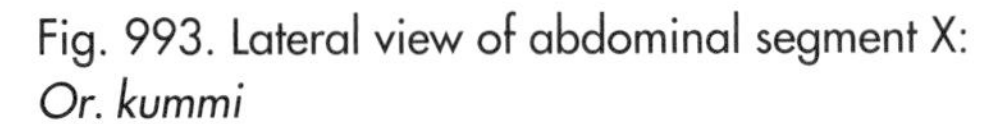

Fig. 993. Lateral view of abdominal segment X: *Or. kummi*

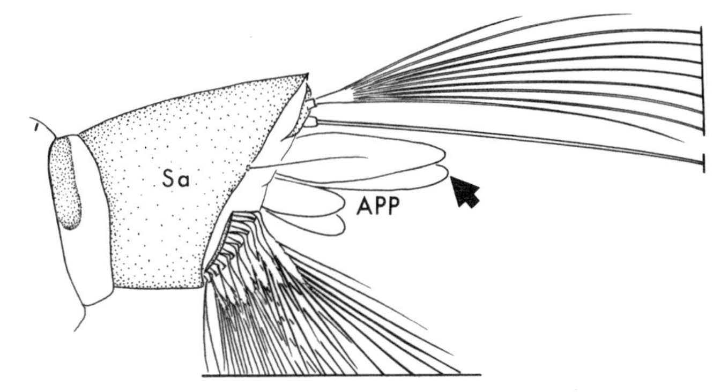


Fig. 995. Lateral view of abdominal segment X: *Or. signifera*

## Key to Fourth Instar Larvae of Genus *Psorophora*

1. Head capsule truncate anteriorly (fig. 996); pecten with 12 or more filamentous spines (fig. 997); antenna small, hardly reaching beyond anterior border of head (fig. 996)(subgenus *Psorophora*) ........................................ 2

   Head capsule rounded anteriorly (fig. 998); pecten with fewer than 10 spines, not produced into filaments (fig. 999); antenna reaching well beyond anterior border of head (fig. 998) ........................................ 3

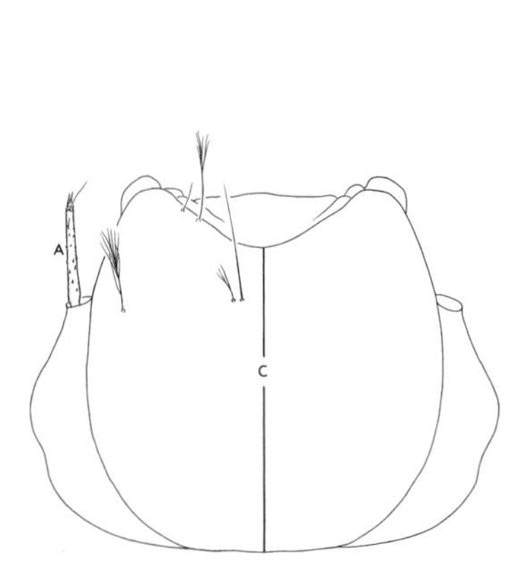


Fig. 996. Dorsal view of head: *Ps. ciliata*

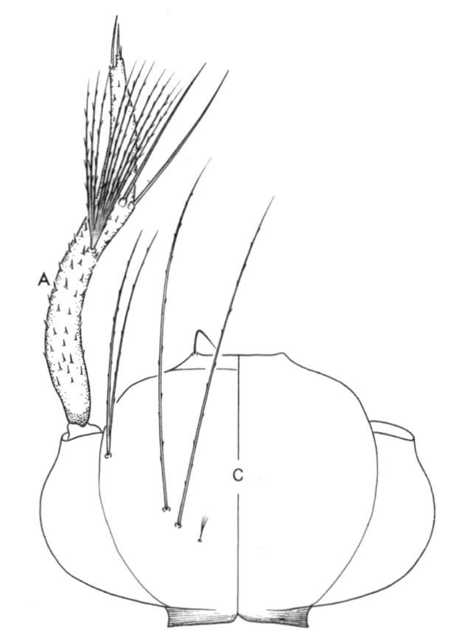


Fig. 998. Dorsal view of head: *Ps. discolor*

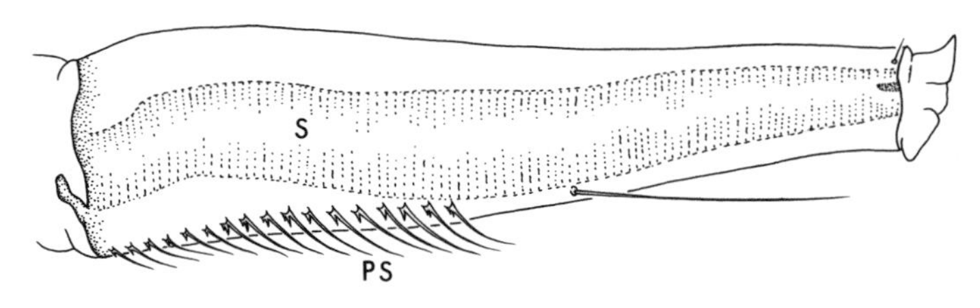

Fig. 997. Lateral view of siphon: *Ps. howardii*

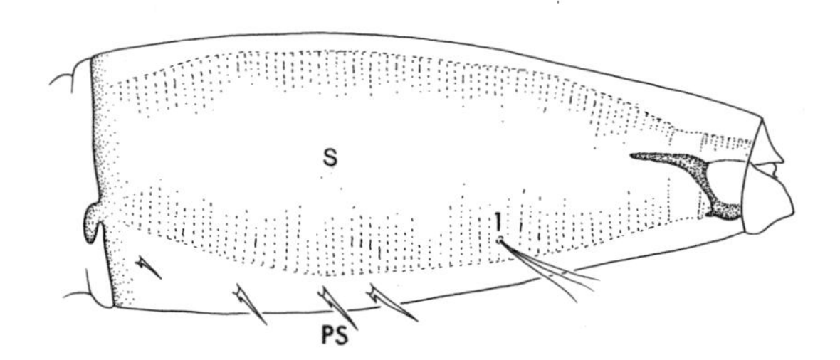


Fig. 999. Lateral view of siphon: *Ps. columbiae*

2(1). Seta 1-X with 3,4 branches from near base (fig. 1000) ....................... *ciliata* (plate 35A)

Seta 1-X single or branched mainly in outer 0.5 (fig. 1001) ........... *howardii* (plate 36B)

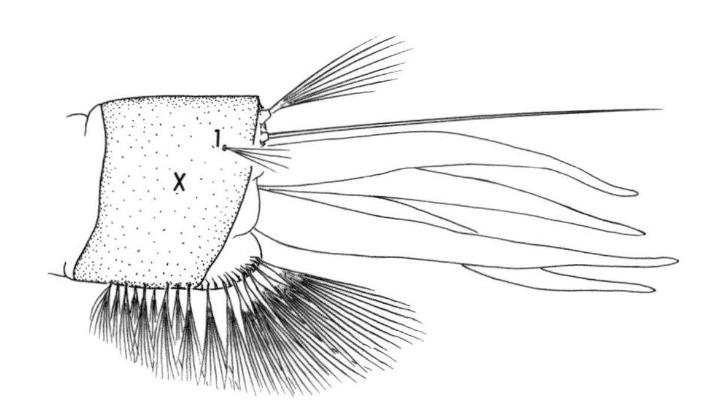


Fig. 1000. Lateral view of abdominal segment X: *Ps. ciliata*

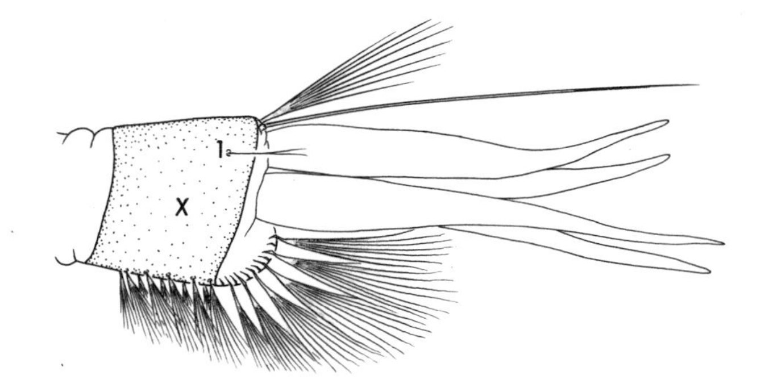


Fig. 1001. Lateral view of abdominal segment X: *Ps. howardii*

3(1). Antenna shorter than head (fig. 1002); if no (*discolor*), then seta 1-S with at least some branches equal to length of siphon; seta 6-S on anterolateral spiracular lobe shorter than apical diameter of siphon (fig. 1003) (subgenus *Grabhamia*) ....................................... 4

Antenna subequal to length of head or longer (fig. 1004), if not (*cyanescens*), then seta 6-S subequal to apical diameter of siphon; seta 1-S much shorter than length of siphon (fig. 1005) (subgenus *Janthinosoma*) ..................................................................................... 7

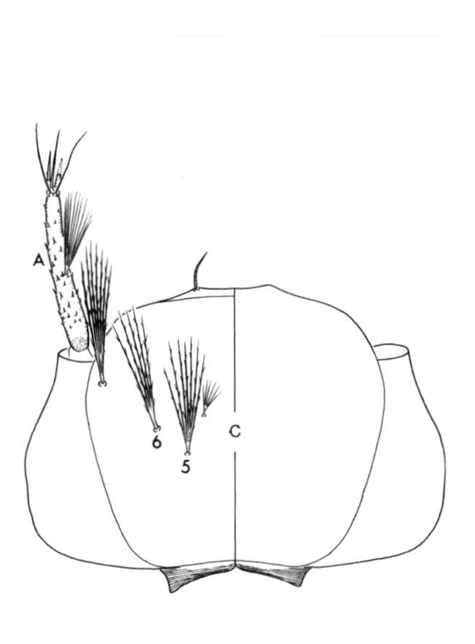


Fig. 1002. Dorsal view of head: *Ps. columbiae*

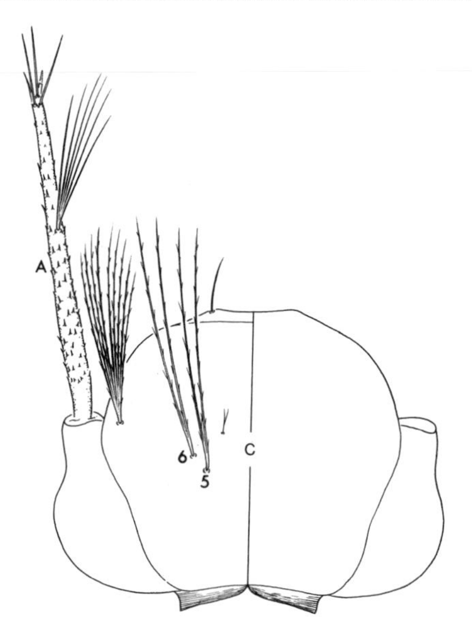


Fig. 1004. Dorsal view of head: *Ps. ferox*

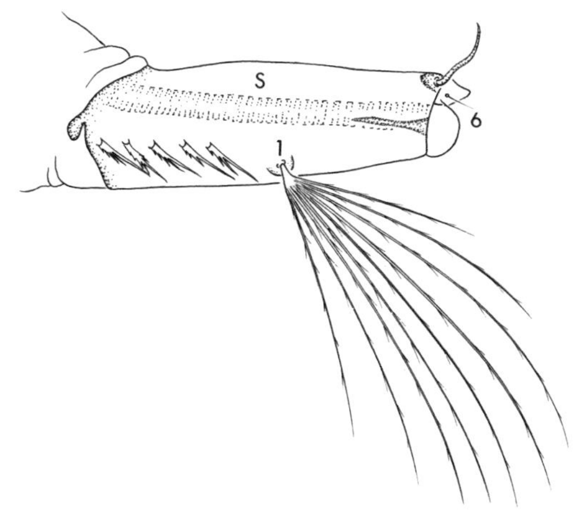


Fig. 1003. Lateral view of siphon: *Ps. discolor*

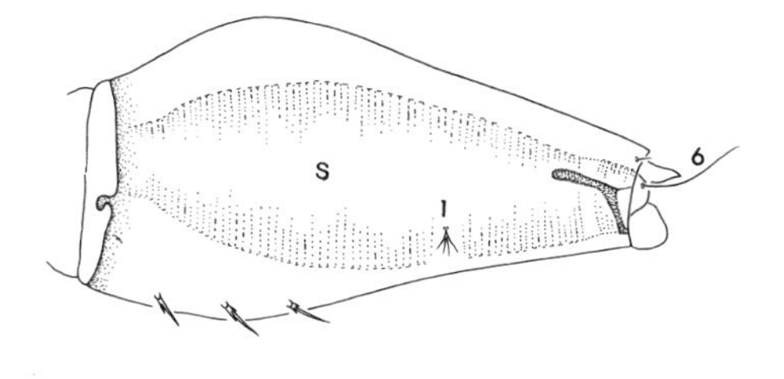


Fig. 1005. Lateral view of siphon: *Ps. cyanescens*

4(3). Antenna longer than head, sinuate, somewhat inflated in distal 0.5 (fig. 1006); seta 1-S very large, with some branches at least equal to length of siphon (fig. 1007) .......................... *discolor* (plate 35C)

Antenna shorter than head, slightly curved, not inflated (fig. 1008); seta 1-S much shorter than length of siphon (fig. 1009) .......................... 5

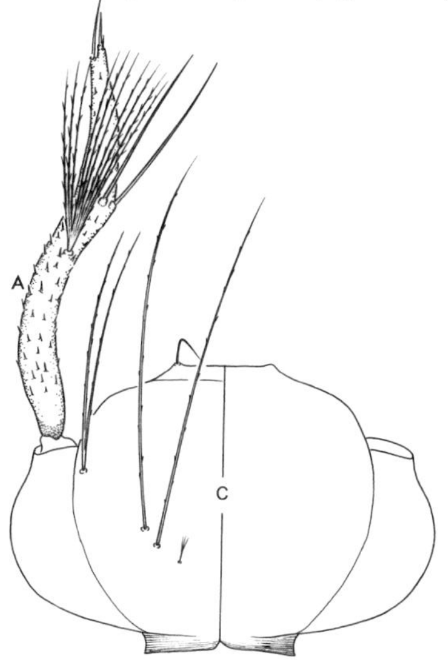


Fig. 1006. Dorsal view of head: *Ps. discolor*

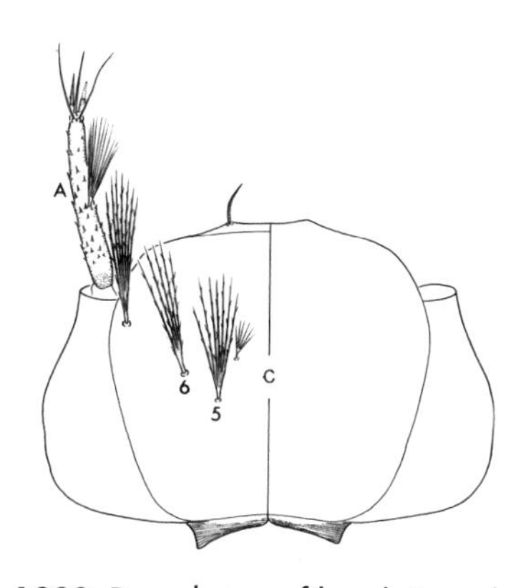


Fig. 1008. Dorsal view of head: *Ps. columbiae*

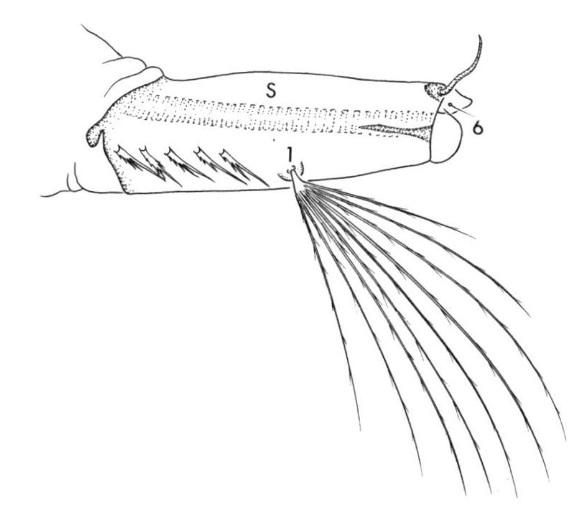


Fig. 1007. Lateral view of siphon: *Ps. discolor*

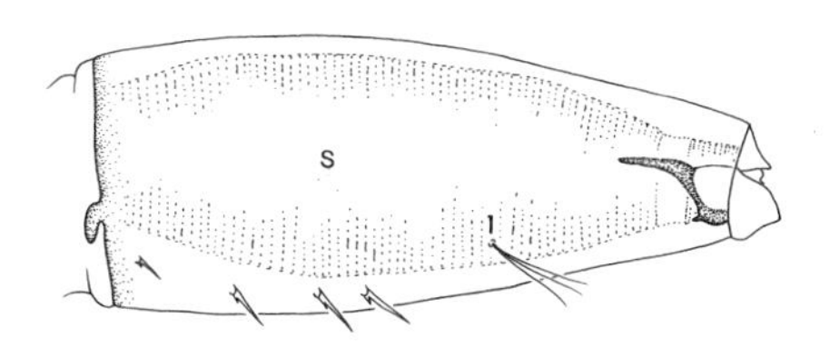


Fig. 1009. Lateral view of siphon: *Ps. columbiae*

5(4). Setae 5,6-C shorter than antenna, with 4 or more branches (fig. 1010) ............ *columbiae* ............................................................................................................. (plate 34D)

Setae 5,6-C about equal to antenna, or longer, single to triple (fig. 1011) ........................ 6

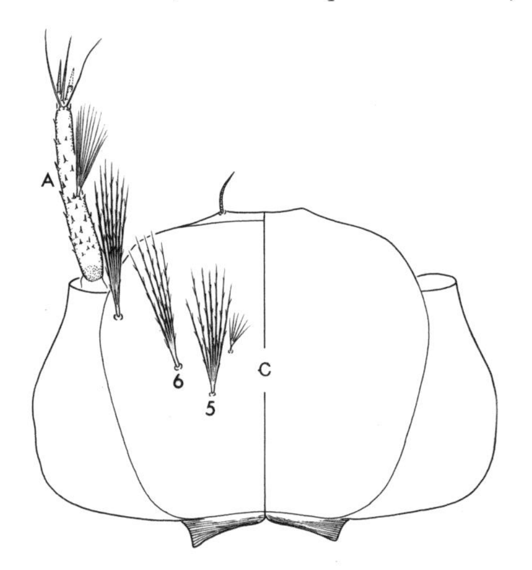


Fig. 1010. Dorsal view of head: *Ps. columbiae*

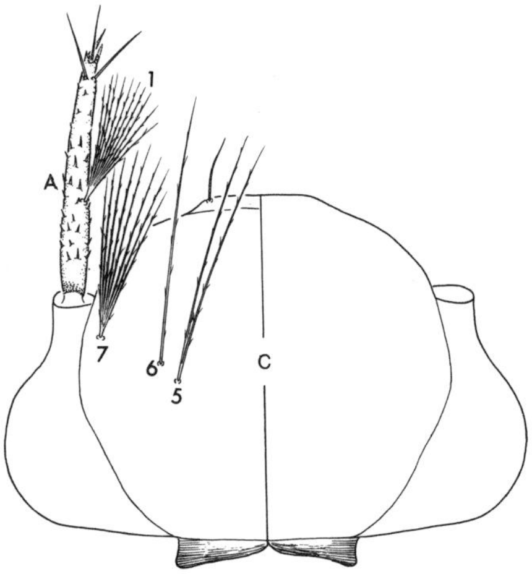

Fig. 1011. Dorsal view of head: *Ps. signipennis*

6(5). Antenna with strong spicules, setae 1-A and 7-C with prominent aciculae, 1-A with 8 or more branches; 7-C with at least 6 branches (fig. 1012) ..........................*signipennis* (plate 37A)

Antenna with weak spicules; setae 1-A and 7-C weakly aciculate, with fewer than 6 branches (fig. 1013) ........................................................................................ *pygmaea* (plate 42C)

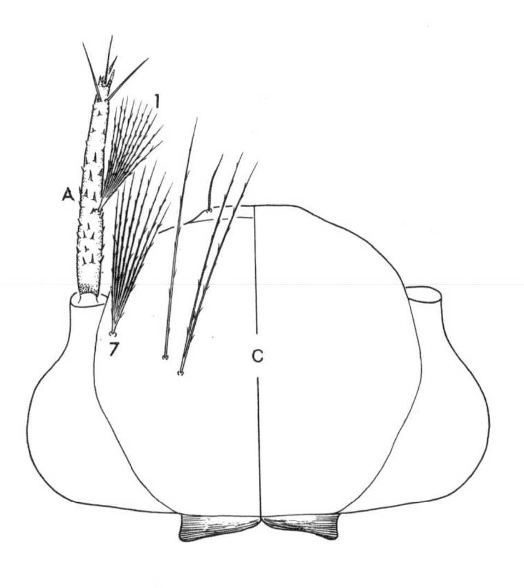


Fig. 1012. Dorsal view of head: *Ps. signipennis*

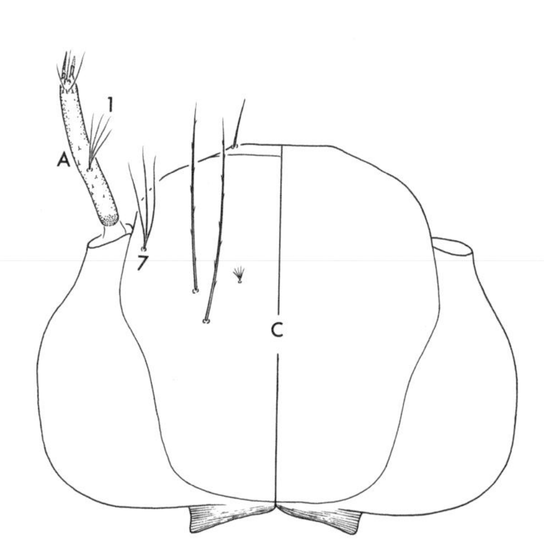


Fig. 1013. Dorsal view of head: *Ps. pygmaea*

7(3). Antenna shorter than head (fig. 1014); seta 6-S on anterolateral spiracular lobe subequal to apical diameter of siphon (fig. 1015) .............................................. *cyanescens* (plate 35B)

Antenna about equal to head, or longer (fig. 1016); seta 6-S much shorter than apical diameter of siphon (fig. 1017) ......................................................................................... 8

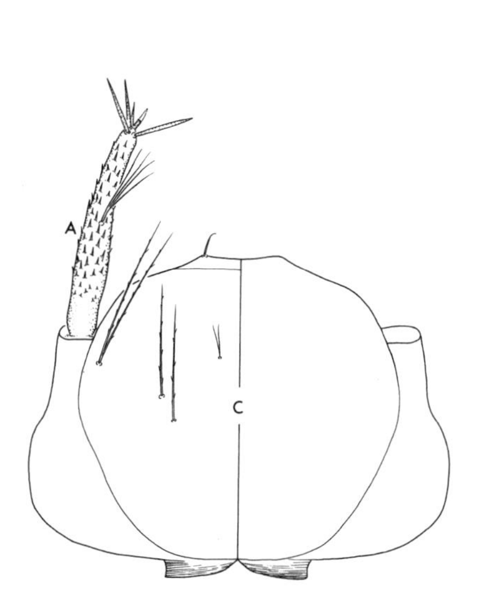


Fig. 1014. Dorsal view of head: *Ps. cyanescens*

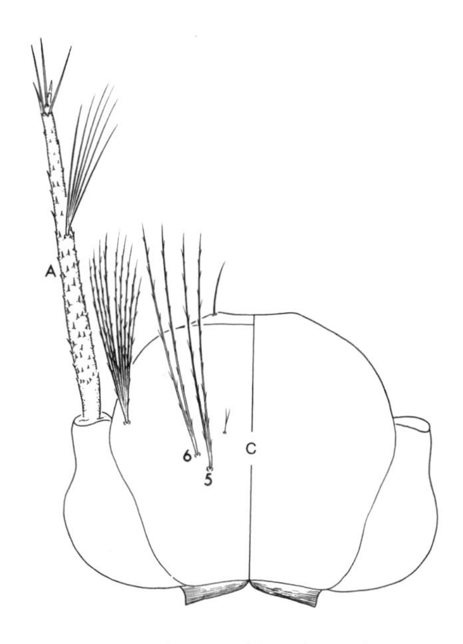


Fig. 1016. Dorsal view of head: *Ps. ferox*

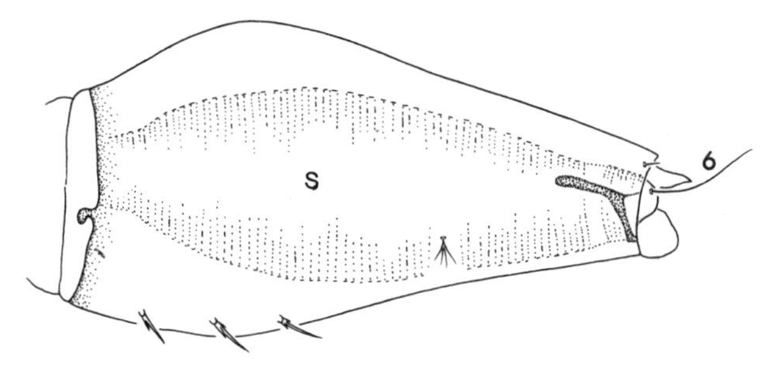


Fig. 1015. Lateral view of siphon: *Ps. cyanescens*

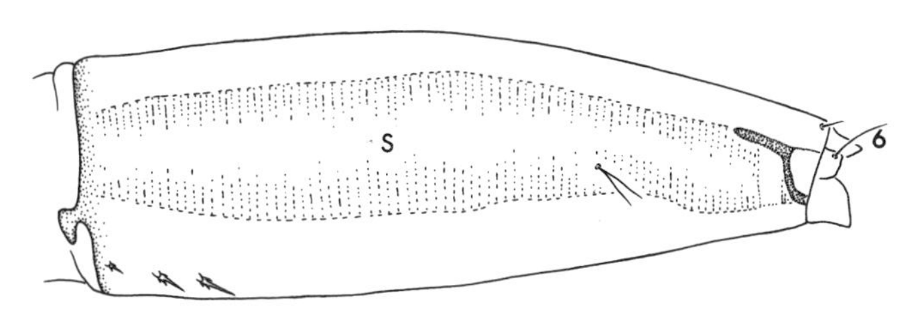


Fig. 1017. Lateral view of siphon: *Ps. ferox*

8(7). Siphon index 2.5–3.0, without subapical narrowed form (fig. 1018); with 4–6 precratal fanlike setae (fig. 1019) ........ *johnstonii* (plate 41C)

Siphon index 3.5 or greater, with distinct subapical narrowing (fig. 1020); with 7 or more precratal fanlike setae (fig. 1021) ........ 9

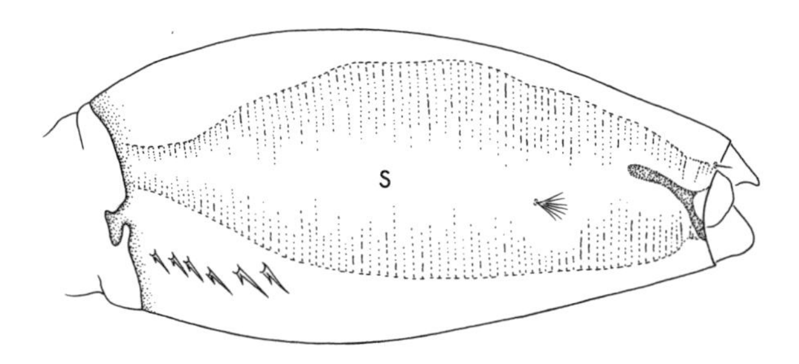


Fig. 1018. Lateral view of siphon: *Ps. johnstonii*

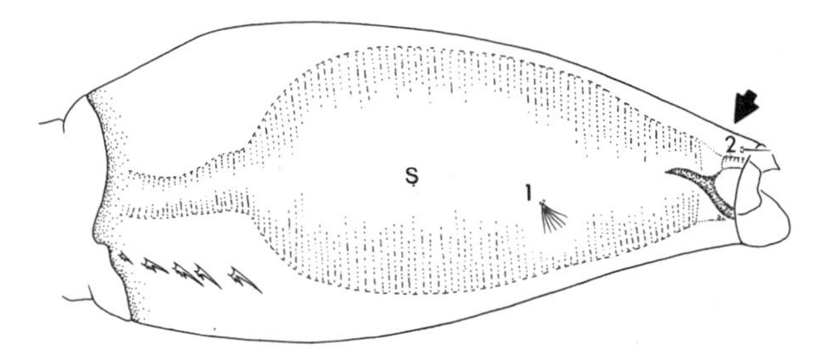


Fig. 1020. Lateral view of siphon: *Ps. horrida*

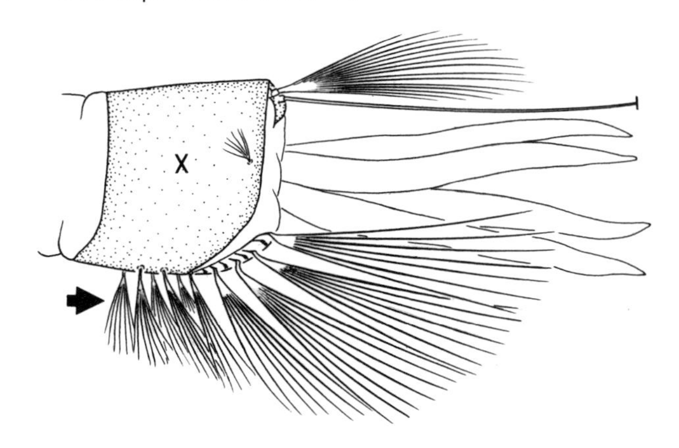


Fig. 1019. Lateral view of abdominal segment X: *Ps. johnstonii*

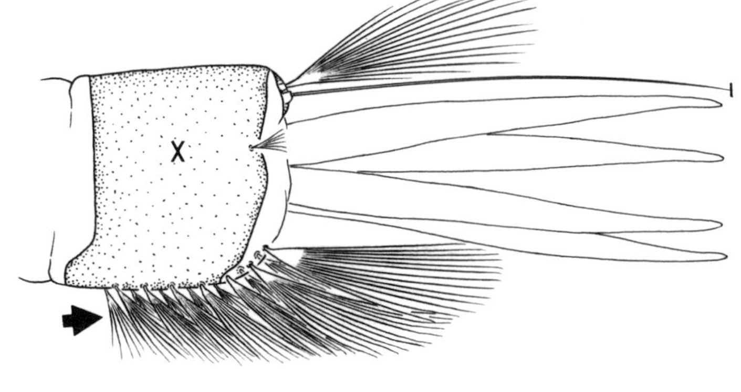


Fig. 1021. Lateral view of abdominal segment X: *Ps. horrida*

9(8). Antenna subequal to median length of head (fig. 1022) ................................................ 10

Antenna distinctly longer than median length of head (fig. 1023) ................................ 11

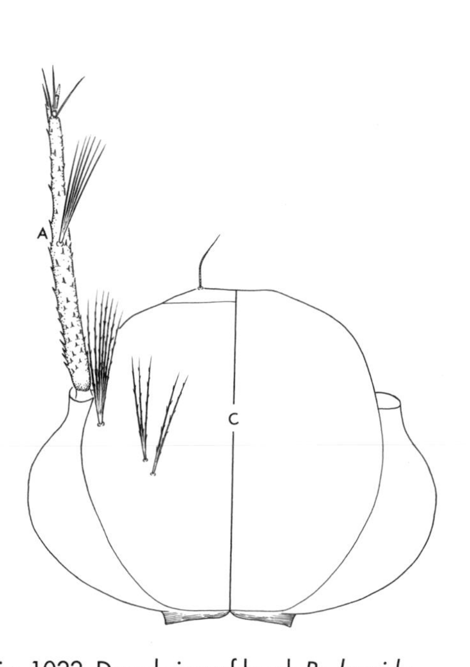

Fig. 1022. Dorsal view of head: *Ps. horrida*

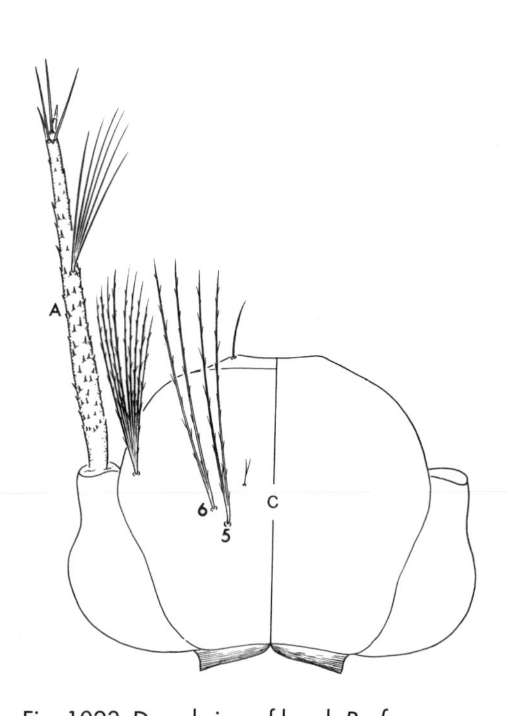


Fig. 1023. Dorsal view of head: *Ps. ferox*

10(9). Setae 6-IV–VI medium sized, not as long as following tergum, double or triple (fig.1024); length of seta 1-S subequal to 2-S (fig. 1025) ....................................... *horrida* (plate 36A)

Setae 6-IV–VI stout, much longer than following tergum, branches variable (fig. 1026); seta 1-S much longer than 2-S (fig. 1027) ............................................. *mathesoni* (plate 36D)

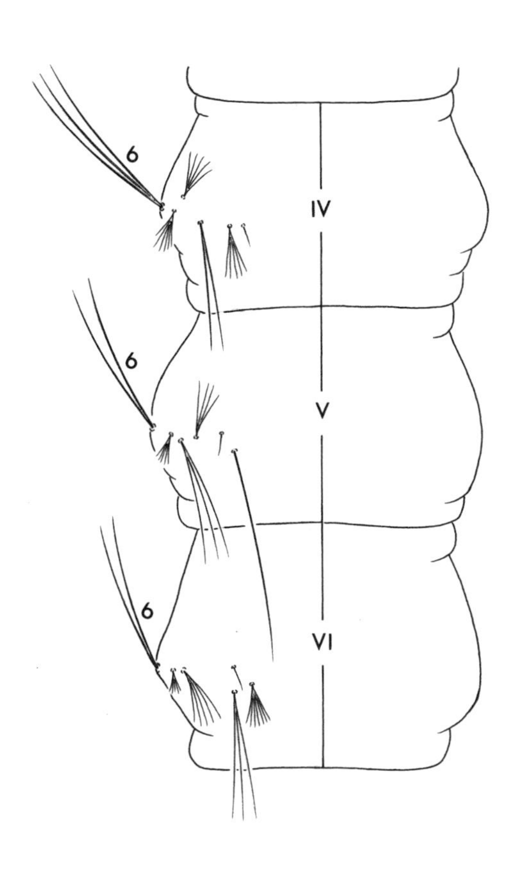


Fig. 1024. Dorsal view of abdominal segments IV–VI: *Ps. horrida*

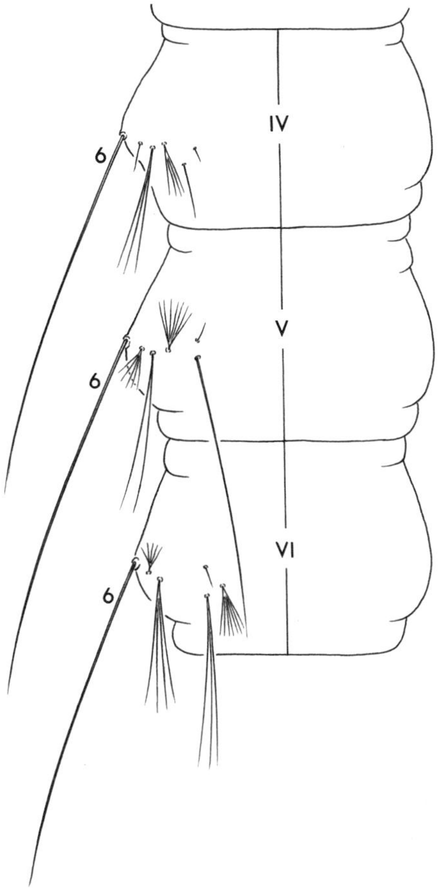

Fig. 1026. Dorsal view of abdominal segments IV–VI: *Ps. mathesoni*

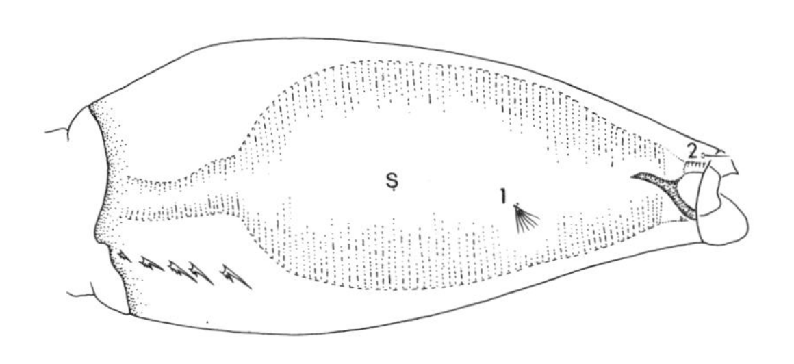


Fig. 1025. Lateral view of siphon: *Ps. horrida*

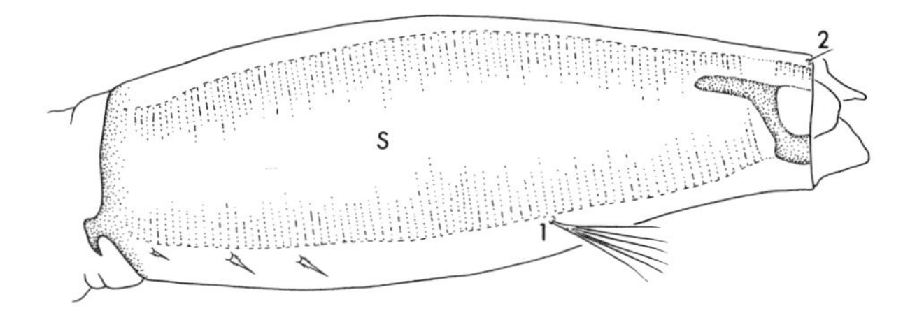


Fig. 1027. Lateral view of siphon: *Ps. mathesoni*

11(9).Setae 6-IV–VI single or double (fig. 1028); individual branches of setae 5,6-C nearly equal in length (fig. 1029) .......................................................................... *ferox* (plate 35D)

Setae 6-IV–VI with 3 or more branches (fig. 1030); individual branches of setae 5,6-C not equal, at least one shorter and weaker (fig. 1031) ......................... *longipalpus* (plate 36C)

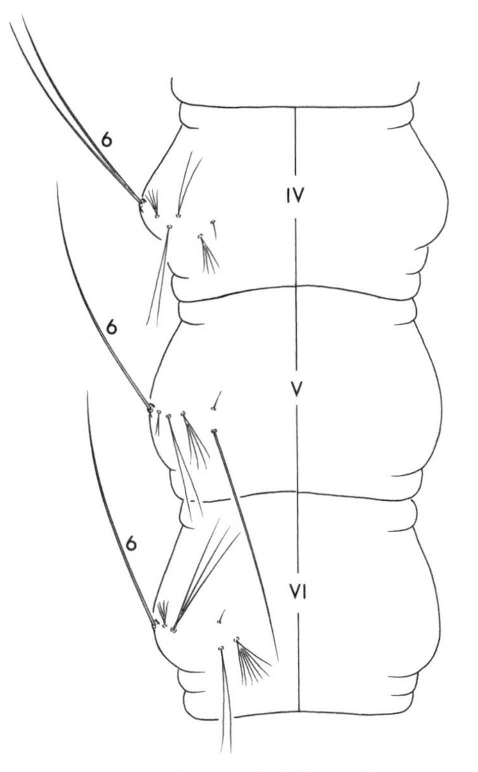


Fig. 1028. Dorsal view of abdominal segments IV–VI: *Ps. ferox*

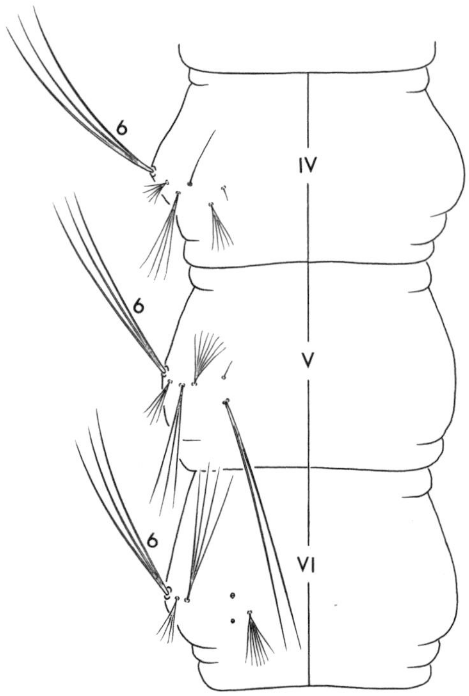

Fig. 1030. Dorsal view of abdominal segments IV–VI: *Ps. longipalpus*

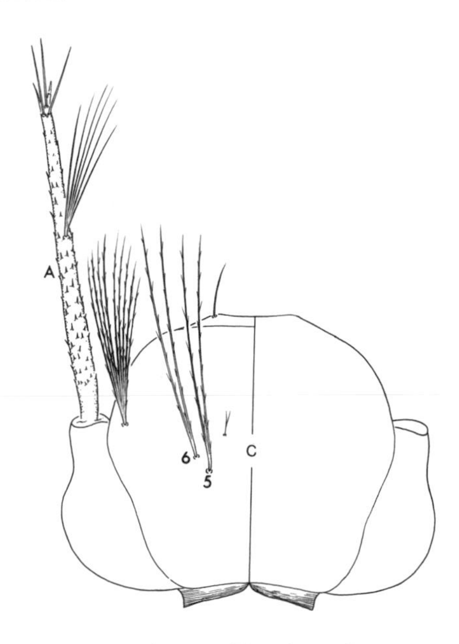


Fig. 1029. Dorsal view of head: *Ps. ferox*

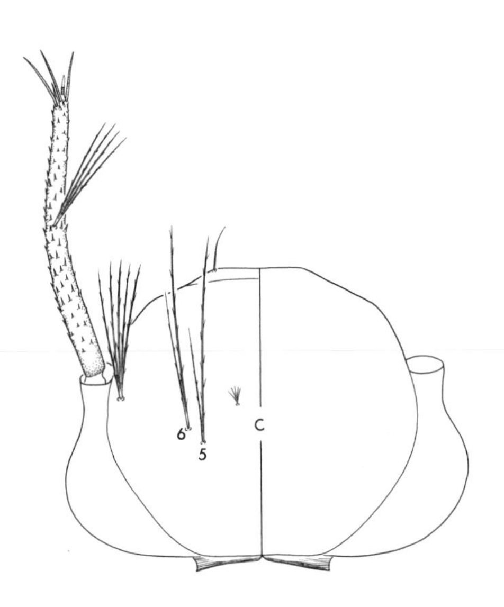


Fig. 1031. Dorsal view of head: *Ps. longipalpus*

## Key to Fourth Instar Larvae of Genus *Uranotaenia*

1.Seta 5-C double or triple, seta 6-C single, both coarse but not spiniform (fig. 1032) (subgenus *Pseudoficalbia*) ............ *a. anhydor* (plate 41C)
*a. syntheta* (plate 37C)

Setae 5,6-C single, stout, spiniform, spinulate (fig. 1033) (subgenus *Uranotaenia*) ........... 2

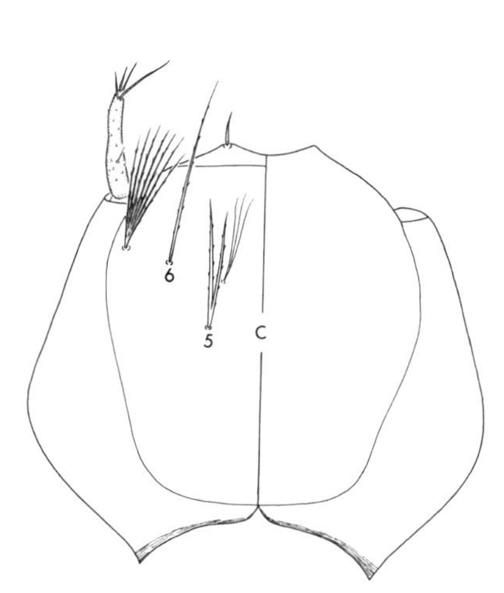


Fig. 1032. Dorsal view of head: *Ur. a. syntheta*

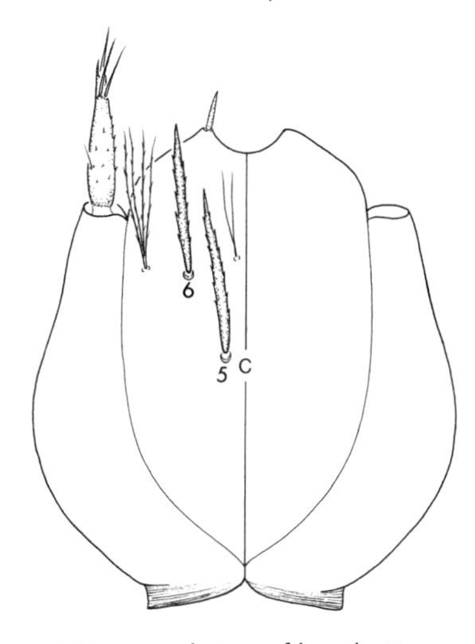


Fig. 1033. Dorsal view of head: *Ur. sapphirina*

2(1). Seta 3-P more than 0.5 length of 1-P, with 4–8 branches (fig. 1034); seta 6-I,II double (fig. 1035) ........................................ *lowii* (plate 37D)

Seta 3-P much shorter than 0.5 length of 1-P, with 8–10 branches (fig. 1036); seta 6-I,II triple (fig. 1037) ........................................ *sapphirina* (plate 38A)

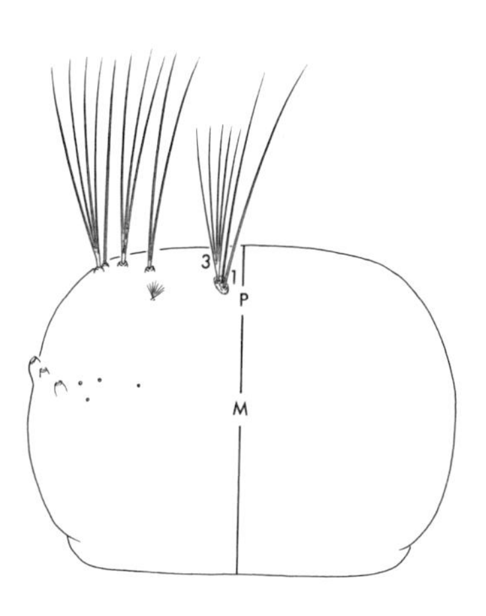


Fig. 1034. Dorsal view of thorax: *Ur. lowii*

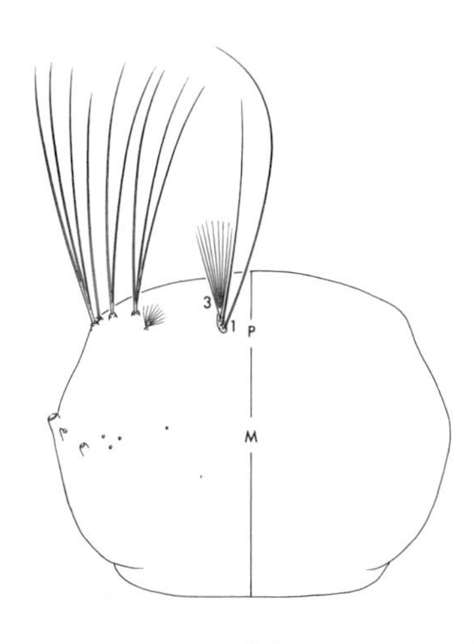


Fig. 1036. Dorsal view of thorax: *Ur. sapphirina*

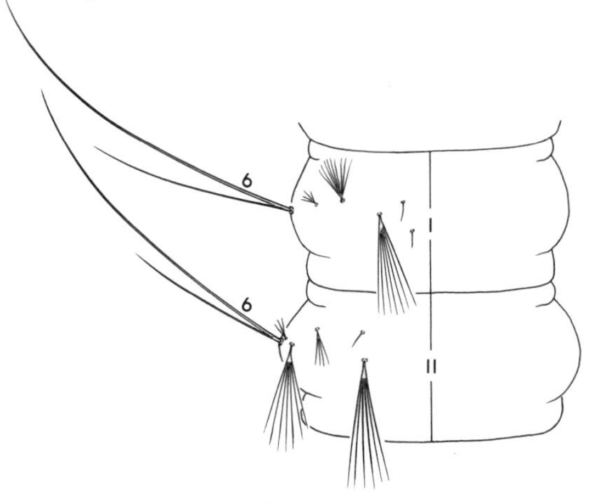


Fig. 1035. Dorsal view of abdominal segments I,II: *Ur. lowii*

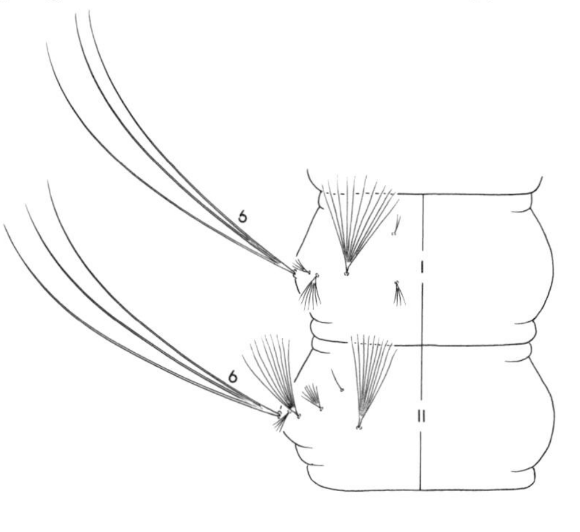


Fig. 1037. Dorsal view of abdominal segments I,II: *Ur. sapphirina*

### Key to Fourth Instar Larvae of Genus *Wyeomyia*

1. Setae 1–3-X single, seta 4-X with 7 or more branches (fig. 1038); seta 5-C with 3,4 branches (fig. 1039) .......... *mitchellii* (plate 37C)

   Setae 1–3-X not all single, seta 4-X with no more than 6 branches (fig. 1040); seta 5-C single (fig. 1041) .......... 2

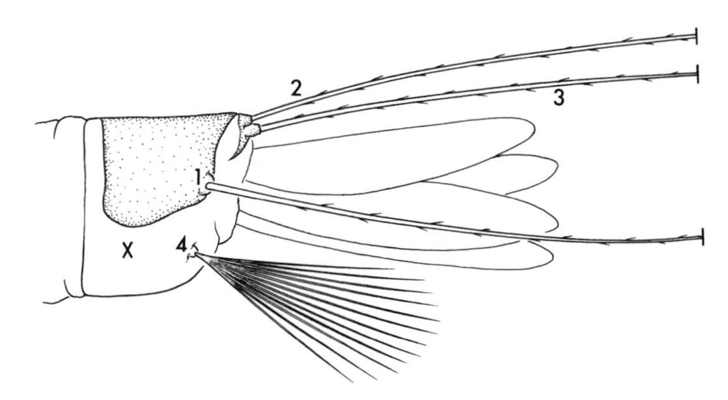


Fig. 1038. Lateral view of abdominal segment X: *Wy. mitchellii*

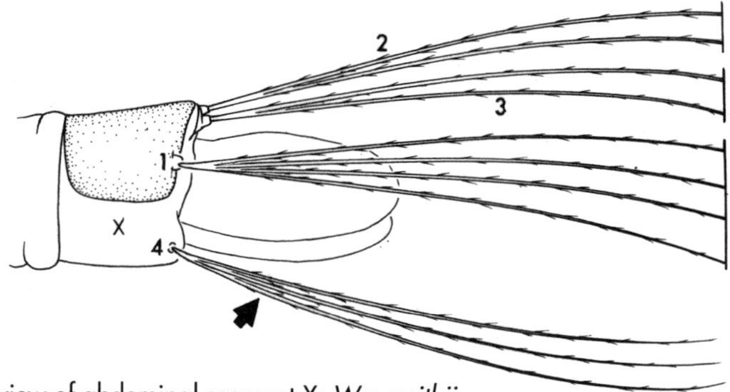


Fig. 1040. Lateral view of abdominal segment X: *Wy. smithii*

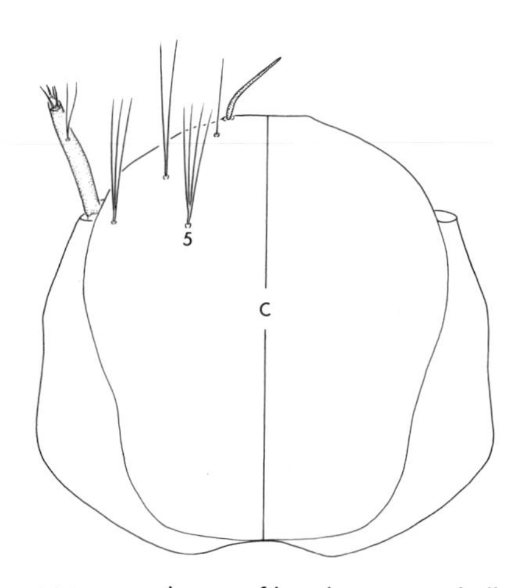


Fig. 1039. Dorsal view of head: *Wy. mitchellii*

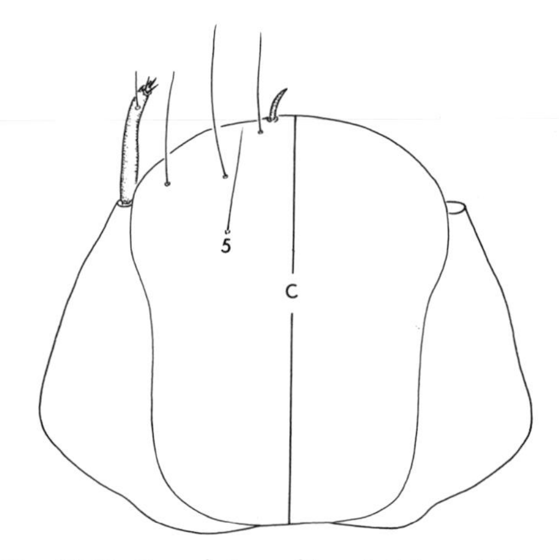


Fig. 1041. Dorsal view of head: *Wy. smithii*

2(1). Siphon index about 6.0 (fig. 1042); seta 4-X with 1,2 long and 3,4 short branches (fig. 1043); several setae on siphon double or triple (fig. 1042) .......... *vanduzeei* (plate 40B)

Siphon index 4.0–5.0 (fig. 1044); seta 4-X with 3 long subequal branches (fig. 1045); setae on siphon all single (fig. 1044) ............................................................ *smithii* (plate 38B)

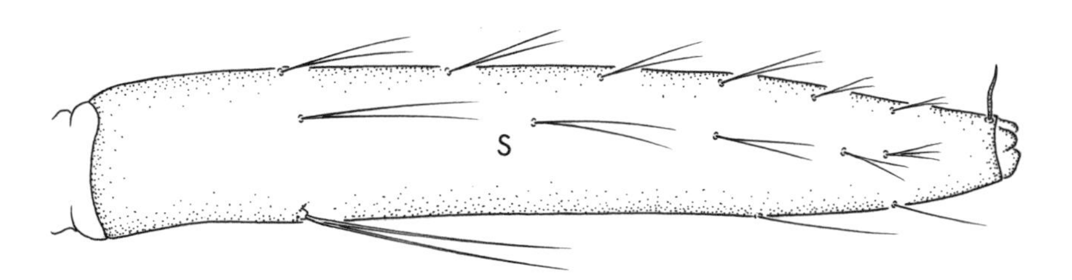


Fig. 1042. Lateral view of siphon: *Wy. vanduzeei*

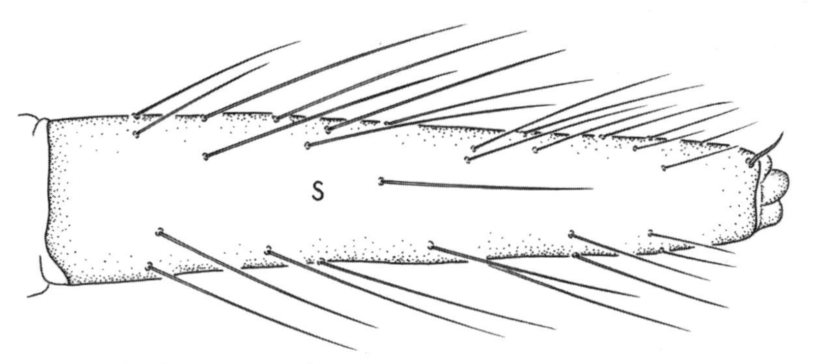


Fig. 1044. Lateral view of siphon: *Wy. smithii*

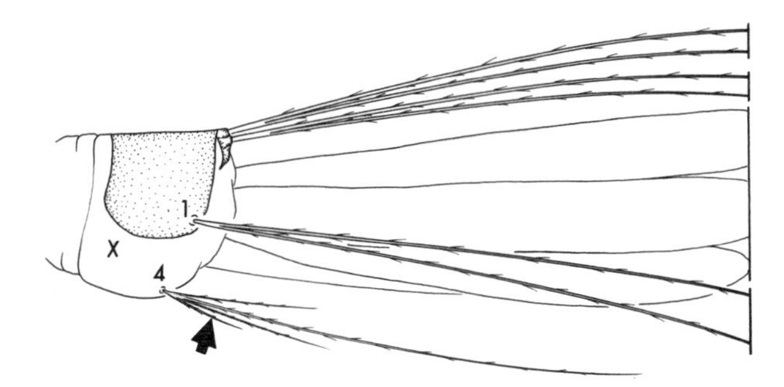


Fig. 1043. Lateral view of abdominal segment X: *Wy. vanduzeei*

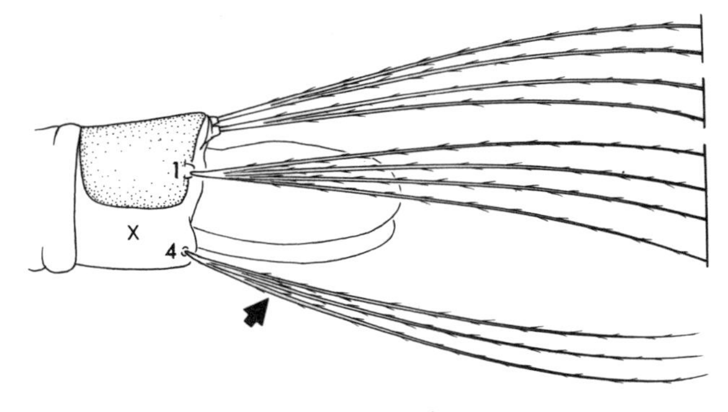


Fig. 1045. Lateral view of abdominal segment X: *Wy. smithii*

# Geographical Distribution of the Culicidae of North America, North of Mexico

Closely associated with the identification of any taxon is its geographical distribution. The process of identification will be greatly enhanced by knowing the limits of dispersion of the fauna with which you are working. Obviously, if you determine that a specimen is a particular species and, by checking its dispersion, discover you collected it outside of its known range, you will check it again. In this publication, identification and distribution have been linked, so the user can check one against the other.

Tables 2–4 list species and subspecies and the states/provinces from which they have been reported. Table 2 registers 108 species and subspecies from 24 eastern states of the United States and the District of Columbia; table 3, 151 species and subspecies from 24 western states, and table 4, 80 species and subspecies from Canada and Alaska. Of the 174 taxa, 77 occur in the conterminous 48 states and Canada/Alaska (the latter two are lumped because they share a similar mosquito fauna); moreover, 95 are found in the 48 states but not in Canada/Alaska and 4 are restricted to the latter area. Comparing Canada with Alaska, each has three species not found in the other area, i.e., Canada: *Oc. churchillensis*, *Oc. rempeli*, and *Oc. nigripes*; Alaska: *Oc. punctodes*, *Oc. ventrovittis*, and *Cs. particeps*. *Oc. churchillensis* is a sibling species of *Oc. communis* (250). *Oc. rempeli* has also been reported from Russia (234), so it has wider distribution than the Canadian records indicate. Therefore, both *Oc. churchillensis* and *Oc. rempeli* could occur in Alaska. All three Alaskan taxa are potential members of the Canadian fauna (784). *Oc. togoi* belongs to the fauna of the Oriental biogeographical region and is apparently a fairly recent introduction into British Columbia that has spread into Washington.

Of the 170 taxa distributed in the 48 states, 97 are disseminated in both eastern and western blocks of states, as listed in tables 2 and 3. Only 17 are restricted to the eastern states, while 3 times as many (61) are confined to the West. In all, 14 (82%) of those found only in the East are limited to peninsular Florida, with a few extending into southern Georgia and South Carolina. The western states, on the other hand, with their plains, high mountains, and deserts, offer a wide variety of weather conditions and habitats, resulting in the development of a diverse mosquito fauna, unique to the West. In fact, three western states, Texas, Arizona, and California, have 7, 4, and 8 species, respectively, not occurring in any other political unit of the region.

There follows in this section maps that depict the distribution in North America, north of Mexico, of all 174 taxa now known in the culicid fauna, except *Ps. varipes* (Coquillett), which was not included in the present treatment since at present its distribution and identity are indefinite (385). Maps have been drawn as accurately as possible within the limits of available information on occurrence in the states and provinces of the region, also referred to as political units. Because of the paucity of information about some, the distribution limits had to be estimated. Some are depicted as having discontinuous distribution based on available records and whether this is real or imagined remains to be determined by further study. No attempt has been made to delineate within political units exactly where widespread species such as *Ae. vexans* have been found.

In studying biogeographical distributional patterns exhibited by the mosquito fauna in North America, north of Mexico, two paths of dispersal are very evident. Southern taxa such as *Oc. thibaulti* and *Ps. ferox* (plates 22, 35) are following the lowlands of the Mississippi Valley and its tributaries and the coastal plain of the Atlantic Ocean northward. On the other hand, northern species such as *Oc. communis* and *Oc. hexodontus* (plates 12, 15) have dispersed southward along the high ranges of the Rocky Mountains. It appears that the eastern Appalachian range lacks either sufficient altitude or, possibly, favorable breeding sites at higher elevations to support the northern fauna; however, the Applachian area has not been studied in depth and its fauna is not well documented.

Another noteworthy dispersal pattern is the avoidance of the southwestern states by such species as *Oc. sticticus*, *Oc. hendersoni*, and *Cx. restuans* (see plates 21, 15, 29). They have been able to colonize large areas of North America, but not the southwest. Apparently they simply cannot tolerate the dry climate and types of habitats found there. On the other hand, widespread species such as *Oc. dorsalis*, *Ae. vexans*, and *Cx. tarsalis* (see plates 13, 24, 30) are well acclimated and have successfully thrived in the dry areas of southwestern United States.

One species, *Ae. aegypti*, presents a peculiar distributional problem. Christophers (165) stated that temperature limits its dispersal, and in North America it is restricted to south of a January isotherm of 1.8°C (35°F) and a July isotherm of 23°C (75°F), which agrees quite well with our limits indicated on plate 9 as the extreme range. Rozeboom (623) spoke of three zones in relation to its limit of distribution: the zone of continuous breeding, the zone of overwintering egg survival, and the temporary summer zone, with incursions during the warm months and complete winter dieoff. On plate 9, then, the usual range marks the limits of continuous breeding, while the extreme range would include the latter two categories. In preparing the map we also considered the records of Morland and Tinker (500) and Tinker and Hayes (725), based on actual surveys.

The study of Wood (784) and our own review of the specimens in the U.S. National Museum have led to the conclusion that *An. occidentalis* does not occur in Canada or Alaska, as depicted by references 146, 199, 284. Gjullin et al. (286) had previously observed that Alaskan records for *An. occidentalis* referred to *An. earlei*.

An analysis of the known distribution of the North American mosquitoes in other parts of the world reveals that only 54 (30.8%) are indigenous, not found outside of the region. The number and proportion in each biogeographical region, area, or specific country are given below.

| Region | No. of Species | Percent of Total |
|---|---|---|
| Indigenous | 54 | 30.8 |
| Palearctic | 29 | 16.5 |
| Neotropical | 33 | 18.8 |
| Oriental | 1 | 0.5 |
| Cosmotropical | 2 | 1.1 |
| Worldwide | 1 | 0.5 |
| Caribbean and Mexico | 5 | 2.8 |
| Caribbean only | 7 | 4.0 |
| Mexico only | 42 | 24.0 |
| Cuba only | 1 | 0.5 |

Three species from the Palearctic region are also distributed elsewhere. *Cx. pipiens* occurs in the southern subsahara Africa and in the southern Neotropical region. *Oc. dorsalis* and *Oc. sticticus* likewise occur in Mexico.

For the general distribution of each taxon outside our region, if applicable, see the systematic index of the Culicidae (table 1, p. 4). The sources for this dispersal information have been principally Knight and Stone (385), Knight (384), and Ward (750, 751).

In all, 85 species found north of Mexico also are distributed in Mexico. Of the total, 33 taxa occur elsewhere in the Neotropical biogeographical region, 2 are cosmotropical, 5 share the Caribbean Islands, 3 are Holoarctic or worldwide and extend into Mexico, and 42 are known only from Mexico outside the target area. The central highlands of Mexico and Baja California are considered part of the Nearctic biogeographical region, while the lowlands are included in the Neotropical zone. Of the 42 taxa, 18 species have Nearctic distribution in Mexico; the other 24 are dispersed in the Neotropical lowlands, although some also may occur in the highlands. In reality, then, 57 taxa (32.5%) occurring in North America, north of Mexico, are also part of the Neotropical fauna. The works of Vargas (735) and Vargas and Martinez Palacios (738) have been helpful in understanding the distribution of the Mexican culicid fauna. Vargas (735) also reports that *Oc. punctor*, *Oc. impiger*, *Oc. s. spencerii*, and *Cs. impatiens* have been collected in Mexico, but their known distributions are so far removed from Mexico that the records need further confirmation (see plates 19, 16, 21, 32).

For 31 of the species, distributional maps have been previously published and used as the basis for those shown here. Their sources are acknowledged in the captions. The other 143 maps are originals, except that the northern extremes of 72 of the 75 Canadian taxa were delimited with the help of maps and information given by Wood et al. (784).

The captions of each of the map plates have been organized in the following manner: first, states of the United States, in alphabetical order of state name, precede the original reference, given in parentheses for each group of states, then provinces of Canada, followed in most instances by taxonomic references.

It must be understood that the starting point for this publication was the monograph by Carpenter and LaCasse (146); the references they cited verifying the occurrence of a species in a given political unit are not repeated here, but the user is referred to their treatise. Additionally, where applicable, there is a notation with substantiating reference if a species was previously reported in a political unit and was subsequently determined not actually occurring there. Listed under the name of each species included in the identification keys is a plate number that refers to the map of its geographical distribution.

In preparing a presentation of geographical distribution a nagging question is the problem of doubtful records. All mosquito specialists who have been responsible for mosquito records in particular political units have an obligation to preserve in an acceptable manner voucher specimens for each species known to occur within its boundaries. Published reports of species found in states/provinces ought to be verified by sample specimens. In the years since 1955 a large number of doubtful records have been settled (see captions on plates 9–42) and those responsible must be commended. Some still remain in doubt and the following 15 records have not been included either because they are quite far removed from the known range of the species or a specimen may have been collected many years ago and no further evidence exists that the species is part of the fauna.

| Species | Political Unit | Reference |
|---|---|---|
| *Oc. aboriginis* | Michigan | 347 |
| *Oc. canadensis mathesoni* | Michigan | 744 |
| *Oc. canadensis mathesoni* | Newfoundland | 784 |
| *Oc. fulvus pallens* | Indiana | 619 |
| *Oc. increpitus* | Manitoba | 784 |
| *Oc. nigromaculis* | Kentucky | 146 |
| *Oc. pullatus* | Michigan | 347 |
| *Oc. trivittatus* | Nova Scotia | 784 |
| *Cx. apicalis* | Illinois | 327 |
| *Cx. pipiens* | Alberta | 784 |
| *Cx. pipiens* | Manitoba | 784 |
| *Cx. territans* | Arizona | 604 |
| *Cs. incidens* | Michigan | 347 |
| *Cs. incidens* | Newfoundland | 263 |
| *Ma. titillans* | Arkansas | 337 |

Table 2. Synopsis of the occurrence of mosquito species in the eastern United States

| Mosquito species | Alabama (55) | Connecticut (44) | Delaware (52) | District of Columbia (37) | Florida (76) | Georgia (61) | Indiana (55) | Kentucky (57) | Maine (37) | Maryland (59) | Massachusetts (46) | Michigan (58) | Mississippi (56) | New Hampshire (44) | New Jersey (63) | New York (59) | North Carolina (57) | Ohio (61) | Pennsylvania (53) | Rhode Island (32) | South Carolina (55) | Tennessee (52) | Vermont (36) | Virginia (52) | West Virginia (26) |
|---|---|---|---|---|---|---|---|---|---|---|---|---|---|---|---|---|---|---|---|---|---|---|---|---|---|
| *Ae. aegypti* | x | | | x | x | x | x | x | | x | | | x | | | x | x | x | | | x | x | | x | |
| *Ae.albopictus* | x | | x | x | x | x | x | x | | x | | | x | | x | | x | x | x | | x | x | | x | x |
| *Ae. cinereus* | x | x | x | x | x | x | x | x | x | x | x | x | x | x | x | x | x | x | x | x | x | x | x | x | |
| *Ae. vexans* | x | x | x | x | x | x | x | x | x | x | x | x | x | x | x | x | x | x | x | x | x | x | x | x | x |
| *Oc. abserratus* | | x | | | | | x | | x | x | x | x | | x | x | x | | x | x | x | | | x | | |
| *Oc. atlanticus* | x | | x | x | x | x | x | x | | x | | | x | | x | x | x | | | | x | x | | x | |
| *Oc. atropalpus* | x | x | | x | | x | x | x | x | x | x | x | | x | x | x | x | x | x | x | x | x | x | x | x |
| *Oc. aurifer* | | x | x | | | | x | | x | x | x | x | | x | x | x | | x | x | x | | | x | x | |
| *Oc. bahamensis* | | | | | x | | | | | | | | | | | | | | | | | | | | |
| *Oc. campestris* | | | | | | | | | | | | x | | | | | | | | | | | | | |
| *Oc. c. canadensis* | x | x | x | x | x | x | x | x | x | x | x | x | x | x | x | x | x | x | x | x | x | x | x | x | x |
| *Oc. c. mathesoni* | | x | x | | x | | | x | x | x | x | x | | | | | | x | | | x | | | x | |
| *Oc. cantator* | | x | x | | | | | x | x | x | x | | | x | x | x | | x | x | x | | | | x | |
| *Oc. communis* | | | | | | | | | x | | x | x | | x | x | x | | | x | | | | x | | |
| *Oc. decticus* | | | | | | | | | x | | x | x | | x | | x | | | x | | | | | | |
| *Oc. diantaeus* | | | | | | | | | x | | x | x | | x | | x | | x | x | | | | x | | |
| *Oc. dorsalis* | | x | x | | | | x | x | | x | x | x | | | x | x | | x | x | | | | | | |
| *Oc. dupreei* | x | | x | | x | x | x | x | | | | x | x | | x | | x | x | | | x | x | | x | |
| *Oc. euedes* | | | | | | | | | | | | x | | | | | | | | | | | | | |
| *Oc. excrucians* | | x | x | | | | x | | x | x | x | x | | x | x | x | | x | x | x | | | x | | |
| *Oc. fitchii* | | x | x | | | | x | | x | x | x | x | | x | x | x | | x | x | x | | | x | | |
| *Oc. flavescens* | | | | | | | x | | | | | x | | x | x | x | | x | | | | | | x | |
| *Oc. fulvus pallens* | x | | | | x | x | x | x | | x | | | x | | | | x | | | | x | x | | x | |
| *Oc. grossbecki* | | x | x | | | | x | x | | x | | | x | | x | x | | x | x | | x | x | x | x | |
| *Oc. hendersoni* | x | x | x | x | x | x | x | x | x | x | x | x | x | x | x | x | x | x | x | | x | x | x | x | x |
| *Oc. impiger* | | | | | | | | | | | | x | | x | | | | | | | | | | | |
| *Oc. implicatus* | | | | | | | | | x | | x | x | | x | x | x | | x | | | | | | | |

*continued*

| Mosquito species | Alabama (55) | Connecticut (44) | Delaware (52) | District of Columbia (37) | Florida (76) | Georgia (61) | Indiana (55) | Kentucky (57) | Maine (37) | Maryland (59) | Massachusetts (46) | Michigan (58) | Mississippi (56) | New Hampshire (44) | New Jersey (63) | New York (59) | North Carolina (57) | Ohio (61) | Pennsylvania (53) | Rhode Island (32) | South Carolina (55) | Tennessee (52) | Vermont (36) | Virginia (52) | West Virginia (26) |
|---|---|---|---|---|---|---|---|---|---|---|---|---|---|---|---|---|---|---|---|---|---|---|---|---|---|
| *Oc. infirmatus* | x | | x | | x | x | x | x | | x | | | x | | x | | x | | | | x | x | | x | |
| *Oc. intrudens* | | x | | | | | | | x | | x | x | | x | x | x | | | x | x | | | x | | |
| *Oc. j. japonicus* | | x | | | | | | | | x | | | | | x | x | | x | x | | | | | | |
| *Oc. mitchellae* | x | | x | x | x | x | | x | | x | | x | x | | x | x | x | x | x | | x | x | | x | |
| *Oc. pionips* | | | | | | | | | x | | | x | | | | | | | | | | | | | |
| *Oc. provocans* | | x | | | | | | | x | | x | x | | x | x | x | | | x | x | | | x | | |
| *Oc. punctor* | | | | | | | x | | x | | x | x | | x | x | x | | | x | | | | x | | |
| *Oc. riparius* | | | | | | | | | | | | x | | | | x | | x | | | | | | | |
| *Oc. scapularis* | | | | | x | | | | | | | | | | | | | | | | | | | | |
| *Oc. sollicitans* | x | x | x | x | x | x | x | x | x | x | x | x | x | x | x | x | x | x | x | x | x | x | | x | |
| *Oc. s. spencerii* | | | | | | | | | | | | x | | | x | x | | x | | | | | | | |
| *Oc. sticticus* | x | x | x | x | x | x | x | x | x | x | x | x | x | x | x | x | x | x | x | | x | x | x | x | x |
| *Oc. stimulans* | | x | | | | x | x | x | x | x | x | x | x | x | x | x | | x | x | x | | | x | x | |
| *Oc. taeniorhynchus* | x | x | x | x | x | x | | | | x | x | | x | x | x | x | x | | x | x | x | | | x | |
| *Oc. thelcter* | | | | | x | | | | | | | | | | | | | | | | | | | | |
| *Oc. thibaulti* | x | x | x | | x | x | x | x | | x | | | x | | x | x | x | x | | | x | x | | x | |
| *Oc. tormentor* | x | | x | | x | x | | x | | x | | | x | | | | x | x | | | x | x | | | |
| *Oc. tortilis* | | | | | x | | | | | | | | | | | | | | | | | | | | |
| *Oc. triseriatus* | x | x | x | x | x | x | x | x | x | x | x | x | x | x | x | x | x | x | x | x | x | x | x | x | x |
| *Oc. trivitatus* | x | x | x | x | | x | x | x | x | x | x | x | | x | x | x | x | x | x | x | x | x | x | x | x |
| *An. albimanus* | | | | | x | | | | | | | | | | | | | | | | | | | | |
| *An. atropos* | x | | | | x | x | | | | x | | | x | | x | | x | | | | x | | | x | |
| *An. barberi* | x | | x | x | x | x | x | x | | x | | x | x | | x | x | x | x | x | | x | x | | x | x |
| *An. bradleyi* | x | | x | | x | x | | | | x | | | x | | x | x | x | | | | x | | | x | |
| *An. crucians* | x | x | x | x | x | x | x | x | | x | x | x | x | | x | x | x | x | x | x | x | x | | x | |
| *An. diluvialis* | | | | | x | | | | | | | | | | | | | | | | | | | | |
| *An. earlei* | | x | | | | | | | x | | x | x | | x | x | x | | | | | | | x | | |
| *An. georgianus* | x | | | | x | x | | | | | | | x | | | | x | | | | x | | | | |

| Mosquito species | Alabama (55) | Connecticut (44) | Delaware (52) | District of Columbia (37) | Florida (76) | Georgia (61) | Indiana (55) | Kentucky (57) | Maine (37) | Maryland (59) | Massachusetts (46) | Michigan (58) | Mississippi (56) | New Hampshire (44) | New Jersey (63) | New York (59) | North Carolina (57) | Ohio (61) | Pennsylvania (53) | Rhode Island (32) | South Carolina (55) | Tennessee (52) | Vermont (36) | Virginia (52) | West Virginia (26) |
|---|---|---|---|---|---|---|---|---|---|---|---|---|---|---|---|---|---|---|---|---|---|---|---|---|---|
| *An. inundatus* | | | | | x | x | | | | | | | | | | | | | | | | | | | |
| *An. maverlius* | | | | | x | x | | | | | | | x | | | | | | | | | x | | | x |
| *An. perplexens* | x | | | | x | x | | x | | | | | | | | | x | x | x | | | x | | | |
| *An. pseudo-punctipennis* | | | | | | | | | | | | | x | | | | | | | | | x | | | |
| *An. punctipennis* | x | x | x | x | x | x | x | x | x | x | x | x | x | x | x | x | x | x | x | x | x | x | x | x | x |
| *An. quadri-maculatus sl*[1] | | | x | x | | | x | | x | x | | x | | x | | | | x | x | x | | | x | x | x |
| *An. quadri-maculatus ss* | x | x | | | x | x | | x | | | x | | x | | x | x | x | | | | x | x | | | |
| *An. smaragdinus* | x | | | | x | x | | x | | | | | x | | | | x | | | | x | x | | | |
| *An. walkeri* | x | x | x | x | x | x | x | x | x | x | x | x | x | x | x | x | x | x | x | x | x | x | x | x | |
| *Cq. perturbans* | x | x | x | x | x | x | x | x | x | x | x | x | x | x | x | x | x | x | x | x | x | x | x | x | x |
| *Cx. atratus* | | | | | x | | | | | | | | | | | | | | | | | | | | |
| *Cx. bahamensis* | | | | | x | | | | | | | | | | | | | | | | | | | | |
| *Cx. biscaynensis* | | | | | x | | | | | | | | | | | | | | | | | | | | |
| *Cx. cedecei* | | | | | x | | | | | | | | | | | | | | | | | | | | |
| *Cx. erraticus* | x | | x | x | x | x | x | x | | x | | x | x | | x | | x | x | x | | x | x | | x | x |
| *Cx. iolambdis* | | | | | x | | | | | | | | | | | | | | | | | | | | |
| *Cx. mulrennani* | | | | | x | | | | | | | | | | | | | | | | | | | | |
| *Cx. nigripalpus* | x | | | | x | x | | x | | | | | x | | | | x | | | | x | x | | | |
| *Cx. peccator* | x | | | | x | x | | x | | | | x | x | | | | x | | | | x | x | | | |
| *Cx. pilosus* | x | | | | x | x | | x | | | | | x | | | | x | | | | x | | | | |
| *Cx. pipiens* | x | x | x | x | | x | x | x | x | x | x | x | x | x | x | x | x | x | x | x | x | x | x | x | x |
| *Cx. quinque-fasciatus* | x | | | x | x | x | x | x | | x | | | x | | | | x | x | | | x | x | | x | x |
| *Cx. restuans* | x | x | x | x | x | x | x | x | x | x | x | x | x | x | x | x | x | x | x | x | x | x | x | x | x |
| *Cx. salinarius* | x | x | x | x | x | x | x | x | x | x | x | x | x | x | x | x | x | x | x | x | x | x | x | x | x |
| *Cx. tarsalis* | x | | | | x | x | x | x | | | | x | x | | x | | | x | x | | x | x | | | |
| *Cx. territans* | x | x | x | | x | x | x | x | x | x | x | x | x | x | x | x | x | x | x | x | x | x | x | x | x |

*continued*

| Mosquito species | Alabama (55) | Connecticut (44) | Delaware (52) | District of Columbia (37) | Florida (76) | Georgia (61) | Indiana (55) | Kentucky (57) | Maine (37) | Maryland (59) | Massachusetts (46) | Michigan (58) | Mississippi (56) | New Hampshire (44) | New Jersey (63) | New York (59) | North Carolina (57) | Ohio (61) | Pennsylvania (53) | Rhode Island (32) | South Carolina (55) | Tennessee (52) | Vermont (36) | Virginia (52) | West Virginia (26) |
|---|---|---|---|---|---|---|---|---|---|---|---|---|---|---|---|---|---|---|---|---|---|---|---|---|---|
| *Cs. impatiens* | | x | | | | | | | x | | x | x | | x | | x | | | x | | | | x | | |
| *Cs. inornata* | x | x | x | x | x | x | x | x | | x | x | x | x | x | x | x | x | x | x | | x | x | | x | x |
| *Cx. melanura* | x | x | x | x | x | x | x | x | x | x | x | x | x | x | x | x | x | x | x | x | x | x | x | x | |
| *Cs. minnesotae* | | x | x | | | | x | | | x | x | x | | x | x | x | | x | | | | | x | | |
| *Cs. morsitans* | | x | x | | | | x | x | x | x | x | x | | x | x | x | | x | x | x | | | x | | |
| *De. cancer* | | | | | x | | | | | | | | | | | | | | | | | | | | |
| *Ma. dyari* | | | | | x | x | | | | | | | | | | | | | | | x | | | | |
| *Ma. titillans* | | | | | x | | | | | | | | | | | | | | | | | | | | |
| *Or. alba* | x | | x | x | x | x | x | x | | x | | x | x | | x | x | x | x | x | | | x | | x | |
| *Or. signifera* | x | x | x | x | x | x | x | x | | x | x | x | x | x | x | x | x | x | x | x | x | x | | x | x |
| *Ps. ciliata* | x | x | x | x | x | x | x | x | | x | x | x | x | x | x | x | x | x | x | x | x | x | x | x | x |
| *Ps. columbiae* | x | | x | x | x | x | x | x | | x | x | | x | | x | x | x | x | x | | x | x | | x | x |
| *Ps. cyanescens* | x | | x | | x | x | x | x | | x | | | x | | x | | x | x | | | x | x | | x | |
| *Ps. discolor* | x | | x | x | x | x | x | x | | x | | | x | | x | | x | x | | | x | x | | x | |
| *Ps. ferox* | x | x | x | x | x | x | x | x | | x | x | x | x | x | x | x | x | x | x | | x | x | x | x | x |
| *Ps. horrida* | x | | x | x | x | x | x | x | | x | | x | x | | | | x | x | x | | x | x | | x | |
| *Ps. howardii* | x | | x | x | x | x | x | x | | x | | | x | | x | | x | x | | | x | x | | x | |
| *Ps. johnstoni* | | | | | x | | | | | | | | | | | | | | | | | | | | |
| *Ps. mathesoni* | x | | x | | x | x | x | x | | x | | | x | | x | x | x | x | | | x | x | | x | |
| *Ps. pygmaea* | | | | | x | | | | | | | | | | | | | | | | | | | | |
| *Ps. signipennis* | | | | | | | | x | | | | | | | | | | | | | | x | | | |
| *Tx. r. rutilus* | | | | | x | x | | | | | | | | | | | | | | | | | | | |
| *Tx. r. septentrionalis* | x | x | x | x | x | x | x | x | | x | x | | x | | x | x | x | x | x | x | x | x | | x | x |
| *Ur. lowii* | x | | | | x | x | | | | | | | x | | | | x | | | | x | | | | |
| *Ur. sapphirina* | x | x | x | x | x | x | x | x | | x | x | x | x | x | x | x | x | x | x | x | x | x | x | x | x |
| *Wy. mitchellii* | | | | | x | x | | | | | | | | | | | | | | | | | | | |
| *Wy. smithii* | | x | x | | x | | x | | x | x | x | x | | x | x | x | x | x | x | x | | | x | | |
| *Wy. vanduzeei* | | | | | x | | | | | | | | | | | | | | | | | | | | |

Table 3. Synopsis of the occurrence of mosquito species in the western United States

| Mosquito species | Arizona (49) | Arkansas (57) | California (56) | Colorado (44) | Idaho (51) | Illinois (60) | Iowa (49) | Kansas (54) | Louisiana (60) | Minnesota (52) | Missouri (57) | Montana (47) | Nebraska (51) | Nevada (36) | New Mexico (55) | North Dakota (31) | Oklahoma (60) | Oregon (45) | South Dakota (40) | Texas (85) | Utah (50) | Washington (46) | Wisconsin (49) | Wyoming (48) |
|---|---|---|---|---|---|---|---|---|---|---|---|---|---|---|---|---|---|---|---|---|---|---|---|---|
| *Ae. aegypti* | x | x | | | | x | | x | x | | x | | | | | | x | | | x | | | | |
| *Ae. albopictus* | | x | x | | | x | | x | x | x | x | | x | | | | x | | | x | | | | |
| *Ae. cinereus* | x | x | x | x | x | x | x | x | x | x | x | x | x | x | x | x | x | x | x | | x | x | x | x |
| *Ae. vexans* | x | x | x | x | x | x | x | x | x | x | x | x | x | x | x | x | x | x | x | x | x | x | x | x |
| *Oc. aboriginis* | | | | | x | | | | | | | | | | | | | x | | | | x | | |
| *Oc. abserratus* | | | | | | x | | | | x | | | | | | | | | | | | | x | |
| *Oc. aloponotum* | | | | | | | | | | | | | | | | | | x | | | | x | | |
| *Oc. atlanticus* | | x | | | | x | | x | x | | x | | | | | | x | | | x | | | | |
| *Oc. atropalpus* | | | | | | x | | x | | x | x | | | | | | | | | | | | x | |
| *Oc. aurifer* | | | | | | x | x | | | x | | | | | | | | | | | | | x | |
| *Oc. bicristatus* | | | x | | | | | | | | | | | | | | | | | | | | | |
| *Oc. bimaculatus* | | | | | | | | | | | | | | | | | | | | x | | | | |
| *Oc. brelandi* | | | | | | | | | | | | | | | | | | | | x | | | | |
| *Oc. burgeri* | x | | | | | | | | | | | | | | | | | | | | | | | |
| *Oc. campestris* | | | x | x | x | | x | | | x | | x | x | x | x | x | | x | x | x | x | x | x | x |
| *Oc. c. canadensis* | | x | | | x | x | x | x | x | x | x | x | x | | | x | x | | x | x | | x | x | x |
| *Oc. cataphylla* | x | | x | x | x | | | | | | | x | x | x | x | | | x | | | x | x | | x |
| *Oc. clivis* | | | x | | | | | | | | | | | | | | | | | | | | | |
| *Oc. communis* | | | x | x | x | | | | | x | | x | | x | x | | | x | | | x | x | x | x |
| *Oc. decticus* | | | | | x | | | | | x | | | | | | | | | | | | | | |
| *Oc. deserticola* | | | x | | | | | | | | | | | | | | | | | | | | | |
| *Oc. diantaeus* | | | | | | | | | | x | | x | | | | | | | | | | | x | x |
| *Oc. dorsalis* | x | x | x | x | x | x | x | x | | x | x | x | x | x | x | x | x | x | x | x | x | x | x | x |
| *Oc. dupreei* | | x | | | | x | x | x | x | | x | | | | | | x | | | x | | | | |
| *Oc. epactius* | x | x | | x | | | | x | x | | x | | | | x | | x | | | x | x | | | |
| *Oc. euedes* | | | | | x | | | | | | | | | | | | | | | | | | x | x |
| *Oc. excrucians* | | | | x | x | x | | | | x | | x | | | x | x | | x | | | x | x | x | x |
| *Oc. fitchii* | x | | x | x | x | x | x | | | x | | x | x | x | x | x | | x | | | x | x | x | x |

*continued*

| Mosquito species | Arizona (49) | Arkansas (57) | California (56) | Colorado (44) | Idaho (51) | Illinois (60) | Iowa (49) | Kansas (54) | Louisiana (60) | Minnesota (52) | Missouri (57) | Montana (47) | Nebraska (51) | Nevada (36) | New Mexico (55) | North Dakota (31) | Oklahoma (60) | Oregon (45) | South Dakota (40) | Texas (85) | Utah (50) | Washington (46) | Wisconsin (49) | Wyoming (48) |
|---|---|---|---|---|---|---|---|---|---|---|---|---|---|---|---|---|---|---|---|---|---|---|---|---|
| *Oc. flavescens* | | | x | x | x | x | x | x | | x | x | x | x | | | x | | x | x | | x | x | x | x |
| *Oc. fulvus pallens* | | x | | | | x | | | x | | x | | | | | | x | | | x | | | | |
| *Oc. grossbecki* | | x | | | | x | | | x | | x | | | | | | | | | x | | | x | |
| *Oc. hendersoni* | | x | | x | x | x | x | x | x | x | x | x | x | | x | | x | x | x | x | x | | x | x |
| *Oc. hexodontus* | | | x | x | x | | | | | | | x | | x | x | | | x | | | x | x | | x |
| *Oc. impiger* | | | | x | x | | | | | | | x | | | | | | x | | | x | x | | x |
| *Oc. implicatus* | x | | | x | x | | x | | | x | | x | x | | x | | | x | | | x | x | | x |
| *Oc. increpitus* | x | | x | x | x | | | | | | | x | x | x | x | | | x | x | | x | x | | x |
| *Oc. infirmatus* | | x | | | | x | | | x | | x | | | | | | | | | x | | | | |
| *Oc. intrudens* | | | | x | x | | | | | x | | x | | | | x | | x | x | | x | x | x | x |
| *Oc. j. japonicus* | | | | | | | | | | | | | | | | | | | | | | x | | |
| *Oc. melanimon* | | | x | x | x | | | | | | | x | x | x | x | | | x | | | x | x | | x |
| *Oc. mercurator* | | | | | x | | | | | | | x | | | | | | | | | | | | x |
| *Oc. mitchellae* | | x | | | | x | | x | x | | x | | | | x | | x | | | x | | | | |
| *Oc. monticola* | x | | | | | | | | | | | | | | x | | | | | | | | | |
| *Oc. muelleri* | x | | | | | | | | | | | | | | x | | | | | x | | | | |
| *Oc. nevadensis* | | | | | x | | | | | | | | | x | | | | x | | | x | x | | x |
| *Oc. nigromaculis* | x | x | x | x | x | x | x | x | x | x | x | x | x | x | x | x | x | x | x | x | x | x | | x |
| *Oc. niphadopsis* | | | x | | x | | | | | | | | | x | | | | x | | | x | | | x |
| *Oc. papago* | x | | | | | | | | | | | | | | | | | | | | | | | |
| *Oc. pionips* | | | | x | x | | | | | x | | x | | | | x | | x | | | | x | | x |
| *Oc. provocans* | | | | | x | | | | | x | | x | | | | | | | | | | x | x | |
| *Oc. pullatus* | x | | x | x | x | | | | | | | x | | x | x | | | x | | | x | x | | x |
| *An. atropos* | | | | | | | | | x | | | | | | | | | | | x | | | | |
| *An. barberi* | | x | | | | x | x | x | x | x | x | | x | | | | x | | x | x | | | x | |
| *An. bradleyi* | | | | | | | | | x | | | | | | | | | | | x | | | | |
| *An. crucians* | | x | | | | x | x | x | x | | x | | | | x | | x | | | x | | | x | |
| *An. earlei* | | | | x | x | | x | x | | x | | x | x | x | | x | | | x | | x | x | x | x |

| Mosquito species | Arizona (49) | Arkansas (57) | California (56) | Colorado (44) | Idaho (51) | Illinois (60) | Iowa (49) | Kansas (54) | Louisiana (60) | Minnesota (52) | Missouri (57) | Montana (47) | Nebraska (51) | Nevada (36) | New Mexico (55) | North Dakota (31) | Oklahoma (60) | Oregon (45) | South Dakota (40) | Texas (85) | Utah (50) | Washington (46) | Wisconsin (49) | Wyoming (48) |
|---|---|---|---|---|---|---|---|---|---|---|---|---|---|---|---|---|---|---|---|---|---|---|---|---|
| *An. franciscanus* | x | | x | x | | | | x | | | | | x | x | x | | x | x | | x | x | | | x |
| *An. freeborni* | x | | x | x | x | | | | | | | x | | x | x | | | x | | x | x | x | | x |
| *An. georgianus* | | | | | | | | | x | | | | | | | | | | | | | | | |
| *An. hermsi* | | | x | x | | | | | | | | | | | x | | | | | | | | | |
| *An. inundatus* | | | | | | | | | x | | | | | | | | | | | | | | | |
| *An. judithae* | x | | | | | | | | | | | | | | x | | | | | x | | | | |
| *An. maverlius* | | | | | | | | | x | | | | | | | | | | | | | | | |
| *An. occidentalis* | | | x | | | | | | | | | | | | | | | x | | | | x | | |
| *An. pseudo-punctipennis* | | x | | | | | | x | x | | x | | | | x | | x | | | x | | | | |
| *An. punctipennis* | x | x | x | x | x | x | x | x | x | x | x | x | x | | x | x | x | x | x | x | | x | x | x |
| *An. quadri-maculatus sl* | | | | | | x | x | x | | | x | | x | x | | | x | | | | | | | |
| *An. quadri-maculatus ss* | | x | | | | | | | x | x | | | | | | | | | | x | | | x | |
| *An. smaragdinus* | | x | | | | | | | x | | | | | | | | | | | x | | | | |
| *An. walkeri* | | x | | | | x | x | x | x | x | x | | x | | | x | | | x | x | | | x | |
| *Cq. perturbans* | x | x | x | x | x | x | x | x | x | x | x | x | x | | x | x | x | x | x | x | x | x | x | x |
| *Cx. abominator* | | | | | | | | | | | | | | | | | | | | x | | | | |
| *Cx. anips* | | | x | | | | | | | | | | | | | | | | | | | | | |
| *Cx. apicalis* | x | | x | | | | | | | | | | | x | x | | x | | | x | x | | | |
| *Cx. arizonensis* | x | | | | | | | | | | | | | | | | | | | x | | | | |
| *Cx. boharti* | | | x | | x | | | | | | | | | x | | | | x | | | | x | | |
| *Cx. chidesteri* | | | | | | | | | | | | | | | | | | | | x | | | | |
| *Cx. coronator* | x | | | | | | | | | | | | | | x | | | | | x | | | | |
| *Cx. declarator* | | | | | | | | | | | | | | | | | | | | x | | | | |
| *Cx. erraticus* | | x | x | | | x | x | x | x | x | x | | x | | x | | x | | x | x | | | x | |
| *Cx. erythrothorax* | x | | x | x | x | | | | | | | | | x | x | | | | | x | x | | | |
| *Cx. interrogator* | | | | | | | | | | | | | | | | | | | | x | | | | |

*continued*

| Mosquito species | Arizona (49) | Arkansas (57) | California (56) | Colorado (44) | Idaho (51) | Illinois (60) | Iowa (49) | Kansas (54) | Louisiana (60) | Minnesota (52) | Missouri (57) | Montana (47) | Nebraska (51) | Nevada (36) | New Mexico (55) | North Dakota (31) | Oklahoma (60) | Oregon (45) | South Dakota (40) | Texas (85) | Utah (50) | Washington (46) | Wisconsin (49) | Wyoming (48) |
|---|---|---|---|---|---|---|---|---|---|---|---|---|---|---|---|---|---|---|---|---|---|---|---|---|
| *Cx. nigripalpus* | x | | | | | | | | x | | | | | | | | x | | | x | | | | |
| *Cx. peccator* | | x | | | | x | | x | x | | x | | | | | | x | | | x | | | | |
| *Cx. pilosus* | | | | | | | | | x | | | | | | | | | | | x | | | | |
| *Cx. pipiens* | | x | x | x | x | x | x | x | | x | x | x | x | | | x | x | x | x | | x | x | x | x |
| *Cx. quinque-fasciatus* | x | x | x | | | x | x | x | x | | x | | x | x | x | | x | | | x | x | | | |
| *Cx. reevesi* | | | x | | | | | | | | | | | | | | | | | | | | | |
| *Cx. restuans* | x | x | x | x | x | x | x | x | x | x | x | x | x | | x | x | x | x | x | x | x | | x | x |
| *Cx. salinarius* | | x | | x | x | x | x | x | x | x | x | | x | | x | x | x | | x | x | | | x | x |
| *Cx. stigmatosoma* | x | | x | | | | | | | | | | | x | x | | x | x | | x | | x | | |
| *Cx. tarsalis* | x | x | x | x | x | x | x | x | x | x | x | x | x | x | x | x | x | x | x | x | x | x | x | x |
| *Cx. territans* | x | x | x | x | x | x | x | x | x | x | x | x | x | x | | | x | x | x | x | x | x | x | x |
| *Cx. thriambus* | x | | x | | | | | | | | | | | x | x | | x | | | x | x | | | |
| *Cs. alaskaensis* | | | | x | x | | | | | | | x | | x | | | | | | | | | | x |
| *Cs. impatiens* | | | x | x | x | | x | | | | x | x | x | x | x | | | x | x | | x | x | x | x |
| *Cs. incidens* | x | | x | x | x | | | | | | | x | x | x | x | x | x | x | x | x | x | x | | x |
| *Cs. inornata* | x | x | x | x | x | x | x | x | x | x | x | x | x | x | x | x | x | x | x | x | x | x | x | x |
| *Cs. melanura* | | x | | | | x | x | x | x | x | x | | x | | | | x | | | x | | | x | |
| *Cs. minnesotae* | | | | | x | x | x | | | x | | x | | | | | | x | | | x | x | x | |
| *Cs. morsitans* | | | | | x | x | x | | | x | | x | | | | x | | | x | | x | | x | |
| *Cs. particeps* | x | | x | | | | | | | | | | | | | | | x | | | | x | | |
| *De. mathesoni* | | | | | | | | | | | | | | | | | | | | x | | | | |
| *De. pseudes* | | | | | | | | | | | | | | | | | | | | x | | | | |
| *Hg. equinus* | | | | | | | | | | | | | | | | | | | | x | | | | |
| *Ma. titillans* | | | | | | | | | | | | | | | | | | | | x | | | | |
| *Or. alba* | | x | | | | x | x | x | x | | x | | x | | x | | x | | | x | | | | |
| *Or. kummi* | x | | | | | | | | | | | | | | x | | | | | | | | | |
| *Or. signifera* | x | x | x | | | x | x | x | x | x | x | | x | | x | | x | x | x | x | x | | x | |
| *Ps. ciliata* | | x | | | | x | x | x | x | x | x | | x | | x | | x | | x | x | | | x | |

| Mosquito species | Arizona (49) | Arkansas (57) | California (56) | Colorado (44) | Idaho (51) | Illinois (60) | Iowa (49) | Kansas (54) | Louisiana (60) | Minnesota (52) | Missouri (57) | Montana (47) | Nebraska (51) | Nevada (36) | New Mexico (55) | North Dakota (31) | Oklahoma (60) | Oregon (45) | South Dakota (40) | Texas (85) | Utah (50) | Washington (46) | Wisconsin (49) | Wyoming (48) |
|---|---|---|---|---|---|---|---|---|---|---|---|---|---|---|---|---|---|---|---|---|---|---|---|---|
| *Ps. columbiae* | | x | x | x | | x | x | x | x | x | x | | x | x | | | x | | x | x | | | | |
| *Ps. cyanescens* | | x | | | | x | | x | x | | x | | x | | x | | x | | | x | | | | |
| *Ps. discolor* | x | x | | | | x | x | x | x | | x | | x | | x | | x | | | x | | | | |
| *Ps. ferox* | | x | | | | x | x | x | x | x | x | | x | | | | x | | x | x | | | x | |
| *Ps. horrida* | | x | | | | x | x | x | x | x | x | | x | | | | x | | x | x | | | x | |
| *Ps. howardii* | x | x | | | | x | x | x | x | | x | | x | | | | x | | | x | | | | |
| *Ps. longipalpus* | | x | | | | | | x | x | | x | | x | | | | x | | x | x | | | | |
| *Ps. mathesoni* | | x | | | | x | | | x | | x | | | | | | x | | | x | | | x | |
| *Ps. mexicana* | | | | | | | | | | | | | | | | | | | | x | | | | |
| *Ps. signipennis* | x | x | x | x | | | x | x | | | x | x | x | x | x | x | x | | x | x | x | | | x |
| *Tx. r. septentrionalis* | | x | | | | x | | x | x | | x | | | | | | x | | | x | | | | |
| *Tx. moctezuma* | x | | | | | | | | | | | | | | | | | | | | | | | |
| *Ur. a. anhydor* | x | | x | | | | | | | | | | | x | | | | | | | | | | |
| *Ur. a. syntheta* | | x | | | | | | | | | | | | | x | | x | | | x | | | | |
| *Ur. lowii* | | x | | | | | | | x | | | | | | | | x | | | x | | | | |
| *Ur. sapphirina* | | x | | | | x | x | x | x | x | x | | x | | x | x | x | | x | x | | | x | |
| *Wy. smithii* | | | | | | x | | | | x | | | | | | | | | | | | | x | |

Table 4. Synopsis of the occurrence of mosquito species in Canada and Alaska

| Mosquito species | Alaska (32) | Alberta (42) | British Columbia (46) | Labrador (26) | Manitoba (46) | New Brunswick (28) | Newfoundland (16) | Northwest Territories (29) | Nova Scotia (26) | Ontario (58) | Prince Edward Island (18) | Quebec (55) | Saskatchewan (40) | Yukon (26) |
|---|---|---|---|---|---|---|---|---|---|---|---|---|---|---|
| *Aedes cinereus* | x | x | x | x | x | x | x | x | x | x | x | x | x | x |
| *Aedes vexans* | x | x | x |  | x | x |  |  | x | x | x | x | x | x |
| *Oc. aboriginis* | x |  | x |  |  |  |  |  |  |  |  |  |  |  |
| *Oc. abserratus* |  |  |  | x | x | x | x |  | x | x | x | x |  |  |
| *Oc. aloponotum* |  |  | x |  |  |  |  |  |  |  |  |  |  |  |
| *Oc. atropalpus* |  |  |  | x |  |  |  |  |  | x |  | x |  |  |
| *Oc. aurifer* |  |  |  |  |  | x |  |  |  | x |  | x |  |  |
| *Oc. campestris* |  | x | x |  | x |  |  |  |  | x |  | x | x | x |
| *Oc. c. canadensis* | x | x | x | x | x | x | x | x | x | x | x | x | x | x |
| *Oc. c. mathesoni* |  |  |  |  |  |  | x |  |  |  |  |  |  |  |
| *Oc. cantator* |  |  |  | x |  | x | x |  | x |  | x | x |  |  |
| *Oc. cataphylla* | x | x | x |  |  |  |  |  |  |  |  |  | x | x |
| *Oc. churchillensis* |  | x |  |  | x |  |  |  |  |  |  |  |  |  |
| *Oc. communis* | x | x | x | x | x | x | x | x | x | x | x | x | x | x |
| *Oc. decticus* | x | x |  | x | x |  |  | x |  | x |  | x |  |  |
| *Oc. diantaeus* | x | x | x | x | x |  |  | x | x | x |  | x | x | x |
| *Oc. dorsalis* |  | x | x |  | x | x |  |  |  | x |  | x | x |  |
| *Oc. euedes* | x | x | x |  | x |  |  | x | x | x |  | x | x |  |
| *Oc. excrucians* | x | x | x | x | x | x | x | x | x | x | x | x | x | x |
| *Oc. fitchii* | x | x | x | x | x | x | x | x | x | x | x | x | x | x |
| *Oc. flavescens* | x | x | x | x | x |  |  | x |  | x |  | x | x | x |
| *Oc. grossbecki* |  |  |  |  |  |  |  |  |  | x |  |  |  |  |
| *Oc. hendersoni* |  |  | x |  | x |  |  |  |  | x |  | x | x |  |
| *Oc. hexodontus* | x | x | x | x | x |  |  | x |  | x |  | x |  | x |
| *Oc. impiger* | x | x | x | x | x |  |  | x |  | x |  | x | x | x |
| *Oc. implicatus* | x | x | x | x | x |  |  | x |  | x | x | x | x | x |

| Mosquito species | Alaska (32) | Alberta (42) | British Columbia (46) | Labrador (26) | Manitoba (46) | New Brunswick (28) | Newfoundland (16) | Northwest Territories (29) | Nova Scotia (26) | Ontario (58) | Prince Edward Island (18) | Quebec (55) | Saskatchewan (40) | Yukon (26) |
|---|---|---|---|---|---|---|---|---|---|---|---|---|---|---|
| *Oc. increpitus* | | x | x | | x | | | | | | | | x | |
| *Oc. intrudens* | x | x | x | x | x | x | | x | x | x | x | x | x | |
| *Oc. j. japonicus* | | | | | | | | | | x | | x | | |
| *Oc. melanimon* | | x | x | | | | | | | | | | x | |
| *Oc. mercurator* | x | x | x | | x | | | x | | x | | x | x | x |
| *Oc. nigripes* | x | | x | x | x | | x | x | | | | x | | x |
| *Oc. nigromaculis* | | x | | | x | | | | | | | | x | |
| *Oc. pionips* | x | x | x | x | x | | | x | | x | | x | x | x |
| *Oc. provocans* | | x | x | | x | x | | x | x | x | x | x | x | |
| *Oc. pullatus* | x | x | x | x | | | x | x | | | | x | | x |
| *Oc. punctodes* | x | | | | | | | | | | | | | |
| *Oc. punctor* | x | x | x | x | x | x | x | x | x | x | x | x | x | x |
| *Oc. rempeli* | | | | | | | | x | | x | | x | | |
| *Oc. riparius* | x | x | x | | x | x | | x | x | x | | x | x | x |
| *Oc. schizopinax* | | x | | | | | | | | | | | | |
| *Oc. sierrensis* | | | x | | | | | | | | | | | |
| *Oc. sollicitans* | | | | | | x | | | x | x | x | x | | |
| *Oc. s. idahoensis* | | | x | | | | | | | | | | | |
| *Oc. s. spencerii* | | x | x | | x | | | | | x | | | x | |
| *Oc. sticticus* | | x | x | x | x | x | | | | x | | x | x | |
| *Oc. stimulans* | | | | | x | x | x | | x | x | x | x | | |
| *Oc. thibaulti* | | | | | | | | | | x | | | | |
| *Oc. togoi* | | | x | | | | | | | | | | | |
| *Oc. triseriatus* | | | | | x | x | | | | x | | x | | |
| *Oc. trivittatus* | | | | | x | | | | x | x | | x | | |
| *Oc. ventrovittis* | x | | | | | | | | | | | | | |
| *An. barberi* | | | | | | | | | | x | | x | | |

*continued*

| Mosquito species | Alaska (32) | Alberta (42) | British Columbia (46) | Labrador (26) | Manitoba (46) | New Brunswick (28) | Newfoundland (16) | Northwest Territories (29) | Nova Scotia (26) | Ontario (58) | Prince Edward Island (18) | Quebec (55) | Saskatchewan (40) | Yukon (26) |
|---|---|---|---|---|---|---|---|---|---|---|---|---|---|---|
| *An. earlei* | x | x | x | x | x | x | | x | x | x | x | x | x | x |
| *An. freeborni* | | | x | | | | | | | | | | | |
| *An. punctipennis* | | | x | | x | x | | | x | x | | x | | |
| *An. quadrimaculatus ss* | | | | | | | | | | x | | x | | |
| *An. walkeri* | | | | | x | x | | | x | x | | x | x | |
| *Cq. perturbans* | | x | x | | x | x | | | x | x | x | x | x | |
| *Cx. pipiens* | | x | x | | x | x | | | x | x | | x | | |
| *Cx. restuans* | | x | | | x | x | | | x | x | | x | x | |
| *Cx. tarsalis* | | x | x | | x | | | x | | x | | x | x | |
| *Cx. territans* | x | x | x | x | x | x | | x | x | x | | x | x | x |
| *Cs. alaskaensis* | x | x | x | x | x | | | x | | | | x | x | x |
| *Cs. impatiens* | x | x | x | x | x | x | x | x | | x | | x | x | x |
| *Cs. incidens* | x | x | x | | | | x | x | | | | | x | x |
| *Cs. inornata* | | x | x | | x | | | x | | x | | x | x | x |
| *Cs. melanura* | | | | | | | | | | x | | x | | |
| *Cs. minnesotae* | x | x | x | | x | | | | | x | | x | x | |
| *Cs. morsitans* | x | x | x | x | x | x | x | x | x | x | x | x | x | x |
| *Cs. particeps* | x | | | | | | | | | | | | | |
| *Or. alba* | | | | | | | | | | x | | x | | |
| *Or. signifera* | | | | | | | | | | x | | | | |
| *Ps. ciliata* | | | | | | | | | | x | | x | | |
| *Ps. columbiae* | | | | | | | | | | x | | | | |
| *Ps. ferox* | | | | | | | | | | x | | | | |
| *Ps. signipennis* | | | | | | | | | | | | | x | |
| *Tx. r. septentrionalis* | | | | | | | | | | x | | | | |
| *Ur. sapphirina* | | | | | | | | | | x | | x | | |
| *Wy. smithii* | | | | x | x | x | x | | x | x | x | x | x | |

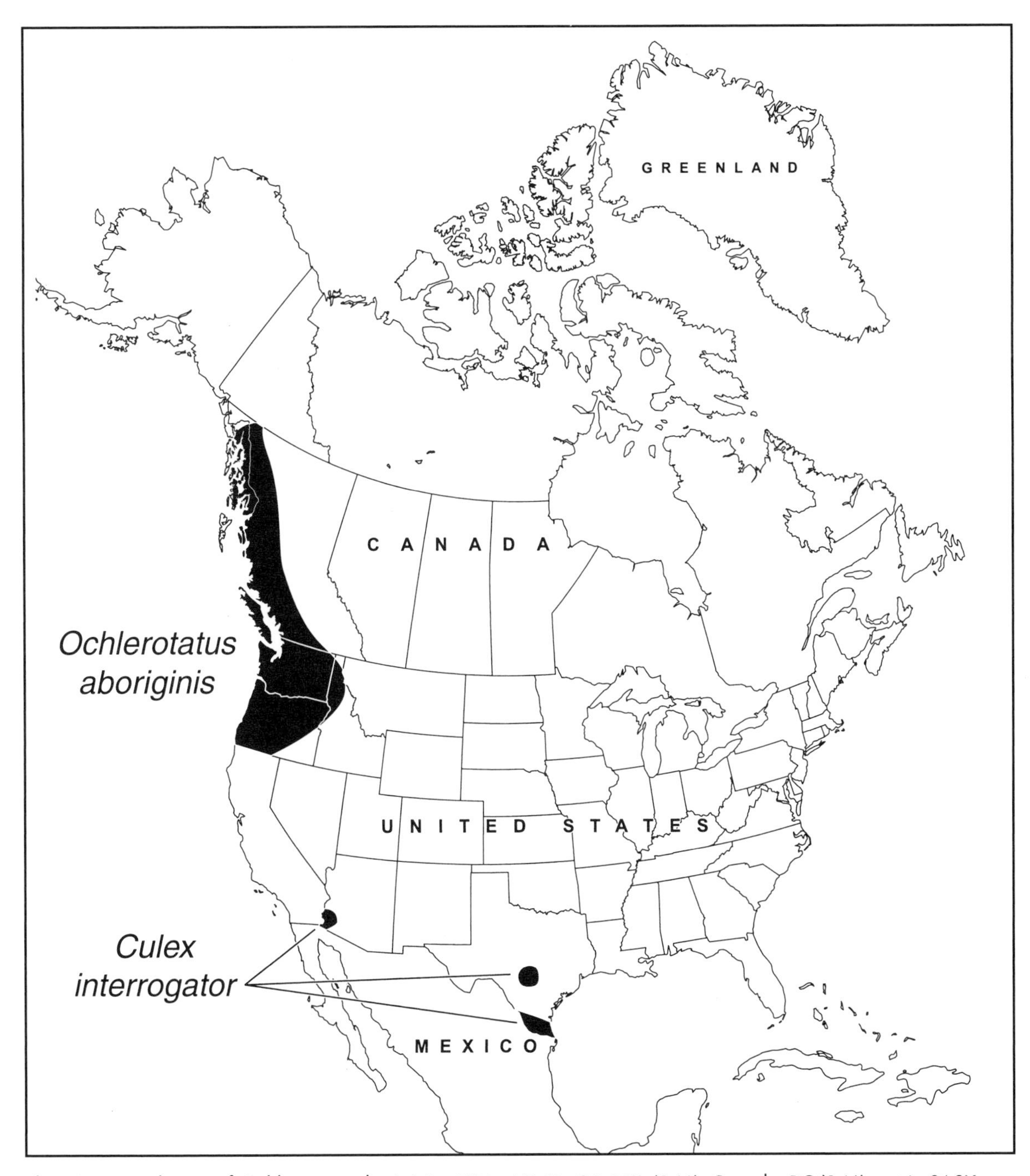


Plate 9 A. Distribution of *Ochlerotatus aboriginis*—USA: AK, ID, OR, WA (146); Canada: BC (146), not in SASK (601). *Culex interrogator*—USA: TX (146), AZ (329); Tax. 87.

Plate 9 B. Distribution of *Aedes aegypti*—USA: AL, AR, DC, FL, GA, IL, IN, KS, KY, LA, MS, MO, NC, OK, SC, TN, TX, VA (146), MD (716), NY (48), NJ (232), AZ (490), OH (Berry & Parsons, pers. comm. 1978); Map modified after Morland & Tinker (500); Tax. 42, 431.

Plate 9 C. Distribution of *Ochlerotatus abserratus*—USA: CT, IL, ME, MA, MI, MN, NH, NJ, NY, OH, PA, RI, VT, WI (146), IN (658), MD (76), WV (371); Canada: LAB, NS, ONT, PEI (146), MAN (107), NB (399), NFLD (739), PQ (442); Tax. 431, 739, 783.

Plate 9 D. Distribution of *Aedes albopictus*—USA: AL, AR, DE, FL, GA, IA, IL, IN, KS, KY, LA, MD, MN, MO, MS, NC, NE, NJ, OH, OK, PA, SC, TN, TX, VA, WV (499); Tax. 325.

Plate 10 A. Distribution of *Ochlerotatus aloponotum*—USA: OR, WA (146); Canada: BC (285); Tax. 285, 431, 783.

Plate 10 B. Distribution of *Ochlerotatus atlanticus*—USA: AL, AR, DE, DC, FL, GA, KS, LA, MD, MS, MO, NJ, NY, NC, OK, SC, TX, VA (146), IL (620), IN (666), KY (184), TN (93); Tax. 431. *Ochlerotatus togoi*—USA: WA (53); Canada: BC (784); Tax. 717.

Plate 10 C. Distribution of *Ochlerotatus atropalpus*—USA: CT, DC, GA, ME, MD, MA, MN, NH, NJ, NY, NC, PA, RI, SC, TN, VT, VA, WV, WI (146), AL, KY (186), IN (602), MI (30), MO (514), OH (Berry & Parsons, pers. comm. 1978), NE (798); Canada: LAB, ONT, PQ (146), NFLD (528); Map modified after Zavortink (792); Tax. 431, 545, 546, 783, 792.

Plate 10 D. Distribution of *Ochlerotatus aurifer*—USA: CT, DE, IL, ME, MD, MA, MI, MN, NH, NJ, NY, OH, RI, VT, WI (146), IN (651), PA (766), VA (Harrison, pers. comm. 2001); Canada: ONT, PQ (146), NB (489); Tax. 431, 784.

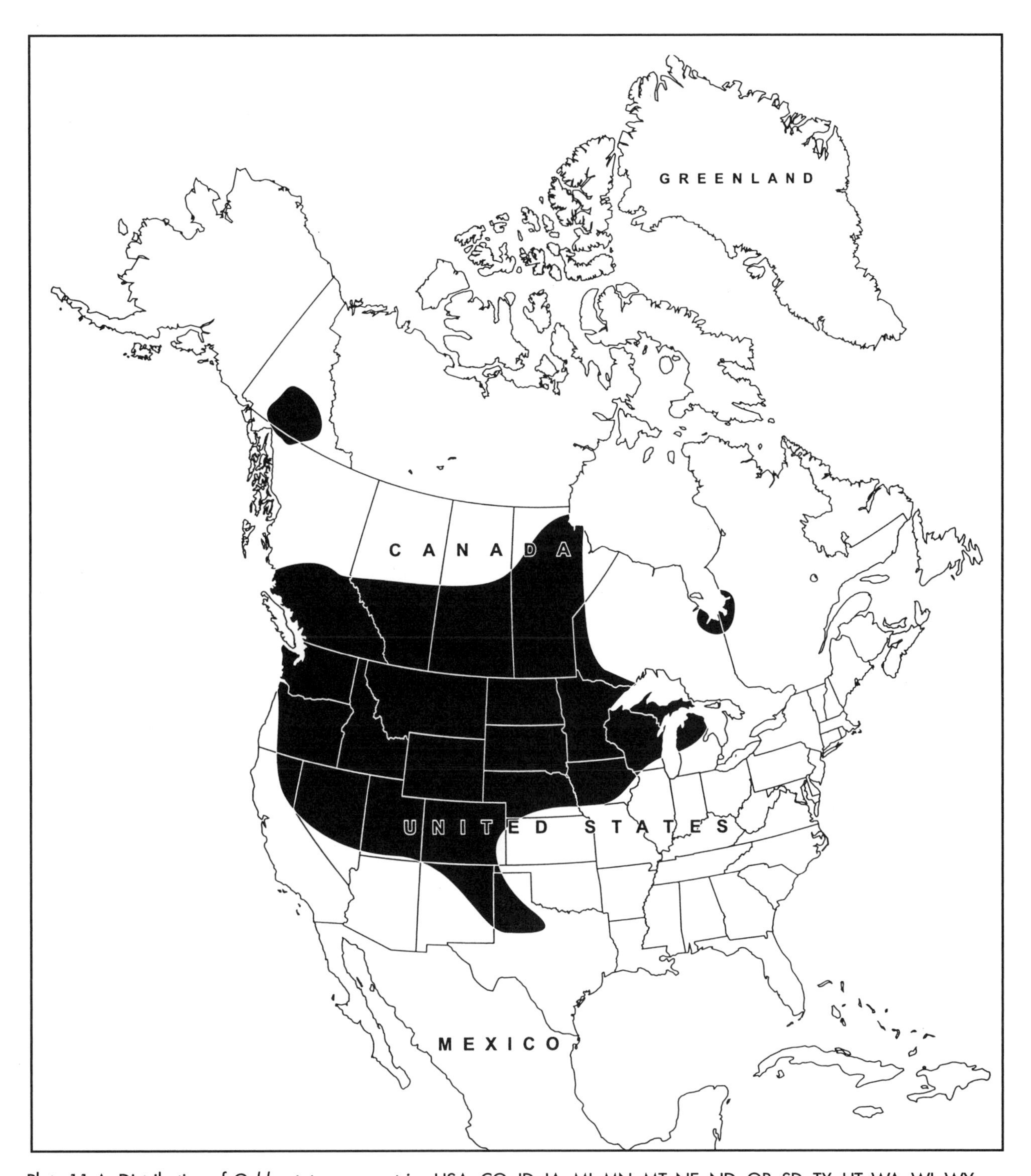


Plate 11 A. Distribution of *Ochlerotatus campestris*—USA: CO, ID, IA, MI, MN, MT, NE, ND, OR, SD, TX, UT, WA, WI, WY (146), CA (229), NV (157), NM (774); Canada: ALTA, BC, MAN, ONT, PQ, SASK, YUK (146); Tax. 431, 739, 784.

Plate 11 B. Distribution of *Ochlerotatus c. canadensis*—USA: AL, AR, CT, DE, DC, FL, GA, ID, IL, IN, IA, KS, KY, LA, ME, MD, MA, MI, MN, MS, MO, MT, NE, NH, NJ, NY, NC, ND, OH, OK, PA, RI, SC, SD, TN, TX, VA, VT, WA, WI, WY (146), AK (188), WV (3), not in NM (777); Canada: ALTA, BC, LAB, MAN, NB, NFLD, NWT, NS, ONT, PEI, PQ, SASK, YUK (146); Tax. 36, 431, 784.

Plate 11 C. Distribution of *Ochlerotatus cantator*—USA: CT, DE, ME, MD, MA, NH, NJ, NY, PA, RI, VA (146), KY (Knapp, pers. comm. 1978), OH (557); Canada: NB, NS, PEI (146), LAB, PQ (446), NFLD (739); Tax. 431, 784.

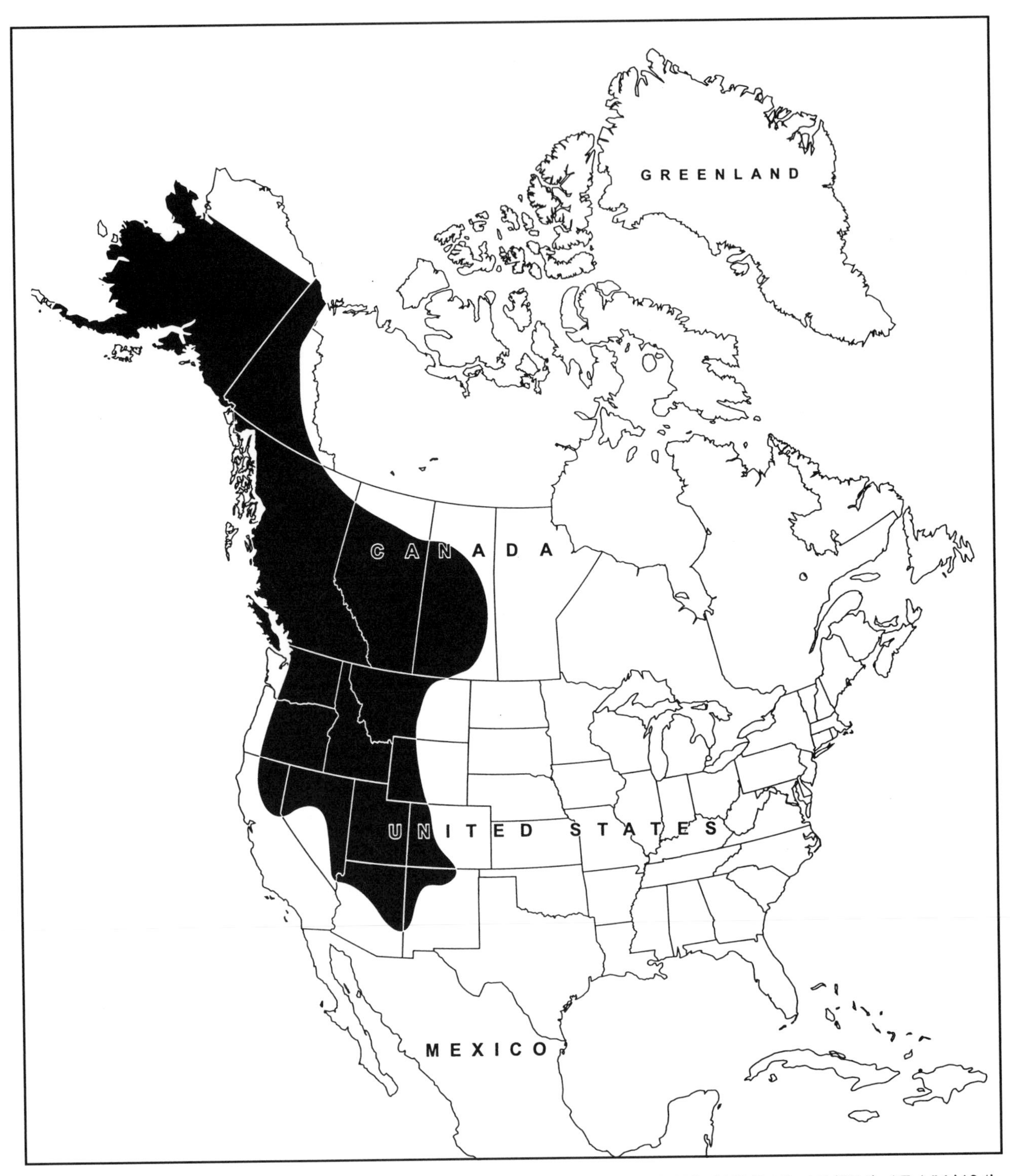


Plate 11 D. Distribution of *Ochlerotatus cataphylla*—USA: CA, CO, ID, MT, OR, UT, WA, WY (146), AK (286), AZ, NV (604), NM (536); Canada: ALTA, BC, SASK, YUK (146); Tax. 431, 777, 784.

Plate 12 A. Distribution of *Ochlerotatus churchillensis*—Canada: ALTA, MAN (250); Tax. 250, 784.

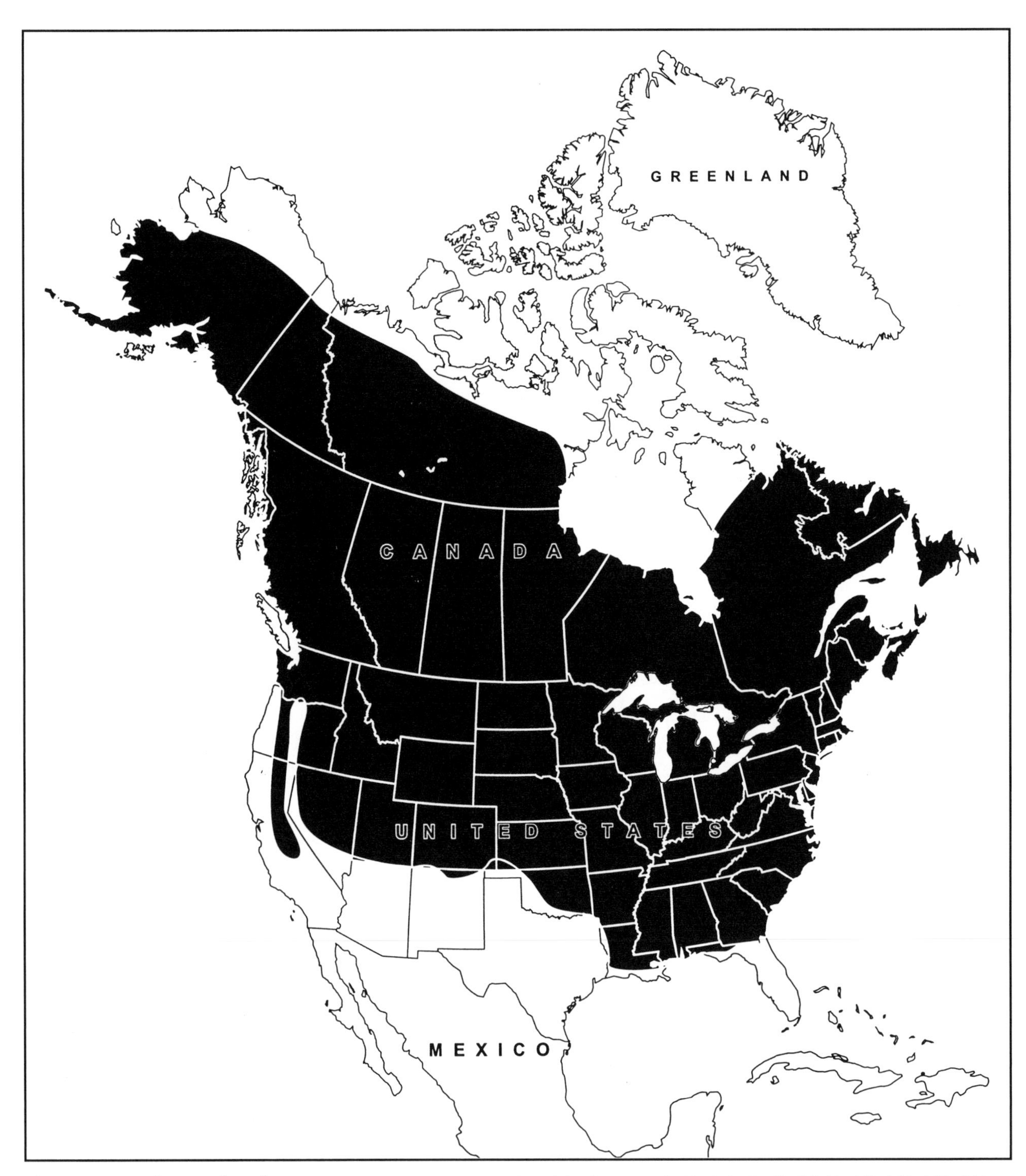


Plate 12 B. Distribution of *Aedes cinereus*—USA: AL, AK, AR, CA, CO, CT, DE, DC, FL, GA, ID, IL, IN, IA, KS, ME, MD, MA, MI, MN, MS, MO, MT, NE, NH, NJ, NY, NC, ND, OH, OK, OR, PA, RI, SC, SD, TN, UT, VT, WA, WI, WY (146), KY (184), LA (158), NV (604), NM (780), VA (287), WV (117); Canada: ALTA, BC, LAB, MAN, NWT, NS, ONT, PEI, PQ, SASK, YUK (146), NB (489), NFLD (784); Tax. 82, 431, 778, 784.

Plate 12 C. Distribution of *Ochlerotatus decticus*—USA: AK, MA, MI, NH, NY (146), ID (Brothers, pers. comm. 2002), ME (641), MN (24), PA (768); Canada: LAB, ONT (146), ALTA, MAN (236), PQ (442), NWT (784), NFLD (528); Map modified after Bourassa et al. (84); Tax. 431, 784.

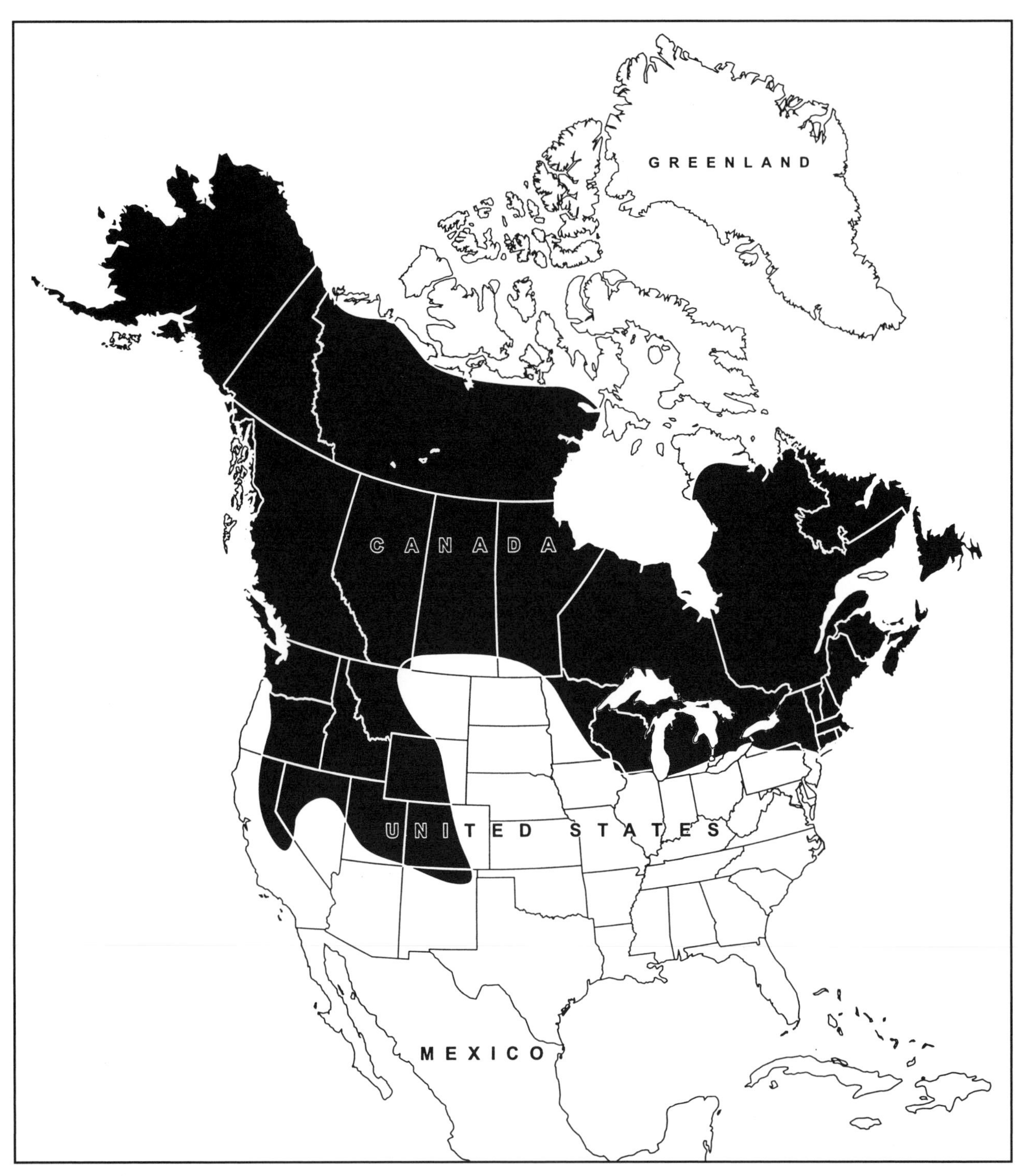


Plate 12 D. Distribution of *Ochlerotatus communis*—USA: AK, CA, CO, ME, MA, MI, MN, MT, NH, NJ, NY, OR, PA, UT, WA, WI, WY (146), ID (503), NV (151), NM (536), VT (291): Canada: ALTA, BC, LAB, MAN, NB, NWT, NS, ONT, PEI, PQ, SASK, YUK (146), NFLD (250); Map modified after Ellis & Brust (250); Tax. 250, 431, 778, 784.

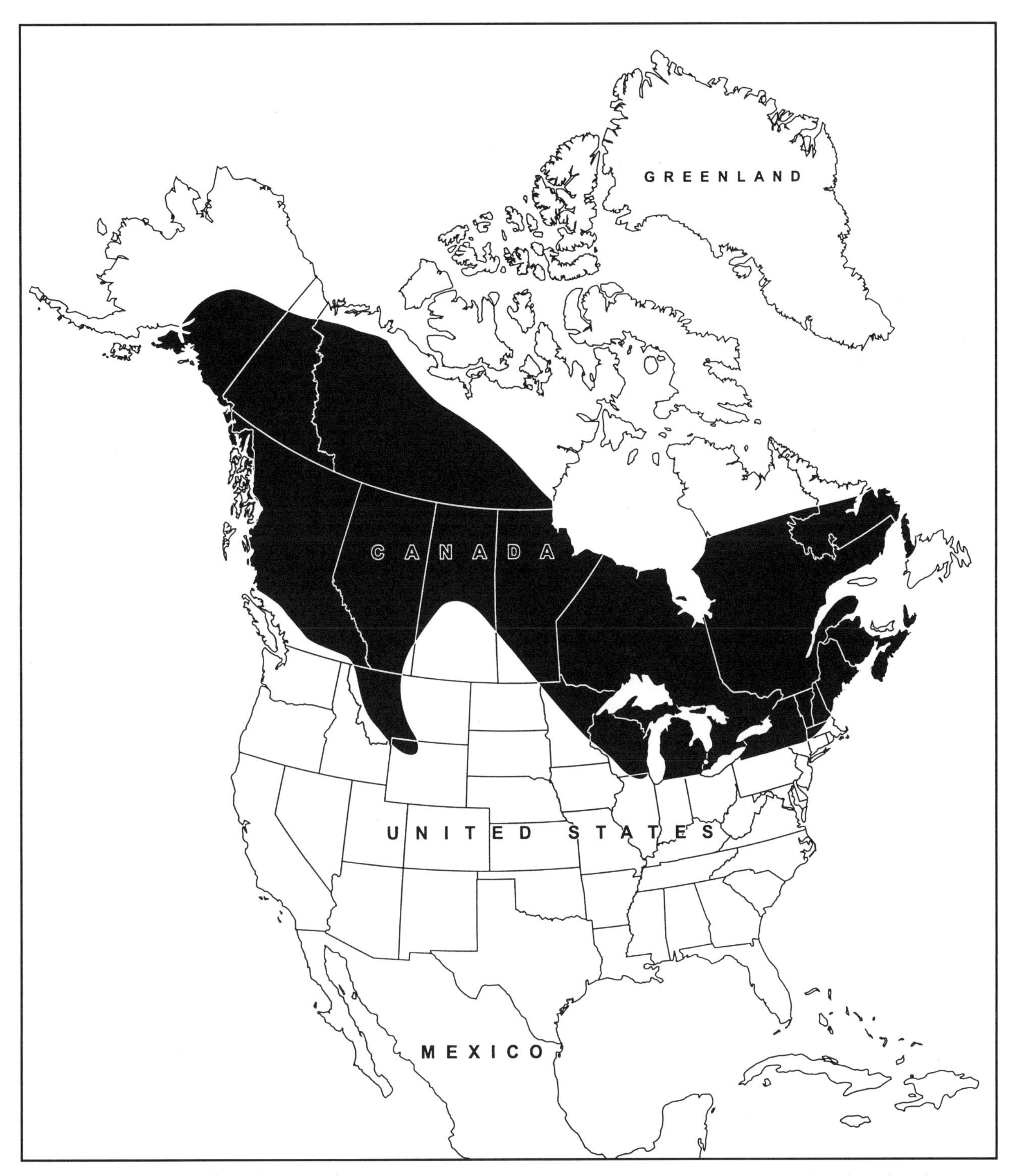


Plate 13 A. Distribution of *Ochlerotatus diantaeus*—USA: AK, ME, MA, MI, MN, NT, NH, NY, VT, WY (146), PA (768), WI (664); Canada: BC, LAB, NWT, NS, ONT, PQ, YUK (146), ALTA (580), SASK (601), MAN (784), NFLD (528); Tax. 303, 431, 784.

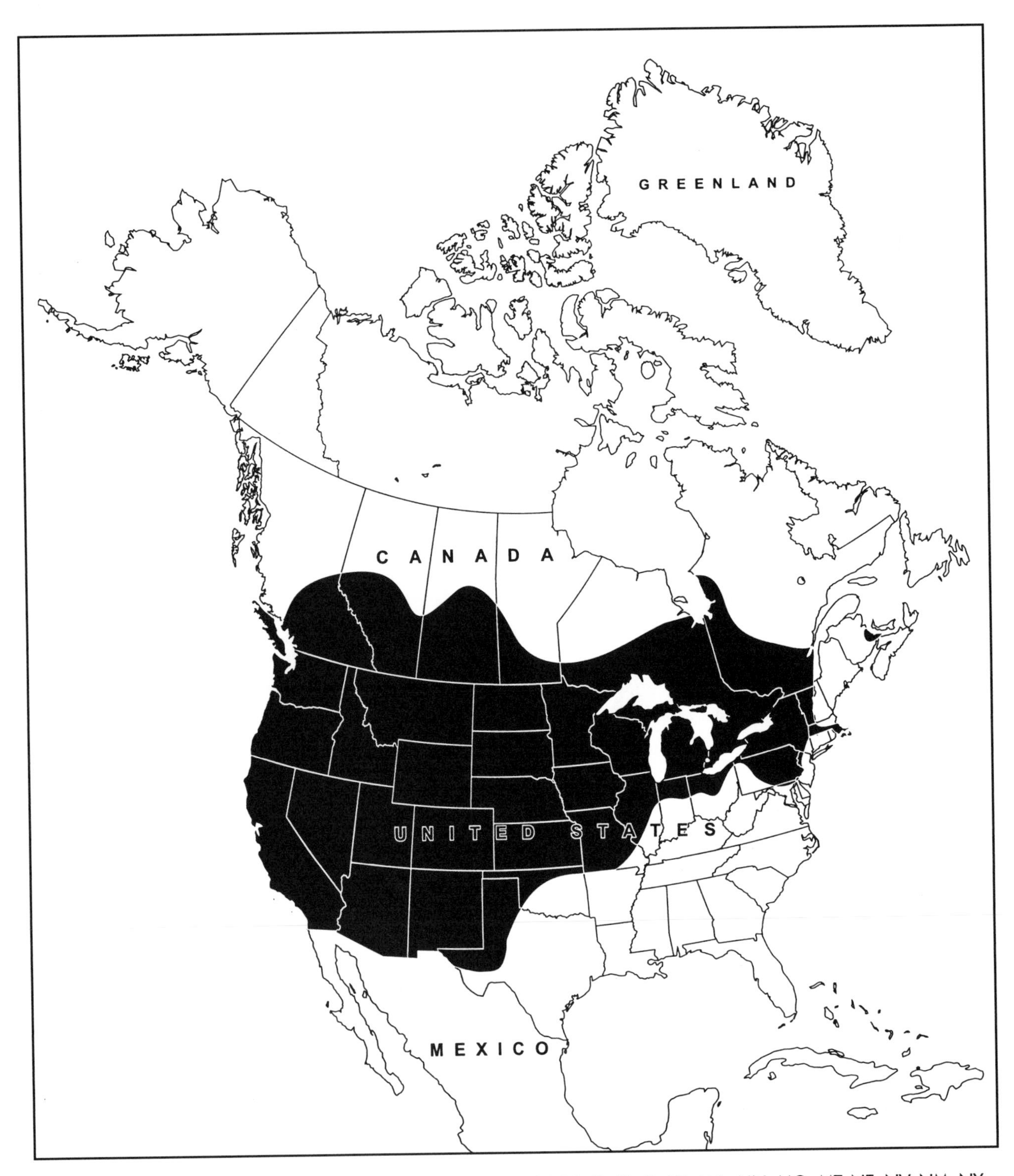


Plate 13 B. Distribution of *Ochlerotatus dorsalis*—USA: CA, CO, CT, DE, ID, IA, IL, KS, MA, MN, MO, MT, NE, NV, NM, NY, ND, OH, OK, OR, PA, SD, TX, UT, WA, WI, WY (146), AZ (604), IN (660), MD (450), MI (30), NJ (110), NH (114); Canada: ALTA, BC, MAN, ONT, PQ, SASK (146), NB (489); Tax. 17, 81, 303, 431, 784.

Plate 13 C. Distrubution of *Ochlerotatus dupreei*—USA: AL, AR, FL, GA, IA, IL, KS, KY, LA, MS, MO, NJ, NC, OK, SC, TN, TX, VA (146), DE (389), IN (660), MI (Newson, pers. comm. 1978), OH (557), Not in MD (67); Tax. 431.

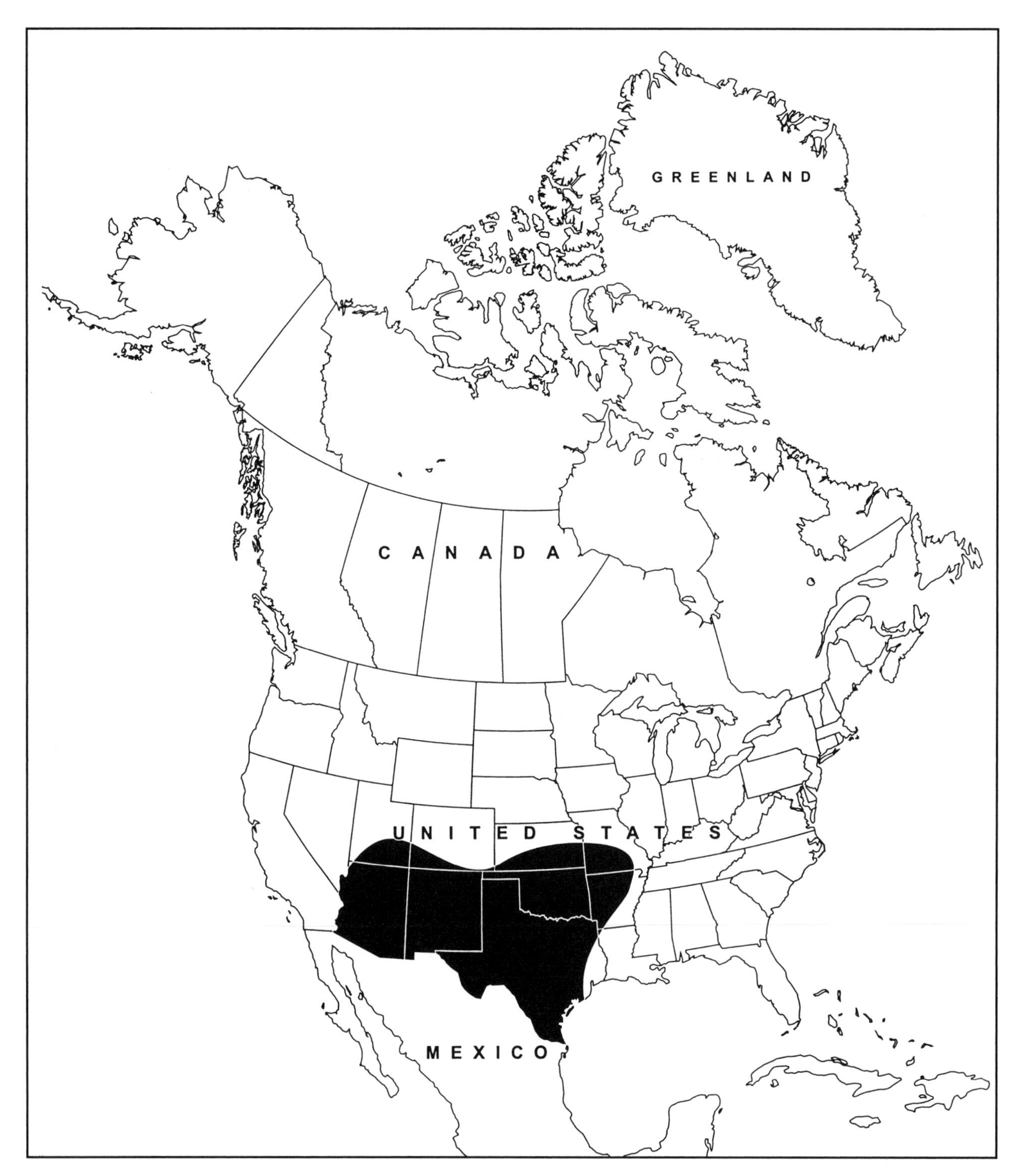


Plate 13 D. Distribution of *Ochlerotatus epactius*—USA: AZ, AR, CO, KS, MO, NM, OK, TX, UT (792), LA (205); Map modified after Zavortink (792); Tax. 545, 546, 792.

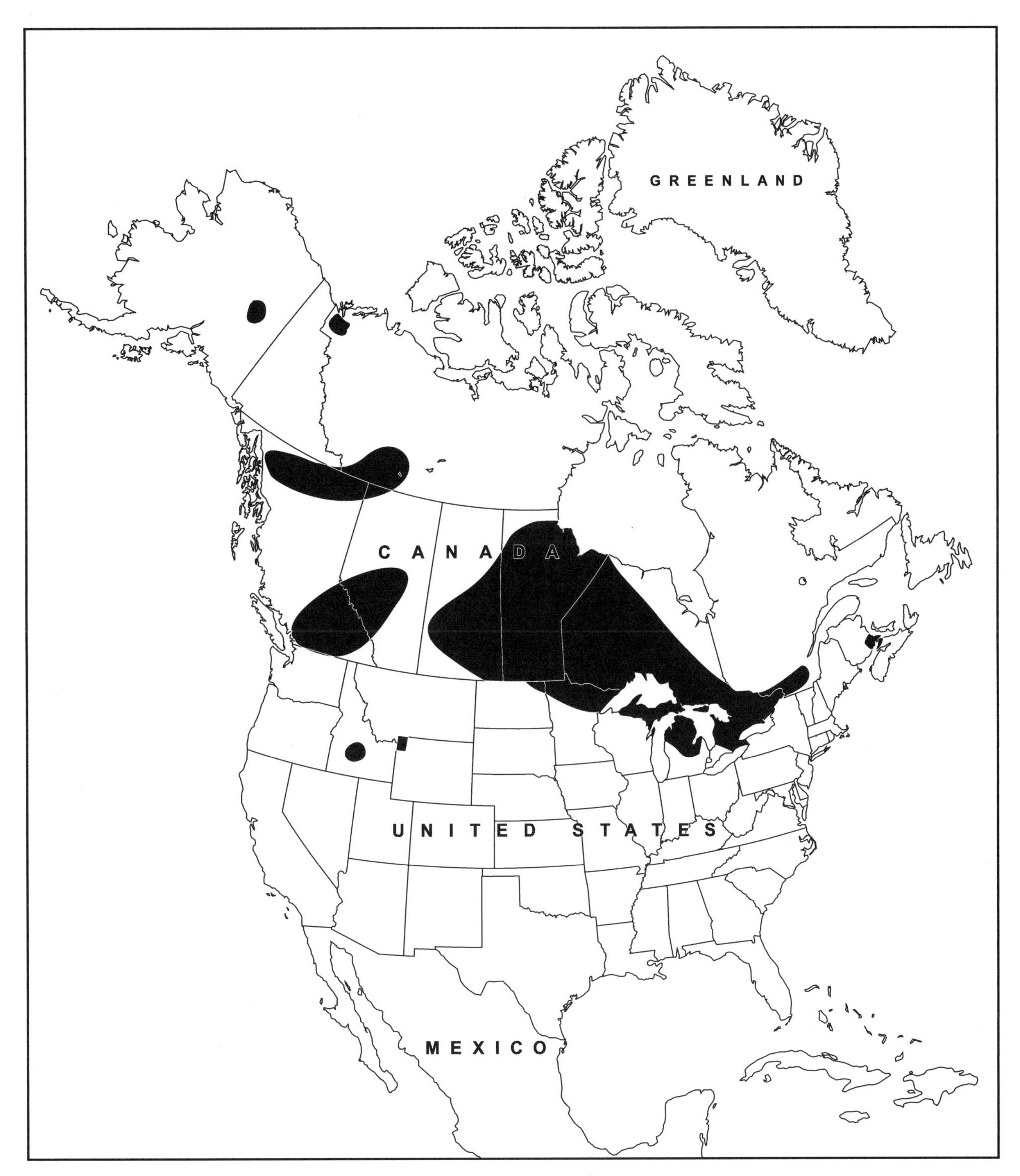


Plate 14 A. Distribution of *Ochlerotatus euedes*—USA: AK (529), ID (Brothers, pers. comm. 2002), MI (771), MN (624), WY (524); Canada: ALTA (252), MAN (107), NS (399), ONT (694), PQ (235), BC, NWT, SASK (784); Tax. 202, 223, 431, 624, 783, 784.

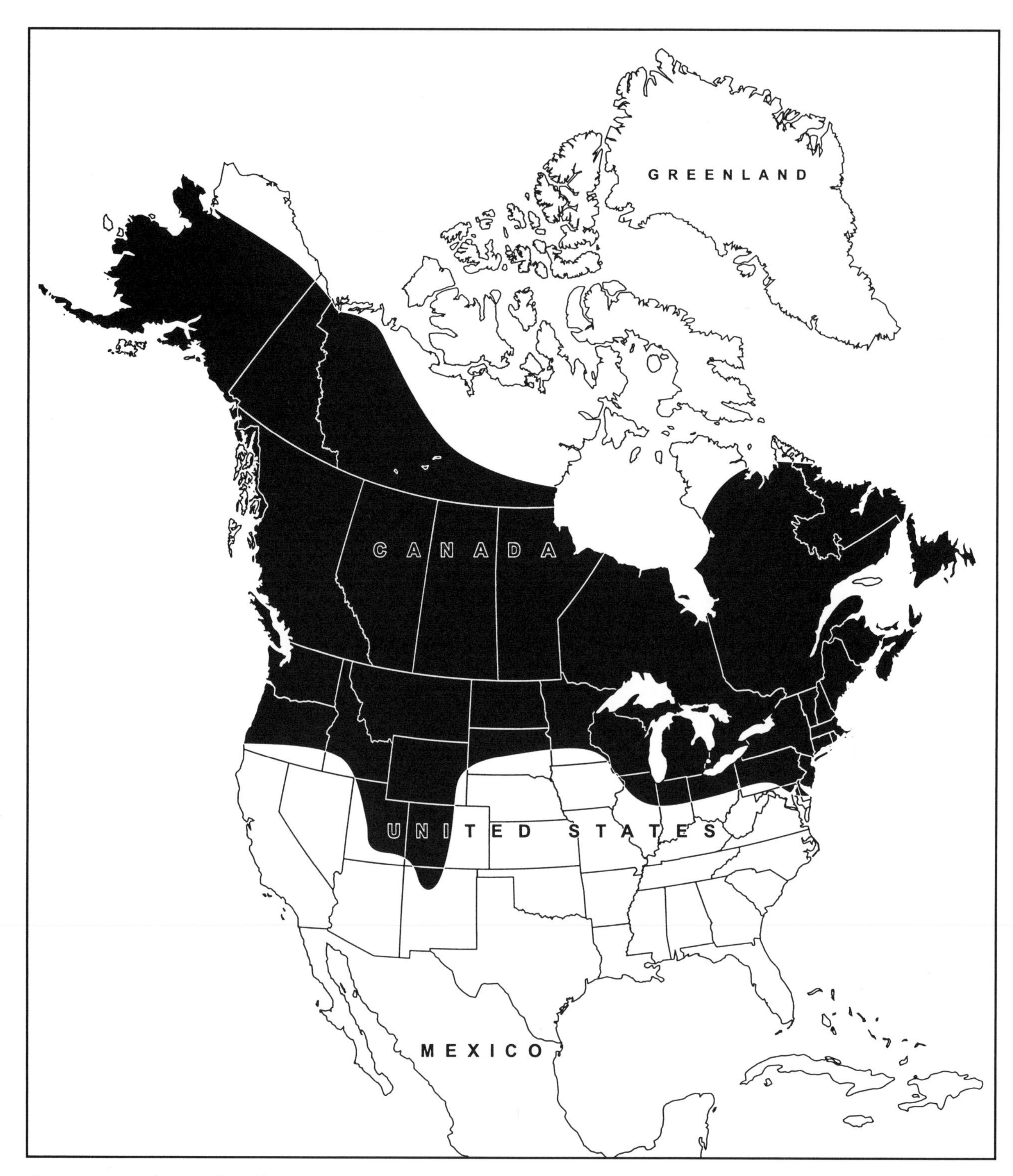


Plate 14 B. Distribution of *Ochlerotatus excrucians*—USA: AK, CO, CT, ID, IL, ME, MA, MI, MN, MT, NH, NJ, NY, ND, OH, OR, PA, RI, UT, VT, WA, WI, WY (146), DE (391), IN (658), MD (76), NM (536); Canada: ALTA, BC, MAN, NFLD, NWT, NS, ONT, PEI, PQ, SASK, YUK (146), LAB (324), NB (489); Tax. 200, 303, 431, 466, 778, 784.

Plate 14 C. Distribution of *Ochlerotatus fitchii*—USA: AK, CA, CO, CT, ID, IA, IL, ME, MI, MN, MT, NE, NH, NJ, NY, ND, OH, OR, RI, UT, VT, WA, WI, WY (146), AZ, NV (604), DE (390), IN (658), MD (67), NM (776), PA (629); Canada: ALTA, BC, LAB, MAN, NFLD, NWT, ONT, PEI, PQ, SASK, YUK (146), NB (399), NS (399); Tax. 431, 778, 784.

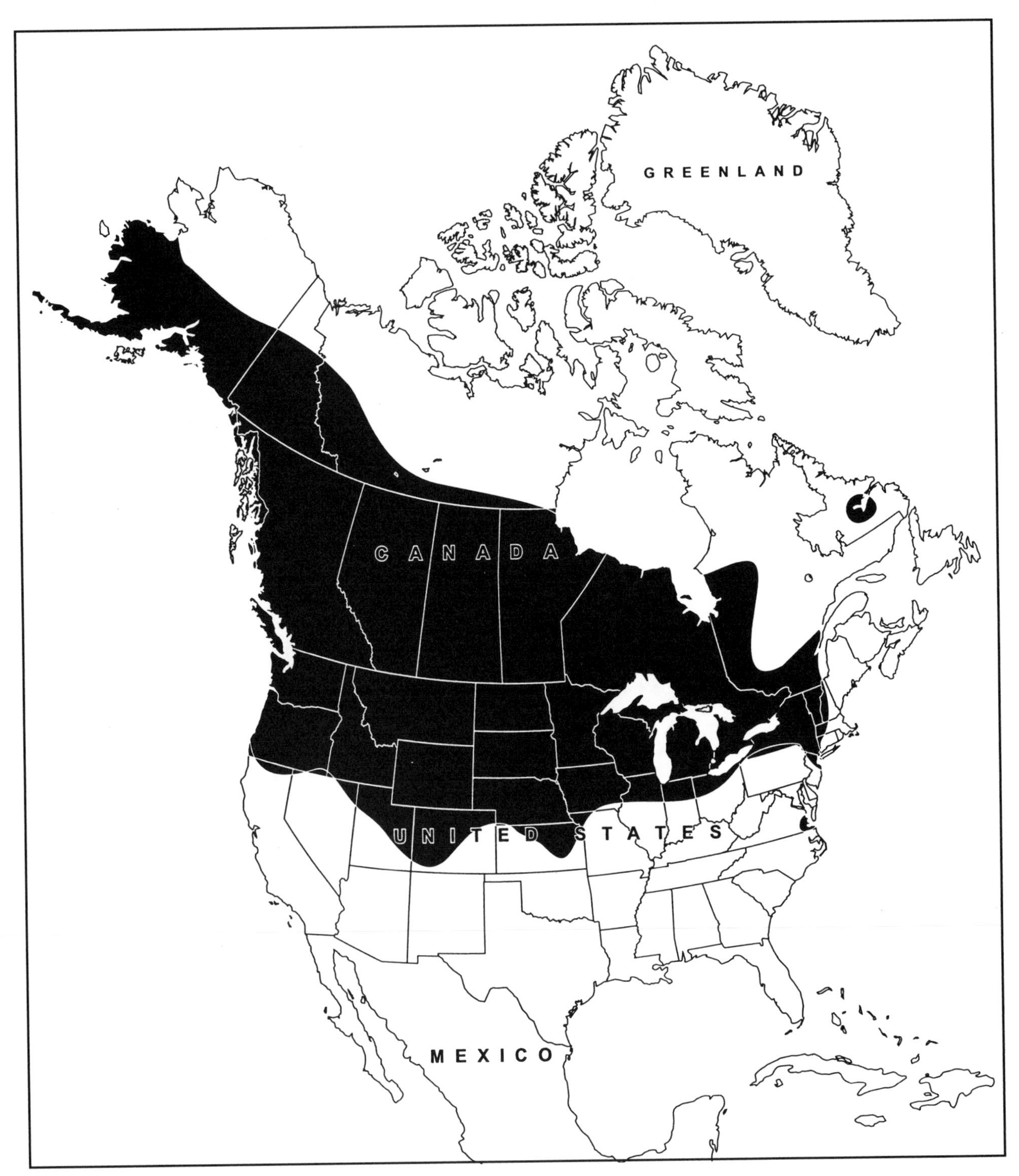


Plate 14 D. Distribution of *Ochlerotatus flavescens*—USA: AK, CA, CO, ID, IL, IA, KS, MI, MN, MO, MT, NE, NY, ND, OR, SD, UT, WA, WI, WY (146), IN (658), NH (114), NJ (191), OH (Berry & Parsons, pers. comm. 1978), VA (*Skeeter* 23:2, 1978); Canada: ALTA, BC, LAB, MAN, NWT, ONT, SASK, YUK (146), PQ (235); Tax. 431, 738, 783.

Plate 15 A. Distribution of *Ochlerotatus fulvus pallens*—USA: AL, AR, FL, GA, IL, KY, LA, MD, MS, NC, OK, SC, TN, TX, VA (146), MO (452); Tax. 431.

Plate 15 B. Distribution of *Ochlerotatus grossbecki*—USA: AR, DE, IL, KY, LA, MD, MS, MO, NJ, NY, OH, SC, TN, VT, VA (146), CT (746), IN (651), PA (768), TX (Harris Co. M.C.D., pers. comm. 1978), WI (724); Canada: ONT (333); Tax. 431, 784.

Plate 15 C. Distribution of *Ochlerotatus hendersoni*—USA: AL, AR, CT, DE, DC, GA, LA, KY, ME, MD, MA, MN, MS, NH, NJ, NY, NC, OR, PA, SC, TN, UT, VA, WV (793), FL (814), CO, TX (98), ID, WY (533), IL (332), IN, MI, OH (730), KS, NE (314), LA (158), MO (673), MT, NM, SD (525), OK (558), WI (419), VT (291); Canada: BC (793), MAN (728), ONT, PQ, SASK (784); Map after Zavortink (793): Tax. 98, 296, 313, 431, 784, 793.

Plate 15 D. Distribution of *Ochlerotatus hexodontus*—USA: AK, CA, CO, ID, MT, OR, WA (146), NV (151), NM (536), UT (592), WY (553); Canada: BC, MAN, NWT, PQ, YUK (146), ALTA (580), LAB, ONT (784), NFLD (528); Tax. 431, 739, 778, 783, 784.

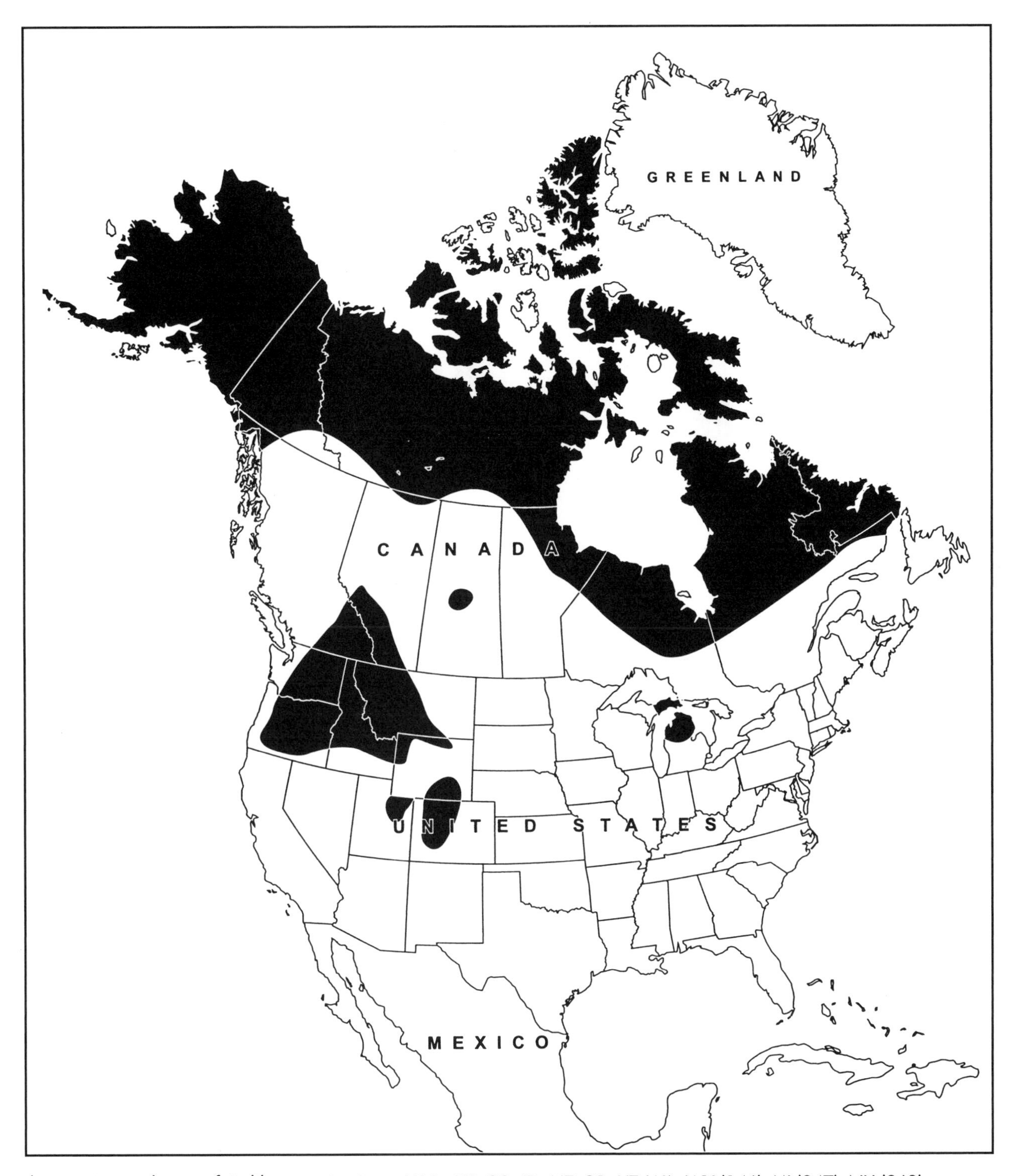


Plate 16 A. Distribution of *Ochlerotatus impiger*—USA: AK, CO, ID, MT, OR, UT, WA, WY (146), MI (347), NH (363); Canada: ALTA, MAN, NWT, ONT, PQ, SASK, YUK(146), BC (263), LAB (784); Tax. 36, 203, 431, 739, 784.

Plate 16 B. Distribution of *Ochlerotatus implicatus*—USA: AK, CO, ID, IA, MA, MI, MN, MT, NE, NH, NY, UT, WA, WY (146), AZ, NM (536), ME (465), NJ (110), OH (557), OR (284), WI (664); Canada: ALTA, BC, MAN, NWT, ONT, PQ, SASK, YUK (146), LAB (784), PEI (399); Tax. 431, 778, 779, 784.

Plate 16 C. Distribution of *Ochlerotatus increpitus*—USA: CA, CO, ID, MT, NV, NM, OR, UT, WA, WY (146), AZ (780), NE (589), SD (280); Canada: BC, SASK (146), ALTA (580); Tax. 431, 777, 783.

Plate 16 D. Distribution of *Ochlerotatus infirmatus*—USA: AL, AR, FL, GA, KY, LA, MS, MO, NC, SC, TN, TX (146), DE (391), IL (326), IN (322), MD (171), VA (11), not in AZ (11); Map modified after Arnell (11).

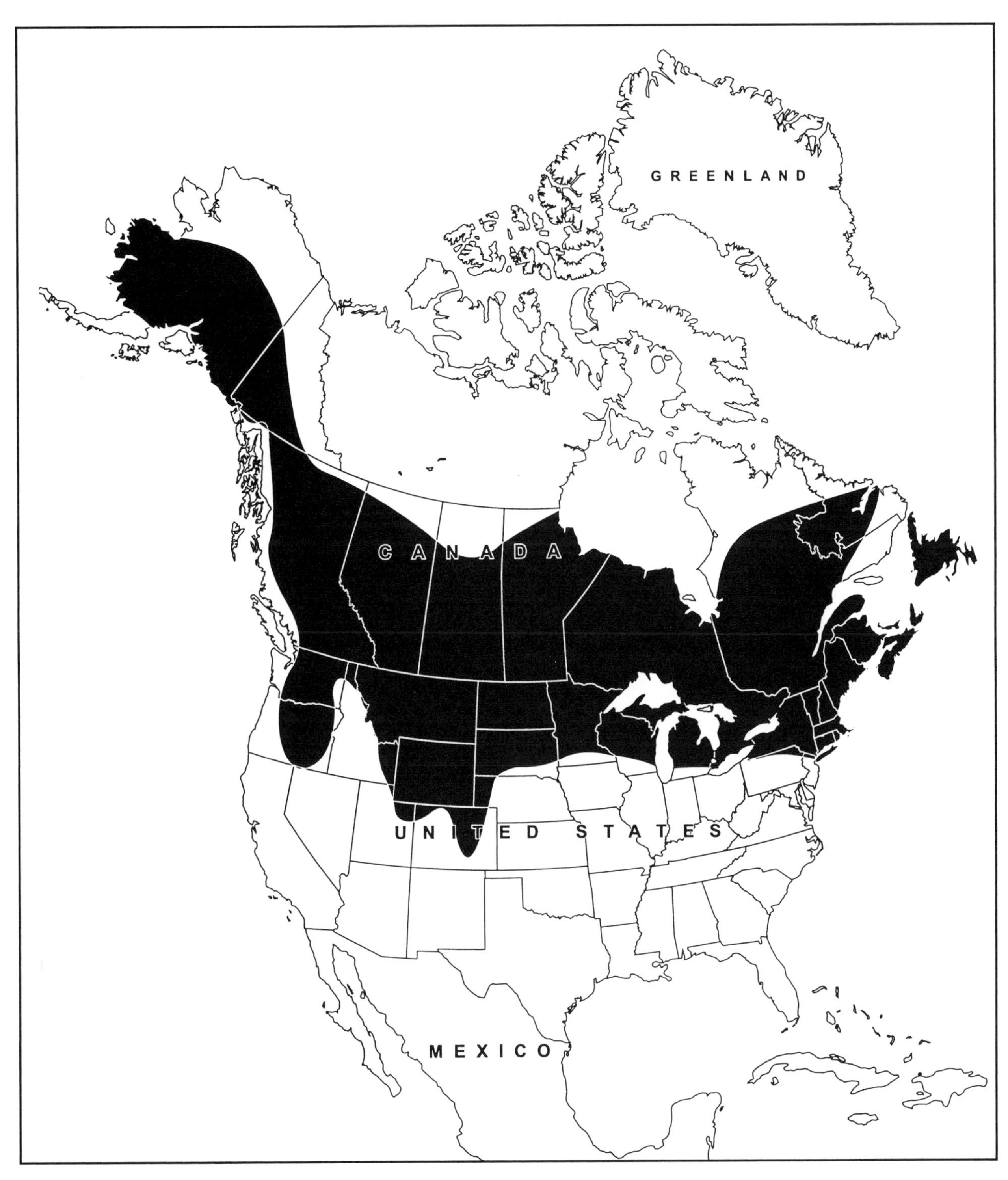


Plate 17 A. Distribution of *Ochlerotatus intrudens*—USA: AK, CO, CT, ID, ME, MA, MI, MN, MT, NH, NY, ND, OR, PA, RI, SD, UT, WA, WI, WY (146), NJ (193), VT (291); Canada: ALTA, BC, LAB, MAN, NB, NS, PEI (146), NFLD (528), ONT (35), PQ (235), SASK (601); Tax. 431, 784.

Plate 17 B. Distribution of *Ochlerotatus melanimon*—USA: CA, CO, MT, NV (146), ID, NE, NM, UT, WA, WY (603), OR (284); Canada: ALTA (115), BC (Belton, pers. comm. 1978), SASK (340); Tax. 17, 81, 431, 603, 784.

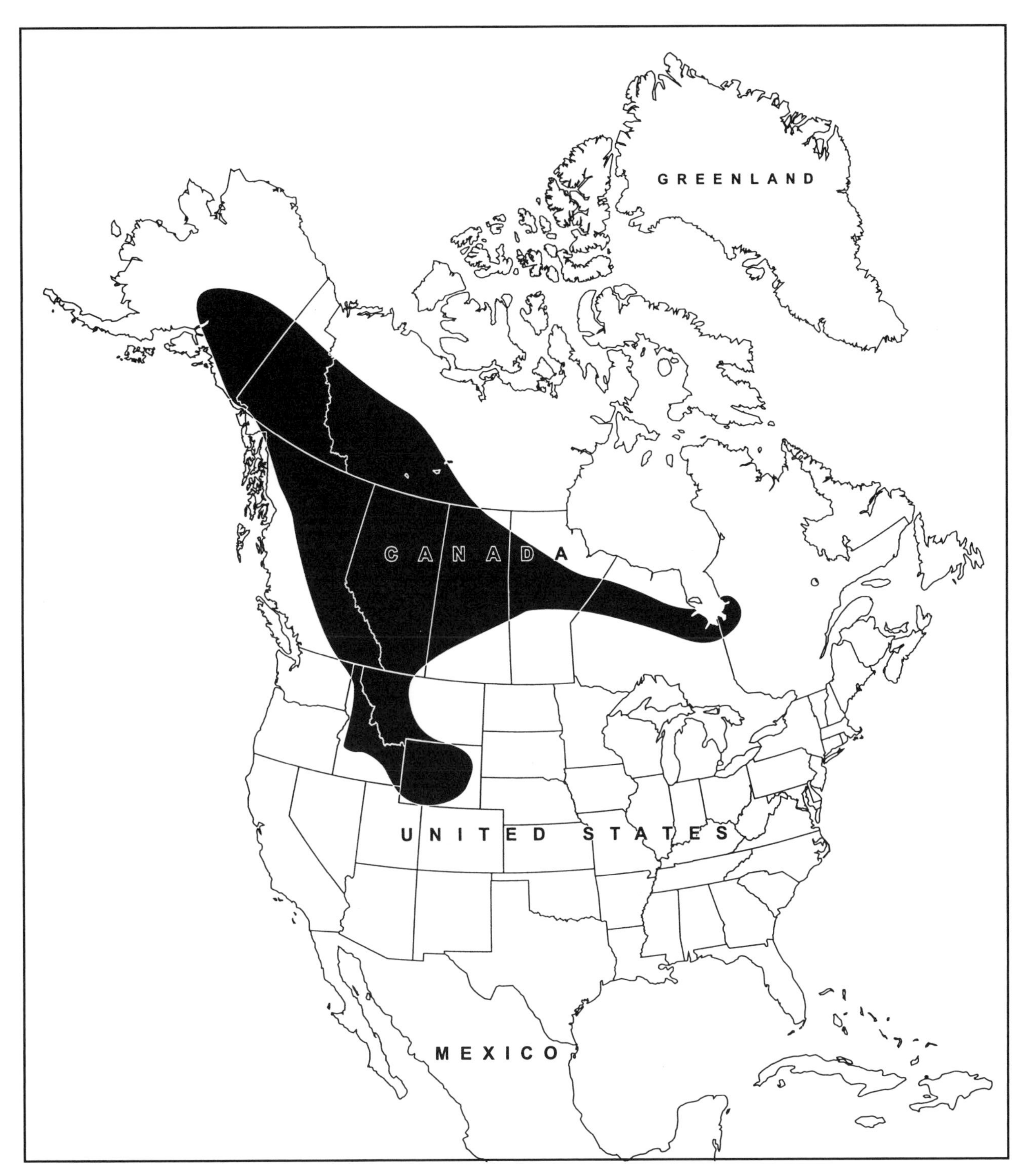


Plate 17 C. Distribution of *Ochlerotatus mercurator*—USA: ID, MT, WY (146 as *Ae. stimulans*), AK (783); Canada: ALTA, MAN, NWT, ONT, SASK, YUK (783), BC (784), PQ (447); Tax. 783, 784.

Plate 17 D. Distribution of *Ochlerotatus mitchellae*—USA: AL, AR, DE, DC, FL, GA, IL, LA, MD, MS, NJ, NM, NY, NC, OK, SC, TN, TX, VA (146), KS (541), KY (184), MI (Newson, pers. comm. 1977), MO (Darsie, personal comm. 1998), OH (557), PA (692); Tax. 431.

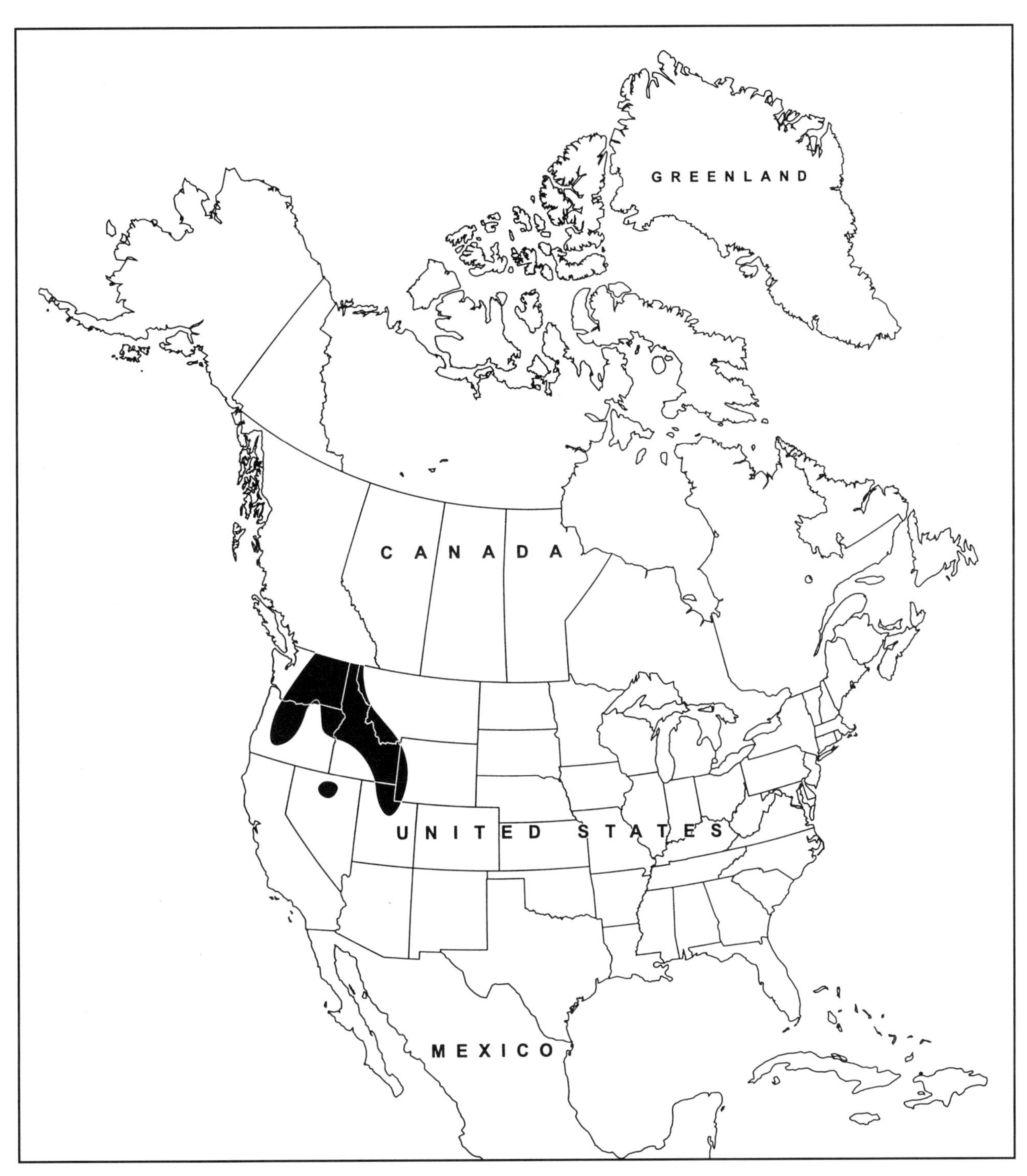


Plate 18 A. Distribution of *Ochlerotatus nevadensis*—USA: ID, OR, WA, (250), NV, UT, WY (159); Map modified after Ellis & Brust (250); Tax. 159, 250, 285, 784.

Plate 18 B. Distribution of *Ochhlerotatus nigripes*—USA: AK (146, 718); Canada: MAN, NWT, PQ, YUK (146), BC, LAB, NFLD (784); Tax. 203, 303, 431, 784.

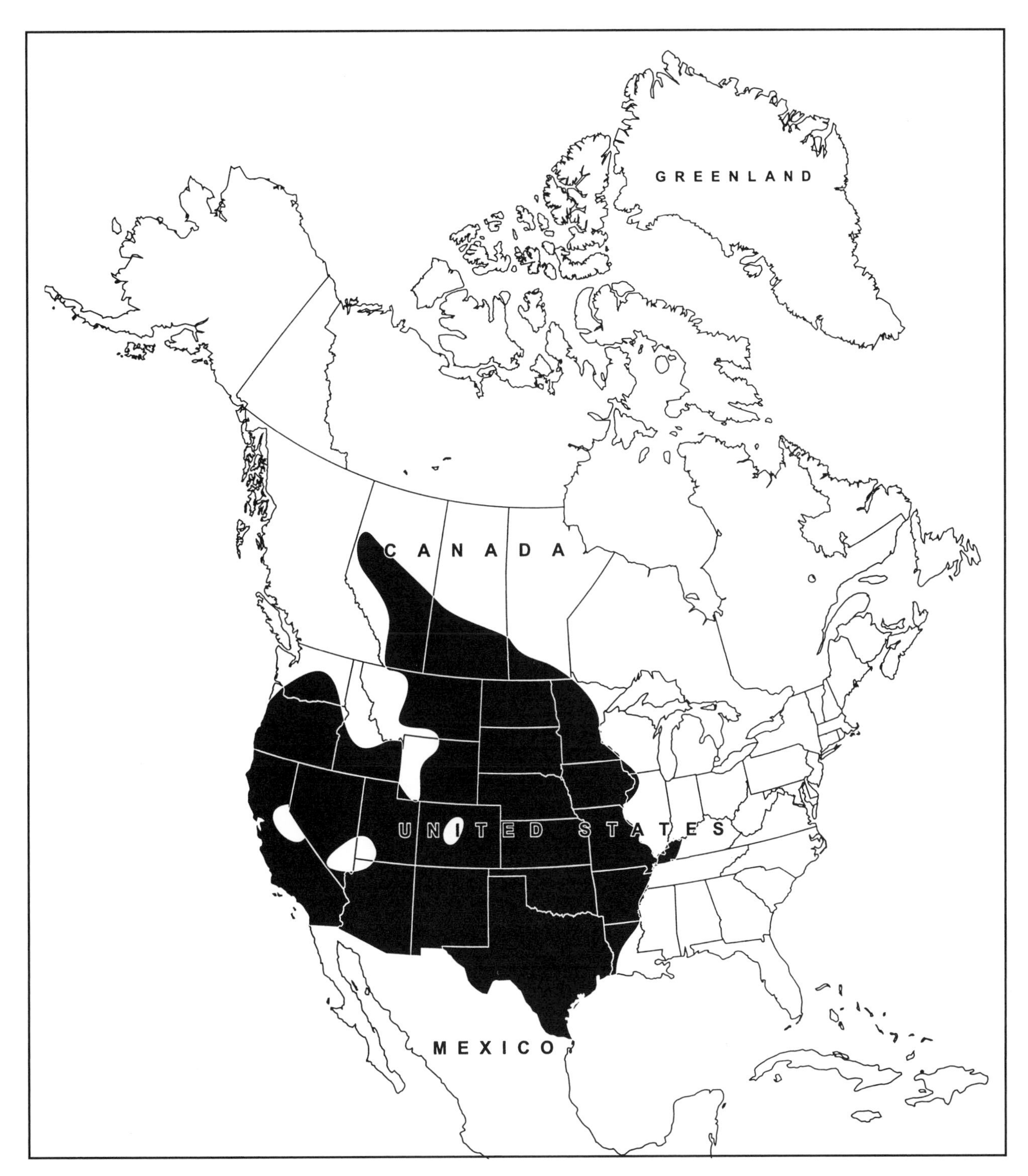


Plate 18 C. Distribution of *Ochlerotatus nigromaculis*—USA: CA, CO, ID, IL, IA, KS, MN, MO, MT, NE, NM, ND, OK, OR, SD, TX, UT, WA, WY (146); AZ (604), AR, LA (162), NV (157); KY (183); Canada: ALTA, MAN, SASK (146); Tax. 431, 784.

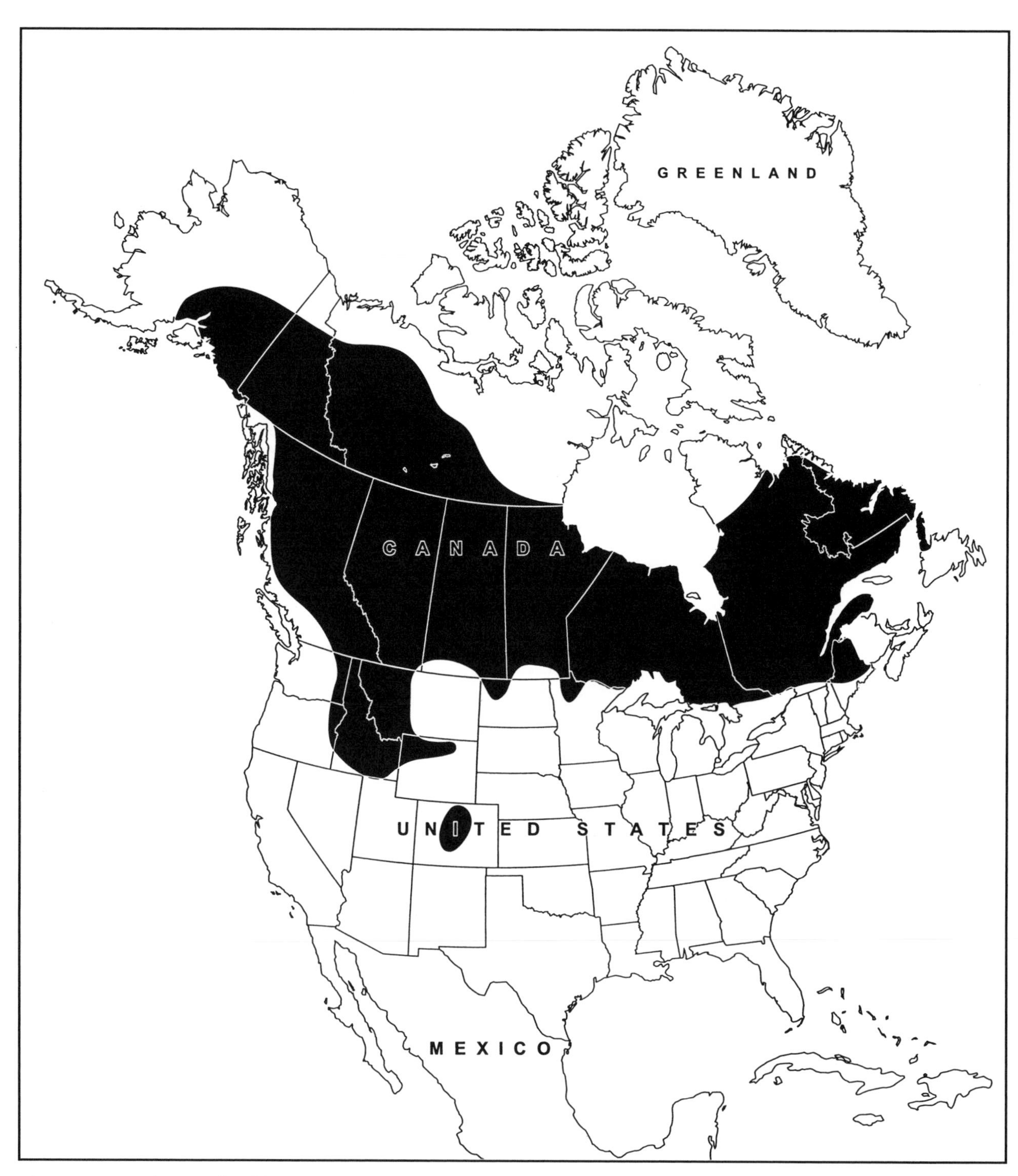


Plate 18 D. Distribution of *Ochlerotatus pionips*—USA: AK, CO, ID, MT, ND, WY (146), ME (465), MI (30), MN (20), OR, WA (284); Canada: ALTA, BC, LAB, MAN, NWT, ONT, PQ, SASK, YUK (146), NFLD (528); Tax. 200, 303, 431, 784.

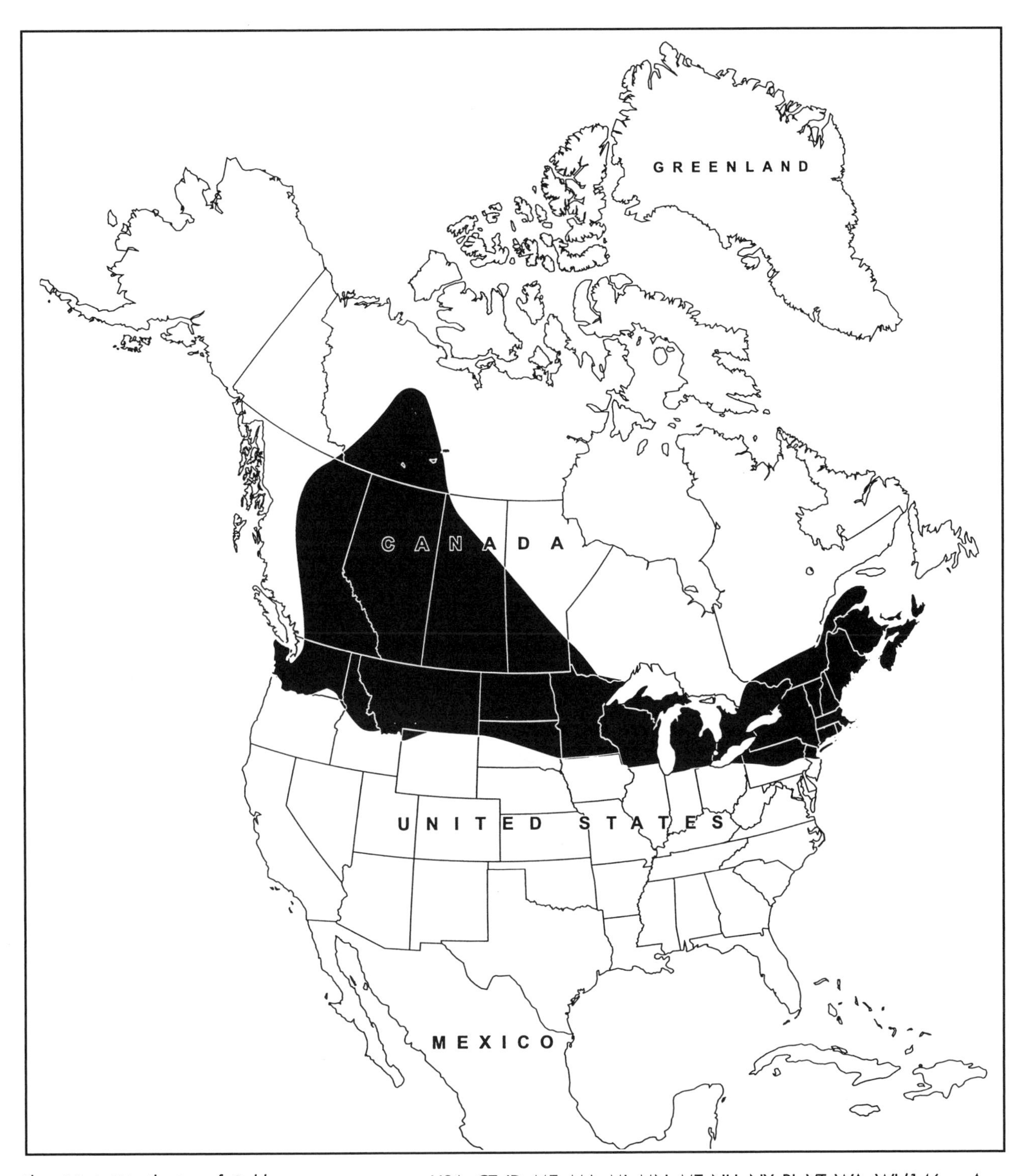


Plate 19 A. Distribution of *Ochlerotatus provocans*—USA: CT, ID, ME, MA, MI, MN, MT, NH, NY, RI, VT, WA, WI (146 as *Ae. trichurus*), NJ (193), PA (639); Canada: ALTA, BC, MAN, NB, NS, ONT, PEI, PQ, SASK (146 as *Ae. trichurus*), NWT (784); Tax. 431, 783, 784.

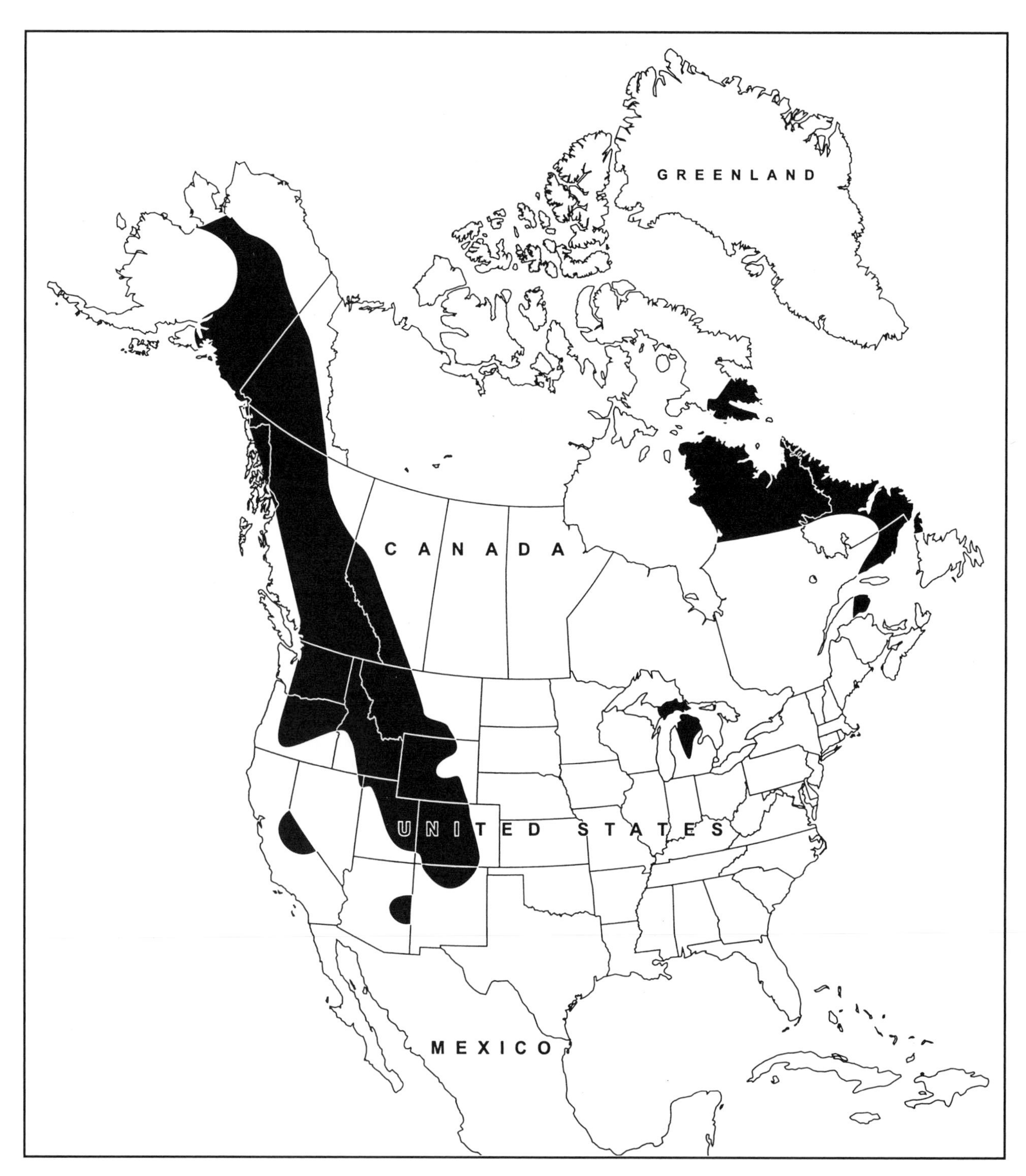


Plate 19 B. Distribution of *Ochlerotatus pullatus*—USA: AK, CA, CO, ID, MT, OR, UT, WA, WY (146), AZ (536), NV (151), NM (495), MI (149); Canada: ALTA, BC, NWT, PQ, YUK (146), LAB, NFLD (784); Tax. 200, 303, 431, 778, 784.

Plate 19 C. Distribution of *Ochlerotatus punctodes*—USA: AK (146); Tax. 200, 431.

Plate 19 D. Distribution of *Ochlerotatus punctor*—USA: AK, CO, IL, ME, MA, MI, MN, MT, NH, NJ, NY, ND, VT, WI, WY (146), ID (284), IN (659), IA (570), PA (105), WA (507), not in MD (67), not in UT (534); Canada: ALTA, BC, LAB, MAN, NB, NWT, NS, ONT, PEI, PQ, SASK, YUK (146), NFLD (263); Tax. 431, 783.

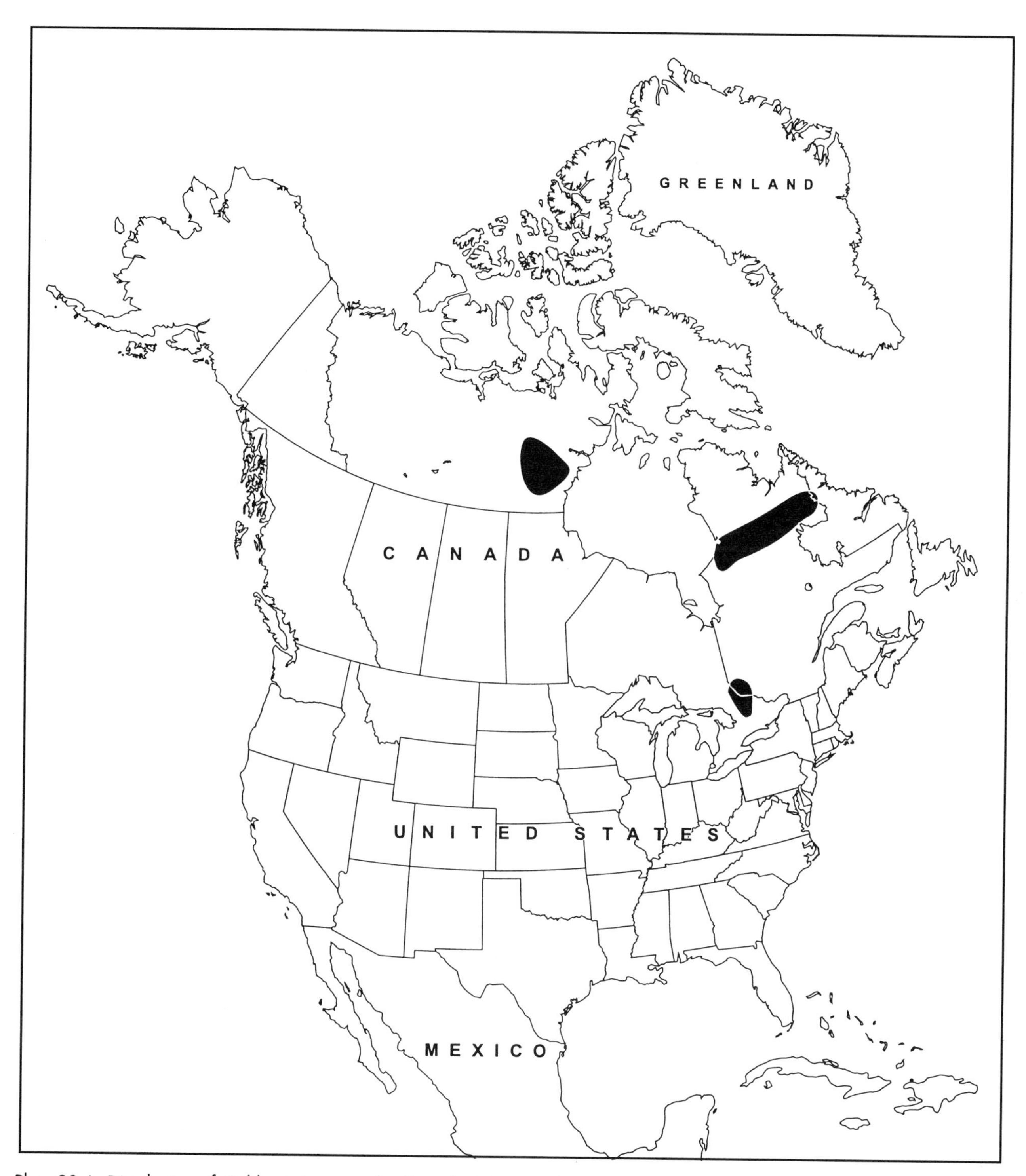


Plate 20 A. Distribution of *Ochlerotatus rempeli*—Canada: NWT, PQ (146), ONT (784); Tax. 234, 431, 681, 784.

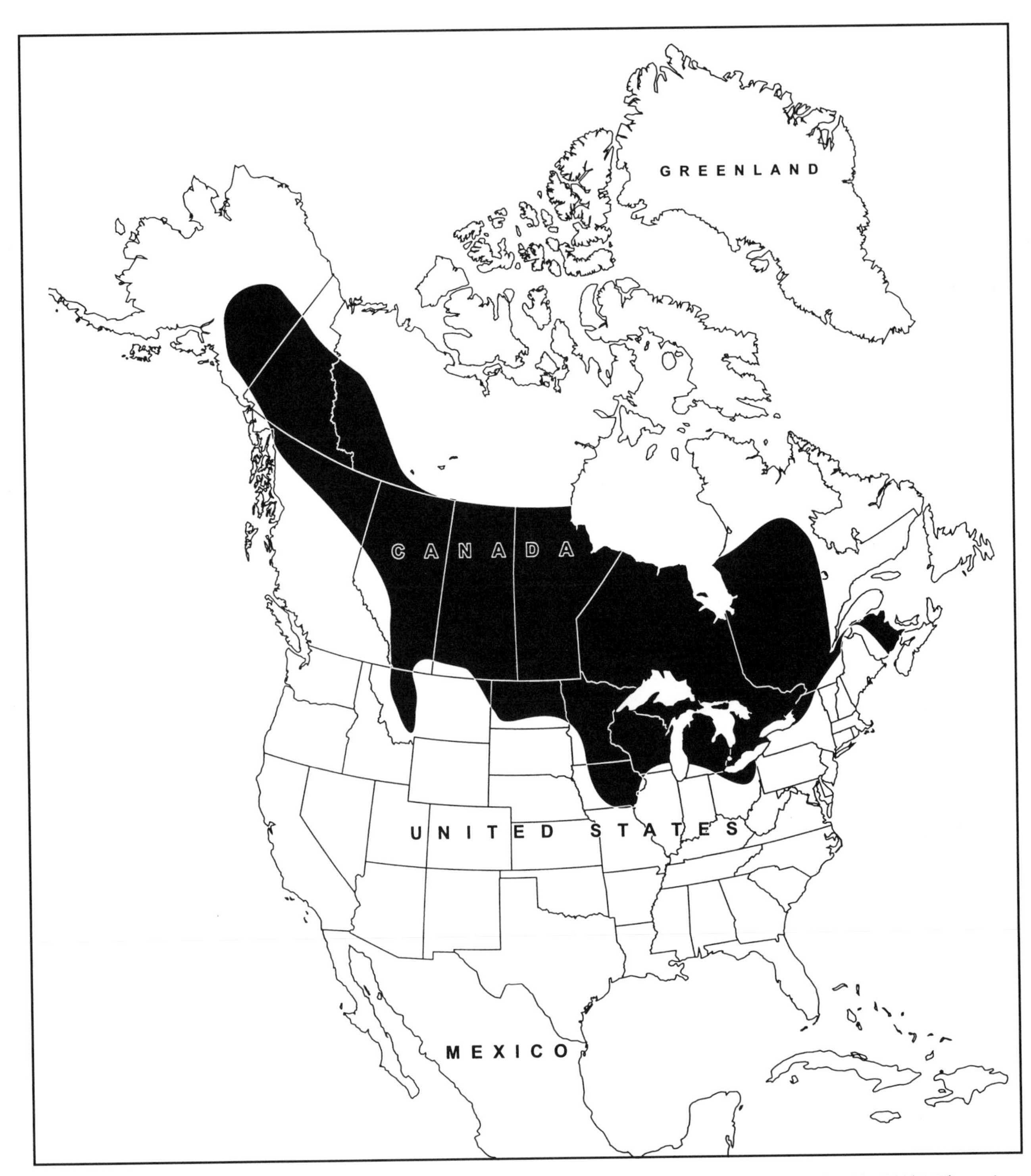


Plate 20 B. Distribution of *Ochlerotatus riparius*—USA: AK, IA, MI, MN, MT, NY, ND, WI (146), MO (673), OH (557), not in CO (314), not in WY (553); Canada: ALTA, BC, MAN, NWT, ONT, SASK, YUK (146), NB (399), NS (784), PQ (235); Tax. 200, 303, 431, 784.

Plate 20 C. Distribution of *Ochlerotatus schizopinax*—USA: MT, WY (146), CA (604), CO, ID, OR (521), NV (151), NM (536), UT (592); Canada: ALTA (252); Tax. 431, 778, 784.

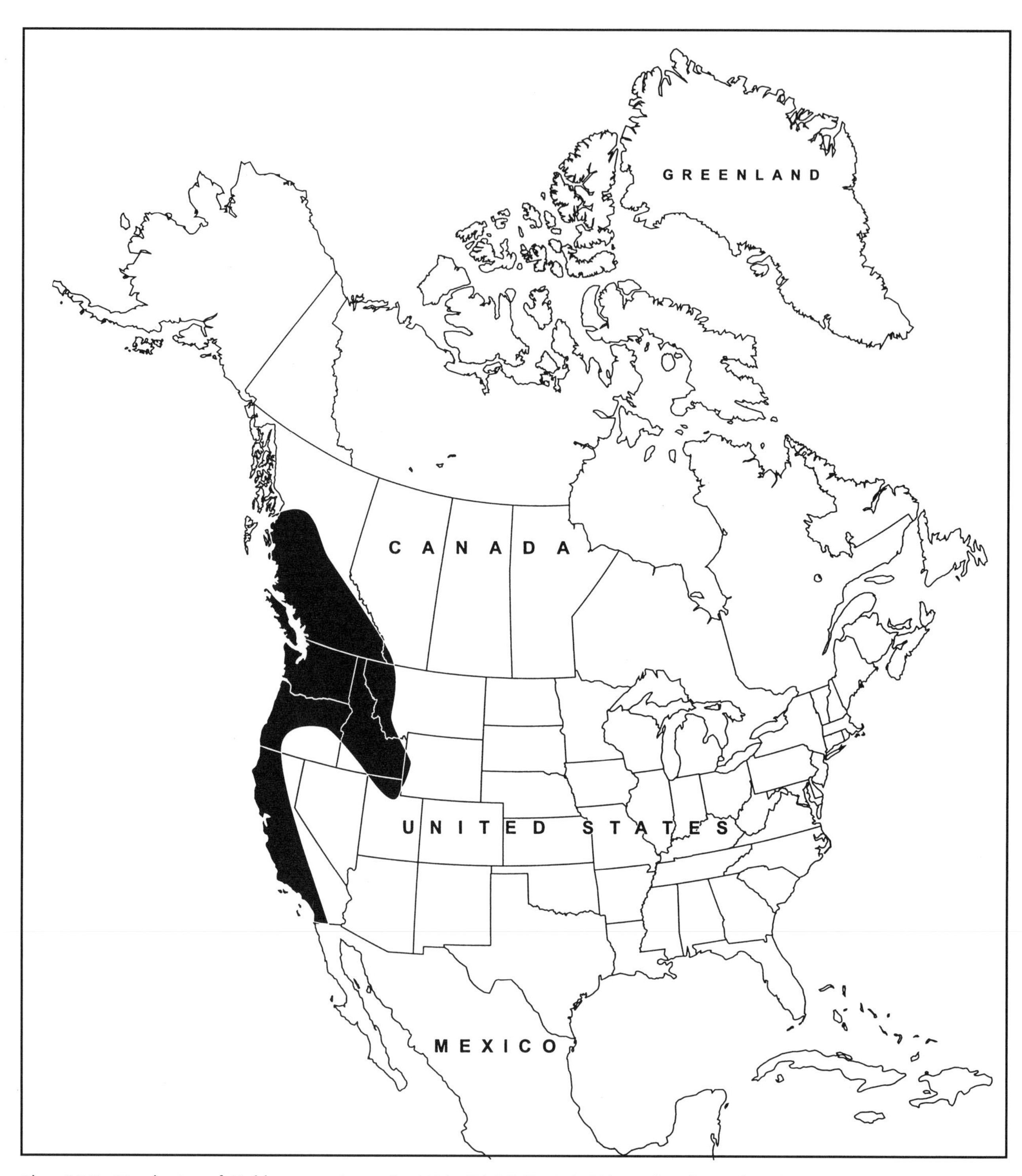


Plate 20 D. Distribution of *Ochlerotatus sierrensis*—USA: CA (418), ID (120), MT (532), NV (154), OR (315), UT (525), WA (507); Canada: BC (199); Map after Arnell & Nielsen (13); Tax. 13, 44, 46, 198, 431, 784.

Plate 21 A. Distribution of *Ochlerotatus sollicitans*—USA: AL, AZ, AR, CT, DE, DC, FL, GA, IL, IN, KS, KY, LA, ME, MD, MA, MS, MO, NE, NH, NJ, NM, NY, NC, ND, OH, OK, PA, RI, SC, TX, VA (146), IA (383), MI (Newson, pers. comm. 1977), SD (280), TN (685), WV (117); Canada: NB, NS, PEI (146), ONT (333); Map modified after Knight (383); Tax. 36, 42, 431, 784.

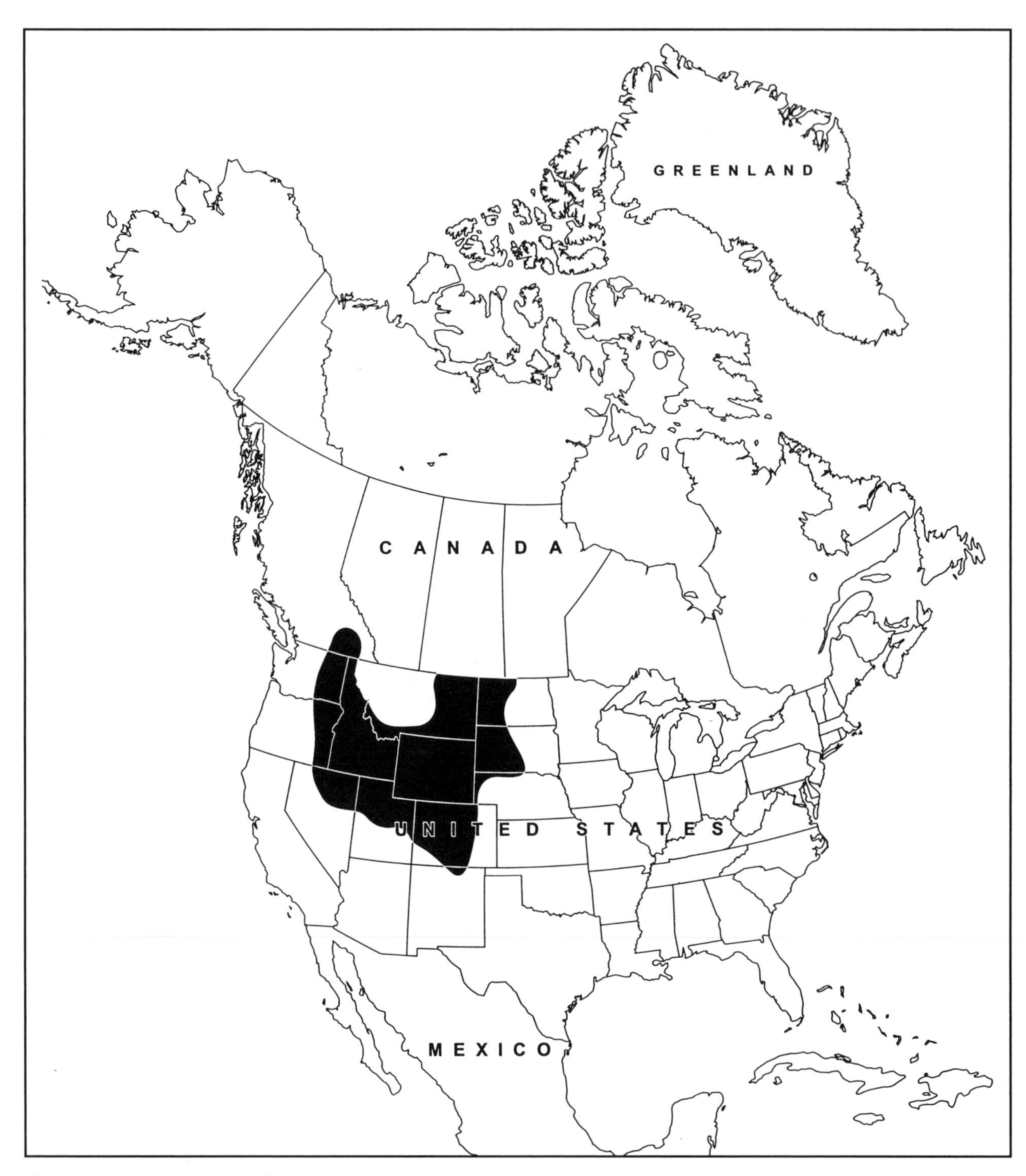


Plate 21 B. Distribution of *Ochlerotatus s. idahoensis*—USA: CO, ID, MT, NE, NV, ND, OR, UT, WA, WY (146), NM (314), SD (280); Canada: BC (146); Tax. 534, 576.

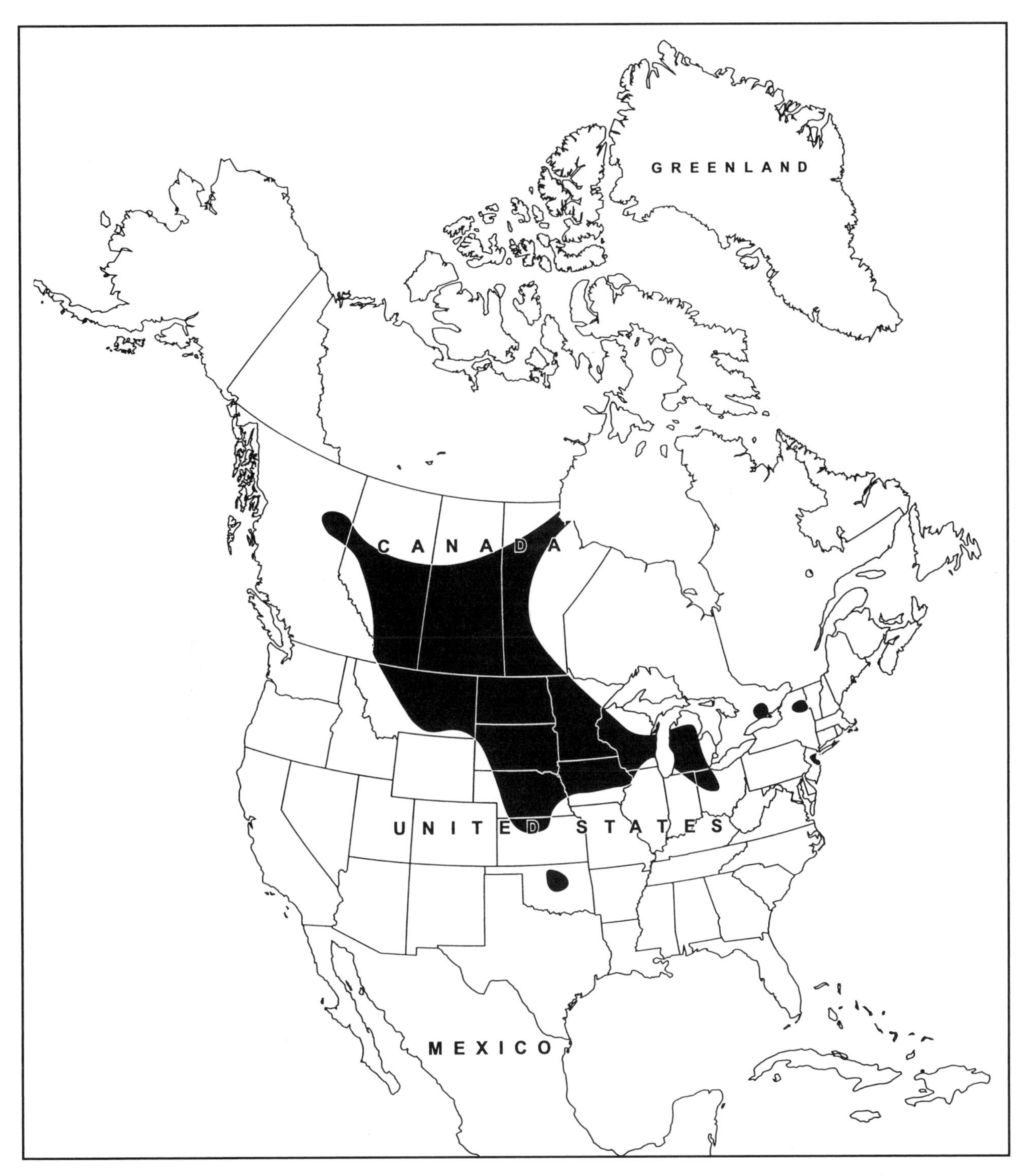


Plate 21 C. Distribution of *Ochlerotatus s. spencerii*—USA: IL, IA, KS, MI, MN, MT, NE, NY, ND, SD, WI, WY (146), OH (135), OK (558); Canada: ALTA, BC, MAN, SASK (146), ONT (784); Tax. 36, 431, 534, 576, 784.

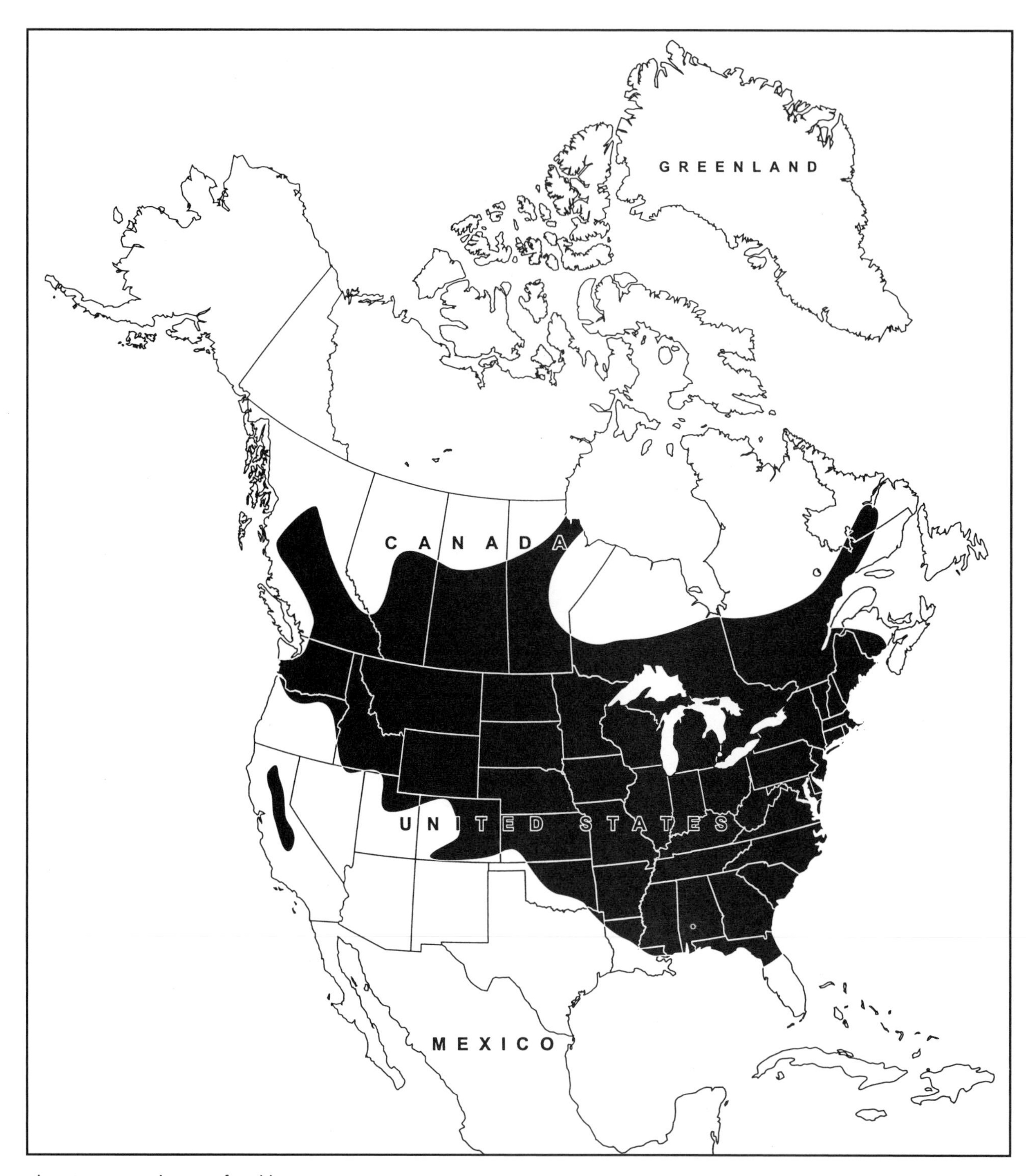


Plate 21 D. Distribution of *Ochlerotatus sticticus*—USA: AL, AR, CA, CO, CT, DE, DC, FL, GA, ID, IL, IN, IA, KS, KY, LA, ME, MD, MA, MI, MN, MS, MO, MT, NE, NH, NJ, NY, NC, ND, OH, OK, OR, PA, SC, SD, TN, TX, UT, VT, VA, WA, WY (146), WV (3), WI (630); Canada: ALTA, BC, MAN, NB, ONT, PQ, SASK (146), LAB (784), NFLD (528); Tax. 34, 303, 431, 784.

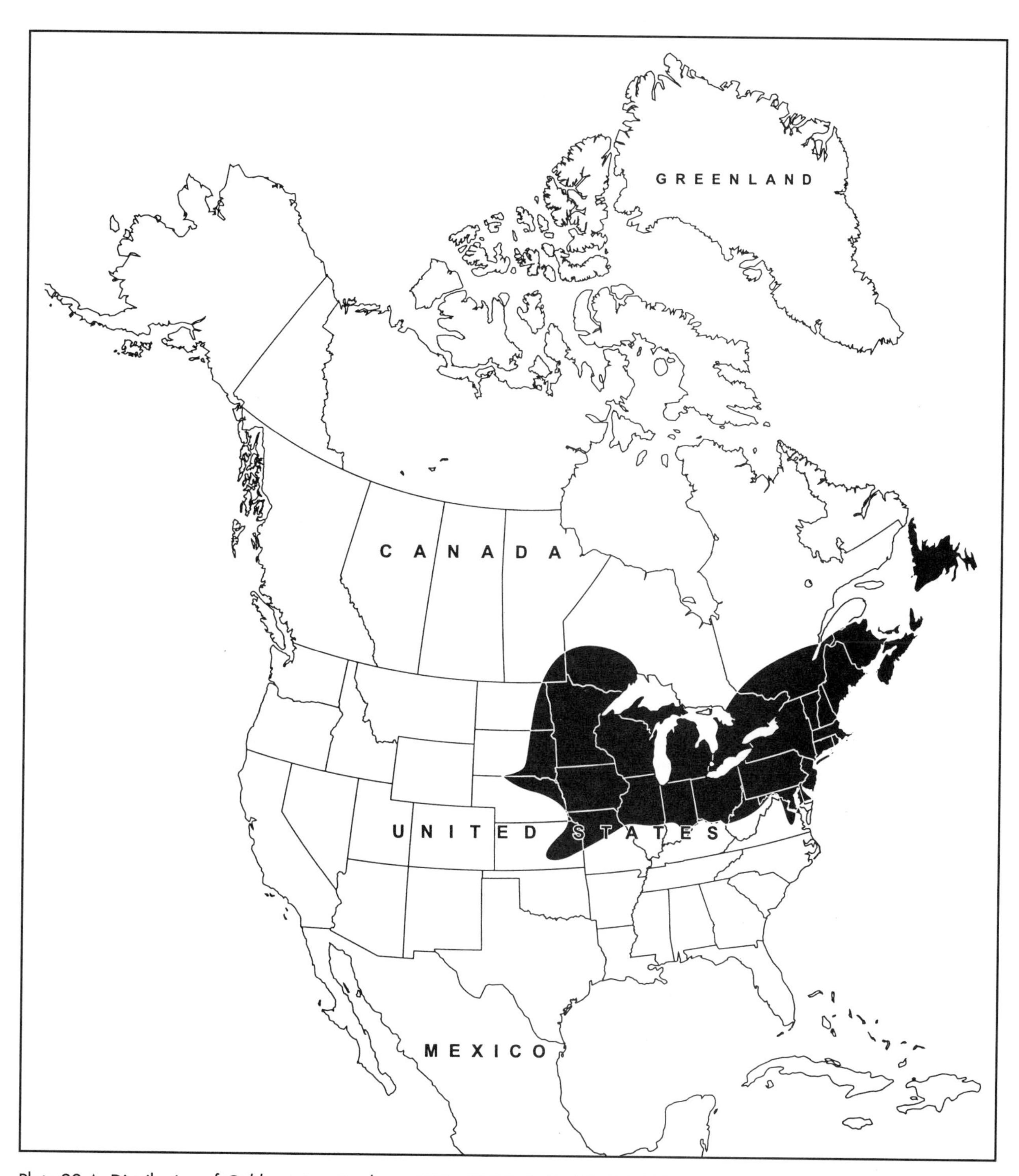


Plate 22 A. Distribution of *Ochlerotatus stimulans*—USA: CT, DE, IL, IA, KS, ME, MA, MI, MN, MS, MO, NE, NH, NJ, NY, OH, PA, RI, SD, VT, WI (146), IN (662), KY (185), MD, VA (67), not in CO (314), not in UT (534); Canada: MAN, NB, NS, ONT, PEI, PQ (146), NFLD (569), not in ALTA, NWT, SASK, YUK (783); Tax. 431, 466, 783, 784.

Plate 22 B. Distribution of *Ochlerotatus taeniorhynchus*—USA: AL, AR, CA, CT, DE, DC, FL, GA, LA, MD, MA, MS, NJ, NY, NC, PA, RI, SC, TX, VA (146), AZ (609), KS (541), NH (114), OK (317, 337), not in NM (778); Map modified after Knight (383); Tax 42, 431.

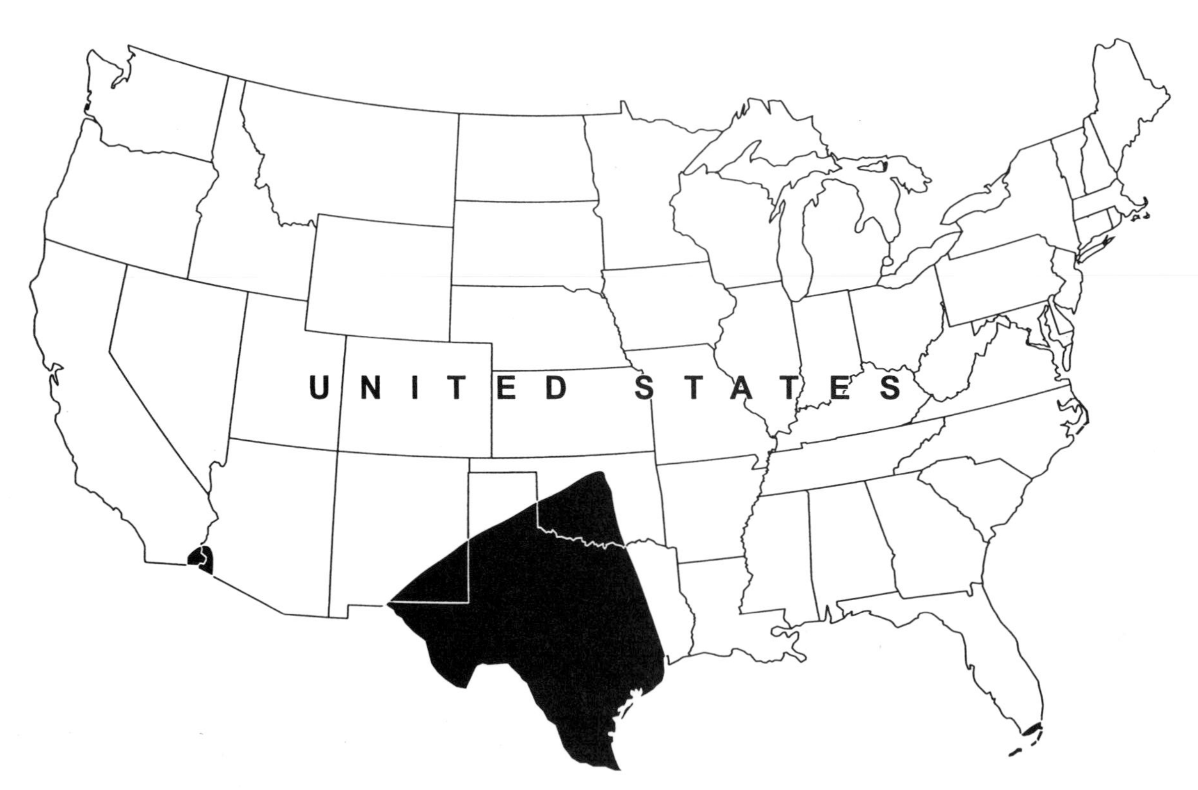


Plate 22 C. Distribution of *Ochlerotatus thelcter*—USA: FL, OK, TX (146), AZ (454), CA (492), NM (495); Map after Arnell (11); Tax. 11, 431.

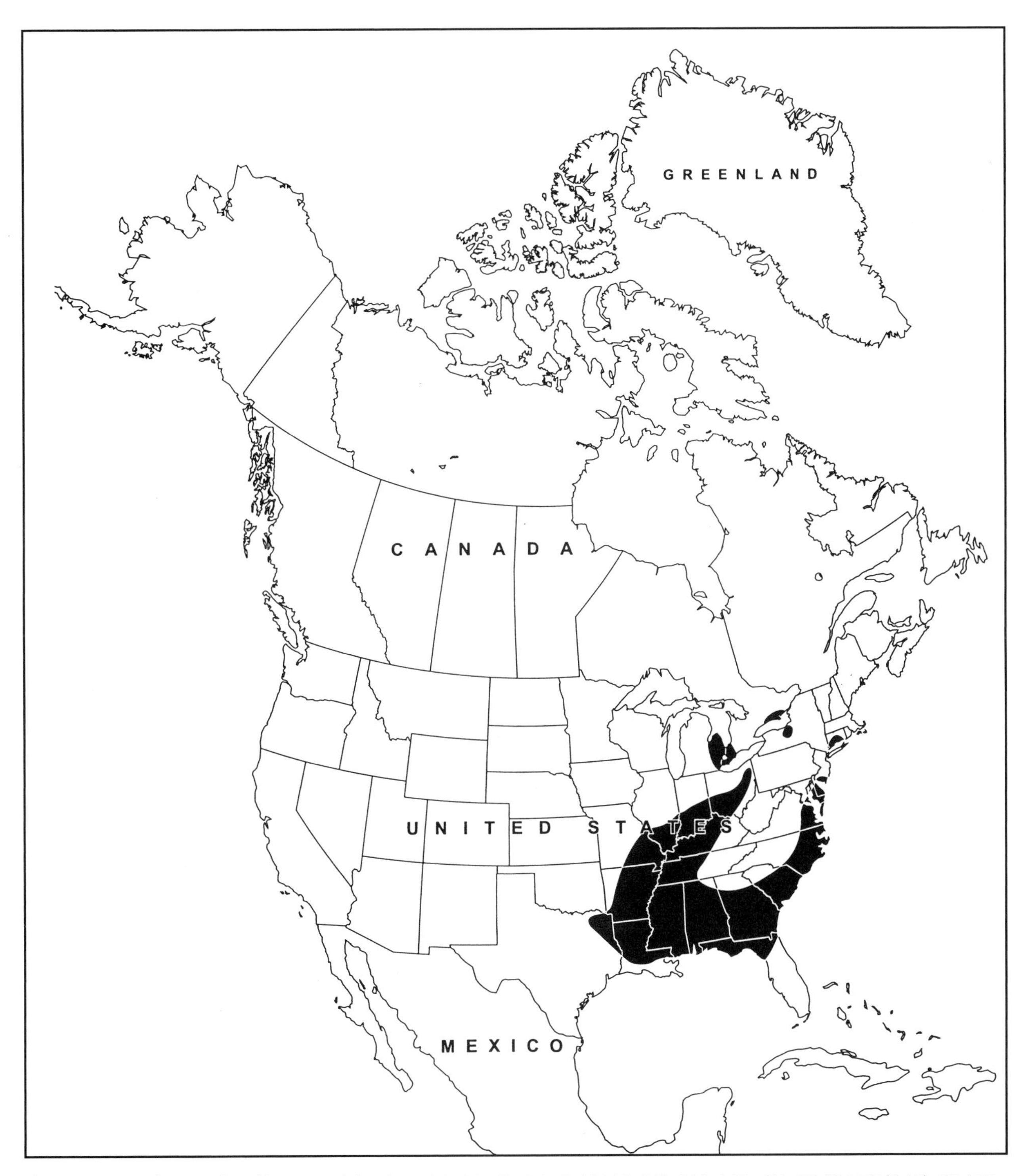


Plate 22 D. Distribution of *Ochlerotatus thibaulti*—USA: AR, FL, GA, IL, KY, LA, MS, MO, NC, OH, SC, TN, TX (146), CT, NY (748), DE (391), IN (652), MD (369), VA (67); Canada: ONT (57); Tax. 431, 784.

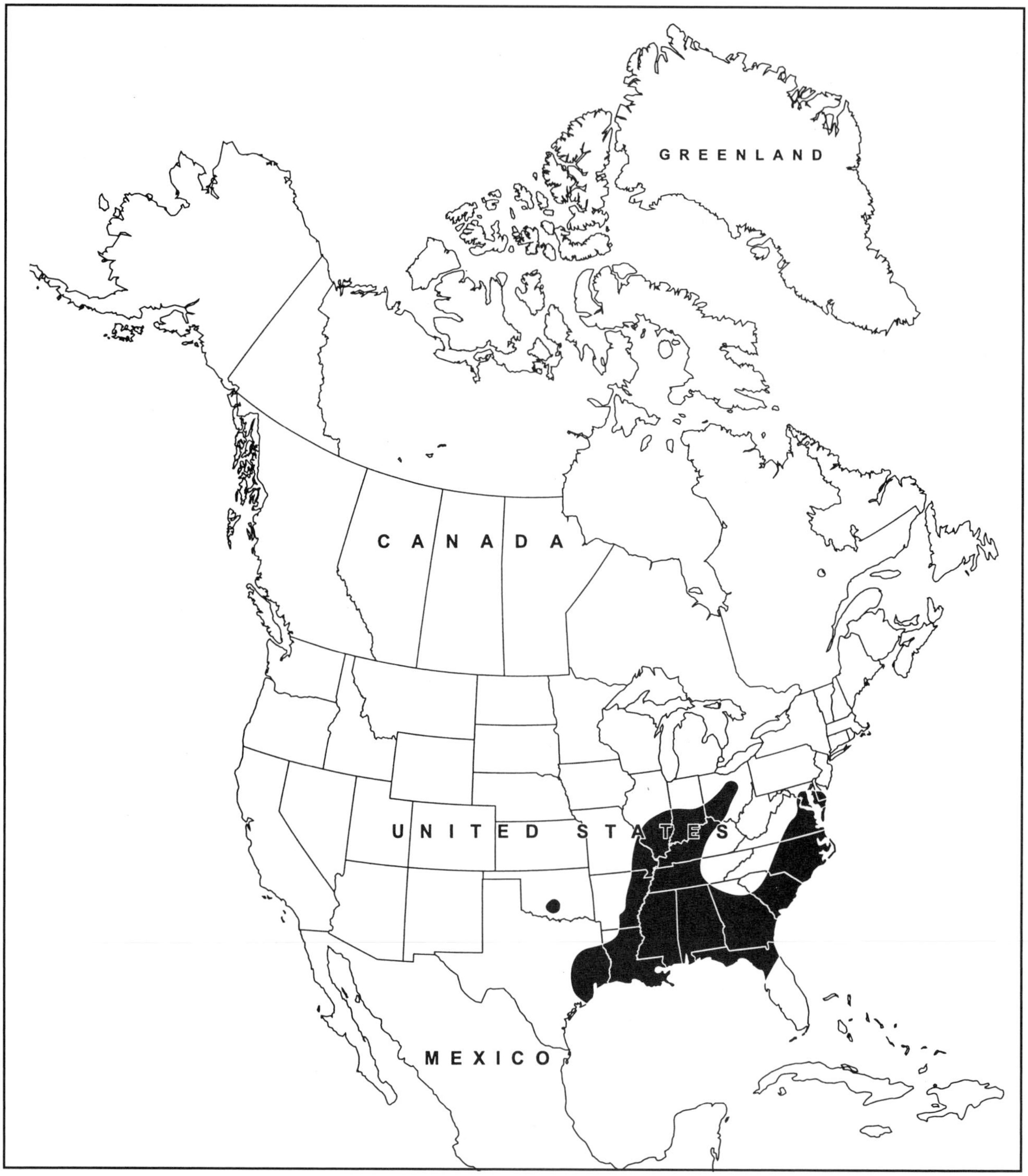

Plate 23 A. Distribution of *Ochlerotatus tormentor*—USA: AL, AR, FL, GA, LA, MS, MO, NC, OH, OK, SC, TX (146), DE (Lake, pers. comm. 1972), IL (620), KY (631), MD (76), TN (93), VA (Harrison, pers. comm. 2001); Tax. 431, 614.

Plate 23 B. Distribution of *Ochlerotatus triseriatus*—USA: AL, AR, CT, DE, DC, FL, GA, IA, IL, IN, KS, KY, LA, ME, MD, MA, MI, MN, MS, MO, NE, NH, NJ, NY, NC, OH, OK, PA, RI, SC, TN, TX, VT, VA, WI (146), WV (3); Canada: ONT, PQ (146), NB (399); Map modified after Zavortink (793); Tax. 98, 296, 313, 431, 784, 793.

Plate 23 C. Distribution of *Ochlerotatus trivittatus*—USA: AR, CO, CT, DE, DC, GA, ID, IL, IN, IA, KS, KY, LA, ME, MD, MA, MN, MO, MT, NE, NJ, NM, NY, NC, ND, OH, OK, PA, RI, SC, SD, TN, TX, VA, WV, WI, WY (146), AL (367), AZ (604), MI (744), NH (78), UT (522); Canada: ONT (146), MAN (728), PQ (416); Map after Arnell (11); Tax. 11, 431, 784.

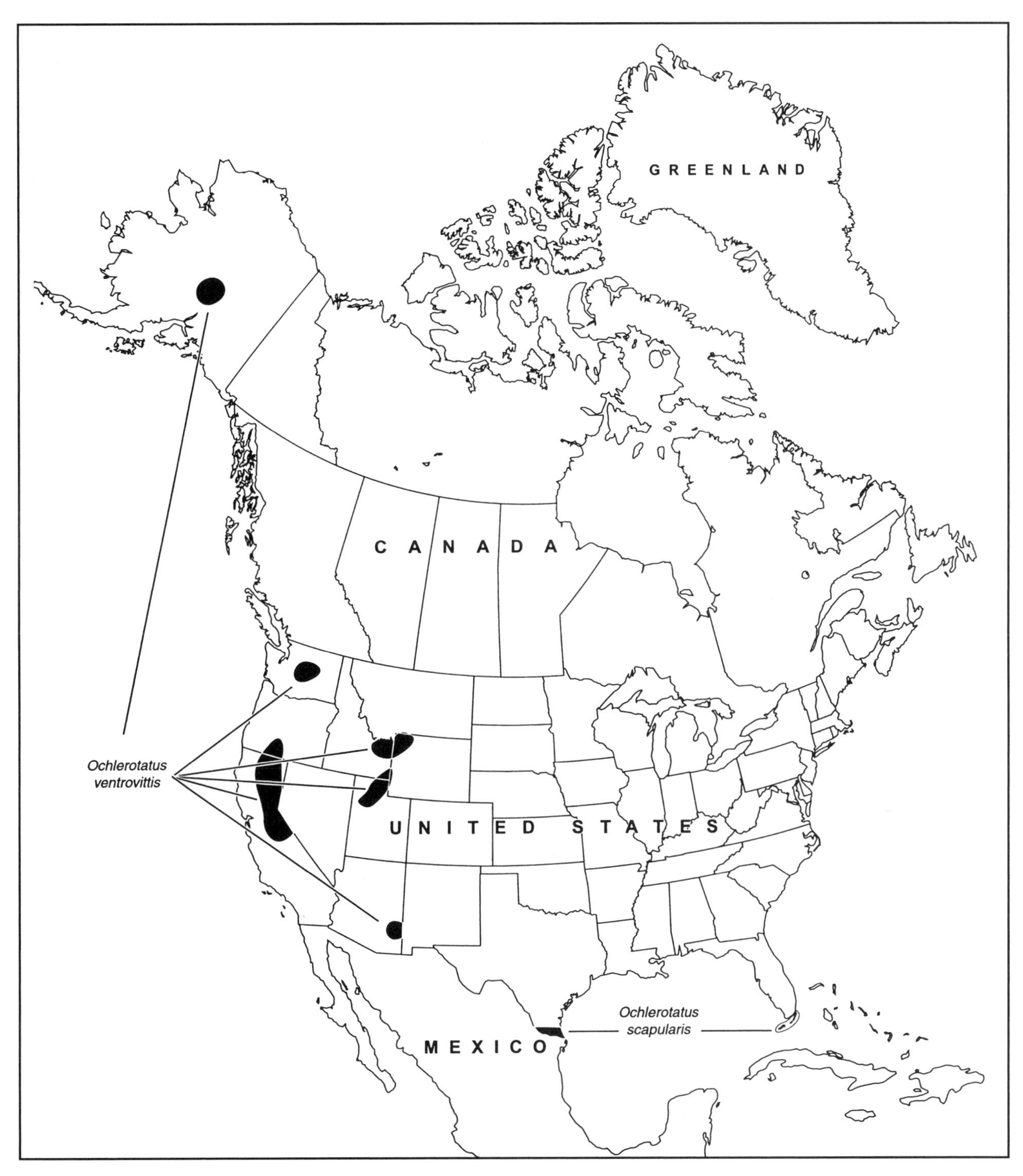


Plate 23 D. Distribution of *Ochlerotatus ventrovittis*—USA: CA, ID, WA (146), AK (69), AZ (468), OR (284), UT (522), WY (523); Tax. 431, 523. *Ochlerotatus scapularis*—USA : TX (146), FL (215), not in LA (135); Tax. 7, 431.

Plate 24 A. Distribution of *Aedes vexans*—USA: AL, AZ, AR, CA, CO, CT, DE, DC, FL, GA, ID, IL, IN, IA, KS, KY, LA, ME, MD, MA, MI, MN, MS, MO, MT, NE, NJ, NM, NY, NC, ND, OH, OK, OR, PA, RI, SC, SD, TN, TX, UT, VT, VA, WA, WY (146), AK (688), NV (604); Canada: ALTA, BC, MAN, NB, NS, ONT, PEI, PQ, SASK, YUK (146); Tax. 431, 784.

Plate 24 B. Distribution of *Ochlerotatus zoosophus*—USA: KS, OK, TX (146), AR (337), LA (364), not in NM (779); Map after Zavortink (793); Tax. 793.

Plate 24 C. Distribution of *Anopheles atropos*—USA: AL, FL, GA, LA, MD, MS, NJ, SC, TX, VA (146); Tax. 42.

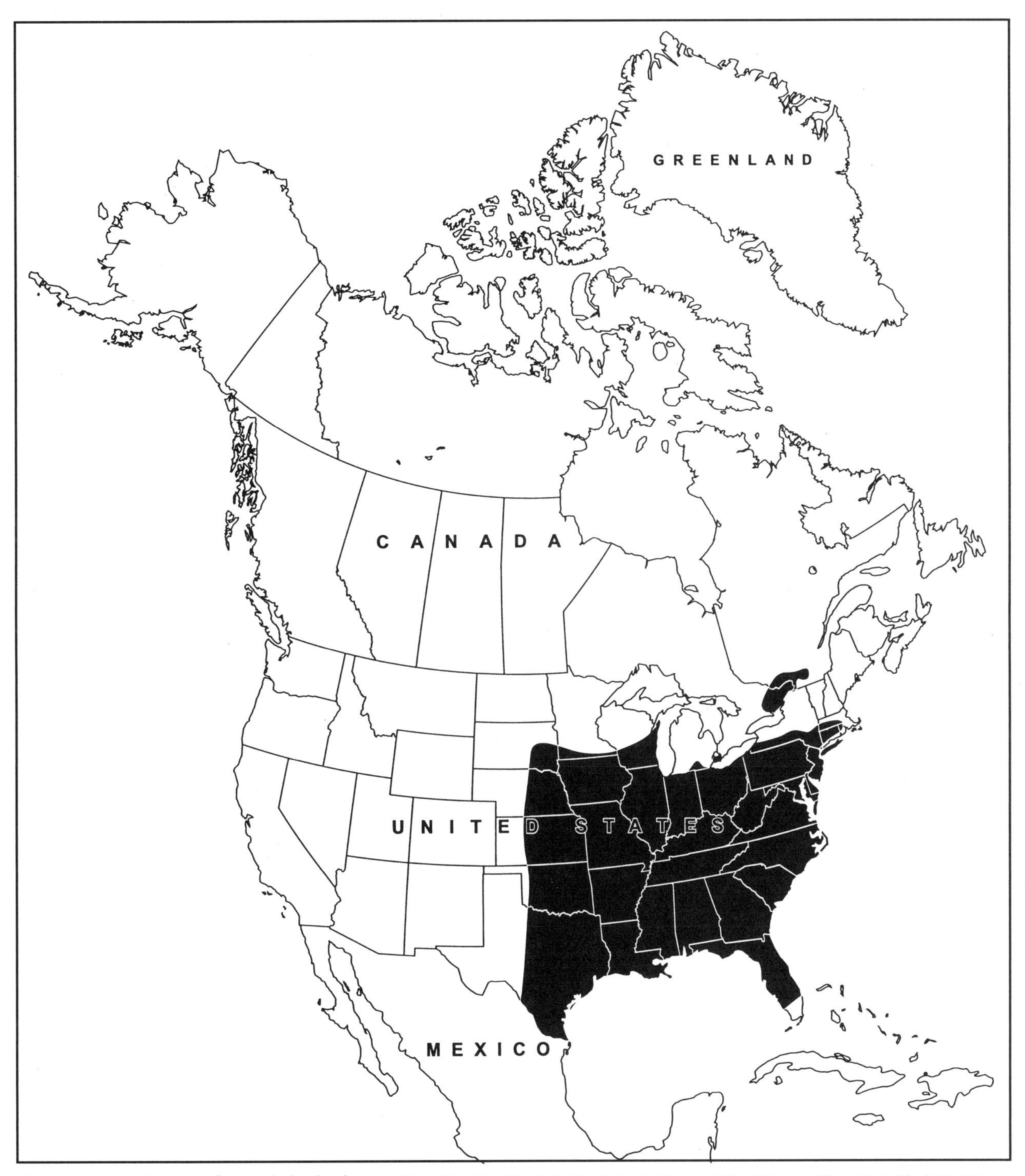


Plate 24 D. Distribution of *Anopheles barberi*—USA: AL, AR, DE, DC, FL, GA, IL, IN, IA, KS, KY, LA, MD, MS, MO, NE, NJ, NY, NC, OH, OK, PA, SC, TN, TX, VA (146), MI (Newson, pers. comm. 1977), MN (579), SD (241), WV (3), WI (573); Canada: ONT (682), PQ (415); Map modified after Zavortink (790); Tax. 784, 788, 790.

Plate 25 A. Distribution of *Anopheles bradleyi*—USA: AL, DE, FL, GA, LA, MD, MS, NJ, NY, NC, SC, TX, VA (146); Map after Floore et al. (258); Tax. 258.

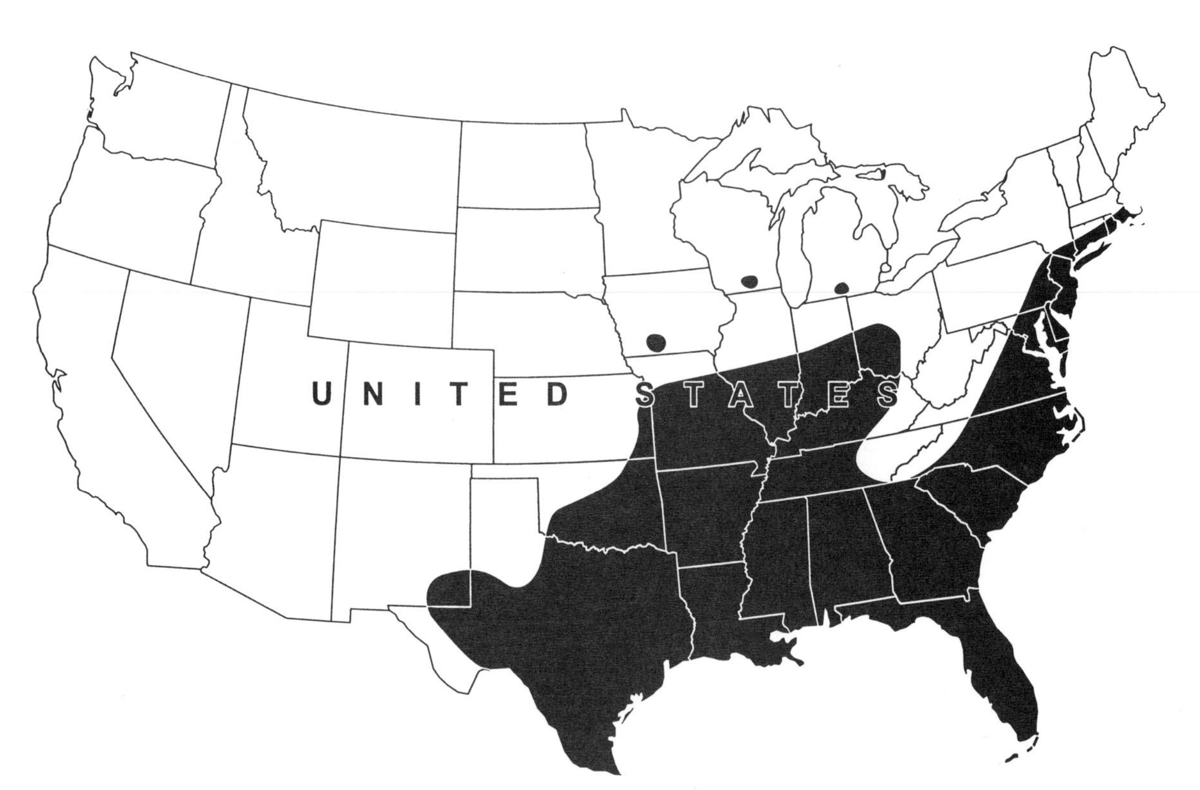


Plate 25 B. Distribution of *Anopheles crucians*—USA: AL, AR, CT, DE, DC, FL, GA, IL, IN, IA, KS, KY, LA, MD, MA, MS, MO, NJ, NM, NY, NC, OH, OK, PA, RI, SC, TN, TX, VA (146), MI (Newson in litt, 1977), WI (Dicke, pers. comm. 1979); Map after Floore et al. (258); Tax 42, 258.

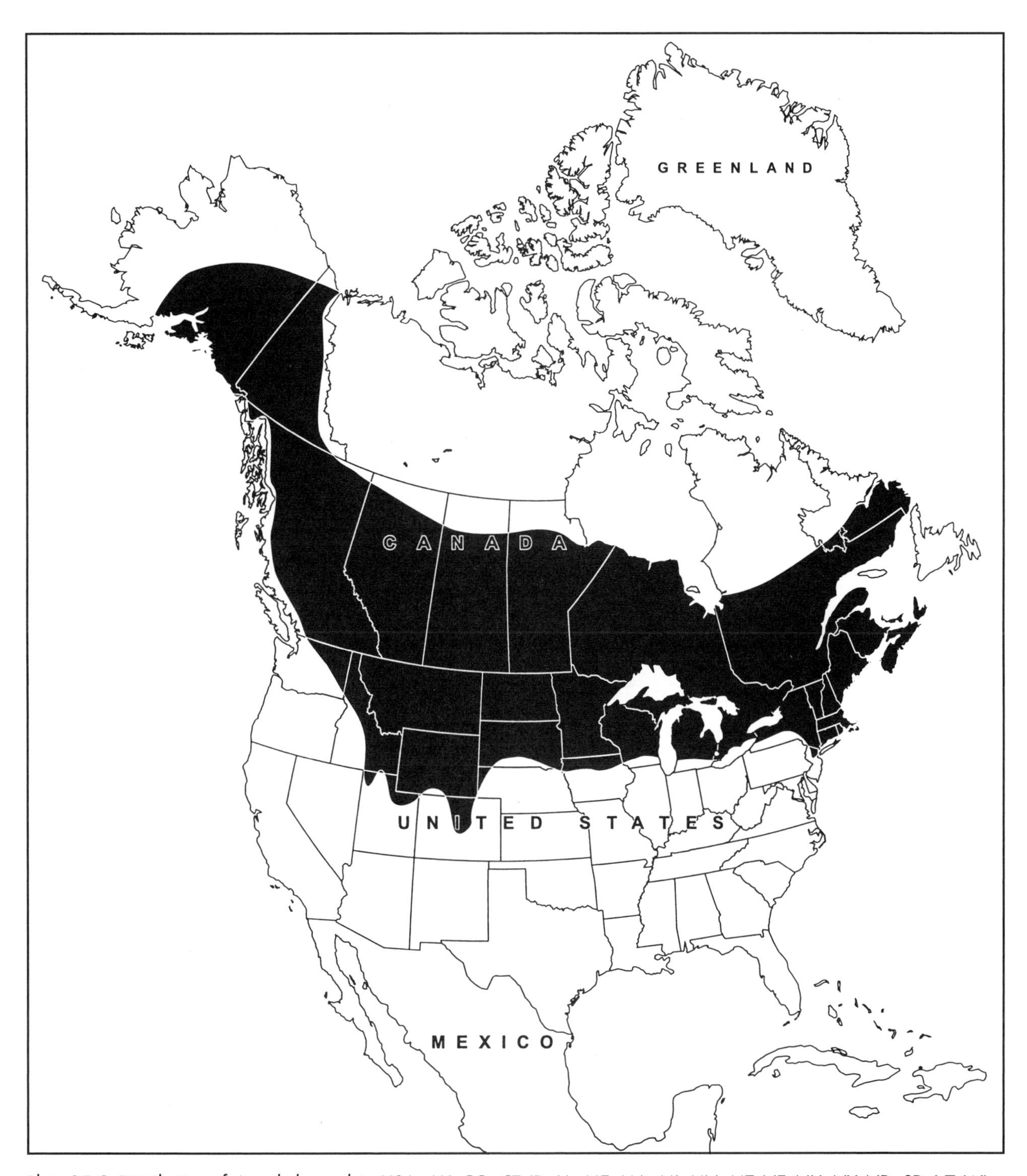


Plate 25 C. Distribution of *Anopheles earlei*—USA: AK, CO, CT, ID, IA, ME, MA, MI, MN, MT, NE, NH, NY, ND, SD, VT, WI, WY (146), KS (541), NV (151), NJ (190), UT (535), WA (284); Canada: ALTA, BC, LAB, MAN, NB, NS, ONT, PQ, SASK (146), NWT, PEI, YUK (784); Tax. 761, 784.

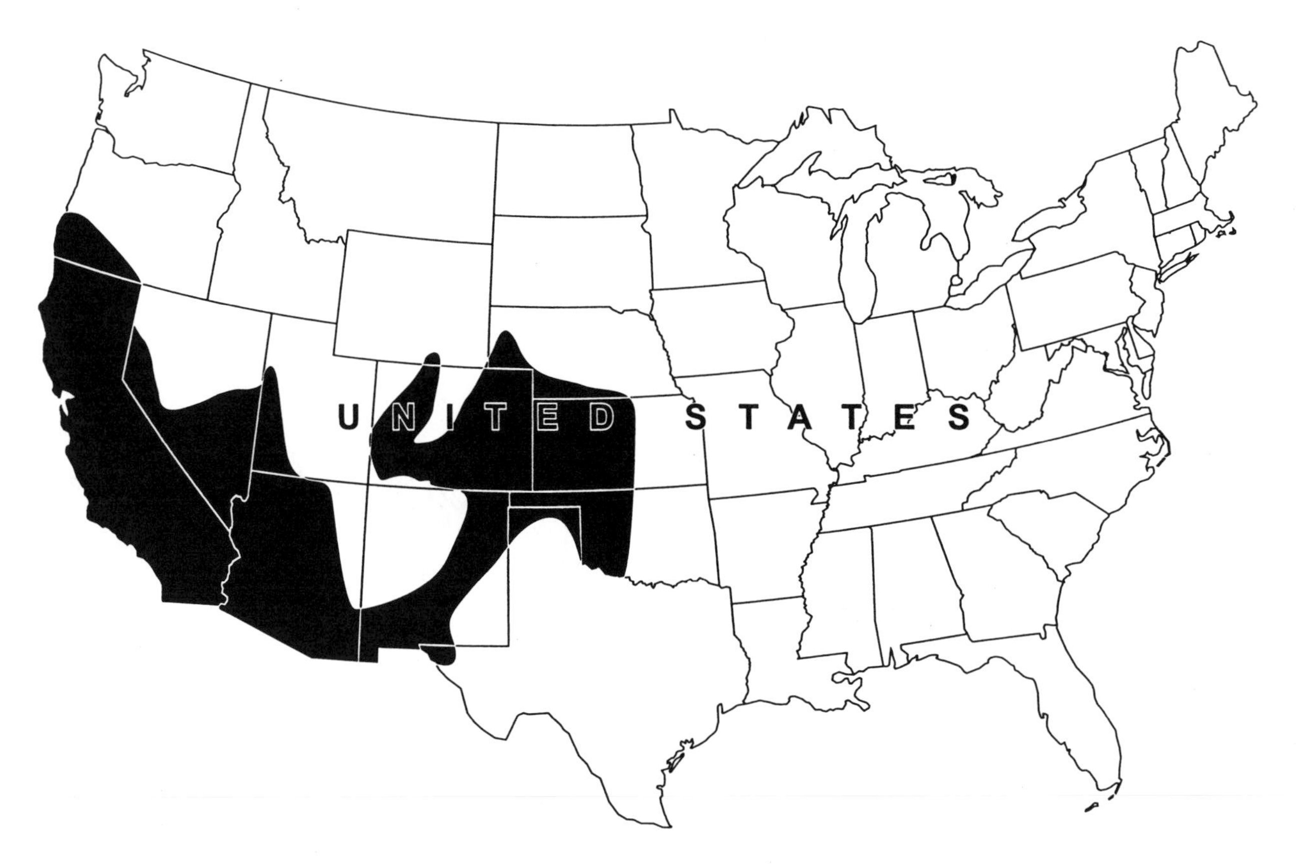


Plate 25 D. Distribution of *Anopheles franciscanus*—USA: AZ, CO, KS, NV, NM, OK, OR, TX, UT, WY (146), NE (587); Tax. 683.

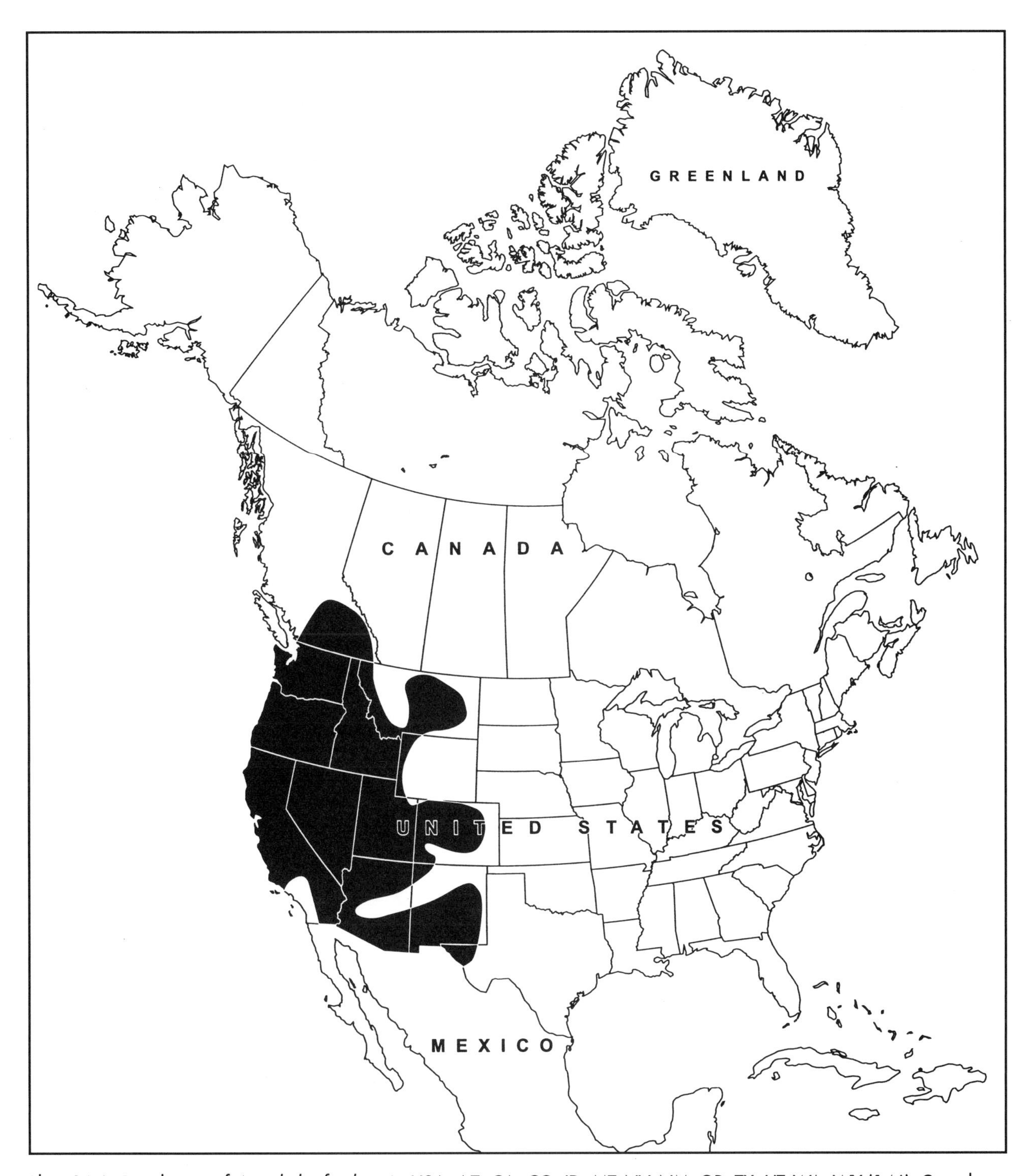


Plate 26 A. Distribution of *Anopheles freeborni*—USA: AZ, CA, CO, ID, MT, NV, NM, OR, TX, UT, WA, WY (146); Canada: BC (146); Tax. 761, 784.

Plate 26 B. Distribution of *Anopheles georgianus*—USA: AL, FL, GA, LA, MS, NC, SC (146); Map after Floore et al. (258); Tax. 258.

Plate 26 C. Distribution of *Anopheles maverlius*—USA: FL, GA, KY, LA, MS, SC (599); Tax. 599.

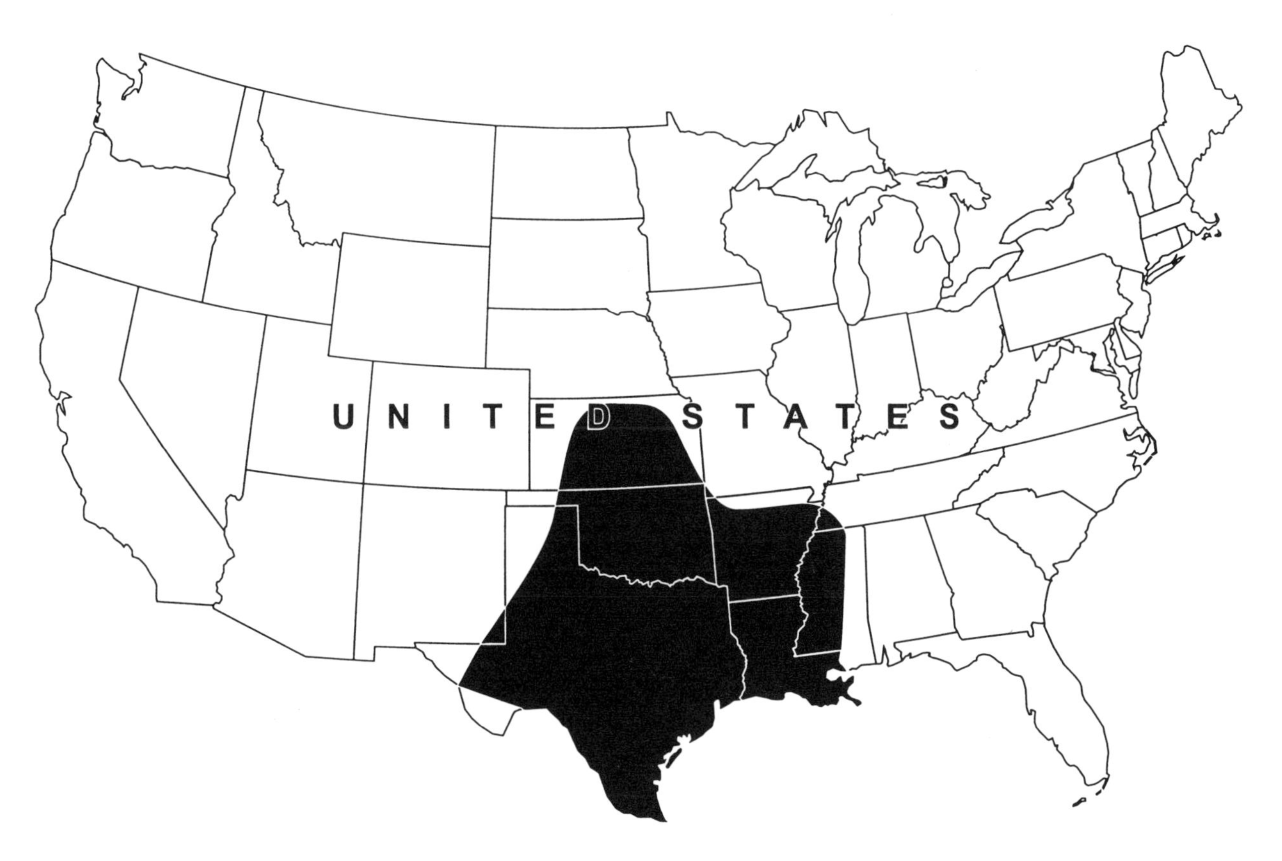


Plate 26 D. Distribution of *Anopheles pseudopunctipennis*—USA: AR, KS, LA, MS, MO, NM, OK, TN, TX (146), not in CO (314); Tax. 683.

Plate 27 A. Distribution of *Anopheles punctipennis*—USA: AL, AR, CA, CO, CT, DE, DC, FL, GA, ID, IL, IN, IA, KS, KY, LA, ME, MD, MA, MI, MN, MS, MO, MT, NE, NH, NJ, NM, NY, NC, ND, OH, OK, OR, PA, RI, SC, SD, TN, TX, VT, VA, WA, WV, WI, WY (146); Canada: BC, MAN, NS, ONT, PQ, (146), NB (731); Tax. 47, 49, 784.

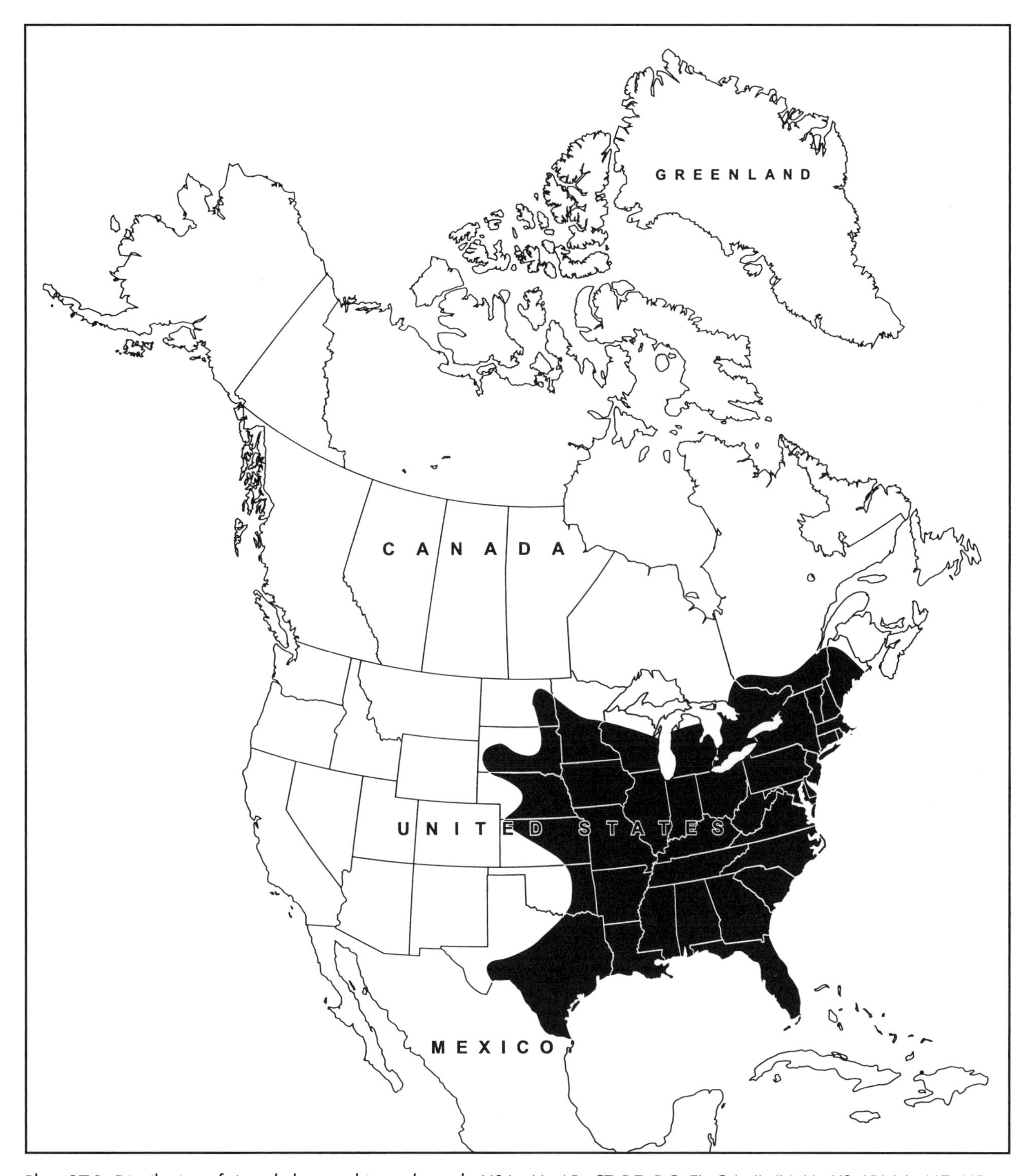


Plate 27 B. Distribution of *Anopheles quadrimaculatus sl*—USA: AL, AR, CT, DE, DC, FL, GA, IL, IN, IA, KS, KY, LA, ME, MD, MA, MI, MN, MS, MO, NE, NH, NJ, NY, NC, ND, OH, OK, PA, RI, SC, SD, TN, TX, VT, VA, WI (146), WV (3); Canada: ONT, PQ (146); Tax. 47, 599.

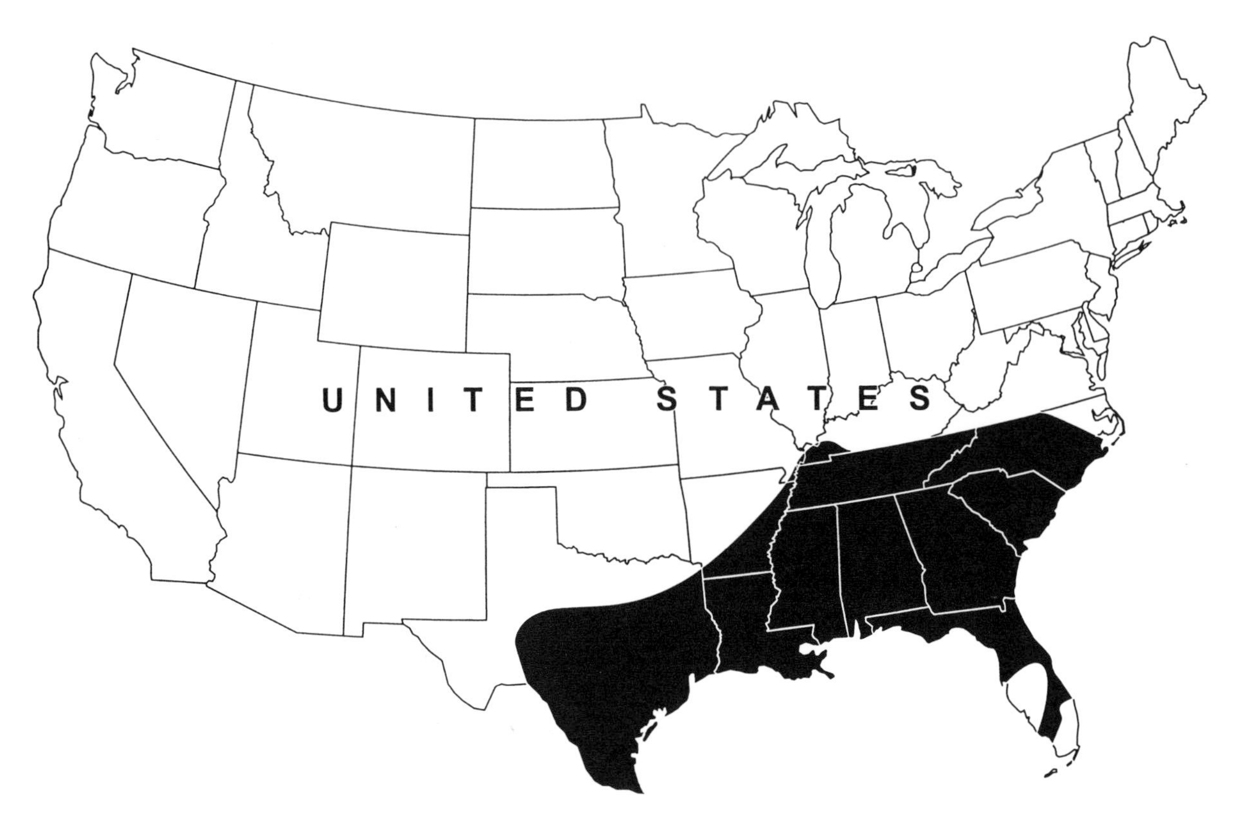


Plate 27 C. Distribution of *Anopheles quadrimaculatus ss*—USA: AL, AR, CT, FL, GA, KY, LA, MA, MI, MN, MS, NC, NJ, NY, S C, TN, TX, WI (599); Tax. 599.

Plate 27 D. Distribution of *Anopheles smaragdinus*—USA: AL, AR, FL, GA, KY, LA (599); Tax. 599.

Plate 28 A. Distribution of *Anopheles walkeri*—USA: AL, AR, CT, DE, DC, FL, GA, IL, IN, IA, KS, KY, LA, ME, MD, MA, MI, MN, MO, NE, NH, NJ, NY, NC, ND, OH, PA, RI, SC, SD, TN, TX, VT, VA, WI (146); Canada: MAN, NB, NS, ONT, PQ, (146), SASK (474), not in BC (688); Tax. 36, 784.

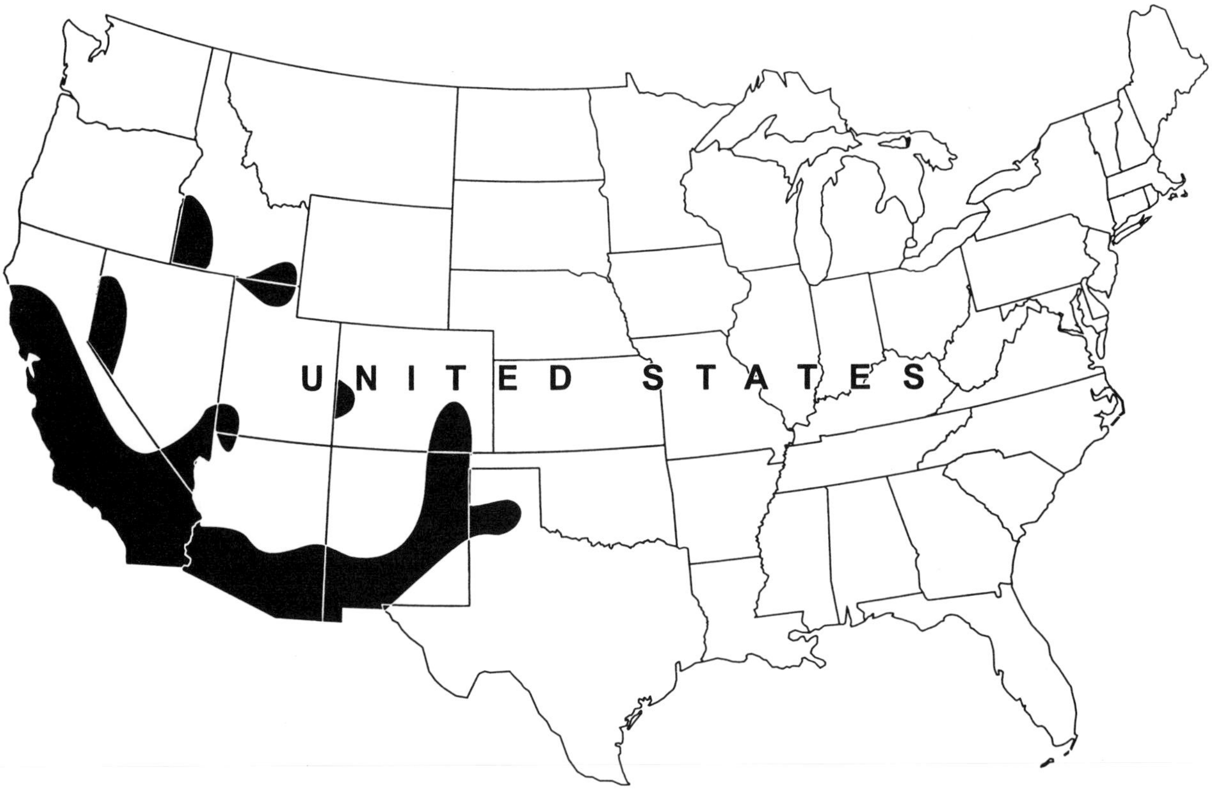

Plate 28 B. Distribution of *Culex erythrothorax*—USA: CA, ID, UT, (146), AZ (604), CO (314), NV (157), NM (712), TX (486); Tax. 87.

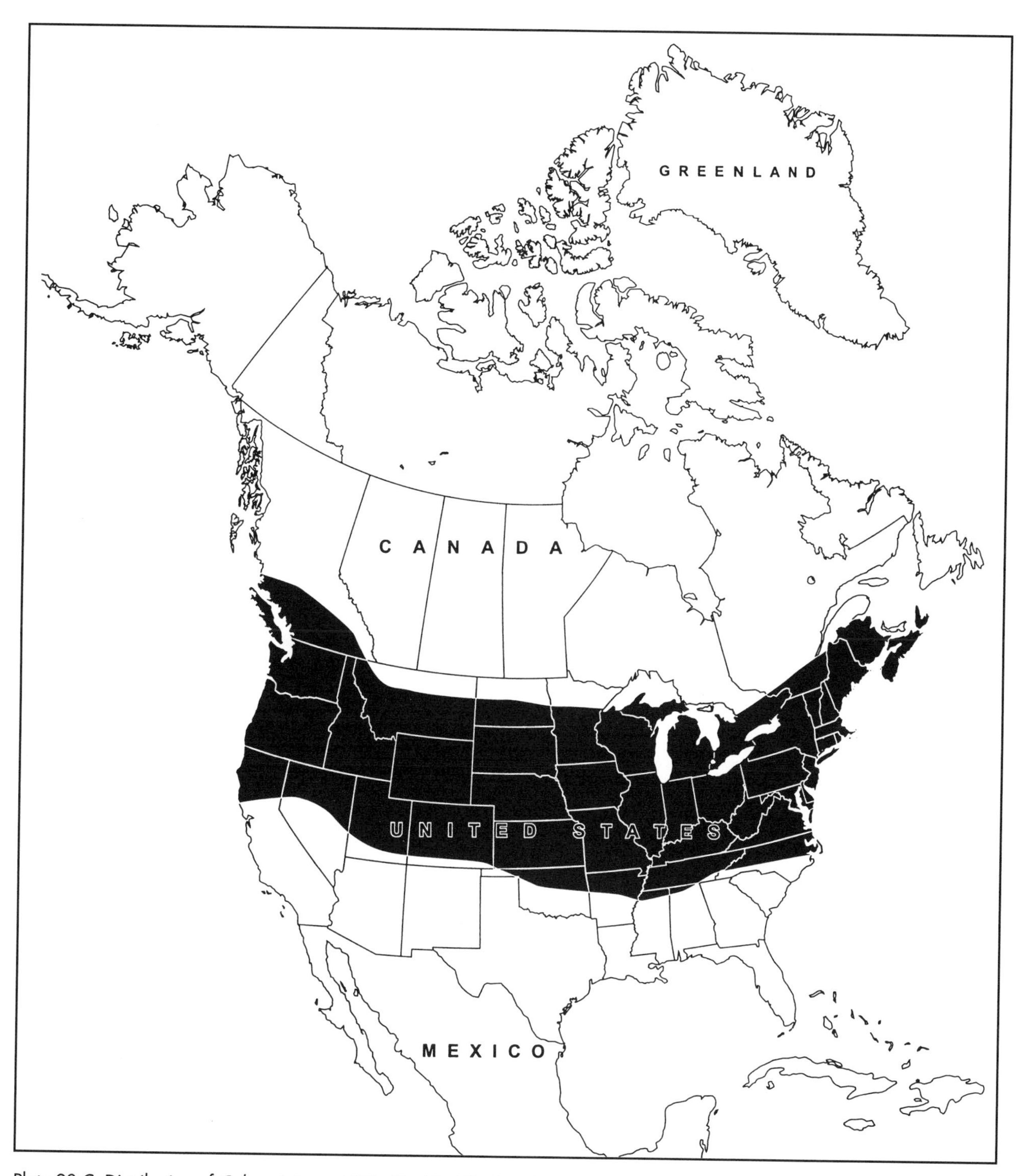


Plate 28 C. Distribution of *Culex pipiens*—USA: AL ,AR, CA, CO, CT, DE, DC, GA, ID, IL, IN, IA, KS, KY, ME, MD, MA, MI, MN, MS, MO, MT, NE, NH, NJ, NY, NC, ND, OH, OK, OR, PA, RI, SC, SD, TN, UT, VT, VA,WA, WI, WY (146), WV (3), not in NM (779); Canada: BC, NB, NS, ONT, PQ (146); Tax. 18, 87, 461, 784.

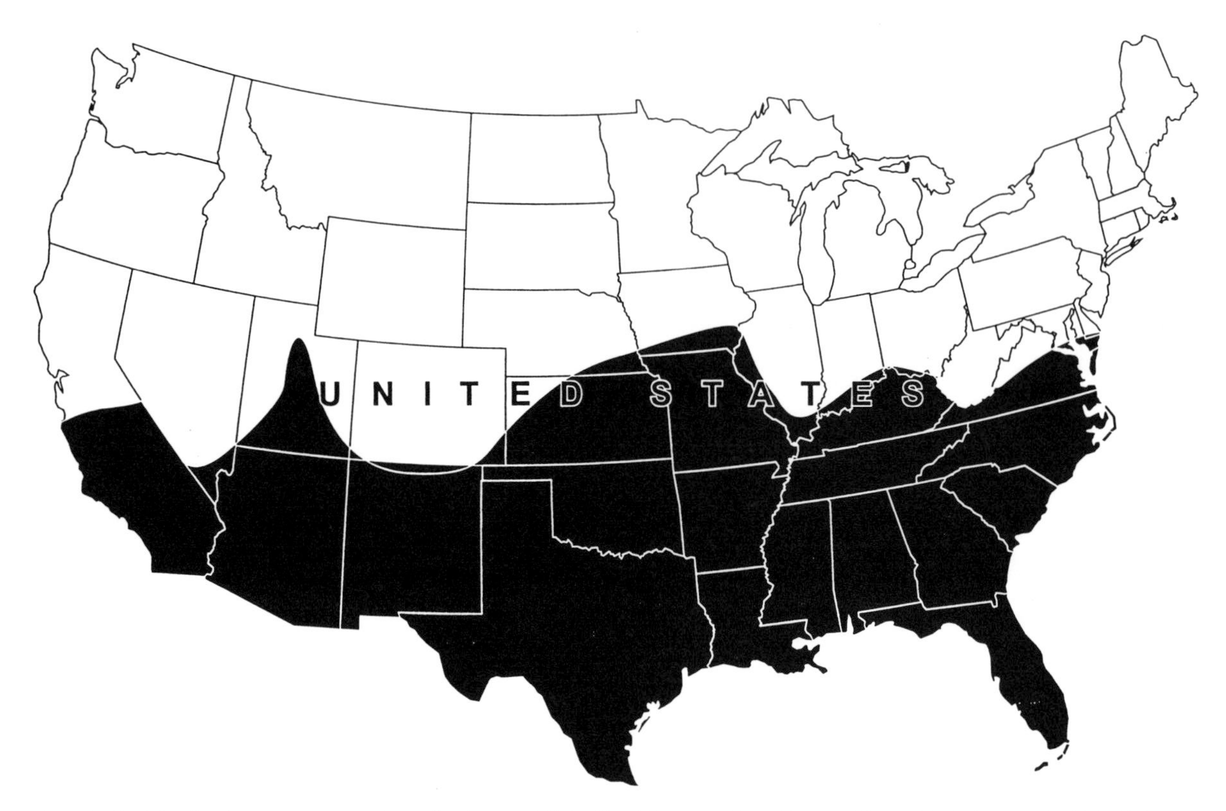


Plate 28 D. Distribution of *Culex quinquefasciatus*—USA: AL, AZ, AR, CA, DC, FL, GA, IL, IA, KS, KY, LA, MS, MO, NE, NM, NC, OH, OK, SC, TN, TX, UT, VA (146), IN (517), MD (18), NV (160), WV (3); Tax. 18, 39, 461, 647, 648.

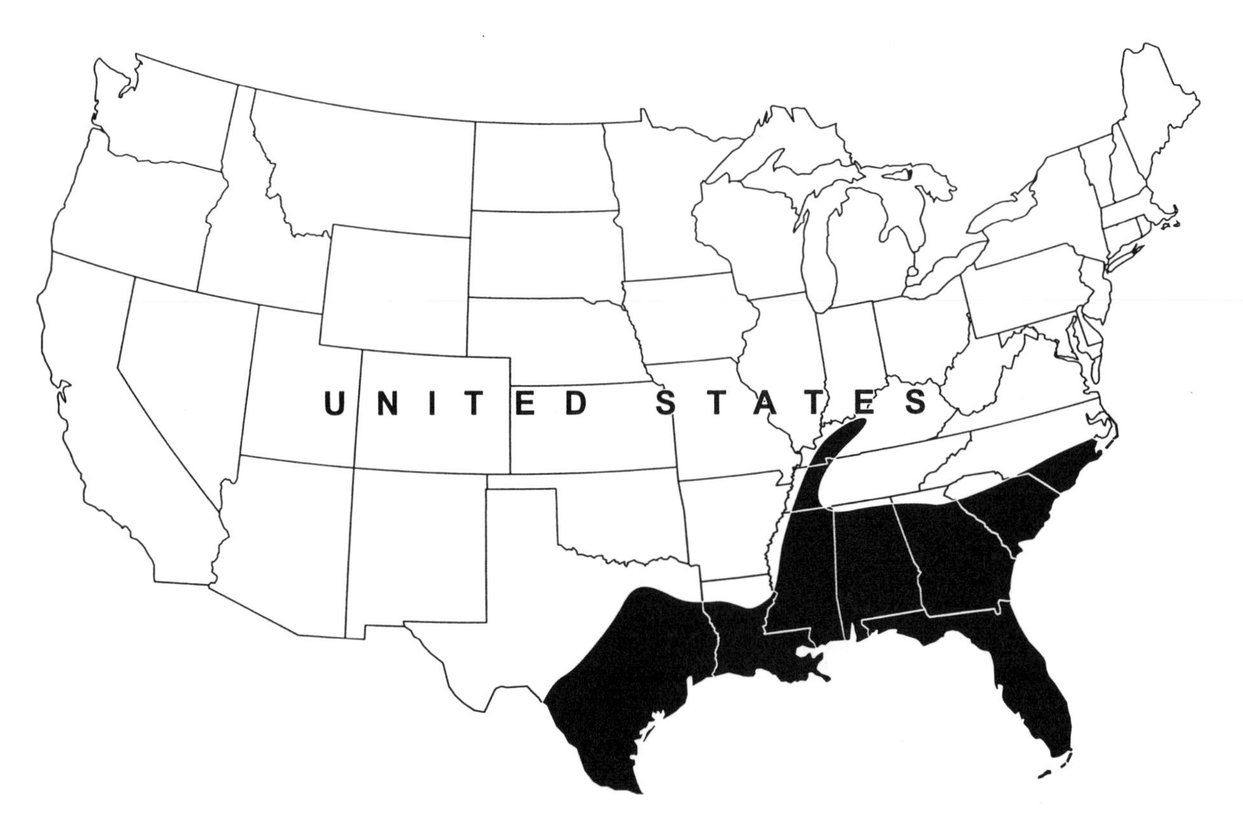


Plate 29 A. Distribution of *Culex nigripalpus*—USA: AL, FL, GA, LA, MS, NC, SC, TN, TX (146), AZ (468), KY (184), OK (337), not in AR, NM (135); Tax. 42, 87.

Plate 29 B. Distribution of *Culex restuans*—USA: AL, AR, CA, CO, CT, DE, DC, FL, GA, ID, IL, IN, IA, KS, KY, LA, ME, MD, MA, MI, MN, MS, MO, MT, NE, NH, NJ, NM, NY, NC, ND, OH, OK, PA, RI, SC, SD, TN, TX, UT, VT, VA, WV, WI, WY (146), AZ (604), OR (315); Canada: MAN, NB, ONT, PQ, SASK (146), ALTA (581), NS (399); Tax. 87, 784.

Plate 29 C. Distribution of *Culex salinarius*—USA: AL, AR, CO, CT, DE, DC, FL, GA, ID, IL, IN, IA, KS, KY, LA, ME, MD, MA, MI, MN, MS, MO, NE, NH, NJ, NM, NY, NC, ND, OH, OK, PA, RI, SC, SD, TN, TX, VT,VA, WI, WY, (146), WV (3), not in UT (522); Canada: not in NS (784); Tax. 87.

Plate 29 D. Distribution of *Culex stigmatosoma*—USA: CA, OK, OR, TX, WA (146), AZ, NV (604), NM (712); Tax. 42, 87, 247.

Plate 30 A. Distribution of *Culex tarsalis*—USA: AL, AZ ,AR, CA, CO, FL, GA, ID, IL, IN, IA, KS, KY, LA, MI, MN, MS, MO, MT, NE, NV, NM, ND, OK, OR, SC, SD, TN, TX, UT, WA, WI, WY (146), NJ (398), OH (557), PA (99); Canada: ALTA, BC, MAN, NWT, SASK (146), ONT (333); Tax. 82, 87, 784.

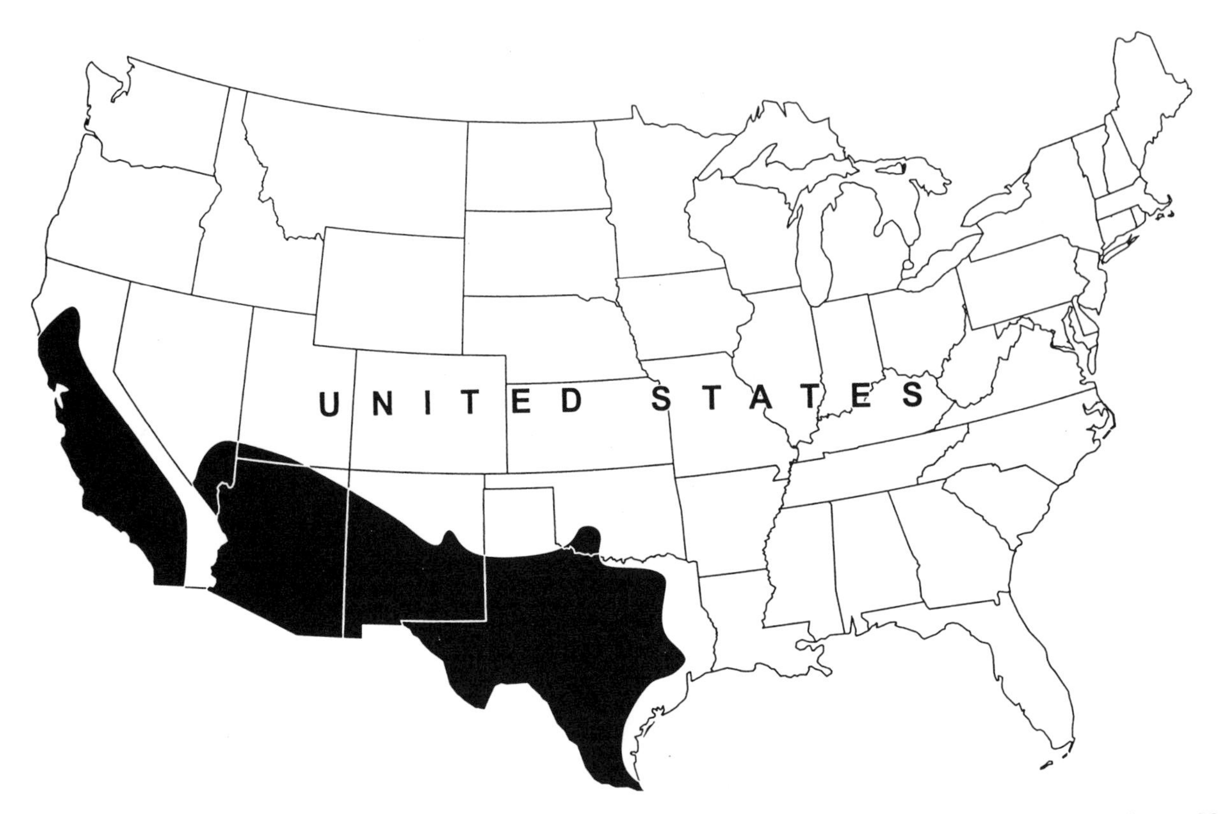


Plate 30 B. Distribution of *Culex thriambus*—USA: CA, OK, TX (146), AZ (609), NV (604), NM (337), UT (530); Tax. 82.

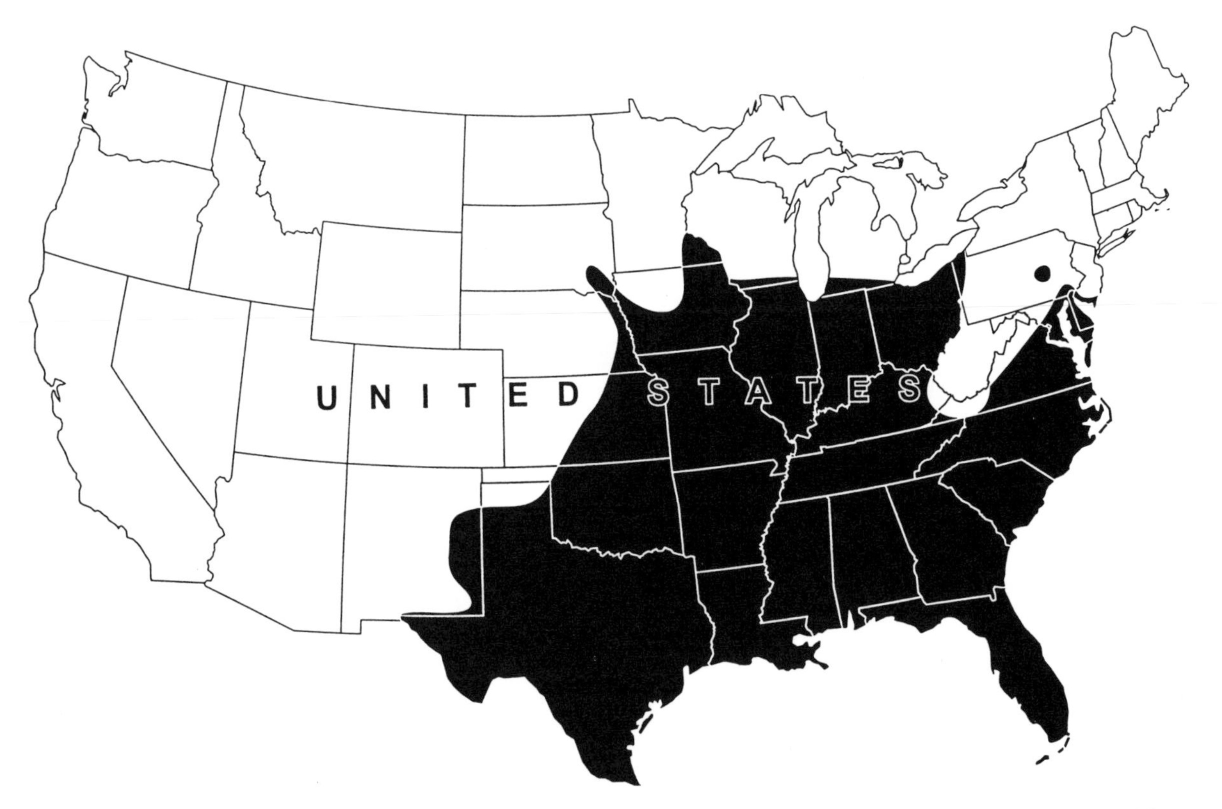


Plate 30 C. Distribution of *Culex erraticus*—USA: AL, AR, DE, DC, FL, GA, IL, IN, IA, KS, KY, LA, MD, MI, MS, MO, NE, NC, OH, OK, SC, SD, TN, TX, VA (146), AZ (329), CA (420), MN (20), NJ (191), NM (779), PA (629), WV (3), WI (Dicke, pers. comm. 1979); Tax. 42, 259, 382.

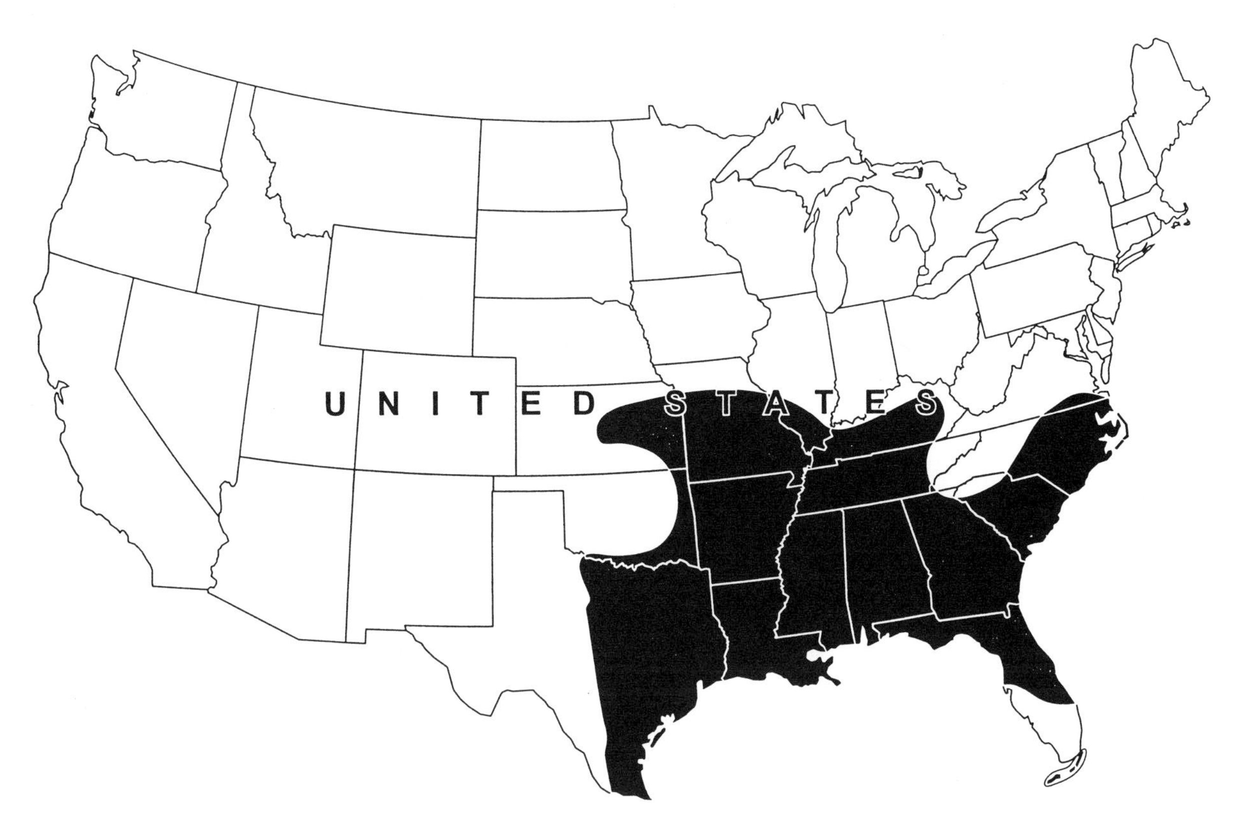


Plate 30 D. Distribution of *Culex peccator*—USA: AL, AR, FL, GA, IL, KS, KY, LA, MI, MS, MO, NC, OK, SC, TN, TX, VA (146), not in DE (Lake, pers. comm. 1972); Tax. 259, 382.

Plate 31 A. Distribution of *Culex pilosus*—USA: AL, FL, GA, KY, LA, MS, NC, SC (146), TX (775); Tax. 42.

Plate 31 B. Distribution of *Culex apicalis*—USA: AZ, CA (146), NV (157), NM (256), OK (558), OR (284), TX (95), UT (535); Tax. 405, 406.

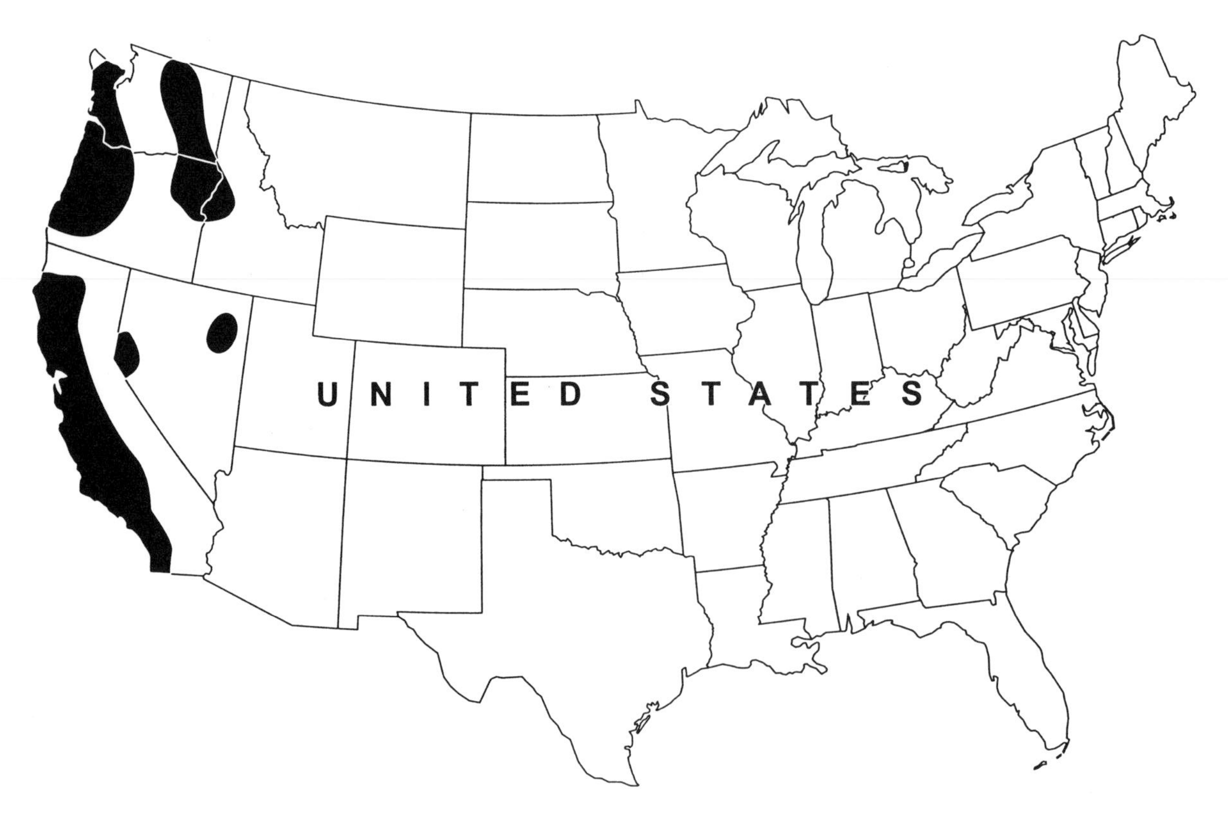


Plate 31 C. Distribution of *Culex boharti*—USA: CA (146), ID, OR, WA (406), NV (604); Tax. 406.

Plate 31 D. Distribution of *Culex territans*—USA: AK, CA, FL, GA, ID, IA, LA, MD, MA, MI, MN, MS, MO, MT, NY, NC, OH, OK, OR, RI, TX, VT, VA, WA (146), AL, SC (380), AZ (604), AR (337), CO (16), CT (746), DE (214), IL (619), IN (658), KS (336), KY (184), ME (465), NE (244), NV (530), NH (78), NJ (109), PA (773), SD (280), TN (685), UT (535), WV (3), WI (222), WY (553); Canada: BC (146), ALTA (580), LAB, NWT (263), MAN (476), NB (489), NS (731), ONT (373), PQ (235), SASK (601), YUK (784); Tax. 36, 47, 303, 406, 784.

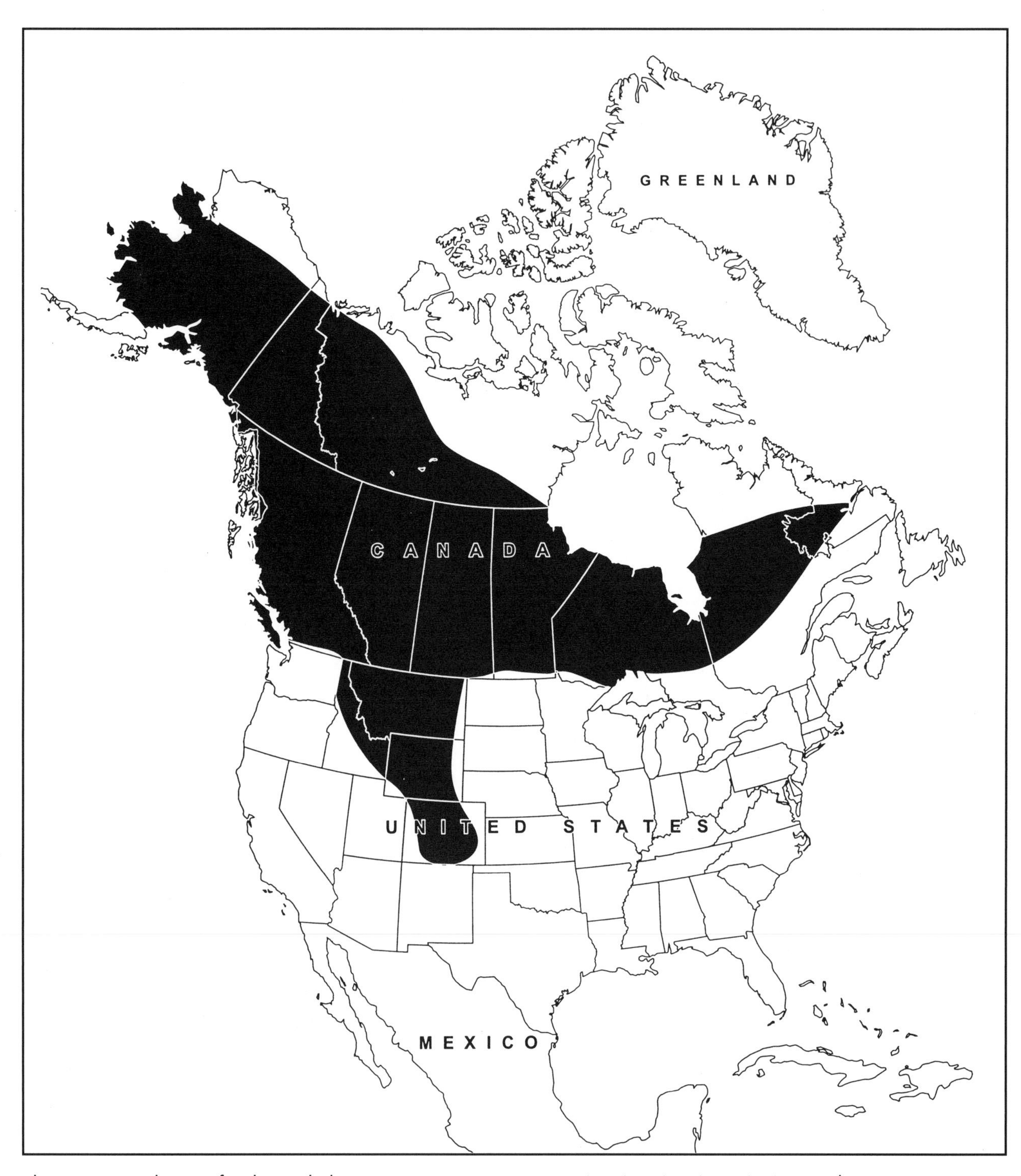


Plate 32 A. Distribution of *Culiseta alaskaensis*—USA: AK, CO, MT, WY (146), ID (219), NV (71); Canada: ALTA, BC, LAB, MAN, NWT, PQ, YUK (146), SASK (601); Map modified after Hopla (341); Tax. 457, 784.

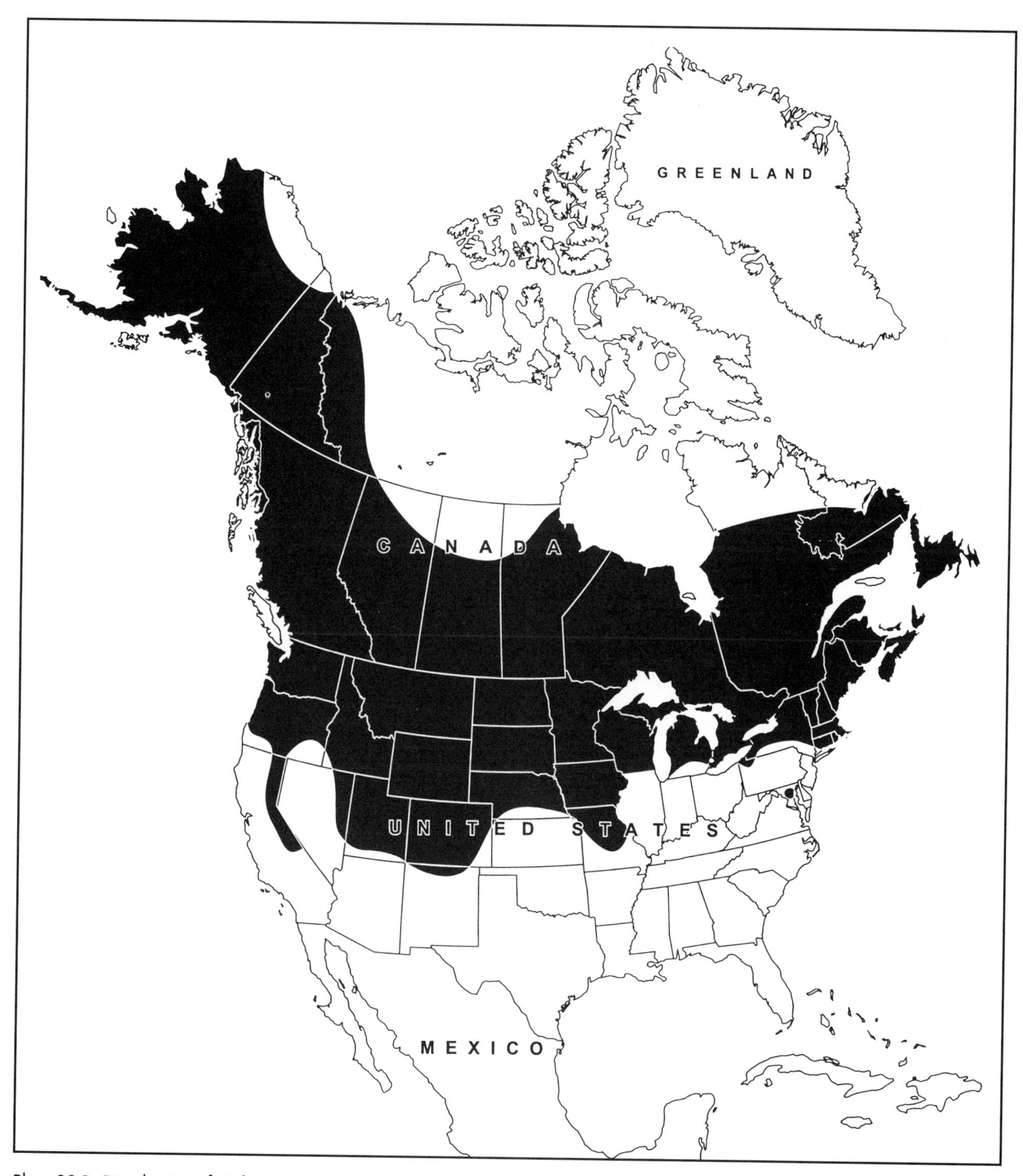


Plate 32 B. Distribution of *Culiseta impatiens*—USA: AK, CA, CO, ID, IA, ME, MA, MI, MO, MT, NE, NH, NY, OR, UT, VT, WA, WI, WY (146), CT (747), NV (604), NM (780), PA (Wills, pers. comm. 1979), SD (280); Canada: ALTA, BC, LAB, MAN, NB, NWT, ONT, PQ, YUK (146), NFLD (569), SASK (601), not in NS (784); Map modified after Hopla (341); Tax. 458, 784.

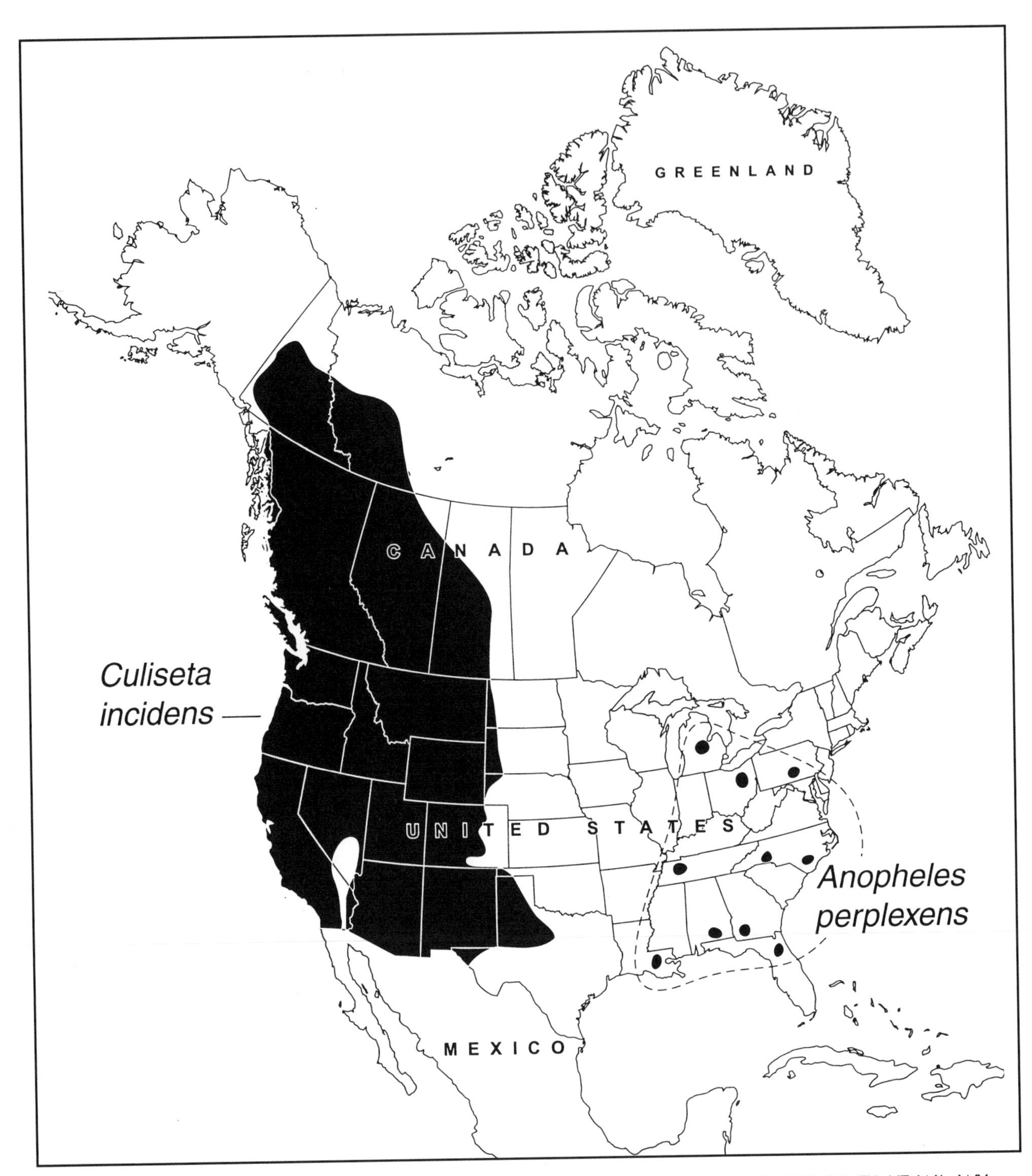


Plate 32 C. Distribution of *Culiseta incidens*—USA: AK, AZ, CA, CO, ID, MT, NE, NV, NM, ND, OK, OR, TX, UT, WA, WY (146), SD (280); Canada: ALTA, BC, NWT, YUK (146), SASK (476), not in NS (263); Tax. 36, 458, 784. *Anopheles perplexens*—USA: AL, FL (380), GA (49), NC, TN (621), OH (557), PA (424); Tax. 49.

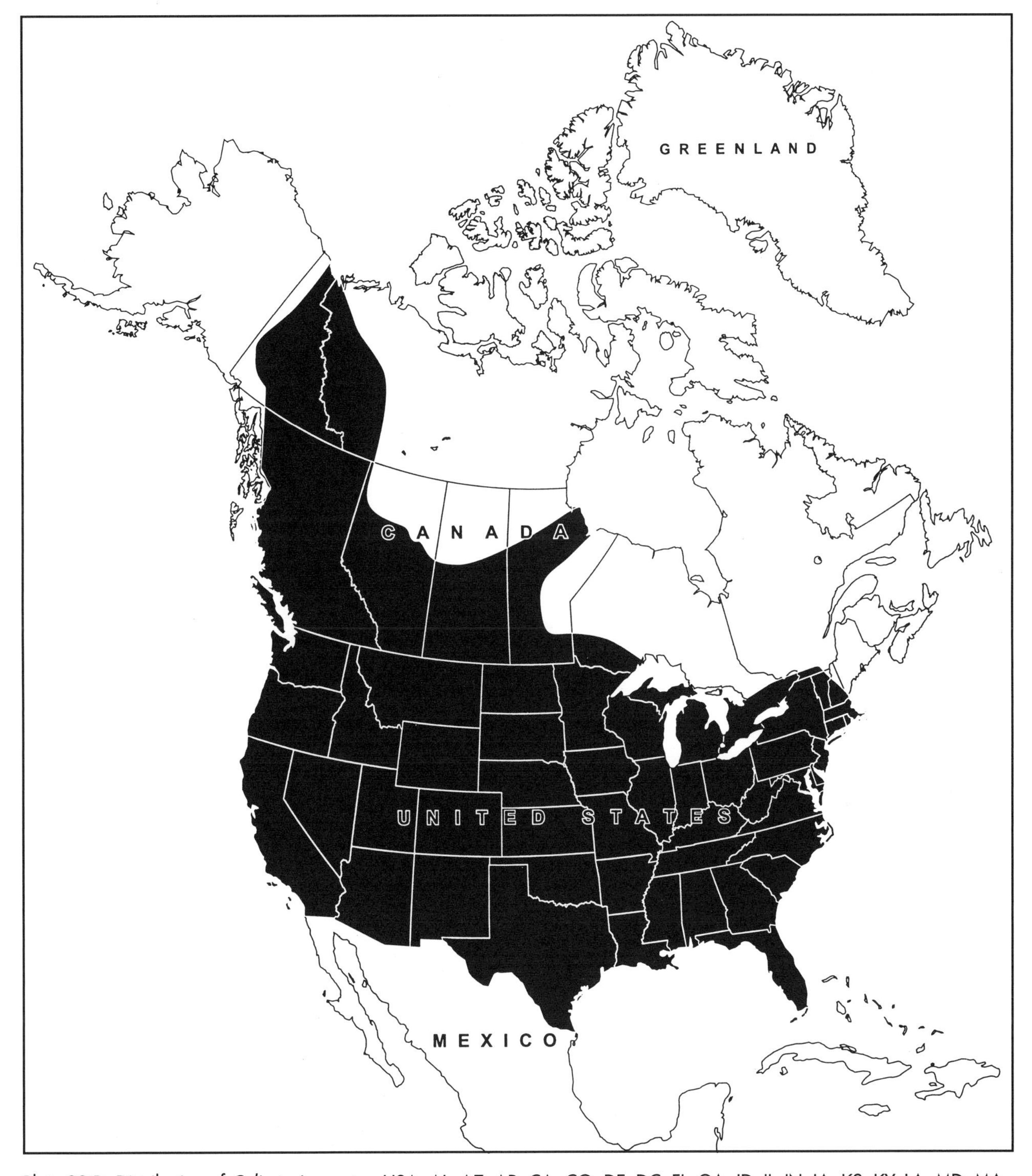


Plate 32 D. Distribution of *Culiseta inornata*—USA: AL, AZ, AR, CA, CO, DE, DC, FL, GA, ID, IL, IN, IA, KS, KY, LA, MD, MA, MI, MN, MS, MO, MT, NE, NV, NH, NJ, NM, NY, NC, ND, OH, OK, OR, PA, SC, SD, TN, TX, UT, VA, WA, WI, WY (146), CT (746), WV (3); Canada: ALTA, BC, MAN, NWT, ONT, SASK, YUK (146), PQ (416); Tax. 458, 784.

Plate 33 A. Distribution of *Culiseta melanura*—USA: AL, AR, DE, DC, FL, GA, IA, KY, LA, ME, MD, MA, MI, MN, MS, MO, NE, NH, NJ, NY, NC, OH, OK, PA, RI, SC, TN, TX, VA, WI (146), CT (745), IL (657), IN (650), KS (541), not in CO (314); Canada: ONT (150), PQ (251, 283); Tax. 784.

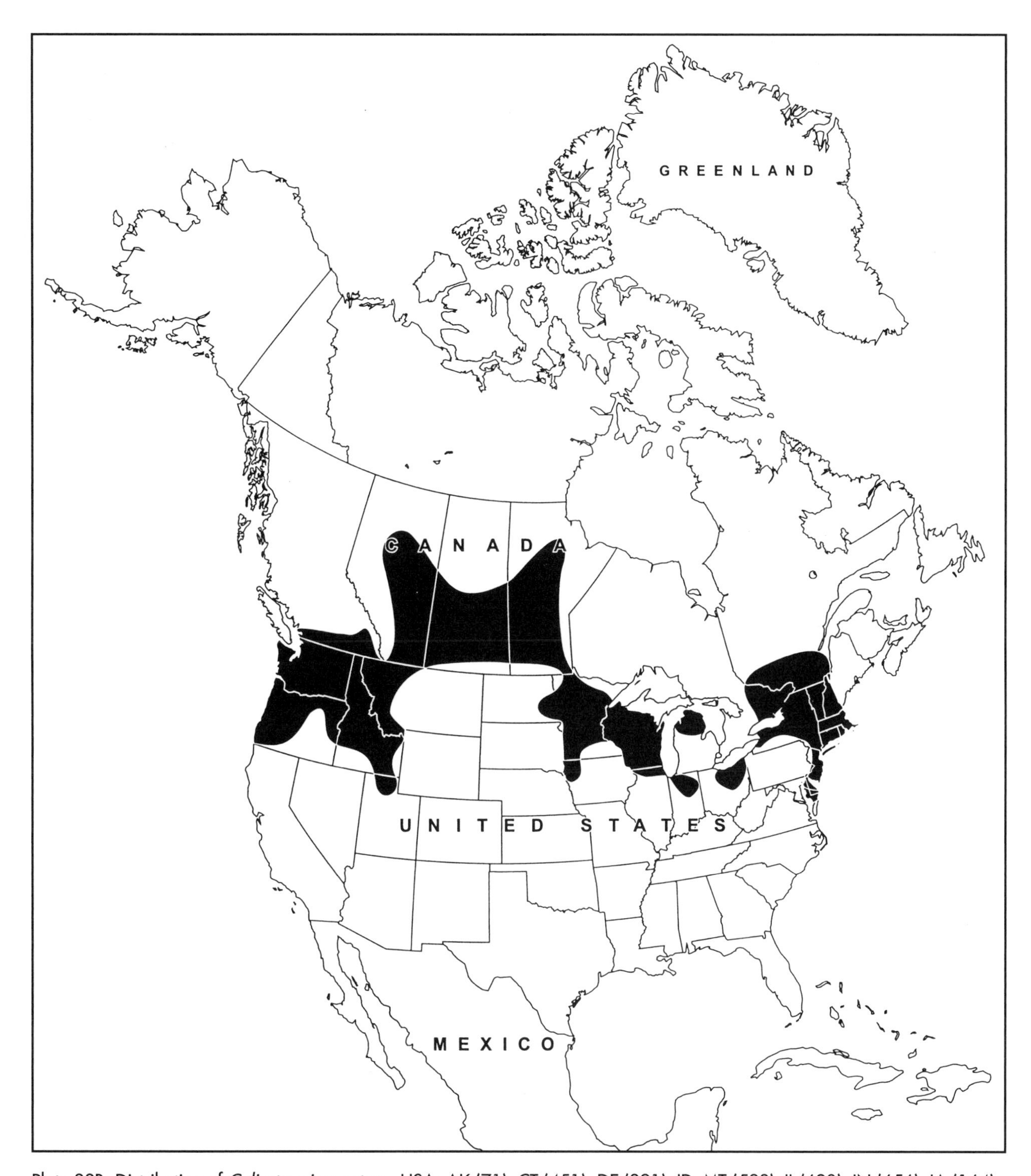


Plate 33B. Distribution of *Culiseta minnesotae*—USA: AK (71), CT (451), DE (391), ID, MT (533), IL (620), IN (656), IA (164), MD (76), MA (328), MI (30), MN (19), NH (113), NJ (110), NY (482), OH (557), OR (284), UT (534), WA (507), WI (630); Canada: ALTA (293), BC (182), MAN (728), ONT (694), PQ (414), SASK (476); Tax. 19, 458, 577, 784.

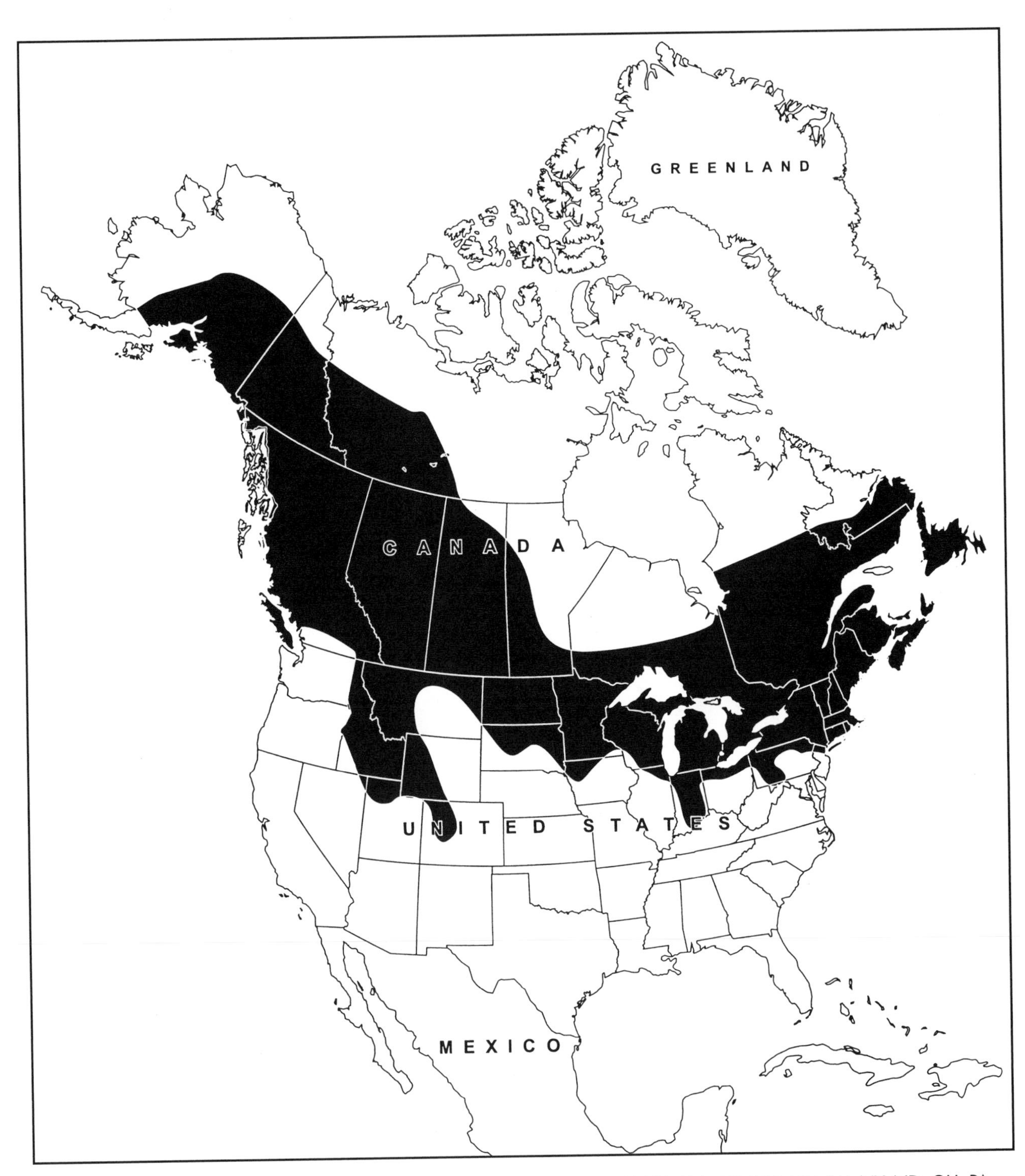


Plate 33 C. Distribution of *Culiseta morsitans*—USA: AK, CT, DE, ID, IL, IA, KY, ME, MA, MI, MN, NH, NJ, NY, ND, OH, PA, RI, SD, WI (146), IN (659), MD (76), MT (533),UT (522), VT (71), not in CO (314); Canada: ALTA, BC, LAB, MAN, NB, NS, NWT, ONT, PEI, PQ, SASK, YUK (146), NFLD (569); Tax. 19, 458, 784.

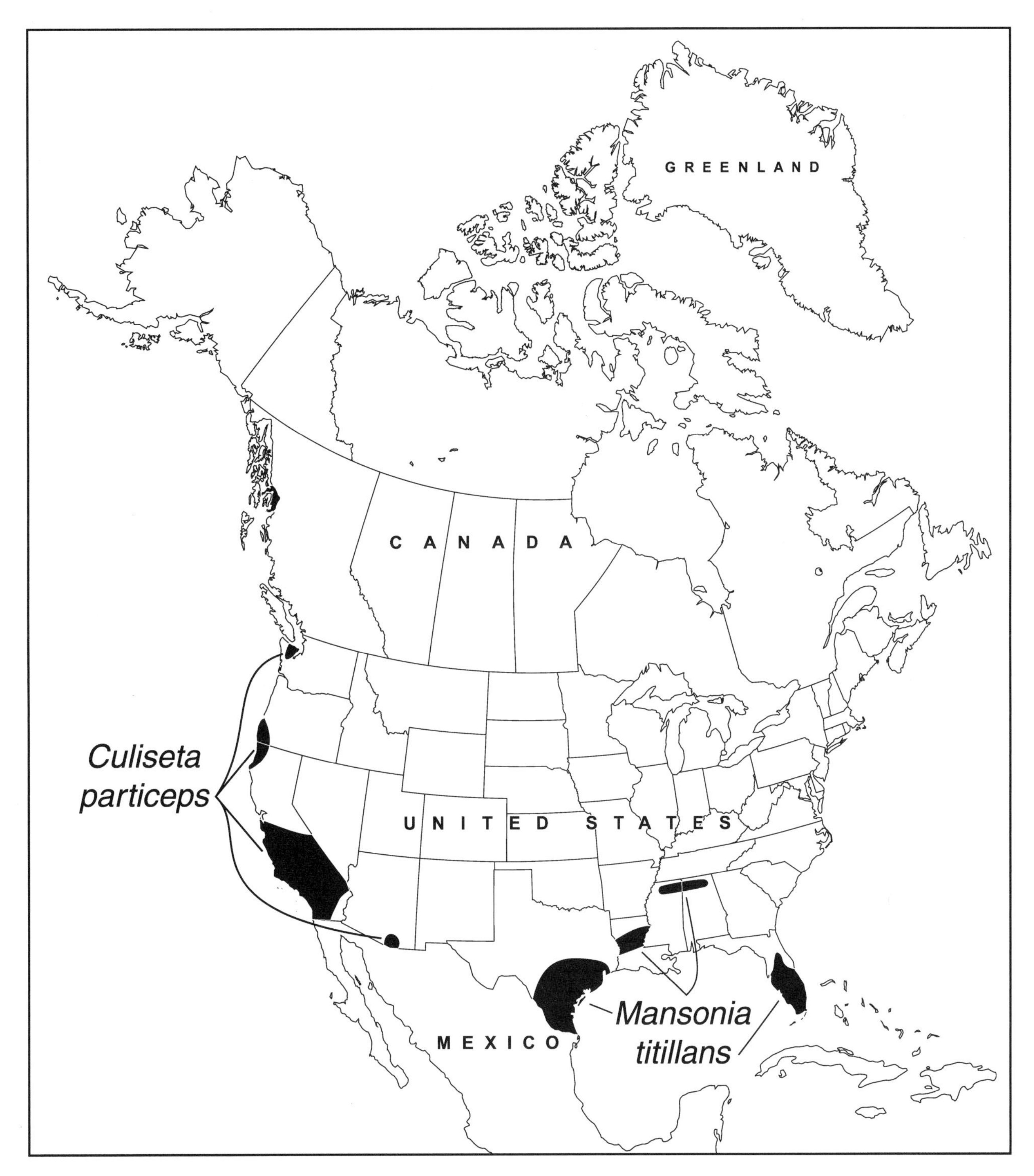


Plate 33 D. Distribution of *Culiseta particeps*—USA: CA, OR (146), AK (69), AZ (604), WA (507); Tax. 458, 698. *Mansonia titillans*—USA: FL, TX (146), LA (E. Milstrey, pers. comm. 1999); Tax. 42, 616, 617.

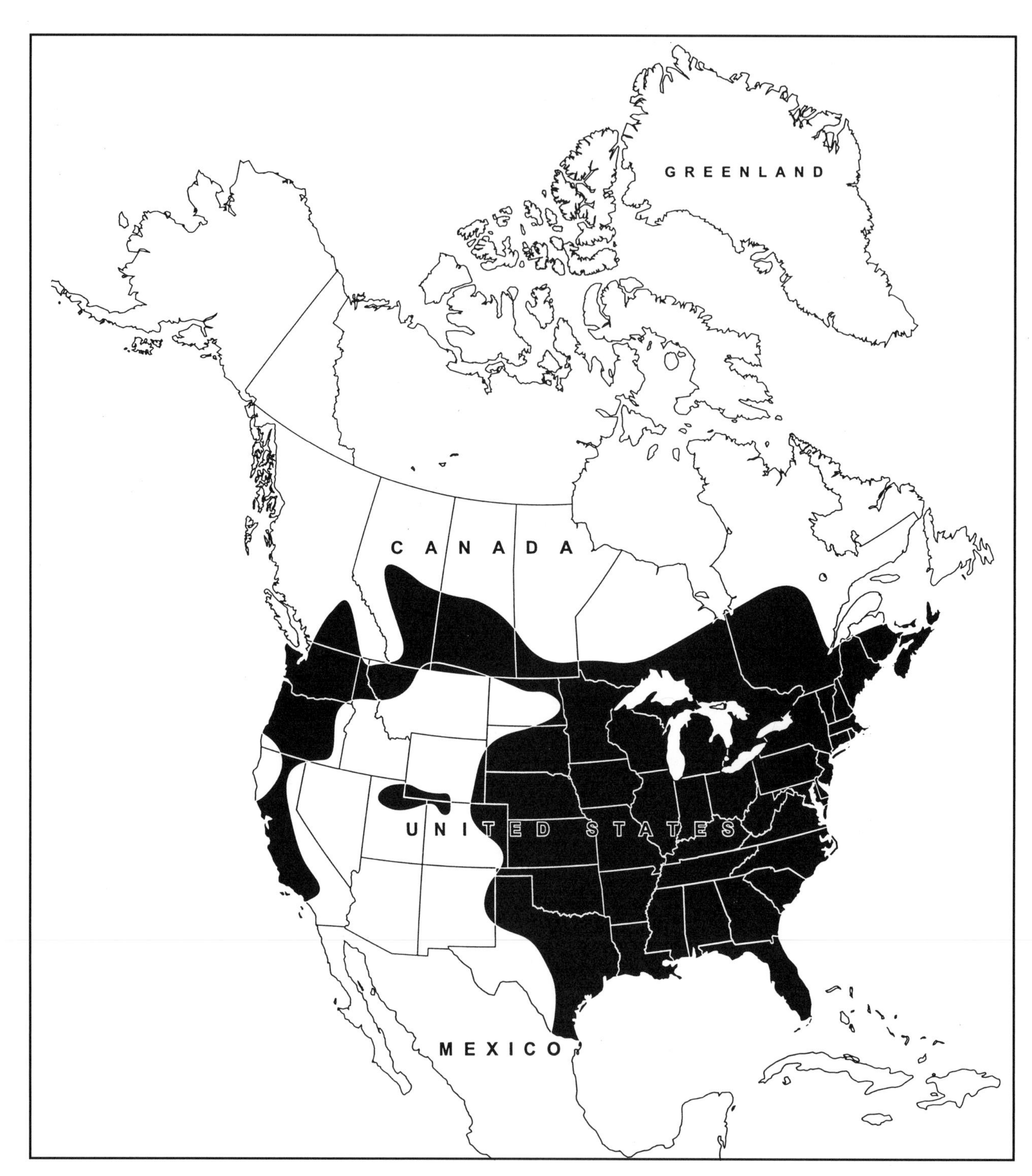


Plate 34 A. Distribution of *Coquillettidia perturbans*—USA: AL, AR, CA, CO, CT, DE, DC, FL, GA, ID, IL, IN, IA, KS, KY, LA, ME, MD, MA, MI, MN, MS, MO, MT, NE, NH, NJ, NY, NC, ND, OH, OK, OR, PA, RI, SC, SD, TN, TX, UT, VA, VT, WA, WI, WY (146), NM (779), WV (3); Canada: BC, MAN, NS, ONT, PEI, PQ, SASK (146), ALTA (306), NB (489); Tax. 47, 616, 617, 784.

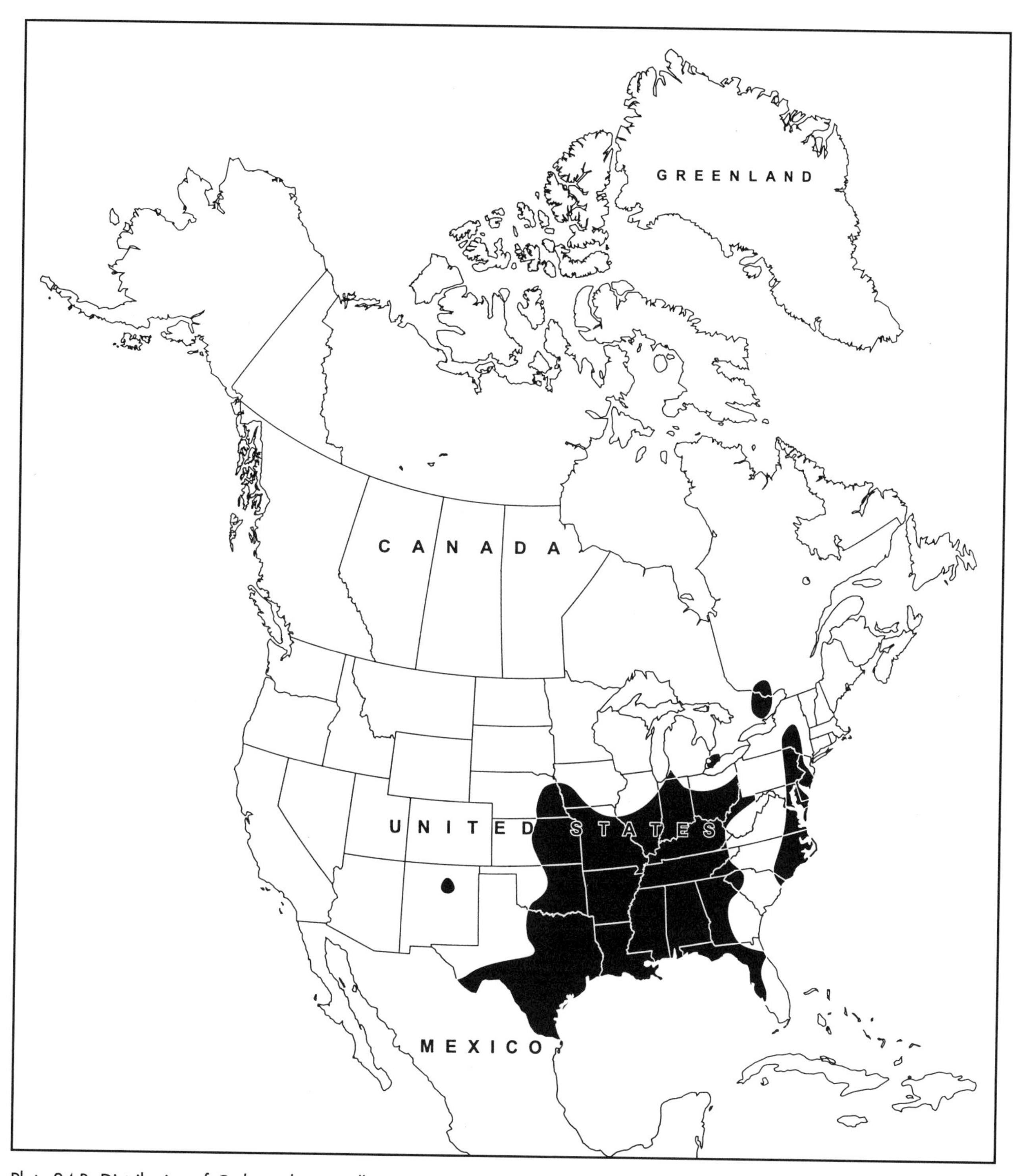


Plate 34 B. Distribution of *Orthopodomyia alba*—USA: AL, IL, KY, LA, MS, MO, NJ, NY, NC, TX, VA (146), AR, DC, MD, OH (787), DE (390), FL (215), GA (714), IN (103), IA (433), KS (544), MI (295), NM (494), OK (337), PA (770), NE (430), TN (90); Canada: ONT (682), PQ (784); Tax. 784, 787.

Plate 34 C. Distribution of *Orthopodomyia signifera*—USA: AL, AR, CT, DE, DC, FL, GA, IL, IN, IA, KS, KY, LA, MD, MA, MS, MO, NE, NJ, NM, NY, NC, OH, OK, PA, RI, SC, TN, TX, VA (146), AZ (604), CA, OR, UT (787), MI (295), MN (579), NH (114), SD (241), WV (3), WI (573); Canada: ONT (682); Tax. 784, 787.

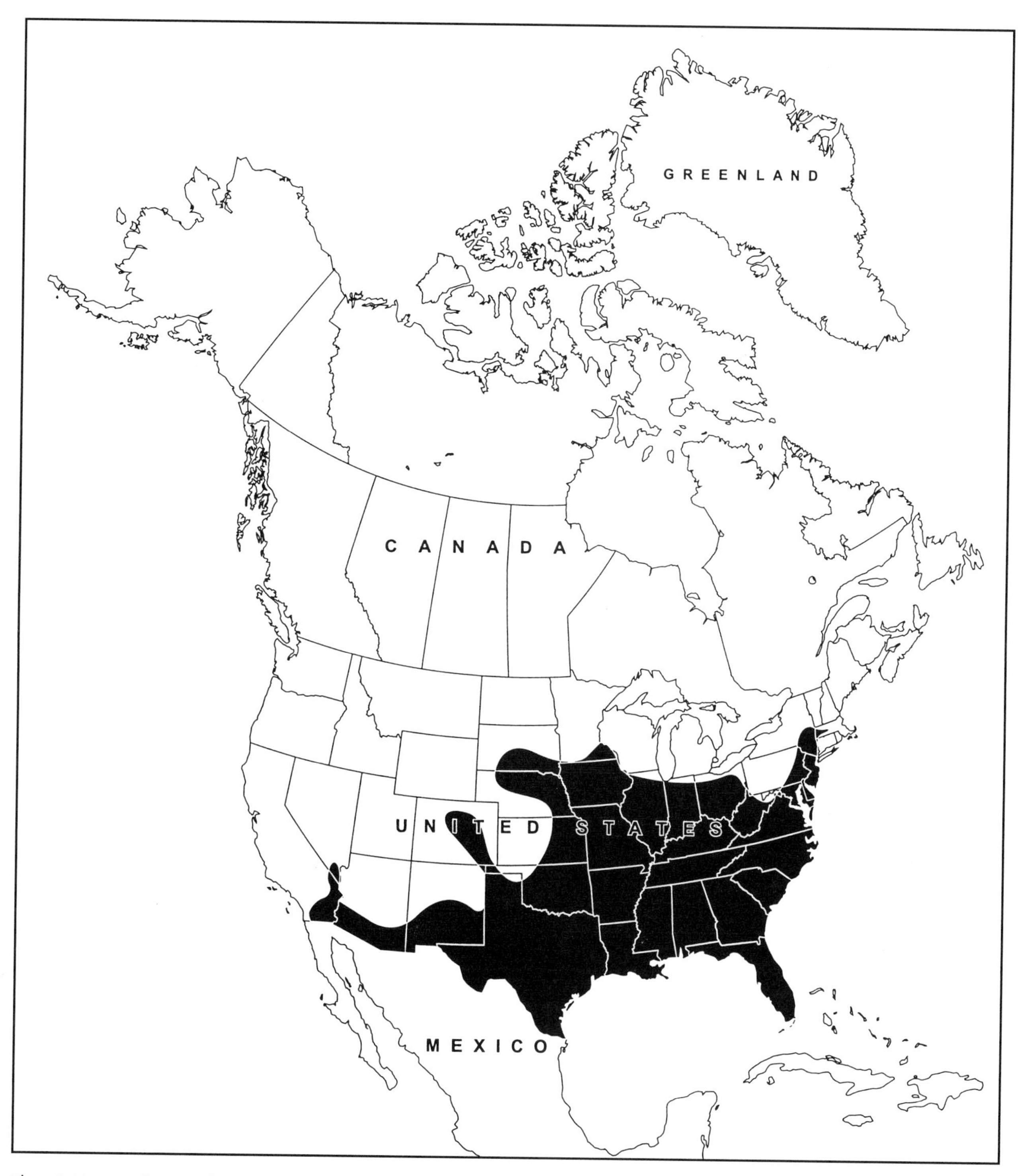


Plate 34 D. Distribution of *Psorophora columbiae*—USA: AL, AR, AZ, CA, CO, DE, DC, FL, GA, IL, IN, IA, KS, KY, LA, MD, MA, MS, MO, NE, NJ, NM. NY, NC, OH, OK, PA, SC, SD, TN, TX, VA, WV (146 as *Ps. confinnis*), CA (82), MN (172), NV (154); Canada: ONT (784); Tax. 42, 70, 206.

Plate 35 A. Distribution of *Psorophora ciliata*—USA: AL. AR, CT, DE, DC, FL, GA, IL, IN, IA, KS, KY, LA, MD, MA, MI, MS, MO, NE, NH, NJ, NY, NC, OH, OK, PA, RI, SC, SD, TN, TX, VA, WV, WI (146), MN (20), NM (495); Canada: ONT, PQ (146); Tax. 36, 42, 47, 784.

Plate 35 B. Distribution of *Psorophora cyanescens*—USA: AL, AR, FL, GA, IL, IN, KS, KY, LA, MS, MO, NE, NM, NC, OH, OK, SC, TN, TX, VA (146), DE (393), MD (370), NJ (191).

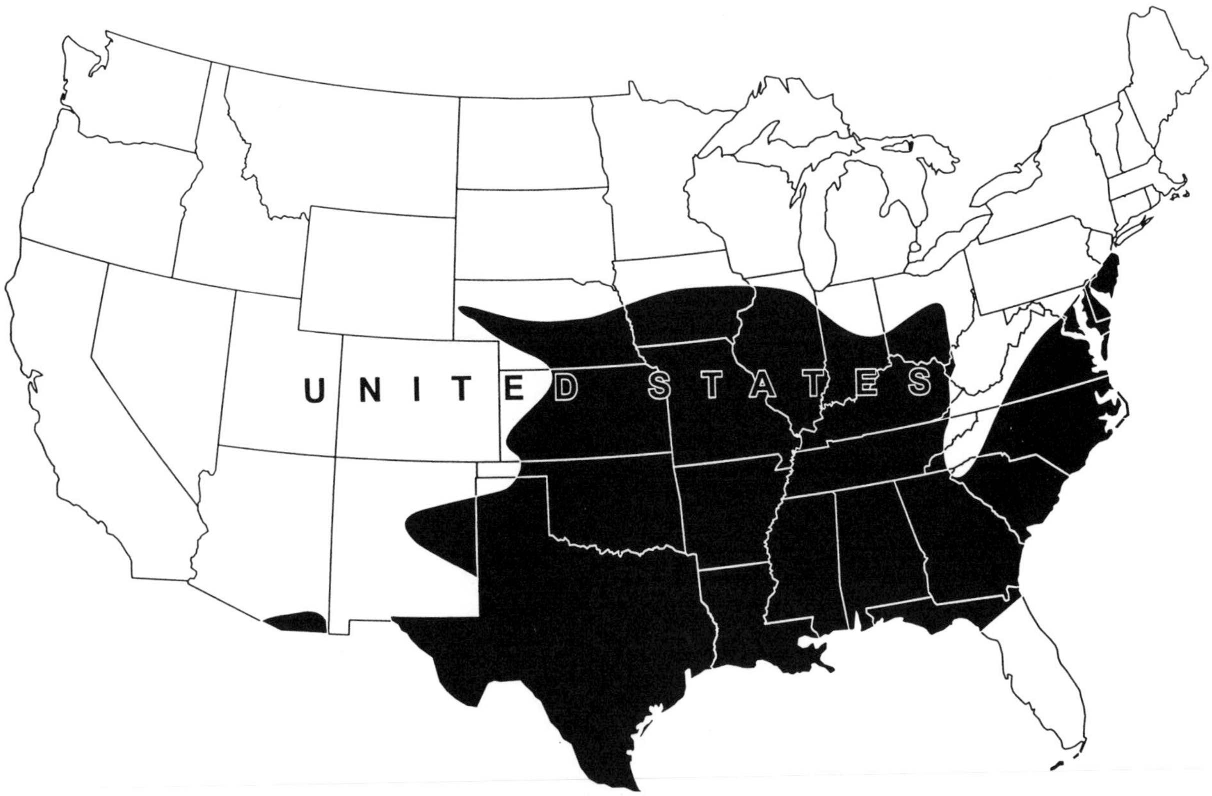

Plate 35 C. Distribution of *Psorophora discolor*—USA: AL, AR, DE, DC, FL, GA, IL, IA, KS, KY, LA, MD, MS, MO, NE, NJ, NM, NC, OH, OK, SC, TN, TX, VA (146), AZ (604), IN (663).

Plate 35 D. Distribution of *Psorophora ferox*—USA: AL, AR, CT, DE, DC, FL, GA, IL, IN, IA, KS, KY, LA, MA, MI, MN, MS, MO, NE, NH, NJ, NY, NC, OH, OK, PA, SC, SD, TN, TX, VA, WI (146), MD (67), WV (3); Canada: ONT (784); Tax. 36, 42, 784.

Plate 36 A. Distribution of *Psorophora horrida*—USA: AL, AR, DC, FL, GA, IL, IN, IA, KS, KY, LA, MD, MS, MO, NE, NC, OH, OK, PA, SC, TN, TX, VA (146), DE (393), MI (Newson, pers. comm. 1977), MN (172), SD(280), WI (Dicke, pers. comm. 1979).

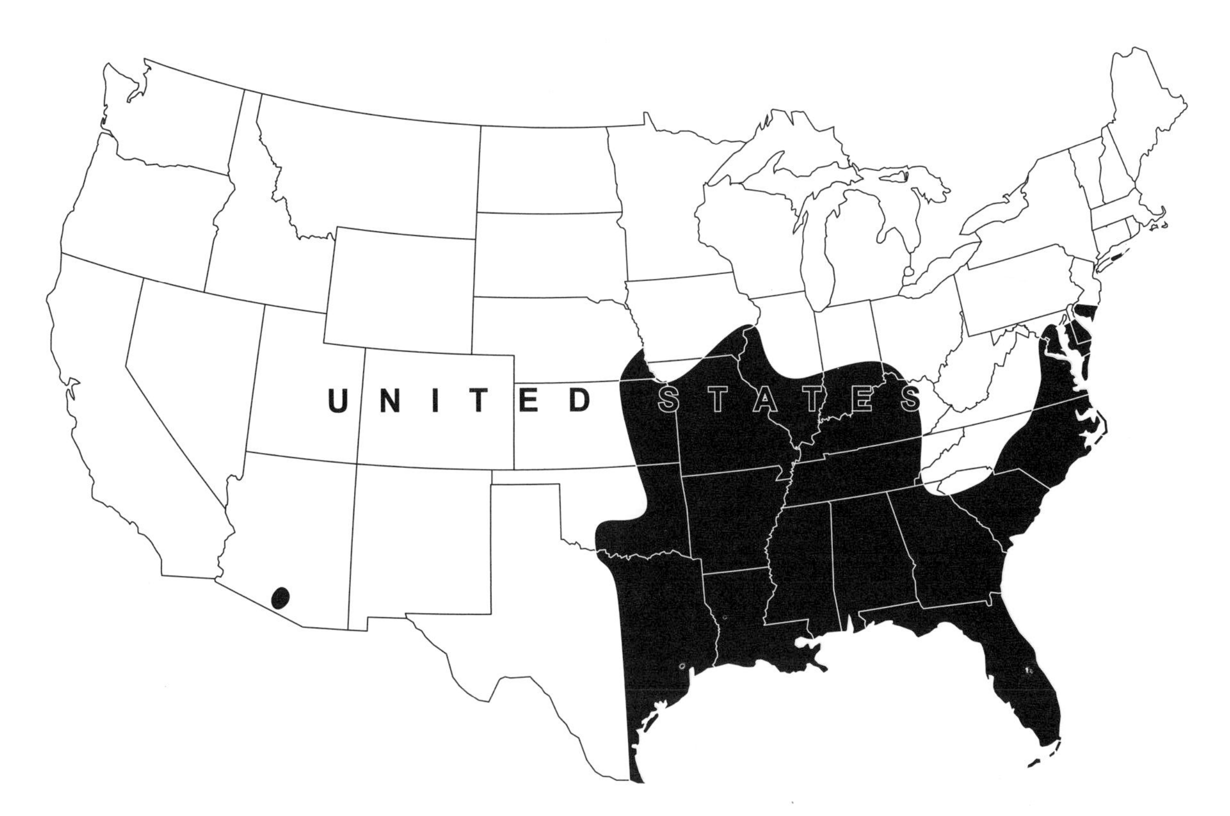


Plate 36 B. Distribution of *Psorophora howardii*—USA: AL, AR, DC, FL, GA, IL, IN, KS, KY, LA, MD, MS, MO, NE, NC, OK, SC, TN, TX, VA (146), AZ (604), DE (389), OH (Berry & Parsons, pers. comm. 1978).

Plate 36 C. Distribution of *Psorophora longipalpus*—USA: AR, KS, LA, MO, OK, SD, TX (146), NE (589); Map after Roth (622); Tax. 622.

Plate 36 D. Distribution of *Psorophora mathesoni*—USA: AL, AR, FL, GA, IL, IN, KY, LA, MS, MO, NY, NC, OH, OK, SC, TN, TX, VA (146), DE (393), MD (67), NJ (110), WI (724); listed in all references as *Ps. varipes;* Tax. 41.

Plate 37 A. Distribution of *Psorophora signipennis*—USA: AZ, AR, CO, IA, KS, KY, MO, MT, NE, NM, ND, OK, SD, TN, TX, WY (146), CA (163), NV (335), UT (604); Canada: SASK (601); Tax. 784.

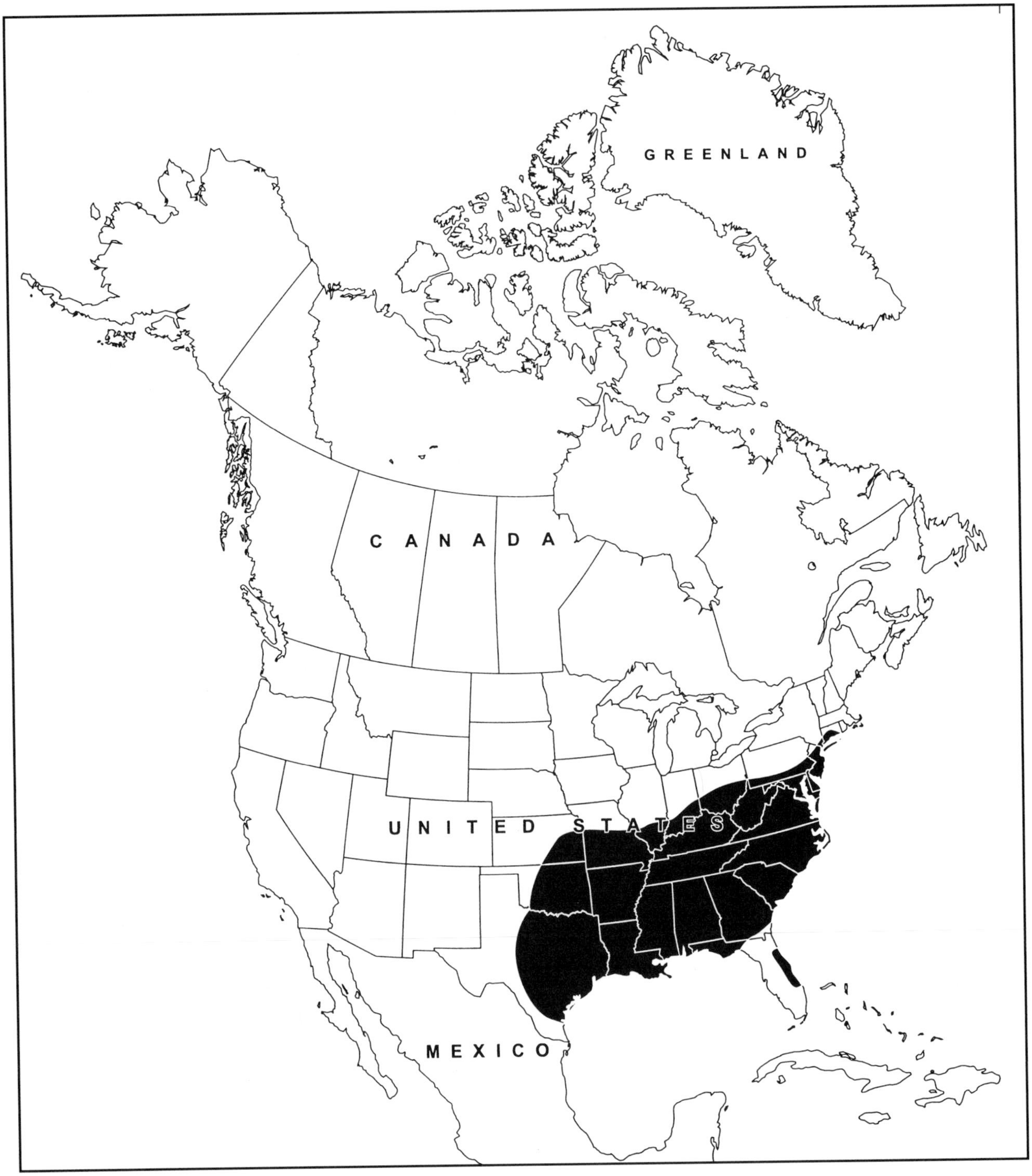

Plate 37 B. Distribution of *Toxorhynchites r. septentrionalis*—USA: AL, AR, DE, DC, FL, GA, IL, KS, KY, LA, MD, MS, MO, NJ, NC, OH, OK, PA, SC, TN, TX, VA, WV (146), CT (439), IN (323), MA (798), NY (381); Canada: ONT (556); Tax. 36, 228, 784.

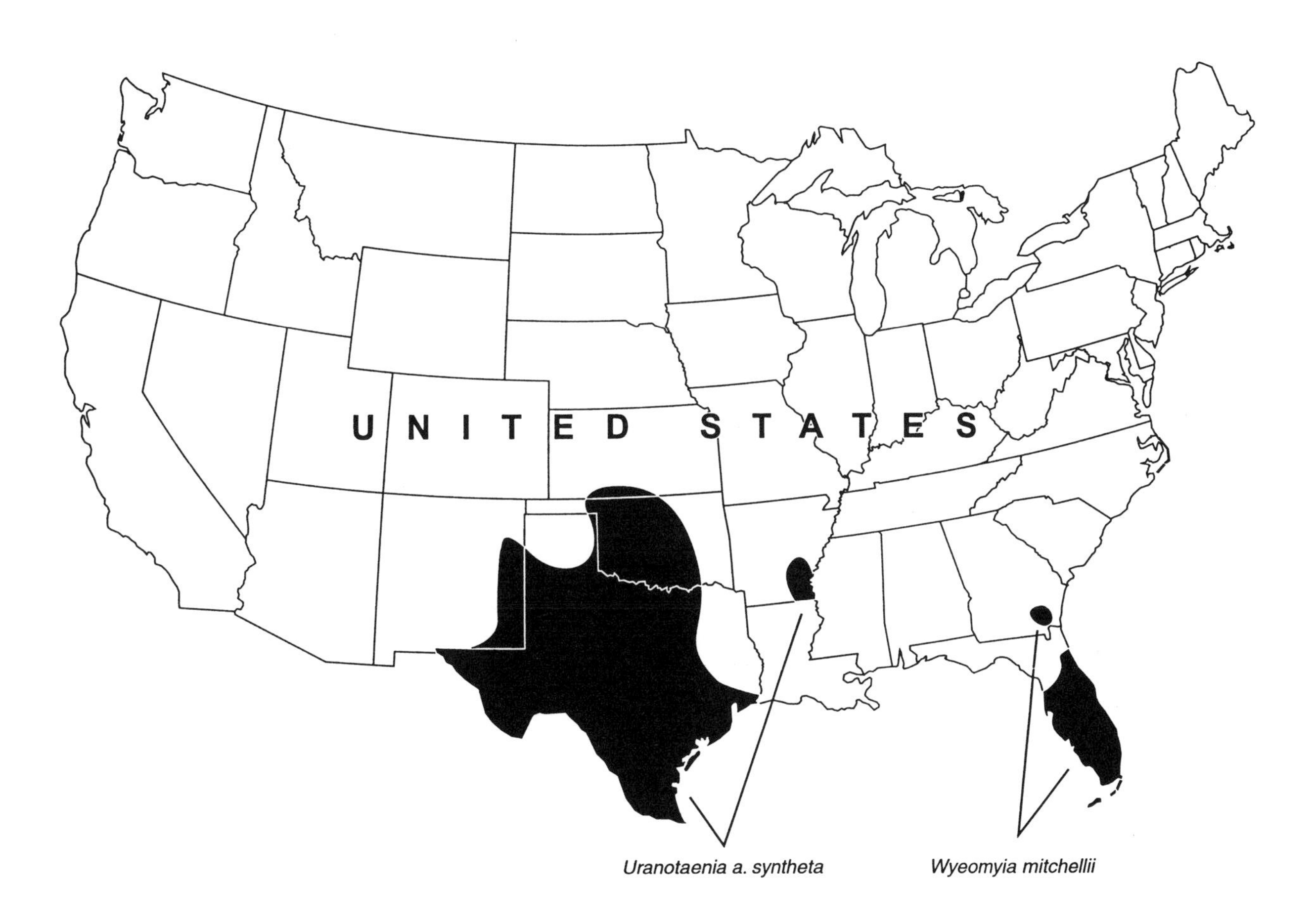


Plate 37 C. Distribution of *Ur. a. syntheta*—USA: NM, OK, TX (146), AR (89); Tax. 45, 94. *Wy. mitchellii*—USA: FL (146), GA (516); Tax. 36, 42, 703.

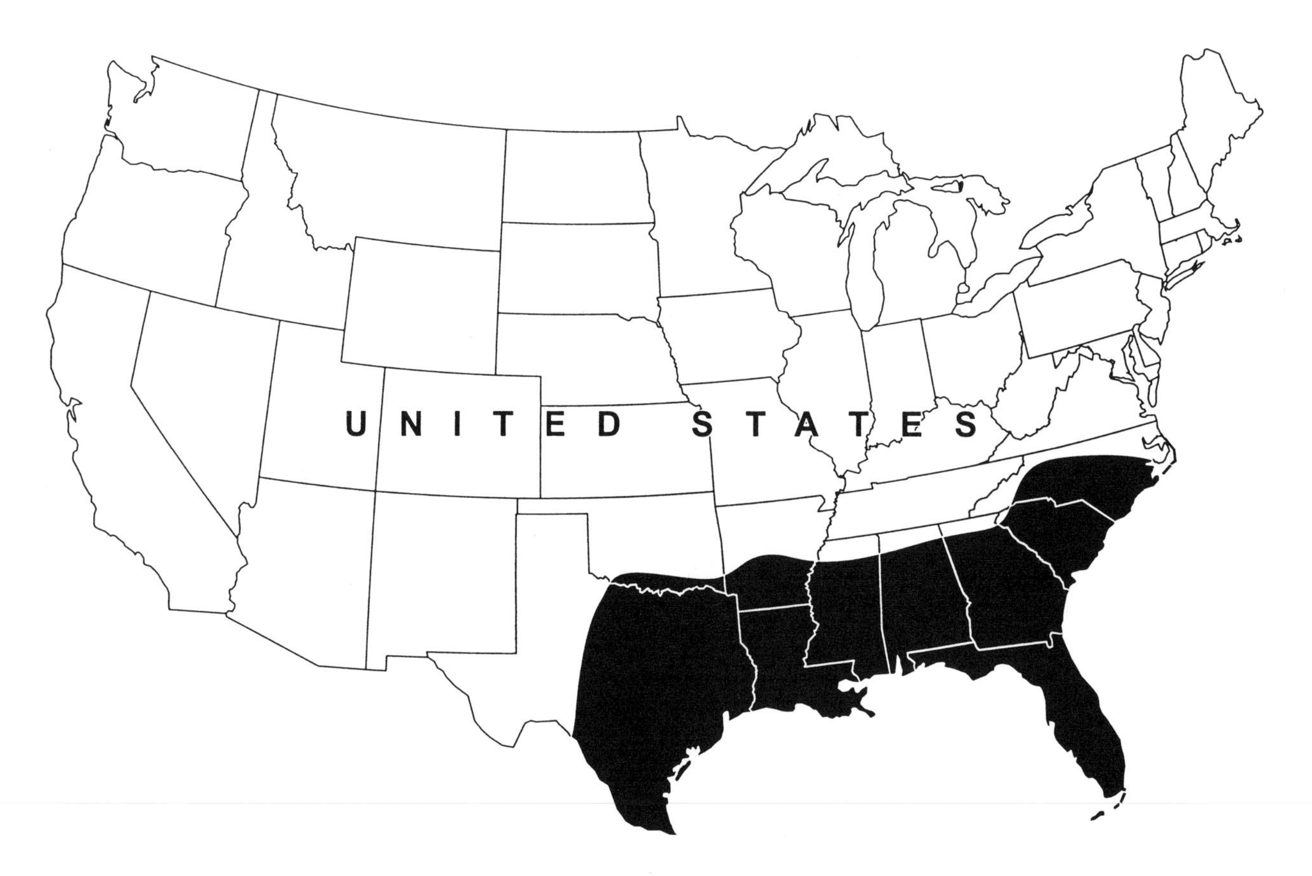


Plate 37 D. Distribution of *Uranotaenia lowii*—USA: AL, AR, FL, GA, LA, MS, NC, SC, TX (146), OK (558); Tax. 36, 42, 275

Plate 38 A. Distribution of *Uranotaenia sapphirina*—USA: AL, AR, CT, DE, DC, FL, GA, IL, IN, IA, KS, KY, LA, MD, MA, MI, MN, MS, MO, NE, NH, NJ, NM, NY, NC, ND, OH, OK, PA, RI, SC, SD, TN, TX, VT, VA, WI (146), WV (3); Canada: ONT, PQ (146); Tax. 42, 784.

Plate 38 B. Distribution of *Wyeomyia smithii*—USA: CT, DE, IL, ME, MA, MI, MN, NH, NJ, NY, OH, RI, WI (146), IN (655), MD (76), PA (767); [AL, NC, SC (146), FL (88), GA (218), MD (77), VA (288) as *Wy. haynei*]; Canada: LAB, MAN, NS, ONT (146), NB (399), NFLD (569), PEI (784), PQ (442), SASK (116); Tax. 218, 784.

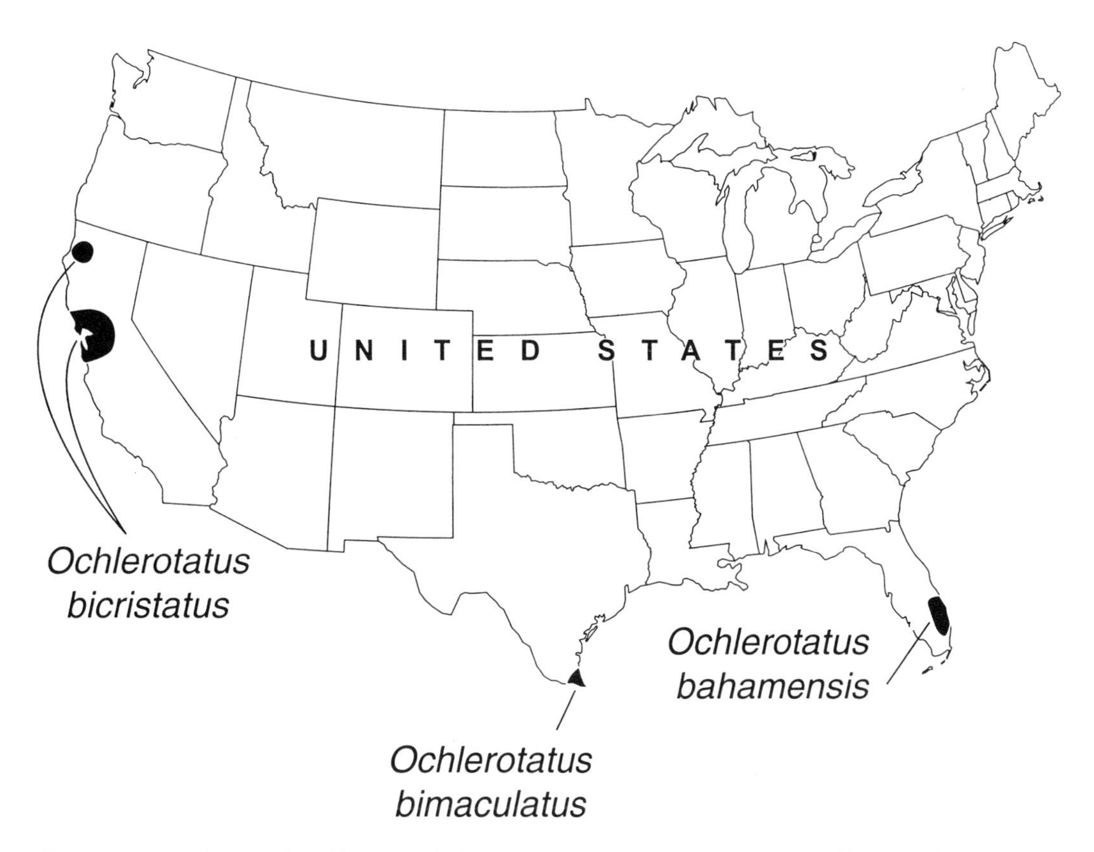


Plate 39 A. Distribution of *Ochlerotatus bahamensis*—USA: FL (554); Tax. 60. *Ochlerotatus bicristatus*—USA: CA (146, 147): Tax. 431. *Ochlerotatus bimaculatus*—USA: TX (146).

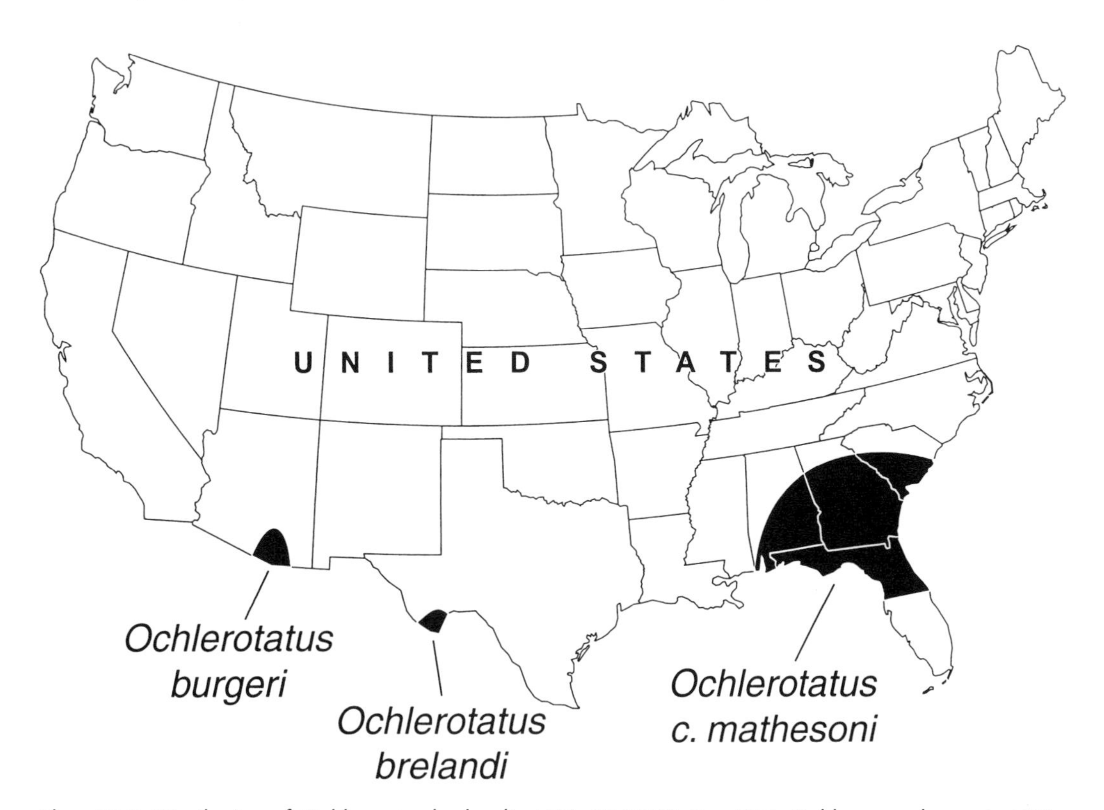


Plate 39 B. Distribution of *Ochlerotatus brelandi*—USA: TX (792); Tax. 792. *Ochlerotatus burgeri*—USA: AZ (793); Map after Zavortink (793); Tax. 111, 793. *Ochlerotatus c. mathesoni*—USA: AL, FL, GA, SC (146), not in OH (Parsons, pers. comm. 1978); Canada: NFLD (569, doubtful 784).

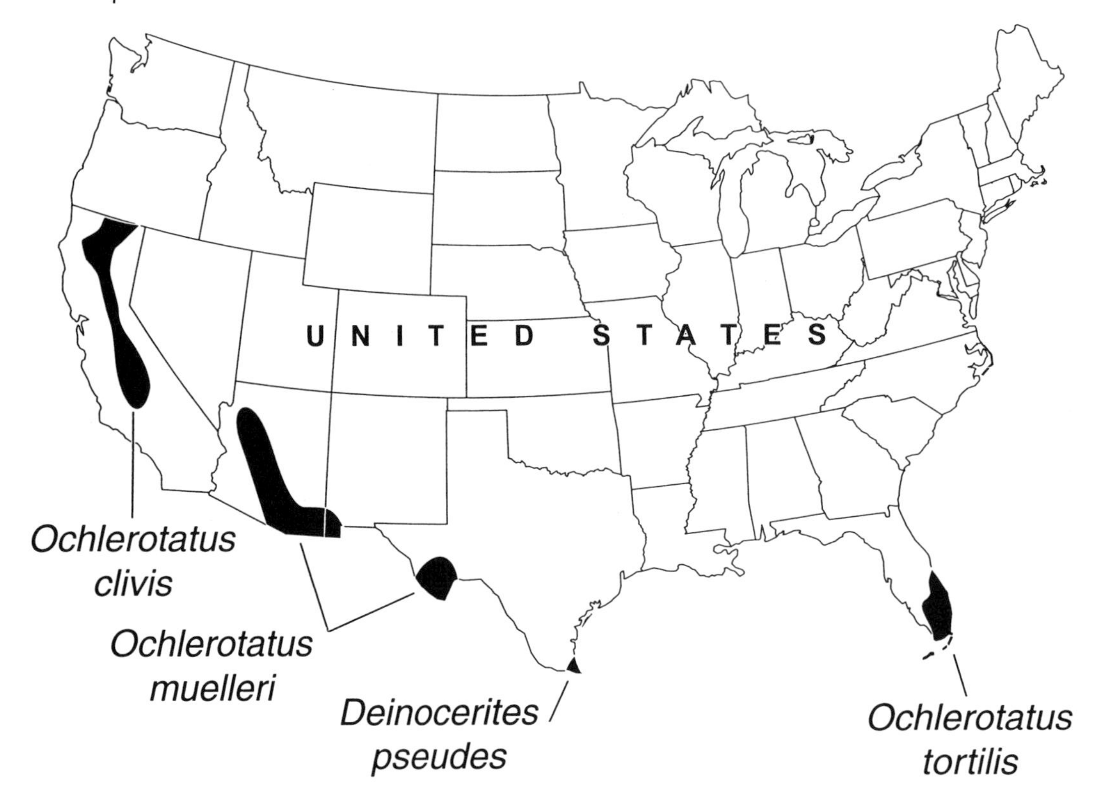


Plate 39 C. Distribution of *Ochlerotatus clivis*—CA (394); Tax. 394. *Ochlerotatus muelleri*—USA: AZ (146), NM (532), TX (97), Map modified after Zavortink (793); Tax. 469, 793. *Ochlerotatus tortilis*—USA: FL (146); Map after Arnell (11); Tax. 11, 36, 42. *Deinocerites pseudes*—USA: TX (43); Tax. 1, 43, 568.

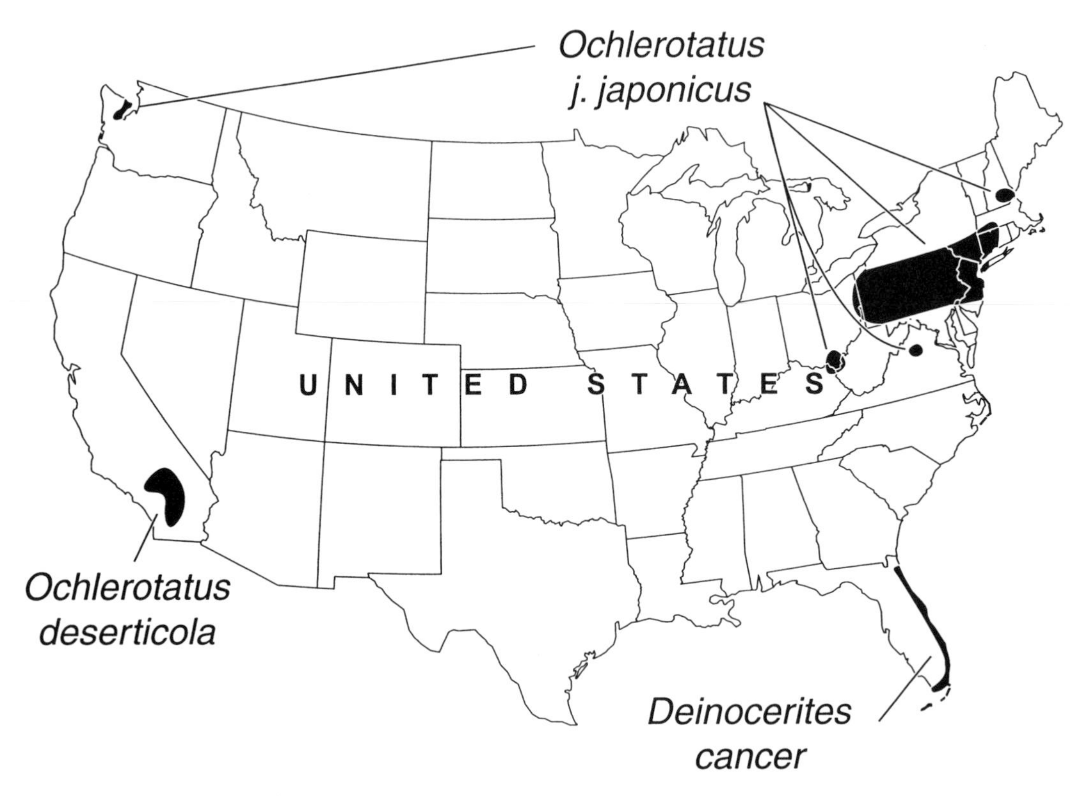


Plate 39 D. Distribution of *Ochlerotatus j. japonicus*—USA: NY, NJ (567), CT (808), OH (Berry, pers. comm. 1999), PA (Spichiger, pers. comm. 2000), WA (Maloney, pers. comm. 2001); Tax. 567. *Ochlerotatus deserticola*—USA: CA (789); Map after Arnell & Nielsen (13); Tax. 13, 789. *Deinocerites cancer*—USA: FL (146); Tax. 1, 42, 43.

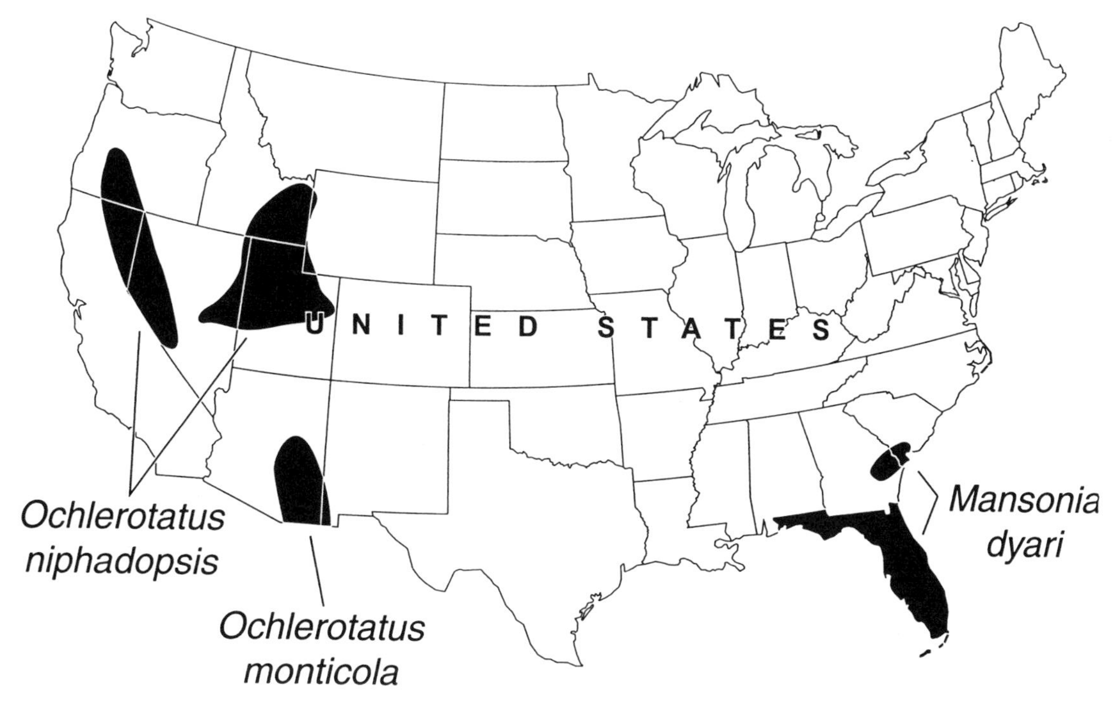


Plate 40 A. Distribution of *Ochlerotatus monticola*—USA: AZ (46), NM (532); Map after Arnell & Nielsen (13); Tax. 13, 46, 431. *Ochlerotatus niphadopsis*—USA: ID, NV, OR, UT (146), CA (137), WY (156): Canada: not in ALTA (Pucat, pers. comm. 1979); Tax. 431, 591. *Mansonia dyari*—USA: FL, GA (146), SC (212); Tax. 42, 616, 617.

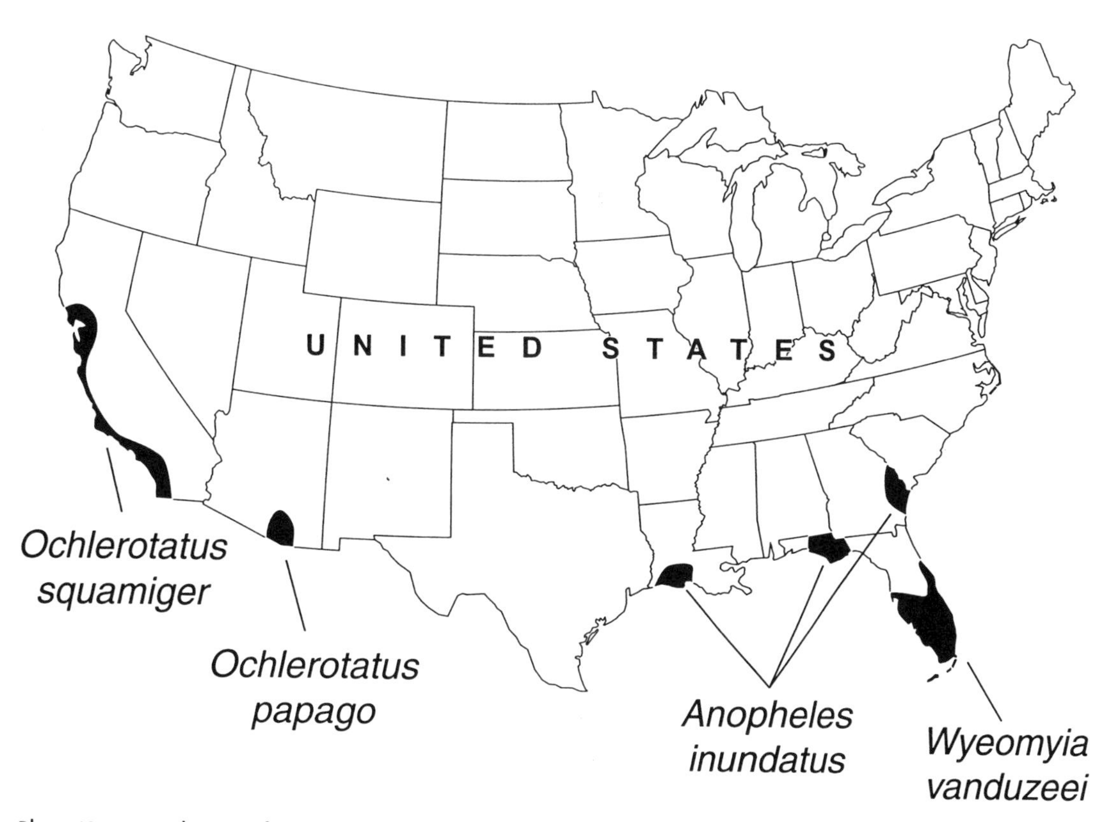


Plate 40 B. Distribution of *Ochlerotatus papago*—USA: AZ (792); Map after Zavortink (792); Tax. 792, 793. *Ochlerotatus squamiger*—USA: CA (146); Tax. 431. *Anopheles inundatus*—USA: FL, GA, LA (599); Tax. 599. *Wyeomyia vanduzeei*—USA: FL (146); Tax. 42.

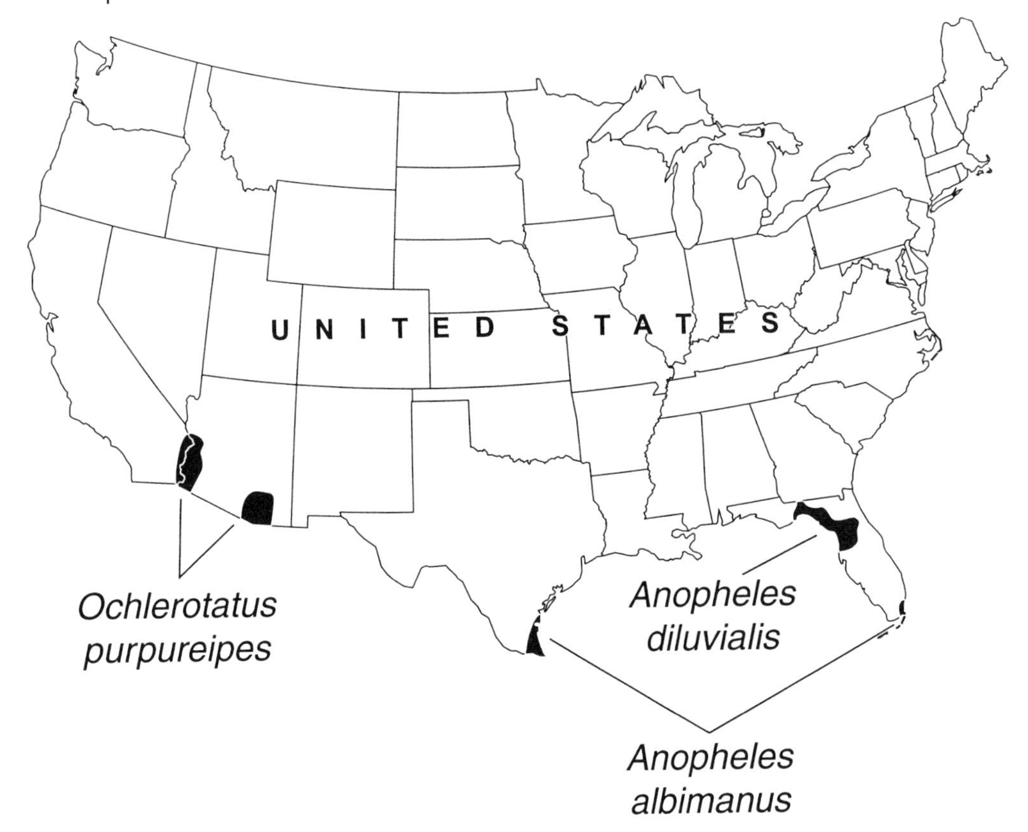


Plate 40 C. Distribution of *Ochlerotatus purpureipes*—USA: AZ (146), CA (491); Tax. 431, 470, 793. *Anopheles diluvialis*—FL (599); Tax. 599. *Anopheles albimanus*—USA: FL, TX (146); Tax. 42.

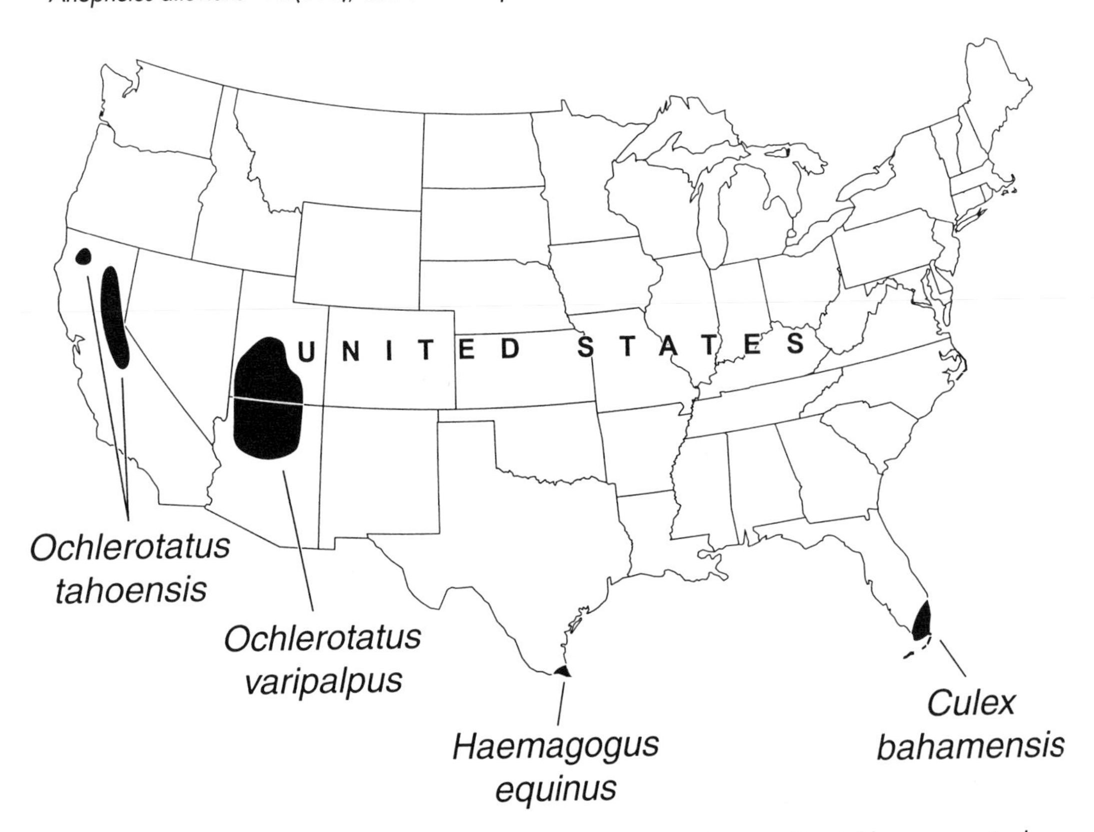


Plate 40 D. Distribution of *Ochlerotatus tahoensis*—USA: CA (108); Tax. 108. *Ochlerotatus varipalpus*—USA: AZ (46), UT (525); Map after Arnell & Nielsen (13); Tax. 13, 44, 46, 431. *Culex bahamensis*—USA: FL (146); Tax. 42, 87. *Haemagogus equinus*—USA: TX (727); Tax. 10, 36, 42.

Plate 41 A. Distribution of *Ochlerotatus washinoi*—USA: CA, OR (394); Tax. 394. *Anopheles judithae*—USA: AZ, NM (788), TX (790); Map after Zavortink (790); Tax. 788, 790. *Culex atratus*—USA: FL (146); Tax. 36, 42, 259, 382.

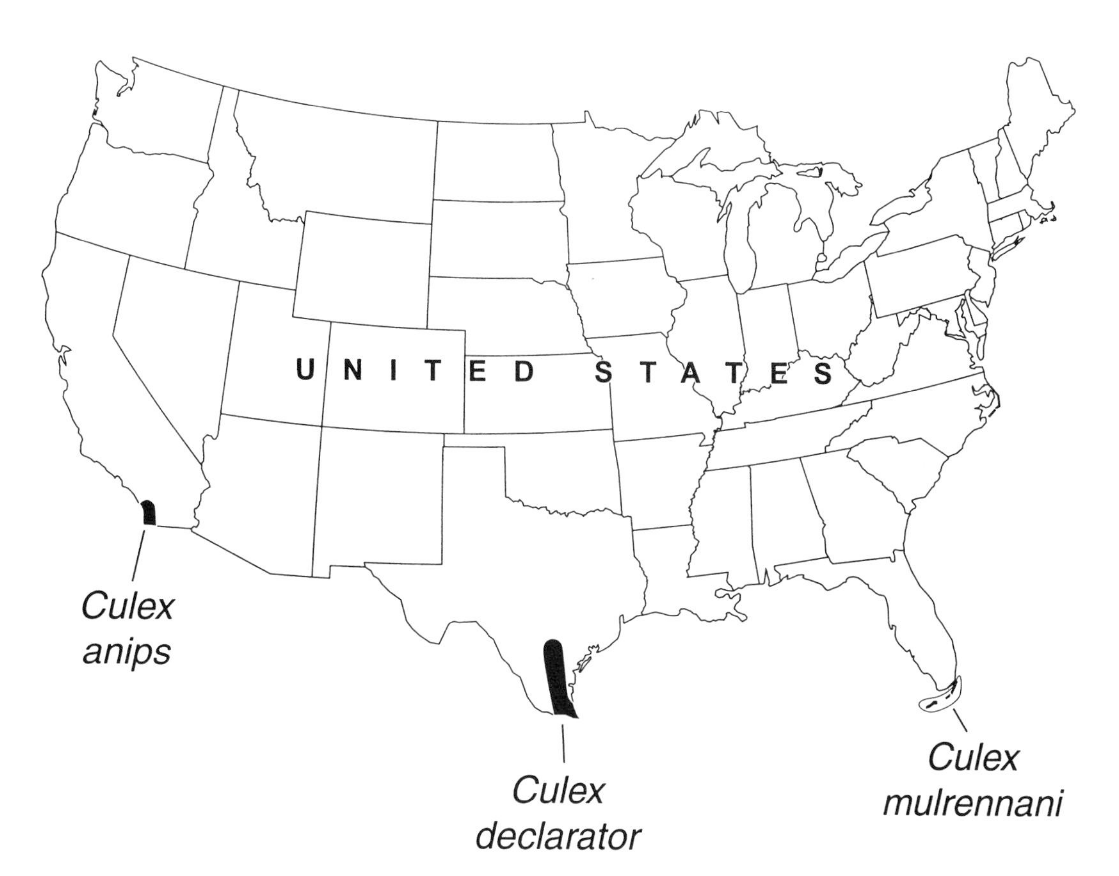


Plate 41 B. Distribution of *Culex anips*—USA: CA (146); Tax. 82, 259. *Culex declarator*—USA: TX (146); Tax. 87, 697. *Culex mulrennani*—USA: FL (146); Tax. 259, 382.

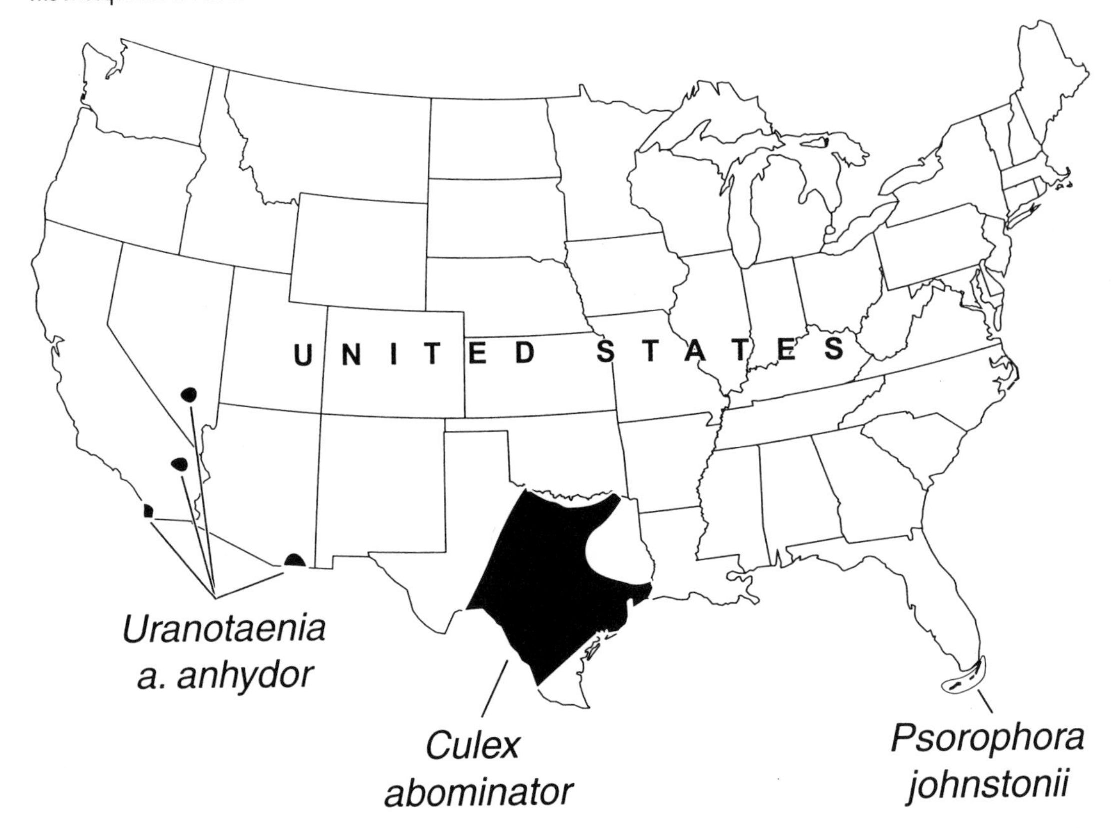


Plate 41 C. Distribution of *Culex abominator*—USA: TX (146), not in LA (135); Map after Fournier & Snyder (261); Tax. 259. *Psorophora johnstonii*—USA: FL (146); Tax. 42. *Uranotaenia a. anhydor*—USA: CA (146), AZ, NV (45); Tax. 45.

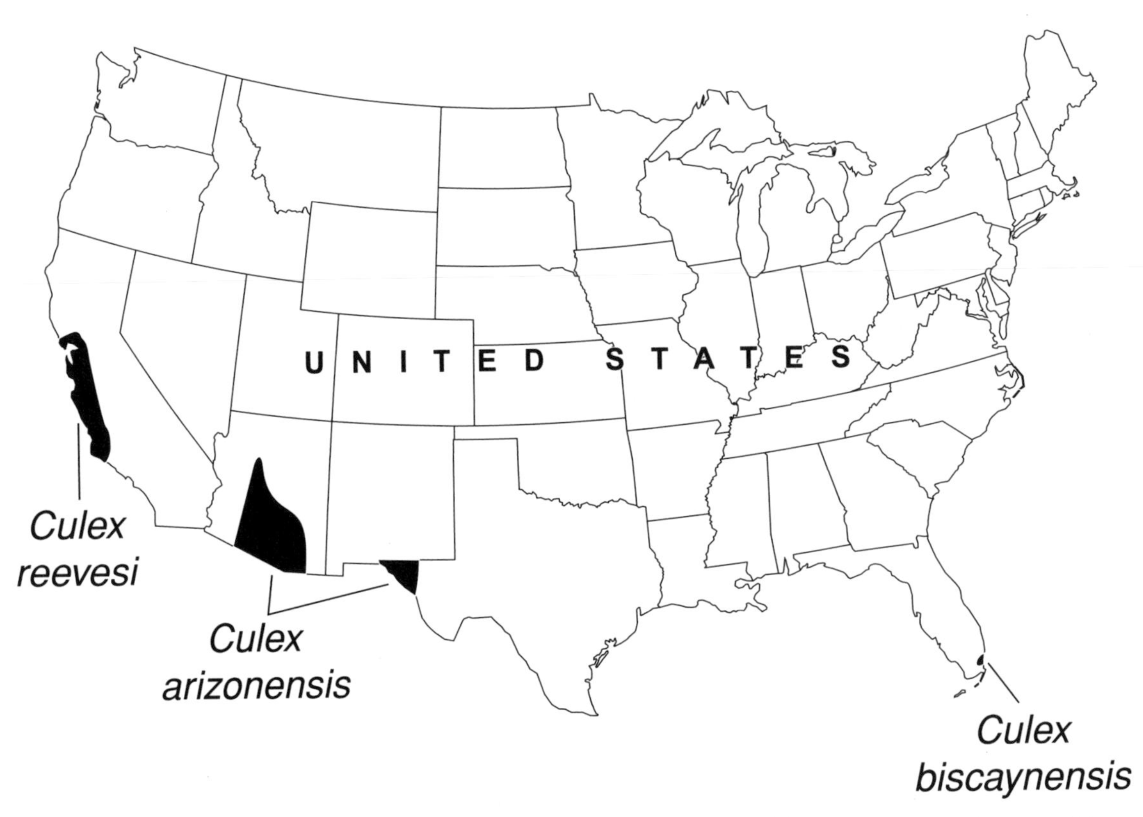


Plate 41 D. Distribution of *Culex biscaynensis*—USA: FL (795); Tax. 547, 795. *Culex arizonensis*—USA: AZ (146), TX (Reeves, pers. comm. 2001); Tax. 406. *Culex reevesi*—USA: CA (146); Tax. 406.

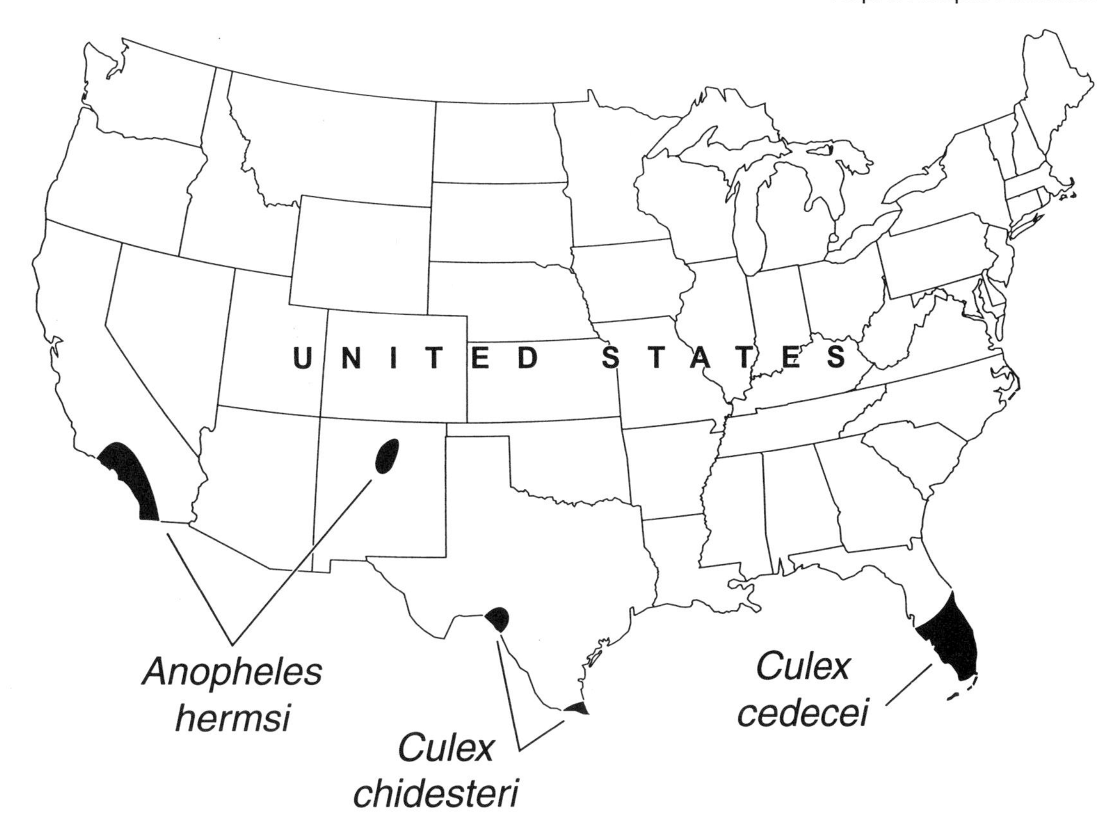


Plate 42 A. Distribution of *Culex cedecei*—USA: FL (146); Tax. 42, 259, 274, 707. *Culex chidesteri*—USA: TX (146); Tax. 42, 87. *Anopheles hermsi*—USA: AZ, CO (806), CA (26), NM (266); Tax. 26.

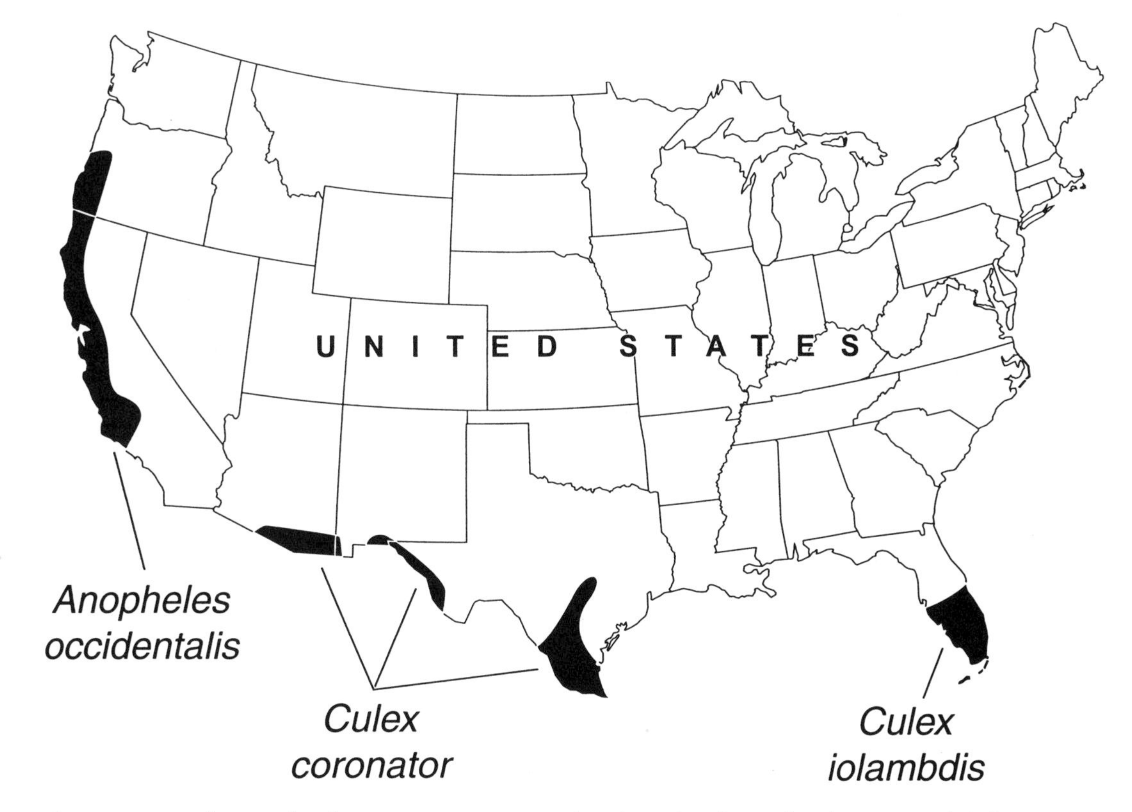


Plate 42 B. Distribution of *Culex coronator*—USA: TX (146), AZ (604), NM (779), not in LA (135); Tax. 87. *Culex iolambdis*—USA: FL (146); Tax. 42, 259. *Anopheles occidentalis*—USA: CA, OR, WA (146), not in AK (286); Canada: not in BC (784), not in YUK (761); Tax. 761.

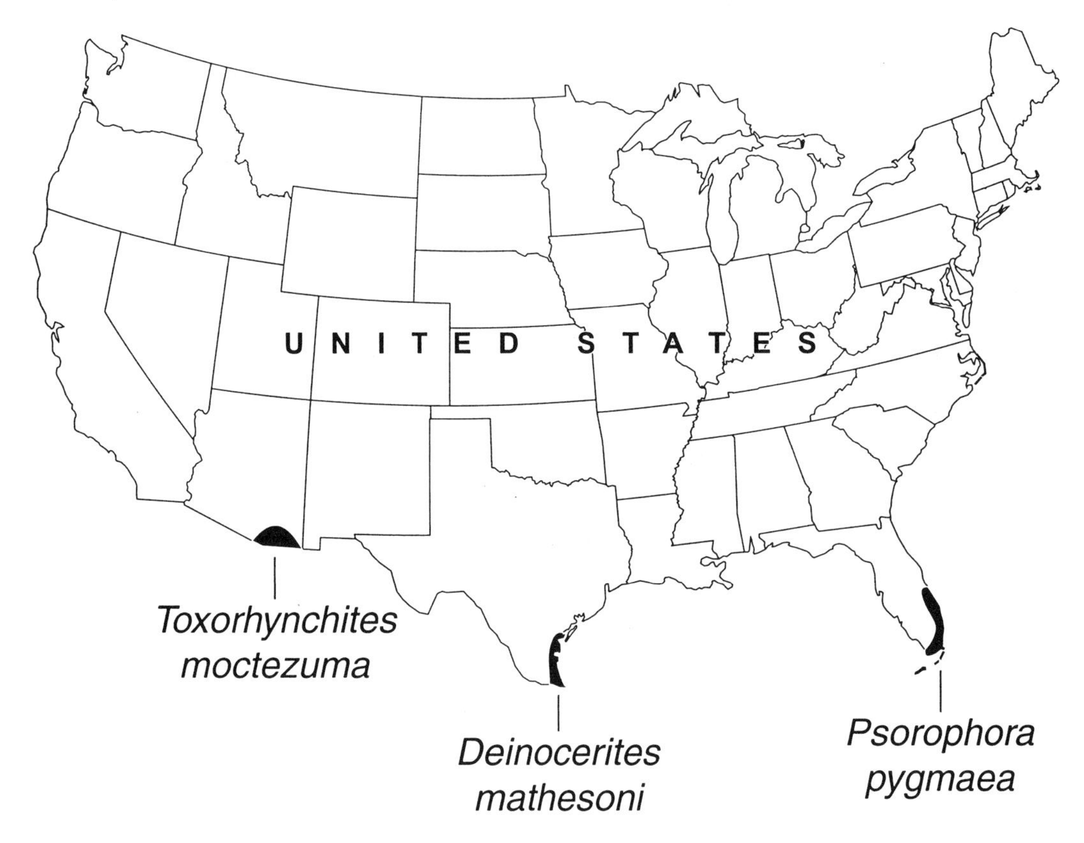


Plate 42 C. Distribution of *Deinocerites mathesoni*—USA: TX (43); Tax. 1, 43, 568. *Psorophora pygmaea*—USA: FL (146), not in MS (135); Tax. 36, 42. *Toxorhynchites moctezuma*—USA: AZ (794).

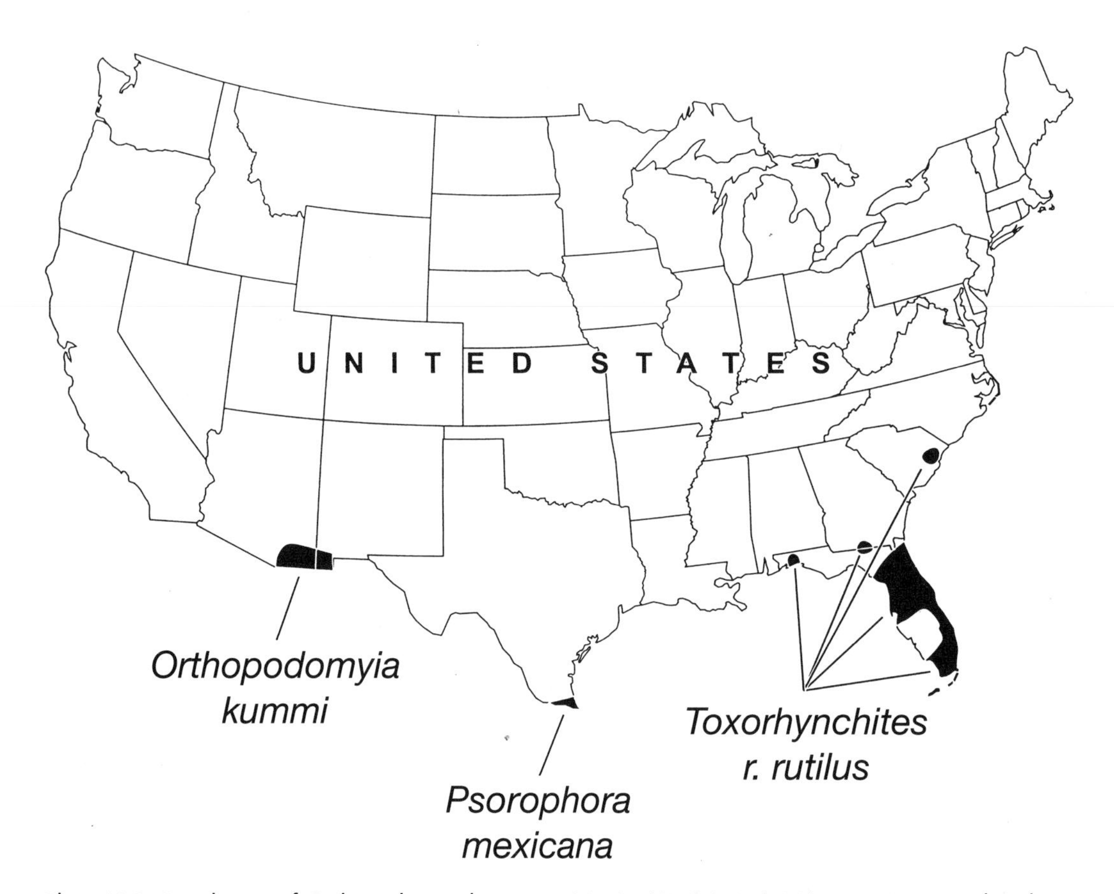


Plate 42 D. Distribution of *Orthopodomyia kummi*—USA: AZ (471), NM (532); Tax. 787. *Toxorhynchites r. rutilus*—USA: FL, GA, SC (146); Tax. 736. *Psorophora mexicana*—USA: TX (372); Tax. 36.

# Selected Bibliography of Mosquito Morphology


Barr, A. R., and C. M. Myers. 1962. Pupae of the genus *Culiseta* Felt. I. The homology of larval and pupal setae (Diptera: Culicidae). *Ann. Entomol. Soc. Am.* 55: 94–98.

Belkin, J. N. 1965. A revised nomenclature for the chaetotaxy of the mosquito larva (Diptera: Culicidae). *Am. Midl. Nat.* 44: 678–698.

———. 1952. The homology of the chaetotaxy of immature mosquitoes and a revised nomenclature for the chaetotaxy of the pupa (Diptera, Culicidae). *Proc. Entomol. Soc. Wash.* 54: 115–130.

———. 1953. Corrected interpretations of some elements of the abdominal chaetotaxy of the mosquito larva and pupa (Diptera, Culicidae). *Proc. Entomol. Soc. Wash.* 55: 318–324.

———. 1954. The dorsal hairless setal ring of mosquito pupae (Diptera, Culicidae). *Pan-Pacific Entomol.* 30: 227–230.

———. 1960. Innervation as a criterion of homology of the elements of the larval and pupal chaetotaxy of mosquitoes (Diptera, Culicidae). *Proc. Entomol. Soc. Wash.* 62: 197.

———. 1962. *The mosquitoes of the South Pacific (Diptera, Culicidae).* Vols. 1 & 2. Berkeley, Univ. Calif. Press, 608 pp. and 412 figures.

Carpenter, S. J., and W. J. LaCasse. 1955. *Mosquitoes of North America (North of Mexico).* Berkeley, Univ. Calif. Press, 360 pp., 127 pl.

Foote, R. H. 1952. The larval morphology and chaetotaxy of the *Culex* subgenus *Melanoconion* (Diptera, Culicidae). *Ann. Entomol. Soc. Am.* 45: 445–472.

Gardner, C. F., L. T. Nielsen, and K. L. Knight. 1973. Morphology of the mouthparts of the larval *Aedes communis* (Diptera: Culicidae). *Mosq. Syst.* 5: 163–182.

Gjullin, C. W., L. F. Lewis, and D. M. Christenson. 1968. Notes on the taxonomic characters and distribution of *Aedes aloponotum* Dyar and *Aedes communis* (Degeer). *Proc. Entomol. Soc. Wash.* 70: 133–136.

Harbach, R. E., and K. L. Knight. 1980. *Taxonomists' glossary of mosquito anatomy.* Marlton, Plexus Publ., 415 pp.

Hochman, R. H., and J. F. Reinert. 1974. Undescribed setae in larvae of Culicidae (Diptera). *Mosq. Syst.* 6: 1–10.

Lunt, S. R., and L. T. Nielsen. 1972. The use of thoracic setae as a taxonomic tool and as an aid in establishing phylogenetic relationship in adult female *Aedes* mosquitoes of North America. Part I. *Mosq. Syst.* 3: 69–98; Part II. *Mosq. Syst.* 3: 102–121.

Reinert, J. F. 1975. Mosquito generic and subgeneric abbreviations (Diptera: Culicidae). *Mosq. Syst.* 7: 105–110.

———. 1976. A ventromedian cervical sclerite of mosquito larvae (Diptera: Culicidae). *Mosq. Syst.* 8: 205–208.

———. 1982. Abbreviations for mosquito generic and subgeneric taxa established since 1975 (Diptera: Culicidae). *Mosq. Syst.* 14: 124–126.

———. 1991. Additional abbreviations of mosquito subgenera. names established since 1982 (Diptera: Culicidae). *Mosq. Syst.* 23: 209–210.

———. 2002. Comparative anatomy of the female genitalia of genera and subgenera in tribe Aedini (Diptera: Culicidae). Part XIII. Genus *Ochlerotatus* Lynch Arribalzaga. Part XIV. Key to genera. *Contrib. Am. Entomol. Inst.* 33(1): 1–117.

Romoser, W. S., and J. G. Stoffolano, Jr. 1998. *The science of entomology.* New York, McGraw-Hill, 720 pp.

Snodgrass, R. E. 1959. The anatomical life of the mosquito. *Smithsonian Misc. Coll.* 139: 1–87.

Wharton, R. H. 1962. The biology of *Mansonia* mosquitoes in relation to the transmission of filariasis in Malaya. *Bull. Med. Res. Fed. Malaya* 11: 1–114.

Wilkerson, R. C., and E. L. Peyton. 1990. Standardized nomenclature for the costal wing spots of the genus *Anopheles* and other spotted-wing mosquitoes (Diptera: Culicidae). *J. Med. Entomol.* 27: 207–224.

Wolff, T. A., and L. T. Nielsen. 1977. A chaetotaxic study of snowpool *Aedes* larvae and pupae with analysis of variance of larvae of eight species. *Mosq. Syst.* 9: 176–236.

# Bibliography of Mosquito Taxonomy and Geographical Distribution


1. Adames, A. J. 1971. Mosquito studies (Diptera, Culicidae). XXIV. A revision of the crabhole mosquitoes of the genus *Deinocerites*. *Contrib. Am. Entomol. Inst.* 7(2): 1–154.
2. Amin, 0. M., and A. G. Hageman. 1974. Mosquitoes and tabanids in southeast Wisconsin. *Mosq. News* 34: 170–177.
3. Amrine, J. W., and L. Butler. 1978. Annotated list of the mosquitoes of West Virginia. *Mosq. News* 38: 101–104.
4. Andreadis, T. G. 1986. New state records for *Aedes communis* and *Aedes punctor* in Connecticut. *J. Am. Mosq. Control Assoc.* 2: 378–379.

   Andreadis, T. G., J. F. Anderson, L. E. Munstermann, R. J. Wolfe, and D. A. Florin. 2001. See ref. 808.
5. Andreadis, T. G., and L. E. Munstermann. 1997. Intraspecific variation in key morphological characters of *Culiseta melanura* (Diptera: Culicidae). *J. Am. Mosq. Control Assoc.* 13: 127–133.
6. Andreadis, T. G., P. M. Capostosto, R. E. Shope, and S. J. Tirrell. 1994. Mosquito and arbovirus surveillance in Connecticut, 1991–1992. *J. Am. Mosq. Control Assoc.* 10: 556–564.
7. Anonymous. 1951. Mosquito records from the Missouri River Basin states. Federal Security Agency, Publ. Hlth. Serv., Surv. Sect., 93 pp. (mimeo).
8. ———. 1978. *Aedes flavescens* in Virginia. *Skeeter* 33:2.
9. Ansell, W. 1994 (1995). *Aedes albopictus* in the Gulf Mosquito Control District—Bay County, Florida. *J. Fla. Mosq. Control Assoc.* 65: 20.
10. Arnell, J. H. 1973. Mosquito studies (Diptera, Culicidae). XII. A revision of the genus *Haemagogus*. *Contrib. Am. Entomol. Inst.* 10(2): 1–174.
11. ———. 1976. Mosquito studies (Diptera, Culicidae). XXXIII. A revision of the *scapularis* group of *Aedes* (*Ochlerotatus*). *Contrib. Am. Entomol. Inst.* 13(3): 1–144
12. Arnell, J. H., and L. T. Nielsen. 1967. Notes on the distribution and biology of tree hole mosquitoes in Utah. *Proc. Utah Mosq. Abatement Assoc.* 20: 28–29.
13. Arnell, J. H., and L. T. Nielsen. 1972. Mosquito studies (Diptera, Culicidae). XXVII. The *varipalpus* group of *Aedes* (*Ochlerotatus*). *Contrib. Am. Entomol. Inst.* 8(2): 1–48.
14. Ashton, A. D., and F. C. Rabalais. 1977. A survey of mosquitoes in Wood County, Ohio. *Mosq. News* 37: 767–770.
15. Axtell, R. C. (ed.). 1974. *Training manual for mosquito and biting fly control in coastal areas.* Univ. N. C. Sea Grant Prog. Publ. UNC-SG-74-08, 249 pp.
16. Baker, M. 1961. The altitudinal distribution of mosquito larvae in the Colorado front range. *Trans. Am. Entomol. Soc.* 87: 231–246.
17. Barr, A. R. 1955. The resurrection of *Aedes melanimon* Dyar. *Mosq. News* 15: 170–172.
18. ———. 1957a. The distribution of *Culex p. pipiens* and *C. p. quinquefasciatus* in North America. *Am. J. Trop. Med. Hyg.* 6: 153–165.
19. ———. 1957b. A new species of *Culiseta* (Diptera: Culicidae) from North America. *Proc. Entomol. Soc. Wash.* 59: 163–167.
20. ———. 1958. *The mosquitoes of Minnesota (Diptera: Culicidae: Culicinae).* Univ. Minn, Agr. Exp. Sta. Tech. Bull. 228, 154 pp.
21. ———. 1960. A review of recent findings in the systematic status of *Culex pipiens*. *Calif. Vector Views* 7: 17–2 1.
22. ———. 1967. Occurrence and distribution of the *Culex pipiens* complex. *WHO Bull.* 37: 293–296.
23. ———. 1988. The *Anopheles maculipennis* complex (Diptera: Culicidae) in western North America, pp. 19–24. In: M. W. Service (ed.), *Biosystematics of haematophagous insects.* Clarendon Press, Oxford.
24. Barr, A. R., and W. V. Balduf. 1965. *Aedes decticus* Howard, Dyar and Knab in Minnesota. *Mosq. News* 25: 344.
25. Barr, A. R., and P. R. Ehrlich. 1958. Mosquito records from the Chukchi Sea coast of northwestern Alaska. *Mosq. News* 18: 12–14.
26. Barr, A. R., and P. Guptavanij. 1988. *Anopheles hermsi* n.sp., an unrecognized American species of the *Anopheles maculipennis* group (Diptera: Culicidae). *Mosq. Syst.* 20: 352–356.

27. Barr, A. R., and H. McMillan. 1956. Preliminary observations on light trap catches of mosquitoes in Kansas. *Proc. Entomol. Soc. Am. N. Cent. Br.* 11: 7.

28. Bartnett, R. E., and B. L. Davis. 1986. *Aedes albopictus* introduction—Texas. DHHS, *Morb. Mort. Weekly Rep.* 35: 141–142.

29. Beadle, L. D. 1957. A review of the mosquito problem in New York City. U.S. Public Health Serv., 25 pp.

30. ———. 1963. The mosquitoes of Isle Royale, Michigan. *Proc. 50th Mtg. N.J. Mosq. Exterm. Assoc.*, 133–139.

31. Beadle, L. D., and F. C. Harmston. 1958. Mosquitoes in sewage stabilization ponds in the Dakotas. *Mosq. News* 18: 293–296.

32. Beck, D. E. 1961. Central Utah County, Utah, mosquito survey studies. *Mosq. News* 21:6–11.

33. Beck, E. C. 1969. The *Culex* (*Melanoconion*) mosquitoes of Florida. *Fla. Anti-Mosq. Assoc. Rept.* 40: 38–43.

34. Beckel, W. E. 1954. The identification of adult female *Aedes* mosquitoes (Diptera, Culicidae) of the black-legged group taken in the field at Churchill, Manitoba. *Can. J. Zool.* 32: 324–330.

35. Beckel, W. E., and H. L. Atwood. 1959. A contribution to the bionomics of the mosquitoes of Algonquin Park. *Can. J. Zool.* 37: 763–770.

36. Belkin, J. N. 1968. Mosquito studies (Diptera, Culicidae). IX. The type specimens of New World mosquitoes in European museums. *Contrib. Am. Entomol. Inst.* 3(4): 1–69.

37. ———. 1969a. *Culex* (*Melanoconion*) *annulipes* invalid. *Mosq. Syst. Newsl.* 1: 68.

38. ———. 1969b. The problem of the identity of the species of *Culex* (*Melanoconion*) related to *opisthopus*. *Mosq. Syst. Newsl.* 1: 26–28.

39. ———. 1977a. *Quinquefasciatus* or *fatigans* for the tropical (southern) house mosquito (Diptera: Culicidae). *Proc. Entomol. Soc. Wash.* 79: 45–52.

40. ———. 1977b *Aedes* (*Ochlerotatus*) *pix* Martini 1935 a synonym of *Aedes* (*0.*) *taeniorhynchus* (Wiedemann 1921). *Mosq. Syst.* 9: 535.

41. Belkin, J. N., and S. J. Heinemann. 1975. *Psorophora* (*Janthinosoma*) *mathesoni* sp. nov. for "*varipes*" of the southeastern U.S. A. *Mosq. Syst.* 7: 363–366.

42. Belkin, J. N., S. J. Heinemann, and W. A. Page. 1970. Mosquito studies (Diptera, Culicidae). XXI. The Culicidae of Jamaica. *Contrib. Am. Entomol. Inst.* 6(l): 1–458.

43. Belkin, J. N., and C. L. Hogue. 1959. A review of the crabhole mosquitoes of the genus *Deinocerites* (Diptera, Culicidae). *Univ. Calif. Publ. Entomol.* 14: 411–458.

44. Belkin, J. N., and W. A. McDonald. 1956a. *Aedes sierrensis* (Ludlow, 1905), a change in name for the western tree-hole mosquito of the Pacific slope. *Proc. Entomol. Soc. Wash.* 58: 344.

45. Belkin, J. N., and W. A. McDonald. 1956b. A population of *Uranotaenia anhydor* from Death Valley, with description of all stages and discussion of the complex (Diptera, Culicidae). *Ann. Entomol. Soc. Am.* 49: 105–132.

46. Belkin, J. N., and W. A. McDonald. 1957. A new species of *Aedes* (*Ochlerotatus*) from tree holes in southern Arizona and a discussion of the *varipa1pus* complex (Diptera: Culicidae). *Ann. Entomol. Soc. Am.* 50: 179–191.

47. Belkin, J. N., R. X. Schick, and S. J. Heinemann. 1966. Mosquito studies (Diptera, Culicidae). VI. Mosquitoes originally described from North America. *Contrib. Am. Entomol. Inst.* 1(6):1–39.

48. Bell, D. D., and J. L. Benach. 1973. *Aedes aegypti* in southeastern New York State. *Mosq. News* 33: 248–249.

49. Bellamy, R. E. 1956. An investigation of the taxonomic status of *Anopheles perplexens* Ludlow, 1907. *Ann. Entomol. Soc. Am.* 49: 515–529.

50. Belton, E. M., and P. Belton. 1990. A review of mosquito collecting in the Yukon. *J. Entomol. Soc. British Columbia* 87: 35–37.

51. Belton, P. 1978a. The mosquitoes of Burnaby Lake, British Columbia. *J. Entomol. Soc. British Columbia* 75: 20–22.

52. ———. 1978b. An erroneous reference to *Aedes aegypti* (L.) in British Columbia. *J. Entomol. Soc. British Columbia* 75: 24.

53. ———. 1980. The first record of *Aedes togoi* (Theo.) in the United States: aboriginal, or ferry passenger? *Mosq. News* 40: 624–626.

54. ———. 1983. *The mosquitoes of British Columbia*. Brit. Columb. Prov. Mus. Handbk. 41, 189 p.

55. Belton, P., and E. M. Belton. 1981. A revised list of the mosquitoes of British Columbia. *J. Entomol. Soc. British Columbia* 78: 55–64.

56. Belton, P., and O. C. Belton. 1990. *Aedes togoi* comes aboard. *J. Am. Mosq. Control Assoc.* 6: 328–329.

57. Belton, P., and D. E. French. 1967. A specimen of *Aedes thibaulti* collected near Belleville, Ontario, Canada. *Can. Entomol.* 99: 1336.

58. Belton, P., and M. M. Galloway. 1965 (1966). Light-trap collections of mosquitoes near Belleville, Ontario, in 1965. *Proc. Entomol. Soc. Ont.* 96: 90–96.

59. Bennett, S. G. 1983. A new record of the treehole mosquito *Aedes sierrensis* (Diptera: Culicidae) from Santa Catalina Island, California. *Bull. Soc. Vector Ecol.* 8: 139–140.

60. Berlin, O. G. W. 1969a. Mosquito studies (Diptera, Culicidae). XII. A revision of the Neotropical subgenus *Howardina* of *Aedes*. *Contrib. Am. Entomol. Inst.* (Ann Arbor) 4(2): 1–190.

61. ———. 1969b. Mosquito studies (Diptera, Culicidae). XVII. The subgenus *Micraedes* of *Culex*. *Contrib. Am. Entomol. Inst.* (Ann Arbor) 5(1): 21–63.

62. Berlin, O. G. W., and J. N. Belkin. 1980. Mosquito studies (Diptera, Culicidae). XXXVI. Subgenera *Aedinus*, *Tinolestes* and *Anoedioporpa* of *Culex*. *Contrib. Am. Entomol. Inst.* (Ann Arbor) 17(2): 1–104.

63. Berry, R. L., E. D. Peterson, and R. A. Restifo. 1988. Records of imported tire-breeding mosquitoes in Ohio. *J. Am. Mosq. Control Assoc.* 4:187–189.

64. Berry, W. J. 1985. Collection of *Aedes atropalpus* from coastal rock holes on the Keweenaw Peninsula, Michigan. *J. Am. Mosq. Control Assoc.* 1:373–374.

65. Berry, W. J., W. A. Rowley, and M. Reynolds. 1986. Collection of *Psorophora howardii* in Scott County, Iowa. *J. Am. Mosq. Control Assoc.* 2: 563.

66. Betts, R. R. 1994 (1995). *Aedes albopictus* and *Aedes aegypti*: species domination in St. Johns County, Florida. *J. Fla. Mosq. Control Assoc.* 65: 17–19.

67. Bickley, W. E. 1957a. Notes on the distribution of mosquitoes in Maryland and Virginia. *Mosq. News* 17: 22–25.

68. ———. 1957b. Note on the occurrence of *Aedes atropalpus* (Coq.) in Western Maryland. *Mosq. News* 17: 318.

69. ———. 1976a. Notes on the distribution of Alaskan mosquitoes. *Mosq. Syst.* 8: 232–236.

70. ———. 1976b. The *Psorophora confinnis* complex. *Mosq. News* 36: 376.

71. ———. 1979. Notes on the geographical distribution of three species of *Culiseta*. *Mosq. News* 39: 392.

72. ———. 1980. Notes on the status of *Aedes cinereus hemiteleus* Dyar. *Mosq. Syst.* 12: 367–369.

73. ———. 1981(1982). Notes on the distribution of *Aedes canadensis mathesoni*. *Mosq. Syst.* 13: 150–152.

74. ———. 1987. Some Maryland mosquitoes are extremists. *Maryland Naturalist* 31: 2–5.

75. Bickley, W. E., and B. A. Harrison. 1989. Separation of variable *Culex territans* specimens from other *Culex* (*Neoculex*) in North America. *Mosq. Syst.* 21: 188–196.

76. Bickley, W. E., S. R. Joseph, J. Mallack, and R. A. Berry. 1971. An annotated list of the mosquitoes of Maryland. *Mosq. News* 31: 186–190.

77. Bickley, W. E., and J. Mallack. 1978. *Wyeomyia haynei* in Maryland. *Mosq. News* 38: 14: 141

78. Blickle, R. L. 1952. Notes on the mosquitoes (Culicinae) of New Hampshire. *Proc. 39th Mtg. N.J. Mosq. Exterm. Assoc.*, 198–202.

79. Blinn, D. W., and M. W. Sanderson. 1989. Aquatic insects in Montezuma Well, Arizona, USA: a travertine spring mound with high alkalinity and dissolved carbon dioxide. *Great Basin Naturalist* 49: 86–88.

80. Bloem, S. C.-G. 1991. Enzymatic variation in the *varipalpus* group of *Aedes* (*Ochlerotatus*) (Diptera: Culicidae). *Ann. Entomol. Soc. Am.* 84: 217–227.

81. Bohart, R. M. 1956. Identification and distribution of *Aedes melanimon* and *Aedes dorsalis*. *Proc. Calif. Mosq. Control Assoc.* 24: 81–83.

82. Bohart, R. M., and R. K. Washino. 1978. *Mosquitoes of California*. 3rd ed. Univ. Calif. Div. Agr. Sci., Berkeley, Publ. 4084, 153 pp.

83. Bosworth, A. B., S. M. Meola, and J. K. Olson. 1983. The chorionic morphology of eggs of the *Psorophora confinnis* complex in the United States. I. Taxonomic considerations. *Mosq. Syst.* 15:285–309.

84. Bourassa, J. P., A. Maire, and A. Aubin. 1976. Nouvelles donnees sur al chorologie et ecologie de quelques especes de culicides (Dipteres) dans le Quebec meridional. *Can. Entomol.* 108: 731–735.

85. Bradley, C. L., and R. L. Post. 1960. Keys to the more common North Dakota mosquitoes with comments on their biology and distribution. *N. Dak. Assoc. Sanitarians Newsl.* Appendix (Mar.), 1–9.

86. Bradshaw, W. E., and L. P. Lounibos. 1977. Evolution of dormancy and its photoperiodic control in pitcher-plant mosquitoes. *Evolution* 31: 546–567.

87. Bram, R. A. 1967. Classification of *Culex* subgenus *Culex* in the New World (Diptera: Culicidae). *Proc. U.S. Natl. Mus.* 120 (3557): 1–122.

88. Branch, N., L. Logan, E. C. Beck, and J. A. Mulrennan. 1958. New distributional records for Florida mosquitoes. *Fla. Entomol.* 41: 155–163.

89. Brandenburg, J. F., and R. D. Murrill. 1947. Occurrence and distribution of mosquitoes in Arkansas. *Ark. Hlth. Bull.* 4: 4–6.

90. Breeland, S. G. 1956. The occurrence of *Orthopodomyia alba* Baker in Tennessee (Diptera: Culicidae). *J. Tenn. Acad. Sci.* 31: 101.

91. ———. 1982. Bibliography and notes on Florida mosquitoes with limited distribution in the United States. *Mosq. Syst.* 14: 53–72.

92. Breeland, S. G., and T. M. Loyless. 1989. Illustrated keys to the mosquitoes of Florida. Adult females and fourth instar larvae. 2nd ed. Fla. Dept. Hlth. Rehab. Serv., 24 pp.

93. Breeland, S. G., W. E. Snow, and E. Pickard. 1961. Mosquitoes of the Tennessee Valley. *J. Tenn. Acad. Sci.* 36: 249–319.

94. Breland, O. P. 1954. Notes on the larvae of *Uranotaenia syntheta* (Diptera: Culicidae). *J. Kans. Entomol. Soc.* 27: 156– 158.

95. ———. 1956a. An eastern extension of the range of the mosquito *Culex apicalis* Adams (Diptera, Culicidae). *Proc. Entomol. Soc. Wash.* 58: 23–24.

96. ———. 1956b. Some remarks on Texas mosquitoes. *Mosq. News* 94–97.

97. ———. 1958. Notes on the *Aedes muelleri* complex (Diptera, Culicidae). *Proc. Entomol. Soc. Wash.* 60: 206.

98. ———. 1960. Restoration of the name *Aedes hendersoni* Cockerell, and its elevation to full specific rank (Diptera: Culicidae). *Ann. Entomol. Soc. Am.* 53: 600–606.

99. Briet, C. G. 1970. New state record. *USDA Coop. Econ. Insect Rept.* 20: 723.

100. Brigham, A. R., W. U. Brigham, and A. Gnilka. 1982. *Aquatic insects and oligochaetes of North and South Carolina.* Mahomet, IL, Midwest Aquatic Enterprises, 824 pp.

101. Brogdon, W. G. 1984a. The siphonal index. I. A method for evaluating *Culex pipiens* subspecies and intermediates. *Mosq. Syst.* 16: 144–152.

102. ———. 1984b. The siphonal index. II. Relative abundance of *Culex* species, subspecies, and intermediates in Memphis, Tennessee. *Mosq. Syst.* 16:153–167.

103. Brooks, I. C. 1947. Tree-hole mosquitoes in Tippecanoe County, Indiana. *Proc. Indiana Acad. Sci.* 56: 154–156.

104. Brothers, D. R. 1971. A check list of the mosquitoes of Idaho. *Tebiwa* 14: 72–73.

105. Brown, W. L., Jr. 1948. Results of the Pennsylvania mosquito survey for 1947. *J. N.Y. Entomol. Soc.* 56: 219–232.

106. Brust, R. A. 1979. Occurrence of *Aedes hendersoni* in Manitoba. *Mosq. News* 39: 395–396.
107. Brust, R. A., and K. S. Kalpage. 1967. New records for *Aedes* species in Manitoba. *Mosq. News* 7: 117–118.
108. Brust, R. A., and L. E. Munstermann. 1992. Morphological and genetic characterization of the *Aedes* (*Ochlerotatus*) *communis* complex (Diptera: Culicidae) in North America. *Ann. Entomol. Soc. Am.* 85: 1–10.
109. Burbutis, P. P. 1958. A new key to the mosquitoes of New Jersey. *Proc. 45th Mtg. N.J. Mosq. Exterm. Assoc.*, 209–212.
110. Burbutis, P. P., and R. W. Lake. 1959. New mosquito records for New Jersey. *Mosq. News* 19: 99–100.
111. Burger, J. F. 1965. *Aedes kompi* Vargas and Downs 1950, new to the United States. *Mosq. News* 25: 396–398.
112. ———. 1977a. New state record. *USDA Coop. Plant Pest Rept.* 2: 674, 676.
113. ———. 1977b. New State record. *USDA Coop. Plant Pest Rept.* 2: 706, 708.
114. ———. 1981. New records of mosquitoes (Diptera: Culicidae) from New Hampshire. *Entomol. News* 92: 49–59.
115. Burgess, L. 1957. Note on *Aedes melanimon* Dyar, a mosquito new to Canada (Diptera: Culicidae). *Can. Entomol.* 89: 532.
116. Burgess, L., and J. G. Rempel. 1971. Collection of the pitcher plant mosquito, *Wyeomyia smithii* (Diptera: Culicidae), from Saskatchewan. *Can. Entomol.* 103: 886–887.
117. Butler, L., and J. W. Amrine. 1980. New state and county records for mosquitoes in West Virginia. *Mosq. News* 40: 347–350.
118. Carpenter, S. J. 1961a. Observations on the distribution and ecology of mountain *Aedes* mosquitoes in California. I. Species and their habitats. *Calif. Vector Views* 8: 49–53.
119. ———. 1961b. Observations on the distribution and ecology of *Aedes* mosquitoes in California. II. *Aedes cataphylla* Dyar. *Calif. Vector Views* 8: 61–63, 65.
120. ———. 1962a. A collection of *Aedes sierrensis* (Ludlow) from Idaho. *Calif. Vector Views* 9:3.
121. ———. 1962b. Observations on the distribution and ecology of mountain *Aedes* mosquitoes in California. III. *Aedes communis* (DeGeer). *Calif. Vector Views* 9: 5–9.
122. ———. 1962c. Observations on the distribution and ecology of mountain *Aedes* mosquitoes in California. IV. *Aedes fitchii* (Felt and Young). *Calif. Vector Views* 9: 17–21.
123. ———. 1962d. Observations on the distribution and ecology of mountain *Aedes* mosquitoes in California. V. *Aedes hexodontus* Dyar. *Calif. Vector Views* 9: 27–32.
124. ———. 1962e. Observations on the distribution and ecology of mountain *Aedes* mosquitoes in California. VI. *Aedes increpitus* Dyar. *Calif. Vector Views* 9: 39–43.
125. ———. 1962f. Observations on the distribution and ecology of mountain *Aedes* mosquitoes in California. VII *Aedes cinereus* Meigen. *Calif. Vector Views* 9: 49–52.
126. ———. 1962g. Observations on the distribution and ecology of mountain *Aedes* mosquitoes in California. VIII. *Aedes schizopinax* Dyar. *Calif. Vector Views* 9: 61–63.
127. ———. 1963. Observations on the distribution and ecologyof mountain *Aedes* mosquitoes in California. IX. *Aedes ventrovittis* Dyar. *Calif. Vector Views* 10: 5–10.
128. ———. 1965. *Culiseta impatiens* (Walker), with keys to the species of *Culiseta* in California. *Calif. Vector Views* 12: 61–66.
129. ———. 1966. Observations on the distribution and ecology of mountain *Aedes* mosquitoes in California. X. Mosquito problems at Sierra Nevada recreational areas. *Calif. Vector Views* 13: 7– 13.
130. ———. 1968a. XI. Observations on the distribution and ecology of mountain *Aedes* mosquitoes in California. XI. *Aedes pullatus* (Coquillett). *Calif. Vector Views* 15: 7–14.
131. ———. 1968b. Review of recent literature on mosquitoes of North America. *Calif. Vector Views* 15: 71–98.
132. ———. 1969a. Observations on the distribution and ecology of mountain *Aedes* mosquitoes in California. XII. Other species found in the mountains. *Calif. Vector Views* 16: 27–32, 34.
133. ———. 1969b. Observations on the distribution and ecology of mountain *Aedes* mosquitoes in California. XIII. Mosquito problems in the Kings Canyon and Sequoia National Parks recreational region. *Calif. Vector Views* 16: 93–98.
134. ———. 1970a. Observations on the distribution and ecology of mountain *Aedes* mosquitoes in California. XIV. Mosquito problems in the Bishop Creek recreational region. *Calif. Vector Views* 17: 13–17.
135. ———. 1970b. Review of recent literature on mosquitoes of North America. Supplement 1. *Calif. Vector Views* 17: 39–65.
136. ———. 1970c. Observations on the distribution and ecology of mountain *Aedes* mosquitoes in California. XV. Mosquito problems in the Yosemite National Park recreational region. *Calif. Vector Views* 17: 101–107.
137. ———. 1971a. *Aedes campestris* and *Aedes niphadopsis* in California. *Calif. Vector Views* 18: 40.
138. ———. 1971b. Observations on the distribution and ecology of mountain *Aedes* mosquitoes in California. XVI. Mosquito problems in the Sonora Pass recreational area in the Sierra Nevada. *Calif. Vector Views* 18: 59–63.
139. ———. 1971c. Observations on the distribution and ecology of mountain *Aedes* mosquitoes in California. XVII. Mosquito problems in the Carson Pass recreational area in the Sierra Nevada. *Calif. Vector Views* 18: 69–74.
140. ———. 1972a. Observations on the distribution and ecology of mountain *Aedes* mosquitoes in California. XVIII. Mosquito problems in the Ebbetts Pass recreational area in the Sierra Nevada. *Calif. Vector Views* 19: 15–19.
141. ———. 1972b. Observations on the distribution and ecology of mountain *Aedes* mosquitoes in California. XIX. Mosquito problems in the Rock Creek recreational area in the Sierra Nevada. *Calif. Vector Views* 19: 81–86.
142. ———. 1973a. Observations on the distribution and ecology of mountain *Aedes* mosquitoes in California. XX. Mosquito problems in the Lake Basin recreational area in the Sierra Nevada. *Calif. Vector Views* 20: 11–17.
143. ———. 1973b. Observations on the distribution and ecology of mountain *Aedes* mosquitoes in California. XXI. Mosquito problems in the Lake Almanor recreational region. *Calif. Vector Views* 20: 19–27.

144. ———. 1974. Review of recent literature on mosquitoes of North America. Supplement II. *Calif. Vector Views* 21: 73–99.

145. Carpenter, S. J., and P. A. Gieke. 1974. Observations on the distribution and ecology of mountain *Aedes* mosquitoes in California. XXII. Mosquito problems in the Lake Tahoe recreational region in the Sierra Nevada. *Calif. Vector Views* 21: 1–8.

146. Carpenter, S. J., and W. J. LaCasse. 1955. *Mosquitoes of North America (North of Mexico).* Berkeley, Univ. Calif. Press, 360 pp., 127 pl.

147. Carpenter, S. J., and D. J. Womeldorf. 1968. Distribution and ecology of *Aedes bicristatus* Thurman and Winkler. *Calif. Vector Views* 15: 37–41.

148. Cassani, J. R., and R. G. Bland. 1978. New distribution records for mosquitoes in Michigan (Diptera: Culicidae). *Great Lakes Entomol.* 11: 51–52.

149. Cassani, J. R., and H. D. Newson. 1980. An annotated list of mosquitoes reported from Michigan. *Mosq. News* 40: 356–368.

150. Chant, G. D., W. F. Baldwin, and L. Forster. 1973. Occurrence of *Culiseta melanura* (Diptera: Culicidae) in Canada. *Can. Entomol.* 105: 1359.

151. Chapman, H. C. 1959a. A list of Nevada mosquitoes, with five new records. *Mosq. News* 19: 155–156.

152. ———. 1959b. Confirmation of *Aedes schizopinax* Dyar in California (Diptera: Culicidae). *Calif. Vector Views* 6: 61.

153. ———. 1960. Observations on *Aedes melanimon* and *A. dorsalis* in Nevada. *Ann. Entomol. Soc. Am.* 53: 706–708.

154. ———. 1961a. Additional records and observations on Nevada mosquitoes. *Mosq. News* 21: 136–138.

155. ———. 1961b. Observations on the snow-water mosquitoes of Nevada. *Mosq. News* 21: 88–92.

156. ———. 1963. Observations on *Aedes niphadopsis* Dyar and Knab and *campestris* Dyar and Knab in Nevada (Diptera: Culicidae) *Pan-Pacific Entomol.* 39: 109–114.

157. ———. 1966. The mosquitoes of Nevada. USDA, ARS, Entomol. Res. Div., Coll. Agric., Univ. Nev., 43 pp.

158. ———. 1968. Some notes on the mosquitoes of Louisiana, including the addition of *Aedes hendersoni* Cockerell. *Mosq. News* 28: 650–651.

159. Chapman, H. C., and A. R. Barr. 1964. *Aedes communis nevadensis*, a new subspecies of mosquito from western North America (Diptera: Culicidae). *Mosq. News* 24: 439–447.

160. Chapman, H. C., and R. C. Bechtel. 1969. Occurrence of *Culex pipiens quinquefasciatus* Say in Nevada. *Mosq. News* 29: 137.

161. Chapman, H. C., and G. Grodhaus. 1963. The separation of adult females of *Aedes dorsalis* (Meigen) and *A. melanimon* Dyar in California. *Calif. Vector Views* 10: 53–56.

162. Chapman, H. C., and E. B. Johnson. 1986. The mosquitoes of Louisiana. Louisiana Mosq. Control Assoc. Tech. Bull. 1 (revised), 17 pp.

163. Chew, R. M., and S. E. Gunstream. 1970. Geographical and seasonal distribution of mosquito species in southeastern California. *Mosq. News* 30: 551–562.

164. Christiansen, M. B., R. R. Pinger, Jr., and W. A. Rowley. 1972. A distributional note for *Culiseta silvestris minnesotae* Barr. *Mosq. News* 32: 637.

165. Christophers, S. R. 1960. Aedes aegypti *(L.), the yellow fever mosquito. Its life history, bionomics and structure.* Cambridge, Cambridge Univ. Press, 739 pp.

166. Christie, G. D., and R. A. LeBrun. 1990. *Culiseta minnesotae* and further notes on *Aedes aegypti* in Rhode Island. *J. Am. Mosq. Control Assoc.* 6: 742.

167. Cilek, J. E., G. D. Moorer, L. A. Delph, and F. W. Knapp. 1989. The Asian tiger mosquito, *Aedes albopictus*, in Kentucky. *J. Am. Mosq. Control Assoc.* 5: 267–268.

168. Clark, G. G., C. L. Crabbs, C. L. Bailey, C. H. Calisher, and G. B. Craig, Jr. 1986. Identification of *Aedes campestris* from New Mexico: with notes on the isolation of Western Equine Encephalitis and other arboviruses. *J. Am. Mosq. Control Assoc.* 2: 529–534.

169. Clover, J. R., E. E. Lusk, and G. Grodhaus. 1973. Additional locality records of mosquitoes from Northeastern California. *Calif. Vector Views* 20: 69–75.

170. Cockburn, A. F., J. Zhang, O. P. Perera, P. Kaiser, J. A. Seawright, and S. E. Mitchell. 1993. A new species of the *Anopheles crucians* complex: detection by mitochondrial DNA polymorphisms. *Host regulated development mechanisms in vector arthropods*, Proc. 3rd Symposium, 32–35.

171. Connell, W. A. 1941. Southern mosquitoes in Maryland. *Mosq. News* 1(3): 14–16.

172. Cook, F. E. 1960. Two new records of mosquito species for Minnesota. *Mosq. News* 20: 318–319.

173. Cook F. E., and W. L. Barton. 1974. *Aedes* (*Protomacleaya*) *hendersoni* Cockerell in Minnesota. *Mosq. News* 34: 232.

174. Cookman, J. E., and R. A. LeBrun. 1986. *Aedes aegypti* larvae in Portsmouth, Rhode Island. *J. Am. Mosq. Control Assoc.* 2: 96– 97.

175. Cookman, J. E., N. E. Scarduzio, and R. A. LeBrun. 1985. *Aedes thibaulti*: A new adult record from Rhode Island. *J. Am. Mosq. Control Assoc.* 1: 251.

176. Cope, S. E. 1989. The distribution of the *Anopheles maculipennis* complex in southern California with special reference to the ecology and biology of *Anopheles hermsi* Barr and Guptavanij (Diptera: Culicidae). Dissertation Abstracts International. B, Sciences and Engineering 50: 140-B.

177. Copeland, R. S. 1984. Occurrence of *Aedes mitchellae* in Indiana. *Mosq. News* 44: 80–81.

178. ———. 1986. The biology of *Aedes thibaulti* in northern Indiana. *J. Am. Mosq. Control Assoc.* 2: 1–6.

179. Copeland, R. S., and G. B. Craig, Jr. 1989. Winter cold influences the spatial and age distributions of the North American treehole mosquito, *Anopheles barberi*. *Oecologia* 79: 287–292.

180. Copps, P. T., G. A. Surgeoner, and B. V. Helson. 1984. Habitat distribution of adult mosquitoes in southern Ontario. *Proc. Entomol. Soc. Ont.* 115: 55–59.

181. Cornel, A. J., C. H. Porter, and F. H. Collins. 1996. Polymerase chain reaction species diagnostic assay for *Anopheles quadrimaculatus* cryptic species (Diptera: Culicidae) based on ribosomal DNA ITS2 sequences. *J. Med. Entomol.* 33: 109–116.

182. Costello, R. A. 1977. The first record of *Culiseta silvestris minnesotae* Barr in British Columbia (Diptera: Culicidae). *J. Entomol. Soc. British Columbia* 74: 9.

183. Courtney, C. C., and B. M. Christensen. 1982. Diversity and seasonal abundance of mosquitoes (Diptera: Culicidae) in Calloway County, Kentucky. *Trans. Ky. Acad. Sci.* 43: 55–59.

184. Covell, C. V., Jr. 1968. Mosquito control and survey in Jefferson County, Kentucky. *Mosq. News* 28: 526–529.

185. ———. 1971. The occurrence of *Aedes stimulans* (Walker) in Kentucky. *Mosq. News* 31:226.

186. Covell, C. V., Jr., and A. J. Brownell. 1979. *Aedes atropalpus* in abandoned tires in Jefferson County, Kentucky. *Mosq. News* 39: 142

187. Coyne, G. E., and L. E. Hagmann. 1970. Distribution of *Wyeomyia* species in New Jersey. *Proc. 57th Mtg. N.J. Mosq. Exterm. Assoc.*, 190–195.

188. Craig, G. B., Jr., and R. L. Pienkowski. 1955. The occurrence of *Aedes canadensis* (Theobald) in Alaska (Diptera, Culicidae). *Proc. Entomol. Soc. Wash.* 57: 268.

189. Craig, G. B., Jr., and W. A. Hawley. 1991. The Asian tiger mosquito, *Aedes albopictus*: whither, whence, and why not in Virginia. Information Series, Virginia Polytechnic Institute and State University, College of Agriculture and Life Sciences no. 91-2, 1–10.

190. Crans, W. J. 1967. *Anopheles earlei* Vargas, an addition to the checklist of New Jersey mosquitoes. *Mosq. News* 27: 430.

191. ———. 1970. The occurrence of *Aedes flavescens* (Muller), *Psorophora cyanescens* (Coquillett) and *Culex erraticus* (Dyar and Knab) in New Jersey. *Mosq. News* 30: 655.

192. ———. 1989. *Psorophora howardii*, a species with an increasing range in New Jersey. *Proc. 76th Mtg. N.J. Mosq. Control Assoc.*: 57–62.

193. Crans, W. J., and L. E. Hagmann. 1965. Two new mosquito records for New Jersey. *Proc. 52nd Mtg. N.J. Mosq. Exterm. Assoc.*: 206–207.

194. Crans, W. J., M. S. Chomsky, D. Guthrie, and A. Acquaviva. 1996. First record of *Aedes albopictus* from New Jersey. *J. Am. Mosq. Control Assoc.* 12: 307–309.

195. Crans, W. J., and D. A. Sprenger. 1996. The blood-feeding habits of *Aedes sollicitans* (Walker) in relation to eastern equine encephalitis in coastal areas of New Jersey. III. Habitat preference, vertical distribution, and diel periodicity of host-seeking adults. *J. Vector Ecol.* 21: 6–13.

196. Crans, W. J., and S. C. Crans. 1998. *Aedes thibaulti* in New Jersey. *J. Am. Mosq. Control Assoc.* 14: 348–350.

197. Crans, W. J., and L. J. McCuiston. 1999. An updated checklist of the mosquitoes of New Jersey *J. Am. Mosq. Control Assoc.* 15: 115–116.

198. Cupp, E. W., and W. R. Horsfall. 1969. Biological bases for placement of *Aedes sierrensis* (Ludlow) in the subgenus *Finlaya* Theobald. *Mosq. Syst. Newsl.* 1: 51–52.

199. Curtis, L. C. 1967. The mosquitoes of British Columbia. Occas. Pap. B.C. Prov. Mus. 15, 9 pp.

200. Dahl, C. 1974. Circumpolar *Aedes* (*Ochlerotatus*) species in North Fennoscandia. *Mosq. Syst.* 6: 57–73.

201. Danilov, V. N. 1974. On the restoration of the name *Aedes* (*0.*) *mercurator* Dyar of a mosquito known in the USSR as *Aedes riparius ater* Gutsevich (Diptera, Culicidae) (in Russian). *Parazitologiya* 8: 322–328.

202. ———. 1975. On the possible identity of the mosquitoes *Aedes* (*Ochlerotatus*) *behlemishevi* Denisova and *A.* (*0.*) *barri* Rueger (Diptera, Culicidae) (in Russian). *Parazitologiya* 9: 61– 63.

203. Danks, H. V., and P. S. Corbet. 1973. A key to all stages of *Aedes nigripes* and *A. impiger* (Diptera: Culicidae) with a description of first instar larvae and pupae. *Can. Entomol.* 105: 367–376.

204. Darsie, R. F., Jr. 1973. A record of changes in mosquito taxonomy in the United States of America 1955–1972. *Mosq. Syst.* 5: 187–193.

205. ———. 1974. The occurrence of *Aedes epactius* Dyar and Knab in Louisiana (Diptera, Culicidae). *Mosq. Syst.* 6: 229–230.

206. ———. 1978. Additional changes in mosquito taxonomy in North America, north of Mexico, 1972–1977. *Mosq. Syst.* 10: 246– 248.

207. ———. 1986. The identification of *Aedes albopictus* in the Nearctic Region. *J. Am. Mosq. Control Assoc.* 2: 336–340.

208. ———. 1989. Keys to the genera, and to the species of five minor genera, of mosquito pupae occurring in the Nearctic Region (Diptera: Culicidae). Mosq. Syst. 21: 1–10.

209. ———. 1992. Key characters for identifying *Aedes bahamensis* and *Aedes albopictus* in North America, North of Mexico. *J. Am. Mosq. Control Assoc.* 8: 323–324.

210. ———. 1995. Identification of *Aedes tahoensis*, *Aedes clivis* and *Aedes washinoi* using Darsie/Ward keys (Diptera: Culicidae). *Mosq. Syst.* 27: 40–42.

Darsie, R. F., Jr. 1996. See ref. 805.

211. Darsie, R. F., Jr., and A. W. Anderson. 1985. A revised list of the mosquitoes of North Dakota, including new additions to the fauna. *J. Am. Mosq. Control Assoc.* 1: 76–79.

212. Darsie, R. F., Jr., and E. J. Haeger. 1993. New mosquito records for South Carolina. *J. Am. Mosq. Control Assoc.* 9: 472– 473.

213. Darsie, R. F., Jr., and D. MacCreary. 1960. The occurrence of *Psorophora discolor* (Coquillett) in Delaware. *Proc. 47th Mtg. N.J. Mosq. Exterm. Assoc.*, 88–92.

214. Darsie, R. F., Jr., D. MacCreary, and L. A. Stearns. 1951. An annotated list of the mosquitoes of Delaware. *Proc. 38th Mtg. N.J. Mosq. Exterm. Assoc.*, 137–146.
215. Darsie, R. F., Jr., and C. D. Morris. 2000. *Keys to the adult females and fourth instar larvae of the mosquitoes of Florida (Diptera, Culicidae).* Tech. Bull. Fla. Mosq. Control Assoc. Vol. 1, 159 pp.
216. Darsie, R. F., Jr., and R. A. Ward. 1981. *Identification and geographical distribution of the mosquitoes of North America, north of Mexico.* Mosq. Syst. Suppl. 1: 1–313.
217. Darsie, R. F., Jr., and R. A. Ward 1989. Review of new Nearctic mosquito distributional records north of Mexico, with notes on additions and taxonomic changes of the fauna. *J. Am. Mosq. Control Assoc.* 5: 552–557.
218. Darsie, R. F., Jr., and R. M. Williams. 1976. First report of *Wyeomyia haynei* in Georgia, with comments on identification of larvae (Diptera, Culicidae). *Mosq. Syst.* 8: 441–444.
219. Davis, T., Jr., and D. M. Rees. 1957. The mosquitoes of Carey and vicinity, Blaine County, Idaho. *Proc. Utah Acad. Sci. Arts Letters* 34: 157.
220. Debboun, M., and R. D. Hall. 1992. Mosquitoes (Diptera: Culicidae) sampled from treeholes and proximate artificial containers in central Missouri. *J. Entomol. Sci.* 27: 19–28.
221. Debrunner-Vossbrinck, B. A., C. R. Vossbrinck, M. H. Vodkin, and R. J. Novak. 1996. Restriction analysis of the ribosomal DNA internal transcribed spacer region of *Culex restuans* and the mosquitoes in the *Culex pipiens* complex. *J. Am. Mosq. Control Assoc.* 12: 477–482.
222. DeFoliart, G. R., M. R. Rao, and C. D. Morris. 1967. Seasonal succession of bloodsucking Diptera in Wisconsin during 1965. *J. Med. Entomol.* 4: 363–373.
223. Denisova, Z. M. 1955. New aspect of *Aedes* (*Ochlerotatus*) (in Russian). *Med. Parasit.* 24: 58–61.
224. Denke, P. M., J. E. Lloyd, and J. L. Littlefield. 1996. Elevational distribution of mosquitoes in a mountainous area of southeastern Wyoming. *J. Am. Mosq. Control Assoc.* 12: 8–16.
Dennehy, J. J., and T. Livdahl. 1999. See ref. 798.
225. Dixon, E. B. 1955. The spread of *Aedes sollicitans* (Walker) in Kentucky. *Mosq. News* 15:42.
226. Dodge, H. R. 1962. Supergeneric groups of mosquitoes. *Mosq. News* 22: 365–368.
227. ———. 1963. Studies on mosquito larvae. I. Later instars of eastern North American species. *Can. Entomol.* 95: 796–813.
228. ———. 1964. Larval chaetotaxy and notes on the biology of *Toxorhynchites rutilus septentrionalis* (Diptera: Culicidae). *Ann. Entomol. Soc. Am.* 57: 46–53.
229. ———. 1966. Studies on mosquito larvae II. The first- stage larvae of North American Culicidae and of world Anophelinae. *Can. Entomol.* 98: 337–393.
230. Doll, J. M. 1970. Notes on the current distribution of *Aedes dorsalis* in central New York 1969. *Mosq. News* 30: 89.
231. Donnelly, J. W. 1992 (1991). Field collections of *Aedes aegypti* in Morris County. *Proc. 79th Mtg. N.J. Mosq. Control Assoc.* pp. 92–95.
232. ———. 1993. *Aedes aegypti* in New Jersey. *J. Am. Mosq. Control Assoc.* 9: 238.
233. Duryea, R. D. 1990. *Aedes trivittatus* in New Jersey. *Proc. 77th Mtg. N.J. Mosq. Control Assoc.*, 73–78.
234. Dubitskii, A. M. 1977. A description of the imago of a little-known species of mosquito, *Aedes* (*Ochlerotatus*) *rempeli* (Culicidae) (in Russian). *Parazitologiya* 11: 72–74.
235. Durand, M., and D. de Oliveira. 1977. Note on Culicidae of the Upper Richelieu, Quebec. *Mosq. News* 37: 423–425.
236. Dyar, H. G. 1919. Westward extension of the Canadian mosquito fauna (Diptera, Culicidae). *Ins. Insc. Menst.* 7: 11–39.
237. ———. 1928. *The mosquitoes of the Americas.* Carnegie Inst. Wash. Publ. 387, 616 pp., 123 pl.
238. Eads, R. B., J. G. Foyle, and R. E. Peet. 1960. Mosquito densities in Orange County, Texas. *Mosq. News* 20: 49–52.
239. Eads, R. B., and L. G. Strom. 1957. An additional United States record of *Haemagogus equinus*. *Mosq. News* 17: 86–89.
240. Easton, E. R., M. A. Price, and O. H. Graham. 1968. The collection of biting flies in West Texas with malaise and animal-baited traps. *Mosq. News.* 28: 465–469.
241. Edman, J. D. 1962. New mosquito records for South Dakota. *J. Kans. Entomol. Soc.* 35: 430–432.
242. ———. 1964. Control of *Culex tarsalis* (Coquillett) and *Aedes vexans* (Meigen) on Lewis and Clark Lake (Gavins Point Reservoir) by water level management. *Mosq. News* 24: 173–185.
243. Edmunds, L. R. 1957. A note on the biology of the mosquito, *Psorophora discolor* (Coquillett), in Mississippi (Diptera: Culicidae). *Ohio J. Sci.* 57: 313–314.
244. ———. 1958. Field observations on the habitats and seasonal abundance of mosquito larvae in Scotts Bluff County, Nebraska (Diptera, Culicidae). *Mosq. News* 18: 23–26.
245. Ehrenberg, H. A. 1983. *Aedes spencerii spencerii* in New Jersey. *Proc. 70th Mtg. N.J. Mosq. Control Assoc.*, 96–97.
246. Elbel, R. E. 1968. Sight identification key for mosquitoes of the Great Salt Lake Basin. *Mosq. News* 28: 167–171.
247. Eldridge, B. F., and R. E. Harbach. 1992. Conservation of the names *Culex stigmatosoma* and *Culex thriambus*. *J. Am. Mosq. Control Assoc.* 8:104–105.
248. Eldridge, B. F., C. L. Bailey, and M. D. Johnson. 1972. A preliminary study of the seasonal geographic distribution and overwintering of *Culex restuans* Theobald and *Culex salinarius* Coquillett (Diptera: Culicidae). *J. Med. Entomol.* 9: 233–238
249. Eldridge, B. F., J. E. Gimmig, K. Lorenzen, K. C. Nixon, and W. C. Reeves. 1998. The distribution of species of the *Aedes increpitus* complex in the western United States. *J. Am. Mosq. Control Assoc.* 14: 173–177.
250. Ellis, R. A., and R. A. Brust. 1973. Sibling species delimitation in the *Aedes communis* (Degeer) aggregate (Diptera: Culicidae). *Can. J. Zool.* 51: 915–959.

251. Ellis, R. A. and D. M. Wood. 1974. First Canadian record of *Corethrella brakeleyi* (Diptera: Chaoboridae). *Can. Entomol.* 106: 221–222.

252. Enfield, M. A. 1977. Additions and corrections to the records of *Aedes* mosquitoes in Alberta. *Mosq. News* 37: 82–85.

253. Evans, E. S., Jr., and L. G. McCuiston. 1971. Preliminary mosquito survey of the Wharton State Forest, summer 1970. *Proc. 58th Mtg. N.J. Mosq. Exterm. Assoc.*, pp. 118–125.

254. Faran, M. E., and C. L. Bailey. 1980. Discovery of an overwintering adult female of *Culiseta annulata* in Baltimore. *Mosq. News* 40: 284–287.

255. Favorite, F. G., and R. Davis. 1958. Some observations on the mosquito fauna of the Okeefenokee Swamp. *Mosq. News* 18: 284–287.

256. Ferguson, F. F., and T. W. McNeel, Sr. 1954. The mosquitoes of New Mexico. *Mosq. News* 14: 30–31.

257. Fletcher, L. W. 1957. The mosquitoes of West Virginia. M.S. thesis, University of West Virginia, 41 pp.

258. Floore, T. G., B. A. Harrison, and B. F. Eldridge. 1976. The *Anopheles* (*Anopheles*) *crucians* subgroup in the United States (Diptera: Culicidae). *Mosq. Syst.* 8: 1–109.

259. Foote, R. H. 1954. The larvae and pupae of the mosquitoes belonging to the *Culex* subgenera *Melanoconion* and *Mochlostyrax*. USDA Tech. Bull. 1091, 126 pp.

Fonseca et al. 2001. See ref. 812.

260. Foster, B. E. 1989. *Aedes albopictus* larvae collected from tree holes in southern Indiana. *J. Am. Mosq. Control Assoc.* 5:95.

261. Fournier, P. V., and J. L. Snyder. 1977. Introductory manual on arthropod-borne disease surveillance. Part 1. Mosquito-borne encephalitis. Texas Dept. Health Resources, Bur. Lab., 92 pp.

262. Frank, J. H., and E. D. McCoy. 1992. Introduction to the behavioral ecology of immigration. The immigration of insects to Florida, with a tabulation of records published since 1970. *Fla. Entomol.* 75: 1–28.

263. Freeman, T. N. 1952. Interim report of the distribution of the mosquitoes obtained in the northern insect survey. Defense Research Board of Ottawa, Environmental Protection Tech. Rept. 1, 2 pp., 43 maps.

264. French, E. W., and B. W. Sweeney. 1971. Mosquitoes recorded in Bucks County, Pennsylvania USA and their relative abundance in the summer of 1970. *Melsheimer Entomol. Ser.* 9: 1–4.

Fritz, G. N., D. Dritz, T. Jensen, and R. K. Washino. 1991. See ref. 811.

265. Fritz, G. N., S. K. Narang, D. L. Kline, J. A. Seawright, R. K. Washino, C. H. Porter, and F. H. Collins. 1991. Diagnostic characterization of *Anopheles freeborni* and *An. hermsi* by hybrid crosses, frequencies of polytene X chromosomes and rDNA restriction enzyme frequencies. *J. Am. Mosq. Control Assoc.* 7: 198–206.

266. Fritz, G. N., and R. K. Washino. 1993. *Anopheles hermsi*, probable vector of malaria in New Mexico. *Am. J. Trop. Med. Hyg.* 49: 419–424.

267. Frohne, W. C. 1954. Mosquito distribution in Alaska with especial reference to a new type of life cycle. *Mosq. News* 14: 10–13.

268. ———. 1955a. Tundra mosquitoes at Naknek, Alaska Peninsula. *Trans. Am. Micro. Soc.* 74: 292–295.

269. ———. 1955b. Characteristic saddle spines of northern mosquito larvae. *Trans. Am. Micro. Soc.* 74: 295–302.

270. ———. 1957. Reconnaissance of mountain mosquitoes in the McKinley Park Region, Alaska. *Mosq. News* 17: 17–22.

271. Frohne, W. C., and S. A. Sleeper. 1951. Reconnaissance of mosquitoes, punkies and blackflies in southeast Alaska. *Mosq. News* 11: 209–213.

272. Fulton, H. R., P. P. Sikorowski, and B. R. Norment. 1974. A survey of north Mississippi mosquitoes for pathogenic microorganisms. *Mosq. News* 34: 86–90.

273. Gaffigan, T. V., and R. A. Ward. 1985. Index to the second supplement to *A catalog of the mosquitoes of the world* with corrections and additions (Diptera: Culicidae). *Mosq. Syst.* 17: 52–63.

274. Galindo, P. 1969. Notes on the systematics of *Culex* (*Melanoconion*) *taeniopus* Dyar and Knab and related species, gathered during arbovirus investigations in Panama. *Mosq. Syst. Newsl.* 1: 82–89.

275. Galindo, P., F. S. Blanton, and E. L. Peyton. 1954. A revision of the *Uranotaenia* of Panama with notes on other American species of the genus (Diptera, Culicidae). *Ann. Entomol. Soc. Am.* 47: 107–177.

276. Gallaway, W. J., and R. A. Brust. 1982. The occurrence of *Aedes hendersoni* Cockerell and *Aedes triseriatus* (Say) in Manitoba. *Mosq. Syst.* 14: 262–263

277. Gebara, A., and D. De Oliviera 1986. Premiere mention de *Culex tarsalis* (Diptera: Culicidae) au Quebec. *Can. Entomol.* 118: 609.

278. Georgia Department of Public Health. 1970. Illustrated key: Adult mosquitoes of Georgia. Pesticide and Vector Control SWM-01-24, 24 pp.

279. ———. 1971. Illustrated key: Larval mosquitoes of Georgia. Pesticide and Vector Control SWM-01-3, 26 pp.

280. Gerhardt, R. W. 1966a. South Dakota mosquito species. *Mosq. News* 26: 37–38.

281. ———. 1966b. South Dakota mosquitoes and their control. *S.D. Agric. Exp. Sta. Bull.* 531, 80 pp.

282. Gilardi, J. W., and W. L. Hilsenhoff. 1992. Distribution, abundance, larval habitats, and phenology of spring *Aedes* mosquitoes in Wisconsin (Diptera: Culicidae). *Trans. Wisconsin Acad. Sci. Arts Let.* 80: 35–50.

283. Gilot, B., G. Pautou, and G. Ain. 1975. Presence au Quebec de *Culiseta* (*Climacura*) *melanura* (Coquillett, 1902). *Ann. Parasit. Hum. Comp.* 50: 649–650.

284. Gjullin, C. M., and C. W. Eddy. 1972. *The mosquitoes of the northwestern United States.* USDA Tech. Bull. 1447, 111 pp.

285. Gjullin, C. M., L. F. Lewis, and D. M. Christenson. 1968. Notes on the taxonomic characters and distribution of *Aedes aloponotum* Dyar and *Aedes communis* (De Geer) (Diptera: Culicidae). *Proc. Entomol. Soc. Wash.* 70: 133–136.

286. Gjullin, C. M., R. W. Sailer, A. Stone, and B. V. Travis. 1961. *The mosquitoes of Alaska.* USDA Agric. Handbook 182, 98 pp.

287. Gladney, W. J., and E. C. Turner, Jr. 1968. Mosquito control on Smith Mountain Reservoir by pumped storage water level management. *Mosq. News* 28: 606–618.

288. ———. 1969. The insects of Virginia. 2. Mosquitoes of Virginia (Diptera: Culicidae). VPI Res. Div. Bull. 49, 24 pp.

289. Gorham, J. R. 1974. Tests of mosquito repellents in Alaska. *Mosq. News* 34: 409–415.

290. ———. 1975. Survey of stored-food insects and other Alaskan insect pests. *Bull. Entomol. Soc. Am.* 21: 113–117.

291. Graham, A. C., J. F. Turmel, and R. F. Darsie, Jr. 1991. New state mosquito records for Vermont including a checklist of the mosquito fauna. *J. Am. Mosq. Control Assoc.* 7: 502–503.

292. Graham, J. E. 1959. The current status of *Aedes nigromaculis* (Ludlow) in Utah. *Proc. Calif. Mosq. Conrol Assoc.* 27: 77–78.

293. Graham, P. 1969a. *Culiseta silvestris minnesotae* Barr and *C. morsitans dyari* (Coquillett) (Diptera: Culicidae) in Alberta. *Mosq. News* 29: 261–262.

294. ———. 1969b. Observations on the biology of the adult female mosquitoes (Diptera: Culicidae) at George Lake, Alberta, Canada. *Quaest. Entomol.* 5: 309–339.

295. Grimstad, P. R. 1977. Occurrence of *Orthopodomyia alba* Baker and *Orthopodomyia signifera* (Coquillett) in Michigan. *Mosq. News* 37: 129–130.

296. Grimstad, P. R., C. E. Garry, and G. R. DeFoliart. 1974. *Aedes hendersoni* and *Aedes triseriatus* (Diptera: Culicidae) in Wisconsin: Characterization of larvae, larval hybrids and comparison of adult and hybrid mesoscutal patterns. *Ann. Entomol. Soc. Am.* 67: 795–804.

297. Grimstad, P. R., and M. J. Mandracchia. 1985. Record of Michigan mosquito species (Diptera: Culicidae) collected in a natural focus of Jamestown Canyon virus in 1984. *Great Lakes Entomol.* 18: 45–49.

298. Grodhaus, G. 1959. Notes on the distribution of *Aedes schizopinax* Dyar in California. *Calif. Vector Views* 6: 67.

299. ———. 1970. Occurrence of *Aedes campestris* in California. *Calif. Vector Views* 17: 108.

300. Guirgis, S. S. 1984. A new record of *Culiseta annulata* with notes on mosquito species in Suffolk County, Long Island, New York. *Mosq. News* 44: 246.

301. ———. 1992. Occurrence of *Psorophora howardii* in Suffolk County, Long Island, New York. *J. Am. Mosq. Control Assoc.* 8: 197.

302. Guirgis, S. S., and J. F. Sanzone. 1978. New records of mosquitoes in Suffolk County, Long Island, New York. *Mosq. News* 38: 200–203.

303. Gutsevich, A. V., A. S. Monchadsky, and A. A. Stackelberg. 1974. *Fauna of the U.S.S.R. Diptera.* Vol. 111, No. 4, *Mosquitoes, Family Culicidae.* Akad. Nauk. SSSR Zool. Inst. N.S. No. 100, 1971, 384 pp. (English translation, 1974)

304. Haeger, J. S., and G. F. O'Meara. 1983. Separation of first-instar larvae of four Florida *Culex* (*Culex*). *Mosq. News* 43: 76– 77.

305. Hanson, S. M., R. J. Novak, R. L. Lampman, and M. H. Vodkin. 1995. Notes on the biology of *Orthopodomyia* in Illinois. *J. Am. Mosq. Control Assoc.* 11: 375–376.

306. Happold, D. C. D. 1965a. Mosquito ecology in central Alberta 1. The environment, the species, and studies of the larvae. *Can. J. Zool.* 43: 795–819.

307. ———. 1965b. Mosquito ecology in central Alberta II. Adult populations and activities. *Can. J. Zool.* 43: 821–846.

Harbach, R. E., and I. J. Kitching. 1998. See ref. 813.

308. Harbach, R. E., and K. L. Knight. 1980. *Taxonomists' glossary of mosquito anatomy.* Marlton, N.J., Plexus Publ., 415 pp.

309. Harbach, R. E., C. Dahl, and G. B. White. 1985. *Culex* (*Culex*) *pipiens* Linnaeus (Diptera, Culicidae): concepts, type designations and descriptions. *Proc. Entomol. Soc. Wash.* 87: 1– 24.

310. Harden, F. W. 1965. Mosquito control at NASA's Mississippi test operation. *Mosq. News* 25: 123–126.

311. Harden, F. W., H. R. Hepburn, and B. J. Ethridge. 1967. A history of mosquitoes and mosquito-borne diseases in Mississippi 1699–1965. *Mosq. News* 27: 60–66.

312. Harden, F. W., and B. J. Poolson. 1969. Seasonal distribution of mosquitoes of Hancock County, Mississippi, 1964–1968. *Mosq. News* 29: 407–414.

313. Harmston, F. C. 1969. Separation of the females of *Aedes hendersoni* Cockerell and *Aedes triseriatus* (Say) (Diptera: Culicidae) by tarsal claws. *Mosq. News* 29: 490–491.

314. Harmston, F. C., and F. A. Lawson. 1967. *Mosquitoes of Colorado.* Bureau of Disease Prevention and Environmental Control, U.S. Public Health Service, Atlanta, 140 pp.

315. Harmston, F. C., L. S. Miller, and R. A. McHugh. 1960. Survey of log pond mosquitoes in Douglas Country, Oregon, during 1956. *Mosq. News* 20: 351–353.

316. Harmston, F. C., G. R. Schultz, R. B. Facts, and G. C. Menzies. 1956. Mosquitoes and encephalitis in the irrigated high plains of Texas. *U.S. Public Health Reports* 71: 759–766.

317. Harrison, B. A., J. F. Reinert, E. S. Saugstad, R. Richardson, and J. E. Farlow. 1973. Confirmation of *Aedes taeniorhynchus* in Oklahoma. *Mosq. Syst.* 5: 157–158.

318. Harrison, B. A., and P. B. Whitt. 1996. Identifying *Psorophora horrida* females in North Carolina (Diptera: Culicidae). *J. Am. Mosq. Control Assoc.* 12: 725–727.

319. Harrison, B. A., P. B. Whitt, E. E. Powell, and E. Y. Hickman, Jr. 1998. North Carolina mosquito records: I. Uncommon *Aedes* and *Anopheles* (Diptera: Culicidae). *J. Am. Mosq. Control Assoc.* 14: 165–172.

320. Harrison, R. J., and G. Cousineau. 1973. Les moustiques au Quebec, leur importance medicale, veterinaire, economique et la necessite d'un programme de demoustication. *Ann. Entomol. Soc. Quebec* 18: 138–146.

321. Harrison, R. J. L. Loiselle, and D. J. LePrince. 1981. Revised list of the mosquitoes (Diptera: Culicidae) of Quebec. *Ann. Entomol. Soc. Quebec* 26: 3–8.

322. Hart, J. W. 1968a. A checklist of the mosquitoes of Indiana with a record of the occurrence of *Aedes infirmatus* D. and K. *Proc. Indiana Acad. Sci.* 78: 257–259.

323. ———. 1968b. Occurrence of *Toxorhynchites rutilus septentrionalis* (Dyar and Knab) in Indiana. *Mosq. News* 28: 118.

324. Haufe, W. O. 1952. Observations on the biology of mosquitoes (Diptera: Culicidae) at Goose Bay, Labrador. *Can. Entomol.* 84: 254–263.

325. Hawley, W. A., and G. B. Craig, Jr. 1989(1990). *Aedes albopictus* in the Americas: future prospects. *Arbovirus research in Australia,* Proc. 5th Symposium, pp. 202–205.

Hayden, C. W., T. M. Fink, F. B. Ramberg, C. J. Mare, and D. G. Mead. 2001. See ref. 806.

326. Hayes, J. 1965. A first report of *Aedes infirmatus* Dyar and Knab in Illinois. *Trans. Ill. State Acad. Sci.* 58: 151.

327. ———. 1970. Check list of the mosquitoes in Johnson and Massac Counties, Illinois. *Trans. Ill. State Acad. Sci.* 63: 109– 112.

328. Hayes, R. O. 1961. Host preference of *Culiseta melanura* and allied mosquitoes. *Mosq. News* 21: 179–187.

329. Hayes, R. O., D. B. Francy, J. S. Lazuick, G. C. Smith, and R. H. Jones. 1976. Arbovirus surveillance in six states during 1972. *Am. J. Trop. Med. Hyg.* 25: 463–476.

330. Heaps, J. W. 1980. Occurrence of *Orthopodomyia alba* in West Virginia. *Mosq. News* 40: 452.

331. Heard, S. B. 1994. Wind exposure and distribution of pitcherplant mosquito (Diptera: Culicidae). *Environ. Entomol.* 23: 1250–1253.

332. Heeden, R. A. 1963. The occurrence of *Aedes hendersoni* Cockerell in northern Illinois. *Mosq. News* 23: 349–350.

333. Helson, B. V., G. A. Surgeoner, R. A. Wright, and S. A. Allan. 1978. *Culex tarsalis*, *Aedes sollicitans*, *Aedes grossbecki*: new distribution records from southwestern Ontario. *Mosq. News* 38: 137–138.

334. Henricksen, K. L., and W. Lundbeck. 1917. Culicidae, pp. 595–596. In: Groenlands Landarthropoder (Insecta et Arachnida Groenlandicae). *Medd. Groenland* 22: 595–596.

335. Hicks, R. C. 1974. The occurrence of *Psorophora signipennis* Coquillett in Nevada. *Mosq. News* 34: 119.

336. Hill, N. D. 1939. Biological and taxonomic observations on the mosquitoes of Kansas. *Trans. Kansas Acad. Sci.* 42: 255–265.

337. Hill, S. 0., B. J. Smittle, and F. M. Phillips. 1958. Distribution of mosquitoes in the Fourth U.S. Army Area. Entomol. Div. 4th U.S. Army Med. Lab., 115 pp.

338. Hinz, E. 1991. Einscheppung und Ausbreitung von *Aedes aegypti* (Diptera: Culicidae) in Amerika. *Tropenmed. Parasit.* 13: 101–110.

339. Hobbs, J. H., E. A. Hughes, and B. H. Eichold II. 1991. Replacement of *Aedes aegypti* by *Aedes albopictus* in Mobile, Alabama. *J. Am. Mosq. Control Assoc.* 7: 488–489.

340. Holmberg, R. G., and D. Trofimenkoff. 1968. *Aedes melanimon* in Saskatchewan. *Mosq. News* 28: 651–652.

341. Hopla, C. E. 1970. The natural history of the genus *Culiseta* in Alaska. *Proc. 57th Mtg. N.J. Mosq. Exterm. Assoc.,* 56–70.

342. Hornby, J. A., and W. R. Opp. 1994(1995). *Aedes albopictus* distribution, abundance and colonization in Collier County, Florida and its effect on *Aedes aegypti*. *J. Fla. Mosq. Control Assoc.* 65: 28–34.

343. Hornby, J. A., and T. W. Miller, Jr. 1994(1995). *Aedes albopictus* distribution, abundance and colonization in Lee County and its effect on *Aedes aegypti*: two additional seasons. *J. Fla. Mosq. Control Assoc.* 65: 21–27.

344. Hornby, J. A., D. E. Moore, and T. W. Miller, Jr. 1994. *Aedes albopictus* distribution, abundance and colonization in Lee County, Florida, and its effect on *Aedes aegypti*. *J. Am. Mosq. Control Assoc.* 10: 397–402.

345. Horsfall, W. R. 1956. *Aedes sollicitans* in Illinois. *J. Econ. Entomol.* 49: 416.

Hribar, L. J. 2001. See ref. 803.

346. Hribar, L. J., and R. R. Gerhardt. 1985. A checklist of the mosquitoes (Diptera: Culicidae) occurring in Knox County, Tennessee. *J. Tenn. Acad. Sci.* 61: 6–7.

Hribar, L. J., and J. J. Vlach. 2001. See. ref. 804.

347. Irwin, W. H. 1941. A preliminary list of the Culicidae of Michigan. Part 1. Culicinae (Diptera). *Entomol. News* 52: 101–105.

348. International Commission on Zoological Nomenclature. 1991. Opinion 1644. *Culex stigmatosoma* Dyar, 1907 and *C. thriambus* Dyar, 1921 (Insecta, Diptera): specific names conserved. *Bull. Zool. Nomencl.* 48: 179–180.

349. Irby, W. S., and C. S. Apperson. 1992. Spatial and temporal distribution of resting female mosquitoes (Diptera: Culicidae). *J. Med. Entomol.* 29: 150–159.

350. Jakob. W. L., and D. B. Francy. 1984. Observations on the DV/D ratio of male genitalia of *Culex pipiens* complex mosquitoes in the United States. *Mosq. Syst.* 16:282–288.

351. Jakob, W. L., D. B. Francy, and R. A. LeBrun. 1986. *Culex tarsalis* in Rhode Island. *J. Am. Mosq. Control Assoc.* 2: 98–99.

352. Jakob, W. L., T. Davis, and D. B. Francy. 1989. Occurrence of *Culex erythrothorax* in southeastern Colorado and report of virus isolations from this and other mosquito species. *J. Am. Mosq. Control Assoc.* 5: 534–536.

353. Jakob, W. L., F. A. Maloney, and F. J. Harrison. 1985. *Aedes purpureipes* in Western Arizona. *J. Am. Mosq. Control Assoc.* 1: 388.

354. James, H. G., G. Wishart, R. E. Bellamy, M. Maw, and P. Belton. 1969 (1970). An annotated list of mosquitoes of southeastern Ontario. *Proc. Entomol. Soc. Ont.* 100: 200–230.

355. Jamieson, D. H., L. A. Olson, and J. D. Whilhide. 1994. A larval mosquito survey in northeastern Arkansas including a new record for *Aedes albopictus*. *J. Am. Mosq. Control Assoc.* 10: 236–239.

356. Jamnback, H. 1961. *Culiseta melanura* (Coq.) breeding on Long Island, N. Y. *Mosq. News* 21: 140–141.
357. ———. 1969. Bloodsucking flies and other outdoor nuisance arthropods of New York State. N.Y. State Mus. Sci. Ser. Mem. 19, 90 pp.
Janousek, T. E., and W. L. Kramer. 1999. See ref. 797.
358. Janovy, J., Jr. 1966. Mosquitoes of the Cheyene Bottoms Waterfowl Management Area, Barton County, Kansas. *J. Kans. Entomol. Soc.* 39: 557–561.
359. Jaynes, H. A., L. Parente, and R. C. Wallis. 1962. Potential encephalitis vectors in Hamden, Connecticut. *Mosq. News* 22: 357– 360.
360. Jensen, T., M. C. Slamecka, and R. J. Novak. 1999. *Culiseta impatiens* (Walker) in Illinois, a new record. *J. Am. Mosq. Control Assoc.* 15: 250.
361. Jensen, T., P. E. Kaiser, T. Fukuda, and D. E. Barnard. 1985. *Anopheles perplexens* from artificial containers and intermittently flooded swamps in northern Florida. *J. Am. Mosq. Control Assoc.* 11: 141–144.
362. Jewell, D., and G. Grodhaus. 1984. An introduction of *Aedes aegypti* into California and its apparent failure to become established, pp. 103–107. In: M. Laird (ed.), *Commerce and the spread of pests and disease vectors.* New York, Praeger Publ.
363. Johnson, C. W. 1925. *Fauna of New England. List of the Diptera or two-winged flies.* Occas. Pap. Boston Soc. Nat. Hist. 7: 1–326.
364. Johnson, E. B. 1959. Distribution and relative abundance of mosquito species in Louisiana. La. Mosq. Control Assoc. Tech. Bull. 1, 18 pp.
365. Johnson, W. E., Jr. 1961. The occurrence of *Orthopodomyia alba* Baker in Oklahoma (Diptera: Culicidae). *Mosq. News* 21: 55–56.
366. ———. 1968. Ecology of mosquitoes in the Wichita Mountains Wildlife Refuge. *Ann. Entomol. Soc. Am.* 61: 1129–1141.
367. Johnson, W. E., Jr., and L. Harrell. 1980. The occurrence of *Aedes trivittatus* in Alabama. *Mosq. News* 40: 296–297.
368. Jordan, S. 1991. Influence of tree trunks on the spatial distribution of *Toxorhynchites r. rutilus* ovipositions in a coastal oak/palm hammock in Florida. *J. Am. Mosq. Control Assoc.* 7: 452–455.
369. Joseph, S. R. 1961. *Aedes thibaulti* in Maryland. *Mosq. News* 21: 251.
370. Joseph, S. R., R. A. Berry, and W. E. Bickley. 1960. A new mosquito record for Maryland (Diptera: Culicidae). *Proc. Entomol. Soc. Wash.* 62: 114.
371. Joy, J. E., C. A. Allman, and B. T. Dowell. 1994. Mosquitoes of West Virginia: an update. *J. Am. Mosq. Control Assoc.* 10: 115–118.
372. Joyce, C. R. 1945. The occurrence of *Psorophora mexicana* (Bellardi) in the United States. *Mosq. News* 5: 86.
373. Judd, W. W. 1957. A study of the population of emerging and littoral insects trapped as adults from tributary waters of the Thames River at London, Ontario. *Am. Midl. Nat.* 58: 394–412.
374. ———. 1962. The mosquito, *Psorophora ciliata* (Fabr.), at London, Ontario. *Mosq. News* 22: 304.
375. Kaiser, P. E., J. A. Seawright, and B. K. Birky. 1988. Chromosome polymorphism in natural populations of *Anopheles quadrimaculatus* Say species A and B. *Genome* 30: 138–146.
376. Kaiser, P. E., S. E. Mitchell, G. C. Lanzaro, and J. A. Seawright. 1988. Hybridization of laboratory strains of sibling species A and B of *Anopheles quadrimaculatus. J. Am. Mosq. Control Assoc.* 4: 34–38.
377. Kaiser, P. E., S. K. Narang, J. A. Seawright, and D. L. Kline. 1988. A new member of the *Anopheles quadrimaculatus* complex, species C. *J. Am. Mosq. Control Assoc.* 4: 494–499.
378. Kalpage, K. S., and R. A. Brust. 1968. Mosquitoes of Manitoba. I. Descriptions and a key to *Aedes* eggs (Diptera: Culicidae). *Can. J. Zool.* 46: 699–718.
379. Kim, S. S., and S. K. Narang. 1989. Restriction site polymorphism of mtDNA for differentiating *Anopheles quadrimaculatus* (Say) sibling species. *Korean J. Appl. Entomol.* 29: 132–135.
380. King, W. V., G. H. Bradley, C. N. Smith, and W. C. McDuffie. 1960. *Handbook of the mosquitoes of the southeastern United States.* USDA Agric. Handbook 173, 188 pp.
381. Klots, A. B. 1961. *Toxorhynchites rutilus* and *Anopheles barberi* in New York City (Diptera: Culicidae). *J. N.Y. Entomol. Soc.* 69: 104.
382. Knight, J. W., and J. S. Haeger. 1971. Key to adults of the *Culex* subgenera *Melanoconion* and *Mochlostyrax* of eastern North America. *J. Med. Entomol.* 8: 551–555.
383. Knight, K. L. 1967. Distribution of *Aedes sollicitans* (Walker) and *Aedes taeniorhynchus* (Wiedemann) within the United States (Diptera: Culicidae). *J. Ga. Entomol. Soc.* 2: 9–12.
384. ———. 1978. *Supplement to a catalog of the mosquitoes of the world (Diptera: Culicidae).* Thomas Say Foundation 6, Suppl., 107 pp.
385. Knight, K. L., and A. Stone. 1977. *A catalog of the mosquitoes of the World (Diptera: Culicidae).* Thomas Say Foundation 6, 611 pp.
386. Knight, K. L., and M. Wonio. 1969. Mosquitoes of Iowa (Diptera: Culicidae). Dept. Zool. and Entomol., Iowa State Univ., Spec. Rept. 61, 79 pp.
387. Kruger, R. M., and R. R. Pinger. 1981. A larval survey of the mosquitoes of Delaware County, Indiana. *Mosq. News* 41: 484–489.
388. Kunz, B. L., S. E. Cope, and R. J. Stoddard. 1989 (1990). Chaetotactic analysis of two species in the *Anopheles maculipennis* complex in California. *Proc. Pap. Ann. Conf. Calif. Mosq. and Vector Control Assoc.* 57: 90–93.
389. Lake, R. W. 1963. The occurrence of *Aedes dupreei* (Coquillett) and *Psorophora howardii* Coquillett in Delaware. *Mosq. News* 23: 160.
390. ———. 1967. Notes on the biology and distribution of some Delaware mosquitoes. *Mosq. News* 27: 324–331.

391. Lake, R. W., and J. M. Doll. 1961. New mosquito distribution records, Delaware, 1960–61. *Proc. 48th Mtg. N.J. Mosq. Exterm. Assoc.*, 191–193.

392. Lake. R. W., F. J. Murphey, and C. J. Stachecki, Jr. 1968a. Distribution and abundance of *Psorophora* species in Delaware 1967. *Proc. 55th Mtg. N.J. Mosq. Exterm. Assoc.*, 139–142.

393. Lake. R. W., F. J. Murphey, and C. J. Stachecki. 1968b. The occurrence of *Psorophora cyanescens* (Coquillett), *P. horrida* (Dyar and Knab) and *P. varipes* (Coquillett) in Delaware. *Mosq. News* 28: 470.

394. Lanzaro, G. C., and B. F. Eldridge. 1992. A classical and population genetic description of two new sibling species of *Aedes* (*Ochlerotatus*) *increpitus* Dyar. *Mosq. Syst.* 24: 85–101.

395. Lawson, D. A., A. D. Gettman, and Y. S. Perullo. 1994. *Toxorhynchites rutilus septentrionalis*: a new adult and larval record from Rhode Island. *J. Am. Mosq. Control Assoc.* 10: 230.

396. LeBrun, R. A., D. Boyes, P. Capotosto, and J. Marques. 1983. Annotated list of the mosquitoes of Rhode Island. *Mosq. News* 43: 435–437.

397. Leprince, D. J., R. J. Harrison, and R. Loiselle. 1978. Nouvelles captures de *Psorophora ciliata* (Fabr.) (Diptera: Culicidae) au Quebec, Canada. *Ann. Entomol. Soc. Quebec* 23: 89– 90.

398. Lesser, F., T. Candeletti, and W. Crans. 1977. *Culex tarsalis* in New Jersey. *Mosq. News* 37: 290.

399. Lewis, D. J., and G. F. Bennett. 1979. Biting flies of eastern Maritime Provinces of Canada II. Culicidae. *Mosq. News* 39: 633–639.

400. Lewis, D. J., and R. A. Webber. 1985. Species composition and relative abundance of anthropophilic mosquitoes in subarctic Quebec. *J. Am. Mosq. Control Assoc.* 1: 521–523.

401. Linam, J. H. 1961. A mosquito survey of Skull Valley, Tooele County, Utah. *Proc. Utah Mosq. Abat. Assoc.* 14: 26–27.

402. ———. 1972 (1973). Distribution of *Aedes hendersoni* Cockerell in Colorado. *Proc. Utah Mosq. Abat. Assoc.* 25: 17–19.

403. Linam, J. H., and L. T. Nielsen. 1963. Notes on the identification of some western *Culex* larvae. *Proc. 50th Mtg. N.J. Mosq. Exterm. Assoc.*, 411–415.

404. Linam, J. H., and L. T. Nielsen. 1964. Utah mosquitoes—their published history: supplement 1. *Proc. Utah Mosq. Abat. Assoc.* 16: 22–23.

405. Linam, J. H., and L. T. Nielsen. 1966. Notes on the distribution, ecology and overwintering habits of *Culex apicalis* Adams in Utah (Diptera: Culicidae). *Proc. Entomol. Soc. Wash.* 68: 136–138.

406. Linam, J. H., and L. T. Nielsen. 1970. The distribution and evolution of the *Culex* mosquitoes of the subgenus *Neoculex* in the New World. *Mosq. Syst. Newsl.* 2: 149–157.

407. Linley, J. R. 1990. Scanning electron microscopy of the eggs of *Aedes vexans* and *Aedes infirmatus* (Diptera: Culicidae). *Proc. Entomol. Soc. Wash.* 92: 685–693.

408. ———. 1992. The eggs of *Anopheles atropos* and *Anopheles darlingi* (Diptera: Culicidae). *Mosq. Syst.* 24: 40–50.

409. Linley, J. R., and D. D. Chadee. 1990. Fine structure of the eggs of *Psorophora columbiae*, *Ps. cingulata* and *Ps. ferox* (Diptera: Culicidae). *Proc. Entomol. Soc. Wash.* 92: 497–511.

410. Linley, J. R., and G. B. Craig, Jr. 1993. The egg of *Aedes hendersoni* and a comparison of its structure with the egg of *Aedes triseriatus* (Diptera: Culicidae). *Mosq. Syst.* 25: 65–72.

411. ———. 1994. Morphology of long- and short-day eggs of *Aedes atropalpus* and *A. epactius* (Diptera: Culicidae). *J. Med. Entomol.* 31: 855–867.

412. Linley, J. R., and P. E. Kaiser. 1994. The eggs of *Anopheles punctipennis* and *Anopheles perplexens* (Diptera, Culicidae). *Mosq. Syst.* 26: 43–56.

413. Linley, J. R., P. E. Kaiser, and A. F. Cockburn. 1993. A description and morphometric study of the eggs of species of the *Anopheles quadrimaculatus* complex (Diptera: Culicidae). *Mosq. Syst.* 25: 124–147.

414. Loiselle, R., and R. J. Harrison. 1977. Trois nouvelles especies de Culicides captures dans les regions de Saint-Hyacinthe et du lac Brome, Quebec. *Ann. Entomol. Soc. Quebec* 22: 143–144.

415. ———. 1978. Presence d'*Anopheles barberi* Coquillett dans la region de Saint-Hyacinthe, Quebec. *Ann. Entomol. Soc. Quebec* 23: 86–88.

416. Loiselle, R., R. J. Harrison, and D. J. Leprince. 1979. Premiere mention d'*Aedes trivittatus* et de *Culiseta inornata* (Diptera: Culicidae) au Quebec. *Can. Entomol.* 111: 39–40.

417. Loomis, L. C. (ed.) 1959. A field guide to common mosquitoes of California. Calif. Mosq. Cont. Assoc., Entomol. Comm., 26 pp.

418. Loomis, L. C., R. M. Bohart and J. N. Belkin. 1956. Additions to the taxonomy and distribution of California mosquitoes. *Calif. Vector Views* 3: 37–45.

419. Loor, K. A., and G. R. DeFoliart. 1970. Field observations on the biology of *Aedes triseriatus*. *Mosq. News* 30: 60–64.

420. Lothrop, B. B., R. P. Meyer, W. K. Reisen, and H. Lothrop. 1995. Occurrence of *Culex* (*Melanoconion*) *erraticus* (Diptera: Culicidae) in California. *J. Am. Mosq. Control Assoc.* 11: 367–368.

421. Love, G. J., and M. H. Goodwin, Jr. 1961. Notes on the bionomics and seasonal occurrence of mosquitoes in southwestern Georgia. *Mosq. News* 21: 195–215.

422. Love, G. J., R. B. Platt, and M. H. Goodwin, Jr. 1963. Observations on the spatial distribution of mosquitoes in southwestern Georgia. *Mosq. News* 23: 13–22.

423. Love, G. J., and W. W. Smith. 1957. Preliminary observations on the relation of light trap collections to mechanical sweep net collections in sampling mosquito populations. *Mosq. News* 17: 9– 14.

424. Ludlow, C. S. 1907. Mosquito notes no. 5—continued. *Can. Entomol.* 39: 129–131.

425. Lungstrom, L. G., and C. A. Scooter. 1961. Mosquito light-trap collections made in conjunction with the encephalitis investigation in southeastern Kansas in 1949 and 1950. *Trans. Kans. Acad. Sci.* 64: 133–143.

426. Lunt, S. R. 1968. A check list of the mosquitoes (Diptera: Culicidae) of Fontenelle Forest. *Proc. Neb. Acad. Sci.* 78: 6.

427. ———. 1969. The occurrence of *Aedes hendersoni* Cockerell in Nebraska (Diptera: Culicidae). *Proc. Neb. Acad. Sci.* 79: 9–10.

428. ———. 1977a. Morphological characteristics of the larvae of *Aedes triseriatus* and *Aedes hendersoni* in Nebraska. *Mosq. News* 37: 654–656.

429. ———. 1977b. The geographical distribution of the sibling mosquito species *Aedes triseriatus* and *Aedes hendersoni* in Nebraska. *Proc. Neb. Acad. Sci.* 87: 19.

430. Lunt, S. R., and L. T. Nielsen. 1968. Setal characteristics and the identification of adult *Aedes* mosquitoes. *Proc. No. Central Branch, Entomol. Soc. Am.* 23: 122–125.

431. ———. 1971. The use of thoracic setae as a taxonomic tool and as an aid in establishing phylogenetic relationships in adult female *Aedes* mosquitoes in North America. Part 1. *Mosq. Syst. Newsl.* 3: 69–98. Part 2. *Mosq. Syst. Newsl.* 3: 102–121.

432. Lunt, S. R., and G. E. Peters. 1974. East-west distribution of tree-hole mosquitoes in Nebraska. *Proc. Pap. 42nd Conf. Calif. Mosq. Contr. Assoc.*, p. 38.

433. ———. 1976. Distribution and ecology of tree-hole mosquitoes along the Missouri and Platte Rivers in Iowa, Nebraska, Colorado, and Wyoming. *Mosq. News* 36: 80–84.

434. Lunt, S. R., and W. F. Rapp, Jr. 1981. An annotated list of the mosquitoes of Nebraska. *Mosq. News* 41: 701–706.

435. Lusk, E. E., and J. R. Clover. 1972. Locality records of *Aedes flavescens* in California. *Calif. Vector Views* 19: 51–52.

436. Lusk, E. E., and C. R. Smith. 1971. Additional collections of *Aedes campestris* and *Aedes niphadopsis* in California. *Calif. Vector Views* 18: 41.

437. Mailhot, Y., and A. Maire. 1978. Caracterisation ecologique des milieux humides: larves de moustiques (Culicides) de la region subarctique continentale d'Opinaca (territoire de la Bale de James, Quebec). *Can. J. Zool.* 56: 2377–2387.

438. Main, A. J., R. O. Hayes, and R. J. Tonn. 1968. Seasonal abundance of mosquitoes in southeastern Massachusetts. *Mosq. News* 28: 619–626.

439. Main, A. J., H. E. Sprance, and R. C. Wallis. 1976. New distribution records for *Toxorhynchites* and *Orthopodomyia* in the northeastern United States. *Mosq. News* 36: 197.

440. Main, A. J., R. J. Tonn, E. J. Randall, and K. S. Anderson. 1966. Mosquito densities at heights of five and twenty-five feet in southeastern Massachusetts. *Mosq. News* 26: 243–248.

441. Maire, A. 1980. Ecologie comparee des especes de moustiques holarctique (Diptera: Culicidae). *Can. J. Zool.* 58: 1582–1600.

442. Maire, A., and A. Aubin. 1976. Inventaire et classification ecologiques des biotopes: larves de moustiques (Culicides) de la region de Radisson (territoire de la Bale de James, Quebec). *Can. J. Zool.* 54: 1979–1991.443.

443. Maire, A., and A. Aubin. 1980. *Les moustiques du Quebec (Diptera: Culicidae). Essai de synthese ecologique.* Mem. Entomol. Soc. Quebec, 107 pp.

444. Maire, A., A. Aubin, and D. M. Wood. 1978. Donnees recentes sur 1'ecologie d'*Aedes rempeli* Vockeroth, 1954 (Diptera: Culicidae). *Ann. Entomol. Soc. Quebec* 23: 182–185.

445. Maire, A., J. P. Bourassa and A. Aubin. 1976. Cartographie ecologique des milieux a larves de moustiques de la region de Trois-Rivieres, Quebec. Doc. de cartographie ecologique. *Lab. Biol. Veg. Univ. Grenoble* 17: 49–71.

446. Maire, A., and Y. Mailhot. 1978. A new record of *Aedes cantator* from the tidal zone of southeastern James Bay, Quebec. *Mosq. News* 38: 207–209.

447. Maire, A., Y. Mailot, C. Tessler, and R. Savignac. 1980. Records of *Aedes mercurator* from eastern James Bay, Quebec. *Mosq. News* 40: 444–445.

448. Maire, A., L. Picard, and A. Aubin. 1979. Presence d'*Aedes (Ochlerotatus) pullatus* (Coquillett) (Diptera: Culicidae) dans les Chic-Chocs, Pare de la Gaspesie, Quebec, implications biogeographiques de cette extension d'aire. *Can. J. Zool.* 57: 1526–1583.

449. Maire, A., C. Tessier, and L. Picard. 1978. Analyse ecologique des populations larvaires de moustiques (Diptera: Culicidae) des zones riveraines de fleuve Saint Laurent, Quebec. *Nat. Can.* 105: 225–241.

450. Mallack, J. 1975. Occurrence of *Aedes hendersoni* and *Aedes dorsalis* in Maryland. *Mosq. News* 35: 412.

451. Mallia, M. J. 1964. A new distribution record for *Culiseta (Culicella) minnesotae* Barr. *Mosq. News* 24: 338–339.

452. Maloney, F. A. 1978. New record for *Aedes fulvus pallens* in Missouri. *Mosq. News* 38:294.

453. ———. 1980. New records for *Uranotaenia sapphirina* in Colorado. *Mosq. News* 40: 451.

454. Maloney, F. A., and B. J. Reid. 1990. New record for *Aedes thelcter* in Arizona. *J. Am. Mosq. Control Assoc.* 6: 138.

455. Maltais, P., and J. Y. Daigle. 1984. Premiere mention d'*Aedes implicatus* et d'*Aedes diantaeus* (Diptera: Culicidae) au Nouveau-Brunswick. *Can. Entomol.* 116: 781–782.

456. Manning, D. L., N. L. Evenhuis, and W. A. Steffan. 1982. Annotated bibliography of *Toxorhynchites* (Diptera: Culcidae). Supplement, *I. J. Med. Entomol.* 19: 429–486.

457. Maslov, A. V. 1964. On the systematics of bloodsucking mosquitoes of the group *Culiseta* (Diptera: Culicidae) (in Russian). *Ent. Obozr.* 43: 193–217. (*Ent. Rev.* 43: 97–107).

458. ———. 1967. *Bloodsucking mosquitoes of the subtribe Culisetina (Diptera: Culicidae) of the world fauna* (in Russian). Akad. Naut. S.S.S.R., Opred. 93: 1–182.

459. Masteller, E. C. 1977. Mosquitoes collected with CDC traps in Erie County, Pennsylvania. *Proc. Pa. Acad. Sci.* 51: 117–121.

460. Matheson, R. 1944. *Handbook of the mosquitoes of North America.* 2nd ed. Ithaca, Comstock, 314 pp.

461. Mattingly, P. F. 1957. Notes on the taxonomy and bionomics of certain filariasis vectors. *Bull. World Health Org.* 16: 686–696.

462. ———. 1961. The culicine mosquitoes of the Indomalayan Area. Part V. Genus *Aedes* Melgen, subgenera *Mucidus* Theobald, *Ochlerotatus* Lynch Arribalzaga and *Neomelanoconion* Newstead. British Museum (Natural History), London, 62 pp.

463. ———. 1971. Contributions to the mosquito fauna of southeast Asia. XII. Illustrated keys to the genera of mosquitoes (Diptera: Culicidae). *Contrib. Am. Entomol. Inst.* 7(4): 1–84.

464. Mayers, P. J. 1983. Recent introduction of *Aedes aegypti* in Bermuda. *Mosq. News* 43: 361–362.

465. McDaniel, I. N. 1975. A list of Maine mosquitoes including notes on their importance as pests of man. *Mosq. News* 35: 232– 233.

466. McDaniel, I. N., and D. L. Webb. 1974. Identification of females of the *Aedes stimulans* group in Maine including notes on larval characters and attempts at hybridization. *Ann. Entomol. Soc. Am.* 67: 915–918.

467. McDonald, J. L., and G. S. Olton. 1974. A list and bibliography of the mosquitoes in Arizona. *Mosq. Syst.* 6: 89–92.

468. McDonald, J. L., T. P. Sluss, J. D. Lang, and C. C. Roan. 1973. Mosquitoes of Arizona. Ariz. Agric. Exp. Sta. Tech. Bull. 205, 21 pp.

469. McDonald, W. A. 1957a. The adults and immature stages of *Aedes muelleri* Dyar (Diptera: Culicidae). *Ann. Entomol. Soc. Am.* 50: 505–511.

470. ———. 1957b. The adults and immature stages of *Aedes purpureipes* Aitken (Diptera: Culicidae). *Ann. Entomol. Soc. Am.* 50: 529–535.

471. McDonald, W. A., and J. N. Belkin. 1960. *Orthopodomyia kummi* new to the United States (Diptera: Culicidae). *Proc. Entomol. Soc. Wash.* 62: 249–250.

472. McHugh, C. P. 1993. Distributional records for *Aedes* mosquitoes from U.S. Air Force ovitrapping program. *J. Am. Mosq. Control Assoc.* 9: 352–355.

473. McHugh, C. P., and P. A. Hanny. 1990. Records of *Aedes albopictus*, *Aedes aegypti* and *Aedes triseriatus* from the U.S. Air Force ovitrapping program. *J. Am. Mosq. Control Assoc.* 6: 549– 551.

474. McLintock, J. 1976. *Anopheles walkeri* Theobald in Saskatchewan and notes on *Culiseta silvestris minnesotae* Barr. *Mosq. News* 36: 308–310.

475. McLintock, J., and J. Iversen. 1975. Mosquitoes and human disease in Canada. *Can. Entomol.* 107: 695–704.

476. McLintock, J., and J. G. Rempel. 1963. Midsummer mosquito abundance in southern Saskatchewan, 1962. *Mosq. News* 23: 242–249.

477. McNelly, J. R. 1984. *Aedes thibauti* in New Jersey. *Mosq. News* 44:247–248.

478. ———. 1989. Occurrence of *Aedes infirmatus* in New Jersey. *J. Am. Mosq. Control Assoc.* 5: 277.

479. McNelly, J., and W. J. Crans. 1983. *Psorophora howardii*, an addition to the checklist of New Jersey mosquitoes. *Mosq. News* 43: 237–239.

480. Means, R. G. 1979. *Mosquitoes of New York Part 1. The genus Aedes Meigen with identification keys to genera of Culicidae.* New York State Mus. Bull. 430a, 219 pp.

481. ———. 1987. *Mosquitoes of New York. Part II. Genera of Culicidae other than Aedes occurring in New York.* New York State Mus. Bull. 430b, 180 pp.

482. Means, R. G., and F. C. Thompson. 1971. A first record of the occurrence of *Culiseta* (*Culicella*) *silvestris minnesotae* Barr (Diptera: Culicidae) in New York. *Mosq. News* 31: 443–445.

483. Meisch, M. V., A. L. Anderson, R. L. Watson, and L. Olson. 1982. Mosquito species inhabiting ricefields in five rice growing regions of Arkansas. *Mosq. News* 42: 341–346.

484. Meisch, M. V., H. Spatz, and W. B. Kottkamp. 1981. Surveillance for yellow fever mosquito in Arkansas. *Ark. Farm Res.* 39: 6.

485. Mekuria, Y., and M. G. Hyatt. 1995. *Aedes albopictus* in South Carolina. *J. Am. Mosq. Control Assoc.* 11: 468–470.

486. Menzies, G. C., R. B. Eads, and F. C. Harmston. 1955. The discovery of *Culex erythrothorax* Dyar in Texas. *Mosq. News* 15: 235–236.

487. Meredith, J., and J. E. Phillips. 1973. Ultrastructure of anal papillae from a seawater mosquito larva *Aedes togoi* (Theobald). *Can. J. Zool.* 51: 349–353.

488. Messersmith, D. H. 1971. Extension of the range of *Aedes triseriatus* (Say) to Greenland. *Mosq. Syst. Newsl.* 3: 7.

489. Meyer, C. L., G. F. Bennett, and C. M. Herman. 1974. Mosquito transmission of *Plasmodium* (*Giovannolata*) *circumflexum* Kikuth, 1931, to waterfowl in the Tantramar Marshes, New Brunswick. *J. Parasitol.* 60: 905–906.

490. Meyer, R. P. 1997. *Aedes aegypti*: yellow fever mosquito in Arizona. Is California next? *Vector Ecol. Newsl.* 28: 8.

491. Meyer, R. P., and S. L. Durso. 1993. Identification of the mosquitoes of California. Calif. Mosq. and Vector Control Assoc., Sacramento, 80 pp.

492. Meyer, R. P., V. M. Martinez, B. R. Hill, and W. K. Reisen. 1988. *Aedes thelcter* from the lower Colorado River in California. *J. Am. Mosq. Control Assoc.* 4: 366–367.

493. Meyer, R. P., W. K. Reisen, and B. R. Hill. 1987. On the occurrence of *Aedes purpureipes* along the lower Colorado River. *J. Am. Mosq. Control Assoc.* 3: 312–313.

494. Miller, B. E. 1962. The occurrence of *Orthopodomyia alba* Baker in New Mexico. *Mosq. News* 22: 309–310.

495. Miller, B. E., J. M. Doll, and J. R. Wheeler. 1964. New records of New Mexico mosquitoes. *Mosq. News* 24: 459–460.

496. Miller, L. S., and R. A. McHugh. 1959. A note on *Mansonia* breeding in Oregon log ponds. *Mosq. News* 19: 198.

497. Mokry, J. 1984. Notes on the *Culiseta* species (Diptera: Culicidae) of Newfoundland, with report of a new record. *Mosq. Syst.* 16: 168–171.

498. Moore, C. G. 1999. *Aedes albopictus* in the United States: current status and prospects for future spread. *J. Am. Mosq. Control Assoc.* 15: 221–227.

499. Moore, C. G., D. B. Francy, D. B. Eliason, R. E. Bailey, and E. G. Campos. 1990. *Aedes albopictus* and other container-inhabiting mosquitoes in the United States: results of an eight-city survey. *J. Am. Mosq. Control Assoc.* 6: 173–178.

Moore, J. P. 1999. See ref. 807.

500. Morland, H. B., and M. E. Tinker. 1965. Distribution of *Aedes aegypti* infestations in the United States. *Am. J. Trop. Med. Hyg.* 14: 892–899.

501. Mullen, G. R. 1971. The occurrence of *Aedes decticus* (Diptera: Culicidae) in central New York. *Mosq. News* 31: 106–109.

502. Mulrennan, J. A., and E. C. Beck. 1955. The distribution of Florida mosquitoes. *Fla. Anti-mosq. Assoc. Rept.* 26: 124–134.

503. Murdoch, W. P. 1956. A preliminary survey of the biting Diptera of the Teton Range. *Proc. 43rd Mtg. N.J. Mosq. Exterm. Assoc.*, 186–191.

504. Murphy, D. R. 1953. Collection records of some Arizona mosquitoes (Diptera: Culicidae). *Entomol. News* 64: 233–238.

505. Myers, C. M. 1964. Identification of *Culex* (*Culex*) larvae in California (Diptera: Culicidae). *Pan-Pacific Entomol.* 40: 13– 18.

506. ———. 1974. A new concept in mosquito identification: the circular mosquito key. *Proc. Pap. Calif. Mosq. Control Assoc.* 42: 167.

507. Myklebust, R. J. 1966. Distribution of mosquitoes and chaoborids in Washington State, by counties. *Mosq. News* 26: 515–519.

508. Narang, S. K., P. E. Kaiser, and J. A. Seawright. 1989a. Dichotomous electrophoretic taxonomic key for identification of sibling species A, B, and C of the *Anopheles quadrimaculatus* complex (Diptera: Culicidae). *J. Med. Entomol.* 26: 94–99.

509. ———. 1989b. Identification of species D, a new member of the *Anopheles quadrimaculatus* species complex: a biochemical key. *J. Am. Mosq. Control Assoc.* 5: 317–324.

510. Narang, S. K., J. A. Seawright, and P. E. Kaiser. 1990. Evidence for microgeographic genetic subdivision of *Anopheles quadrimaculatus* species C. *J. Am. Mosq. Control Assoc.* 6: 179– 187.

511. Narang, S. K., J. A. Seawright, S. E. Mitchell, P. E. Kaiser, and D. A. Carlson. 1993. Multiple-technique identification of sibling species of the *Anopheles quadrimaculatus* complex. *J. Am. Mosq. Control Assoc.* 9: 463–464.

512. Narang, S. K., S. R. Toniolo, J. A. Seawright, and P. E. Kaiser. 1989. Genetic differentiation among sibling species A, B and C of the *Anopheles quadrimaculatus* complex (Diptera: Culicidae). *Ann. Entomol. Soc. Am.* 82: 508–515.

513. Nasci, R. S., D. B. Taylor, and L. Munstermann. 1983. Record of the mosquitoes *Aedes dupreei*, *Psorophora horrida* and *Psorophora mathesoni* (Diptera: Culicidae) in St. Joseph County, Indiana. *Great Lakes Entomol.* 16:33.

514. Nawrocki, S. J., and G. B. Craig, Jr. 1989. Further extension of the range of the rock pool mosquito, *Aedes atropalpus*, via tire breeding. *J. Am. Mosq. Control Assoc.* 5: 110–114.

515. Nemjo, J., and M. Slaff. 1984. Head capsule width as a tool for instar and species identification of *Mansonia dyari*, *M. titillans*, and *Coquillettidia perturbans* (Diptera: Culicidae). *Ann. Entomol. Soc. Am.* 77: 633–635.

516. Newhouse, V. F., R. W. Chamberlain, J. G. Johnston, and W. D. Sudia. 1966. Use of dry ice to increase mosquito catches of the CDC miniature light trap. *Mosq. News* 26: 30–35.

517. Newhouse, V. F., and R. E. Siverly. 1965. The *Culex pipiens* complex in southern Indiana. *Mosq. News* 25: 489–490.

518. Niebylski, M. L., and G. B. Craig, Jr. 1994. Dispersal and survival of *Aedes albopictus* at a scrap tire yard in Missouri. *J. Am. Mosq. Control Assoc.* 10: 339–343.

519. Nielsen, E. T., and H. T. 1966. Observations on mosquitoes in Greenland. *Medd. Gronland* 170: 5–9.

520. Nielsen, L. T. 1959. Seasonal distribution and longevity of Rocky Mountain snow mosquitoes of the genus *Aedes*. *Proc. Utah Acad. Sci. Arts Letters* 36: 83–87.

521. ———. 1961. *Aedes schizopinax* Dyar in the western United States. *Proc. Calif. Mosq. Cont. Assoc.* 29: 21–24.

522. ———. 1968. A current list of mosquitoes known to occur in Utah with a report of new records. *Proc. Utah Mosq. Abat. Assoc.* 21: 34–37.

523. ———. 1969. *Aedes cacothius* Dyar, a synonym of *Aedes ventrovittis* Dyar (Diptera: Culicidae). *Proc. Entomol. Soc. Wash.* 71: 530.

524. ———. 1982. *Aedes euedes* H. D. and K.: a report of a new record from Wyoming with notes on the species. *Mosq. Syst.* 14: 133–134.

525. Nielsen, L. T., J. H. Arnell, and J. H. Linam. 1967. A report on the distribution and biology of tree hole mosquitoes in the western United States. *Proc. Calif. Mosq. Cont. Assoc.* 35: 72–76.

526. Nielsen, L. T., and M. S. Blackmore. 1996. The mosquitoes of Yellowstone National Park (Diptera: Culicidae). *J. Am. Mosq. Control Assoc.* 12: 695–700.

527. Nielsen, L. T., and J. E. Mokry. 1982a. *Culiseta melanura* in Newfoundland. *Mosq. News* 42: 274–275.

528. ———. 1982b. Mosquitoes of the island of Newfoundland: a report of new records and notes on the species. *Mosq. Syst.* 14: 34–40.

529. Nielsen, L. T., and W. R. Horsfall. 1973. The occurrence of *Aedes barri* Rueger in Alaska with notes on its distribution. *Mosq. News* 33: 243.

530. Nielsen, L. T., and J. H. Linam. 1963. New distributional records for the mosquitoes of Utah. *Proc. Utah Acad. Sci. Arts Letters* 40: 193–196.

531. ———. 1964. Additional distributional records for Utah mosquitoes with notes on biology. *Proc. Utah Mosq. Abat. Assoc.* 17: 29–31.

532. Nielsen, L. T., J. H. Linam, J. H. Arnell, and T. J. Zavortink. 1968. Distributional and biological notes on the tree hole mosquitoes of the western United States. *Mosq. News* 28: 361–365.

533. Nielsen, L. T., J. H. Linam, and D. M. Rees. 1963. New distribution records for mosquitoes in the Rocky Mountain states. *Proc. 50th Mtg. N.J. Mosq. Exterm. Assoc.*, 424–428.

534. Nielsen, L. T., and D. M. Rees. 1959. The mosquitoes of Utah: a revised list. *Mosq. News* 19: 45–47.

535. ———. 1961. An identification guide to the mosquitoes of Utah. Univ. Utah Biol. Ser. 12(3): 1–58.

536. Nielsen, L. T., T. A. Wolff, and J. H. Linam. 1973. New distribution records for snowpool *Aedes* mosquitoes in the mountains of Arizona and New Mexico. *Mosq. News* 33: 378–380.

537. Novak, R. J., and J. H. Linam. 1970. The *Aedes* mosquitoes of the front range of Custer County, Colorado. *Proc. Utah Mosq. Abat. Assoc.* 23: 42–46.

538. Novak, R. J., B. A. Steinly, and D. W. Webb. 1990. *Aedes albopictus* in Illinois. *Proc. Ill. Mosq. Vector Control Assoc.* 1: 25–30.

539. Obrecht, C. B. 1967. New distribution records of Michigan mosquitoes, 1948–1963. *Mich. Entomol.* 1: 153–158.

540. Ochoa, O., Jr., and T. L. Biery. 1978. Distribution of mosquitoes in the continental United States. USAF Sch. Aerospace Med. Rept. SAM-TR-78-28, 54 pp.

541. Oldham, T. W. 1977. Distributional records of mosquitoes in Kansas. Tech. Publ. State Biol. Surv. Kansas 4: 51–62.

542. Olinger, L. D. 1957. Observations on the mosquito, *Toxorhynchites rutilus rutilus* (Coquillett) in Alachua County, Florida. *Fla. Entomol.* 40: 51–52.

543. Olson, J. K., R. E. Elbel, and K. L. Smart. 1968. Mosquito collections by CDC miniature light traps and livestock-baited stable traps at Callao, Utah. *Mosq. News* 28: 512–516.

544. Olson, T. A., and H. L. Keegan. 1944. New mosquito records from the Seventh Service Command Area. *J. Econ. Entomol.* 37: 847– 848.

545. O'Meara, G. F., and G. B. Craig, Jr. 1970a. A new subspecies of *Aedes atropalpus* (Coquillett) from southwestern United States (Diptera: Culicidae). *Proc. Entomol. Soc. Wash.* 72: 475–479.

546. ———. 1970b. Geographical variation in *Aedes atropalpus* (Diptera: Culicidae). *Ann. Entomol. Soc. Am.* 63: 1392–1400.

547. O'Meara, G. F., and L. F. Evans, Jr. 1997. Discovery of a bromeliad-inhabiting *Culex* (*Micraedes*) sp. in south Florida. *J. Am. Mosq. Control Assoc.* 13: 208–210.

548. O'Meara, G. F., L. F. Evans, Jr., A. D. Gettman, and J. P. Cuda. 1995. Spread of *Aedes albopictus* and decline of *Ae. aegypti* (Diptera: Culicidae) in Florida. *J. Med. Entomol.* 32: 554–562.

549. O'Meara, G. F., A. D. Gettman, L. F. Evans, Jr., and G. A. Curtis. 1993. The spread of *Aedes albopictus* in Florida. *Am. Entomol.* 39: 163–172.

550. O'Meara, G. F., V. L. Larson, D. H. Mook, and M. D. Latham. 1989. *Aedes bahamensis*: its invasion of south Florida and association with *Aedes aegypti*. *J. Am. Mosq. Control Assoc.* 5: 1–5.

551. O'Meara, G. F., L. F. Evans, Jr., and M. L. Womack. 1997. Colonization of rock holes by *Aedes albopictus* in the southeastern United States. *J. Am. Mosq. Control Assoc.* 13: 270–274.

552. Osmun, J. 1967. Mosquitoes of the general Great Lakes area. Their bionomics and discussion of major problems. *Pap. Ohio Mosq. Control Assoc.* 19–20: 36–43.

553. Owen, W. B., and R. W. Gerhardt. 1957. The mosquitoes of Wyoming. *Univ. Wyo. Publ.* 21: 71–141.

554. Pafume, B. A., E. G. Campos, D. B. Francy, E. L. Peyton, A. N. Davis, and M. Nelms. 1988. Discovery of *Aedes* (*Howardina*) *bahamensis* in the United States. *J. Am. Mosq. Control Assoc.* 4: 380.

555. Pagac, B. B., Jr., H. J. Harlan, S. D. Doran, and M. A. Brosnihan. 1992. New state record for *Culiseta impatiens* in Maryland. *J. Am. Mosq. Control Assoc.* 8: 196.

556. Parker, D. J. 1977. The biology of the tree-holes of Point Pelee National Park, Ontario. II. First record of *Toxorhynchites rutilus septentrionalis* in Canada (Diptera: Culicidae). *Can. Entomol.* 109: 93–94.

557. Parsons, M. A., R. L. Berry, M. Jalil, and R. A. Masterson. 1972. A revised list of the mosquitoes of Ohio with some new distribution and species records. *Mosq. News* 32: 223–226.

558. Parsons, R. E., and D. E. Howell. 1971. A list of Oklahoma mosquitoes. *Mosq. News* 31: 168–169.

559. Peacock, B. E., J. P. Smith, P. G. Gregory, T. M. Loyless, J. A. Mulrennan, Jr., P. R. Simmonds, L. Padgett, Jr., E. K. Cook, and T. R. Eddins. 1988. *Aedes albopictus* in Florida. *J. Am. Mosq. Control Assoc.* 4: 362–365.

560. Pecor, J. E., V. L. Mallampalli, R. E. Harbach, and E. L. Peyton. 1992. Catalog and illustrated review of the subgenus *Melanoconion* of *Culex* (Diptera: Culicidae). *Contrib. Am. Entomol. Inst.* 27(2): 1–228.

561. Pennington, R. G., and J. E. Lloyd. 1975. Mosquitoes captured in a bovine-baited trap in a Wyoming pasture subject to river and irrigation flooding. *Mosq. News* 35: 402–408.

562. Perera, O. P., S. E. Mitchell, A. K. Cockburn, and J. A. Seawright. 1995. Variation in mitochondrial and ribosomal DNA of *Anopheles quadrimaculatus* species A (Diptera: Culicidae) across a wide geographic range. *Ann. Entomol. Soc. Am.* 88: 836–845.

563. Pest Control. 1961. Pictorial key to U.S. genera of mosquito larvae. *Pest Control* 29: 36.

564. Peus, F. 1972. Uber clas subgenus *Aedes* sensu stricto in Deutschland (Diptera: Culicidae). *Zeitsch. Angewandte Entomol.* 72: 177–194.

565. Peyton, E. L. 1972. A subgeneric classification of the genus *Uranotaenia* Lynch Arribalzaga, with a historical review and notes on other categories. *Mosq. Syst.* 4: 16–40.

566. ———. 1973. Notes on the Genus *Uranotaenia*. *Mosq. Syst.* 5: 194–196.

567. Peyton, E. L., S. R. Campbell, T. M. Candeletti, M. Romanowskiand, and W. J. Crans. 1999. *Aedes* (*Finlaya*) *japonicus japonicus* (Theobald), a new introduction into the United States. *J. Am. Mosq. Control Assoc.* 15: 238–241.

568. Peyton, E. L., J. F. Reinert, and N. E. Peterson. 1964. The occurrence of *Deinocerites pseudes* Dyar and Knab in the United States, with additional notes on the biology of *Deinocerites* species of Texas. *Mosq. News* 24: 449–458.
569. Pickavance, J. R., G. F. Bennett, and J. Phipps. 1970. Some mosquitoes and blackflies from Newfoundland. *Can. J. Zool.* 48: 621–624.
570. Pinger, R. R., Jr., and W. A. Rowley. 1970. A distributional note for *Aedes punctor* (Kirby). *Mosq. News* 30: 649–650.
571. ———. 1972. Occurrence and seasonal distribution of Iowa mosquitoes. *Mosq. News* 32: 234–241.
572. Porter, C. H., and F. H. Collins. 1991. Species-diagnostic differences in a ribosomal DNA internal transcribed spacer from the sibling species *Anopheles freeborni* and *Anopheles hermsi* (Diptera: Culicidae). *Am. J. Trop. Med. Hyg.* 45: 271–279.
573. Porter, C. H., and W. L. Goimerac. 1970. Mosquitoes of Point Beach State Forest. University of Wisconsin, Coll. Agric. Life Sci. Res. Rept. 53, 15 pp.
574. Porter, J. L. 1964. *Deinocerites cancer* Theobald recovered from tree holes at Miami, Florida. *Mosq. News* 24: 222.
575. Portman, R. F. 1957. *Mansonia perturbans* in Butte County. *Calif. Vector Views* 4: 5.
576. Pratt, H. D. 1956. A checklist of the mosquitoes (Culicinae) of North America (Diptera: Culicidae). *Mosq. News* 16: 4–10.
577. Price, R. D. 1958. A description of the larva and pupa of *Culiseta* (*Culicella*) *minnesotae* Barr. *J. Kans. Entomol. Soc.* 31: 47–53.
578. ———. 1963. Frequency of occurrence of spring *Aedes* (Diptera: Culicidae) in selected habitats in northern Minnesota. *Mosq. News* 23: 324–329.
579. Price, R. D., and L. R. Abrahamsen. 1958. The discovery of *Orthopodomyia signifera* (Coquillett) and *Anopheles barberi* Coquillett in Minnesota (Diptera, Culicidae). *J. Kans. Entomol. Soc.* 31: 92.
580. Pucat, A. 1964. Seven new records of mosquitoes in Alberta. *Mosq. News* 24: 419–421.
581. ———. 1965. List of mosquito records from Alberta. *Mosq. News* 25: 300–302.
582. Quickenden, K. L. 1972. Montana mosquitoes Part 1. Identification and biology. Vector Control Bull. No. 1, Mont. State Dept. Hlth. Environ. Sci., 34 pp.
583. Quickenden, K. L., and V. C. Jamison. 1979. Montana mosquitoes Part 1: Identification and biology. Montana Vector Control Bull. No. 1 (Revised), 35 pp.
584. Rai, K. S. 1991. *Aedes albopictus* in the Americas. *Ann. Rev. Entomol.* 36: 459–484.
585. Rapp, W. F., Jr. 1956. Notes on the mosquitoes (Culicinae) of the Crete (Nebraska) region. *J. Kans. Entomol. Soc.* 29: 55–57.
586. ———. 1958. The mosquitoes (Culicidae) of the Missouri Valley region of Nebraska. *Mosq. News* 18: 27–29.
587. ———. 1959. A distributional check-list of Nebraska mosquitoes. *J. Kans. Entomol. Soc.* 32: 128–133.
588. ———. 1985. The distribution and natural history of *Culex tarsalis* in the Great Plains Region. *Proc. West Central Mosq. Vector Control Assoc.* 10:29–33.
589. Rapp, W. F., Jr., and F. C. Harmston. 1961. New mosquito records from Nebraska, I. *J. Kans. Entomol. Soc.* 34: 86–87.
590. ———. 1965. Notes on the mosquitoes (Culicinae) of northwestern Nebraska. *Mosq. News* 25: 302–306.
591. Rees, D. M., and G. C. Collett. 1954. The biology of *Aedes niphadopsis* Dyar and Knab (Diptera, Culicidae). *Proc. Entomol. Soc. Wash.* 56: 207–214.
592. Rees, D. M., and L. T. Nielsen. 1955. Additional mosquito records from Utah (Diptera: Culicidae). *Pan-Pacific Entomol.* 31: 31–33.
593. Reinert, J. F. 1975. Mosquito generic and subgeneric abbreviations (Diptera: Culicidae). *Mosq. Syst.* 7: 105–110.
594. ———. 1982. Abbreviations of mosquito generic and subgeneric taxa established since 1975 (Diptera: Culicidae). *Mosq. Syst.* 14: 124–126
595. ———. 1991. Additional abbreviations of mosquito subgenera: names established since 1982 (Diptera: Culicidae). *Mosq. Syst.* 23: 209–210.
596. ———. 1997 (1998). Bibliography of *Anopheles quadrimaculatus* Say *sensu lato* (Diptera: Culicidae). *J. Am. Mosq. Control Assoc.* 13 (Suppl.): 112–161.
597. ———. 1999a. Morphological abnormalities in species of the *quadrimaculatus* complex of *Anopheles* (Diptera: Culicidae). *J. Am. Mosq. Control Assoc.* 15: 8–14.
598. ———. 1999b. Separation of fourth instar larvae of *Culex nigripalpus* from *Culex salinarius* in Florida using the spiracular apodeme. *J. Am. Mosq. Control Assoc.* 15: 84–85.
———. 1973. See ref. 799.
———. 2000a. See ref. 800.
———. 2000b. See ref. 801.
———. 2000c. See ref. 802.
———. 2001. See ref. 815.
599. Reinert, J. F., P. E. Kaiser, and J. A. Seawright. 1997 (1998). Analysis of the *Anopheles quadrimaculatus* complex of sibling species (Diptera: Culicidae) using morphological, cytological, molecular, genetic, biochemical, and ecological techniques in an integrated approach. *J. Am. Mosq. Control Assoc.* 13 (Suppl.): 1–102.
600. Reiter, P., and R. F. Darsie, Jr. 1984. *Aedes albopictus* in Memphis, Tennessee (USA): an achievement of modern transportation? *Mosq. News* 44: 393–399.
601. Rempel, J. G. 1953. The mosquitoes of Saskatchewan. *Can. J. Zool.* 31: 433–509.
602. Restifo, R. A., and G. C. Lanzaro. 1980. The occurrence of *Aedes atropalpus* (Coquillett) breeding in tires in Ohio and Indiana. *Mosq. News* 40: 292–294.

603. Richards, C. S. 1956. *Aedes melanimon* Dyar and related species. *Can. Entomol.* 88: 261–269.

604. Richards, C. S., L. T. Nielsen, and D. M. Rees. 1956. Mosquito records from the Great Basin and the drainage of the Lower Colorado River. *Mosq. News* 16: 10–17.

605. Richardson, J. H., W. E. Barton, and D. C. Williams. 1995. Survey of container-inhabiting mosquitoes in Clemson, South Carolina, with emphasis on *Aedes albopictus*. *J. Am. Mosq. Control Assoc.* 11: 396–400.

606. Richtor, J. A., B. R. Farmer, and J. L. Clarke, Jr. 1987. *Aedes albopictus* in Chicago, Illinois. *J. Am. Mosq. Control Assoc.* 3: 657.

607. Rigby, P. T. 1968. Occurrence of *Aedes infirmatus* D. and K. in Arizona. *Mosq. News* 28: 239.

608. Rigby, P. T., and H. Ayers. 1961. Occurrence of *Orthopodomyia californica* in Arizona. *Mosq. News* 21: 56.

609. Rigby, P. T., T. E. Blakeslee, and C. E. Forehand. 1963. The occurrence of *Aedes taeniorhynchus* (Wiedemann), *Anopheles barberi* (Coquillett), and *Culex thriambus* (Dyar) in Arizona. *Mosq. News* 23: 50.

610. Riley, J. A., and R. A. Hoffman. 1963. Observations on the meteorological–mosquito population relationship at Stoneville, Miss., 1959–1960. *Mosq. News* 23: 36–40.

611. Rings, R. W., and E. A. Richmond. 1953. Mosquito survey of Horn Island, Mississippi. *Mosq. News* 13: 252–255.

612. Ritchie, S. A., and W. A. Rowley. 1980. A new distribution record for *Psorophora cyanescens* (Coquillett) in Iowa. *Mosq. News* 40: 118.

613. Roberts, D. R., and J. E. Scanlon. 1975. The ecology and behavior of *Aedes atlanticus* D. and K. and other species with reference to Keystone virus in the Houston area, Texas. *J. Med. Entomol.* 12: 537–546.

614. Roberts, D. R., and J. E. Scanlon. 1979. An evaluation of entomological characters for separating females of *Aedes* (*Ochlerotatus*) *atlanticus* Dyar and *Aedes* (*Ochlerotatus*) *tormentor* Dyar and Knab (Diptera: Culicidae). *Mosq. Syst.* 11: 203–208.

615. Roch, J. F. 1990. Liste annotee de dipteres recoltes a Granby, division de recensement de Shefford, Quebec. *Fabreries* 15: 43–47.

616. Ronderos, R. A., and A. O. Bachman. 1962(1963). A proposito del complejo *Mansonia* (Diptera, Culicidae). *Rev. Soc. Entomol. Argentina* 25: 43–51.

617. ———. 1963. Mansonini neotropicales I (Diptera, Culicidae). *Rev. Soc. Entomol. Argentina* 26: 57–65.

618. Rosay, B., and L. T. Nielsen. 1973. The *Culex pipiens* complex in Utah. *Proc. Utah Mosq. Abat. Assoc.* 22: 30–35.

619. Ross, H. H. 1947. The mosquitoes of Illinois (Diptera, Culicidae). *Bull. Ill. Nat. Hist. Surv.* 24: 1–96.

620. Ross, H. H., and W. R. Horsfall. 1965. A synopsis of the mosquitoes of Illinois (Diptera: Culicidae). *Ill. Nat. Hist. Surv. Biol. Notes* 52, 50 pp.

621. Roth, L. M. 1945a. Aberrations and variations in anopheline larvae of the southeastern United States (Diptera, Culicidae). *Proc. Entomol. Soc. Wash.* 47: 257–278.

622. ———. 1945b. The male and larva of *Psorophora* (*Janthinosoma*) *horrida* (Dyar and Knab) and a new species of *Psorophora* from the United States (Diptera, Culicidae). *Proc. Entomol. Soc. Wash.* 47: 1–23.

623. Rozeboom, L. E. 1940. The overwintering of *Aedes aegypti* L. in Stillwater, Oklahoma. *Proc. Okla. Acad. Sci.* 19: 81–82.

624. Rueger, M. E. 1958. *Aedes* (*Ochlerotatus*) *barri*, a new species of mosquito from Minnesota (Diptera, Culicidae). *J. Kans. Entomol. Soc.* 31: 34–46.

625. Rupp, H. R. 1990. New Jersey *Psorophora*. *Proc. 77th Mtg. N.J. Mosq. Control Assoc.*, 78–90.

626. Rutledge, C. R., A. J. Cornel, C. L. Meek, and F. H. Collins. 1996. Validation of a ribosomal DNA-polymerase chain reaction species diagnostic assay for the common malaria mosquito (Diptera: Culicidae). *J. Med. Entomol.* 33: 952–954.

627. Rutledge, C. R., and C. L. Meek. 1994. Record of *Anopheles quadrimaculatus* species C in Louisiana. *J. Am. Mosq. Control Assoc.* 10: 585–586.

628. ———. 1998. Distribution of the *Anopheles quadrimaculatus* sibling species complex in Louisiana. *Southwest. Entomol.* 23: 161–167.

629. Rutschky, C. W., T. C. Mooney, Jr., and J. P. Vanderberg. 1958. Mosquitoes of Pennsylvania. Pa. Agric. Exp. Sta. Bull. 630, 26 pp.

630. Ryckman, R. E. 1952. Ecological notes on mosquitoes of Lafayette County, Wisconsin (Diptera: Culicidae). *Am. Midl. Nat.* 47: 469–470.

Sallum, M. A., and O. P. Forattini. 1996. See ref. 809.

631. Saugstad, E. S. 1977. Initial record of *Aedes tormentor* in Kentucky. *Mosq. News* 37: 298.

632. Savage, H. M., and G. C. Smith. 1994. Identification of damaged adult female specimens of *Aedes albopictus* and *Aedes aegypti* in the New World. *J. Am. Mosq. Control Assoc.* 10: 440–442.

633. Savage, H. M., L. T. Nielsen, and B. R. Miller. 1994. First record of *Culiseta morsitans* from Wyoming. *J. Am. Mosq. Control Assoc.* 10: 462.

634. Savage, H. M., and G. C. Smith. 1995. *Aedes albopictus* y *Aedes aegypti* en las Americas: implicaciones para la transmission de arbovirus e identificacion de hembras adultas danadas. *Bol. Oficina Sanit. Panam.* 118: 473–478.

635. Savignac, R., and A. Maire. 1981. A simple character for recognizing second and third instar larvae of five Canadian mosquito genera (Diptera: Culicidae). *Can. Entomol.* 113: 13–20.

636. Scholefield, P. J., and J. McIntosh. 1984. A further addition to the mosquitoes of Alberta. *Mosq. News* 44: 423–424.

637. Scholefield, P. J., G. Pritchard, and M. A. Enfield. 1981. The distribution of mosquito (Diptera, Culicidae) larvae in southern Alberta, 1976–1978. *Quaest. Entomol.* 17: 147–167.

638. Schutz, S. J., and B. F. Eldridge. 1993. Biogeography of the *Aedes* (*Ochlerotatus*) *communis* complex (Diptera: Culicidae) in the western Unites States. *Mosq. Syst.* 25: 170–176.

639. Schuyler, K. 1978. The occurrence of *Aedes provocans* in Pennsylvania. *Mosq. News* 38: 286–287.

640. Seawright, J. A., P. E. Kaiser, S. K. Narang, K. J. Tennessen, and S. E. Mitchell. 1992. Distribution of sibling species A, B, C and D of the *Anopheles quadrimaculatus* complex. *J. Agric. Entomol.* 9: 289–300.
641. Shaw, F. R. 1959. New records and distribution of the biting flies of Mt. Desert Island, Maine. *Mosq. News* 19: 189–191.
642. Shaw, F. R., and S. A. Maisey. 1961. The biology and distribution of the rockpool mosquito, *Aedes atropalpus* (Coq.) *Mosq. News* 21: 12–16.
643. Shemanchuk, J. A. 1959. Mosquitoes (Diptera, Culicidae) in the irrigated areas of southern Alberta and their seasonal changes in abundance and distribution. *Can. J. Zool.* 37: 899–912.
644. Shipp, J. L., and R. E. Wright. 1978. A new northern limit for the distribution of *Orthopodomyia signifera*. *Mosq. News* 38: 286.
645. Shipp, J. L., R. E. Wright, and D. H. Pengelly. 1978. Distribution of *Aedes triseriatus* (Say) and *Aedes hendersoni* Cockerell in southwestern Ontario, 1975–76. *Mosq. News* 38: 408– 412.
646. Shroyer, D. A., and R. W. Meyer. 1973. New distribution records of mosquitoes in Indiana, 1973 (Diptera, Culicidae). *Proc. Indiana Acad. Sci.* 83: 218–219.
647. Sirivanakarn, S. 1976. Medical entomology studies—III. A revision of the subgenus *Culex* in the Oriental Region (Diptera: Culicidae). *Contrib. Am. Entomol. Inst.* 12(2): 1–272.
Sirivavakarn, S., and J. N. Belkin. 1980. See ref. 810.
648. Sirivanakarn, S., and G. B. White. 1978. Neotype designation of *Culex quinquefasciatus* Say (Diptera, Culicidae). *Proc. Entomol. Soc. Wash.* 80: 360–372.
649. Sibal, I. H. 1994 (1995). *Aedes albopictus* and *Aedes aegypti* in Polk County (Florida, USA). *J. Fla. Mosq. Control Assoc.* 65: 15–16.
650. Siverly, R. E. 1957 (1958). Occurrence of *Culiseta melanura* (Coquillett) in Indiana. *Proc. Indiana Acad. Sci.* 67: 137.
651. ———. 1958 (1959). Occurrence of *Aedes grossbecki* Dyar and Knab and *Aedes aurifer* (Coquillett) in Indiana. *Proc. Indiana Acad. Sci.* 68: 149
652. ———. 1960 (1961). Occurrence of *Aedes thibaulti* Dyar and Knab in Indiana. *Proc. Indiana Acad. Sci.* 70: 137.
653. ———. 1961 (1962). Occurrence of *Culex territans* Walker in Indiana. *Proc. Indiana Acad. Sci.* 71: 115.
654. ———. 1962 (1963). Occurrence of *Aedes excrucians* (Walker) in Indiana. *Proc. Indiana Acad. Sci.* 72: 140.
655. ———. 1963 (1964). Occurrence of *Wyeomyia smithii* (Coquillett) in Indiana. *Proc. Indiana Acad. Sci.* 73: 144–145.
656. ———. 1965 (1966). Occurrence of *Culiseta minnesotae* Barr in Indiana. *Proc. Indiana Acad. Sci.* 75: 108.
657. ———. 1966a. Occurrence of *Culiseta melanura* (Coquillett) in Illinois. *Mosq. News* 26: 95–96.
658. ———. 1966b. Mosquitoes of Delaware County, Indiana. *Mosq. News* 26: 221–229.
659. ———. 1967. Occurrence of *Aedes abserratus* (Felt and Young) and *Culiseta morsitans* (Theobald) in Indiana. *Mosq. News* 27: 116.
660. ———. 1969. Occurrence of *Aedes dorsalis* (Meigen), *A. dupreei* (Coquillett), and *A. punctor* in Indiana. *Mosq. News* 29: 689.
661. ———. 1972. *Mosquitoes of Indiana.* Indianapolis, Indiana State Board of Health, 126 pp.
662. ———. 1973. Distribution of *Aedes stimulans* (Walker) in east central United States. *Proc. Indiana Acad. Sci.* 82: 227.
663. Siverly, R. E., and R. W. Burkhardt, Jr. 1964 (1965). Occurrence of *Psorophora discolor* (Coquillett) in Indiana. *Proc. Indiana Acad. Sci.* 74: 195.
664. Siverly, R. E., and G. R. DeFoliart. 1968a. Mosquito studies in northern Wisconsin I. Larval studies. *Mosq. News* 28: 149–154.
665. ———. 1968b. Mosquito studies in northern Wisconsin II. Light trapping studies. *Mosq. News* 28: 162–167.
666. Siverly, R. E., and J. W. Hart. 1971. Occurrence of *Aedes atlanticus* Dyar and Knab in Indiana. *Mosq. News* 31: 224.
667. Siverly, R. E., and D. A. Shroyer. 1974. Illustrated key to the genitalia of male mosquitoes of Indiana. *Mosq. Syst.* 6: 167–200.
668. Slaff, M., and C. S. Apperson. 1989. A key to the mosquitoes of North Carolina and the mid-Atlantic states. N.C. Agric. Ext. Ser. Publ. AG-412, vi + 38 pp.
669. Smith, J. P., T. M. Loyless, and J. A. Mulrennan, Jr. 1990. An update on *Aedes albopictus* in Florida. *J. Am. Mosq. Control Assoc.* 6: 318–320.
670. Smith, L. W., Jr. 1969a. History of mosquito occurrence in Missouri. *Mosq. News* 29: 220–222.
671. ———. 1969b. The relationship of mosquitoes to oxidation lagoons in Columbia, Missouri. *Mosq. News* 29: 556–563.
672. Smith, L. W., Jr., and W. R. Enns. 1967. Laboratory and field investigations of mosquito populations associated with oxidation lagoons in Missouri. *Mosq. News* 27: 462–466.
673. ———. 1968. A list of Missouri mosquitoes. *Mosq. News* 28: 50–51.
674. Smith, M. E. 1958. The *Aedes* mosquitoes of New England. Part 1: Key to adult females. *Bull. Brooklyn Entomol. Soc.* 53: 39–47.
675. ———. 1965a. Instar recognition in *Aedes* larvae (Diptera, Culicidae). *Proc. 12th Int. Congr. Entomol.*, London, 762–763.
676. ———. 1965b. Larval differences between *Aedes communis* (DeG.) and *A. implicatus* Vock., (Diptera, Culicidae) in a Colorado community. *Mosq. News* 25: 187–191.
677. ———. 1966. Mountain mosquitoes of the Gothic, Colorado, area. *Am. Midl. Nat.* 76: 125–150.
678. ———. 1969a. The *Aedes* mosquitoes of New England (Diptera, Culicidae). II. Larvae: keys to instars, and to species exclusive of first instar. *Can. Entomol.* 101: 41–51.

679. ———. 1969b. The *Aedes* mosquitoes of New England. III. Saddle hair position in 2nd and 3rd instar larvae, with particular reference to instar recognition and species relationships. *Mosq. Syst. Newsl.* 1: 57–62.
680. Smith, S. M., and R. A. Brust. 1970. Autogeny and stenogamy of *Aedes rempeli* (Diptera, Culicidae) in arctic Canada. *Can. Entomol.* 102: 253–256.
681. Smith, S. M., and R. M. Trimble. 1973. The biology of tree-holes of Point Pelee National Park, Ontario. 1. New mosquito records for Canada (Diptera, Culicidae). *Can. Entomol.* 105: 1585– 1586.
682. Smith, S. M., and R. M. Trimble 1994. Nectar feeding by early-spring mosquito, *Aedes provocans. Med. Vet. Entomol.* 8: 201–213.
683. Smithson, T. W. 1972. Species rank for *Anopheles franciscanus* based on failure of hybridization with *Anopheles pseudopunctipennis pseudopunctipennis. J. Med. Entomol.* 9: 501– 505.
684. Sollers-Riedel, H. 1972. 1970 world studies on mosquitoes and diseases carried by them. *Proc. 58th Mtg. N.J. Mosq. Exterm. Assoc.* 1971 Suppl., 52 pp.
685. Snow, W. E., and E. Pickard. 1956. Seasonal history of *Culex tarsalis* and associated species in larval habitats of the Tennessee Valley region. *Mosq. News* 16: 143–148.
686. Snow, W. E., and G. E. Smith. 1956. Observations on *Anopheles walkeri* Theobald in the Tennessee Valley. *Mosq. News* 16: 294–298.
687. Sommerman, K. M. 1966. True-false key to species of Alaskan biting mosquitoes. *Mosq. News* 26: 540–543.
688. ———. 1968. Notes on Alaskan mosquito records. *Mosq. News* 28: 233–234.
689. Spadom, R. D., and R. 0. Hayes. 1970. Mosquitoes on the offshore islands in California. *Proc. Calif. Mosq. Control Assoc.* 38: 97.
690. Spielman, A. 1964. Swamp mosquito, *Culiseta melanura*: occurrence in an urban habitat. *Science* 143: 361–362.
691. Sprenger, D., and T. Wuithiranyagool. 1986. The discovery and distribution of *Aedes albopictus* in Harris County, Texas. *J. Am. Mosq. Control Assoc.* 2: 217–219.
692. Stabler, R. M. 1945. New Jersey light-trap versus human bait as a mosquito sampler. *Entomol. News* 56: 93–99.
693. Steffan, W. A., N. L. Evenhuis, and D. L. Manning. 1980. Annotated bibliography of *Toxorhynchites* (Diptera: Culicidae). *J. Med. Entomol.* Suppl. 3, 180 pp.
694. Steward, C. C., and J. W. McWade. 1961. The mosquitoes of Ontario (Diptera, Culicidae) with keys to the species and notes on distribution. *Proc. Entomol. Soc. Ont.* 91: 121–188.
695. Stojanovich, C. J. 1960. Illustrated key to common mosquitoes of southeastern United States. Atlanta, 36 pp.
696. ———. 1961. Illustrated key to common mosquitoes of northeastern North America. Atlanta, 49 pp.
697. Stone, A. 1956. Corrections in the taxonomy and nomenclature of mosquitoes (Diptera, Culicidae). *Proc. Entomol. Soc. Wash.* 58: 333–344.
698. ———. 1958. Types of mosquitoes described by C. F. Adams in 1903 (Diptera, Culicidae). *J. Kans. Entomol. Soc.* 31: 235–237.
699. ———. 1961a. A correction in mosquito nomenclature (Diptera: Culicidae). *Proc. Entomol. Soc. Wash.* 63: 246.
700. ———. 1961b. A synoptic catalog of the mosquitoes of the world, Supplement I (Diptera: Culicidae). *Proc. Entomol. Soc. Wash.* 63: 29–52.
701. ———. 1963. A synoptic catalog of the mosquitoes of the world, Supplement II (Diptera: Culicidae). *Proc. Entomol. Soc. Wash.* 65: 117–140.
702. ———. 1965. Family Culicidae, p. 105–120. *In* A. Stone, C. W. Sabrowsky, W. W. Wirth, R. H. Foote, and J. R. Coulson. *A catalog of the Diptera of America north of Mexico.* USDA Handbook 276, 1696 pp.
703. ———. 1967. A synoptic catalogue of the mosquitoes of the world, Supplement III (Diptera, Culicidae). *Proc. Entomol. Soc. Wash.* 69: 197–224.
704. ———. 1968. A new mosquito record for the United States (Diptera: Culicidae). *Proc. Entomol. Soc. Wash.* 70: 384.
705. ———. 1969. Breedin-Archbold-Smithsonian biological survey of Dominica: The mosquitoes of Dominica (Diptera, Culicidae). Smithsonian Contrib. Zool. 16, 8 pp.
706. ———. 1970. A synoptic catalog of the mosquitoes of the world, Supplement IV. (Diptera: Culicidae). *Proc. Entomol. Soc. Wash.* 72: 137–171.
707. Stone, A., and J. A. Hair. 1968. A new *Culex* (*Melanoconion*) from Florida (Diptera, Culicidae). *Mosq. News* 28: 39–41.
708. Stone, A., K. L. Knight, and H. Starcke. 1959. *A synoptic catalog of the mosquitoes of the world (Diptera, Culicidae).* Thomas Say Foundation 6, 358 pp.
709. Strickman, D. 1988a. *Culex stigmatosoma* and *Cx. peus*: identification of female adults in the United States. *J. Am. Mosq. Control Assoc.* 4: 555–556.
710. ———. 1988b. Redescription of the holotype of *Culex* (*Culex*) *peus* Speiser and taxonomy of *Culex* (*Culex*) *stigmatosoma* Dyar and *thriambus* Dyar (Diptera: Culicidae). *Proc. Entomol. Soc. Wash.* 90:484–494.
711. Stryker, R. G., and W. W. Young. 1970. Effectiveness of carbon dioxide and L(+) lactic acid in mosquito light traps with and without light. *Mosq. News* 30: 388–393.
712. Sublette, M. S., and J. E. Sublette. 1970. Distributional records of mosquitoes on the southern high plains with a checklist of species from New Mexico and Texas. *Mosq. News* 30: 533–538.
713. Sudia, W. D., R. W. Emmons, V. F. Newhouse, and R. F. Peters. 1971. Arbovirus-vector studies in the Central Valley of California, 1969. *Mosq. News* 31: 160–168.
714. Sudia, W. D., and R. H. Gogel. 1953. The occurrence of *Orthopomyia alba* Baker in Georgia (Diptera: Culicidae). *Bull. Brooklyn Entomol. Soc.* 48: 129–131.
715. Swales, D. E. 1966. Species of insects and mites collected at Frobisher Bay, Baffin Island, 1964, and Inuvik, N. W. T., 1965, with brief ecological and geographical notes. *Ann. Entomol. Soc. Quebec* 11: 189–199.

716. Sweeney, K. J., M. A. Cantwell, and J. Dorothy. 1988. The collection of *Aedes aegypti* and *Ae. albopictus* from Baltimore, Maryland. *J. Am. Mosq. Control Assoc.* 4: 381.

717. Tanaka, K., K. Mizusawa, and E. S. Saugstad. 1979. A revision of the adult and larval mosquitoes of Japan (including the Ryukyu Archipelago and the Ogasawara Islands) and Korea (Diptera: Culicidae). *Contrib. Am. Entomol. Inst.* 16: 1–987.

718. Tanimoto, R. M. 1971. Introductory survey of adult mosquitoes in the Yukon-Kuskokwim Delta of Alaska. *Mosq. News* 31: 544–551.

719. Tawfik, M. S., and R. H. Gooding. 1970. Observations on mosquitoes during 1969 control operations at Edmonton, Alberta. *Quaest. Entomol.* 6: 307–310.

720. Taylor, D. B. 1983. New distribution records for mosquitoes (Diptera: Culicidae) in St. Joseph County, Indiana. *Proc. Indiana Acad. Sci.* 90: 274–280.

721. Taylor, S. A., R. F. Darsie, Jr., and W. L. Jakob. 1984. *Anopheles crucians*: a new adult record from Michigan. *Mosq. News* 44: 69–70.

722. Tessier, C., A. Maire, and A. Aubin. 1981. Productivite en larves de moustique (Diptera: Culicidae) demi milieux aquatiques peu profonds d'un secteur du Moyen-nord Quebecois (LG-1, Territoire de la Baie de James). *Can. J. Zool.* 59: 738–749.

723. Thompson, G. A. 1965. An invasion of the Gulf Coast by saltmarsh mosquitoes. *Mosq. News* 15: 164–165.

724. Thompson, P. H., and G. R. Defoliart. 1966. New distribution records of biting Diptera from Wisconsin. *Proc. Entomol. Soc. Wash.* 68: 85.

725. Tinker, M. E., and G. R. Hayes, Jr. 1959. The 1958 *Aedes aegypti* distribution in the United States. *Mosq. News* 19: 73–78.

726. Tipton, V. J., and R. C. Saunders. 1971. A list of arthropods of medical importance which occur in Utah, with a review of arthropod-borne diseases endemic in the state. Brigham Young Univ. Sci. Bull., Biol. Ser. 15(2): 1–31.

727. Trapido, H., and P. Galindo. 1956. Genus *Haemagogus* in the United States. *Science* 123:634.

728. Trimble, R. M. 1972. Occurrence of *Culiseta minnesotae* and *Aedes trivittatus* (Diptera: Culicidae) in Manitoba, including a list of mosquitoes from Manitoba. *Can. Entomol.* 104: 1535–1537.

729. Trimble, R. M., and S. M. Smith. 1975. A bibliography of *Toxorhynchites rutilus* (Coquillett) (Diptera: Culicidae). *Mosq. Syst.* 7: 115–126.

730. Truman, J. W., and G. B. Craig, Jr. 1968. Hybridization between *Aedes hendersoni* and *Aedes triseriatus*. *Ann. Entomol. Soc. Am.* 61: 1020–1025.

731. Twinn, C. R. 1949. Mosquitoes and mosquito control in Canada. *Mosq. News* 9: 35–41.

732. U.S. Department of Agriculture. 1971. Cooperative Economic Insect Report 21: 780.

732. ———. 1978. Cooperative Plant Pest Report 3: 420–425.

733. Urbanelli, S., F. Silvestrini, W. K. Reisen, E. de Vito, and L. Bullini. 1997. Californian hybrid zone between *Culex pipiens pipiens* and *Cx. p. quinquefasciatus* revisited (Diptera: Culicidae). *J. Med. Entomol.* 34: 116–127.

734. Vargas, J. A., and Z. Prusak. 1994 (1995). The status of *Aedes albopictus* within the Reedy Creek Improvement District, Orange County, Florida. *J. Fla. Mosq. Control Assoc.* 65: 12–14.

735. Vargas, L. 1956a. Especies y distribución de mosquitos mexicanos no anofelinos. *Rev. Inst. Salubr. Enferm. Trop.* (Mex.) 16: 19–36.

736. ———. 1956b. Algunas diferencias morfológicas entre *Toxorhynchites rutilus* y *T. septentrionalis*. *Rev. Inst. Salubrid. Enferm. Trop.* (Mex.) 16: 33–36.

737. ———. 1974. Bilingual key to the New World genera of mosquitoes (Diptera: Culicidae) based upon the fourth stage larvae. *Calif. Vector Views* 21: 15–18.

738. Vargas, L., and A. Martinez Palacios. 1956. Anofelinos mexicanos, taxonomía y distribución. Mexico, D. F., Sec. Salubr. Y Asist., Com. Nacional Errad. Palud., 81 pp.

739. Vockeroth, J. R. 1954. Notes on the identities and distributions of *Aedes* species of northern Canada, with a key to the females (Diptera, Culicidae). *Can. Entomol.* 86: 241–255.

740. Wada, Y. 1965. Population studies on Edmonton mosquitoes. *Quaest. Entomol.* 1: 187–222.

741. Wagner, V. E., and H. D. Newson. 1975. Mosquito biting activity in Michigan state parks. *Mosq. News* 35: 217–222.

742. Walker, E. D. 1983. Occurrence of *Anopheles barberi* in Massachusetts. *Mosq. News* 43: 73.

743. Walker, E. D., G. F. O'Meara, and W. T. Morgan. 1996. Bacterial abundance in larval habitats of *Aedes albopictus* in a Florida cemetery. J. Vector Ecol. 21: 173–177.

744. Wallace, R. C. 1960. Mosquitoes collected in the vicinity of Marquette, Michigan, during the summer of 1959. *Trans. Ill. State Acad. Sci.* 53: 46–47.

745. Wallis, R. C. 1954. Notes on the biology of *Culiseta melanura* (Coquillett). *Mosq. News* 14: 33–34.

746. ———. 1960. Mosquitoes in Connecticut. Conn. Agric. Exp. Sta. Bull. 632, 30 pp.

747. Wallis, R. C., and L. Whitman. 1968. Mosquitoes of the genus *Culiseta* in Connecticut (Diptera: Culicidae). *Proc. Entomol. Soc. Wash.* 70: 187–188.

748. Wallis, R. C., and L. Whitman. 1971a. First report of *Aedes thibaulti* Dyar and Knab in Connecticut and New York. *Mosq. News* 31: 111.

749. ———. 1971b. New collection records of *Psorophora ciliata* (Fabricius), *Psorophora ferox* (Humboldt) and *Anopheles earlei* Vargas in Connecticut (Diptera: Culicidae). *J. Med. Entomol.* 8: 336–337.

750. Ward, R. A. 1984. A second supplement to *A catalog of the mosquitoes of the world* (Diptera: Culicidae). *Mosq. Syst.* 16: 227–270.

751. ———. 1992. Third supplement to *A catalog of the mosquitoes of the world* (Diptera: Culicidae). *Mosq. Syst.* 24: 177–230.

752. Ward, R. A., and R. F. Darsie, Jr. 1982. Corrections and additions to the publication *Identification and geographical distribution of the mosquitoes of North America, north of Mexico*. *Mosq. Syst.* 14: 209–219.

753. Weathersbee, A. A., and F. T. Arnold. 1947. A resume of the mosquitoes of South Carolina. *J. Tenn. Acad. Sci.* 22: 210–229.

754. Weaver, S. C., W. F. Scherer, C. D. Taylor, D. A. Castello, and E. W. Cupp. 1986. Laboratory vector competence of *Culex* (*Melanoconion*) *cedecei* for sympatric and allopatric Venezuelan equine encephalomyelitis viruses. *Am. Trop. Med. Hyg.* 35: 619– 623.

755. Weber, R. M., and R. G. Weber. 1985. The egg raft seam as an indicator of species in *Culex pipiens* and *Culex restuans*. *Mosq. Syst.* 17: 363–370.

756. Welch, J. B., and J. D. Long. 1984. *Aedes aegypti* collections in rural southeast Texas. *Mosq. News* 44: 544–547.

757. Wesson, D., W. Hawley, and G. B. Craig, Jr. 1990. Status of *Aedes albopictus* in the midwest: LaCrosse belt distribution, 1988. *Proc. Ill. Mosq. Vector Control Assoc.* 1: 11–15.

758. West, A. S., and A. Hudson. 1960. Notes on mosquitoes of eastern Ontario. *Proc. 47th Mtg. N.J. Mosq. Exterm. Assoc.*, 68–74.

759. West, D. F., C. F. Bosio, and W. C. Black IV. 1994. New state record for *Culiseta morsitans* in Colorado. *J. Am. Mosq. Control Assoc.* 10: 588.

760. White, D. J., and C. P. White. 1980. *Aedes atropalpus* breeding in artificial containers in Suffolk County, New York. *Mosq. News* 40: 106–107.

761. White, G. B. 1978. Systematic reappraisal of the *Anopheles maculipennis* complex. *Mosq. Syst.* 10: 13–44

762. White, M. S. 1956. *Aedes bicristatus* occurrence. *Calif. Vector Views* 3: 17.

763. Whitlaw, J. T., Jr., W. E. Bickley, and E. N. Cory. 1956. Mosquitoes in farm ponds in Maryland. *J. Econ. Entomol.* 49: 273.

764. Wilkerson, R. C., and E. L. Peyton. 1990. Standardized nomenclature for the costal wing spots of the genus *Anopheles* and other spotted-wing mosquitoes (Diptera: Culicidae). *J. Med. Entomol.* 27: 207–224.

765. Williams, R. W. 1956. A new distribution record for *Culex salinarius* Coq.: the Bermuda Islands. *Mosq. News* 16:29–30.

766. Wills, W., and R. L. Beaudoin. 1966. Distribution of mosquitoes in Pennsylvania. *Proc. Pa. Acad. Sci.* 39: 166–169.

767. Wills, W., and V. McElhattan. 1963. *Aedes aurifer* (Coquillett) and *Wyeomyia smithii* (Coquillett) in Pennsylvania (new state record). *Mosq. News* 23: 264.

768. ———. 1968. Additions to the list of *Aedes* species in Pennsylvania. *Mosq. News* 28: 108–109.

769. Wills, W., and D. Steinhart. 1966. Inland records for salt marsh mosquitoes in Pennsylvania. *Mosq. News* 26: 254–255.

770. Wills, W., and G. Whitmyre, Jr. 1970. New Pennsylvania record of *Orthopodomyia alba* Baker. *Mosq. News* 30: 472.

771. Wilmot, T. R., J. M. Henderson, and D. W. Allen. 1987. Additional collection records for mosquitoes of Michigan. *J. Am. Mosq. Control Assoc.* 3: 318.

772. Wilmot, T. R., D. S. Zeller, and R. W. Merritt. 1992. A key to container-breeding mosquitoes of Michigan (Diptera: Culicidae). *Great Lakes Entomol.* 25: 137–148.

773. Wilson, C. A., R. C. Barnes, and H. L. Fellton. 1946. A list of the mosquitoes of Pennsylvania with notes on their distribution and abundance. *Mosq. News* 6: 78– 84.

774. Wilson, W. T. 1959. A study of the medically important mosquitoes at Holloman Air Force Base, New Mexico. *Mosq. News* 19: 17–19.

775. Wiseman, J. S. 1965. A list of mosquito species reported from Texas. *Mosq. News* 25: 58–59.

776. Wolff, T. A. 1970. The presence of *Aedes fitchii* (Felt and Young) in New Mexico. *Mosq. News* 30: 472.

777. Wolff, T. A., and L. T. Nielsen. 1976. The distribution of snowpool *Aedes* mosquitoes in the southwestern states of Arizona and New Mexico with notes on biology and past dispersal patterns. *Mosq. Syst.* 8: 413–439.

778. ———. 1977. A chaetotaxic study of snowpool *Aedes* larvae and pupae with analysis of variance of the larvae of eight species. *Mosq. Syst.* 9: 176–236.

779. Wolff, T. A., L. T. Nielsen and R. O. Hayes. 1975. A current list and bibliography of the mosquitoes of New Mexico. *Mosq. Syst.* 7: 13–18.

780. Wolff, T. A., L. T. Nielsen and J. H. Linam. 1974. Additional records of culicine and chaoborine mosquitoes from the mountains of Arizona and New Mexico. *Proc. Pap. 42nd Conf. Calif. Mosq. Cont. Assoc.*, 41–42.

781. Womack, M. L. 1993. Distribution, abundance and bionomics of *Aedes albopictus* in southern Texas. *J. Am. Mosq. Control Assoc.* 9: 367–369.

782. Womack, M. L., T. S. Thuma and B. R. Evans. 1995. Distribution of *Aedes albopictus* in Georgia, USA. *J. Am. Mosq. Control Assoc.* 11: 237.

783. Wood, D. M. 1977. Notes on the identities of some common Neartic *Aedes* mosquitoes. *Mosq. News* 37: 71–81.

784. Wood, D. M., P. T. Dang, and R. A. Ellis. 1979. *The mosquitoes of Canada (Diptera: Culicidae)*. Series: The insects and arachnids of Canada. Part 6. Biosystematics Res. Inst., Canada Dept. Agric. Publ. 1686, 390 pp.

785. Yamaguti, S., and W. J. LaCasse. 1951. *Mosquito fauna of North America*. Parts I-V. Office of the Surgeon, Hq. Japan Logistical Command, 629 pp.

786. Zaim, M., H. D. Newson, and G. D. Dennis. 1977. *Psorophora horrida* in Michigan. *Mosq. News* 37: 763.

787. Zavortink, T. J. 1968. Mosquito studies (Diptera, Culicidae). VIII. A prodrome of the genus *Orthopodomyia*. *Contrib. Am. Entomol. Inst.* 3(2): 1–221.

788. ———. 1969a. Mosquito studies (Diptera, Culicidae). XV. A new species of treehole breeding *Anopheles* from the southwestern United States. *Contrib. Am. Entomol. Inst.* 4(4): 27–38.

789. ———. 1969b. Mosquito studies (Diptera, Culicidae). XVI. A new species of treehole breeding *Aedes* (*Ochlerotatus*) from southern California. *Contrib. Am. Entomol. Inst.* 5(l): 1–7.

790. ———. 1969c (1970). Mosquito studies (Diptera: Culicidae). XIX. The treehole *Anopheles* of the New World. *Contrib. Am. Entomol. Inst.* 5(2): 1–35.

791. ———. 1969d. New species and records of treehole mosquitoes from the southwestern United States. *Mosq. Syst. Newsl.* 1: 22.

792. ———. 1970. Mosquito studies (Diptera, Culicidae). XXII. A new subgenus and species of *Aedes* from Arizona. *Contrib. Am. Entomol. Inst.* 7(1): 1–11.

793. ———. 1972. Mosquito studies (Diptera, Culicidae). XXVIII. The New World species formerly placed in *Aedes* (*Finlaya*). *Contrib. Am. Entomol. Inst.* 8(3): 1–206.

794. ———. 1985. Observations on the ecology of treeholes and treehole mosquitoes in the southwestern United States, pp. 473–487. In: L. P. Lounibos, J. R. Rey, and J. H. Frank (eds), *Ecology of mosquitoes.* Florida Medical Entomology Laboratory, Vero Beach.

Zavortink, T. J., and J. N. Belkin. 1979. See ref. 814.

795. Zavortink, T. J., and G. F. O'Meara. 1999. A new species of *Culex* (*Micraedes*) from Florida (Diptera: Culicidae). *J. Am. Mosq. Control Assoc.* 15: 263–270.

796. Zingmark, R. G. (ed.) 1978. *An annotated checklist of the biota of the coastal zone of South Carolina.* Baruch Inst. Marine Biol. Coastal Res. 364 pp.

## Addendum

797. Janousek, T. E., and W. L. Kramer. 1999. Seasonal incidence and geographical variation of Nebraska mosquitoes, 1994–95. *J. Am. Mosq. Control Assoc.* 15: 253–262.

798. Dennehy, J. J., and T. Livdahl. 1999. First record of *Toxorhynchites rutilus* (Diptera: Culicidae) in Massachusetts. *J. Am. Mosq. Control Assoc.* 15: 423–424.

799. Reinert, J. F. 1973. Contributions to the mosquito fauna of Southeast Asia—XVI. Genus *Aedes* Meigen, subgenus *Aedimorphus* Theobald in southeast Asia. *Contrib. Am. Entomol. Inst.* 9(5): 1– 218.

800. ———. 2000a. Assignment of two North American species of *Aedes* to subgenus *Rusticoidus*. *J. Am. Mosq. Control Assoc.* 16: 42–43.

801. ———. 2000b. Separation of trap-collected adults of *Anopheles atropos* from species of the *quadrimaculatus* complex. *J. Am. Mosq. Control Assoc.* 16: 44.

802. ———. 2000c. New classification for the composite genus *Aedes* (Diptera: Culicidae: Aedini), elevation of subgenus *Ochlerotatus* to generic rank, reclassification of the other subgenera, and notes on certain subgenera and species. *J. Am. Mosq. Control Assoc.* 16: 175–188.

803. Hribar, L. J. 2001. Uncommonly collected mosquitoes from the Florida Keys. *Entomol. News* 112: 123.

804. Hribar, L. J., and J. J. Vlach. 2001. Mosquito (Diptera: Culicidae) and biting midges (Diptera: Ceratopogonidae) collections in Florida Keys state parks. *Fla. Sci.* 64: 219–223.

805. Darsie, R. F., Jr. 1996. A survey and bibliography of the mosquito fauna of Mexico (Diptera: Culicidae). *J. Am. Mosq. Control Assoc.* 12: 298–306.

806. Hayden, C. W., T. M. Fink, F. B. Ramberg, C. J. Mare, and D. C. Mead. 2001. Occurrence of *Anopheles hermsi* (Diptera: Culicidae) in Arizona and Colorado. *J. Med. Entomol.* 38: 341–343.

807. Moore, J. P. 1999. Mosquitoes of Fort Campbell, Kentucky (Diptera: Culicidae). *J. Am. Mosq. Control Assoc.* 15: 1–3.

808. Andreadis, T. G., J. F. Anderson, L. E. Munstermann, R. J. Wolfe, and D. A. Florin. 2001. Discovery, distribution and abundance of the newly introduced mosquito *Ochlerotatus japonicus* (Diptera: Culicidae) in Connecticut, USA. *J. Med. Entomol.* 38: 774–779.

809. Sallum, M. A., and O. P. Forattini. 1996. Revision of the *spissipes* section of *Culex* (*Melanoconion*) (Diptera: Culicidae). *J. Am. Mosq. Control Assoc.* 12: 517–600.

810. Sirivanakarn, S., and J. N. Belkin. 1980. The identity of *Culex* (*Melanoconion*) *taeniopus* Dyar and Knab and related species with notes on the synonymy and description of a new species (Diptera, Culicidae). *Mosq. Syst.* 12: 7–28.

811. Fritz, G. N., D. Dritz, T. Jensen, and R. K. Washino. 1991. Pattern variation in *Anopheles punctipennis* (Say). *Mosq. Syst.* 23: 81–86.

812. Fonseca, D. M., S. Campbell, W. J. Crans, M. Mogi, I. Miyagi, T. Toma, M. Bullians, T. G. Andreadis, R. L. Berry, B. Pagac, M. R. Sardelis, and R. C. Wilkerson. 2001. *Aedes* (*Finlaya*) *japonicus* (Diptera, Culicidae), a newly recognized mosquito in the United States and putative source populations. *J. Med. Entomol.* 38: 135–146.

813. Harbach, R. E., and I. J. Kitching. 1998. Phylogeny and classification of the Culicidae (Diptera). *Syst. Entomol.* 23: 327–370.

814. Zavortink, T. J., and J. N. Belkin. 1979. Occurrence of *Aedes hendersoni* in Florida (Diptera, Culicidae). *Mosq. News* 39: 673.

815. Reinert, J. F. 2001. Revised list of abbreviations for genera and subgenera of Culicidae (Diptera) and notes on generic and subgeneric changes. *J. Am. Mosq. Control Assoc.* 17: 51–55.

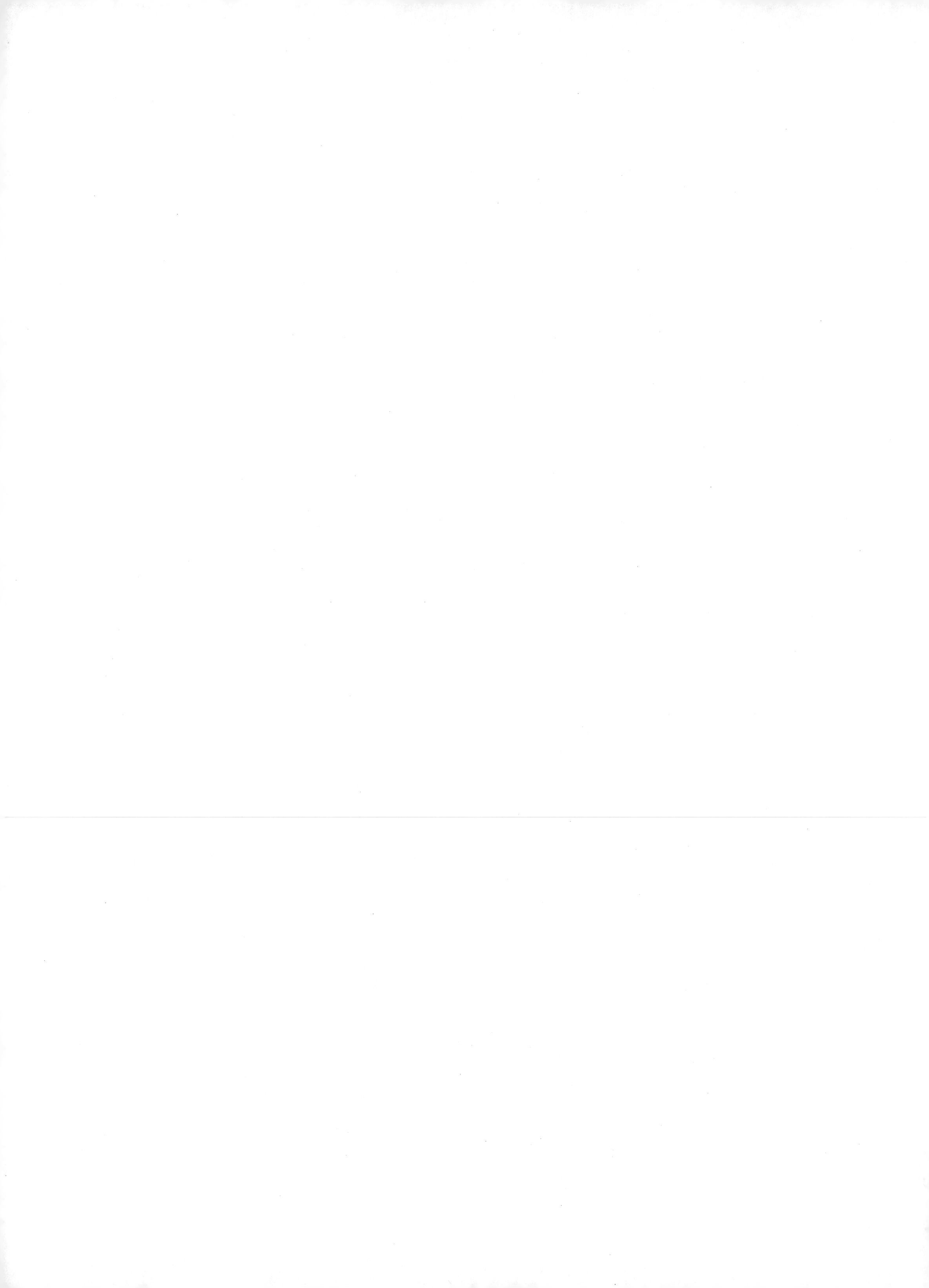

# Appendix: Locality Data for Mosquito Specimens Used to Prepare Illustrations for Keys

For the benefit of mosquito taxonomists and other scientists interested in the localities from which the specimens were collected, the following list presents locality data. Ninety-six percent of the specimens are from the United States and Canada, but for some species that are neotropical in distribution and are found only in the extreme southern parts of the United States, it was necessary to select specimens from the Caribbean Islands, Mexico, Central America, Panama, and Colombia. Specimens were utilized from all states of the continental United States, except Iowa, Indiana, New Hampshire, South Dakota, and West Virginia, and all provinces of Canada, except New Brunswick, Newfoundland, Nova Scotia, Prince Edward Island, and Quebec. In all, adult females were selected from 38 states of the United States, 7 provinces of Canada, and 8 foreign countries, while larvae were from 35 states of the United States, 4 provinces of Canada, and 9 foreign countries.

Since the mosquito fauna is better known in some regions than others, it is not surprising that specimens from only 11 states or provinces were used to prepare 50 percent of the adult illustrations, while larvae from 10 states or provinces accounted for 64.7 percent of the drawings used in the keys to immatures.

**Locality Data for Mosquito Specimens Used to Prepare Illustrations for Keys**

| Figure Number | Species | State/ Country | Prov. | County | Locality |
|---|---|---|---|---|---|
| 1,2 | Tx. r. septentrionalis | USA | DE | Kent | Bombay Hook |
| 3,4 | Ae. vexans | Canada | ONT | Kenora | Dryden |
| 5,6 | An. quadrimaculatus | USA | NC | Robeson | Maxton |
| 7,8 | Ae. vexans | Canada | ONT | Kenora | Dryden |
| 9,10 | Wy. smithii | USA | MA | Hampden | Westfield |
| 11 | Ae. vexans | Canada | ONT | Kenora | Dryden |
| 12 | Ae.vexans | USA | ND | Grand Forks | Grand Forks |
| 13,14 | Ur. sapphirina | USA | VA | Fairfax | Falls Church |
| 15 | Cx. pipiens | USA | NJ | Middlesex | Nixon |
| 16 | Ae. vexans | USA | ND | Grand Forks | Grand Forks |
| 17 | Ps. ciliata | USA | VA | Accomack | Chincoteague |
| 18 | Cs. inornata | USA | OR | Portland | Portland |
| 19,20 | Ma. titillans | Mexico | | Tamaulipas | Tampico |
| 21,22 | Ae. vexans | Canada | ONT | Kenora | Dryden |
| 23 | Ps. ciliata | USA | VA | Accomack | Chincoteague |

| Figure Number | Species | State/ Country | Prov. | County | Locality |
|---|---|---|---|---|---|
| 24 | Ps. cyanescens | USA | TX | Dallas | Dallas |
| 25 | Ae. vexans | USA | ND | Grand Forks | Grand Forks |
| 26 | Ae. vexans | Canada | ONT | Kenora | Dryden |
| 27,28 | Cs. inornata | USA | OR | Middlesex | Nixon |
| 31 | Hg. equinus | USA | TX | Cameron | Brownsville |
| 32 | Cx. pipiens | USA | NJ | Middlesex | Nixon |
| 33,34 | Or. signifera | USA | FL | Indian River | Vero Beach |
| 35,36 | Cx. pipiens | USA | NJ | Middlesex | Nixon |
| 37 | Cq. perturbans | USA | NY | Wayne | Fair Haven |
| 38,40 | Cx. pipiens | USA | NJ | Middlesex | Nixon |
| 39 | De. pseudes | USA | TX | Cameron | Brownsville |
| 41,42 | Oc. purpureipes | USA | AZ | Santa Cruz | Madera Canyon |
| 43 | Cx. pipiens | USA | NY | Brooklyn | Ft. Hamilton |
| 44 | Cx. pipiens | USA | NJ | Middlesex | Nixon |
| 45,47 | Oc. excrucians | USA | MT | Glacier | Glacier Natl Pk |
| 46 | Oc. triseriatus | USA | KY | Jefferson | Louisville |
| 48 | Oc. c. canadensis | USA | MN | Roseau | Warroad |
| 49 | Oc. sollicitans | USA | FL | Dade | Miami |
| 50 | Ae. vexans | Canada | ONT | Kenora | Dryden |
| 51,52 | Oc. taeniorhynchus | USA | FL | Palm Beach | Palm Beach |
| 53,54 | Oc. sollicitans | USA | FL | Dade | Miami |
| 55,56 | Oc. mitchellae | USA | FL | | |
| 57–59 | Oc. sollicitans | USA | FL | Dade | Miami |
| 60 | Oc. sollicitans | USA | FL | Dade | Miami |
| 61,62 | Oc. nigromaculis | USA | ID | Ada | Boise |
| 63,64 | Oc. papago | USA | AZ | Pima | Mendoza Canyon |
| 65 | Oc. taeniorhynchus | USA | FL | Palm Beach | Palm Beach |
| 66 | Ae. vexans | USA | ND | Grand Forks | Grand Forks |
| 67 | Ae. aegypti | USA | FL | | |
| 68 | Oc. c. canadensis | USA | MN | Roseau | Warroad |
| 69,70 | Oc. j. japonicus | USA | CT | Middlesex | Portland |
| 71 | Oc. j. japonicus | USA | CT | Middlesex | Portland |
| 72,73 | Ae. aegypti | USA | FL | | |
| 74 | Ae. aegypti | USA | FL | | |
| 75 | Oc. zoosophus | USA | TX | Frio | |
| 76 | Oc. epactius | USA | TX | Travis | |
| 77,79 | Ae. vexans | Canada | ONT | Kenora | Dryden |
| 78 | Oc. excrucians | USA | MT | Glacier | Glacier Natl Pk |
| 80 | Ae. vexans | USA | ND | Grand Forks | Grand Forks |
| 81,82 | Oc. cantator | USA | NY | Long Island | |
| 83 | Oc. c. canadensis | USA | MN | Roseau | Warroad |
| 84 | Ae. albopictus | USA | FL | Indian River | Vero Beach |
| 85 | Oc. stimulans | Canada | ONT | Carleton | Ottawa |
| 86 | Oc. excrucians | USA | MT | Glacier | Glacier Natl Pk |
| 87,88 | Oc. bahamensis | USA | FL | Dade | |
| 89,90 | Ae. albopictus | USA | FL | Indian River | Vero Beach |
| 90 | Oc. grossbecki | USA | LA | Rapides | Alexandria |
| 92 | Oc. stimulans | Canada | ONT | Carelton | Ottawa |
| 93 | Oc. squamiger | USA | CA | Orange | Huntington |
| 94 | Oc. squamiger | USA | CA | San Diego | San Diego |
| 95,96 | Oc. grossbecki | USA | LA | Rapides | Alexandria |
| 97 | Oc. nigromaculis | USA | ND | Ramsey | Devil's Lake |
| 98 | Oc. nigromaculis | USA | ID | Ada | Boise |
| 99,100 | Oc. increpitus | USA | UT | Cache | River Heights |

| Figure Number | Species | State/ Country | Prov. | County | Locality |
|---|---|---|---|---|---|
| 101 | Oc. flavescens | Canada | SASK | | Oxbow |
| 100,102 | Oc. increpitus | USA | UT | Cache | River Heights |
| 103 | Oc. excrucians | USA | MT | Glacier | Glacier Natl Pk |
| 104 | Oc. increpitus | USA | UT | Cache | River Heights |
| 105,107 | Oc. riparius | Canada | ALTA | | Red Deer |
| 106 | Oc. stimulans | Canada | ONT | Carleton | Ottawa |
| 108 | Oc. fitchii | Canada | ONT | Algoma | White River |
| 109,110 | Oc. riparius | Canada | ALTA | | Red Deer |
| 111,112 | Oc. aloponotum | USA | OR | Marion | Idanha |
| 113–116 | Oc. euedes | USA | MN | Clearwater | Itasco St. Pk. |
| 117–120 | Oc. fitchii | Canada | ONT | Algoma | White River |
| 121–122 | Oc. increpitus | USA | UT | Cache | River Heights |
| 123,124 | Oc. stimulans | Canada | ONT | Carleton | Ottawa |
| 125 | Oc. fitchii | Canada | ONT | Klondike | Dawson |
| 127 | Oc. stimulans | Canada | ONT | Carleton | Ottawa |
| 128 | Oc. euedes | USA | MN | Clearwater | Itasca St. Pk. |
| 129–131 | Oc. mercurator | Canada | YUK | Klondike | Dawson |
| 132–134 | Oc. fitchii | Canada | ONT | Algoma | White River |
| 135,136 | Oc. stimulans | Canada | ONT | Carleton | Ottawa |
| 137,138 | Oc. euedes | USA | MN | Clearwater | Itasca St. Pk. |
| 139,140 | Oc. dorsalis | USA | OR | Klamath | Klamath Falls |
| 141 | Oc. c. canadensis | USA | MN | Roseau | Warroad |
| 142 | Oc. atropalpus | USA | MA | Essex | |
| 143,144 | Oc. melanimon | USA | MT | Hill | Havre |
| 145,146 | Oc. dorsalis | USA | OR | Klamath | Klamath Falls |
| 147,148 | Oc. dorsalis | USA | OR | Klamath | Klamath Falls |
| 149,150 | Oc. campestris | USA | NV | Elko | Carlin |
| 151 | Oc. togoi | Taiwan | | | |
| 152–154 | Oc. c. canadensis | USA | MN | Roseau | Warroad |
| 155,156 | Oc. atropalpus | USA | Ma | Essex | |
| 157,158 | Oc. c. canadensis | USA | MN | Roseau | Warroad |
| 159,160 | Oc. c. mathesoni | USA | FL | Clay | Camp Blanding |
| 161–163 | Oc. atropalpus | USA | MA | Essex | |
| 164–166 | Oc. sierrensis | USA | WA | Mason | Lake Cushman |
| 167–169 | Oc. epactius | USA | TX | Travis | |
| 170–172 | Oc. atropalpus | USA | MA | Essex | |
| 173 | Oc. monticola | USA | AZ | Pima | Sabino Basin |
| 174 | Oc. sierrensis | USA | WA | Mason | Lake Cushman |
| 175 | Oc. varipalpus | USA | AZ | Coconino | Williams |
| 176,177 | Oc. sierrensis | USA | CA | Los Angeles | Pearblossom |
| 178 | Oc. sierrensis | USA | WA | Mason | Lake Cushman |
| 179,180 | Oc. deserticola | USA | CA | Riverside | Joshua Tree Natl Monument |
| 181 | Oc. fulvus pallens | USA | LA | Rapides | Alexandria |
| 182 | Oc. triseriatus | USA | KY | Jefferson | Louisville |
| 183,184 | Oc. fulvus pallens | USA | LA | Rapides | Alexaandria |
| 185,186 | Oc. bimaculatus | USA | TX | Cameron | Brownsville |
| 187 | Oc. purpureipes | USA | AZ | Santa Cruz | Madera Canyon |
| 188 | Oc. hendersoni | USA | CO | Weld | Kuner |
| 189 | Oc. atlanticus | USA | NC | Brunswick | Wilmington |
| 190,192 | Oc. triseriatus | USA | KY | Jefferson | Louisville |
| 191 | Oc. pullatus | USA | CO | Grand | Grand Lake |
| 193 | Oc. atlanticus | USA | NC | Brunswick | Wilmington |
| 194,195 | Oc. triseriatus | USA | KY | Jefferson | Louisville |

| Figure Number | Species | State/ Country | Prov. | County | Locality |
|---|---|---|---|---|---|
| 196 | Oc. hendersoni | USA | OH | Portage | Ravenna |
| 197 | Oc. hendersoni | USA | CO | Weld | Kuner |
| 198,199 | Oc. hendersoni | USA | OH | Portage | Ravenna |
| 200,201 | Oc. brelandi | USA | TX | Brewster | Big Bend Natl Pk |
| 202 | Oc. trivittatus | USA | MO | Clay | Kansas City |
| 203,205 | Oc. atlanticus | USA | NC | Brunswick | Wilmington |
| 204 | Oc. infirmatus | USA | FL | Gulf | |
| 206,207 | Oc. scapularis | USA | TX | Hidalgo | Mission |
| 208,209 | Oc. infirmatus | USA | FL | Gulf | |
| 210,211 | Oc. burger | USA | AZ | Santa Cruz | Bodie Canyon |
| 212,213 | Oc. atlanticus | USA | NC | Brunswick | Wilmington |
| 214,215 | Oc. muelleri | USA | AZ | Santa Cruz | Bodie Canyon |
| 216,217 | Oc. atlanticus | USA | NC | Brunswick | Wilmington |
| 218 | Oc. dupreei | USA | LA | East Baton | Baton Rouge |
| 219 | Oc. atlanticus | USA | NC | Brunswick | Wilmington |
| 220 | Oc. niphadopsis | USA | UT | Salt Lake | Salt Lake City |
| 221,223 | Oc. s. idahoensis | USA | UT | Uintah | Ouray |
| 222 | Oc. pullatus | USA | CO | Grand | Grand Lake |
| 224 | Oc. niphadopsis | USA | UT | Salt Lake | Salt Lake City |
| 225,226 | Oc. s. spencerii | USA | UT | Uintah | Ouray |
| 229,230 | Oc. ventrovittis | USA | WY | Teton | Yellowstone |
| 231 | Oc. bicristatus | USA | CA | Lake | Lower Lake |
| 232 | Oc. cataphylla | Canada | BC | | Cranbrook |
| 233,234 | Oc. niphadopsis | USA | UT | Salt Lake | Salt Lake Cuty |
| 235,236 | Oc. cataphylla | Canada | BC | | Cranbrook |
| 237,238 | Oc. bicristatus | USA | CA | Lake | Lower Lake |
| 239,240 | Oc. cataphylla | Canada | BC | | Cranbrook |
| 241 | Oc. pullatus | USA | CO | Grand | Grand Lake |
| 242 | Oc. diantaeus | USA | MI | Keweenaw | Copper Harbor |
| 243,244 | Oc. pullatus | USA | CO | Grand | Grand Lake |
| 245,246 | Oc. implicatus | USA | ID | Kootenai | Athol |
| 247,248 | Oc. intrudens | USA | ME | Washington | Crawford |
| 249,250 | Oc. pullatus | USA | CO | Grand | Grand Lake |
| 251,252 | Oc. implicatus | USA | ID | Kootenai | Athol |
| 253,254 | Oc. provocans | USA | MN | Roseau | Warroad |
| 255 | Oc. diantaeus | USA | MI | Keweenaw | Copper Harbor |
| 256 | Oc. intrudens | USA | ME | Washington | Crawford |
| 257,258 | Oc. aurifer | USA | DE | New Castle | Glasgow |
| 259–261 | Oc. thibaulti | USA | AL | Lauderdale | Wilson Dam |
| 262–264 | Oc. decticus | USA | MA | Hampshire | Belchertown |
| 265,266 | Oc. diantaeus | USA | MI | Keweenaw | Copper Harbor |
| 267 | Oc. sticticus | USA | MA | Hampshire | Northhampton |
| 268 | Oc. punctor | USA | MA | Hampshire | Chesterfield |
| 269 | Oc. thelcter | USA | TX | Bexar | San Antonio |
| 270–272 | Oc. intrudens | USA | ME | Washington | Crawford |
| 273,274 | Oc. sticticus | USA | MA | Hampshire | Northampton |
| 275,276 | Ae. cinereus | USA | MN | Roseau | Warroad |
| 277–279 | Oc. intrudens | USA | ME | Washington | Crawford |
| 280 | Oc. tortilis | Bahamas | | | |
| 281,282 | Oc. rempeli | Canada | NWT | | Baker Lake |
| 283–286 | Oc. sticticus | USA | MA | Hampshire | Northampton |
| 287–289 | Oc. communis | USA | MI | Keweenaw | Copper Harbor |
| 290,291 | Oc. nevadensis | USA | NV | Elko | Lamoille Canyon |
| 292 | Oc. churchillensis | Canada | MAN | | Churchill |

| Figure Number | Species | State/ Country | Prov. | County | Locality |
|---|---|---|---|---|---|
| 293,294 | Oc. ventrovittis | USA | WY | Teton | Yellowstone Pk |
| 295 | Oc. implicatus | USA | ID | Kootenai | Athol |
| 296 | Oc. punctor | USA | MA | Hampshire | Chesterfield |
| 297,298 | Oc. impiger | USA | AK | | Nome |
| 299 | Oc. pionips | Canada | ONT | Algoma | White River |
| 300 | Oc. implicatus | USA | ID | Kootenai | Athol |
| 301,302 | Oc. impiger | USA | AK | | Nome |
| 303,304 | Oc. nigripes | Canada | MAN | | Churchill |
| 305,306 | Oc. schizopinax | USA | CA | Nevada | Boca |
| 307,308 | Oc. punctor | USA | MA | Hampshire | Chesterfield |
| 309 | Oc. implicatus | USA | ID | Kootenai | Athol |
| 310 | Oc. hexodontus | Canada | BC | | Prince Rupert |
| 311–313 | Oc. implicatus | USA | ID | Kootenai | Athol |
| 314–316 | Oc. punctor | USA | MA | Hampshire | Chesterfield |
| 317,318 | Oc. pionips | Canada | ONT | Algoma | White River |
| 319–321 | Oc. hexodontus | Canada | BC | | Prince Rupert |
| 322 | Oc. pionips | Canada | ONT | Algoma | White River |
| 323 | Oc. punctor | USA | MA | Hampshire | Chesterfield |
| 324 | Oc. hexodontus | Canada | BC | | Prince Rupert |
| 325 | Map | | | | |
| 326 | An. crucians | USA | GA | Baker | Newton |
| 327 | An. quadrimaculatus | USA | AR | Arkansas | Stuttgart |
| 328 | An. earlei | USA | MN | Ramsey | St. Paul |
| 329 | An albimanus | Panama | | | Canal Zone |
| 330 | An. punctipennis | USA | CT | Fairfield | Redding |
| 331 | An. crucians | USA | GA | Baker | Newton |
| 332–334 | An. punctipennis | USA | CT | Fairfield | Redding |
| 335 | An. pseudopunctipennis | USA | TX | Cameron | Ft. Brown |
| 336 | An. pseudopunctipennis | USA | TX | Travis | |
| 337 | An. punctipennis | USA | CT | Fairfield | Redding |
| 338 | An. perplexens | USA | FL | | |
| 339 | An. pseudopunctipennis | USA | TX | Travis | |
| 340 | An. pseudopunctipennis | USA | TX | Cameron | Brownsville |
| 341,342 | An. franciscanus | USA | NM | Eddy | Artesia |
| 343,345 | An. earlei | USA | MN | Ramsey | St. Paul |
| 34 | An. quadrimaculatus | USA | AR | Arkansas | Stuttgart |
| 346 | An. occidentalis | USA | CA | Alameda | Palo Alto |
| 347,348 | An. barberi | USA | DE | Kent | Bombay Hook |
| 349 | An. quadrimaculatus | USA | AR | Arkansas | Stuttgart |
| 350 | An. freeborni | USA | CA | Stanislaus | Modesto |
| 351 | An. barberi | USA | DE | Kent | Bombay Hook |
| 352 | An. judithae | USA | AZ | Santa Cruz | Nogales |
| 353 | An. quadrimaculatus | USA | AR | Arkansas | Stuttgart |
| 355 | An. walkeri | USA | MI | Livingston | |
| 356 | An. atropos | USA | LA | Plaquemines | Buras |
| 357 | An. freeborni | USA | CA | Nevada | Auburn |
| 358 | An. quadrimaculatus | USA | AR | Arkansas | Stuttgart |
| 359,360 | An. quadrimaculatus | USA | FL | Dixie | Cross City |
| 361,362 | An. inundatus | USA | FL | Walton | Bruce |
| 363,364 | An. quadrimaculatus | USA | FL | Dixie | Cross City |
| 365,366 | An. smaragdinus | USA | FL | Levy | Manatee Springs |
| 367 | An. diluvialis | USA | FL | Dixie | Bear Bay Swamp |
| 368 | An. inundatus | USA | FL | Walton | Bruce |
| 369–371 | An. maverlius | USA | MS | Tishomingo | High Point |

| Figure Number | Species | State/ Country | Prov. | County | Locality |
|---|---|---|---|---|---|
| 372–374 | An. inundatus | USA | FL | Walton | Bruce |
| 375–377 | An. walkeri | USA | MI | Livingston | |
| 378–380 | An. atropos | USA | LA | Plaquemines | Buras |
| 381,382 | Cx. pipiens | USA | OR | Multnomah | Portland |
| 383,384 | Cx. erraticus | USA | FL | Dade | Miami |
| 385,388 | Cx. restuans | USA | WI | Dane | |
| 386 | Cx. territans | USA | VA | Fairfax | Falls Church |
| 387,389 | Cx. tarsalis | USA | TX | Victoria | Victoria |
| 388 | Cx. restuans | USA | TX | Victoria | Victoria |
| 389 | Cx. tarsalis | USA | CA | | |
| 390 | Cx. pipiens | USA | NJ | Middlesex | Nixon |
| 391 | Cx. bahamensis | USA | FL | Monroe | Key Largo |
| 392,393 | Cx.tarsalis | USA | TX | Victoria | Victoria |
| 394 | Cx. tarsalis | USA | CA | | |
| 395 | Cx. stigmatosoma | USA | CA | Mariposa | |
| 396 | Cx. stigmatosoma | USA | CA | | |
| 397 | Cx. thriambus | USA | TX | Kerr | Kerrville |
| 398,399 | Cx. coronator | USA | TX | Cameron | Weslaco |
| 400 | Cx. declarator | Costa Rica | | | |
| 401,402 | Cx. erythrothorax | USA | CA | San Luis Obispo | San Luis Obispo |
| 403–405 | Cx. nigripalpus | USA | FL | | |
| 406 | Cx. restuans | USA | WA | Dane | |
| 407,408 | Cx. biscaynensis | USA | FL | Dade | Miami |
| 409,410 | Cx. pipiens | USA | NJ | Middlesex | Nixon |
| 411,412 | Cx. nigripalpus | USA | FL | | |
| 413–415 | Cx. salinarius | USA | MD | Calvert | Chesapeake City |
| 416 | Cx. chidesteri | USA | TX | Cameron | Brownsville |
| 417 | Cx. pipiens | USA | TN | Campbell | Loyston |
| 418 | Cx. pipiens | USA | NJ | Middlesex | Nixon |
| 419–422 | Cx. restuans | USA | WI | Dane | Madison |
| 423,424 | Cx. interrogator | Panama | Canal Zone | | |
| 425 | Cx. reevesi | Mexico | Baja California Norte | | |
| 426–428 | Cx. territans | USA | VA | Fairfax | Falls Church |
| 429 | Cx. arizonensis | USA | AZ | Yavapai | Prescott |
| 430 | Cx. apicalis | USA | AZ | Cochise | Portal |
| 431,432 | Cx. territans | USA | VA | Fairfax | Falls Church |
| 433 | Cx. boharti | USA | CA | San Diego | San Diego |
| 434 | Cx. boharti | USA | CA | Placer | Lake Tahoe |
| 435 | Cx. apicalis | USA | AZ | Cochise | Portal |
| 436 | Cx. arizonensis | USA | AZ | Yavapai | Prescott |
| 437 | Cx. erraticus | USA | TX | Kinnney | Brackettsv'e |
| 438,439 | Cx. peccator | USA | LA | La Salle | Olla |
| 440 | Cx. atratus | Cuba | | | Havana |
| 441 | Cx. cedecei | USA | FL | Dade | Miami |
| 442,444 | Cx. peccator | USA | LA | La Salle | Olla |
| 443 | Cx. abominator | USA | TX | Bexar | |
| 445 | Cx. iolambdis | USA | FL | Monroe | Key Largo |
| 446,447 | Cx. atratus | Cuba | | | Havana |
| 448,449 | Cx. pilosus | USA | FL | Broward | Ft. Lauderdale |
| 450 | Cx. mulrennani | USA | FL | Monroe | Big Pine Key |
| 451 | Cs. melanura | USA | IL | | |
| 452,453 | Cs. morsitans | USA | MI | Livingston | |
| 454 | Cs. impatiens | USA | CO | Grand | Grand Lake |

| Figure Number | Species | State/ Country | Prov. | County | Locality |
|---|---|---|---|---|---|
| 455,456 | Cs. particeps | USA | CA | Humboldt | Arcata |
| 457 | Cs. morsitans | USA | MI | Livingston | |
| 458 | Cs. impatiens | USA | CO | Grand | Grand Lake |
| 459 | Cs. particeps | USA | CA | Humboldt | Arcata |
| 460 | Cs. alaskaensis | USA | AK | Anchorage | |
| 461 | Cs. incidens | USA | WA | Whatcom | Bellingham |
| 462 | Cs. impatiens | USA | CO | Grand | Grand Lake |
| 463 | Cs. minnesotae | USA | MN | St. Louis | Virginia |
| 464 | Cs. morsitans | USA | MI | Livingston | |
| 465–466 | Cs. inornata | USA | MO | St. Louis | W.St.Louis |
| 467,468 | Cs. impatiens | USA | CO | Grand | Grand Lake |
| 469 | De. pseudes | Panama | | | |
| 470,471 | De. cancer | USA | FL | Indian River | Vero Beach |
| 472 | De. mathesoni | USA | TX | Cameron | Brownsville |
| 473 | Ma. titillans | Cuba | | | |
| 474 | Ma. titillans | USA | FL | | |
| 475 | Ma. dyari | USA | FL | Indian River | Vero Beach |
| 476 | Ma. dyari | USA | FL | Okeechobee | Ocheechobee |
| 477 | Or. kummi | Costa Rica | | | |
| 478,479 | Or. kummi | Panama | | | El Volcan |
| 480 | Or. alba | USA | MA | Anne Arundel | Patuxent |
| 481 | Or. signifera | USA | LA | Chicot | Kilbourne |
| 482 | Or. alba | USA | TX | Travis | Austin |
| 483 | Or. signifera | USA | LA | Orleans | Camp Planche |
| 484 | Or. signifera | USA | LA | Chicot | Kilbourne |
| 485 | Or. alba | USA | MA | Anne Arundel | Patuxent |
| 486 | Or. alba | USA | TX | Travis | Austin |
| 487,488 | Ps. columbiae | USA | TX | Cameron | Brownsville |
| 489 | Ps. ciliata | USA | SC | Beaufort | Parris Isl. |
| 490 | Ps. cyanescens | USA | NJ | Cumberland | Fairton |
| 491 | Ps. pygmaea | USA | FL | Monroe | Key West |
| 492–494 | Ps. columbiae | USA | TX | Cameron | Brownsville |
| 495,496 | Ps. discolor | USA | GA | Fulton | Ft. McPherson |
| 497 | Ps. signipennis | USA | TX | Sutton | Sinora |
| 498 | Ps. discolor | USA | GA | Fulton | Ft. McPherson |
| 499 | Ps. ciliata | USA | SC | Beaufort | Parris Island |
| 500 | Ps. cyanescens | USA | NJ | Cumberland | Fairton |
| 501 | Ps. ferox | USA | NC | Columbus | Lake Waccamaw |
| 502 | Ps. ciliata | USA | SC | Beaufort | Parris Island |
| 503' | Ps. ciliata | USA | SC | Charleston | McClellanville |
| 504,505 | Ps. howardii | USA | DE | New Castle | Newport |
| 506 | Ps. cyanescens | USA | NJ | Cumberland | Fairton |
| 507 | Ps. cyanescens | USA | TX | Dallas | Dallas |
| 508,509 | Ps. ferox | USA | NC | Columbus | Lake Waccamaw |
| 510 | Ps. mathesoni | USA | DE | Sussex | Thompsonville |
| 511 | Ps. ferox | USA | NC | Columbus | Lake Waccamaw |
| 512 | Ps. johnstonii | USA | FL | Indian River | Vero Beach |
| 513 | Ps. mathesoni | USA | DE | Sussex | Thompsonville |
| 514 | Ps. mexicana | USA | TX | Cameron | Brownsville |
| 515–517 | Ps. ferox | USA | NC | Columbus | Lake Waccamaw |
| 518–521 | Ps. horrida | USA | LA | E. Baton Rouge | Baton Rouge |
| 522,523 | Ps. longipalpus | USA | TX | Cameron | Brownsville |
| 524 | Ur. lowii | USA | FL | Highlands | |
| 525,526 | Ur. sapphirina | USA | DC | Washington | |

| Figure Number | Species | State/ Country | Prov. | County | Locality |
|---|---|---|---|---|---|
| 527,528 | Ur. a. anhydor | USA | CA | Bernardino | Saratoga Springs |
| 529 | Ur. a. syntheta | USA | TX | Cameron | |
| 530–531 | Wy. vanduzeei | USA | FL | | |
| 532,533 | Wy. smithii | USA | NJ | Union | Rahway |
| 534,535 | Wy. mitchellii | USA | FL | Indian River | Vero Beach |
| 536,537 | Wy. smithii | USA | NJ | Union | Rahway |
| 538 | An. quadrimaculatus | USA | NC | | Camp Sutton |
| 539,541 | Cx. pipiens | USA | PA | Allegheny | Turtle Creek |
| 540 | Ma. dyari | USA | FL | Palm Beach | W. Palm Beach |
| 542,543 | Ma. dyari | USA | FL | Palm Beach | W. Palm Beach |
| 544 | Cq. perturbans | USA | MN | Clearwater | |
| 545 | Cq. perturbans | USA | FL | Palm Beach | W. Palm Beach |
| 546 | Or. signifera | USA | GA | Fulton | Atlanta |
| 547 | Ae. aegypti | USA | GA | Chatham | Savannah |
| 548 | Tx. r. septentrionalis | USA | GA | Richmond | Augusta |
| 549 | Tx. r. septentrionalis | USA | FL | Palm Beach | Boca Raton |
| 550 | Cx. pipiens | USA | MO | St. Louis | St. Louis |
| 551 | Ae. aegypti | USA | GA | Chatham | Savannah |
| 552 | Wy. smithii | USA | MD | Worchester | |
| 553 | Or. signifera | USA | GA | Fulton | Atlanta |
| 554 | Ur. sapphirina | USA | FL | Palm Beach | Camp Murphy |
| 555 | Ur. sapphirina | USA | GA | Bryan | Ft. Stewart |
| 556,557 | Ps. columbiae | USA | DE | New Castle | Summit Bridge |
| 558,559 | De. pseudes | USA | TX | Cameron | Brownsville |
| 560 | Ps. columbiae | USA | DE | New Castle | Summit Br. |
| 561,563 | Ae. aegypti | USA | GA | Chatham | Savannah |
| 562 | Cs. inornata | USA | LA | Rapides | Esler Field |
| 564,566 | Cx. pipiens | USA | PA | Allegheny | Turtle Creek |
| 565 | Ae. aegypti | USA | GA | Chatham | Savannah |
| 567 | Oc. provocans | USA | NY | Tompkins | Ithaca |
| 568 | Ps. columbiae | USA | DE | New Castle | Summit Br. |
| 569,570 | Oc. atlanticus | USA | GA | Fulton | Atlanta |
| 571,573 | Ae. aegypti | USA | GA | Chatham | Savannah |
| 572 | Hg. equinus | Guatemala | | | |
| 574 | Oc. provocans | USA | NY | Tompkins | Ithaca |
| 575 | Ae. aegypti | USA | GA | Chatham | Savannah |
| 576 | Ae. cinereus | USA | CA | El Dorado | |
| 577,578 | Oc. provocans | USA | NY | Tompkins | Ithaca |
| 579 | Oc. bicristatus | USA | CA | Lake | |
| 580 | Oc. atlanticus | USA | GA | Fulton | Atlanta |
| 581 | Ae. aegypti | USA | GA | Chatham | Savannah |
| 582 | Oc. nigromaculis | USA | CA | Tulare | Visalia |
| 583 | Oc. abserratus | USA | NY | Tompkins | Ringwood |
| 584,585 | Oc. nigromaculis | USA | CA | Tulare | Visalia |
| 586,587 | Oc. f. pallens | USA | LA | Rapides | |
| 588 | Oc. nigromaculis | USA | CA | Tulare | Visalia |
| 589 | Oc. nigripes | Canada | MAN | | |
| 590 | Oc. f. pallens | USA | LA | Rapides | |
| 591 | Oc. f. pallens | USA | SC | Horry | Myrtle Beach |
| 592,593 | Oc. thelcter | USA | TX | Cameron | Brownsville |
| 594 | Oc. tormentor | USA | KY | Bullitt | Ft. Knox |
| 595 | Oc. abserratus | USA | NY | Tompkins | Ringwood |
| 596 | Oc. bimaculatus | USA | TX | Cameron | Brownsville |

| Figure Number | Species | State/ Country | Prov. | County | Locality |
|---|---|---|---|---|---|
| 597 | Oc. tormentor | USA | KY | Bullitt | Ft. Knox |
| 598 | Oc. atlanticus | USA | GA | Fulton | Atlanta |
| 599 | Oc. sollicitans | USA | IL | St. Clair | Dupo |
| 600 | Oc. taeniorhynchus | USA | FL | Palm Beach | Camp Murphy |
| 601 | Oc. taeniorhynchus | USA | FL | Highlands | Avon Park |
| 602,603 | Oc. abserratus | USA | NY | Tompkins | Ringwood |
| 604 | Oc. taeniorhynchus | USA | FL | Hillsborough | MacDill Field |
| 605 | Oc. taeniorhynchus | USA | FL | Palm Beach | Camp Murphy |
| 606 | Oc. dupreei | USA | GA | Fulton | Atlanta |
| 607,608 | Oc. atlanticus | USA | GA | Fulton | Atlanta |
| 609 | Oc. sollicitans | USA | IL | St. Clair | Dupo |
| 610 | Oc. atlanticus | USA | GA | Fulton | Atlanta |
| 611 | Oc. hexodontus | USA | CA | Tuolumne | Yosemite Pk. |
| 612 | Oc. punctor | USA | ME | Penobscot | Orono |
| 613 | Oc. sollicitans | USA | IL | St. Clair | Dipo |
| 614,615 | Oc. mitchellae | USA | MS | Harrison | Gulfport |
| 616,617 | Oc. sollicitans | USA | IL | St. Clair | Dupo |
| 618,620 | Oc. infirmatus | USA | FL | Highlands | Avon Park |
| 619 | Oc. taeniorhynchus | USA | FL | Palm Beach | Cp. Murphy |
| 621 | Oc. trivittatus | USA | IL | Champaign | Champaign |
| 622,623 | Oc. rempeli | Canada | NWT | | |
| 624 | Oc. taeniorhynchus | USA | FL | Hillsborough | MacDill Field |
| 625 | Oc. scapularis | Guatemala | | | |
| 626 | Oc. taeniorhynchus | USA | FL | Hillsborough | Tampa |
| 527 | Oc. taenioehynchus | USA | FL | Hillsborough | MacDill Field |
| 628,629 | Oc. scapularis | Guatemala | | | |
| 630 | Oc. scapularis | Dominican Republic | | | |
| 631 | Oc. scapularis | Guatemala | | | |
| 632,633 | Oc. tortilis | St. Lucia | | | |
| 634 | Oc. excrucians | USA | MA | Hampden | Springfield |
| 635 | Oc. melanimon | USA | CA | Merced | |
| 636 | Oc. cataphylla | USA | OR | Grant | Dixie Pass |
| 637 | Oc. excrucians | USA | MA | Hampden | Springfield |
| 638 | Oc. j. japonicus | USA | CT | | |
| 639 | Oc. atropalpus | USA | MD | Montgomery | Bethesda |
| 640,641 | Oc. cataphylla | USA | CA | Mono | |
| 642 | Oc. atropalpus | USA | ME | Hancock | Mt.Desert Is. |
| 643–645 | Oc. atropalpus | USA | MD | Montgomery | Bethesda |
| 646,647 | Oc. epactius | USA | TX | Comal | New Braunsfel |
| 648,650 | Oc. diantaeus | USA | VT | Windham | Jacksonville |
| 649 | Ae. vexans | USA | LA | Rapides | |
| 651 | Oc.diantaeus | USA | MI | Keweenaw | Isle Royale |
| 652 | Oc. aurifer | USA | DE | New Castle | New Castle |
| 653 | Oc. aurifer | USA | MD | Prince Georges | |
| 654 | Oc. s. spencerii | USA | MN | Ramsey | |
| 655 | Oc. campestris | USA | NV | Churchill | |
| 656,657 | Oc. s. spencerii | USA | MN | Ramsey | |
| 658 | Oc. s. idahoensis | USA | UT | Summit | Oakley |
| 659 | Oc. s. idahoensis | USA | CO | Grand | Grand Lake |
| 660 | Oc. excrucians | USA | MD | Cecil | Elkton |
| 661 | Oc. intrudens | USA | NY | Tompkins | Ringwood |
| 662 | Oc. excrucians | USA | MA | Hamoden | Springfield |

| Figure Number | Species | State/ Country | Prov. | County | Locality |
|---|---|---|---|---|---|
| 663 | Oc. excrucians | USA | NY | Tompkins | MacLean |
| 664–667 | Oc. campestris | USA | NV | | Churchill |
| 668,669 | Oc. flavescens | Germany | | | Spandau |
| 670 | Oc. flavescens | USA | AK | | Anchorage |
| 671 | Oc. flavescens | Germany | | | Spandau |
| 672,673 | Oc. aloponotum | USA | OR | Marion | Idanha |
| 674 | Oc. intrudens | USA | NY | Tompkins | Ringwood |
| 675 | Oc. niphadopsis | USA | UT | Tooele | Grantsville |
| 676 | Ae. vexans | USA | GA | Fulton | Atlanta |
| 677 | Oc. euedes | USA | MN | Clearwater | Itasca St.Pk. |
| 678 | Oc. intrudens | USA | AK | | Steese Hwy. |
| 679 | Oc. intrudens | USA | NY | Tomkins | Ringwood |
| 680,681 | Oc. euedes | USA | MN | Clearwater | Itasca St.Pk. |
| 682 | Oc. decticus | USA | MA | Hampshire | Belchertown |
| 683–685 | Oc. niphadopsis | USA | UT | Tooele | Grantsville |
| 686 | Oc. niphadopsis | USA | UT | | |
| 687–689 | Oc. riparius | USA | MN | Ramsey | |
| 690 | Oc. euedes | USA | MN | Clearwater | Itasca St.Pk. |
| 691 | Oc. ventrovittis | USA | CA | Alpine | |
| 692,693 | Oc. riparius | USA | MN | Ramsey | |
| 694,695 | Oc. ventrovitttis | USA | CA | Alpine | |
| 696 | Oc. triseriatus | USA | OH | Portage | Ravenna |
| 697 | Oc. fitchii | USA | NY | Tompkins | Ringwood |
| 698 | Oc. purpureipes | USA | AZ | Santa Cruz | |
| 699 | Oc. triseriatus | USA | OH | Portage | Ravenna |
| 700 | Ae. aegypti | USA | GA | Chatham | Savannah |
| 701 | Oc. papago | USA | AZ | Pima | Mendoza Canyon |
| 702 | Oc. muelleri | USA | AZ | Santa Cruz | Madera Canyon |
| 703 | Oc. purpureipes | USA | AZ | Santa Cruz | |
| 704,705 | Oc. papago | USA | AZ | Pima | Mendoza Canyon |
| 706 | Oc. triseriatus | USA | LA | Calcasie | Lake Charles |
| 707–710 | Ae. aegypti | USA | GA | Chatham | Savannah |
| 711–713 | Ae. albopictus | USA | FL | Indian River | Vero Beach |
| 714,715 | Oc. purpureipes | USA | AZ | Santa Cruz | |
| 716 | Ae. aegypti | USA | GA | Chatham | Savannah |
| 717 | Oc. muelleri | USA | AZ | Santa Cruz | Madera Canyon |
| 718,720 | Oc. sierrensis | USA | CA | San Diego | |
| 719 | Oc. zoosophus | USA | TX | Pecos | Sheffield |
| 721,723 | Oc. monticola | USA | AZ | Santa Cruz | |
| 722 | Oc. deserticola | USA | CA | Los Angeles | |
| 724–725 | Oc. monticola | USA | AZ | Santa Cruz | |
| 726 | Oc. varipalpus | USA | AZ | Coconino | |
| 727 | Oc. varipalpus | USA | UT | Kane | |
| 728,729 | Oc. bahamensis | USA | FL | Dade | |
| 730,731 | Oc. triseriatus | USA | OH | Portage | Ravenna |
| 732,733 | Oc. burgeri | USA | AZ | Santa Cruz | |
| 734,735 | Oc. triseriatus | USA | OH | Portage | Ravenna |
| 736 | Oc. zoosophus | USA | TX | Pecos | Sheffield |
| 737–739 | Oc. triseriatus | USA | OH | Portage | Ravenna |
| 740–742 | Oc. hendersoni | USA | CO | Boulder | Boulder |
| 743 | Oc. brelandi | USA | TX | Brewster | |
| 744 | Oc. impiger | USA | AK | Liberty Falls | |
| 745 | Oc. cantator | USA | RI | Washington | Westerly |

| Figure Number | Species | State/ Country | Prov. | County | Locality |
|---|---|---|---|---|---|
| 746 | Oc. fitchii | USA | NY | Tompkins | Ringwood |
| 747 | Oc. c. canadensis | USA | MA | Hampshire | Belchertown |
| 748 | Oc. impiger | USA | AK | | Liberty Falls |
| 749 | Oc. stimulans | USA | MN | Clearwater | |
| 750 | Oc. punctodes | USA | AK | | Anchorage |
| 751 | Oc. impiger | USA | AK | | Umiat |
| 752 | Oc. stimulans | USA | MN | Clearwater | |
| 753 | Oc. aboriginis | USA | OR | Columbia | Vernonia |
| 754 | Oc. melanimon | USA | CA | Kern | Bakersfield |
| 755 | Oc. sticticus | USA | GA | Bibb | Macon |
| 756,757 | Oc. melanimon | USA | CA | Merced | |
| 758,759 | Oc. stimulans | USA | MN | Clearwater | |
| 760,761 | Oc. nevadensis | USA | NV | Elko | Lamoille Cy. |
| 762,763 | Oc. stimulans | USA | MN | Clearwater | |
| 764,765 | Oc. mercurator | Canada | YUK | | Dawson |
| 766,767 | Oc. sticticus | USA | GA | Bibb | Macon |
| 768,769 | Oc. flavescens | USA | AK | | Anchorage |
| 770,771 | Oc. sticticus | USA | GA | Bibb | Macon |
| 772,773 | Oc. schizopinax | USA | CA | Nevada | |
| 774,775 | Oc. aboriginis | USA | OR | Columbia | Vernonia |
| 776 | Oc. pullatus | USA | CO | Latimer | Rocky Mt.Pk. |
| 777 | Oc. dorsalis | USA | KS | Stafford | |
| 778,779 | Oc. pullatus | USA | CO | Latimer | Rocky Mt. Pk. |
| 780–781 | Oc. c. canadensis | USA | MA | Hampshire | Belchertown |
| 782,783 | Oc. pionips | USA | MI | Keweenaw | Isle Royale |
| 784 | Oc. pullatus | USA | AK | | Eklutna |
| 785,786 | Oc. pullatus | USA | CO | Latimer | Rocky Mt.Pk. |
| 787 | Oc. pullatus | USA | AK | | Eklutna |
| 788 | Oc. cantator | USA | MD | Ann Arundel | Shelby on Bay |
| 789 | Oc. cantator | USA | RI | Washington | Westerly |
| 790 | Oc. togoi | Canada | BC | | |
| 791 | Oc. c. canadensis | USA | GA | Rabun | |
| 792,793 | Oc. thibaulti | USA | DE | Sussex | Redden Park |
| 794,795 | Oc. c. canadensis | USA | GA | Rabun | |
| 796 | Oc. squamiger | USA | CA | Marin | Richmond |
| 797 | Oc. communis | USA | MN | Clearwater | |
| 798,799 | Oc. dorsalis | USA | KS | Stafford | |
| 800,801 | Oc. increpitus | USA | CA | Mariposa | |
| 802 | Oc. campestris | USA | NV | Churchill | |
| 803–805 | Oc. dorsalis | USA | KS | Stafford | |
| 806,807 | Oc. grossbecki | USA | LA | Rapides | Alexandria |
| 808,809 | Oc. communis | USA | AK | | Umiat |
| 810,811 | Oc. melanimon | USA | CA | Merced | |
| 812 | Oc. tahoensis | USA | CA | | |
| 813 | Oc. communis | USA | MI | Clearwater | Churchill |
| 814,815 | Oc. churchillensis | Canada | MAN | | |
| 816,817 | Oc. communis | USA | MN | Clearwater | |
| 818,819 | Oc. melanimon | USA | CA | Merced | |
| 820,821 | Oc. increpitus | USA | CA | Mono | |
| 822–824 | Oc. implicatus | USA | MN | Clearwater | |
| 825–827 | Oc. increpitus | USA | CA | Mariposa | |
| 828 | Oc. clivis | USA | CA | | |
| 829 | Oc. washinoi | USA | CA | | |

| Figure Number | Species | State/ Country | Prov. | County | Locality |
|---|---|---|---|---|---|
| 830,831 | An. judithae | USA | AZ | Cochise | Portal |
| 832 | An. albimanus | USA | FL | Monroe | |
| 833 | An. albimanus | USA | TX | Cameron | Cosmas |
| 834 | An. barberi | USA | OH | Stark | Canton |
| 835 | An. barberi | USA | MD | Montgomery | Cabin John |
| 836 | An. judithae | USA | AZ | Cochise | Portal |
| 837 | An. judithae | USA | AZ | Santa Cruz | Patagonia |
| 838 | An. albimanus | USA | FL | Monroe | |
| 839 | An. quadrimaculatus | USA | TN | Dyer | Dyersburg |
| 840 | An. albimanus | USA | TX | Cameron | Cosmos |
| 841 | An. albimanus | USA | FL | Monroe | |
| 842,843 | An. pseudopunctipennis | USA | TX | Bell | Temple |
| 844 | An. pseudopunctipennis | Dutch WI | Curacao | | |
| 845 | An. pseudopunctipennis | USA | TX | Hidalgo | Edinburg |
| 846,847 | An. franciscanus | USA | NM | Eddy | Artesia |
| 848 | An. atropos | USA | FL | Monroe | Key Largo |
| 849 | An. quadrimaculatus | USA | TN | Dyer | Dyersburg |
| 850 | An. crucians | USA | LA | Calcasi | Lake Charles |
| 851 | An. punctipennis | USA | LA | Rapides | Esler Field |
| 852 | An. walkeri | USA | TN | Obion | Walnut Log |
| 853 | An. walkeri | USA | TN | Dyer | Dyersburg |
| 854,855 | An. quadrimaculatus | USA | TN | Dyer | Dyersburg |
| 856,858 | An. bradleyi | USA | MS | Harrison | Kessler Fd |
| 857 | An. quadrimaculatus | USA | LA | St. Charles | Narco |
| 859 | An. bradleyi | USA | AL | Mobile | Mobile |
| 860,861 | An. georgianus | USA | GA | Bibb | Macon |
| 862 | An. earlei | USA | MN | Beltrami | Bemidji |
| 863,864 | An. quadrimaculatus | USA | TN | Dyer | Dyersburg |
| 865 | An. punctipennis | USA | LA | Rapides | Esler Field |
| 866,867 | An. maverlius | USA | MS | Tishomingo | High Point |
| 868–871 | An. quadrimaculatus | USA | FL | Levy | Chiefland |
| 872,873 | An. inundatus | USA | FL | Walton | Bruce |
| 874–876 | An. quadrimaculatus | USA | FL | Levy | Chiefland |
| 877–879 | An. smaragdinus | USA | FL | Levy | Manatee Sp. |
| 880 | An. diluvialis | USA | FL | Dixie | Bear Bay Sw. |
| 881 | An. inundatus | USA | FL | Walton | Bruce |
| 882 | An. occidentalis | USA | CA | S.L.Obispo | Pismo Beach |
| 883 | An. punctipennis | USA | LA | Rapides | Esler Fld |
| 884 | An. freeborni | USA | UT | Salt Lake | Salt L. City |
| 885 | An. freeborni | USA | UT | Weber | Ogden |
| 886 | An. punctipennis | USA | LA | Rapides | Alexandria |
| 887 | An. punctipennis | USA | CA | Shasta | Tower Hse |
| 888 | Cx. pipiens | USA | MO | St. Louis | St. Louis |
| 889 | Cx. territans | USA | GA | Fulton | Atlanta |
| 890,891 | Cx.bahamensis | USA | FL | Monroe | Matecumbe Key |
| 892 | Cx. pipiens | USA | PA | Allegheny | Turtle Creek |
| 893 | Cx. pipiens | USA | MO | St. Louis | St. Louis |
| 894 | Cx. interrogator | USA | TX | Cameron | Harlingen |
| 895,897 | Cx. tarsalis | USA | CA | Contra Costa | Pittsburg |
| 896 | Cx. restuans | USA | SC | Richland | Columbia |
| 898 | Cx. restuans | USA | NC | Robeson | Maxton |
| 899 | Cx. thriambus | USA | CA | Riverside | Coachella V. |
| 900 | Cx. coronator | USA | TX | Cameron | Brownsville |

| Figure Number | Species | State/ Country | Prov. | County | Locality |
|---|---|---|---|---|---|
| 901,902 | Cx. tarsalis | USA | CA | Contra Costa | Pittsburg |
| 903 | Cx. pipiens | USA | PA | Allegheny | Turtle Creek |
| 904 | Cx. tarsalis | USA | CA | Contra Costa | Pittsburg |
| 905 | Cx. chidesteri | USA | TX | Cameron | Brownsville |
| 906 | Cx. declarator | USA | TX | Caldwell | Luling |
| 907,908 | Cx. pipiens | USA | PA | Allegheny | Turtle Creek |
| 909 | Cx. salinarius | USA | MD | Ann Arundel | Shelbyon Bay |
| 910 | Cx. stigmatosoma | USA | CA | Sacramento | Sacramento |
| 911 | Cx. stigmatosoma | USA | CA | Marin | |
| 912 | Cx. pipiens | USA | MO | St. Louis | St. Louis |
| 913 | Cx. pipiens | USA | NE | Otoe | Dunbar |
| 914,915 | Cx. nigripalpus | USA | FL | Palm Beach | Gulf Stream |
| 916,920 | Cx. salinarius | USA | MD | Ann Arundel | Selbyonbay |
| 917,921 | Cx. salinarius | USA | KS | Douglas | |
| 918,919 | Cx. biscaynensis | USA | FL | Dade | Miami |
| 922 | Cx. erythrothorax | USA | CA | San Luis Obispo | |
| 923 | Cx. salinarius | USA | KS | Douglas | |
| 924 | Cx. territans | USA | GA | Richmond | Ft.Gordon |
| 925 | Cx. peccator | USA | FL | | |
| 926,927 | Cx. arizonensis | USA | AZ | Yavapai | Prescott |
| 928 | Cx. territans | USA | GA | Richmond | Ft. Gordon |
| 929 | Cx territans | USA | GA | Richmond | Ft.Gordon |
| 930 | Cx. apicalis | USA | TX | Brewster | BigBendPk |
| 931 | Cx. territans | USA | GA | Richmond | Ft. Gordon |
| 932 | Cx. reevesi | USA | CA | San Luis Obispo | |
| 933 | Cx. territans | USA | GA | Fulton | Atlanta |
| 934,935 | Cx. territans | USA | GA | Fulton | Atlanta |
| 936,937 | Cx. boharti | USA | CA | Benito | |
| 938 | Cx. pilosus | USA | LA | Orleans | Cp.Villere |
| 939 | Cx. atratus | Puerto Rico | | | |
| 940 | Cx. pilosus | USA | GA | Fulton | Atlanta |
| 941 | Cx. erraticus | USA | GA | Baker | |
| 942,944 | Cx. cedecei | USA | FL | Broward | Ft.Lauderdale |
| 943 | Cx. peccator | USA | FL | | |
| 945 | Cx. cedecei | USA | FL | Dade | |
| 946,947 | Cx. atratus | USA | FL | Monroe | Vaca Key |
| 948 | Cx. abominator | USA | TX | Comal | |
| 949 | Cx. iolambdis | USA | FL | Martin | Jensen |
| 950 | Cx. peccator | USA | GA | Fulton | Atlanta |
| 951 | Cx. iolambdis | USA | FL | Palm Beach | Jupiter |
| 952 | Cx. anips | Mexico | Baja California | | Tijuana |
| 953 | Cx. peccator | USA | FL | | |
| 954 | Cx. iolambdis | USA | FL | Palm Beach | Jupiter |
| 955 | Cx. mulrennani | USA | FL | Monroe | Big Pine Key |
| 956 | Cs. melanura | USA | FL | Okaloosa | Baker |
| 957 | Cs. inornata | USA | CO | Larimer | Estes Pk. |
| 958,959 | Cs. morsitans | USA | MN | Clearwater | |
| 960,961 | Cs. inornata | USA | CO | Larimer | Estes Pk. |
| 962,963 | Cs. minnesotae | USA | MN | Clearwater | |
| 964,965 | Cs. morsitans | USA | MN | Clearwater | |
| 966 | Cs. impatiens | USA | AK | | Ketchikan |
| 967 | Cs. inornata | USA | CO | Larimer | Estes Pk. |
| 968,969 | Cs. particeps | USA | CA | Kern | Kernville |

| Figure Number | Species | State/ Country | Prov. | County | Locality |
|---|---|---|---|---|---|
| 970 | Cs. inornata | USA | CO | Larimer | Estes Pk. |
| 971,972 | Cs. inornata | USA | DE | New Castle | Newark |
| 972 | Cs. incidens | USA | ID | Valley | McCall |
| 974,975 | Cs. alaskaensis | USA | AK | | Glen Allen |
| 976,977 | Cs. incidens | USA | CA | Madera | |
| 978,979 | De. mathesoni | USA | TX | Cameron | Brownsville |
| 980–982 | De. pseudes | USA | TX | Cameron | Brownsville |
| 983,985 | De. cancer | USA | FL | Palm Beach | Boca Raton |
| 984 | De. pseudes | USA | TX | Cameron | Brownsville |
| 986,987 | Ma. titillans | Jamaica | | | |
| 988,989 | Ma. dyari | USA | FL | Palm Beach | W. Palm Beach |
| 990 | Or. alba | USA | GA | Fulton | Atlanta |
| 991 | Or. signifera | USA | MS | Forrest | Cp. Shelby |
| 992,993 | Or. kummi | USA | AZ | Santa Cruz | |
| 994,995 | Or. signifera | USA | MS | Forrest | Cp. Shelby |
| 996 | Ps. ciliata | USA | DE | New Castle | Summit Br. |
| 997 | Ps. howardii | USA | MA | Prince Georges | College Pk |
| 998 | Ps. discolor | USA | GA | Baker | Newton |
| 999 | Ps. columbiae | USA | DE | New Castle | Delaware Cty. |
| 1000 | Ps. ciliata | USA | TX | Hidalgo | Mission |
| 1001 | Ps. howardii | USA | DE | New Castle | Newport |
| 1002 | Ps. columbiae | USA | DE | New Castle | Summit Br. |
| 1003 | Ps. discolor | USA | GA | Baker | Newton |
| 1004 | Ps. ferox | USA | GA | Worth | |
| 1005 | Ps. cyanescens | USA | LA | Rapides | Alexandria |
| 1006,1007 | Ps. discolor | USA | GA | Baker | Newton |
| 1008,1009 | Ps. columbiae | USA | DE | New Castle | Delaware Cty. |
| 1010 | Ps. columbiae | USA | DE | New Castle | Summit Br. |
| 1011,1012 | Ps. signipennis | USA | KS | Reno | Hutchinson |
| 1013 | Ps. pygmaea | Puerto Rico | Juan Diaz | | |
| 1014,1015 | Ps. cyanescens | USA | LA | Rapides | Alexandria |
| 1016,1017 | Ps. ferox | USA | GA | Worth | |
| 1018,1019 | Ps. johnstonii | USA | FL | Monroe | Long Key |
| 1020–1022 | Ps. horrida | USA | GA | Chattahoochee | Ft. Benning |
| 1023 | Ps. ferox | USA | GA | Worth | |
| 1024,1025 | Ps. horrida | USA | GA | Chattahoochee | Ft. Benning |
| 1026 | Ps. mathesoni | USA | LA | | |
| 1027 | Ps. mathesoni | USA | DE | Sussex | Thompsonville |
| 1028,1029 | Ps. ferox | USA | GA | Worth | |
| 1030 | Ps. longipalpus | USA | TX | Grayson | Denison |
| 1031 | Ps. longipalpus | USA | OK | Tulsa | Tulsa |
| 1032 | Ur. a. syntheta | USA | TX | Bexar | San Antonio |
| 1033 | Ur. sapphirina | USA | GA | Bryan | Ft. Stewart |
| 1034,1035 | Ur. lowii | USA | FL | Palm Beach | Boca Raton |
| 1036,1037 | Ur. sapphirina | USA | LA | Rapides | Alexandria |
| 1038,1039 | Wy. mitchellii | USA | FL | Palm Beach | Boca Raton |
| 1040 | Wy. smithii | USA | MA | Prince Geo. | Suitl'd |
| 1041 | Wy. smithii | USA | MN | Clearwater | |
| 1042 | Wy. vanduzeei | USA | FL | Palm Beach | Boca Raton |
| 1043 | Wy. vanduzeei | USA | FL | Dade | Miami |
| 1044 | Wy. smithii | USA | MN | Clearwater | |
| 1045 | Wy. smithii | USA | MA | Prince Geo. | Suitl'd |

# Index to Scientific Names

This is an alphabetical listing of the scientific names mentioned in the text, including the adult and larval morphology, the keys, the illustrations of the key couplets, and the map plate captions. Abbreviations of the genera follow Reinert (815). Page numbers for scientific names in the couplet illustrations are in italics; all others are in roman type. The map numbers are parenthetically in boldface. A page number in roman type to the left of a slant line refers to the adult key; to the right of the line, the larval key. In a similar manner, the italicized numbers to the left and right of the second slant line pertain to the figure numbers of the species used to illustrate the adult and larval keys, respectively.

Richard F. Darsie Jr. is a research entomologist. He was on the faculty of the Department of Entomology, University of Delaware, for twelve years, and has worked in Nepal, the Philippines, El Salvador, and Guatemala as well as Fort Collins, Colorado, and South Carolina in the field of medical entomology and vector-borne disease control. He retired from the Centers for Disease Control and Prevention in 1987 after more than 20 years' service but continued at the Center until 1996, when he moved to his present position as courtesy professor, University of Florida, Florida Medical Entomology Laboratory, Vero Beach. He has authored 125 scientific articles and is a member of the Entomological Society of America, Entomological Society of Washington, American Mosquito Control Association, and American Society of Tropical Medicine and Hygiene.

Ronald A. Ward has been a medical entomologist with the Walter Reed Army Institute of Research, Washington, D.C., since 1958, after serving three years as a biology instructor at Gonzaga University, Spokane. He has conducted field studies in Southeast Asia, the Middle East, and Africa on malaria and African trypanosomiasis. He has served on the Armed Forces Pest Management Board and is an honorary research associate in entomology at the Smithsonian Institution. He has published more than 65 papers in medical entomology with emphasis on vector-parasite relationships. He is an active member of many scientific societies, including the American Mosquito Control Association, and is the local Washington, D.C., secretary of the Royal Society of Tropical Medicine and Hygiene.